PROGRESS IN BIOMETEOROLOGY

:DITOR-IN-CHIEF: S. W. TROMP EXECUTIVE EDITOR: JANNEKE J. BOUMA

BIOMETEOROLOGICAL RESEARCH CENTRE, LEIDEN
AN STEENLAAN 3, OEGSTGEEST (LEIDEN)
THE NETHERLANDS

DIVISION B

PROGRESS IN ANIMAL BIOMETEOROLOGY

The Effect of Weather and Climate on Animals

VOLUME 1 PART I PERIOD 1963-1973

EFFECT OF TEMPERATURE ON ANIMALS

(Including Effects of Humidity, Radiation and Wind)

EDITED BY H. D. JOHNSON
DEPARTMENT OF DAIRY HUSBANDRY
UNIVERSITY OF MISSOURI
COLUMBIA (MO) U.S.A.
IN COOPERATION WITH 87 CONTRIBUTORS

SWETS & ZEITLINGER B.V. AMSTERDAM 1976

Printed by Drukkerij F. E. Mac Donald, Nijmegen

ISBN 90 265 0196 x

LIST OF CONTRIBUTORS OF DIVISION B, VOLUME 1

NAME	ADDRESS	CHAPTER and SECTION*
Y. ALARIE	Dept of Occupational Health, Graduate School of Public Health, Univ. of Pittsburgh, Pittsburgh, Pa 15261, U.S.A.	7.4
T. E. ALLEN	Div. of Animal Genetics, C.S.I.R.O., McMaster Field Station, Private Mail Bag, Liverpool, N.S.W., 2170, Australia.	2.2a
C. W. ALLISTON	Dept of Animal Sciences, The Roy W. Herrick Lab., Lilly Hall, Purdue Univ., West Lafayette, Indiana 47907, U.S.A.	1.15b(1)
P. D. ALTLAND	Physiological Section, Lab. of Chemical Physics, National Inst. of Arthritis, Metabolism and Digestive Diseases, National Inst. of Health, Bethesda, Maryland 20014, U.S.A.	3.2
A. ANTHONY	Dept of Biology, College of Science, The Pennsylvania State Univ., 208 Life Sciences I, Univ. Park, Pa 16802, U.S.A.	4.0
E. BEDRAK	Dept of Biology, Univ. of the Negev and Research and Development Authority, P.O.B. 1025, Beer-Sheva 84110, Israel	1.15d
S. C. BHATIA	Dept of Biochemistry, Vallabhohdi Patel Chest Inst., Univ. of Delhi, Delhi 110007, India Temporarily: Chef des Travaux at the Dept de Biochimie Médicale, Ecole de Médecine, Univ de Génève, 20, rue de l'Ecole de Médecine, 1211 Genève 4, Switzerland.	7.5
W. BIANCA	Institut f. Tierzucht, ETH, Universitätstr. 2, 8006 Zürich, Switserland	1.17
J. BLIGH	ARC Institute of Animal Physiology, Babraham, Cambridge, Great Britain.	1.4a
W. G. BRADLEY	Dept of Biological Sciences, Univ. of Nevada, Las Vegas, Nevada 89154, U.S.A.	1.18a
J. R. BRETT	Physiological Investigation Lab., Pacific Biological Station, Dept of the Environment, Nanaimo, B.C., Canada V9R 5K6	1.12c

* 1.5.b, 2.1.g means that the Contributor wrote Chapter 1, Section 5, Sub-Section b and Chapter 2, Section 1, sub-Section g.

G. D. BROWN	Div. of Wildlife Research, C.S.I.R.O., P.O.B. 84, Lyneham, A.C.T. 2602, Australia.	2.2a
W. E. CALDWELL	Dept of Psychology, The George Washington Univ., Washington, D.C. 20006, U.S.A.	5.2a
D. P. CARDINALI	Instituto Latino-Americano de Fisiología de la Reproducción (ILAFIR), Univ del Salvador, Casilla de Correo 10, San Miguel (P.B.A.), Argentina.	2.2b(2)
K. CENA	Academy of Physical Education, ul. Banacha 11, 51-617 Wrocław, Poland.	1.13a
R. R. CHAFFEE	Dept of Ergonomics, Univ. of California, Santa Barbara, Cal. 93106, U.S.A.	1.1a
J. W. DEATON	South Central Poultry Research Lab., Agricultural Res. Service, P.O.B. 5367, State College, Mississippi 39762, U.S.A.	1.12d
R. W. DOUGHERTY	National Animal Disease Lab., P.O.B. 70, Ames, Iowa 50010, U.S.A.	1.20.a
D. F. DOWLING	Dept of Animal Husbandry, Univ. of Queensland, St Lucia., Brisbane 4067, Australia.	1.12.a
T. W. DUKES	Animal Diseases Res. Inst., P.O.B. 11300, Station H, Ottawa K2H 8P9, Canada.	2.2c
G. E. FOLK Jr	Dept of Physiology and Biophysics, University of Iowa, College of Medicine, Iowa City, Iowa 52242, U.S.A.	1.11.b(1)
F. E. J. FRY	Ramsay Wright Zoological Labs, Univ. of Toronto, Harbord Str. 25, Toronto, Ontario, Canada, M5S 1A1.	1.4.e
M. GEBCZÝNSKI	Mammals Research Inst., 17-230 Bialowicza, Woj. Bialystok, Poland.	2.3
R. F. GROVER	Cardiovascular Pulmonary Research Lab., Div. of Cardiology, Dept of Medicine, Univ. of Colorado Medical Center, 4200 East-ninth Av., Denver, Col. 80220, U.S.A.	3.1
D. E. GRUBBS	Dept of Biology, California State Univ., Fresno, Calif., U.S.A.	1.3.c
G. L. HAHN	Dept of Agricultural Engineering, Building T-12, Univ. of Missouri-Columbia, Columbia, Mo, 65201, U.S.A.	1.21a
J. R. S. HALES	C.S.I.R.O., Div. of Animal Physiology, Ian Clunies Ross Animal Res. Lab., P.O.B. 239, Blacktown, N.S.W. 2148, Australia.	1.5.a 1.11.a(1)
C. L. HAMILTON	Dept of Physiology, Univ. of Pennsylvania, School of Medicine, Philadelphia, Pa 19104, U.S.A.	1.7
P. C. HARRISON	Dept of Animal Science, 124 Animal Sciences Lab., College of Agriculture, Univ. of Illinois, Urbana, Ill. 61801, U.S.A.	1.11.a(3) 1.11.b(2)

W. O. HAUFE	Veterinary and Medical Entomological Section, Canada Agriculture Research Station, Lethbridge, Alb. T1J-4B1, Canada.	1.20.d
F. L. HAYS	Eidgenössische Techn. Hochschule, Lehr- und Versuchsgut Chamau, Unterhünenberg (Zug), Bahnstation Cham, 6331 Unterhünenberg, Switzerland	7.1
J. C. HOFFMANN	Dept of Anatomy and Reproductive Biology, Univ. of Hawaii, School of Medicine, 1960 East-West Road, Honolulu, Hawaii 96822, U.S.A.	2.2.b(1)
M.E. HUGH-JONES	Pathological and Epidemiology Dept, Veterinary Lab., New Haw, Weybridge, Surrey, Great Britain.	1.20.b
J. C. D. HUTCHINSON	Division of Animal Physiology, Ian Clunies Ross Animal Research Lab., Prospect, CSIRO, P.O.B. 239, Blacktown, New South Wales 2148, Australia.	2.1.g 2.2.a
D. L. INGRAM	Agricultural Research Council, Institute of Animal Physiology, Babraham, Cambridge, Great Britain.	1.5.d 1.18.c(2)
G. L. JACKSON	Dept of Veterinary Physiology and Pharmacology, College of Veterinary Medicine, Univ. of Illinois at Urbana-Champaign, Ill. 61801, U.S.A.	2.1.c
L. JANSKÝ	Dept of Comparative Physiology, Charles University, Prague, Czechoslovakia.	1.8.b(4)
Elisabeth JOHNSON	Dept of Zoology, Univ. of Reading, Whiteknights, Reading RG6-2AJ, Great Britain.	2.1.f
H. D. JOHNSON	Dept of Dairy Husbandry, University of Missouri, Columbia, Missouri 65201, U.S.A.	1.2.a 1.14
K. G. JOHNSON	ARC Institute of Animal Physiology, Babraham, Cambridge, Great Britain.	1.5.c
T. H. KAMAL	Bioclimatology Unit, Radiobiology Division, Atomic Energy Establishment, Cairo, Egypt.	1.19
K. L. KNOX	Dept of Nutritional Sciences, College of Agriculture, The Univ. of Connecticut, Storrs, Conn. 06268, U.S.A.	1.8.a(2)
A. P. KRUEGER	School of Public Health, Earl Warren Hall, Univ. of Cal., Berkeley, Cal. 94720, U.S.A.	5.1
M. A. LESSLER	Dept of Physiology, Ohio State Univ., 333 W.10th Av., Columbus, Ohio 43210, U.S.A.	7.2
W. V. MACFARLANE	Dept of Physiology, Waite Institute, University of Adelaide, Private Bag 1, Glen Osmond, Adelaide (S.A.), Australia.	1.3.a 1.18.b

R. E. MACMILLEN	Dept of Population and Environmental Biology, Univ. of California, Irvine, Cal., U.S.A.	1.3.c
G. M. O. MALOIY	Dept of Animal Physiology, Univ. of Nairobi, P.O.B. 30197, Nairobi, Kenya, E. Africa.	1.3.b
E. J. MASORO	Dept of Physiology, The Univ. of Texas Health Center at San Antonio, 7703 Floyd Curl Drive, San Antonio, Texas 78284, U.S.A.	1.1.b
W. F. Y. MAWSON	Development Planning Branch, Dept of Primary Industries, Bribane, Qld 4000, Australia.	1.21.b
J. A. MCLEAN	Hannah Dairy Institute, Ayr, Scotland Great Britain	1.8.a(1)
J. L. MONTEITH	Dept of Physiology and Environmental Studies, Univ. of Nottingham, School of Agriculture, Sutton Bonington, Loughborough, Leics. LE12-5RD, Great Britain	1.13.a
P. R. MORRISON	Dept of Zoophysiology, Institute of Arctic Biology, Univ. of Alaska, Fairbanks, Alaska 99701, U.S.A.	3.3
L. E. MOUNT	Dept of Applied Biology, ARC Institute of Animal Physiology, Babraham, Cambridge, Great Britain.	1.8.b(3)
L. D. OLSON	Dept of Veterinary Pathology, University of Missouri, Columbia, Mo. 65201, U.S.A.	1.20.c
J. OYAMA	Aimes Research Center, NASA, Moffett Field, Cal. 94035, U.S.A.	6
Alice Y. S. P'AN	School of Public Health, Univ. of Montreal, 2375 Chemin Côte Ste-Catherine, Montreal 101, Quebec, Canada.	7.3
Pamela R. PENNYCUIK	Div. of Animal Genetics, CSIRO, P.O.B. 90, Epping, N.S.W. 2121, Australia.	1.18.c(3)
M. A. PERSINGER	Environmental Psychophysiological Lab., Dept of Psychology, Laurentian University, Sudbury, Ont., Canada.	5.2.b
V. M. PETROVIC	Inst. of Physiology and Biochemistry, Fac. Science, Univ. of Belgrado, Studentski/trg 16, Belgrado, Yugoslavia.	1.2.b(2)
W. S. PLATNER	Dept of Physiology, Univ. of Missouri, School of Medicine, Columbia, Missouri 65201, U.S.A.	1.9
R. S. POGRUND	Dept of Occupational and Environmental Medicine, School of Public Health, Univ. of Illinois at the Medical Centre, P.O.B. 6998, Chicago, Ill. 60680, U.S.A.	5.3
R. O. RAWSON	John B. Pierce Foundation, 290 Congress Av., New Haven, Connecticut 06519, U.S.A.	1.4.b

F. N. REECE	South Central Poultry Research Lab., Agricultural Research Service, P.O.B. 5367, Mississippi State College, 39762, U.S.A.	1.12.d
R. J. REITER	Dept of Anatomy, The Univ. of Texas, Health Science Center at San Antonio, 7703 Floyd Curl Drive, San Antonio, Texas 78284, U.S.A.	2.1.d
J. D. RISING	Ramsay Wright Zoological Lab., Univ. of Toronto, 25 Hardford Street, Toronto M5S-1A1, Ont., Canada.	1.4.f
D. ROBERTSHAW	Indiana Univ. School of Medicine, Medical Sciences Program, Meyers Hall, Bloomington, Ind. 47401, U.S.A.	1.5.b
F. RUSSO	Psychological Section, Behavioral Sciences Dept, Naval Medical Research Inst., Bethesda, Md 20014, U.S.A.	5.2.a
M. L. SCOTT	Dept of Poultry Science and Animal Nutrition, Rice Hall, New York State College of Agriculture, Cornell Univ. Ithaca, New York 14850, U.S.A.	1.10
H. S. SIEGEL	South East Poultry Research Lab., Univ. of Georgia, 934 College Station Road, Athens, Georgia 30601, U.S.A.	1.8.a(4) 1.8.b(5)
E. SIMENSEN	Dept of Veterinary Pathology, Univ. of Missouri, Columbia, Mo. 65201, U.S.A. Temporarily: Dept of Animal Husbandry and Genetics, Vet. College of Norway, P.O. Box 8146, Oslo-Dep, Oslo, Norway.	1.20.c
U. A. SOD-MORIAH	Dept of Biology, Univ. of the Negev and Research and Development Authority, P.O.B. 1025, Beer-Sheva 84110, Israel.	1.15.d
J. STEINBACH	Dept of Animal Science, Univ. of Ibadan, Ibadan, Nigeria.	1.12.b 1.15.c
M. B. TER HAAR	A.R.C. Inst. of Animal Physiology, Babraham, Cambridge, Great Britain.	2.1.b
G. E. THOMPSON	The Hannah Dairy Research Inst., Ayr, Scotland KA6 5HL, Great Britain.	1.8.b(1)
C. J. THWAITES	Dept of Livestock Production, Univ. of New England, Armidale, New South Wales 2351, Australia.	1.13.b
M. VAN KAMPEN	Lab. for Veterinary Physiology, Alex. Numankade 93, State Univ., Utrecht, The Netherlands.	1.5.e
H. A. W. VENTER	Dept of Animal Science, Univ. of Pretoria, Pretoria, South Africa.	1.15.a

J. E. VERCOE	C.S.I.R.O., Div. of Animal Genetics, Cattle Research Lab., Rockhampton, Qld, Australia.*	1.18.c(1)
A. J. F. WEBSTER	Rowett Research Inst., Bucksburn, Aberdeen, Scotland, AB2-9SB, Great Britain	1.2.b(1) 1.8.b(2)
G. C. WEST	Inst. of Arctic Biology, Lab. of Zoophysiology, Univ. of Alaska, Fairbanks, Alaska 99701, U.S.A.	1.18.d(2)
G. C. WHITTOW	Dept of Physiology, School of Medicine, Univ. of Hawaii, Honolulu, Hawaii 96822, U.S.A.	1.4.d 1.11.a(2)
W. O. WILSON	Dept of Avian Sciences, Univ. of California, Davis, Calif. 95616, U.S.A.	1.16
G. L. WRIGHT	Lab. of Cell Physiology, Dept of Physiology, 4196 Medical Science Building, 333 West Tenth Av., Columbus, Ohio 43210, U.S.A.	1.6 7.2
N. T. M. YEATES	Dept of Livestock Production, Univ. of New England, Armidale, New South Wales, Australia.	1.15.b(2)
M. K. YOUSEF	Dept of Biological Sciences, Univ. of Nevada, Las Vegas, Nevada 89154, U.S.A. and Lab. of Environmental Patho-Physiology, Desert Research Inst., Univ. of Nevada System, Boulder City, Nevada, U.S.A.	1.4.g 1.8.a(3) 1.18.a

* Temporary address: Animal Production and Health Section, Joint FAO/I.A.E.A. Division, Kärtnerring 11, P.O.B. 590, A–1101 Vienna, Austria.

PART II

PREFACE

Since 1963 Biometeorology has been developing rapidly due to the considerable increase in interest of the whole scientific world in man's physical environment. Also the recent developments in Space Research in connection with the landings on the moon and studies of the atmosphere of the planets have stimulated many research workers to enter this field of research.

The present book "PROGRESS IN ANIMAL BIOMETEOROLOGY" represents the first Volume of Division B in the series PROGRESS IN BIOMETEOROLOGY, following Volume 1 of Division A, PROGRESS IN HUMAN BIOMETEOROLOGY and Volume 1 of Division C, PROGRESS IN PLANT BIOMETEOROLOGY. The book PROGRESS IN HUMAN BIOMETEOROLOGY was prepared as a continuation of a previous textbook "MEDICAL BIOMETEOROLOGY" (Weather, Climate and the Living Organism) which was published by Elsevier Publishing Company, Amsterdam, 1963. It was written by the present Editor-in-Chief of the series PROGRESS IN BIOMETEOROLOGY.

Swets & Zeitlinger Publishing Company agreed to publish a continuation of "Medical Biometeorology" which covers the period 1963-1973. As the effects both on healthy man and on diseased subjects would be reviewed, the original title was changed to "PROGRESS IN HUMAN BIOMETEOROLOGY" with the subtitle "THE EFFECTS OF WEATHER AND CLIMATE ON MAN AND HIS LIVING ENVIRONMENT".

However, it was realized that not only "Human Biometeorology", needed a new up to date textbook, but also the recent developments in plant and animal biometeorology made it highly necessary to cover these developments.
For this reason it was decided to publish a completely new series which was called PROGRESS IN BIOMETEOROLOGY. It consists of three divisions: Division A: PROGRESS IN HUMAN BIOMETEOROLOGY; Division B: PROGRESS IN ANIMAL BIOMETEOROLOGY and Division C: PROGRESS IN PLANT BIOMETEOROLOGY. Each division has its own Editor. Division A is being prepared by Dr S. W. TROMP, Director of the Biometeorological Research Centre, Leiden, The Netherlands; Division B is being prepared by Prof. H. D. JOHNSON, Director of the Department of Dairy Husbandry, University of Missouri, Columbia, U.S.A.; Division C by Mr L. P. SMITH, former President of the Commission for Agricultural Meteorology of the World Meteorological Organization, Meteorological Office, Bracknell, Great Britain.
General Coordinator and Editor-in-Chief of the three divisions is Dr S. W. TROMP. Executive Editor of all volumes is Mrs Janneke J. BOUMA, Research Associate of the Biometeorological Research Centre, Leiden, The Netherlands. Depending on the amount of new information received by the Editors in their specific divisions, new volumes will be prepared every two to three years. For this reason the present book is called DIVISION B, VOLUME I (PERIOD 1963-1973).

Instead of giving a list of references after each Chapter or Section the same method was followed, previously used in MEDICAL BIOMETEOROLOGY and in Division A and C of PROGRESS IN HUMAN BIOMETEOROLOGY, which proved to be very satisfactory. All references are given at the end of each book arranged in the same sequence as the Chapters and Sections. At the end of each Section we briefly indicated the pages where the references of this particular Section can be found. An extensive Author and Subject Index will enable the reader to find each Section in the book and the Contributors of these Sections.

The Volumes 1 of Division A, B and C represent in a way the standard type of a Progress Series but differ from the usual publications in which a number of authors review only certain aspects of a whole science. In this Progress Series we have tried to prepare up to date textbooks on Human, Animal and Plant Biometeorology.

It is fully realized that the task ahead is a very difficult one, but we trust that the efforts of all the cooperating experts in the field of Biometeorology will enable us to keep abreast of the fast developments of this modern branch of science and will help future students in Meteorology, Biology, Medicine, Veterinary, Agricultural and many other sciences, in finding their way in this dense forest of rapidly growing scientific knowledge.

S. W. Tromp
Editor-in-Chief
Progress in Biometeorology

INTRODUCTION

PROGRESS IN ANIMAL BIOMETEOROLOGY was designed to provide a concise, up-to-date text of biometeorological effects on the mechanisms of physiological and behavioural responses of terrestrial vertebrates. Emphasis is on domesticated and wild mammalian and avian species; though, when possible other vertebrates and invertebrates are included. Most Sections will not cover all species; however, as information and authors become available, additional Sections will be added upon revision of the textbook.

This book will serve as a basic text for students of animal biometeorology and environmental physiology and related fields. The description and measurement of the physical and chemical environments and intensive ecological relationships must be provided from other sources of information.

Each Section has endeavored to include a concise concensus by the author of the present state of knowledge. When possible, figure(s) or diagram(s) have been used to illustrate the known or possible mode of action of the environmental effect and responses of the organisms.

Volume 1 consists of two Parts:
Part I comprises only Chapter 1, dealing with environmental temperature effects on animal functions and processes. It is the most extensive due to the greater availability of research on this subject.

Cell metabolism, endocrine functions, water and electrolyte balance, temperature regulation, evaporative and non-evaporative heat loss, energy metabolism, mineral and vitamin requirements and cardiovascular physiology have received major emphasis. Also the environmental effects on the process of growth, lactation, reproduction, fibre growth, egg production and aging are covered.
Adaptability and processes of engineering and management to alleviate or modify effects of heat and cold receive attention. Finally the effects of heat and cold on animal diseases are discussed.

In Part II a current evaluation is presented on the effects of light, high altitude, noise, electric, magnetic and electro-magnetic fields, ionization, gravity and chemical air pollution on animals.

Harold D. Johnson
Editor
Progress in Animal Biometeorology

DEFINITION AND CLASSIFICATION OF BIOMETEOROLOGY

Various definitions and classifications have been suggested since the Foundation of the International Society of Biometeorology (I.S.B.) in 1956. Biometeorology can be defined briefly as the study of the biological effects of weather and climate on living organisms, plants, animals and man and of their direct physico-chemical environment.

In 1956, during the first Biometeorological Symposium of the International Society of Biometeorology, at UNESCO Headquarters in Paris, the following definition was accepted by the members of ISB:

"Biometeorology comprises the study of the direct and indirect interrelations between the geophysical and geochemical environment of the atmosphere and living organisms, plants, animals and man. The term "environment" is broadly conceived and includes micro, macro and cosmic environments in so far as they affect the earth's atmosphere and the diverse physical and chemical factors which comprise these environments. The investigations in these disciplines should be conducted in nature or in the laboratory under as rigidly controlled conditions as possible to describe measurable and reproducible physical, chemical and biological factors which shows a sufficiently high statistical correlation with measurable physiological and pathological processes to suggest a valid cause and effect relationship between organism and environment."

Since 1969 the members of ISB were not satisfied anymore with this very long definition and the greatest majority of the members from 53 countries decided to use for official purposes the short definition only, as indicated above.

However, this short definition is insufficient if the entire field of biometeorological activities should be sharply defined. For this reason another slightly longer definition was suggested indicating the various types of phenomena studied by biometeorology. According to this definition biometeorology comprises the study of the direct and indirect effects (of an irregular, fluctuating or rhythmic nature) of the physical, chemical and physico-chemical micro- and macro-environments of both the earth atmosphere, and of similar extra-terrestrial environments, on physico-chemical systems in general and on living organisms (plants, animals and man) in particular.

The biometeorological sciences have been divided into six main groups: phytological, zoological, human, cosmic, space and paleo-biometeorology.

1. PHYTOLOGICAL BIOMETEOROLOGY studies the influence of weather and climate on the development and distribution of plants for general phytological, agricultural and forestry purposes, and also the effects on small plant organisms responsible for the development of piant, animal and human diseases.

2. ZOOLOGICAL BIOMETEOROLOGY studies the influence of weather and climate on animals in general, on milk production, reproduction and coating of cattle, etc. (*Veterinary biometeorology*), on egg-laying rhythm, etc., of birds (*Avian biometeorology*), but also the effect of weather and climate on insects and terrestrial arthropoda, their physiological activities and development and the outbreak of plant, animal and human diseases as a result of these developments, such as the abundant development of locusts, malaria mosquitoes, etc. (*Entomological biometeorology*).
3. HUMAN BIOMETEOROLOGY studies the influence of weather and climate on man. It is usually divided into five important sections:
   a. PHYSIOLOGICAL BIOMETEOROLOGY: the influence of weather and climate on physiological processes in normal healthy man;
   b. SOCIAL BIOMETEOROLOGY: use of favorable meteorological factors for the treatment of large population groups as preventive or curative measures;
   c. PATHOLOGICAL BIOMETEOROLOGY: the influence of weather and climate on the various physiological and pathological phenomena associated with the diseases of man, the period of outbreak, intensity and geographical distribution;
   d. ARCHITECTURAL AND URBAN BIOMETEOROLOGY: the influence of the microclimates in houses and cities on the health of man and the effect of architectural constructions and town planning on these microclimates;
   e. NAUTICAL BIOMETEOROLOGY: the influence of weather and climate on animals and man living on moving ships.

Of special importance for the general practitioner are the results obtained in *Acclimatization biometeorology* (a branch of physiological biometeorology which studies the various physiological processes involved during adaptation to extreme conditions of temperature, humidity, etc., e.g., in the tropics or polar regions); *Air pollution biometeorology* (influence of pollen, spores, dust and chemical air pollution on man) and the various methods applied in *Climatotherapy*.

4. COSMIC BIOMETEOROLOGY studies the influence of extraterrestrial factors such as variations in solar activity, cosmic radiation, etc., on living organisms on earth.
5. SPACE BIOMETEOROLOGY is closely related to space research. It studies the microclimatic conditions in space vehicles and the biological effects (both observed and expected) of extra terrestrial physical environments on man and animals on the moon and the planets.
6. PALEO-BIOMETEOROLOGY studies the influence of climatic conditions in the past (millions of years ago) on the evolutional development and geographic distribution of plants, animals and man.

Table 1. Classification of HUMAN BIOMETEOROLOGY

1. PHYSIOLOGICAL BIOMETEOROLOGY
   a. General physiological biometeorology
   b. Geographical biometeorology
   c. Ethnological biometeorology
   d. Acclimatization biometeorology
2. SOCIAL BIOMETEOROLOGY
   a. General social biometeorology
   b. Psychological biometeorology (including aestheto-biometeorology)
   c. Archaelogical biometeorology
3. PATHOLOGICAL BIOMETEOROLOGY
   a. General pathological biometeorology
   b. Meteorological pathology
   c. Climatological pathology
   d. Air pollution biometeorology
      (1) Pollution with organic particles (pollen, spores, etc.)
      (2) Pollution with inorganic particles (dust, etc.)
      (3) Chemical pollution
      (4) Aerosol biometeorology
   e. Geographical climatopathology
   f. Climatotherapy (or therapeutic biometeorology)
      (1) General climatotherapy
      (2) Thalassotherapy (climatological)
      (3) Heliotherapy
      (4) Thermotherapy
      (5) Aerosol (and ionization) therapy
      (6) Socio-climatotherapy
      (7) Climatic health resorts
      (8) Climatic chamber treatments
   g. Pharmacological biometeorology
4. ARCHITECTURAL AND URBAN BIOMETEOROLOGY
   a. General urban biometeorology
   b. Architectural biometeorology
   c. Sanatorium biometeorology
5. NAUTICAL BIOMETEOROLOGY
   a. General nautical biometeorology
   b. Cargo biometeorology

S. W. TROMP

CHAPTER 1

EFFECTS OF TEMPERATURE ON ANIMALS
(Including Effects of Humidity, Radiation and Wind)

SECTION 1: CELLULAR METABOLISM, ENZYMES AND OTHER CELLULAR CHANGES
- (a) Effects of Heat on Cellular Metabolism: Enzymes and Other Cellular Changes (R. R. J. CHAFFEE)
- (b) Effects of Cold on Cellular Metabolism: Enzymes and Other Cellular Changes (E. J. MASORO)

SECTION 2: ENDOCRINE FUNCTIONS IN DOMESTIC ANIMALS
- (a) Effects of Heat
  - (1) Effects of Heat on Endocrine Responses of Domestic and Laboratory Animals (H. D. JOHNSON)
- (b) Effects of Cold
  - (1) Effects of Cold on Endocrine Responses of Ruminants and Other Domestic Animals (A. J. F. WEBSTER)
  - (2) Effects of Acute and Chronic Exposure to Cold on the Endocrine System of Laboratory Animals (V. M. PETROVIC)

SECTION 3: WATER AND ELECTROLYTE BALANCE
- (a) Water Metabolism in Sheep in Relation to Heat and Cold (W. V. MACFARLANE)
- (b) Water and Salt Metabolism in the Camel (G. M. C. MALOIY)
- (c) Water Metabolism in Rodents (R. E. MACMILLEN and D. E. GRUBBS)

SECTION 4: THERMOREGULATORY MECHANISMS
- (a) Temperature Regulation in Cattle (J. BLIGH)
- (b) Temperature Regulation in Sheep (R. O. RAWSON)
- (c) Temperature Regulation in Pigs (D. L. INGRAM) (see Chapter 1, Sect. 5d and 18c(2))
- (d) Temperature Regulation in Marine Mammals (G. C. WHITTOW)
- (e) Temperature Regulation in Fish (F. E. J. FRY)
- (f) Temperature Regulation in Birds (Effects of Cold on Metabolism and Endocrinal Reactions in Birds) (J. D. RISING)
- (g) Thermoregulation in Wild Ungulates (M. K. YOUSEF)

P.H.B. = Progress in Human Biometeorology (Ed. S. W. TROMP)

SECTION 5: EVAPORATIVE (both Respiratory and Skin) TEMPERATURE REGULATION IN HOT AND COLD ENVIRONMENTS

(a) Interactions between Respiratory and Thermoregulatory Systems of Domestic Animals in Hot Environments (J. R. S. HALES)
(b) Evaporative Temperature Regulation in Ruminants (D. ROBERTSHAW)
(c) Evaporative Temperature Regulation in Sheep (K. G. JOHNSON)
(d) Evaporative Temperature Regulation in Pigs (D. L. INGRAM)
(e) Evaporative Temperature Regulation in Birds (M. VAN KAMPEN)

SECTION 6: POSSIBLE MECHANISMS INVOLVED IN DEATH IN LABORATORY ANIMALS DUE TO THERMAL STRESSES (HEAT AND COLD) (G. L. WRIGHT)

SECTION 7: ENVIRONMENTAL TEMPERATURES AND FEEDING BEHAVIOUR (C. L. HAMILTON)

SECTION 8: ENERGY METABOLISM (INCLUDING EXERCISE AND TEMPERATURE RELATED ENDOCRINAL EFFECTS)

(a) Effects of Heat
  (1) Effects of Heat and Exercise on Energy Metabolism of Cattle (J. A. McLEAN)
  (2) Effects of Heat on Energy Metabolism of Sheep (K. L. KNOX)
  (3) Effects of Heat on Energy Metabolism and Related Endocrine Functions in Rodents (M. K. YOUSEF)
  (4) Effects of Heat on Energy Metabolism in Birds (H. S. SIEGEL)
(b) Effects of Cold
  (1) Effects of Cold on Energy Metabolism of Cattle (G. E. THOMPSON)
  (2) Effects of Cold on Energy Metabolism of Sheep (A. J. F. WEBSTER)
  (3) Effects of Heat and Cold on Energy Metabolism of the Pig (L. E. MOUNT)
  (4) Effects of Cold and Exercise on Energy Metabolism of Small Mammals (L. JANSKÝ)
  (5) Effects of Cold on Energy Metabolism in Birds (H. S. SIEGEL)

SECTION 9: ELECTROLYTE METABOLISM

Effects of Cold on Electrolyte Metabolism in Rodents and Laboratory Animals (W. S. PLATNER)

SECTION 10: VITAMIN METABOLISM

Effects of Heat on Vitamin Metabolism (M. L. SCOTT)

SECTION 11: CARDIOVASCULAR PHYSIOLOGY

(a) Effects of Heat
  (1) The Redistribution of Cardiac Output in Animals During Heat Stress (J. R. S. HALES)

Effects of Heat on Cellular Metabolism: Enzymes and Other Cellular Changes

by

R. R. J. Chaffee

INTRODUCTION

In nature homeotherms face numerous environmental variables which challenge their abilities to maintain a constant body temperature. In such situations most mammals respond by the activation of compensatory physiological thermoregulatory mechanisms and by seeking a more sheltered environment. In studies concerning the changes which occur in response to a given environmental temperature, the animal is placed in a controlled laboratory situation where all other environmental variables are held as nearly constant as possible and behavioral avoidance is prevented. Although there are many different specific thermoregulatory responses to changes in environmental temperature in mammals, these can be classified into three major divisions of physiological modulation: (1) heat conservation, (2) heat loss, and (3) heat gain. Classically physiological changes which modify heat conservation or loss have been considered "physical regulation" because chemical (metabolic) change is minimal: responses which result in the modification of heat gain have been considered "chemical regulation" because they primarily involve metabolic (thermogenic) changes (Rubner, 1902).

One of the advantages of a high metabolic rate (MR) in a cool environment is that it yields heat for the maintenance of body temperature. On the other hand, the generation of metabolic heat in a hot environment can present a problem in that it adds an internal heat load to the existing environmental heat load. This may force the animal to activate heat loss mechanisms, which in itself requires additional metabolic thermogenesis. Therefore, since nature tends to be conservative, it is not surprising that in many animals acclimation to heat involves a suppression of the metabolism of many tissues not directly involved with heat loss activities (Chaffee and Roberts, 1971) and the net result is a lowering of the animal's MR (Collins and Weiner, 1968).

In this Section we will focus our attention on some of the cellular metabolic changes that accompany, and may be the basis for, the heat-induced depression of the MR. Furthermore, because many studies have involved comparisons of not only heat acclimated and control animals, but also cold acclimated animals as a means of magnifiying temperature effects, and because many studies made on cellular metabolic control mechanisms in cold acclimated animals may shed light on what is occurring in heat acclimation, certain fragments of such information will be woven directly into the fabric of this discussion.

EFFECTS OF CHEMICAL THERMOSUPPRESSION ON TEMPERATURE TOLERANCE (SURVIVAL)

In heat acclimation there is evidence of decreased skeletal muscle contractile thermogenesis (Collins and Weiner, 1968), and in many species there is a decline in mass and oxidative enzymatic activity of certain organs and tissues which has been referred to as chemical thermosuppression (CTS) (Cassuto and Chaffee, 1966; Chaffee and Roberts, 1971).

Although heat acclimated animals have a lower tolerance to cold than controls, they survive extreme heat better (Cassuto and Chaffee, 1966; Collins and Weiner, 1968). This increased heat tolerance has been attributed at least in part to CTS (Chaffee and Roberts, 1971).

Weeks of heat exposure are required before CTS develops fully (Chaffee and Roberts, 1971) and it is now generally accepted that CTS depends not only on initial rapid responses associated with the sympathetic adrenomedullary system but also on slower changes involving the thyroid and adrenal cortex, among other endocrines (Fregly, 1968; Yousef and Johnson, 1968; Collins and Weiner, 1968; Cassuto and Amit, 1968; Chaffee and Roberts, 1971).

GENERAL CONSIDERATIONS IN CELLULAR METABOLIC CONTROL AND THERMOGENESIS

In all but a few mammalian cells most of the heat generated comes from oxidative reactions in the electron transport system (ETS) of the mitochondria and the exothermic expenditure of ATP. Since it is essential in so many intracellular processes, it is not surprising that the rates of production and expenditure of ATP are closely correlated with the overall metabolic activity of the cell.

Most of the biochemical research in temperature acclimation has involved two major approaches in the investigation of cellular changes which could be of importance in explaining changes in the MR. One approach involves assays of the activity of enzymes associated with intermediary metabolism. The other involves studies of possible biochemical mechanisms which could alter cellular levels of ADP by influencing: the rate of expenditure of ATP and the efficiency of its production, and ultimately the rate of flow of electrons through the ETS.

Assays of the activity of oxidative enzymes have been made in an attempt to learn something about the THERMOGENIC CAPACITY of tissues. These studies are based on the concept that the level of activity or concentration of enzymes in a cell is representative of the amount of use of those enzymes (Smith and Hoijer, 1962; Chaffee and Roberts, 1971). Thus if the in vitro measurements on a tissue preparation from a heat acclimated animal indicate that the activity of a number of its oxidative enzymes is below control values, it is assumed that these enzymes are doing less catalytically in vivo and that the maximal thermogenic capacity of that tissue in the heat acclimated animal is thus limited.

Although the levels of certain enzymes may be indicative of the METABOLIC POTENTIAL of a cell, under most conditions other factors CONTROL THE METABOLIC RATE of the cell. This makes sense since rapid changes in cellular thermogenesis are frequently necessary, and enzymatic syntheses and degradations are comparatively slow processes. Thus much attention has been given to the exploration of other biochemical mechanisms which might be involved in the control of cellular respiration which could respond more quickly to the thermogenic requirements of the animal.

It has long been known from in vitro studies that the respiration rate of intact mitochondria in the presence of adequate oxygen and substrates is limited by the availability of ADP and inorganic phosphate (P_i) (Lardy and Wellman, 1952) and since there is an adequate supply of P_i in vivo, it is thought that ADP is the rate-limiting factor. In shivering, where sarcomeric ATPase is activated to catalyze the reaction ATP $\rightarrow$ ADP + P_i, the ADP-stimulated muscle respiration is easy to understand. However, how ADP levels may be increased to account for the elevated nonshivering thermogenesis (NST) in cold acclimation, and decreased to restrict cellular respiration in heat acclimation, is not well understood.

Recently some excellent biochemical models have been proposed to explain how ADP replenishment from ATP in the cell may be controlled. Some of these deal with the activity of phosphate transfer reactions which would be thermogenic primarily because they supply ADP, while others deal with changes in phosphate hydrolytic reactions (ATPases) which would be calorigenic not only because of the formation of ADP, but because of the heat of hydrolysis of the terminal high energy phosphate bond of ATP.

One important factor which influences the rate of both dephosphorylation and hydrolysis of ATP is 3′, 5′-cyclic adenosine monophosphate (CAMP) which is derived from ATP under the catalytic activation of the enzyme adenyl cyclase. CAMP modulates the actions of a number of hormones which affect the metabolism of carbohydrates, fats and proteins (Sutherland, 1972), and stimulates the plasma membrane sodium-potassium pump (Na^+/K^+ ATPase) (Horwitz, 1974). In these ways then, among others, it affects the rate of utilization of ATP and the replenishment of ADP and thus the metabolic rate of the cell.

Interestingly, the levels of CAMP have been shown to increase in certain tissues in response to epinephrine and norepinephrine (NE), presumably due to an activation of adenyl cyclase (Sutherland, 1972); and since these catecholamines have long been associated with rapid thermogenic responses, considerable research is being conducted to learn more about the precise mechanisms by which they produce their effects in temperature acclimated animals (Therriault et al., 1969; Horwitz and Smith, 1972; Horwitz, 1974). One question now being investigated is whether the increased rate of ADP replenishment is possibly due also to a direct depolarization of the cell membrane by NE which would increase the rate of entrance of sodium into the cell and activate the Na^+/K^+ ATPase (Horwitz, 1974).

Another mechanism which could be significant in the increased replenishment of ADP and thus stimulate thermogenesis is the increased synthesis and degradation of complex molecules which involve the conversion of many ATP molecules to ADP. Of particular interest in this respect is that there is evidence that protein molecules may be built up and degraded at a more rapid rate in cold acclimation (Chaffee etal., 1968a; Chaffee, 1968; Yousef and Chaffee, 1970; Himms-Hagen et al., 1972) and at a slower rate in heat acclimation (Chaffee, 1968; Chaffee et al., 1969a), and that in protein synthesis ATP is converted either directly to ADP or to AMP which yields ADP by the reaction $ATP + AMP \rightleftarrows 2\,ADP$ (catalyzed by adenylate kinase). Since thyroid activity is elevated in the cold (Hoch, 1971) and decreased in the heat (Collins and Weiner, 1968) and protein synthesis is stimulated by thyroid hormone (Sokoloff et al., 1964; Hoch, 1971), the thyroid could regulate this cellular thermogenic mechanism according to the thermogenic requirements of the animal.

The above mentioned increase in protein (i.e., amino acid) recycling in the cold would be of particular thermogenic significance if it involved preferential synthesis of thermogenesis-promoting enzymes or enzyme systems at the expense of amino acids from nonthermogenic cellular proteins. It is known that the specific activity (activity per mg protein) of many oxidative enzymes increases in the cold (Chaffee and Roberts, 1971) and these effects may be associated with increased thyroid activity (ibid). Also of interest is the fact that the plasma membrane Na^+/K^+ ATPase (a protein), which might serve as a major ADP replenishing system in NST, is increased in specific activity under the influence of the thyroid hormone T_3 (Ismail-Beigi and Edelman, 1971). Thus there may be any number of such preferential syntheses at the expense of other proteins in the cold acclimation-induced biochemical transformation of the cell to better serve the thermogenic needs of the animal. Conversely, in the heat if the degradation of these thermogenic proteins occurs without the resynthesis of others, it would result in a decline of the thermogenic activity of the cell and it could contribute to the well known negative nitrogen balance of heat exposure. Furthermore, in light of the increased consumption of food (which is partially protein) in the cold and the characteristic anorexia in the heat, the rate of recycling of proteins could possibly be influenced by the availability of amino acids.

Some researchers have proposed that in addition to changes in ADP replenishment, there are also changes in the mitochondrial respiration due to variations in levels of factors which alter the coupling between oxidative and phosphorylative reactions in the mitochondria (Smith, 1960; Hoch, 1971). A change in the efficiency of this coupling would not only affect the rate of electron flow in the ETS, and thus the oxidative heat evolved, but also would affect the rate of removal of ADP in the synthesis of ATP. Since injections of thyroxin can produce changes in mitochondrial oxidative phosphorylation efficiency, and it has long been known that the thyroid activity is a prime factor in the regulation of the basal metabolic rate, there is considerable interest in learning more about the precise mechanisms involved (Hoch, 1971).

We will have more to say about both extramitochondrial factors which modify intracellular levels of ADP, and mitochondrial oxidative phosphorylation efficiency in our consideration of the effects of temperature acclimation on specific tissues and organs.

Although enzyme levels and release of electron flow through the ETS are important considerations, one must always keep in mind that availability of substrate, not only from intracellular stores, but also from extracellular sources, may not always be adequate to maintain high levels of cellular respiration. Thus thermogenesis in one organ can be influenced by activities of other organs or tissues such as the liver and white adipose tissue (WAT) that dispense circulating carbohydrates and fatty acid fuels respectively. Therefore, although we will be primarily concerned with changes which modify the thermogenic output of tissues, we will also consider some tissue changes which may alter the systemic fuel supply.

BROWN ADIPOSE TISSUE

Although the role of BAT in NST has been studied in great detail in cold acclimated rats (Smith and Horwitz, 1969; Himms-Hagen, 1972; Horwitz, 1974), only a few studies have been made on this tissue in heat acclimation (Chaffee and Roberts, 1971). In the first study on the effects of heat exposure on BAT it was found that significant decreases occur in its mass and oxidative enzyme capacity (Chaffee and Smith, 1963). Similar changes have since been reported in other rodents and species of a variety of mammalian orders (Chaffee et al., 1969b). In some of these animals if the environmental heat is sufficiently high and the time of exposure long enough, it was not possible to find any distinguishable BAT at all (Chaffee, 1968).

The color of the BAT from heat acclimated animals is a much more pale brown than that of the controls which may be due to infiltration by WAT and a decline in such colored substances as flavoproteins and electron transport elements, since there are far less mitochondria and a lower specific activity of oxidative enzymes in those present (Chaffee and Roberts, 1971). In general, changes in the BAT in heat acclimation are opposite to those of cold acclimation and this antithetical effect of heat vs. cold acclimation on the chemical potential of BAT is particularly striking in *Macaca mulatta* (Chaffee et al., 1974). When these monkeys are kept for six months at two extremes in environmental temperature, 35 ± 1°C and 5 ± 1°C, the axillary BAT of the heat exposed animals is less than 1/3 wet weight of that of the cold acclimated animal and has less than one tenth the protein content. Its total calculated enzymatic activity associated with succinate is 1/50th, with α ketogluterate 1/16th, with β-hydroxybutyrate 1/19th, and with α glycerophosphate 1/60th that of the axillary BAT of the animal kept at 5°C. In control monkeys kept at a temperature of 24 ± 1°C the axillary BAT is somewhat intermediate with respect to weight, protein content, and oxidative enzymatic activity.

In considering the importance of BAT as a thermogenic tissue, its strategic anatomical distribution for efficient heat production and rich content of oxidative enzymes is impressive (Smith and Horwitz, 1969). A model has recently been developed which gives considerable insight into how norepinephrine might influence the respiratory rate of this tissue (Horwitz, 1974) and thus cause it to contribute whatever thermogenic potential it might have directly to the well known NE-induced increase in the MR of cold acclimated animals described by Hsieh and Carlson (1957). NE stimulates BAT respiration by causing an increase in the cellular availability of ADP in two different ways. It depolarizes the BAT cell membrane (Girardier et al., 1968; Horwitz et al., 1969) and thus changes its sodium permeability, and it activates adenyl cyclase so that there is an increase in CAMP. Both these effects of NE cause an activation of Na^+/K^+ ATPase (Horwitz, 1974). The resulting increase in ADP formation coupled with a NE-induced lipolysis in BAT and WAT (Himms-Hagen, 1972; Horwitz, 1974), which increases intracellular availability as well as the circulating levels of fatty acids, suggests a mechanism whereby the high BAT respiratory rate induced by cold stress or NE stimulation may be sustained. Adding to this cellular metabolic release is the fact that fatty acids can directly loosen the oxidative phosphorylation of BAT mitochondria (Himms-Hagen, 1972).

To date no comparable studies of catecholamines have been done on BAT in heat acclimation. It is known however that injected NE has no effect on the MR of diverse species of heat acclimated animals (Cassuto et al., 1970; Chaffee and Allen, 1973), and there is considerable evidence that fats are not mobilized as an energy source (Collins and Weiner, 1968). On the basis of the many previously mentioned antithetical responses of BAT in heat versus cold acclimation, one might predict that the membrane potential is less subject to depolarization by NE and that ionic permeability and Na^+/K^+ ATPase activity is lower. However, to date no one has investigated these factors. Furthermore, no one has investigated the lipase activity of the BAT of heat acclimated animals, but in light of the fact that β-hydroxybutyric acid dehydrogenase and α glycerophosphate dehydrogenase as well as TCA cycle enzymes show decreased activity and the BAT is engorged with white lipid material, it is probable that lipolysis as well as the oxidation of fatty acids and other substrates is suppressed (Chaffee and Roberts, 1971).

In thinking about the BAT as a principal biothermic effector in temperature acclimation one must always keep in mind the fact that the mass of this tissue in comparison to the body mass is very low in most animals (Jansky, 1973). Therefore, just as it is difficult to defend the hypothesis that an increase in the thermogenic activity of this tissue directly contributes to survival in the cold, it is difficult to defend the hypothesis that a decline in its thermogenic output could account for the depressed MR of heat acclimated animals (Chaffee and Roberts, 1971).

It has been suggested that perhaps in cold acclimation BAT plays an important controlling influence on the metabolic response to sympathetic stimulation by secreting an endocrine factor (or factors) which modifies the ability of other tissues such as skeletal muscle to respond calorigenically to catecholamines (Himms-Hagen, 1969; Leduc and Revist, 1969). Conversely, in the heat acclimated animal the decrease in the amount of BAT and perhaps any secretions it produces, might thus result in a reduction of the metabolic responsive capacity of other tissues such as skeletal muscle, and thus be of importance in CTS. Obviously, the question of whether or not BAT does produce such a secretory factor, is a very important one which demands further investigation.

THE LIVER

In spite of the fact that the liver represents only a small percentage of the body weight it has been estimated to account for more than 20% of the MR in the rat (Jansky, 1966; Hoch, 1971). Thus a decline in its metabolism in heat acclimation could be a significant factor in lowering the internal heat load of the animal.

OXIDATIVE ENZYME LEVELS

There are striking decreases in the activity of oxidative enzymes in the liver of rodents (Cassuto and Chaffee, 1966; Chaffee and Roberts, 1971) which are opposite to those which occur in the cold (Smith and Hoijer, 1962; Hoch, 1971). Thus it is generally agreed that in some animals there are temperature induced changes in the metabolic potential of the liver. On the other hand there is considerable lack of agreement concerning which biochemical alterations are most important in its metabolic regulation in temperature acclimation.

MITOCHONDRIAL OXIDATIVE PHOSPHORYLATION EFFICIENCY

It has been postulated that in temperature acclimation there is a change in the high energy bond phosphate formed per gram-atom of oxygen consumed (P:O ratio) in the mitochondria and that this alters cellular thermogenesis (Smith and Fairhurst, 1958). In studies of liver mitochondria from cold acclimated laboratory rodents some investigators found no change in the P:O ratio (Chaffee, 1957; Chaffee et al., 1961) whereas others found a significant decrease (Smith and Fairhurst, 1958; Pangos et al., 1958). This discrepancy was soon resolved when it was found that if the mitochondria were washed thoroughly the P:O ratios were not different and it was concluded that the extra washing might remove some factor which increases the quantity of electrons which bypass phosphorylating sites of the mitochondrial oxidative-phosphorylating system (Smith, 1960). It has been suggested that this factor may be of thyroid origin (Smith, 1960). In defense of this hypothesis is the fact that circulating thyroid hormone levels increase in the cold and injections of thyroid hormones into a euthyroid animal cause a demonstrable lowering of liver mitochondrial P:O ratios (Hoch, 1971). However, the quantities which must be injected to cause this lowering far exceeds physiological levels of circulating thyroid hormones, and this fact has somewhat dampened interest in the mitochondrial uncoupling hypothesis in cold acclimation (ibid; Chaffee and Roberts, 1971). Nevertheless, there is the possibility that a natural accumulation of this thyroid factor occurs in the mitochondria over a period of time in cold exposed animals and that it produces an in vivo uncoupling of oxidative phosphorylation which cannot be demonstrated in vitro because there is a given percentage loss of it in the isolation washing procedure. In the artificial injection situation the mitochondria may become superloaded with the thyroid derivative so that in spite of the percentage loss in the isolation procedure, there is still enough of it left in the mitochondria to loosen oxidative phosphorylation coupling.

In heat acclimation where thyroid activity is depressed one might expect an increase in the P:O ratio, but again there is no demonstrable difference in this ratio between mitochondria of heat acclimated and the control animals (Cassuto and Chaffee, 1966).

RESPIRATORY CONTROL

In mitochondria from heat acclimated hamsters, there is a greater release of respiration caused by the addition of ADP (i.e., increased respiratory control) than in the mitochondria of controls (Cassuto and Chaffee, 1966). Here again this is different from the case in cold acclimated rodents where there is either no change, or a decrease in respiratory control (Chaffee, 1957; Lianides and Beyer, 1960).

A notable phenomenon in heat acclimation is the effect of pharmacological uncouplers on respiratory rates of liver mitochondria. After 3 to 4 weeks of exposure to heat the responses of the liver mitochondria to DNP and dicoumarol were 50% greater than that in control mitochondria (Cassuto, 1968). This is further evidence of a greater respiratory control. This sensitivity to uncouplers was also exhibited in in vivo studies. Cassuto (1968) concludes that some modification of the respiratory chain develops in the liver mitochondria of heat acclimated animals such that uncouplers act to relieve the mitochondria from some kind of inhibition which does not affect oxidative phosphorylation efficiency. Thus under conditions in which tighter control of energy metabolism is demanded, a modification of some coupling factors in the mitochondria takes place and therefore a higher percentage of the flow of electrons is channeled through an oxidative phosphorylating pathway rather than a non-phosphorylating by-pass which decreases energy leaks and reduces the total heat output (Cassuto, 1968).

CALORIGENIC SHUNTING SYSTEMS

Another closely related hypothesis to explain the increased thermogenesis by liver mitochondria in vivo in cold acclimation concerns the calorigenic shunting systems of microsomal NADPH and NADH-cytochrome c reductase, which increase in liver microsomes in cold acclimated rodents (Reynafarje and Chaffee, 1960). It has been hypothesized that these microsomal shunting systems would be calorigenic in that they are thought to involve a transfer of hydrogen reducing equivalents from microsomal NADPH and NADH to cytochrome c in the mitochondrial ETS via certain non-phosphorylating pathways (Reynafarje and Chaffee, 1960; Beyer, 1963). This transfer, however, would be dependent on the presence of an as yet unidentified factor which would make possible electron transfer from the endoplasmic reticulum to the mitochondria. There are two possible nonstationary mitochondrial substances which are electron acceptors whose reoxidation would not involve the first phosphorylative step: cytochrome c and coenzyme Q. Both the former (Klain, 1963; Depocas, 1966) and the latter (Beyer et al., 1962) have been found to increase in the liver of cold acclimated rats.

In heat acclimated rodents there is a striking depression in the microsomal NADPH-cytochrome c reductase (Cassuto and Chaffee, 1963), and a decline in the cytochrome c (Chaffee, unpublished). Whether there is a corresponding decrease in the level of coenzyme Q in heat acclimated animals remains to be determined. If indeed this calorigenic shunting system does account for a significant thermogenesis, a decline in its component parts and thus its thermogenic activity in heat acclimation could be an asset in reducing the internal heat load.

LIVER ADP REPLENISHING SYSTEMS

There is evidence that increased thyroid activity can increase the activity of liver

Na^+/K^+ ATPase (Edelman and Ismail-Beigi, 1972), which is presumably what occurs in cold acclimation. Conversely, since there is so much evidence that thyroid activity is depressed in heat acclimation, there may well be a decline in the activity of this ATPase so that ADP replenishment by this mechanism would be reduced.

Since the rate of protein synthesis is dependent on thyroid activity (Hoch, 1971), one might predict that there would be a depressed protein synthesis in heat acclimation. One piece of evidence indicating this is that liver ribosomes from heat acclimated animals incorporate amino acids into protein at a slower rate than do ribosomes from controls (Chaffee, 1968). Furthermore, there is a lowered whole body protein turnover rate in heat acclimated animals (Chaffee et al., 1969a) which is opposite to the increase that occurs in whole body and liver protein recycling in cold acclimation (Chaffee, 1968; Yousef and Chaffee, 1970). Himms-Hagen et al. (1972) also have reported evidence of increased protein recycling in cold acclimated rats. They found increased recycling in mitochondria from the BAT and skeletal muscle, but as was the case in studies by Lusena and Depocas (1967), there was no change in the recycling rate in liver mitochondria. Therefore, the faster protein recycling in the liver in cold acclimated rats (Chaffee, 1968; Yousef and Chaffee, 1970) must involve extramitochondrial proteins.

LIVER GLYCOLYSIS

In addition to the depression of the thyroid there is a decreased adrenal cortical output of glucocorticoids in the heat (Yousef and Johnson, 1968; Collins and Weiner, 1968). Thus it is not surprising that there is a depressed rate of glucose production and glycogen breakdown (Cassuto, 1967). In keeping with the lowered conversion of glucose-6-phosphate to glucose is the lowered glucose-6-phosphatase activity (Cassuto et al., 1970). These findings are in accord with the fact that the liver supplies other tissues with glucose, and fuel requirements are lowered in the heat. There is also evidence that the liver itself may utilize less glucose in the heat since a metabolite of glucose degradation such as phospho-enol-pyruvate is in lower concentration in heat acclimation even though its catabolism via the pyruvate-Krebs cycle oxidation pathway is much reduced (Cassuto and Chaffee, 1966). Thus in some ways these liver glycolytic responses are opposite to those found in cold acclimated rodents (Hannon, 1963a; Morrison et al., 1966). But whether the development of CTS represents a decline in whole body gluconeogenesis has not been determined, whereas the development of NST in cold acclimation is apparently not related to changes in whole body gluconeogenesis (Pennor and Himms-Hagen, 1968).

In other species studied, the liver seems to show few if any indications of CTS in heat acclimation. For example in the tree shrew (*Tupaia chinensis*) and in *Macaca mulatta* we failed to find evidence of hepatic CTS, and in the squirrel monkey *Saimiri sciureus* the only indication was a decrease in liver weight of about 20% (Chaffee and Roberts, 1971). On the other hand, it is possible that with further studies on different enzyme systems indications of CTS in the liver will be found in these mammals as in the rodents.

WHITE ADIPOSE TISSUE

Little is known concerning the effects of heat acclimation of WAT. In cold acclimation this tissue is the most important fuel supply in mammals (Masoro, 1973) and lipolysis is increased whereas in heat acclimation fats are not mobilized as a major energy source (Collins and Weiner, 1968; Hluszko et al., 1969). In the mobilization of fats, triglycerides are hydrolyzed yielding blood-borne fatty acids and glycerol. Thus in the heat the lowered mobilization of fats would not only lower blood-borne fatty acids, but also glycerol. Glycerol passes readily into cells and can be converted into α glycerophosphate. In many tissues of cold acclimated animals α glycerophosphate dehydrogenase is increased (Chaffee and Roberts, 1971) and there is a direct correlation between thyroxin levels and mitochondrial α glycerophosphate dehydrogenase in many tissues of the rat (Ruegemer et al., 1964; Lee and Lardy, 1965). Since thyroid function is depressed it is not surprising that in many tissues the level of α glycerophosphate dehydrogenase is lowered in heat acclimation (Arine et al., 1973). The importance of this enzyme in linking cytoplasmic glycolytic reductive reactions involving NADH to mitochondrial oxidation has been established (Estabrook and Sacktor, 1958) and it has been hypothesized that a change in the levels of α glycerophosphate dehydrogenase might be a significant factor in thermogenesis in temperature acclimated animals (Chaffee et al., 1964).

Another factor favoring the thermogenic role of α glycerophosphate dehydrogenase is that this enzyme in the mitochondria may catalyze a calorigenic shunting reaction involving the direct delivery of electrons from NADH to an acceptor in the ETS which represents a by-passing of a phosphorylating step (Estabrook and Sacktor, 1958; Bucher and Kingenberg, 1958).

THE KIDNEY

In a great many animals the kidney shows a decline in mass as a response to prolonged heat exposure (Chaffee and Roberts, 1971). Here again this response is antithetical to the increase in kidney mass which occurs in animals during cold acclimation (Chaffee and Roberts, 1971). The decline in kidney mass in the heat varies in different animals between 10 and 35%. In rodents there is considerable evidence that the drop in kidney mass is accompanied not only by a lower total enzymatic activity per kidney but also by a decline in the specific enzymatic activity associated with a number of substrates, for example succinate (Chaffee et al., 1963), glutamate (Chaffee et al., 1969b), and α glycerophosphate (Arine et al., 1973). Also there is a significant decline in NADH-cytochrome c reductase (Chaffee et al., 1963). However, as is the case with BAT in many animals, the percent organ-to-body weight is so low that it would seem that a change in the kidney metabolic rate could not have a very significant direct effect in accounting for the observed decline in the MR of the animal during heat acclimation. On the other hand, some have argued (Arine et al., 1973) that since the blood flow to the kidney is about 25% of the cardiac output, and the respiration rate per mg protein of the kidney cortex greatly exceeds that of liver, it may be a small but significant thermogenic source. If the kidney is producing considerable heat, it is possible that it may warm the suprarenals and thus indirectly affect the MR of the animal. Thus if renal thermogenesis decreases in the heat, this might result in a slight decrease in suprarenal activity. Whether this is true remains to be demonstrated.

THE HEART

In most rodents heat acclimation results in a decline in heart mass which is yet another response opposite to that which occurs in the cold (Chaffee et al., 1969b). This smaller heart in heat acclimation as opposed to an enlarged heart in the cold, has also been found in the shrew *Suncus murinus*, the protoprimate, *Tupaia chinensis*, the squirrel monkey *Saimiri sciureus*, and in rhesus, *Macaca mulatta* (Chaffee et al., 1969b). To date most of the enzymatic studies on the effects of heat acclimation on the heart have involved succinoxidase which some consider representative of mitochondrial oxidative enzyme activity. Succinoxidase shows no change in specific activity, but because of the reduction of heart mass, the enzyme shows a decrease in total activity which may indicate a decline in total cardiac thermogenic potential. Of more significance however, is the fact that the MR is lower in the heat, and the heart rate under most conditions correlates well with the MR. Thus if the heart is slower, its sarcomeric ATPase is activated less frequently. Hence, there would be less ADP available to sustain metabolism in the heart mitochondria of the heat acclimated animals. Furthermore, there would be a slower overall rate of ATP hydrolysis which is exothermic in itself. Consequently, it would seem that the heart produces less heat in heat acclimated animals.

SKELETAL MUSCLE

Because of the large mass of skeletal muscle and its great metabolic potential, it is most probable that its metabolic activity decreases in heat acclimation. However, this decrease must involve a change in metabolic control mechanisms rather than in metabolic potential, since oxidative enzyme levels of skeletal muscle tissue seem to be unaffected by temperature acclimation in most of those species studied (Chaffee et al., 1969b).

P:O RATIOS

Most studies indicate that there is no change in oxidative phosphorylation efficiency in skeletal muscle in response to heat or cold acclimation (Chaffee et al., 1969b), except possibly in rats (Beyer, 1971), and even in the latter study the author points out that none of these changes can be viewed as a major basis for the observed differences in the MR. Thus the mechanisms for skeletal metabolic control probably depend primarily on the modulation of cellular levels of ADP.

ADP REPLENISHING SYSTEMS

It has been reported that muscle tonus, which is dependent upon a pulsed stimulation of sarcomeric ATPase, decreases in the heat (Collins and Weiner, 1968). Such a response in itself would depress the rate of hydrolysis of ATP to ADP, and hence metabolic thermogenesis. In addition, other metabolic changes which would depress the rate of ADP production in the muscle cell are suggested by the fact that thyroid activity is depressed in the heat. The activity of Na^+/K^+ ATPase of skeletal muscle is known to be dependent upon the level of circulating thyroid hormones (Edelman and Ismail-Beigi, 1972). Thus in the heat where thyroid activity is depressed, it is probable that the level of activity of this ATPase declines. Also since it is known that the NE-stimulated metabolic response in heat acclimated animals is dependent upon the levels of thyroid hormone (Cassuto and Amit, 1968; Chaffee and Roberts, 1971), and that NE can directly depolarize skeletal muscle (Teskey et al., 1974), it is probable that this depolarizing affect decreases in the heat, and thus the skeletal muscle membrane potential would be stabilized. If this is indeed the case, it also could contribute to a decline in the activation of those Na^+/K^+ ATPase pumping units present.

Another factor which is influenced by the levels of thyroid activity, and was mentioned before in the case of the liver, is the protein turnover rate. Himms-Hagen et al. (1972) have reported that in cold acclimation protein recycling increases in muscle mitochondria and represents a significant component in the nonshivering thermogenic metabolic contribution of skeletal muscle to the MR; and as mentioned previously, mitochondrial protein synthesis is probably dependent upon thyroid activity (Hoch, 1971). Conversely, the decline in thyroid activity in heat acclimation should result in the depression of this ADP replenishing system and thus contribute to CTS in skeletal muscle tissue. Other evidence in accord with the hypothesis that protein recycling may decline in skeletal muscle in the heat, is that skeletal muscle makes up a very large percentage of the active tissue mass of the body, and there is evidence that total body protein recycling is markedly depressed in the heat (Chaffee et al., 1971).

SOME METHODOLOGICAL PROBLEMS INHERENT IN IN VITRO STUDIES OF TEMPERATURE INDUCED CELLULAR BIOCHEMICAL CHANGES

In the study of the cellular biochemical aspects of temperature acclimation the major goal has been and is to study changes at the cellular, subcellular and biochemical levels and relate these changes to modifications in the thermogenesis of the tissue being studied, and hence to the overall metabolism of the animal. Since presently there are few methods indeed which make such studies possible in vivo, almost all of these studies have been made using in vitro techniques. The justification for such studies is based on the premise that changes seen in vitro reflect what must be occurring in vivo. Unfortunately, the defense of this assumption is not always easy, but no one can deny that this approach will serve as a valuable starting point for future in vivo research as technology advances.

One serious problem in the study of enzyme levels in temperature acclimation is that some enzymes change in activity early in the acclimation process and then return to preexposure (or near preexposure) levels, whereas others may change gradually and then stabilize (Hannon, 1963b). To complicate matters the timing of these changes varies in different species and is dependent upon the extent of the temperature challenge.

Another major problem in trying to determine what enzymatic changes actually occur in temperature acclimation is that the assay techniques are subject to criticism because in vitro versus in vivo conditions are so different. What an enzyme can do in vivo depends on many different factors: substrate supply, cofactor availability, organelle and cellular compartmental pH conditions, levels of ionic enzyme activators, levels of metabolites which may act as allosteric inhibitors or activators, the effects of the intracellular milieu on the tertiary and quaternary enzyme structure, and levels of cyclic AMP, to mention only a few. In vitro many if not all of these factors in the reaction mixture are artificially changed so that the enzyme will show maximum activity. The philosophy behind this approach is that if the in vitro conditions are the same for the heat acclimated and control enzymatic preparations, both will perform under the same handicaps or advantages afforded by the conditions in the reaction mixture. This may be true with respect to certain assays, but not in others. For example, when there is a temperature acclimation-induced change in the isozymic (i.e., subunit) constitution of an enzyme, such as lactic dehydrogenase (Allen and Chaffee, 1964; Blatt et al., 1965), the in vitro conditions for maximum activity of the enzyme from the control may differ from those from the temperature acclimated animal. Also in situations where the fatty acid constituents have been altered, and hence possibly the structure of the membrane of the mitochondria, as occurs in temperature acclimation (Chaffee et al., 1968b; Patton and Platner, 1970), the permeability of the mitochondria to substrate may be changed so that the availability of a substrate to the enzyme in the mitochondria may be rate limiting. This situation would mimic an enzyme limited reaction in that the more mitochondria added the greater the speed of reaction. This may explain the frequent conflicting findings when studies are made on thoroughly homogenized preparations versus those made on intact mitochondria.

Another source of error in the in vitro approach is that the conditions for maximal enzymatic activity are almost always determined using control preparations. This is no doubt because enzyme assay procedures are most frequently borrowed by physiologists from biochemists who use animals living in an animal colony where the standard temperature is maintained at about 24°C. Therefore it should always be kept in mind that the problem encountered in in vitro assays are legion, so that until the technology for in vivo (cellular) biochemical assays improves, advances will probably not be very rapid in this field.

SUMMARY

Chemical thermosuppression (CTS) occurs in the cells of various organs and tissues during heat acclimation. Included among fundamental changes which may occur in CTS are: (1) the lowering of the mass and oxidative enzymatic capacity of certain tissues and organs and (2) a restriction of cellular metabolism by lowering levels of available ADP or possibly other metabolic rate limiting factors. The first type of change is caused rather gradually and represents a slow-acting thermoregulatory modulation. The second type of change involves more rapid-acting modulations of the levels of factors which influence the rate of dephosphorylation or hydrolysis of ATP, its conversion rate to cyclic AMP, the efficiency of its production, and ultimately the rate of flow of electrons through the electron transport system. There is evidence that CTS may play a significant role in survival in exposure to extreme heat.

Evidence is presented which supports the concept that cellular metabolic changes of many organs and tissues in heat acclimation are antithetical to those which occur in cold acclimation.
Some of the technological limitations which exist in the in vitro biochemical methods used in temperature acclimation research are discussed.

REFERENCES: see pp. 513-515.

Effects of Cold on Cellular Metabolism: Enzymes and Other Cellular Changes

by

E. J. Masoro

INTRODUCTION

Homeotherms maintain body temperature by means of a broad spectrum of physiologic reactions, the predominant response relating to the species and to other coexisting environmental circumstances. Although a variety of specific responses is involved in the reactions to cold environments, they all can be classified within two broad and obvious categories: heat conservation mechanisms and heat production mechanisms (Brengelmann, 1973).

Heat conservation involves behavioral, hemodynamic, and insulative phenomena. The seeking of shelter by most animals, the wearing of clothes by man, and central heating are examples of behavioral responses. In the case of man, so highly developed has he become at "coming in from the cold" that it is his major way of dealing with cold environments. However, man and other mammals also have many physiologic mechanisms to call upon to meet a cold environment when they are not able to circumvent the problem by behavioral responses. Cutaneous vasoconstriction, directed by the central nervous system, permits the temperature of the skin to fall and in this way decreases heat loss by radiation, conduction and convection between the environment and the body surface. Important changes in insulation occur in some species on cold exposure e.g., increased pelage thickness, piloerection, and increased thickness of the subcutaneous adipose tissue.

In some species of mammals, an increased rate of heat production is often the predominant response to cold exposure. This response involves a variety of enzymatic and metabolic changes.

Although without question changes in cell physiology are the bases for responses involved in heat conservation, little or no information about the cellular events is yet available. Therefore, our attention will be focused in the one area concerning which data are available i.e., the cellular mechanisms underlying the increase in heat production in response to cold environments.

GENERAL CHARACTERISTICS OF METABOLIC RESPONSE TO COLD

If an animal that has been living for some time at what is often called a thermoneutral environment (i.e., an environment causing the animal to make no metabolic response to maintain body temperature nor any sudomotor response for that matter) is exposed to a cold environment, there is an immediate increase in the rate of heat production brought about primarily by shivering (Brengelmann, 1973). This response is called, therefore, SHIVERING THERMOGENESIS and is the result of increased metabolic activity of skeletal muscle.

However, if exposure to cold is continued for a prolonged period, animals become cold acclimated, a term covering a wide variety of physiological responses that better enable them to survive the deleterious effects of cold environments. In some species (but not all) a most important component of this acclimation process is the development of significant ability to increase the rate of heat production by mechanisms other than shivering i.e., by a process or processes called NONSHIVERING THERMOGENESIS (Chaffee and Roberts, 1971). Both the brown adipose tissue and the skeletal muscle are sites of nonshivering thermogenesis (Jansky and Hart, 1963; Smith and Horwitz, 1969). There is some evidence that liver and other viscera may also participate but this is still in debate.

CELLS INVOLVED IN SHIVERING THERMOGENESIS

In this section, only those cells clearly identified as either the sites of thermogenesis or sources of fuel for thermogenesis will be discussed.

The muscle cells of the skeletal muscles are the sole site of shivering thermogenesis. The biochemical mechanism of this thermogenic activity is readily explained on the basis of the currently accepted concepts of the control of rate of electron transport and oxygen utilization (Hoch, 1971). Mitochondria of most cells are limited in their rate of electron transport by the availability of adenosine diphosphate (ADP) and inorganic orthophosphate (Pi) because of the concomitant occurrence of oxidative phosphorylation. Cells in which electron transport is so controlled are said to possess "respiratory control"; presumably mitochondria of skeletal muscle under in vivo conditions in normal animals possess "respiratory control". When cold exposure induces shivering an increased frequency of muscular contractile activity causes an increased rate of adenosine triphosphate (ATP) hydrolysis to ADP and Pi in the muscle cell which increases the amount of available ADP and Pi and thus elevates the rate of electron transport in muscle mitochondria. In turn, this stepped-up rate of electron transport provides an increased rate of generation of ATP for the heat producing hydrolysis of ATP by contractile elements and an increased rate of heat production directly because of the inefficiency of electron transport reactions in terms of ATP generation. On removal from the cold environment, shivering stops and thus cessation of shivering thermogenesis occurs with a resultant fall in the metabolic rate to that which occurred prior to cold exposure. All foodstuffs (proteins, carbohydrates, and fats) can support cold-induced shivering thermogenesis (Masoro, 1966). The extent to which fat may be utilized preferentially relates to the fact that it is the major energy reserve of the body and is called upon preferentially whenever caloric intake is insufficient to meet caloric expenditure. Since acute exposure to cold environments usually imposes a negative caloric balance, the fat stored in the adipose tissue reservoir becomes a prime fuel for shivering thermogenesis.

This fuel, of course, must be mobilized from the adipocytes of the white adipose tissue. Indeed, this mobilization during acute cold exposure is brought about by the activity of the sympathetic nervous system (Himms-Hagen, 1972), its neurotransmitter norepinephrine promotes the activity of the hormone-sensitive lipase of adipocytes thus causing triglycerides to be broken down into FFA and glycerol. These hydrolysis products are released into the circulation, the whole process being termed fat mobilization. Because of this, the concentration of FFA in the blood rises during cold exposure and in this way FFA becomes a prime fuel for the skeletal muscle for shivering thermogenesis.

Since the animal undergoing shivering thermogenesis is usually in the negative caloric balance state, it is also to be expected that the hepatocytes of the liver will exhibit an increased gluconeogenic activity (Masoro, 1973). There is strong evidence that this is indeed the case. For example, the activities of key gluconeogenic enzymes in the rat liver markedly increase within a short time after exposure to cold. The following have been shown to be increased in activity: Phosphoenolpyruvate carboxykinase, serine dehydratase, tyrosine-amino transferase, glutamic-pyruvic transaminase, glutamic-oxalacetic transaminase, urea cycle enzymes. Presumably the increased activity of these enzymes is part of the process involved in supplying glucose as fuel for the body.

CELLS INVOLVED IN NONSHIVERING THERMOGENESIS

Brown adipose tissue, a site of nonshivering thermogenesis, responds to prolonged cold exposure by growth with an increase of the number of cells, and by an increase in the number of mitochondria per cell. Moreover, each mitochondrion has more cristae than was the case before cold exposure (Smith and Horwitz, 1969).

Cold-induced nonshivering thermogenesis in brown adipose tissue cells is brought about by the following mechanism (Himms-Hagen, 1972). Triglyceride stored in the brown adipose tissue cells is converted to FFA and glycerol at a rapid rate because cold-exposure brings into action the sympathetic nervous system supplying the brown adipose tissue thus releasing norepinephrine to the cells which in turn activates the hormone-sensitive lipase. The FFA so formed are used primarily within the brown adipose tissue cell. These FFA have the dual function of (a) altering the mitochondria so that they become loosely coupled and are thus able to respire at high rates independently of the supply of ADP and Pi and of (b) providing the substrate in the form of FFA to support this elevated metabolic rate. The amount of heat produced by brown adipose tissue is relatively small compared to the total heat production of the organism but it is nevertheless important because of its local distribution particularly to the thoracic region. Indeed, an important function of the heat produced by the brown adipose tissue in the thoracic region may be the suppression of shivering via the warming of receptors for shivering in the spinal cord.

Besides being a site of nonshivering thermogenesis, brown adipose tissue has a second function in this regard. It appears the brown adipose tissue plays a role in the development and maintenance of the nonshivering thermogenic activity of the skeletal muscle during cold acclimation. There is evidence that a protein produced by brown adipose tissue mitochondria is the mediator of this action of brown adipose tissue on skeletal muscle nonshivering thermogenesis (Himms-Hagen, 1972).

It is also of interest that during cold acclimation the activities of several enzymes increase in brown adipose tissue (Chaffee and Roberts, 1971). These include glucose phosphate dehydrogenase, glycerophosphate dehydrogenase, malic enzyme, succinic dehydrogenase and cytochrome oxidase.

Quantitatively it is clear that skeletal muscle is the major site of nonshivering thermogenesis. However, unlike the brown adipose tissue, the biochemical mechanism of nonshivering thermogenesis in skeletal muscle is not known. There is strong evidence to indicate that norepinephrine is the trigger for nonshivering thermogenesis by skeletal muscle (Carlson, 1966) and that in some way the brown adipose tissue is required for the development and maintenance of nonshivering thermogenesis in skeletal muscle (Himms-Hagen, 1971). In rodents, oxidative enzymes in the skeletal muscle are elevated during acclimation to cold (Chaffee and Roberts, 1971). Also the carnitine concentration of skeletal muscle is increased in cold acclimated animals which fits well with the evidence that there is an increased capacity to oxidize fatty acids in skeletal muscle of cold-acclimated mammals. There also is an increased rate of turnover of certain proteins in the mitochondria of the skeletal muscle of cold acclimated rats. The relationship, if any, of this alteration in mitochondrial protein turnover to nonshivering thermogenesis is at present unknown. However, this is a clear demonstration of a biochemical difference between the skeletal muscle of cold-acclimated animals and the skeletal muscle of animals not so acclimated and thus may open up a new approach to the study of the biochemical nature of the nonshivering thermogenesis process (Himms-Hagen, 1972).

Even though the liver does not appear to be an important site of nonshivering thermogenesis, the effect that cold acclimation has on the enzymatic machinery of liver has been the subject of many studies, possibly because of the role that the liver plays in supplying fuel (via gluconeogenesis) to those sites that are involved in nonshivering thermogenesis but more probably because the liver is a tissue that is easy to study enzymologically. In rodents, the level of oxidative enzyme activity in liver increases during cold acclimation and coenzyme Q also increases in concentration in the liver of cold acclimated animals (Hannon, 1963). It has also been claimed that oxidative phosphorylation is uncoupled in liver mitochondria from cold acclimated rats, but this is an area of debate since it is now clear that if the isolated mitochondria are very well washed before being assayed that mitochondria isolated from cold-acclimated animals are not uncoupled (Chaffee and Roberts, 1971). However, contamination with other cellular components uncouples oxidative phosphorylation in the case of material derived from cold-acclimated rats and this may be physiologically relevant information.

Many hepatic enzymes involved in the intermediary metabolism of carbohydrate are increased in activity in cold acclimated rats (Chaffee and Roberts, 1971). In particular, the enzymes involved in gluconeogenesis are increased in activity which fits with the role of the liver as an important supplier of glucose fuel. The enzymes involved in glycogenolysis, glycogenesis, or glycolysis have not been found to be increased in activity during cold acclimation.

There are also metabolic adaptive changes in the white adipose tissue induced by cold acclimation. This is not surprising when it is recognized that white adipose tissue serves as the major reservoir of fuel supply in mammals (Masoro, 1973). Indeed, the lipolytic response to norepinephrine is enhanced in white adipose tissue in cold acclimated animals (Hannon and Larson, 1962), a reaction basic to the mobilization of fuel from adipocytes. This probably relates to the finding that the specific activity of the norepinephrine-stimulated adenylcyclase in white adipose tissue is increased in cold acclimated animals (Therriault et al., 1969).

ANALYSIS OF THE METHODOLOGY OF THE FIELD

At the moment, further understanding in this field is limited by the methodology currently available to approach such questions as to how norepinephrine brings about nonshivering thermogenesis in the cold acclimated animal. The techniques that have been used are those designed to study the enzymes in tissue homogenates or the enzymatic and metabolic characteristics of organelles isolated from tissues or the metabolic characteristics of whole tissues isolated from animals. In such assay systems, of course, conditions are utilized for isolating the cellular component or for assaying the enzyme that yields the greatest activity for the reaction under study. However, such "optimal conditions" may not be those that are occurring within the cell in regard to the physiologic state being investigated (e.g., cold acclimation). A case in point is the extensive literature on the effect of cold acclimation on oxidative phosphorylation of mitochondria. Clearly, if these mitochondria are optimally isolated and thoroughly washed, there is no difference in the oxidative phosphorylation efficiency between the mitochondria obtained from cold-acclimated rats and those obtained from appropriate control animals. However, if washing of the mitochondria is less than thorough then uncoupling

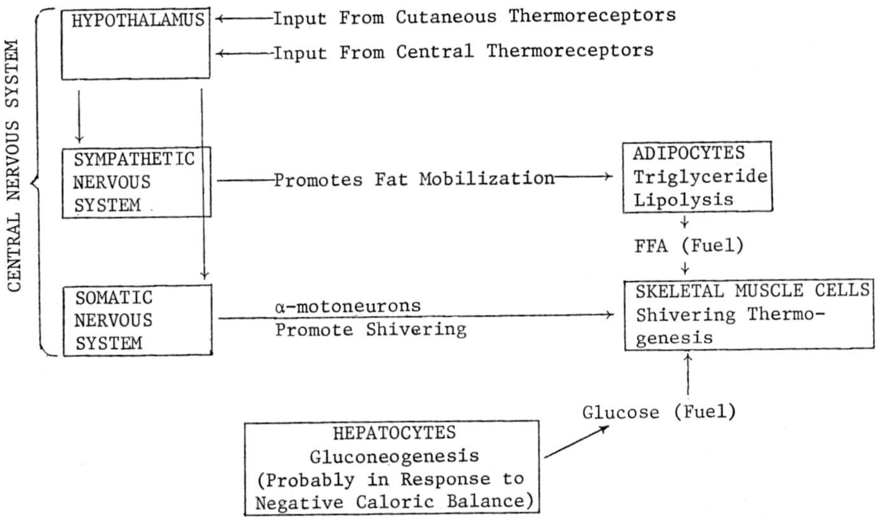

Fig. 1. Overview of cellular responses in shivering thermogenesis.

of oxidative phosphorylation is seen in the mitochondria isolated from cold acclimatized rats, but not in those isolated from the appropriate control animals. From such experiments it is not possible to know if oxidative phosphorylation is affected by cold acclimation. Clearly this field awaits methodology that permits the study in vivo of many of the phenomena discussed in this paper which were uncovered by in vitro experiments.

OVERVIEW OF CELLULAR RESPONSES DURING THERMOGENESIS

Figure 1 is a schematic overview of the cellular responses involved in shivering thermogenesis. The hypothalamus is apprised of the thermal stress by information sent to it from cutaneous thermoreceptors and from central thermoreceptors located in the hypothalamus. The hypothalamus integrates this information and directs the nervous

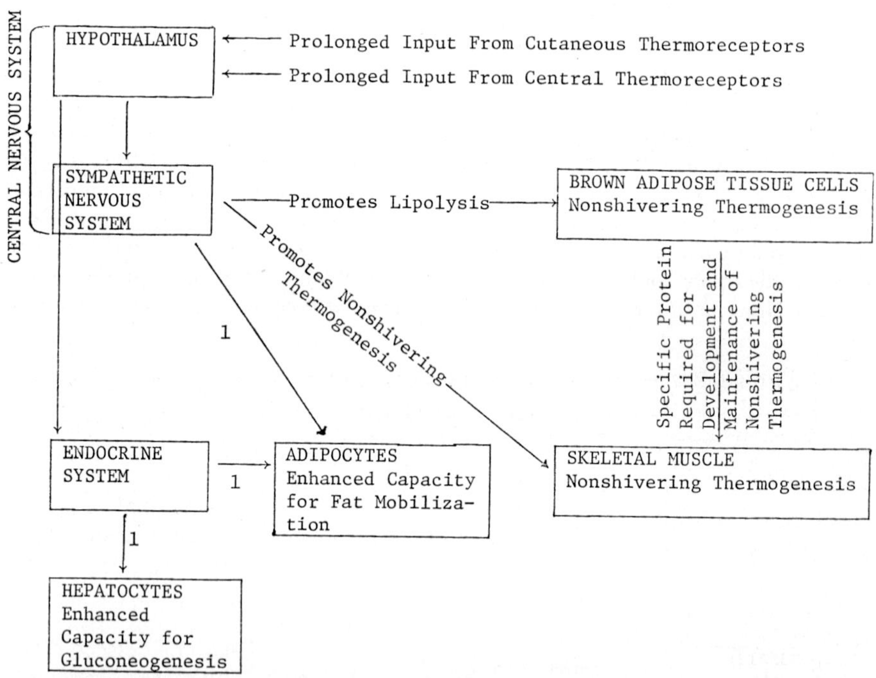

Fig. 2. Overview of cellular responses in nonshivering thermogenesis.

[1] When acclimatized to a given cold environment, animals are in a state of caloric balance and therefore do not require high rates of fat mobilization or gluconeogenesis. Nevertheless, the capacity of the adipose tissue and liver to carry out these respective processes is greatly enhanced by cold acclimation which becomes evident when the need for these processes arises (e.g., if deprived of food or placed in a significantly colder environment). Presumably the enhanced capacity of the adipose tissue to carry out fat mobilization is brought about by chronic influences from the sympathetic nervous system and/or endocrine system. The increased capacity of the hepatocytes for gluconeogenesis again probably involves chronic influences from the endocrine system.

system into appropriate efferent action. This includes activity over α-motoneurons to the skeletal muscle causing the muscle response we call shivering and which generates heat by the biochemical mechanisms discussed above. It also involves increased sympathetic nervous activity to the adipocytes causing triglyceride stores of these cells to be mobilized as FFA and glycerol by the mechanisms described above. This FFA serves as an important fuel for shivering thermogenesis since in acute cold exposure increased food intake is not sufficient to meet the increased caloric expenditure. It is probable that it is this negative caloric balance state involving falling insulin levels and rising glucagon levels in the body fluids that is responsible for the increased gluconeogenesis by hepatocytes. The glucose so generated also serves importantly as a fuel.

Figure 2 is a schematic overview of the cellular responses involved in nonshivering thermogenesis. Nonshivering thermogenesis develops significantly only after prolonged exposure to cold; presumably, therefore, it involves prolonged processing of this thermal information by the hypothalamus. Using the sympathetic nervous system as the efferent outflow, nonshivering thermogenesis occurs in the brown adipose tissue cells by the biochemical mechanism detailed above. In addition the brown adipose tissue appears also to generate and secrete a protein required for the development and maintenance of nonshivering thermogenesis in skeletal muscle. Since this thermogenic activity of skeletal muscle is triggered by norephinephrine, presumably the sympathetic nervous system is involved in this function of skeletal muscle; but the biochemical mechanisms are not known. The adipocytes of white adipose tissue develop an increased capacity to mobilize fat in response to sympathetic nervous activity, presumably this cellular response involves actions on it by the nervous system or the endocrine system or both. The hepatocytes also develop an increased gluconeogenic capacity and again the endocrine system is suspected as being involved. This increased capacity of both of these cell types makes it possible for them to supply much fuel should the need arise.

SUMMARY

The cellular mechanisms underlying shivering thermogenesis are rather well understood. The biochemical basis of the increased heat production by the skeletal muscle cells (the site of this thermogenesis) is established as is the mechanism by which adipocytes make fuel available. The molecular basis of the increased gluconeogenic activity of hepatocytes is not unequivocally established.

In contrast, nonshivering thermogenesis is poorly understood at a cellular and biochemical level. A reasonable scheme, suggested by in vitro evidence, is currently accepted in regard to the biochemical basis of nonshivering thermogenesis in brown adipose tissue cells but there is no molecular understanding of nonshivering thermogenesis in skeletal muscle which quantitatively is the major site of this heat production. The field awaits the methodology for the in vivo exploration of this biochemistry before it is likely that further understanding about this process in skeletal muscle will be forthcoming. The capacity of adipocytes to mobilize fat and for hepatocytes to carry on gluconeogenesis is enhanced under conditions where nonshivering thermogenesis is well developed, but the biochemical mechanisms involved are not fully understood probably again because of the limitations of current methodology. Finally, there is suggestive evidence that the brown adipose tissue secretes a protein which is involved in the development and maintenance of nonshivering thermogenesis in skeletal muscle; much more work is needed to establish the biochemical mechanism of action of this brown adipose tissue nonshivering thermogenic factor.

REFERENCES: see p. 516.

Effects of Heat on Endocrine Responses of Domestic and Laboratory Animals

by

H. D. Johnson

Animal biometeorologists and environmental physiologists are gradually gaining a better understanding of how the primary physical environmental factors such as temperature, humidity, wind, and radiation influence the homeostatic and homeothermic capacities of various species of animals (see A Guide to Environmental Research on Animals, 1971). However, to be able to characterize a particular climate or physical environment specifically and in meaningful terms to environmental physiologists requires much more intensive investigations by biometeorologists. But major questions immediately confront the biometeorologist, namely, what physiological or combination of physiological factors characterize optimal animal acclimation or acclimitization to the specific environmental conditions? These physiological factors of a particular species or breed of animal which constitute the physiological profile are influenced by age, level of activity, lactation, rate of growth, and reproductive state. Thus I am suggesting the present state of knowledge or understanding of what constitutes an environmental profile for a specific geographic region or shelter area requires not only specific environmental characterization but the development of a meaningful physiological profile for different species or breeds of animals in various physiological states. Thus the physiological needs – to be optimally served by the environment – must have a physiologically desirable environmental profile. The fitting of an animal to an environment or fitting the environment to the animal is of major importance for domestic animals of all types which are continuously being transported all over the world to varied physical, chemical, and biological environments and on many occasions are not successful in their new environmental surrounding. This is especially true in the cattle industry where animals and semen are introduced into tropical areas with only meager knowledge of the "physiological profile" of the species, breed, or particular individual. Also the new Animal Welfare Acts in many regions of the world characterize the environments needed by a laboratory animal with a minimum amount of physiological justification for some of the requirements and regulations.

A number of climatic classifications have been proposed. The early work of Köppen and Geiger (1936) has been followed by schemes of Thornthwaite (1948) and Trewartha (1954) applicable to plant systems. Lee (1953), Williamson and Payne (1965), and Webster and Wilson (1966) have proposed general climatic classifications for use by animal production specialists in tropics. However, the need to provide a concise climatic index based on animal response data was furthered by the temperature, humidity index (THI) (Johnson et al., 1963). But, wind and radiation are important climatic factors for animal comfort and must be included in an environmental profile. McDowell (1972) has an excellent discussion of the environmental profile. However, little attention has been given to the concept of a physiological profile nor to the development of a biometeorological index based on the merger of environmental and physiological profiles.

In this Section the author will restrict discussions to the thermal effect of environmental profile on a mammal with special attention to laboratory and domestic animals. The hormones, an important aspect of the physiological profile, that are known to respond to thermal stimuli will receive attention since the Sections 12, 14 and 15 of Chapter 1 will refer to thermal and other environmental factors effects on lactation, reproduction, and growth.

A major objective is to summarize the present state of knowledge and focus on immediate and long range studies necessary to attain an understanding of environmental heat effects on hormonal change and involved metabolic acclimation and adaptation.

Since the endocrine system and neuroendocrine system with their acknowledged roles in thermal, water, energy, and hormonal balance of an animal are important aspects of physiological adaptability to an environment, the author will describe the present state of knowledge in the area of environmental heat effects – that is, the resultant effects of temperature, humidity, wind, radiation, (T, H, W, R), or combination thereof on known endocrine and neuroendocrine functions of animals. The endocrines and neuroendocrines may provide important indicators of the environmental needs or indices of adaptability of laboratory and domestic animals.

Temperature, humidity, and radiation conditions above the thermoneutral initiate peripheral thermal stimuli in the skin which project by sensory fibers the lateral spinal-thalamic tract to synaptic relays in the reticular system of the medulla and to the posterolateral ventralmedial nuclei of the thalamus. Neural projections into cortical areas presumably mediate conscious thermal sensations, and ascending noradrenergic tracts connect the brain stem with the hypothalamus (an excellent review on neuroendocrine thermoregulation is by Gale, 1973, and climatic physiology by Thompson, 1973).

Thus, the environmental heat may affect the hypothalamus by skin neural receptors, spinal CNS, and hypothalamic neural processes and functions responsive to this peripheral neural stimulation, or rises in core temperature may directly affect hypothalamic neuroendocrine functions.

It is generally known that the hypothalamus controls the release of the pituitary hormones via neurotransmitters and neurohormones called "releasing" or "inhibiting" factors and/or "hormones" (see recent reviews by Blackwell and Guillemin, 1973 and Convey, 1973). The neuroendocrine hormones are listed in Figure 1.

In the hypothalamic-pituitary neuroendocrine systems, thermal and chemical receptors in the preoptic anterior hypothalamus (POAH) influence secretion from the anterior pituitary of thyroid stimulating hormone (TSH), adrenalcorticotropin (ACTH), and growth hormone (GH). These actions are presumably mediated by hypothalamic

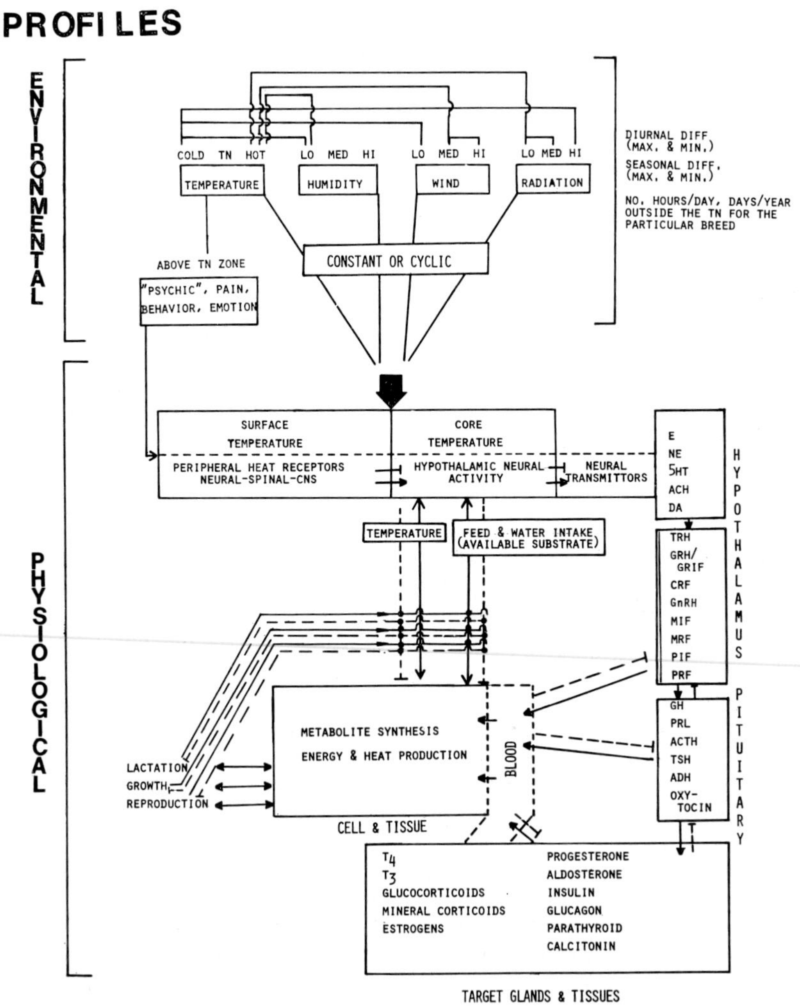

Fig. 1. General guide to enumerate factors of the environmental profile (THWR) and physiological profile (neuroendocrine) and major avenues of possible interaction. This interaction is proposed as the basis for a biometeorological index.
Connecting lines suggest the importance of the levels of T, H, W, R in determining the severity of a "hot" or "cold" environment. These lines suggest various more optimal combinations of the four factors. The genetic differences, acclimation or acclimitization, level of lactation, rate of growth, reproductive stage influence the energy, thermal, water, hormonal, mineral balances of the animal and the relative activity, plasma levels and rates of utilization of the various hormones. The arrows (→) or inhibition (⊣) indicate interactions between the hormones, cell activity, heat production, and including interactions of body temperature, feed and water intake, lactation, growth, and reproduction.

releasing hormones for thyrotrophin (TRH), corticotropin (CRH), and growth hormone (GH). Recent evidence suggests that ADH may be influenced by heat (Seif, 1973; Hayward and Baker, 1968). The thermal effects on prolactin (releasing factors or gonadotrophic releasing factors) have not been quantitated as yet.

This hypothalamic control of the pituitary is modulated by pituitary hormone output which may alter hormonal releasing and inhibiting factors, or change in pituitary responsiveness to the hypothalamic neurohormones. The release of the pituitary hormones also are modified by hormonal feedback from target glands or tissues. The level of utilization of these target gland hormones and such pituitary hormones as prolactin and growth hormone by the tissue also plays a role in level of neuroendocrine activity. Figure 1 suggests the influence of the level of metabolite synthesis energy production and heat production on hormonal utilization (See also Chapter 1, Section 1a by Chaffee). Also the level of growth, lactation, and reproductive state will similarly influence hormonal usage. There are many other aspects of this overall hormonal metabolism scheme that may be discussed, but this does illustrate in a general manner the possible relative influences of an environmental factor such as heat on hormonal functions and the possible influences of the level of lactation, growth, and reproductive state.

Although it is not intended to further complicate this environmental-physiological profile interaction concept, there is a time factor, a time response in level of hormone following an environmental stimuli, or stimuli resulting from greater or less energy production (Johnson, 1974). It is typical for a hormone to initially "overshoot" or increase its plasma level, followed in some instances by an "undershoot" and then stabilization at a higher or lower than normal plasma level unless the animal becomes exhausted, in which case the animal may succumb to the stress. Thus the time of exposure to the stressful environment is critical in proper assessment and interpretation of the plasma hormonal levels or functional hormonal activity of the animal and may be generally regarded as the influence of acclimation or acclimitization (Johnson, 1974).

Short and long term environmental heat exposure effects on plasma hormone levels are now becoming well documented for some of the pituitary and target tissue hormones due to the recent development of competitive binding and radioimmunoassay techniques for hormone assays. As antibodies and purified hormones become available for the various hormones, rapid advances will be seen in the knowledge of the thermal effects on hypothalamic neurotransmitters, releasing factors, or hormones, pituitary, and target gland.

At present the author is not aware of any environmental heat data on releasing hormones, however, thermal data on pituitary and target gland suggest the responsiveness of TRH, PRF/PIF, GnRH, and CRF which will be discussed later.

Acute heat exposures of cattle (Mitra and Johnson, 1972) increase the plasma levels of growth hormone with 15 minutes, whereas, long term exposure causes a decline in growth hormone levels (Mitra et al., 1972). Plasma cortisol similarly increases immediately upon acute exposure to heat (Christison et al., 1970) or declines following days of environmental heat exposure (Christison and Johnson, 1972). Since plasma prolactin increases due to various stress stimuli (Rand et al., 1971), it is conceivable that prolactin would increase due to acute thermal stress. Longer term summer seasonal effects on plasma prolactin indicated higher plasma values although response of prolactin to stimuli of milking was less in the summer. More data are needed on prolactin to fully resolve these environmental effects.

Insufficient data are presently available to indicate whether TSH increases or decreases on acute exposure to heat, however, numerous publications (Yousef and Johnson, 1965) describe a decline in ^{131}I-thyroxine disappearance and lowered PBI levels after 72 hours of exposure in cattle. The time response in total plasma thyroxine, free thyroxine, and triiodothyronine should be determined with the newer radioimmunoassay techniques available. ADH shows a moderate increase in Zebu and Highlander cattle (Seif, 1973) due to acute heat exposure, however, the long term effects are only significantly depressed for the Zebu as there were no changes in the Highlander. Surprisingly little information is available on plasma insulin and thermal heat with the exception of the report by Kamal et al. (1970).

Plasma levels of epinephrine and norepinephrine respond differently than most other hormones in that an immediate increase in both epinephrine and norepinephrine occurs upon heat exposure, and this elevated level persists for many days (Alvarez and Johnson, 1973).

Environmental heat effects on reproduction will be discussed in Section 15, but a brief summary here would indicate that pituitary LH peaks as well as base levels on female cattle were depressed by moderate environmental heat exposure (Madan and Johnson, 1973). Progesterone levels are elevated by environmental heat (Stott et al., 1967). Since radioimmunoassay procedures for estrogens are underway, data should be expected very soon which will aid environmental physiologists to better ascertain the hormonal basis for poor reproductive performance normally associated with female domestic animals during temperate summer heat and in the semitropics and tropical regions of the world.

The reproductive performance of male domestic animals has long been observed in hot seasons and semitropical and tropical regions of the word. Recent studies by Rhynes and Ewing (1973) showed that environmental heat lowered blood testosterone levels to 43% of control values in Hereford bulls.

Ulberg (1967) has produced a wealth of fundamental data on environmental heat effects on embryonic death in sheep. At this time, the author is not aware of information on heat effects on parathyroid, calcitonin, glucagon, or oxytocin plasma levels or functions.

To gain a greater understanding of the longer term or adaptative changes due to environmental heat effects and related tropical environmental factors on physiological and hormonal functions which are associated with reproduction, lactation, and growth requires measurements on domestic animals in the tropics under laboratory and field conditions.

The long term effects of acclimitization and adaptation of domestic mammals to adverse low tropical conditions are a lowered heat production, poor reproductive status, and low levels of growth rate and milk production. The further complication of the typical lower energy, lower protein and higher fiber tropical rations makes it difficult to compare plasma hormonal levels of "high energy-fed" temperate animals. Additionally the feed intake may be reduced due to thermal heat (Fig. 1), therefore, the hormonal levels may even be high in the plasma due to a lower utilization of the hormones. To quantitate this overall problem of environmental heat effects on feed intake reduction (including climate effects on feed quality), possible reduction in most hormonal usage with the exception of epinephrine and norepinephrine, and interactions due to reduction in heat production, growth, and lactation, is an immediate challenge to environmental physiologists. These plasma hormone values are needed to scientifically assess the reasons for lowered productivity in hot climates and make valid recommendations for improvement of lactation, growth, and reproductive performance. That is, use the physiological tools available for improvement of management during temperate summer heat and in semitropical and tropical environments. As emphasized in this section, the development of environmental profiles, physiological profiles, and resultant biometeorological indexes for a specific breed in a specific environment will rapidly advance our scientific technology in this important area of animal biometeorology.

CONCLUSION

In order to be able to quantitate the major physical factors of the climatic environment a characterization of an environmental profile is required for the environment in which the animal is expected to pursue its daily activities. This will require intensive investigations by biometeorologists in the laboratory and field. An index readily applicable to this environmental profile will enable the environmental physiologist to quantitate the environmental thermal effects on physiological compensations necessary for successful animal acclimitization or adaptation to varied natural or sheltered climatic conditions throughout the world. For domestic animals in which compensations in growth and lactation should be minimal, this physiological profile or understanding of the physiological mechanisms that respond to known environmental factors may allow either shelter modification of the environmental profile factors or physiological profile modifications by physiological, nutritional, or hormonal therapy and other management practices.

REFERENCES: see pp. 516-517.

Effects of Cold on Endocrine Responses of Ruminants and Other Domestic Animals

by

A. J. F. Webster

INTRODUCTION

Ruminant animals range wild, or are herded domestically in all the climatic regions that are considered habitable by man, including the Arctic and Boreal regions where man can suivive only by cultural adaptation; the creation of a microclimate far removed from that of the outdoor environment. Ruminants have little ability to create for themselves a favourable microclimate (they do not build nests!) and since they depend for their food on what they can graze and browse, the winter period presents the combined stresses of chronic cold and chronic food shortage. Chronic cold and undernutrition both deplete the energy reserves of the body and the effects of the combined stresses are more than additive, each serves to exacerbate the effects of the other.

It is, however, a fact that domestic and wild reminants and even the relatively hairless pig can thrive in climates intolerably cold for man. It is also true that adaptation to cold in large mammals like the ox or sheep (Webster, 1974b) takes a very different form from that observed in small mammals like the laboratory rat on which so many investigations have been made. Small mammals that inhabit cold regions do modify their thermal environment by building burrows or nests. However, it is probable that they live at an air temperature which is below their critical temperature (i.e., elicits a thermogenic response) for most of the year. When they leave their burrows, they may briefly require a very large elevation in heat production in order not to succumb to cold. In small mammals, therefore, it is not surprising that adaptation to cold should take the form of an enhanced capacity to elevate and sustain metabolic rate due in large part to noradrenaline mediated regulatory non-shivering cold thermogenesis (Jansky, 1970). In large mammals the likelihood of a cold stress so severe that it will overwhelm their capacity to sustain homeothermy by cold thermogenesis is remote. The problem of exhaustion of energy reserves or of those hormones which regulate the release of the energy reserves is real.

Endocrine responses to cold are concerned with the maintenance of homeothermy in the face of a challenge from the external environment. These responses may be considered in terms of their physiological consequences which are:

1. Increased metabolic rate (cold thermogenesis) – direct effect of hormones on exergonic reactions in body tissues;
2. Mobilization of energy rich substrates from reserves, especially in fat, for export to actively catabolizing tissues;

3. Diminution in the intensity of the cold stimulus – changes in the peripheral circulation, hair growth and habituation to the sensation of cold;
4. Maintenance of the integrity of the tissues of the extremities (ears and feet) by regulation of the peripheral circulation.

The role of the endocrine system in mediating the first two responses has been written about at great length, although it can only be reviewed briefly in this Section. Endocrinological aspects of the latter two responses are largely unexplored. The work which has been done suggests however, that more intensive study would be worthwhile.

METHODOLOGY: MEASUREMENT OF THE STIMULUS AND RESPONSE

The effect of cold on a homeotherm is a readily definable stimulus-response situation. When the thermal demand of the environment exceeds the thermoneutral metabolic rate of an animal it must either elevate heat production or permit body temperature to fall. The first law of thermodynamics requires that in this case a constant stimulus must achieve a constant response although the way in which the response is mediated by hormonal and other factors may change as a consequence of adaptation to cold. A reduction in the metabolic response to cold can only mean that the intensity of the stimulus has been reduced to the same degree.

It would seem, at first sight, that exposing an animal to a controlled low air temperature in a refrigerated room would provide a constant stimulus. This is a dangerous assumption particularly when endocrine factors affecting the metabolic response to cold are being considered. An extreme example is the metabolic response of a sheep on exposure to $-20°$ which may vary by a factor of four depending on the depth of its fleece. Other, less obvious reasons why the metabolic response to an apparently constant intensity of cold may differ include habituation to the sensation of cold at the skin surface and changes in the blood supply through the superficial tissues, both of which affect tissue insulation. Provision of a constant air temperature is, therefore, not enough. An intensity of cold stress that produced a constant mean skin temperature would be better, although this would not escape changes in response due to habituation. Cooling of the hypothalamus has proved a useful and inexpensive (compared with the cost of climatic chambers) way of eliciting a metabolic response (Andersson et al., 1967). However, the hypothalamus is not the only thermosensitive area in the body and the physiological response that is evoked is abnormal in that the temperature of other regions of the body rises to an unnatural degree. The best approach is probably to regulate the intensity of an external cold stimulus in order to produce a constant metabolic response as assessed from a continuous record of respiratory exchange. At air temperatures below 0°C this is best achieved using open circuit respiration calorimetry (Webster, 1974a; and Chapt. 1, Sect. 8b(2)).

Regulation of temperature in the exposed extremities of the body during exposure to cold is synonymous with regulation of blood flow, since in regions like the ears and the shanks incoming blood provides the only significant source of heat. Regional variations in blood flow are brought about by varying the cross-sectional area of the main resistance vessels, the precapillary arterioles and arteriovenous anastomoses. Studies of the effects of vasoactive agents on the smooth muscle of the resistance vessels have traditionally been made using isolated arteriolar strips or isolated perfused limbs. More recently measurements of regional blood flow have been made using radioactive inert gases (Sejrsen, 1969) or microspheres (Hales, 1973).

ENDOCRINOLOGY

Endocrine changes accompanying the response to cold may be grouped as follows:

1. Catabolic hormones e.g., the sympathoadrenal system, the thyroid and the adrenal cortex;
2. Anabolic hormones e.g., insulin and growth hormone;
3. Hormones regulating the peripheral circulation e.g., catecholamines, vasoactive polypeptides;

The role of the catecholamines within the central nervous system will not be considered here. The tools for studying the roles of the sympathoadrenal system in the metabolic and circulatory response to cold are many and have been well described (Carlson, 1966; Himms-Hagen, 1967). They include the surgical techniques of stimulation and ablation of sympathetic tissue, immunosympathectomy and an impressive array of pharmacological agents which can stimulate or inhibit the release of catecholamines, prolong their life in the circulation and more or less selectively block the alpha and beta receptor sites on the target cells. The life of catecholamines released into the blood stream is very short and they may experience any of several fates. Measurements of blood concentrations of noradrenaline and adrenaline do not therefore give any good indication of the activity of the sympathoadrenal system. A better, though incomplete, picture of the synthesis and secretion of catecholamines can be obtained by measuring the urinary excretion of noradrenaline, adrenaline and their major metabolites (Shum, et al., 1969).

The involvement of the thyroid and adrenal cortex in the response to cold were established by the classical techniques of ablation and replacement therapy. More searching questions as to the quantitative nature of their roles require more demanding techniques. The most successful modern approach to the measurement of thyroxine, the adrenal glucocorticords, principally cortisol and corticosterone, insulin and growth hormone is that of radioimmunoassay (London, 1968). This technique, in its simplest form, is based on the ability of an unlabelled hormone competitively to inhibit the binding of a radioactively labelled hormone to specific antibodies. The concentration of hormone in an unknown sample is determined by comparing the degree to which it inhibits the binding of a labelled hormone by antibody with the degree of inhibition produced by known standards.

Measurement of the concentration (C) of a hormone in blood or plasma does not directly give any indication of the rate of secretion of that hormone into the circulation ($\dot{q}i$) or of its rate of uptake by target organs ($\dot{q}o$).

If the volume of distribution of a hormone is V then,

$$V \frac{dC}{dt} = \dot{q}i - \dot{q}o$$

Changes in C following, for example, acute exposure to cold, only indicate the extent and duration of imbalance between $\dot{q}i$ and $\dot{q}o$. At equilibrium (i.e. over any extended period of time) dC/dt must be close to zero so that $\dot{q}i = \dot{q}o$, which may be called the rate of turnover of the hormone. This may be measured by determining the fractional rate of clearance per unit time (K) of a small quantity of radioactively labelled hormone added to the endogenous pool (Yousef & Johnson, 1967).

$$\text{Turnover } (\dot{q}o) = K \times V \times C$$

An alternative approach that has been used with sheep has involved the transplantation of either a lobe of the thyroid (Falconer, 1963) or an adrenal gland (McDonald et al. 1958) into a loop of skin in the neck in such a way that samples can be taken of blood entering and leaving the gland. If this can be combined with a sufficiently precise measurement of blood flow through the gland, then secretion rate ($\dot{q}i$) can be determined directly. The technique is more demanding than that of isotope turnover, but it is potentially more rewarding.

Effects of cold on the secretion of anabolic hormones such as insulin and growth hormone have not been studied in detail. The isotope clearance and radioimmunoassay techniques work well with these hormones, but the endocrine glands do not lend themselves to transplantation. The pituitary gland can be transplanted into, for example, the anterior chamber of the eye but once broken from its unique vascular link to the hypothalamus its function is quite abnormal.

Table 1. Plasma concentrations and secretion rates of hormones regulating the metabolism of ruminants.

Hormone	Species	Environment	Plasma concentration (μg/l)	Secretion rate μg/kg body wt. per day
Thyroxine	Cattle	18°C	64	3.2
		1°C	(64)	(4.4)
	Sheep	Thermoneutral	54	2.9
		Moderate cold	72	4.5
Cortisol	Sheep	Thermoneutral	9	326
		Very cold	31	1500
	Reindeer	Summer	22	276
		Winter	34	452
Insulin	Sheep	Thermoneutral, fed	1.1†	16
		Thermoneutral, fasted	0.3†	3
Growth hormone	Cattle	18°C	15 (variable)	43

† Assuming 24 mU per μg.

SECRETION OF HORMONES IN RESPONSE TO COLD

In sheep, goats and probably in all domestic animals urinary excretion of catecholamines increases in the cold to a degree proportional to the intensity of the stimulus (Andersson et al., 1967; Webster et al., 1969). The increase in catecholamine excretion begins at an air temperature about 5-10° higher than the lower critical temperature determined from measurements made of metabolic rate. More precise quantification of this response is required.

Changes in the plasma concentration and secretion rate of other hormones involved in the regulation of energy metabolism are shown in Table 1. Thyroxine secretion rate relative to its plasma concentration is relatively slow, the biological half life of released thyroxine being 30-40 h. It has long been an article of faith that slow adaptation to cold involves an increase in the secretion rates of thyroid hormones. This assumption naturally followed the observation that cold and hyperthyroidism have similar effects on metabolic rate. Evidence for an increased thyroxine secretion rate in domestic animals exposed to cold is rather slender and requires further study. Héroux (1969) has shown that elevated plasma thyroxine concentrations in rats are not necessarily related to thyroxine secretion rate, can frequently be attributed to an increased intake of diets rich in iodine, and when this is prevented and thyroxine turnover remains constant, acclimation to cold proceeds without impediment.

The secretion rate of cortisol is undoubtedly markedly increased during exposure either to acute cold in the laboratory or to the fluctuating and unpredictable stress of the winter (Panaretto and Vickery, 1970; Yousef et al., 1971). It is surprising, but reassuring to observe in Table 1 that plasma concentrations of cortisol, and of the other hormones listed, are rather closely associated with estimated turnover rates. There is, of course, an element of autocorrelation here since the latter value is, in part, derived from the former. However, one may infer from the common observation that plasma concentrations of glucocorticords tend to return to normal values following prolonged exposure to cold that secretion rates do the same, although as indicated earlier it is seldom clear in the literature whether adaptation to cold means a change in the mediation of the metabolic response or merely a diminution in the intensity of the stimulus.

The secretion rates of insulin and growth hormone are reduced by starvation, by increased sympathetic activity and probably by exposure to cold (Trenkle, 1971, 1972). There is evidence for an inverse relationship in rats between release of growth hormone and adrenocorticotropic hormone from the anterior pituitary (Kokka et al., 1972). A meat producing animal exposed to the combined, if intermittent, stresses of cold and shortage of food may exhibit an abnormal pattern of growth. This problem is largely unexplored, but its practical importance is obvious.

HORMONAL REGULATION OF COLD THERMOGENESIS

CATECHOLAMINES

The principal function of catecholamines (Fig. 1) in domestic animals exposed to cold is to mobilize energy rich substrates for catabolism in shivering muscles (Himms-Hagen, 1967). In the main energy depot, white adipose tissue, catecholamines acting via beta receptors stimulate lipolysis by causing adenyl cyclase to increase the intracellular concentration of cyclic AMP and thereby activate triglyceride lipase. Free fatty acids are liberated into the plasma and exported to the sites of thermogenesis. Catecholamines also stimulate glycogenolysis in the liver and increase catabolism and gluconeogenesis from amino acids in muscle. They also selectively inhibit glucose uptake in muscle by inhibiting insulin release (an alpha adrenergic mechanism) and thereby spare glucose

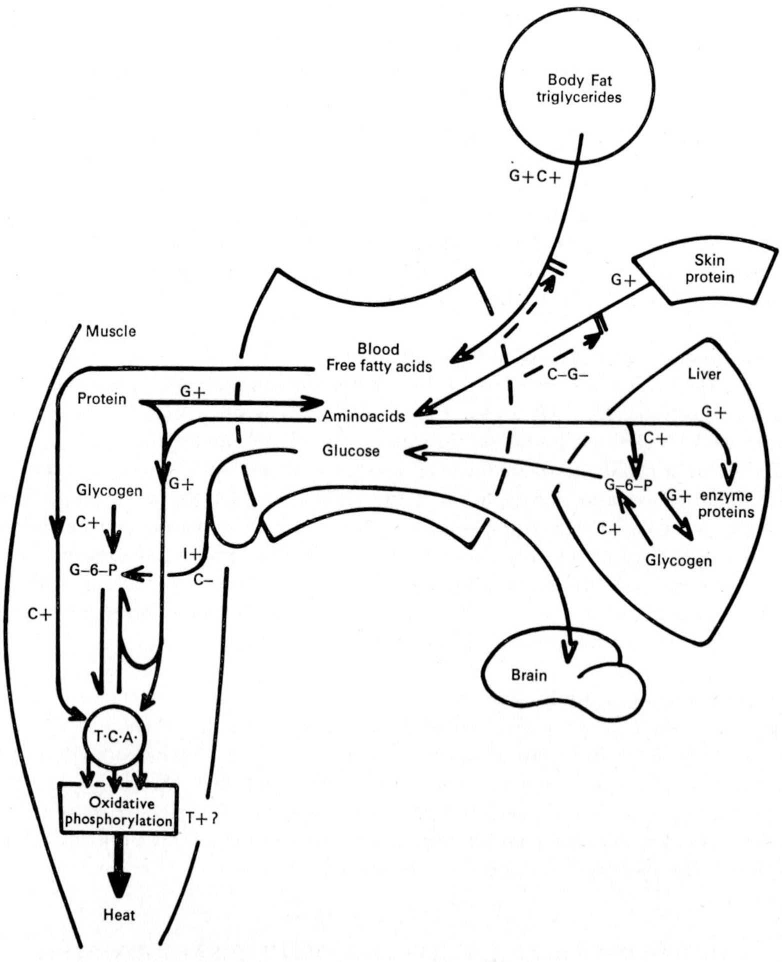

Fig. 1. Sites of action of hormones involved in the elevation of energy metabolism during exposure to cold. C = catecholamines; G = glucocorticoids; T = thyroxine; I = insulin; G-6-P = glucose 6 phosphate. + = stimulated; — = inhibited.

for vital tissues such as the brain. They probably also inhibit the release of growth hormone although this is less certain.

There is now reasonable agreement that noradrenaline induced regulatory nonshivering thermogenesis only occurs to a significant degree in those species which possess brown adipose tissue. Among the common farm animals brown adipose tissue has only been observed in the newborn lamb. In cattle, pigs, chickens and adult sheep therefore noradrenaline has no significant direct thermogenic effect (Webster, 1974a).

THYROID HORMONES

The work of Héroux (1969) has cast serious doubts as to the exact role of the thyroid hormones in the response of animals to prolonged cold. Thyroidectomized goats and sheep can elevate metabolic rate in the cold, but this is accompanied by an abnormally high increase in the excretion of catecholamines (Andersson et al., 1967). This indicates only that a euthyroid condition is essential to ensure a normal response to cold. Synergism between the thyroid and sympathoadrenal systems is well known. It is less certain whether hyperthyroidism is involved in the response to cold, and if so, whether it would be of any advantage to the animal. The effects of thyroxine are slow in onset and decay, and therefore are not particularly valuable to a precise control system. Domestic animals are not continuously below their lower critical temperatures even in severe winters, and a mechanism such as hyperthyroidism which produced an elevation in metabolic rate that persisted when the animals were not cold would deplete the body's reserves of energy unnecessarily. In view of this, studies of effects of thyroxine on oxidative phosphorylation (Fig. 1) for example probably have little relevance to the endocrinology of the response to cold.

ADRENAL STEROIDS

The steroid hormones most concerned in the metabolic response to cold are the glucocorticords cortisol and corticosterone. The actions of glucocorticoids are many and various (Ashmore and Morgan, 1967) (Fig. 1). Their best documented effect is to increase nitrogen excretion i.e., increase net protein catabolism. More precisely, glucocorticoids stimulate the breakdown of protein, and inhibit its synthesis in muscle, skin and lymphatic tissue, which might be considered as protein depots. In this way amino acids are mobilized for catabolism, gluconeogenesis and synthesis of new enzyme proteins in the liver. Glucocorticoids act in some way with the catecholamines to facilitate the mobilisation and probably the beta oxidation of free fatty acids. Their effect on carbohydrate metabolism is complex. Briefly, they enhance gluconeogenesis in the liver from amino acids, glycogen synthesis also in liver; and they may inhibit glucose uptake in tissues such as muscle, which can use other substrates, so as to spare it for tissues like the central nervous system for which it is essential. The ruminant obtains very little glucose direct from dietary carbohydrate. Thus, when glucose requirement is high, as in pregnancy and lactation, and the animal is faced by the stress of cold or undernutrition, ruminants are particularly susceptible to hyperglycaemia and ketosis. Many cases respond to therapy with glucocorticoids or synthetic steroids, which indicates that they are due to exhaustion of the adrenal cortex rather than to exhaustion of the substrates available for gluconeogenesis.

Glucocorticoids probably act in the liver by increasing the synthesis and activity of the enzymes involved in the reactions described above. Whether their other actions, such as their permissive effect on catecholamine stimulated mobilisation of free fatty acids, are similarly mediated is uncertain. If so, then these effects are surprisingly rapid compared for example with those of thyroxine.

INSULIN AND GROWTH HORMONE

These hormones are principally involved in the normal processes of energy and nitrogen metabolism whereby the substrates made available by digestion are directed towards the maintenance and growth of body tissues. During cold stress, catecholamines and corticosteroids probably act both to inhibit the release and the activity of insulin and growth hormone so that accumulated deposits of energy and nitrogen might be withdrawn rapidly in order to meet the environmental stress. The uptake of glucose by the brain is probably not insulin dependent and is therefore not inhibited by catecholamines.

ENDOCRINE PHYSIOLOGY OF THE SKIN AND BODY SHELL

Adaptation to cold in domestic animals, with the possible exception of the pig, has over successive generations involved changes in the coat of hair or wool, and probably changes in the skin as well. Cattle that are chronically exposed to cold have longer coats than cattle kept indoors. The evidence suggests that this is not because cold stimulates hair growth, but that it reduces shedding (Webster, 1974b). In sheep, which do not spontaneously shed their fleece, cold has apparently no effect on wool yield. However, the growth of hair or wool must be to some degree under hormonal control. More attempts have been made to exploit this phenomenon than to understand it. Ryder and Stephenson (1968) describe several experiments in which large doses of thyroxine were given to sheep. Wool growth was usually increased, at least temporarily, to the considerable detriment of the sheep underneath. It is unlikely that thyroid hormones play more than a permissive role in the maintenance of a heavy winter coat in cold exposed animals.

Exposure of sheep to very severe cold or starvation, or the administration of ACTH, have a markedly inhibitory effect on wool growth (Ryder and Stephenson, 1968). In extreme cases wool growth is stopped altogether; and several weeks later when the break in growth reaches the surface, the sheep may shed almost its entire fleece with catastrophic consequences. The reason why moderate cold apparently retards hair shedding in cattle is not known. An investigation into the hormonal mechanisms that underlie this phenomenon would be most interesting and might yield some surprising fringe benefits. Temperature regulation in regions such as the ears, the shanks and feet which have a very limited capacity to produce heat in situ is achieved entirely by variations in blood flow. Cold stress presents two conflicting challenges. The initial response to cold is usually a reduction in the rate of convection of heat to these extremities achieved by sympathetic vasoconstriction. At air temperatures below 0°C, however, this vasoconstrictor effect must be overcome or relaxed in order that the tissues do not freeze or suffer from avascular necrosis. The phenomenon of cold induced vasodilatation has been recognised for 40 years but it is still unexplained (Webster, 1974a). On balance, the evidence suggests that cold induced vasodilatation is an active process i.e., the vasoconstrictor effects of noradrenaline are not so much relaxed as overwhelmed by some powerfull vasodilator substance or substances. What these substances are, how their action changes during adaptation to cold and whether they fall into our traditional concepts of what constitutes a hormone remains to be seen.

REFERENCES: see pp. 517-518.

Effects of Acute and Chronic Exposure to Cold on the Endocrine System of Laboratory Animals

by

V. M. Petrović

INTRODUCTION

Immediate increase of the heat production in laboratory animals exposed to cold is a product of shivering and mechanism of this increased metabolism is linked with the sympathetic nervous system and the adrenal medulla (see Jansky, 1973). In the experiments in which the animals were prevented from shivering, a great portion of the heat production was found to be due to the activity of the adrenal medulla. So Thibault (1949) registered that about 35–40% of the cold thermogenesis of the white rat, in the absence of shivering, resulted from a secretion of medullary hormones. It seems that both hormones of the adrenal medulla, adrenaline and noradrenaline contribute to the heat production during cold exposure of laboratory animals. Urinary output of adrenaline and noradrenaline under these conditions was increased (Cottle, 1960b; Le Blanc and Nadeau, 1961; Leduc, 1961a,b). Kventansky et al. (1971) showed that adrenal epinephrine level was significantly lower after one day of exposure to cold, but not different from that in control animals after 7, 21 or 42 days. It was also found that the activity of catecholamine-synthesizing enzymes in the adrenals of the rat and ground squirrel exposed to cold was significantly higher than in the controls (Kventnansky et al., 1971; Petrović et al., 1973c). During cold acclimation noradrenaline becomes more calorigenic and this response becomes much greater than that of adrenaline. Noradrenaline is considered to be a true mediator of nonshivering thermogenesis, because of the magnitude of the response and of the absence of hyperglycaemia, unlike that found when adrenaline was injected (Carlson, 1960; Depocas, 1960). In the time course of acclimation to cold the capacity of rats to respond to noradrenaline becomes greater, shivering heat production disappears and nonshivering heat production dominates. Rats and guinea pigs adapted to 30°C do not respond to noradrenaline (see Jansky, 1973), but contrary to that adaptation to 30°C and even to 34°C does not prevent the sensitive reaction to noradrenaline in the ground squirrel (Petrović and Marković-Giaja, 1973).

In addition to the sympathetic nervous system and to the hormones of the adrenal medulla during acute exposure to cold also some other endocrine glands have very significant role in the metabolic alterations. These glands are the adrenal cortex, thyroid, and pituitary. Hypothalamopituitary system in general plays a fundamental role in the activity of other endocrine glands involved in the acclimation to cold.

ADRENAL CORTEX

There is much evidence that the activity of adrenal cortex of rats and guinea pigs considerably increases during the initial phase of cold exposure. The following parameters were analysed: cholesterol and ascrobic acid content in the adrenals, circulating glucocorticoids, incorporation of P^{32}, in vivo measurements of steroids production, fall in eosinophils, incorporation of labelled amino acids pool into adrenal proteins or orotic acid C^{14} into nucleic acids. It was found that cold exposure resulted in depletion of adrenal ascorbic acid and cholesterol content and in a decrease of eosinophils as well as in an increased incorporation of P^{32} (Levin, 1945; Héroux and Hart, 1954a; Nichols and Rossiter, 1956). In the rat and guinea pig as well as in the ground squirrel or golden hamster, an increase of circulating glucocorticoids during cold exposure was found (Boulouard, 1966; Schoenbaum et al., 1959; Schoenbaum, 1960; Petrović and Janić, 1964; Petrović et al., 1964). Furthermore a high level of circulating glucocorticoids found in the active ground squirrel coming from a natural habitat did not prevent an additional increase under the conditions of cold exposure (Petrović and Janić, 1964). Cold exposure resulted in an increase in the adrenal weight, protein and RNA but not in DNA content and also an initial increase in the incorporation of labelled amino acids into the total and soluble proteins (Petrović et al., 1966; Petrović and Rajčić, 1969; Petrović et al. 1973a,b). This initial increase was followed by a return towards the control level in the cold adapted rats. Many indications were found that adrenocortical activity of rats tends to return also towards normal control level after 3-4 weeks of cold exposure (Nichols and Rossiter, 1955, 1956; Rositer et al., 1959; Héroux and Hart,

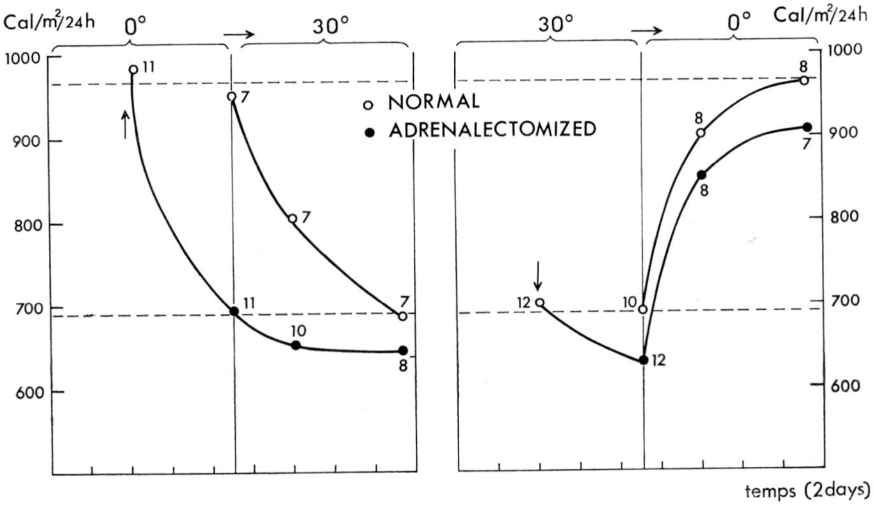

Fig. 1. Resting metabolism of normal and adrenalectomized rats before and after changes of ambient temperatures with numbers of experimental animals. The time is plotted on abscissa, each section representing 2 day periods. Broken lines: the mean values of resting metabolism of the controls adapted to cold (0°C) or to 30°C; horizontal arrows: the direction of changes of thermal environment; vertical arrows: the time of adrenalectomy. (From Petrović et al., 1961).

1954b; Schoenbaum, 1960). Some proofs for such changes in adrenocortical activity and related influence of glycocorticoids on the liver proteins were recently obtained in the rat (Petrović et al., 1973a,b). In fact the incorporation of labelled amino acids into the total and soluble proteins in rats being exposed to cold for 7 days was increased but in cold adapted animals was about the same level as in the controls. It seems that adrenal cortex is very important for the first reaction to cold but not indispensable for cold adaptation. It was found that cold-adapted adrenalectomized rats could be maintained on a constant but relatively small amount of cortical extracts daily (Héroux, 1955). Furthermore, it seems that a considerably amount of cold adaptation in the rat is possible without adrenals (Héroux, 1955; Petrović and Andjus, 1960; Petrović et al. 1961).

In the guinea pig the response of the pituitary-adrenal system is particularly intense, judging by the increase in plasma corticoids level, urinary corticoids excretion and ACTH concentration in the pituitary (D'Angelo, 1960b; Boulouard, 1966). Chowers (1970) found that in dexamethasone treated rabbits an increase in 17-hydroxicorticosteroids did not occur and there was a marked drop in body temperature, contrary to the control animals in which an increase in adreno-cortical activity and as a consequence, no change in body temperature was found.

The role of ascorbic acids seems to be very important for cold adaptation of laboratory animals (Desmarais, 1957). The positive correlation of growth in the guinea pig in the cold with ascorbic acid level in adrenals has been shown. In the rat the level of ascorbate in the tissues as well as its excreted amount, increased progressively with prolonged exposure to cold this animal being capable of synthetizing its own ascorbic acid. In the mouse ascorbic acid also has a beneficial effect for the survival in the cold. The effect was comparable to that found when cortisone was administered (Booker, 1960). It seems that the enhanced survival of mice in the cold by ascorbic acid is largely dependent on the adrenals or on cortisone administration. Ascorbic acid also increases metabolic rate of adrenocortical hormones in incubated liver slices and facilitates the conversion of C^{14} acetate to cholesterol in the presence of ACTH. It was indicated that ascorbic acid can stimulate the transport of electrons to increase oxidation of substrate without increasing phosphorilation (Hoijer, 1957).

THYROID

The increase in thyroid secretion during acute or chronic exposure to cold was shown in rats, guinea pigs, rabbits, hamsters and ground squirrels (Woods and Carlson, 1956; Stevens et al., 1955; Brown-Grant et al., 1954; Knigge, 1960; Fountain and Lachiver, 1955; Lachiver and Petrović, 1960; Cottle, 1960a; Cottle and Carlson, 1956; Smith, 1960). The amplitude and rapidity of the thyroid response to cold depends on the degree to which thermogenesis is stimulated. If the external temperature is very low, histological changes in thyroid occur within a few hours; but if temperature is moderate low, in mice, rats, rabbits, and guinea pigs, the thyroid activation is not apparent in 6 to 24 hours, but is apparent in 3 or 7 days, with a maximal effect in about three weeks of cold exposure (see Fontain and Lachiver, 1955). This direct involvement of thyroid in

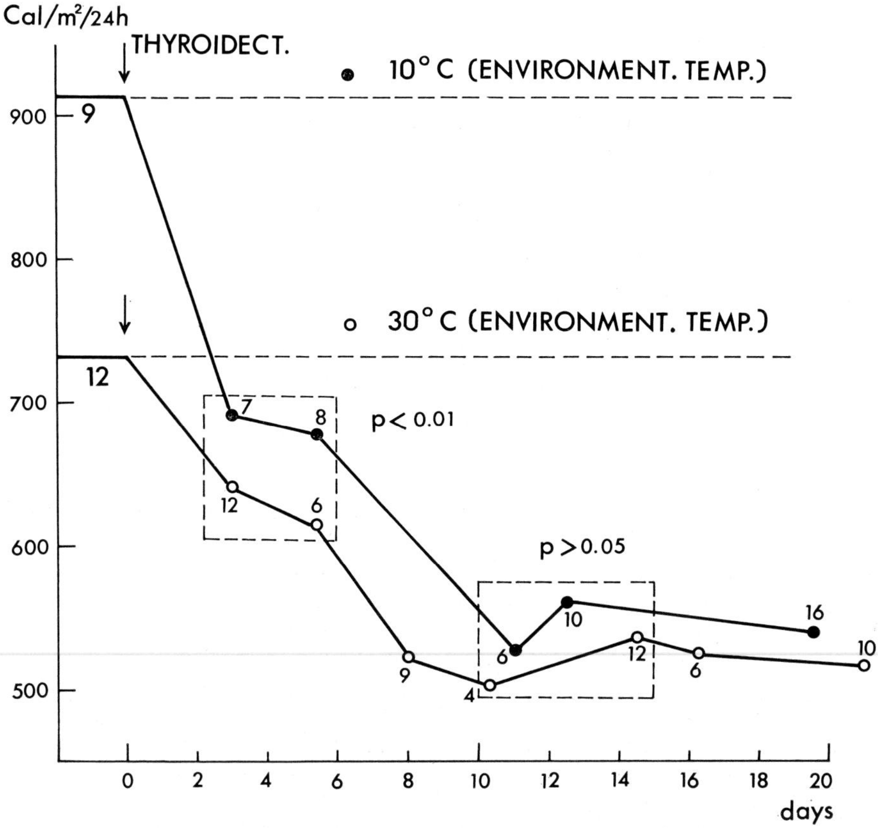

Fig. 2. Resting metabolism of normal and thyroidectomized rats adapted to cold (10°C) and to 30°C with numbers of experimental animals. Broken lines: the mean value of resting metabolism of the controls adapted to cold (10°C) or to 30°C; vertical arrows: the time of thyroidectomy. (From Petrović et al., 1961).

the cold adaptation by some authors became questionable when evidence of a normal or even reduced thyroid activity was found in animals adapted outdoors to winter conditions (Héroux, 1960; Héroux and Brauer, 1965). It seems that adaptation to outdoors conditions differs from that of laboratory conditions. Besides, in laboratory animals there is almost no evidence that the thyroid activity is increased during the cold period of the year (Héroux, 1960). This occurred in spite of the development of a metabolic acclimatization to winter conditions, of increased sensitivity to noradrenaline and of increased metabolism. However, it is to be pointed out that many other factors besides temperature may be operating in natural habitat; and it is usually not possible to say whether temperature, light or other physiological factors are responsible for some of the observed changes. Seasonal and circadian endogenous rhythms may also be very important. Outdoor conditions differ from laboratory conditions, and the obtained results concerning indoor and outdoor adapted rats cannot be always compared completely. We suppose that outdoor conditions may be also simulated indoors, at least in part, by including many factors taking place in natural habitats (humidity, wind, changeable cold, light, radiation). But, such experiments could be done only in collaboration with meteorologists and by using a special equipment.

Cadot et al. (1969) estimated thyroid function in rats exposed or adapted to 5 or 30°C, after an isotopic equilibration with J^{125} and found the similar level of thyroid function in rats fully adapted to 5 and 30°C as judged by total iodine turnover in the gland. Moreover, when the daily iodine intake was higher (75 μg/day), rats at 5°C had lower total thyroidal iodine turnover than rats at 30°C, but these authors concluded from their results that some additional experimenting should be done. Pretreatment with thyroxine and especially with thyroxine and noradrenaline, increased the resistance to cold in the rat and a chronic treatment increased the sensitivity to noradrenaline (LeBlanc and Villemaire, 1970). On the other hand, changes in thyroid activity in cold acclimated animals as well as in heat acclimated ones appear to partially depend on food intake (Héroux and Petrović, 1969; Yousef and Johnson, 1968).

Many facts showed that the action of various hormones cannot be considered independently but in the contest of their interactions. Already the studies of Thibault (1949) have shown that the thyroid hormones have a pronounced influence on the calorigenic effect of adrenaline which may be reduced without thyroid secretion. Swanson (1957) found that the response to 60μg of adrenaline was greater in coldacclimated rats to an extent that was consistent with the increased turnover of thyroxine observed by Cottle and Carlson (1956). Hsieh and Carlson (1957) have shown that cold induced nonshivering thermogenesis progressively decreases with the time following thyroidectomy. Recent experiments of Lutherer et al. (1969) also support the concept that thyroid hormones affect tissue sensitivity to catecholamines. Sellers et al. (1970) demonstrated that the reciprocal relationship between thyroid hormones and catecholamines in heat production during cold adaptation is only partial and that both are essential.

HYPOTHALAMUS AND ADENOHYPOPHYSIS

The increased thyroid activity in the cold exposed animals is brought about by an increased secretion of thyroid-stimulating hormone of the adenohypophysis. In rats adapted to cold, besophil cells which are responsible for TSH production, are more numerous than in the nonadapted controls, and serum TSH is more easily detected. In laboratory animals, such as the rat, guinea pig, and hamster, ventral hypothalamic lesions, which produce persistant oestrus, the basal pituitary-thyroid function appears normal, but thyroid activation does not occur when the animals are placed at 5°C (Knigge and Bierman, 1958; Knigge, 1960; D'Angelo, 1960). Cold stimulus acts on the hypothalamus and gives rise to a neurohormone at the level of the median eminence which is transported to the cells of the Pars Distalis and there activates the secretion of TSH (see Harris, 1955; Harris and Donovan, 1966). Exogenous thyroxine is known to have a depressing effect on TSH secretion and on thyroid activity, but the exact site of

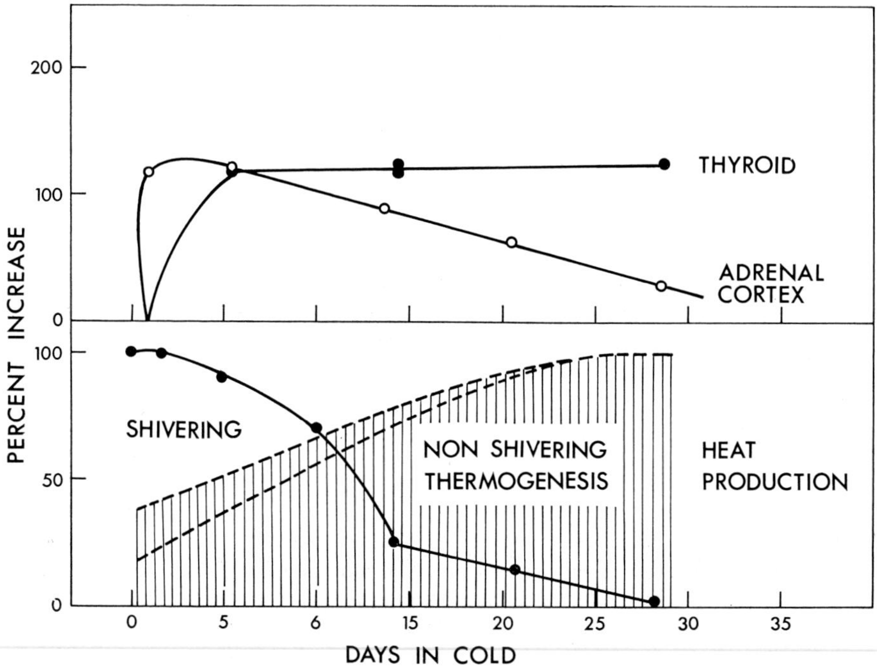

Fig. 3. Summary of endocrinological changes and thermogenesis during acclimation to cold. (Summarized by Hart, 1963).

thyroxine action in regulating TSH secretion is not completely defined. It seems that both sites, hypothelamus and pituitary, are involved in the regulation of TSH secretion.

Cold exposure seems to activate neural mechanisms activating TSH discharge within increase in short time, being 0.5 and 3 hrs for the rabbit and hamster respectively. This relatively TSH secretion is intense but transient. As a consequence a marked decrease in the pituitary TSH content was observed (Itoh et al., 1969).

The increased activity of the adrenal cortex during the first few hours of cold exposure is due to the effect of ACTH stimulation, the concentration of which in the pituitary augmented considerably during cold exposure of laboratory animals. Later phase of adrenocortical activity in the cold conditions is probably related with the increased thyroid function. Thiouracil treatment caused a reduced P^{32} incorporation into the cortex which was reversed by thyroxine in animals at room temperature (Hart, 1963; Fortier, 1966; Chaffee and Roberts, 1971; Mallory et al., 1959).

Because of the fact that in laboratory animals hypothalamo-pituitary adrenocortical system reacts immediately after cold exposure it is probably that feed back mechanism of activation in this case does not play a significant role at least in the conditions of acute exposure. It is possible to conclude that under these conditions blood corticoid levels do not appear to play a central role in the initiation or control of ACTH secretion. The neural drive seems to be capable of inducing and maintaining a sustained increase in blood corticoid levels (see Fortier, 1966).

In the hypothalamic disconnected rats, contrary to the controls, (except those with anterior hypothalamic disconnection) a significant inhibition of corticosterone release upon exposure to cold was found (Chowers et al., 1972). Dupont et al. (1971) have recently demonstrated that hippocampal stimulation in the rat inhibits the pituitary-adrenocortical response to stress. The similar effect has been previously foond in the cat (Endröczi and Lissak, 1962) and rabbit (Kawakami et al., 1968). These results suggest that the role of the hippocampus and of other components of the limbic system in the control of ACTH and TSH secretion should be also taken into consideration for futher systematic studies on neurohumoral mechanism involved in short-term exposure to cold as well as in adaptation and seasonal acclimatization of laboratory animals.

SUMMARY AND CONCLUSION

The first reaction to cold is linked with the sympathetic nervous system and the adrenal medulla. Both hormones of the adrenal medulla, adrenaline and noradrenaline, contribute to the heat production of laboratory animals under cold conditions.

Hypothalmo-pituitary-adrenocortical system reacts immediately after cold exposure and neural drive is capable of inducing and maintaining a sustained increase in blood corticoid levels. Some results indicate that adrenocortical activity in the rat tends to return towards the control level after 3-4 weeks of cold exposure. Furthermore, a considerable amount of cold adaptation is possible also without adrenals.

Cold exposure affects neural mechanism activating TSH discharge in a relatively short time. The increased thyroid function is persistent in the cold conditions with a maximal value in about three weeks of cold exposure. The thyroid seems to be indispensable for cold adaptation of laboratory animals. This conclusion became questionable, however, when a normal thyroid activity in animals adapted outdoors to winter conditions was found. Besides, food intake seems to play a significant role in the increased thyroid activity in the cold.

Many facts showed that the action of various hormones cannot be considered independently but in the context of their interactions.

Recent data suggest that thyroid activity in adaptation and seasonal acclimatization of laboratory animals should be studied more precisely. The role of the involved neurohumoral mechanisms including that of hippocampus has to be also taken into consideration.

REFERENCES: see pp. 518-521.

Water Metabolism in Sheep in Relation to Heat and Cold

by

W. V. Macfarlane

INTRODUCTION

In the field several situations inducing differential effects of heat and cold on water handling by sheep have been investigated. These are (1) studies of the seasonal flux in water turnover and content between summer and winter, (2) removal of insulating fleece in summer and winter, (3) translocation of animals between 21°S (summer rainfall area) and latitude 35°S (winter rainfall area).

SEASONAL WATER METABOLISM

Among Merino sheep grazing mixed pasture in a Mediterranean environment, body composition and water spaces change seasonally (Fig. 1). Winter minimum temperatures reach −5°C and summer maxima 45°, while the rainfall ranges between 350 and 550 mm, more than two thirds of the precipitation being in winter and spring. At the beginning of summer there is dry pasture but by the end of summer there is very little food since no growth of pasture takes place until the end of autumn. During the drought period sheep eat little, drink relatively little and lose body weight as their fat reserves are used. During this period (Fig. 1) in spite of higher temperatures which induce respiratory evaporative cooling and some sweating, total water turnover is less than 3 l a day (60 ml/kg/24 hr). In this period of undernutrition on dry grass little drinking is undertaken beyond that needed for urine and respiratory water losses (Macfarlane and Howard, 1970). The extracellular space expands relative to the body weight and the total body water increases as fat is displaced. In sheep partly starved during summer drought the strain is rather less than in cold starvation, since the metabolic rate is depressed and there is no need for shivering. Fleece provides protection from the sun by reflection or by reradiating long-wave energy from its surface – which may reach 93°C in the sun (Macfarlane et al., 1956, 1964; Priestley, 1957).

When rain falls (usually in April) there is a latency of 6 weeks before sufficient vegetation has grown to alter the nutritional status of the sheep. Initially there is an increase in water turnover as they nibble fresh grown grass of high water content. As the supply of this improves, both food and water intakes increase and at the same time wool growth accelerates (Macfarlane and Howard, 1970). During summer the rate of wool growth is less than one third that achieved in the winter. Growth of wool is thus a direct function of both nutrient intake and of water turnover. Some time after wool synthesis increases there is sufficient excess energy to allow deposition of fat to replace that lost during the summer drought. By middle to late spring, the daily food intake is between 2 and 4 kg of

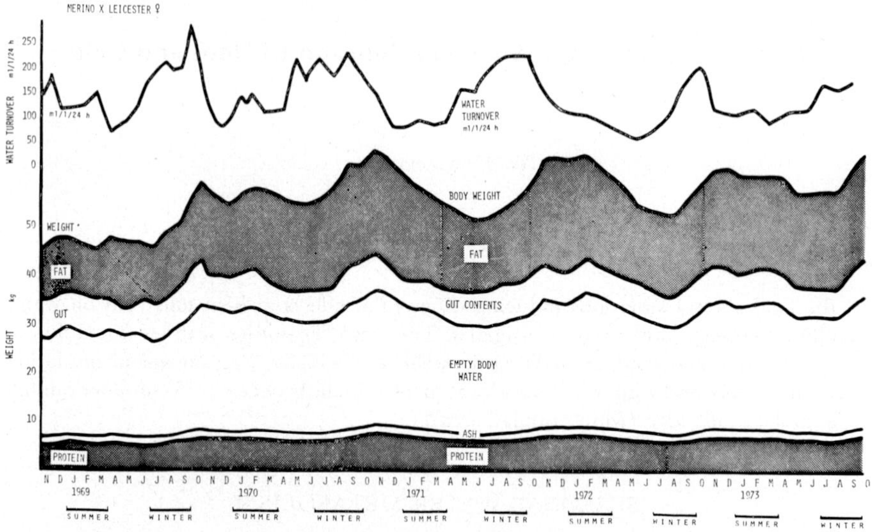

Fig. 1. Body composition of a grazing ewe during 5 years, measured by TOH dilution (lat. 35°S, winter rainfall zone). When rain falls in late autumn, grass grows and intake of food with its contained water increases. Water turnover is proportional to food intake. After 2 months body weight rises and fat is deposited. When spring grass fails (1971) fat is drawn on through late spring to autumn. But in years with summer rain (1970) there is little change in fat content over summer. Protein content of the body as calculated is greater in spring than autumn but the flux is smaller than that of water and fat.

dry matter and since virtually all of this food carries 3 parts of water to 1 part dry matter, the water turnover reaches 8 to 12 l daily (160-250 ml/kg/24 hr). This is 4 to 6 times more than in summer. More food is eaten when there is more water and the extra metabolizable energy allows both tissue growth and storage of fat for the reproductive period, which is optimal after the longest day. As summer approaches, the grasses dry off but retain nutrient value. As they dry the water turnover decreases. The total dry matter intake also decreases (Fig. 1).

Towards mid-summer the sheep are mated and lambs are born at the end of autumn usually as the new grass grows. Gestation begins while sheep live on fat reserves and any dry food available. Lactation – which increases water turnover by nearly 50% (Macfarlane and Howard, 1972) and energy turnover by about the same increment – is undertaken on new-grown pasture but even so the ewes lose fat more than dry sheep, and the extracellular and total body waters increase.

There is thus a cyclicity of water intake and body fluid content in reproduction which runs parallel with the rates at which energy is consumed seasonally by grazing sheep.

SHEARING AND WATER TURNOVER

A. WINTER SHEARING

The removal of up to 6 clo of insulation from a sheep in winter instantly subjects it to thermal strain. The time of year chosen for shearing is tactically important since lack of insulation can lead to rapid death in a hot, or a cold environment. There are 3 major forms of insulation in a sheep. There may be subcutaneous fat to assist in the reduction of heat flow. In some skin areas there is vasoconstriction, which is confined essentially to legs, ears and face. There is almost no effective vasoconstriction in the woolcovered skin of sheep. When the wool is removed heat continues to flow rapidly between blood and the environment in the trunk region. The third insulation is of course the fleece which is effective when dry, but it loses insulation value when wet. The thermoneutral point of a sheep without wool is 25° to 27° (Blaxter, 1962) and the balance point drops proportionally to the amount of fleece grown. With full insulation a sheep is in balance at about −30°C in still air, but in wind and rain this may rise to +6°C (Webster and Park, 1967). Winter shearing may have a death rate of 2 to 36% (Brown and Hutchinson, 1973). Lambs are even more affected by cold than adults (Alexander, 1970). Energy balances are discussed also by Porter and Gates (1969).

Table 1. Effects of summer and of winter shearing on fluid spaces and body composition of Merino sheep

		Winter shearing		Summer shearing
	Sown pasture lat. 35°S	Well fed 15/ha	Poorly fed 22.5/ha	*Astrebla* pasture lat. 27°S
Water turnover ml/kg/24 hr	Before	63	32	81
	After	91	53	159
	Δ%	+44.5%	+65.5%	+96.3%
Body water ml/kg	Before	562	561	686
	After	573	666	670
	Δ%	+ 1.9%	+18.7%	− 2.3%
Extracellular volume ml/kg	Before	242	246	256
	After	277	288	287
	Δ%	+11.5%	+17.1%	+12.1%
Plasma volume ml/kg	Before	52	49	63
	After	58	60	74
	Δ%	+11.5%	+22.4%	+17.5%

Summer maximum temperatures were 35-38°C in the sub tropics; winter 8-20°C in the cool temperature zone. The tropical sheep had higher body water content and turnover than the lat. 35°C sheep. Water turnover increased more because of evaporative cooling in the tropical sheep when shorn.

In winter water turnover was proportional to food supply, and it was greater in the 15/ha than in the overstocked 22.5/ha sheep. The replacement of fat by fluid was greatest in the 22.5/ha animals, and extracellular fluid expanded most in this group.

Water turnover in the cold rose to meet increased energy expenditure, but in the heat, water turnover and extracellular fluid increased with the demands of evaporative cooling.

When sheep are shorn in winter the consequences in metabolism and water economy are a function of the prior nutritional state. Two groups (Table 1) of 4-yr old Merino wethers were studied before and after shearing in winter in an environment where the daily maximum temperatures averaged 10°C and the minima −2 to −7°C with considerable wind and some rain (Morris et al., 1962). Fluid spaces were measured with thiocyanate for extracellular volume, Evans' blue for plasma volume and tritiated water for total body water and water turnover. All the sheep were initially well supplied with fat, since body water comprised 56% of the body weight. Sheep grazing at 15/ha turned over water at 63 ml/kg/24 hr. Overgrazed pasture (22.5/ha) provided only 32 ml/kg/24 hr (in the cool weather neither sweating nor drinking was required and water turnover derived from food). In winter, food intake (Macfarlane and Howard, 1970) is directly proportional to the water turnover. The food available for the 22.5/ha sheep was only half that taken by the better fed 15/ha sheep, as confirmed by control estimates from oesophagostomy sheep. Rectal and skin temperatures were the same in the two groups.

After shearing in the coldest month with exposure to wind, rain and frosts, both groups of sheep shivered continuously for more than a month. During this time the water intake of the 15/ha sheep increased steeply to 91 ml/kg/24 hr as they ate more food (Table 1). The water turnover of the 22.5/ha group rose from 32 to 53 ml/kg/24 hr with less available food. The rectal temperatures of the 22.5/ha sheep were at least 1° lower than those of the better fed animals, and over the coldest period they fell to 34°C. About 1 month after shearing the 22.5/ha sheep had increased total body water to 67% of body weight (from 56% thus they lost body solids) and they had gained 2.1 l of water (distributed between extracellular, vascular and cell and gut water spaces). On the other hand the better fed sheep increased relative water space by only 1% and they lost 1.3 l of fluid. So the 15/ha sheep lost 7.5% of body solids while the 22.5/ha sheep lost 30.7% of body solids (essentially fat). Since the absolute weight changes of the groups were similar it was apparent that as fat was burnt by the poorly fed sheep, it was replaced by water, which concealed the tissue loss. The 15/ha sheep lost 14.3% of cell and gut water, but gained extracellular volume, while the 22.5/ha sheep gained 10.7% of cell and gut water as well as extracellular fluid. This finding was reinforced in other circumstances where undernourished sheep, even in warm weather, were found not to have lost as much weight as would be expected, because extracellular volume had increased and twice as much fluid was present in the rumen as in rumena of well fed sheep at the same time of the year. It appears that starved animals take water which accumulates in the rumen presumably as a compensation for the lack of dry matter.

The findings illustrate several points in relation to water and nutrition.

1. Fluid intake is proportional to dry matter intake. The more food eaten the more water is turned over, and the more water is retained in the body spaces.
2. As fat reserves are used they are in part replaced by increased water in extracellular spaces.
3. Undernourished sheep increase the volume of fluid in cell and gut space as well as in the extracellular volume, an adjustment which conceals the real losses of tissue weight.

Removal of insulation in the cold is a way of inducing acute starvation since metabolic rates may be raised 3 and 4 times above the normal level. The 22.5/ha sheep were probably burning over 18,000 kJ each day, relative to the normal 6,000 kJ for sheep of this size. With this high energy turnover and inadequate food intake, a starvation type of water distribution was induced. Panaretto and Ferguson (1969) have shown also that cortisol is mobilised in the cold and as metabolic damage occurs transaminases appear in the plasma of sheep.

B. SUMMER SHEARING

The removal of fleece in summer is also traumatic (Table 1) in its consequences. Ten Merino wethers were shorn in January, the hottest month, in the dry sub-tropics on latitude 27° with maximum air temperatures between 36° and 38°C. The *Astrebla* (Mitchell grass) pasture was dry and low in protein but the sheep were in good nutritional condition with an average total body water of 686 ml/kg. The surface of the wool reached 85°C under the afternoon sun but the wool protected the skin from solar heating by long-wave radiation from the wool surface (Macfarlane, 1956, 1964). Rectal temperatures averaged 39.9°C while sheep were in the sun. Between 12.00 hr and 15.00 hr the respiratory rates averaged 137/min (Macfarlane et al., 1966).

After shearing the sheep changed in behaviour. They no longer grazed during the day but sought any minimal shade that was available, panting at rates up to 200/min. The average rate while standing in the sun was 178/min relative to 125/min for unshorn sheep in the same environment. In the sun the surface of the clipped wool (6-10 mm long) was 46°C and the skin 42°C, with rectal temperatures averaging 40.5°C, a rise of 0.6°C. During the day the skin became red from sunburn and the animals doubled their water intake. This rose from 81.5 ml/kg/24 hr to 159 ml/kg in the first 3 days, falling to 140 ml/kg/24 hr in the next 5 days. At the same time the extracellular volume rose from 256 ml/kg to 287 ml/kg and the plasma space rose proportionally (Table 1). There was a 17% rise in plasma volume and the plasma proteins were diluted 17% during the time of heat strain brought about by shearing. From calculations relating to body solids loss, and water turnover increment, it may be estimated that the heat input rose by some 1550 kcal/24 hr (6500 kJ/24 hr) after shearing. This additional heat equals the standard metabolic rate of Merinos of this size. Shearing, therefore, more than doubled energy and water turnover and increased extracellular spaces. The circulatory compensation for the extra heat and water flow was an increase of plasma and interstitial volumes, without which a state of dehydration and shock would be induced.

Sunburn is a hazard in summer shearing since ultra-violet radiation penetrates to the skin, which becomes necrotic, and plasma ooze is added to the fluid losses, so that death is not an uncommon sequel. There is no pigmentary protection in most sheep skin.

The tactics of shearing, therefore, can considerably modify the fluid spaces and water turnover of sheep shorn in winter or in summer. The judicious choice is clearly spring or autumn when it is less likely that there will be extremes of temperature. Probably the most dangerous combinations are high air temperature with solar radiation on the one side, and low air temperature with wind and rain (particularly wind) on the other side. Both are clearly to be avoided.

TRANSLOCATION BETWEEN LATITUDES 21°S AND 35°S

The effects of summer and winter rainfall patterns on the same types of animal and their fluid dynamics were investigated by two reciprocal transfers of Merino sheep between 21° and 35° latitude in Australia. In tropical arid grassland rain falls in summer, usually about 350 mm for the year. Pasture comprises sparse clumps of *Astrebla* spp. (Mitchell grass), a low protein food. There is no shade on the dry plains. At 35° latitude there is a Mediterranean climate with winter rains of between 400 and 600 mm for the year. Mixed rye-grass and clover pastures grow in winter and spring. Sheep from each environment were transferred to the other environment in autumn, and other sheep in spring. The sheep from one environment rapidly took on the body fluid distribution pattern and wool growth rate of sheep in the new environment, but differences between local and translocated animals remained in each environment. Temperate zone sheep increased in body water content and lost fat on transfer to the tropics, but they did not fall as low in body solid content as the local sheep. Wool growth rate decreased but remained above the production level of the local sheep. When tropical sheep were moved to the Mediterranean environment in autumn they slowly increased in body solid content. It was, however, nearly a year before the rate of wool growth rose from that in the tropics towards that in the cooler climate in spite of large amounts of green feed. Even when adaptation had taken place the tropical sheep maintained a lower growth rate of wool than the local sheep. Reproduction was initially poor in all the translocated animals. The amount of milk produced by the Mediterranean sheep in the tropics was lower than that of their congeners in the south but greater than that of tropical animals. The tropical sheep in the cooler climate produced less milk than the local sheep.

Table 2. Water content and turnover of similar Merino ewes grazing in or translocated from the tropics lat. 21°S, to the temperate zone, lat. 35°S.

	Body solids, % B.wt.		Water turnover/24 hr			
			l		ml/kg	
	Summer	Winter	Summer	Winter	Summer	Winter
Lat. 21°S	23.8 (W)	31.0(D)	6.0 (W)	2.4 (D)	190 (W)	62 (D)
Lat. 35°S	36.0 (D)	26.3 (W)	2.9 (D)	8.5 (W)	69 (D)	250 (W)

A group of 40 ewes 15 months old was divided so that half remained in the tropics, half grazed on temperate pastures. The relative effects of heat and cold, compared with wet and dry seasons are clear, so that water in the pasture determines most of the water turnover. W = wet; D = dry

Water turnover and water content were greatest (Table 2) in the wet season of either environment. In the summer wet of the tropics twice as much water went through the animals eating succulent pasture as in the hot dry summer of the 35° latitude animals on dry grass. Similarly in winter when it is dry in the tropics only one quarter as much water passed through the tropical sheep as was turned over by their congeners on cool temperate pastures. This of course was due to the high water content (70-80%) of the winter pastures of the temperate area. So in sheep, total water content, extracellular volume and water turnover are as much products of the food available and its water content, as of the ambient temperature. During the dry winter season in the tropics much the same water picture is found as in the dry season of the temperate summer. These water-energy relationships were reflected also in the amounts of wool grown, in the potential fecundity and in milk production of the animals.

ELECTROLYTES AND WATER METABOLISM

All ruminants are exposed to high potassium intakes since normally the plants they eat are made up of cells containing potassium and they have little sodium content. Tropical pastures are even lower in sodium than temperate zone grasses (by a factor of 10 to 20) since sodium is leached from tropical (and mountain) soils by the high rainfall. Potassium concentration is usually greater in the temperate zone grasses than in tropical pasture. Potassium acts as a diuretic in sheep and it moves rapidly through the kidneys, partly by direct passage and partly by secretion from the distal tubule. If there is no renal damage, potassium is normally fully excreted and the animal balances readily with this cation (Macfarlane, 1971).

In mountains and tropics where there is little sodium, aldosterone is secreted in order to retain renal sodium and thus maintain the extracellular volume (Blair-West et al., 1965). In the desert, however, there are many plants which accumulate sodium and some that store potassium (*Kochia*). Leaves of the saltbush family (*Atriplex*) may contain over 15% of their dry weight as sodium salts (3 mol/kg) with potassium at 1 mol/kg. When such food is eaten, sheep compensate by drinking extra water proportionally to the salt load, which is thus excreted. If water is not available either sheep do not eat the saltbush or the nervous system suffers damage from the sodium. In desert conditions salt concentrates in troughs, pools or lakes in summer. Sheep have been poisoned by 2% (234 mmol/l) of sodium chloride in drinking water but they can live satisfactorily on 1.3 to 1.5% sodium chloride (Moule, 1945; Potter, 1963; Peirce, 1968). Equatorial goats tolerate somewhat more salt in their drinking fluid than sheep, reaching 257 mmol/l without apparent ill effect. Most sheep begin losing appetite at 150 mmol/l.

The salt tolerance of Leicester and Merino wethers was compared in the field, when their exclusive grazing was *Atriplex nummularia*, a form of saltbush which contains about 1 mol potassium and 3 mol sodium per kg of dry matter (Macfarlane et al., 1967). When grazing grassland (*Danthonia* spp.) the Leicesters used some 30% more water (compensated for size difference) during summer than the Merinos. The *Atriplex* intake was enough to maintain weight in the sheep although it is not their favored food. There was a rapid increase of water turnover when the sheep ate *Atriplex*, and total body water increased. The greater the water turnover the lower plasma sodium concentration became, falling from an average of 154 down to 136 mequiv/l. Urinary concentrations did not rise above 1000 mosmol/l. Merinos produced 924 $\pm$ 173 mosmol/l although these sheep can readily achieve 3000 mosmol/l when deprived of water (Macfarlane et al., 1961). Evidently the salt intake induced drinking to maintain a relatively low electrolyte content of urine and the salt intake was washed through without accumulation – in terms of concentration of body fluids – but extracellular volume increased (Macfarlane et al., 1967; Jones et al., 1970). There was a 100% scatter between the responses of individuals to salt in each breed. Some drank nearly twice as much as others while eating saltbush but invariably the Leicester sheep took more than the Merinos and maintained the same overall ratio shown on grassland (Table 3). Wilson and Hindley (1968) also found that if sheep have restricted access to water while eating saltbush, Leicesters took more food and more water per $kg^{0.82}$ than the Merinos. When water was restricted, however, food intake was reduced proportionally to water lack. Thus limits to the amount of saltbush that could be grazed are imposed by the supply of water available. The other limitation comes from the total water cost imposed by the environment. If radiation and air temperatures are high, water will be required for cooling, rather than for salt excretion. When food is desiccated, the demand for drinking water is increased; and if the water is salty lower limits to survival in saltbush associations are set (Wilson, 1966, 1968).

Table 3. Water intake of Border Leicester and Merino wethers grazing grass pasture and saltbush in summer.

Breed	Border Leicester		Merino	
Grazing	Grass*	Saltbush†	Grass	Saltbush
Body solids %	22.8	22.6	28.0	25.5
Water intake				
1./24 h	6.9 ± 1.0	13.7 ± 2.6	5.9 ± 1.5	9.4 ± 2.9
ml/kg/24 h	173	350	111	196
$ml/kg^{0.82}/24$ h	337 ± 40	677 ± 126	227 ± 53	385 ± 117

* *Danthonia* spp.
† *Atriplex nummularia*

Leicester sheep have higher water and energy turnover rates than Merinos. On saltbush they used more water than Merinos, and were more heterogeneous than Merinos in response to salt. Saltbush increased the variance of water intake. Means ± S.D.

When sheep ingest more salt than they can rapidly excrete (either by drinking salt water or by eating saltbush without access to water) there are pathological sequelae. Slow salt poisoning leads to loss of weight, wasting, reduced wool growth and poor lactation. Neurones are intoxicated by excess sodium. The sheep lose appetite, become weak, incoordinated and may have epileptic convulsions. There is diarrhoea and the intestine is inflamed. Adaptation to sodium excess may occur, more in some sheep than others. Part of this may be associated with increase of Na K ATPase activity in the intestine, liver and kidney (unpublished observation on rats), and there is increase in adrenal size (Potter, 1968) of sheep chronically ingesting 220 mequiv/l sodium chloride. Both cellular pump activity and hormone output may contribute to salt tolerance, on a genetic background.

SUMMARY

The water content of sheep responds to seasonal heat or cold mainly through changes of pasture composition. Lush pasture increases water content. Hot climates increase water content also, mainly in the extracellular compartments.

When water turnover is increased by the succulence of pasture, by the demands of evaporative cooling or by additional intake of salt, the total body water increases.

The effects of heat and cold are exaggerated by removal of the fleece. This leads to greater water intake, enlarged total water space and increases of interstitial and plasma volume. Winter shearing is the equivalent of starvation, since energy equal to several times the normal metabolic rate is lost through the skin. This produces a water metabolism similar to that of starvation, in which fluid spaces are expanded and fat is replaced by extracellular and gut fluid.

When the same type of sheep are exposed to hot wet, cool dry, as against hot dry and cool wet seasons (by translocation) there is some persistence of the initial fluid distribution pattern. But there is also high water turnover in the wet season and lower turnover in the dry whatever the temperature, though more water is used by the temperate zone than by tropical sheep.

REFERENCES: see p. 521.

Water and Salt Metabolism in the Camel

by

G. M. O. Maloiy

INTRODUCTION

The ability of the camel to cross hot waterless tracted country for days or even weeks without drinking water has long been appreciated by traders and travellers in the desert. Yet it was only recently that physiological and behavioural mechanisms which enable this animal to survive and thrive in these hostile environments have come to be studied by biologists. This paper will review and summarize the current knowledge on the physiology of this desert animal.

Camels are well distributed in most deserts of the world where they are herded by desert people like the Bedouin. They are used as pack and saddle animals, as well as being kept for milk, meat and hides. Physiological studies of K. Schmidt-Nielsen and his associates have shown how a small desert rodent, the kangaroo, rat, can live in the desert without drinking water. The secret of this success lies as much in its behaviour as in its physiological adaptation, for it manages to escape the heat of the day by retiring to its underground burrow. The kangaroo rat has also got a low water trunover and a remarkable kidney which is capable of excreting a high osmotically concentrated urine.

Studies carried out by Schmidt-Nielsen and associates (1956, 1957a and b), Macfarlane et al. (1963, 1972) and Maloiy (1972) on temperature regulation, metabolism, renal function, water and salt metabolism, clearly demonstrated that the camel is physiologically adapted for withstanding the harsh conditions imposed by the desert environment. The camel weighs over 500 kg and it cannot of course escape the heat of the day, though by appropriate behaviour, it does reduce the amount of heat it receives from its surroundings. In addition we now know that the camel keeps cool by sweating following heat exposure. In the desert, therefore, the camel uses water for evaporative cooling. Unlike small desert rodents, the camel cannot be independent of drinking water. In the Sahara, for example, Bedouin still require to water their herds of camels regularly.

WATER METABOLISM OF THE CAMEL

WATER BALANCE

The water metabolism of the one-humped camel (*Camelus dromedarius*) has been studied in the field in the Sahara and Central Australia (Schmidt-Nielsen et al., 1956), Macfarlane et al., 1963, 1964; Siebert and Macfarlane, 1971) and in Kenya under controlled conditions inside a climatic chamber (Maloiy, 1972). Before discussing these results in detail it is advantageous to consider the question of the avenues of water intake and water loss in the camel.

The camel acquires water by drinking, through preformed water in the food it eats, and by formation of water as it oxidizes its food. It loses water through evaporation (mostly sweating) and in its urine and faeces. Female camels may also lose small amounts of water to the foetus during pregnancy and large amounts in their milk during lactation. The camel is said to be in water balance when the avenues of water intake equal those of water loss.

In the laboratory, large metabolism cages were built for experiments on camels. These cages permitted separate collections of urine and faeces as well as the measurements necessary for a complete water balance. Facilities were also available to expose the camels to environmental temperatures of either 22°C or 40°C. These data thus enabled quantifications of each of these avenues of water intake and loss in both normally hydrated and dehydrated camels at differing degrees of heat stress. The water expenditure of camels fed on dry hay and exposed to a moderately cool environment (22°C) and in a simulated desert environment (alternating 22-40°C) is shown in Figure 1. When camels had free access to water, the estimated daily intake from all sources averaged 2.06 l/100 kg body weight in the cool environment and 3.84 l/100 kg body weight in the simulated desert environment. Comparable figures for another desert herbivore, the Somali donkey, were 7.45 and 8.82 l/100 kg body weight, respectively, for the same environmental conditions. The camel then is found to be economical in its water expenditure compared with other wild and domestic desert herbivores studied by Taylor (1968), Maloiy (1970), and Maloiy and Taylor (1971). Studies of Macfarlane (1964) and Macfarlane and Howard (1972) seem to indicate that there are differences between the water metabolism results obtained in the field and the laboratory.

Schmidt-Nielsen et al. (1957 and 1967) have made a study of body temperature of camels in relation to the water economy and metabolic rate. When exposed to high air temperatures (40-45°C) and solar radiation, the camel dissipates excess heat by increasing cutaneous evaporation of water (sweating). At the same time, at these high temperatures and especially when dehydrated, the camel allowed its body temperature to increase by as much as 3-5°C, from below 35°C in the early morning to 40°C after 8 or more hours of exposure to a heat load and solar radiation in the desert. The physiological significance of these temperature fluctuations in the body of the camel has been discussed elsewhere (Schmidt-Nielsen et al., 1957b, 1967).

Figure 2 depicts the amounts of water lost in evaporation, urine and faeces by the camel and the donkey in a simulated desert environment. In both these desert herbivores, evaporation accounts for a large fraction of the total water lost. When exposed to heat and dehydrated, the camel reduced evaporative water loss. This reduction of evaporative water loss during dehydration in the heat is especially important in the overall water conservation of the camel in the desert.

It has also been observed that following dehydration in the desert, the camel reduces its faecal water loss. Thus a camel exposed to the desert heat and deprived of water for

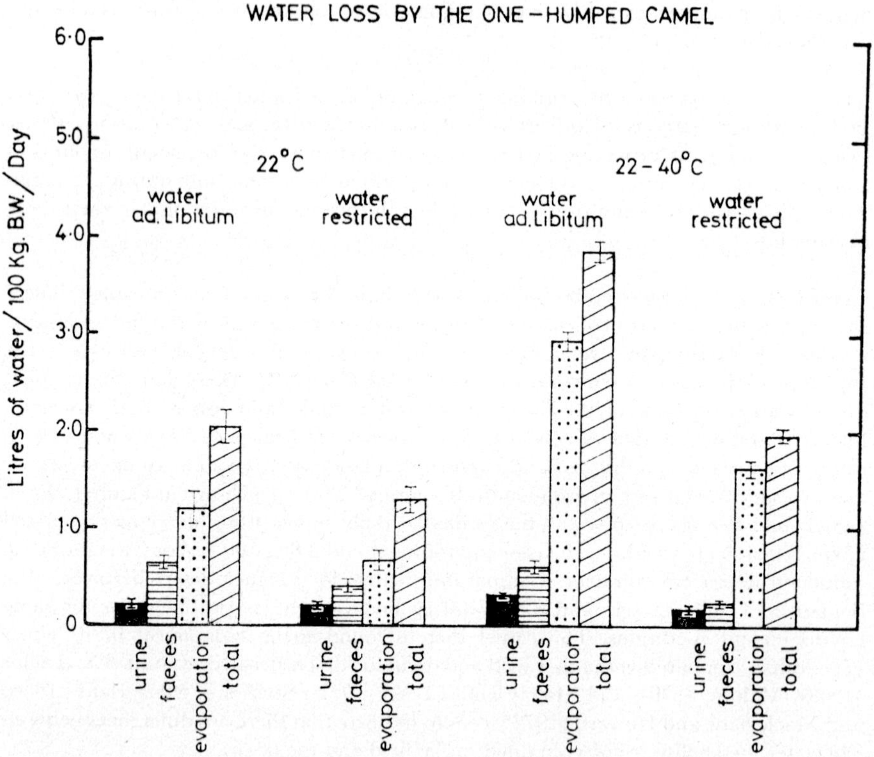

Fig. 1. Amounts of water loss in faeces, urine and evaporation when the camel was watered *ad libitum* and when water was restricted to near minimal levels at ambient temperatures of 22°C and 22-40°C.

long periods will excrete small dry faecal pellets which bounce on hitting the bare desert ground. There is greater water and salt absorption from the large intestine (colon) following severe dehydration in the camel (Maloiy, unpublished observations) than in other large herbivores.

Another interesting feature of camel water metabolism is the remarkable extent of its ability to reduce urinary water loss following dehydration. The camel reduces the volume and increases the osmolar concentration of the excreted urine (Maloiy, 1972; Siebert and Macfarlane, 1971), both by decreasing glomerular filtration rate and increasing tubular reabsorption of water. Under these conditions, substances normally excreted are retained in the extracellular fluid (Macfarlane, 1964).

TOLERANCE TO DEHYDRATION AND SUDDEN REHYDRATION

The camel can tolerate extreme water depletion. Camel, donkey and sheep tolerate without ill effects a loss of water corresponding to 30-31% of the initial body water. Man

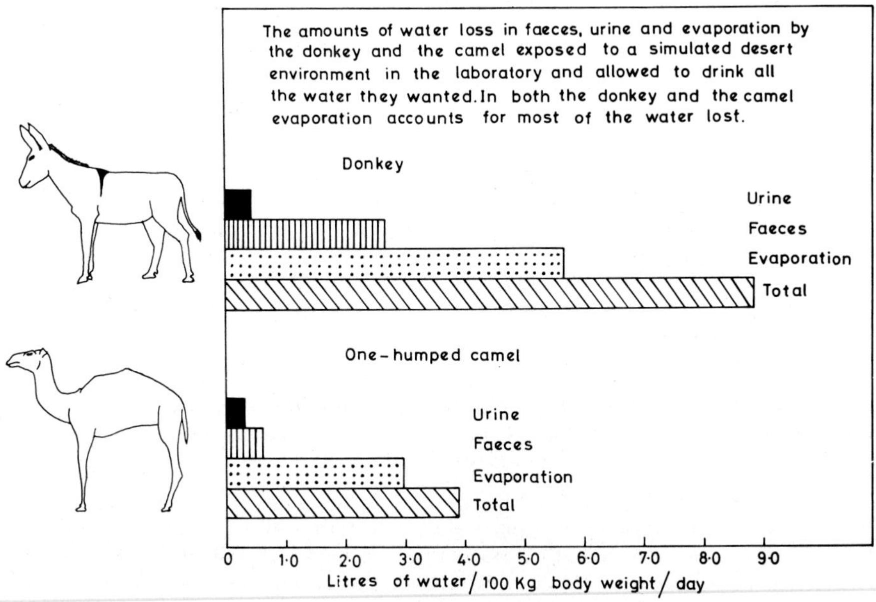

Fig. 2. The amounts of water loss in faeces, urine and evaporation by the donkey and the camel exposed to a simulated desert environment.

is only able to withstand a loss of water corresponding to 10% of the body weight and he will collapse and die after a loss which exceeds 12%.

Coupled with this high degree of tolerance to water loss, the camel, like other ruminants, rehydrates itself quickly following dehydration. A dehydrated camel can drink 40-60 l of water or 60-80% of its loss of weight in 10-15 min to restore the water deficit. This aspect of the water economy of the camel is of importance in survival amongst predators who normally wait around water holes. The rate of water absorption across the rumen wall seems to be slow in camels (Macfarlane and Howard, 1972). The camel's red blood cells are interesting in that during the period of rehydration the red blood cells do not undergo haemolysis as has been observed in other animals such as cattle. This helps the camel to ingest large quantities of water without any threat to red blood cell rupture.

BODY FLUID, TOTAL BODY WATER AND WATER TURNOVER

Macfarlane and his associates used dilution of tracer substances to estimate water turnover and have made detailed studies on seasonal changes in body fluid compartments, total body water and water turnover rates in both hydrated and dehydrated camels grazing freely in the field in Central Australian desert. They also made similar measurements in lactating and non-lactating camels. Siebert and Macfarlane (1971) working in Australia, found that daily water turnover of camels in the dry season is double that during the cool rainy season; and it is higher in lactating than in non-lactating animals (89 ml/kg vs 62 ml/kg).

Table 1. Comparison of water content and turnover of some East African herbivores in the equatorial desert in the Northern region of Kenya (NFD). (Macfarlane, 1964).

Species	N	Body solids	Daily water turnover	
		% body weight	ml/Kg	ml/Kg$^{0.82}$
Boran cattle	4	23	135	347
Somali fat-tailed sheep	12	32	107	197
Goats	10	31	96	185
Camels	4	30	61	188

N = number of animals used.

Table 1 shows that in a tropical semi-desert, Boran cattle (*Bos indicus*) turned over more water than goats, sheep or camels. Goats use less water than sheep, and camels have water turnover rates near to or below those of goats.

Macfarlane (1964) working on camels that had been without water for 9 days in Central Australia made interesting observations on body composition, water loss, and changes in plasma volume. It was observed that as camels become dehydrated, the major part of the water loss occurred from the cell and gut water compartments. Under these conditions, the reduction in plasma volume was very small.

The distribution of the water loss in a camel that had been without water for 9 days is shown in Figure 3. Initially, this animal weighed 288 kg. After 9 days without water the camel lost 49 l of water (17% of initial body weight). Of this total water loss only 1 l was lost from the plasma volume; 16 l were removed from the interstitial fluid, while the remaining 32 l of water were lost from gut and cell water compartments.

BEHAVIOURAL FACTORS THAT CONTRIBUTE TO THE WATER METABOLISM OF CAMELS

Certain behavioural factors are important in the water economy of a desert herbivore. A camel, for example, will seek the shade so as to avoid the high solar heat load. Other important factors listed and discussed by Schmidt-Nielsen (1964) are given in Table 2.

Table 2. Behavioural characteristics that contribute to the water economy of camels. (From Schmidt-Nielsen, 1964).

A. Seeking shade.
B. Orienting lengthwise to the sun's rays.
C. Sitting with legs folded under body.
D. Exposing minimal surface.
E. Remaining sitting on same spot of ground thus avoiding contact with hot surface.

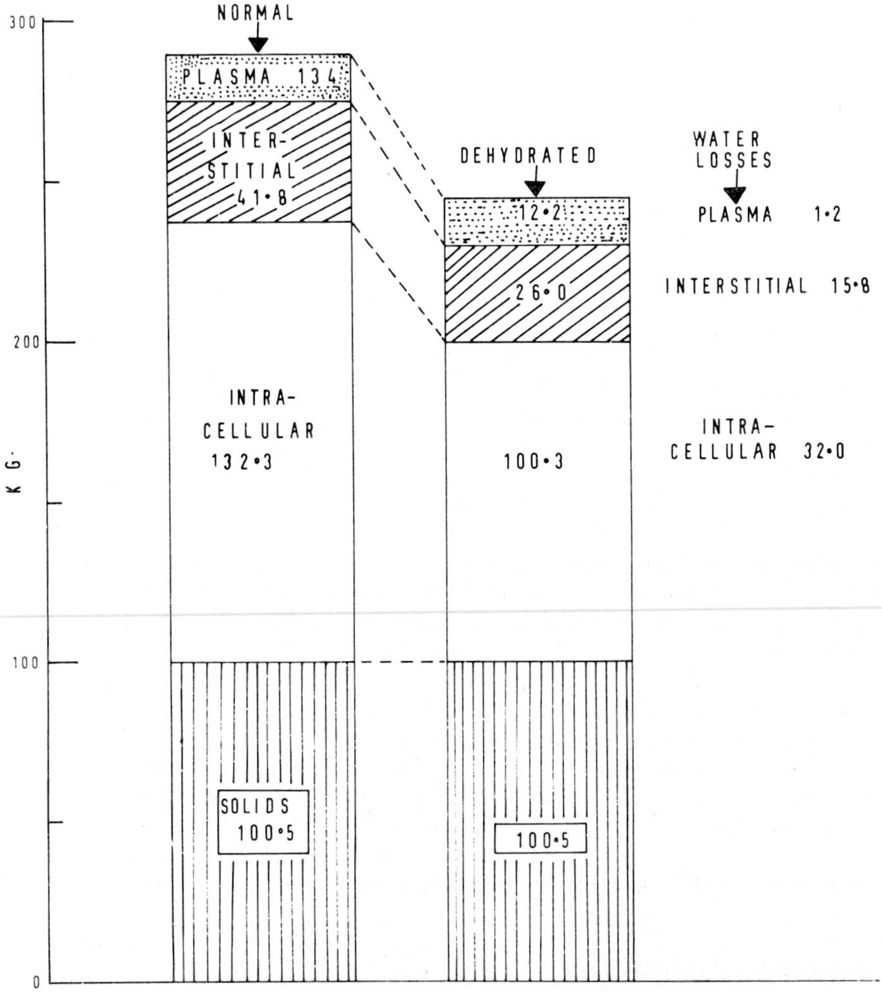

Fig. 3. Changes in body fluid compartments normally hydrated and dehydrated camel. (Macfarlane, 1964).

RENAL CONTROL OF SALT AND ELECTROLYTE EXCRETION IN THE CAMEL

SALT TOLERANCE

Herbivores will tolerate considerable amounts of NaCl in their drinking water. Maloiy investigated this problem by increasing the NaCl concentrations in the drinking water of the donkey, camels, sheep, goats and some wild herbivores (unpublished observations). Somali fat-tailed sheep tolerated only 220 mmol/NaCl/l, while Turkana goats ate and maintained weight on 257 mmol/l salt in the drinking water. The one-humped camel at rest and at environmental temperature of 22°C on the other hand maintained both appetite and body weight while drinking 1850 mmol/l (5.5% NaCl). The donkey reduced food intake and lost weight while drinking 0.75-1.00% NaCl. The camel also tolerates raised extracellular Na, since a plasma Na of 200 meq/l does not seem to either disturb circulation or nerve and muscle function (Siebert and Macfarlane, 1971). When drinking NaCl solutions of varying concentrations (0.25-5.50%) the camel increased fluid and food intake. Similarly, the daily volume of urine excreted increased up to 9 l/day. Urinary Cl concentrations rose to above 1000 meq/l.

GFR, RENAL TUBULAR FUNCTION AND ELECTROLYTE EXCRETION

Glomerular filtration rate (GFR) and renal tubular function have been studied (Macfarlane et al., 1962, 1967; Siebert and Macfarlane, 1971; Maloiy, 1972) in camels subjected to various physiological conditions. There is a general agreement in all these results that dehydration in the camel is accompanied by a decrease in GFR, in such a way that low urine flow rates (0.5-1.5 ml/min) are reached with urine concentrations of only 2,000-2,500 mosm/l. Maloiy (1972) found that salt loading increased GFR and urine flow by about 52% and 103%, respectively, as in other herbivores.

In dehydrated and salt-loaded camels, respectively 99.8% and 99.2% of the water filtered at the glomerulus was reabsorbed. The observed changes of urine flow in the camel appeared to be brought about by alteration in GFR and renal tubular water reabsorption.

Urea and electrolyte excretion in camels has been studied (Schmidt-Nielsen et al., 1957; Maloiy, 1972). Potassium, because of its concentration in the vegetation, is the principal cation excreted by the camel. Urinary urea concentrations can be either high, up to 950 mM/l, or very low. Data in the literature convincingly shows that in the camel, the renal excretion of urea is an actively regulated process (Schmidt-Nielsen et al., 1957a). Table 3 shows the urine volume and concentrations in hydrated and dehydrated camels. From these results, urea accounted for a substantial fraction of the osmotic concentration.

ACTION OF VASOPRESSIN

Studies by Macfarlane and associates have shown that the action of vasopressin on camels is similar to that on sheep. The water-loaded camels are, however, very sensitive to the antidiutetic action of vasopressin (responding to 1 μU/kg or 0.5 mU for a 500 kg camel). As in other herbivores studied, Macfarlane found out that an injection of vasopressin increases urinary electrolyte, particularly K, excretion. During dehydration also, K^+ is excreted, although Na^+ tends to be retained in the extracellular fluid and is used later to restore fluid volume.

Although an intravenous infusion of 750 mU of vasopressin (in 60 min) into a hydrated camel, with urine flow rates of 0.9-1.3 ml/min, increased urine flow to 3.3 ml/min, Macfarlane (1964) concluded from these and other studies that vasopressin is not intimately concerned with water regulation in the camel.

Table 3. Urine volume and concentration of its constituents in the one-humped camel subjected to heat and dehydration in a climatic chamber.

Ambient temperature and experimental conditions	Urine volume ml/day	Sodium m.equiv/l	Potassium m.equiv/l	Chloride m.equiv/l	Urea m.mole/l	Osmolality m.osmole/kg H_2O
22°C Water ad lib.	702.0 ± 43.4	2.3 ± 0.48	778.0 ± 30.7	250.0 ± 20.8	538.0 ± 88.6	2094 ± 79.6
22°C Minimum water	442.0 ± 29.9	9.0 ± 1.0	987.0 ± 66.3	278.0 ± 32.6	608.0 ± 107.5	2581.0 ± 99.4
22-40°C Water ad lib.	857.0 ± 80.5	2.2 ± 0.27	423.0 ± 76.8	130.0 ± 28.3	618.0 ± 108.9	1473.0 ± 212.1
22-40°C Minimum water	205.0 ± 7.4	6.0 ± 0.75	630.0 ± 28.6	153.0 ± 22.6	954.0 ± 66.8	2230.0 ± 68.5

n = 3 camels; values are means ± S.E.; 18 samples were taken from each animal when water was available freely and 10 samples when the animals were dehydrated.

SUMMARY

The well-known ability of the camel to survive in desert environments is due to a combination of behavioural adaptations, which reduce its exposure to heat load, and physiological adaptations, which allow it to conserve water, to resist the effects of prolonged dehydration and to rehydrate without ill-effect. The responses of the camel to heat load and water deprivation are reviewed, and particular emphasis is given to recent studies of kidney function in camels subjected to dehydration or salt loading.

In the camel, the gross mechanisms involved in water conservation have now been described; but the actual regulatory and cellular mechanisms involved in the responses of dehydration now require investigation. Little is also known about the water metabolism of the camel during pregnancy, lactation, and exercise performance. For example, will load-carrying in the desert heat increase water intake of the camel? Considerably more research is required to solve these problems.

ACKNOWLEDGEMENT: The unpublished observations of the author reported here were made during laboratory and field experiments financed by the Wellcome Trust, Nuffield Foundation, and the University of Nairobi research and publication grants.

REFERENCES: see p. 522.

Water Metabolism in Rodents

by

R.E. MacMillen and D. E. Grubbs

INTRODUCTION

Among mammals rodents comprise the most diverse, the most numerous, and the most ubiquitous order. As such they are spectacularly successful in nearly every terrestrial habitat on earth and, indeed, in some aquatic ones as well. While many live in equable habitats, many others live where surface conditions are potentially extremely demanding with respect to heat or cold, and paucity of water of low salinity. Among such extreme habitats where rodents flourish are arctic-alpine situations of high latitude or altitude, deserts with low rainfall, alkaline sinks, and salt marshes.

Being typically small and fossorial, rodents are infrequently exposed to the most demanding of surface conditions, avoiding temporary climatic extremes in the more stable microclimates of burrows. Thus behavioral flexibility allows temporary escape from conditions which could otherwise exceed physiological tolerances. However, it is the long-term or average climatic conditions encountered during periods of surface activity to which are geared the physiological adaptations with respect to heat and cold; the degree of such physiological exposure is carefully regulated by behavior.

DISCUSSION

Critical to the understanding of an organism's water balance are the avenues of water input and output. In rodents water input is by imbibition (succulent foods, foods with a high yield of oxidative water, drinking), and output is by lactation, defecation, urination, and evaporation from skin and lungs. Evaporative water loss represents by far the greatest source of output (70 per cent of the total in the desert kangaroo rat, *Dipodomys merriami*; Schmidt-Nielsen, 1972), yet its absolute importance in the water economy of rodents is improperly understood.

Evaporative water loss is misunderstood (particularly the respiratory component) because with each amount lost in respiration there is an amount of water gained as foodstuffs are oxidized (oxidative or metabolic water). The amounts lost are indirectly related to the absolute humidity of the inspired air and directly related to the temperature of the expired air. The amount of oxidative water gained depends upon the composition of the foodstuff and the metabolic rate of the rodent which, below thermal neutrality, is indirectly related to air temperature. Schmidt-Nielsen and colleagues (Jackson and Schmidt-Nielsen, 1964; Schmidt-Nielsen et al., 1970) elegantly showed that rodents possess a counter-current heat exchange system in the nasal passages which results in cooling of exhalent air to below the inhalent temperature, thereby promoting condensation and recovery of water prior to expiration; the magnitude of recovery is indirectly related to air temperature. Thus the rate of evaporative water loss (of which a very large proportion is respiratory water loss) in virtually all rodents is directly related to ambient temperature (MacMillen, 1972). It is commonly known that below thermal neutrality in normothermic rodents, metabolic rate (and hence metabolic water production) is indirectly related to ambient temperature. MacMillen (1972) postulated and demonstrated that there is a predictable ambient temperature for each of four species of desert rodents at which oxidative water production should equal evaporative water loss; this temperature was between 15 and 20°C.

A more thorough analysis, incorporating most of the existent data on simultaneous measurements of evaporative water loss and oxygen consumption in rodents demonstrates very aptly the direct relationship between ambient temperature and evaporative water loss (Fig. 1). This analysis also reveals, contrary to Schmidt-Nielsen's (1964) suggestion, that the rate of evaporative water loss is independent of habitat, with virtually the same rates for both desert and non-desert forms at the same temperature. A regression analysis of the data for desert rodents yields the equation:

$\log EWL = -0.663 + 0.027\, t_a$ (EWL is evaporative water loss in mg H_2O/Cm^3O_2 and t_a is ambient temperature in °C).

A similar analysis of the data for non-desert rodents yields the equation:

$\log EWL = -0.570 + 0.025\, t_a$.

Lines fitted to these data by these equations do not differ statistically from each other either in slope or in Y intercept. Since there was no statistical difference between rates of evaporative water loss in desert and non-desert rodents, the data were pooled to describe the relationship between evaporative water loss and ambient temperature for rodents in general. This relationship is described by the equation:

$\log EWL = -0.624 + 0.026\, t_a$

The line fitted by this equation to these pooled data is shown in Fig. 1.

By making certain assymptions about the kinds of foodstuffs being oxidized these data lend themselves to further analysis, which reveals the relationships between evaporative water loss, metabolic water production, and ambient temperature (Fig. 2; see caption for full explanation of assumptions) for desert and non-desert rodents. Again the analysis shows no difference between rodents from the two habitats, and demonstrates that, ignoring for the moment avenues of water loss other than evaporative, rodents in general produce more water metabolically than they lose by evaporation at ambient temperatures below 16.6°C; at still lower temperatures metabolic water production even further exceeds evaporative water loss. Although these data relating metabolic water production (MWP) to evaporative water loss (EWL) were measured at rest, Raab and Schmidt-Nielsen (1972) have shown that basically the same relationship persists during activity: activity increases both MWP and EWL proportionately. A line fitted to the pooled data relating MWP/EWL to ambient temperature is shown in Figure 2, and is described by the regression equation:

$\log MWP/EWL = 0.430 - 0.026\, t_a$

The line indicates that MWP = EWL at t_a = 16.6°C. This is a very moderate environmental temperature, commonly exceeding those encountered on the surface in most

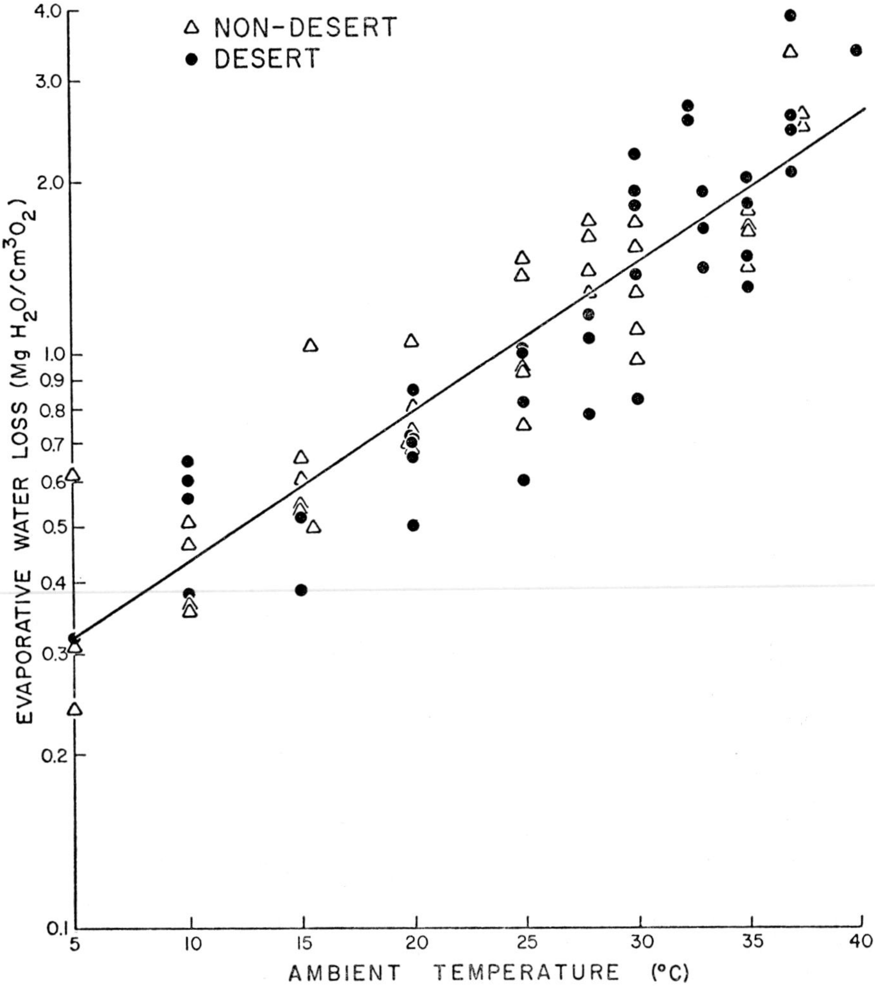

Fig. 1. The relationship between evaporative water loss (EWL, expressed as a function of oxygen consumption) and ambient temperature in a variety of rodents. Non-desert rodents are indicated by △ and desert rodents by ●. The diogonal line is fitted to all of the rodent data by the method of last squares, and is described by the equation: log EWL = —0.624+0.026 t_a. The non-desert and desert rodents on which these data are based and the sources of the data are indicated in Table 1. Data for rodents which salivated profusely in response to high temperatures were not included, as this represented a component of evaporative water loss that did not occur across the entire temperature range.

habitats and seasons by nocturnally-active rodents. Thus nocturnality imbues rodents automatically with a very favorable relationship between metabolic water production and evaporative water loss. Provided with this relationship, the maintenance of water balance is a relatively simple matter for nocturnal rodents through the prudent choice

Table 1. Species of Rodents, Habitats, and Sources Employed in Figs. 1 & 2.

SPECIES	SOURCE
DESERT	
Family Heteromyidae	
Dipodomys merriami	Carpenter, 1966
Dipodomys merriami	Raab & Schmidt-Nielsen, 1972
Family Muridae	
Acomys cahirinus	Shkolnik & Borut, 1969
Acomys russatus	Shkolnik & Borut, 1969
Leggadina hermannsburgensis	MacMillen, et al., 1972
Notomys alexis	MacMillen & Lee, 1970
Notomys cervinus	MacMillen & Lee, 1970
Family Sciuridae	
Citellus leucurus	Hudson, 1962
Citellus tereticaudus	Hudson, 1964
NON-DESERT	
Family Cricetidae	
Microtus californicus	Church, 1966
Peromyscus eremicus	MacMillen, 1965
Reithrodontomys megalotis	Coulombe, 1970
Family Heteromyidae	
Dipodomys agilis	Carpenter, 1966
Dipodomys venustus	Church, 1969
Liomys irroratus	Hudson & Rummel, 1966
Liomys salvani	Hudson & Rummel, 1966
Family Muridae	
Pseudomys australis	Macfarlane, et al., 1971
Rattus lutreolus	Collins, 1973a
Rattus norvegicus	Raab & Schmidt-Nielsen, 1972
Rattus rattus	Collins, 1973b
Rattus villosissimus	Collins, 1973b
Family Sciuridae	
Eutamias merriami	Wunder, 1970
Spermophilus beecheyi	Baudinette, 1972

of foodstuff (either with a high yield of metabolic water, or preformed water), and appropriate control of urinary and fecal water loss. Consequently, one finds a complete spectrum of species of nocturnal rodents (often in the same habitat) varying from those that are independent of exogenous sources of water (other than that absorbed in their foods) with greatly reduced rates of urinary and fecal loss, to those that have elected a highly succulent diet (either insectivorous/carnivorous or herbivorous) and can readily afford the luxury of rather high rates of urinary and fecal water loss. Diurnally-active rodents have far less option, at least if they are active at ambient temperatures exceeding 17°C. They, in general, are dependent on exogenous water sources, with rates of evaporative water loss that exceed metabolic water production. In addition, these relationships reveal that in order for laboratory studies of responses of rodents to water deprivation to be ecologically meaningful, they must be conducted at the temperatures to which the animals are exposed while active on the surface.

All of the data in Figure 2 were collected in dry air, simulating surface conditions for desert rodents. For rodents in humid burrows or active under more mesic conditions, metabolic water production would be on the same order as the relationship predicts, but evaporative water loss would be reduced, depending upon the humidity of the inspired air. Kenagy (1973) has shown that at least certain desert rodents spend most of the 24 hrs in burrows, foraging for as little as one hour on the surface where conditions approximate the experimental situation. Hence the relationship between MWP/EWL and t_a as depicted in Figure 2 must be considered a conservative estimate.

For granivorous nocturnally-active rodents one can conclude from Figure 2 that under most circumstances while active, evaporative water loss is overcompensated for by metabolic water gain. Complete water balance can be achieved without additional water input by voiding relatively dry feces and adjusting urine concentrations consistent with the temperature-induced difference between metabolic water production and evaporative water loss. Thus, under natural circumstances in water-independent rodents, one would expect to find minimal urine concentrations during the colder winter months and maximal concentrations during the warmer summer months. This is precisely what MacMillen (1972) has demonstrated in the water-independent desert heteromyids: *Dipodomys merriami*, *Perognathus fallax*, and *P. longimembris* (Table 2).

Table 2. Summer and winter urine concentrations of sympatric desert rodents at Joshua Tree, San Bernardino County, California.

	Urine concentrations			
	Summer		Winter	
	$\bar{X}$ mOsm/l	N*	$\bar{X}$ mOsm/l	N*
Water independent				
Dipodomys merriami	3936	13	1065	15
Perognathus fallax	3797	4	1307	3
Perognathus lonbimembris	3716	9	1546	16
Water dependent				
Onychomys torridus	3259	1	3188	5
Neotoma lepida	1484	3	757	2

* N = number of animals

Another strategy for achieving water balance by nocturnal rodents is to rely on succulent foods (either animal or vegetable) of low salinity, augmenting rather than depending upon metabolic water production. Such a strategy does not require the development of extreme renal water-conserving capacities, but rather the kidneys are used equally to

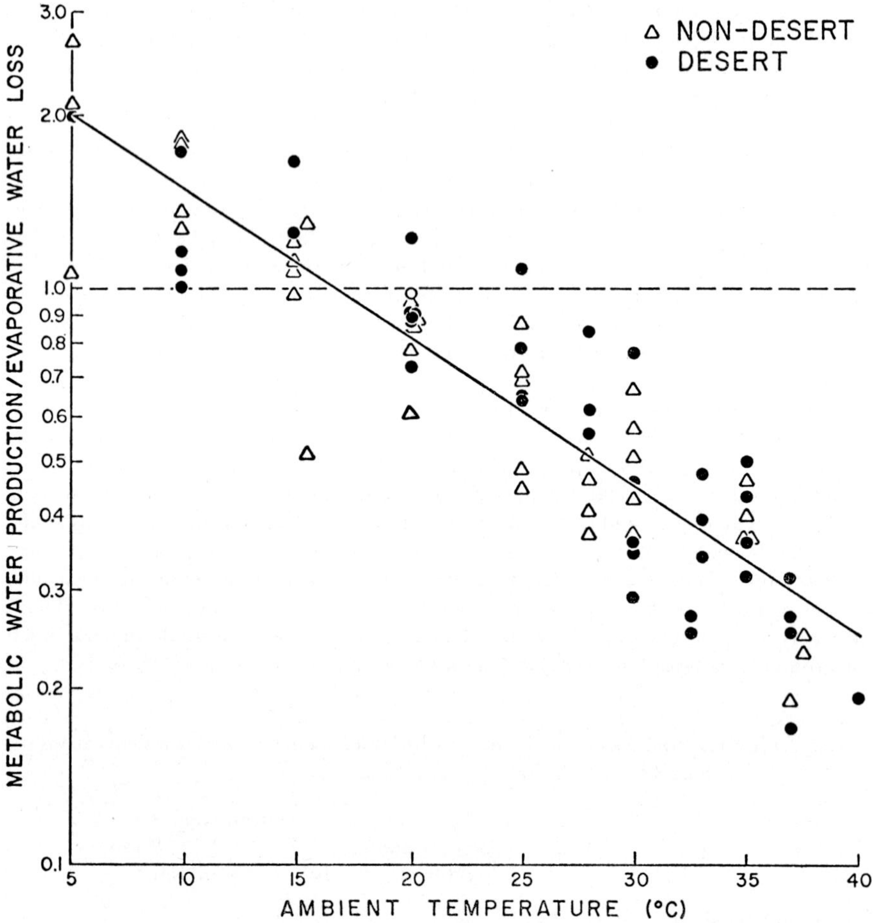

Fig. 2. The relationship between metabolic water production (MWP), evaporative water loss (EWL) and ambient temperature in a variety of rodents. Non-desert rodents are indicated by △, and desert rodents are indicated by ●. The values for MWP/EWL were derived from the values in Fig. 1 by assuming all rodents were oxidizing pearled barley which yields 0.64 mg of metabolic water for each 1.0 CM^3O_2 consumed (after Schmidt-Nielsen, 1964, and MacMillen, 1972). The horizontal dashed line represents equality between MWP and EWL. The diagonal line is fitted to the data by the method of least squares, and is described by the equation: log MWP/EWL $= 0.430—0.026\ t_a$. The least squares line crosses the line of equality at $t_a = 16.6°C$.

void excess water and solute loads imposed by the diet. In such rodents while active on the surface one would expect more uniform urine concentrations throughout the year, with higher concentrations imposed by urea loads in insectivorous/carnivorous rodents and lower concentrations in herbivores. Again, MacMillen (1972) has demonstrated just such patterns in the insectivorous/carnivorous desert rodent *Onychomys torridus* and the herbivorous desert rodent *Neotoma lepida*. Table 2 depicts representative summer (high temperature) and winter (low temperature) field-collected urine concentrations for sympatric "water-independent" and "water-dependent" desert rodents from the vicinity of Joshua Tree, San Bernardino County, California. This site is transitional between The Sonoran and Mojave Deserts, with a mean annual rainfall of 105 mm and elevation of 970 m.

Except at high latitudes and altitudes diurnally-active rodents operate typically at ambient temperatures in which evaporative water loss exceeds metabolic water production, precluding the possibility of "water-independence". Although insufficient data are available one would predict that while active on the surface these rodents must rely either on succulent foods or frequent drinking or both. For diurnally-active rodents maximal osmotic stress would be expected under hot, dry desert circumstances where water availability is minimal and water loss is maximal, especially in the summer. Even with access to relatively succulent vegetation, evaporative water loss rates would be so much in excess of metabolic water production that the requirement of renal water conservation through production of highly concentrated urine would be a necessity. Hence, Hudson (1962) reports urine concentrations as high as 3900 mOsm/l in the diurnally-active desert antelope ground squirrel *Citellus leucurus*; such values exceed those of some "water-independent" nocturnally active desert rodents, even though *C. leucurus* cannot tolerate prolonged periods of water deprivation.

SUMMARY AND CONCLUSIONS

This relationship between evaporative water loss and metabolic water production in rodents accentuates the intimate relationship between water metabolism and energy metabolism in homeotherms. Although no differences between desert and non-desert rodents are revealed, we believe there are differences in rodents in patterns of energy metabolism that relate to water metabolism; these differences have been alluded to but not elucidated. The differences which we suggest herein, and which will await verification are that water independent rodents will have generally narrow to no thermal neutral zones, and will show a maximal increase of metabolic rate (and metabolic water production) as ambient temperature decreases below thermal neutrality. This will, theoretically, increase metabolic water production below thermal neutrality, while not affecting the rate of evaporative water loss. In contrast, water dependent rodents will be expected to have broader thermal neutral zones.

REFERENCES: see pp. 522-523.

Temperature Regulation in Cattle

by

J. Bligh

INTRODUCTION

The inclusion in this series of papers on progress in animal biometeorology of a Section on temperature regulation in cattle might be taken to imply that there are particular features in the thermoregulation of this species which are of fundamental physiological significance. Essentially, this does not seem to be so, and the recent introductory discussion of the physiology of temperature regulation by Richards (1973), and the more detailed treatment of the subject by Bligh (1973) will serve as well for the understanding of the fundamentals of bovine thermoregulation as for the other mammalian species. Review articles by Hensel (1973), Bligh and Hensel (1974) and Hales (1974) summarize the most recent researches into mammalian thermoregulation, and that by Thompson (1973) is a summary of the studies on the climatic physiology of cattle since Bianca's (1965) review of the literature on cattle and heat.

The significance of the bovine animal in relation to its thermal environment is not in the peculiarities of its thermoregulatory processes, but in its occurrence as a domesticated species in many environmental circumstances. Man has settled over much of the earth's surface, and can endure a wide range of macro-climates because of his particularly well-developed ability to provide himself with a comfortable and physiologically tolerable microclimate. Thus his permanent colonization of many inhospitable regions of the world is not determined so much by the direct effects of climate, which he can easily avoid, as on the availability of an adequate and continuous supply of nutrients. The role of the domesticated herbivore is as a converter of plant matter which is inedible by man, or of low nutritional value to him, into animal matter which is edible by man and of greater nutritional value. With the aid of his domesticated herbivores, man has been able to migrate across the land surfaces of the earth, and develop techniques of land management and permanent farming.

Although these domestic herbivores are commonly provided with some degree of shelter against the more extreme seasonal cold stresses, similar protection against heat stresses has not been a feature of tropical animal husbandry, and in general, cattle have not been as well buffered against climatic stresses as their masters have been. Consequently, they have had greater need to become adapted to the conditions in which they are being husbanded. In each animal there is some degree of phenotypic adaptation to the general climate and to its seasonal variations, but there has also been a great deal of genotypic adaptation which, by both natural and man-directed selection, has produced breeds which are able to survive to maturity and reproduce themselves in the regions in which they have been settled.

HEAT TOLERANCE IN NATURE AND IN MODERN AGRICULTURE

In nature, adaptation to and tolerance of the natural environment means no more than that the species can survive. The primary requisite of the domesticated herbivores which migrated with man into thermally stressful environments was only that they should survive. As long as they could do this, they were available as converters of plant matter into animal matter for human consumption. Thus it may be assumed that the first characteristics of breeds of cattle indigenous to different regions was their capacity to survive in different environmental circumstances. This implies the existence of physiological differences between breeds in their capacities to cope with the direct and indirect effects of the climate. The particular interest in studies of the responses of cattle to heat stresses and cold stresses resides more in the differences between breeds of cattle genotypically adapted to the one extreme and to the other, than in the differences between the bovine animal and other mammals.

The commercial motives of profit, and the social and political motives of matching the growth of population with the growth of food supplies, compel the selection of cattle for the maximum level of productivity in the prevailing environmental circumstances. For reasons of both technical competence and educational organisation, and because of the generally temperate climate, improvements in milk and meat production have been greatest in European breeds of cattle. The contrast between the productivity of European cattle (*Bos taurus*) and tropical cattle (*Bos indicus*), coupled with the fact that in general the standards of living in the tropical areas of the world are well below those of Europe and North America, have inspired efforts to raise the productivity of cattle in tropical zones by the importation of European cattle, by cross-breeding between European and native breeds, or by the up-grading by selection of the native breeds.

The underlying problem is that high productivity involves a high rate of heat production which is incompatible with the toleration of external heat stress. Obviously only those individual European animals or breeds which are the most efficient converters of plant matter into body tissues (i.e., with the lowest rate of heat production per unit of conversion), and/or those animals with the greatest ability to shed heat despite a lowered thermal gradient between animal and environment by evaporative heat loss, are likely to sustain anything like their maximum productivity when transported to tropical zones, or when used in the production of cross-bred progeny with improved body weight gain (meat production) or milk yield. Currently the search is for breeds of cattle which can maintain a relatively high level of productivity in tropical or semi-tropical environments, and heat tolerance has come to be related to productivity, or sustained agricultural performance, rather than to survival.

THE SEARCH FOR AN INDEX OF HEAT TOLERANCE

The selection of cattle which can contribute to an improvement in meat and milk production in thermally stressful environments can be best achieved by direct measurement of productivity in the stressful environment, as is done in progeny testing schemes. While this may be the surest means of selection, it is also a slow and laborious one. If there were some simple and reliable physiological index of heat tolerance which correlated well with the ultimate record of sustained productivity, the processes of selection both of the sires and dams, and of their progeny, could be greatly speeded up.

During the last 20-30 years there have been a great number of studies of "heat tolerance" in cattle, both on individual animals within breeds, and of breeds and cross-breeds. Some of these studies have been made under field conditions, and others have been conducted in simulated thermal environments in climatic chambers. Various thermoregulatory and other effects of heat on the animals have been measured, and correlated with the environmental conditions and with body temperature. Few, if any, of these studies have established a useful correlation between the various responses of the animals to climatic stress and their agricultural productivity when they are more permanently exposed to climatic stress.

Progress is frequently hindered by the presence of intellectual barriers which artificially separate areas of applied biology that have been given different names. An inexplicable separation between the progeny testers and the environmental physiologists seems to be such an instance. The progeny testers are making empirically sound but physiologically blind decisions on the agricultural worth of animals in particular environmental circumstances. The physiologists use a measure of heat tolerance not necessarily correlating with productivity. A useful initial approach to the problem of finding a physiological index of heat tolerance, defined in the agricultural sense of sustained productivity in a tropical environment, might be to take the best of the animals selected by the progeny testers, and the worst of the animals rejected by them, to look for physiological differences between them, and to see if these differences correlate with the measured differences in performance as meat or milk producers.

A second stage might be to make a nominal rating of the animals submitted to a progeny selection scheme, on the basis of a possible physiological index, and to see if there is any correlation between the measured performance and the physiological index.

CORE TEMPERATURE AS AN INDEX OF HEAT TOLERANCE

The majority of heat tolerance studies on cattle are not based on the criterion of productivity, but on the stability of core temperature. A relation between the ability of an animal to remain thermostable in a particular thermal environment and its ability to maintain its productivity if kept permanently in that thermal environment is assumed. The basis of this assumption is that the bovine animal, like many other mammals, maintains a fairly constant body temperature so long as it is physiologically and physically possible for it to do so. A deviation beyond the nychthemeral variation of body temperature in a temperate climate indicates that the animal is unable to lose heat at the same rate as it is producing it except by raising its body temperature so that the gradient from the organism to the environment is increased. This ability, or inability, to equate heat production and heat loss at the level of body temperature at which it is normally regulated is obviously a reflection of the rates of heat production and/or the ability to activate and sustain heat loss. Since hyperthermic animals lose their appetite, hypophagia results in a lowered heat production, and as productivity depends on appetite, the presumed correlation between body temperature and productivity seem reasonable even though unestablished.

This presumed relation between body temperature and productivity was the basis of the IBERIA HEAT TOLERANCE TEST (Rhoad, 1944) which was subsequently abandoned as an index of heat tolerance in beef cattle in the same research station at which it had originated (Vernon et al., 1951). However, the heat tolerance of cattle continues to be judged on the basis of rectal temperature records (for example, Moran 1973). Chquiloff (1964) reported that the Gir breed, unlike the European breeds, voluntarily stayed exposed to solar radiation, and body temperature rose, but this was unaccompanied by thermoregulatory responses. This thermolability seemed to be passively tolerated, but no evidence was offered on whether this rendered the Gir breed more productive.

Although, with notable exceptions, most mammals do seem to be fairly thermostable, species vary in the extent of the nychthemeral variation on core temperature (for examples see Bligh and Harthoorn, 1964). Schmidt-Nielsen et al. (1957) confirmed earlier reports that the camel is somewhat thermolabile, and supposed that this lability enables the camel to inhabit hot arid environments unsuited to more thermostable species such as man, which maintain their thermostability by the evaporation of body fluids. The possibility of a wider occurrence of thermolability in species adapted to hot arid environments was appreciated by Schmidt-Nielsen (1964) who suggested that this might equally apply to agriculturally important animals. However, when Bligh and Lampkin (1964) recorded the core temperature patterns of pure Hereford and pure Zebu cows under the same non-thermally-stressful conditions, they found no significant differences, and Berman (1971) failed to find any evidence of increased thermolability of cattle during summer heat in Israel. These findings indicate that at least some breeds of both *Bos taurus* and *Bos indicus* are thermostable when ambient conditions are suitable, and this leaves unexplained the failure to establish a positive correlation between thermostability and sustained productivity in cattle in general, even if there are exceptions.

Bianca (1971) made the interesting suggestion that an effect of producing a high degree of uniformity within a breed, as is generally the case where agricultural practices are fairly well-developed, has been to greatly diminish the extent of the gene pool.

From a recent study of the core temperature patterns of South American camelidae – which are fairly near relatives to the dromedary camels studied by Schmidt-Nielsen et al. (1957), Bligh et al. (1975) have tentatively concluded that the ancestors of the camels were probably fairly good thermoregulators as are the modern llamas and alpacas. When the ancestors of the present-day camels were pushed by some environmental pressure into the inhospitable environments of hot dry deserts, possibly only those individual animals which were more thermolabile were able to avoid dehydration and death, and to reproduce their kind. This could account for the emergence of this thermolabile species. If, as Bianca suggests, already highly selected breeds of cattle have a much reduced gene pool, there may be little hope of finding bovine "sports" which are more thermolabile than the rest of the breed, or with other characteristics which render them more suited to thermally adverse environments. It is possible, of course, that larger gene pools still exist in some more "primitive" breeds which have not been subjected to intensive breeding for particular qualities, and in the creation of new breeds of cattle more suited to thermally stressful environments, it might be profitable to extend the search to those areas in which there has been less development of breeds with high degrees of uniformity in form and function.

FITTING CATTLE TO THE NATURAL ENVIRONMENT OR THE ENVIRONMENT TO THE CATTLE

The emphasis in the last section was on the selection of individual animals, or of breeds, which are genetically suited to thermally stressful natural environments. There is, however, the alternative approach of selecting animals that are highly productive in a temperate climate and then creating a local environment within which this high productivity can be sustained. There is nothing in the literature to suggest that these highly productive animals will be other than fairly fine regulators of core temperature. On the assumption that in such breeds an upward displacement of core temperature is evidence of a thermal stress which will result in a decline in productivity, one can establish the range of environmental circumstances in which thermostability can be maintained, and then engineer a micro-climate accordingly. Obviously, the maintenance of productivity will depend also on adequate standards of feeding and watering and other aspects of good management, and on adequate immunity to local infection hazards.

The provision of a non-stressful micro-climate in a thermally stressful macro-climate is costly in both construction and maintenance, and generally speaking the provision of non-stressful micro-climates for the production of meat and dairy produce in stressful macroclimates cannot be contemplated. More modest compromises such as the provision of protection from direct solar radiation and perhaps also of a "self-service" water sprinkler are more general. Such partial amelioration of the thermal stress of the natural environment is unlikely to be sufficient to maintain the productivity of high performance cattle, but may be of some value when used in conjunction with a suitable breed of cattle.

BEHAVIOURAL AMELIORATION OF HEAT STRESS

Until quite recently the literature on mammalian responses to heat stress emphasised the involvement of the autonomic processes (e.g., panting, sweating, shivering and peripheral vasomotor tone), but there is now an increasing realization of the importance of behavioural responses to heat stress. Indeed, the primary responses against heat and cold stress in most mammals, including man and cattle, are behavioural ones. When it is available cattle seek protection against high intensity solar radiation, and thereby spare the autonomic processes of evaporative heat loss by panting and sweating. While this conservation of body fluids may aid production, especially of milk, shade-seeking alters the grazing pattern and may therefore have an adverse effect on production. Thus in some circumstances it may be better to keep animals permanently under cover and to provide them with harvested feed than to provide optional shelter for freegrazing animals. However, for this alternative, the fodder must be readily harvestable, and the cost of harvesting and feeding must be economic.

CONCLUSIONS

The prime motivation of studies of the environmental and thermoregulatory physiology of cattle is to maximise the effectiveness of this species as a converter of inedible or nutritionally poor plant matter into edible or nutritionally better animal matter for human consumption.

In nature the tolerance of a thermally stressful environment is manifest simply by the capacity of an organism to survive and reproduce. For cattle in modern agriculture, however, tolerance of a thermally stressful environment relates not just to the survival of the species, but to the maintenance of the highest yield of milk and meat that can be achieved by good management and the selection and breeding of suitable animals.

A thermally stressful environment, through its effects on the availability and utilization of food and water, has indirect as well as direct effects on the health, weight gain and milk yield of cattle, and so laboratory studies in which only the direct thermal stress alone is imposed on the animals does not necessarily give information of practical value in forecasting the performance of animals under field conditions.

The most reliable basis for the selection of cattle for a thermally stressful environment is the process of progeny testing in that environment, as the basis is then some measure of productivity. However, the process is slow, and efforts are being made to find some simple measurement which is a reliable index of performance.

In studies of the environmental physiology of cattle, much emphasis has been given to core temperature as an index of heat tolerance or intolerance. There is, however, doubt whether this can be used as a reliable indication of the ability or inability of an individual animal or breed to maintain productivity during exposure to heat stress.

It is possible that, in hot arid environments at least, a thermolabile animal may be better able to maintain productivity than a thermostable animal. However, because of the restricted gene pool in highly selected European breeds, such mutants might exist only in less highly selected populations.

The best hope of establishing a physiological index of heat tolerance might be to look for the differences between the individual animals selected and rejected by progeny testers, for here the differences relate directly to the productivity of the animals.

An alternative to matching the animal to the environment is to match the environment to the animal. Environmental physiological studies can be used to determine the range of ambient conditions in which an animal can remain thermostable and/or productive, but the provision of an ideal micro-climate in a thermally stressful macro-climate is seldom an economic proposition. Partial amelioration of the macro-climate by the provision of shade and other simple facilities may be inadequate for the husbanding of highly productive European cattle in thermally stressful climates, but may afford some improvement in the performance of more suited breeds of cattle.

REFERENCES: see pp. 523-524.

Temperature Regulation in Sheep

by

R. O. Rawson

INTRODUCTION

In a hot environment the sheep is capable of rapidly reducing its brain temperature by panting, while the remainder of the body core is more slowly reduced. This phenomenon has been described by Baker and Hayward (1968a) as a protective mechanism in the sheep which enables this animal to cool its brain selectively in a heat stress situation. The cooling of cerebral arterial blood by means of the carotid rete located in the cavernous sinus, as described by Daniel et al. (1953) and Baker and Hayward (1968b), has been suggested as a means of preventing cerebral overheating, and to be partly responsible for the high heat tolerance of the sheep.

In spite of a lowered brain temperature, the sheep continues to pant until the core (deep body) temperature is reduced. From this observation it is apparent that until this is accomplished, a significant heat loss drive still remains; and it is probable that nervous structures outside the brain are sending important temperature information to a central integrator.

It will be mentioned later that the experimental procedure of additional cooling of the hypothalamus is capable of inducing a decline in panting rate even while the animal remains under heat stress. Under these circumstances, the sheep hypothalamus still appears to be operating as a regulator.

From these two observations in sheep, that panting selectively reduces brain temperature, and further reduction of brain temperature reduces panting, one concludes that hypothalamic temperature receptors and temperature receptors outside the hypothalamus are at one time or another involved in body temperature control and probably exhibit some definite hierarchy with respect to their influence on the effector responses of panting, shivering, and vasomotor activity.

It has been well documented that temperature receptors outside the hypothalamus that are important to mammalian body temperature control can be found in the skin and spinal cord. Evidence of this will be found in other chapters of the text and will not be presented here. It will be the intention in the discussion to follow to describe a third major system of extra-hypothalamic temperature receptors that has been quite recently discovered in the sheep.

Although these receptors are monitors of the internal thermal environment, they are nonetheless important to the study of the effect on the animal of the external thermal environment, since it can be shown that their activity does modify the animal's response to heat or cold exposure.

Since the sheep hypothalamus can be cooled by panting while the temperature of the rest of the core remains elevated, then one should logically look for temperature receptors in areas outside the brain and spinal cord that could sustain a drive for heat loss. Spinal neurones could serve this function, but they cannot be stimulated by heat generated in the abdominal viscera until the heat has been transferred to the blood and pumped to the spinal cord.

Several years ago our laboratory was studying the physiological effects of long term warming of the abdominal viscera of sheep. It was during this procedure that we noted a sudden onset of panting that could not be accounted for by any change in body temperatures that were being measured.

This phenomenon is demonstrated in Figure 1 which depicts some data from one of our early experiments. During period A heat was introduced into the abdomen of a 41 kg ewe by means of electrical heating devices implanted against the inner surface of the abdominal wall.*

The respiratory evaporative heat loss (E) immediately increased in spite of a marked decline in hypothalamic temperature (T_{hy}). The temperature of the skin directly over the heater implants (T_{skin}) rose 0.7°C during heating. In period B heaters with the same surface area applied to the skin directly over the implants heated the skin temperature 1.5°C for better than 20 min without increasing E. Repeating internal heating in period C again stimulated panting.

When this procedure was repeated in a cold environment, metabolic heat production was markedly depressed by internal heating but not by the local external heating of the skin (Rawson and Quick 1970).

It was evident from these results that temperature sensors were being stimulated that were different from those known to exist in the brain, spinal cord or skin. Bligh (1961, 1963) has demonstrated that when cold saline was infused into the vena cava of sheep via a femoral or jugular catheter, there was a reduction of respiratory frequency in the absence of a decline in brain temperature. The walls of large veins or the vena cava were, therefore, considered to be possible locations of the thermoreceptors he was stimulating.

There is experimental evidence which shows, without discounting Bligh's hypothesis, that the thermoreceptors that were stimulated in the intra-abdominal heating experiments did not lie in the walls of the vena cava.

A study by Rawson and Quick (1972) showed that the response to the intra-abdominal

* Four electrical heat sources each measuring 5.5 cm wide, 16 cm long and 1 cm thick, were sutured side by side to the right and left abdominal wall, and powered through externalized wire leads. Details of construction, calibration and operation of these devices have been described elsewhere, along with a detailed description of basic experimental procedures used in the investigations reported here, (Rawson and Quick, 1970).

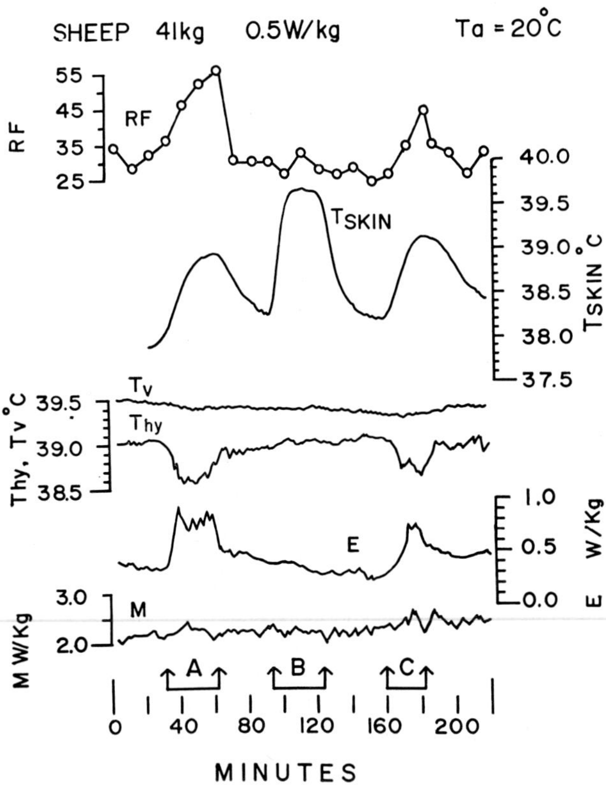

Fig. 1. The effect of intra-abdominal heating on respiratory frequency and hypothalamic temperature. Plotted from top to bottom are: RF, respiratory frequency; T_{skin}, the temperature of skin directly overlying the intra-abdominal heaters; T_v, vaginal temperature; T_{hy}, hypothalamic temperature: E, respiratory evaporative heat loss in watts per kilogram of body weight; and M, metabolic heat production calculated from oxygen consumption. Periods A and C denote internal heating, and D external skin heating.

heating of sheep was independent of vena cava blood temperature. In these experiments thermocouples in a sealed catheter in the posterior vena cava measured blood temperatures at four equally spaced points along the vena cava, the most anterior thermocouple lying at the entrance to the right atrium. The animal was also prepared with a gastric fistula permitting insertion of a large rubbeı balloon into the rumen to serve as a water perfused thermode.

With the sheep in a 28°C environment and with 16°C water being perfused through the thermode at one liter per minute, the deep body temperature rapidly declined 0.5°C as did vena cava blood temperature. Metabolic heat production rose rapidly to about 1.0 w/kg of body weight higher than the control level. When the intra-abdominal heaters were turned on (0.35 w/kg), the metabolic heat production was depressed by approximately 0.6 w/kg. During the 40 min heating period, there was no increase in any body temperature measured that could account for the depression of heat production, including all vena cava blood temperatures.

In other experiments vena cava temperatures also showed no increase when panting was stimulated by intra-abdominal heating alone (Rawson and Quick, 1972). These results were interpreted as evidence that the intra-abdominal thermoreceptors being stimulated by the electrical heat sources did not lie in the walls of the vena cava.

AFFERENT PATHWAYS FOR INTRA-ABDOMINAL THERMORECEPTORS

Denervation studies have been conducted to investigate the afferent neural pathways serving thermoreceptors. Rawson and Quick (1971) found that unilateral vagotomy did not appear to affect the responses to intra-abdominal heat. However, when the splanchnic nerves were sectioned on one side where they entered the adrenal plexus, both depression of metabolic heat production in the cold, and increase in respiratory heat loss in a warm environment, were completely absent when the operated side was warmed. However, sectioning of the nerves had no effect whatever on the response to stimulation of the contralateral side.

These results clearly demonstrated that the splanchnic nerves are the path for afferent impulses from the intra-abdominal thermoreceptors. At the same time these data provided conclusive evidence that the receptors responsible for the heat stimulus response do not lie in the spinal cord, skin, or muscle, since those tissues are not served by the splanchnics. The unilateral nature of the abolition of the response also discounted the possibility that thermal receptors in the liver were being stimulated by warm blood returning through the hepatic-portal system, since all blood from the gut eventually flows together at the liver.

THE ANATOMICAL LOCATION OF INTRA-ABDOMINAL THERMORECEPTORS

With the evidence so far available, no specific location for these receptors within the abdomen can be designated. Wood (1970) reported finding temperature sensitive neurons in Auerbach's plexus in the gut of the cat. Although not suggested by Wood, it can be interpreted that he was recording from temperature sensitive afferents in the intestine that may affect more central components involved in thermoregulation. It is appropriate to note that, similar to the sheep, the cat is also capable of selectively lowering its brain temperature by panting (Baker, 1972). Wood's finding supports the point of view that intra-abdominal thermoreceptors probably lie in the wall of the gut. This is an especially attractive hypothesis when one considers that in sheep, fermentation of the ruminal contents can raise its temperature 1.0°C or more. A monitoring system close to this source of additional endogenous heat would be most appropriate.

However, the results of the unilateral splanchnotomy studies described earlier tend to conflict with the idea of receptors in the gut wall, unless one revises a classical anatomical concept. The gut is believed to be innervated along its length by splanchnic nerves from both the right and left sides. Cutting the nerves on one side should reduce but not completely abolish the heating response, as happened in the experiments reported. Since the heaters on one side were in contact with only the rumen and those on the other with only the intestine, the unilateral abolition of response could be interpreted to mean that the rumen is innervated primarily by splanchnic nerves from one side, while the intestine is innervated by those from the contralateral side. It is believed, however, there have been no anatomical studies made to support such a concept.

Riedel et al. (1973) have thermally stimulated the dorsal wall of rabbits within the abdomen. They found that there was an optimum dorsal location where panting and skin vasodilation could be maximally elicited. Thermal stimulation of the stomach wall or intestine produced no response. In the sheep experiments that have been discussed, the level of response was the same whether the dorsally or ventrally located heaters were turned on. It was not possible, however, to stimulate abdominal viscera and abdominal wall separately. One cannot therefore discount the possibility that the intra-abdominal thermoreceptors of sheep are situated in the peritoneum of the abdominal wall. This concept would agree well with the observations made following unilateral splanchnotomy. Until more evidence is found, the question of the exact location of these thermoreceptors remains unresolved.

INTERACTION BETWEEN INTRA-ABDOMINAL AND HYPOTHALAMIC THERMORECEPTORS

In recent studies of intra-abdominal thermoreceptors in sheep, some observations have been made about their behavior with respect to certain other well documented thermosensitive areas.

A powerful drive to lose or conserve body heat can be elicited by heating or cooling the hypothalamus. When one plots body heat loss against average hypothalamic temperature, a linear regression line is obtained similar to those shown in Figure 2. In the upper right corner of the graph are two regression lines showing the effect of warming the preoptic hypothalamic area on respiratory evaporative heat loss in a 30°C environment. The closed circles depict the response when no intra-abdominal heating was done. The second line (open squares) shows the effect of warming the same hypothalamic area during the dissipation of 0.3 w/kg of heat within the abdomen. The intra-abdominal receptor stimulation lowered the threshold for stimulation of E, without changing the slope of the regression. The absence of a change in slope is interpreted to mean that the input from these receptors to some central integrating system did not change preoptic area sensitivity to warming.

In the upper left corner are data from similar experiments conducted in an environment just 6°C cooler. The regressions are plotted together in the lower part of the figure.

Note that in the cooler environment the slope of the regression has increased. This of course means that for a given rise in preoptic temperature, a greater panting response

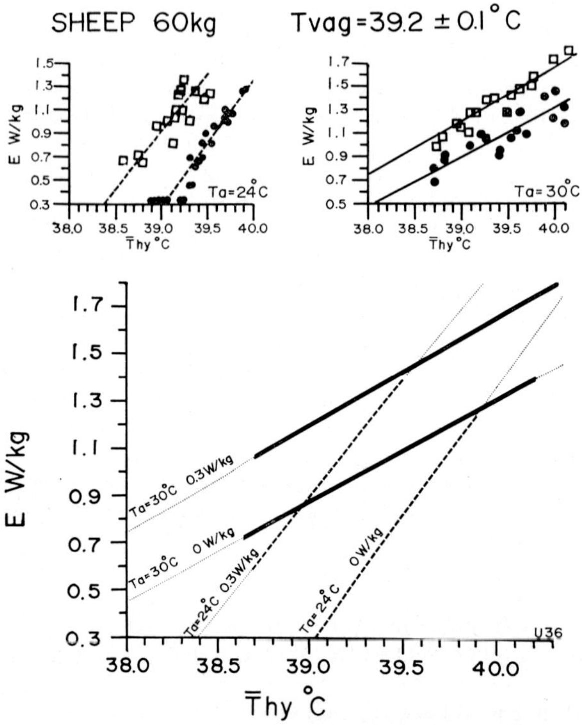

Fig. 2. Respiratory evaporative heat loss (E) as a function of preoptic hypothalamic temperature (T_{hy}) at two ambient temperatures and two levels of intra-abdominal heating. All points were taken when vaginal temperature measured 39.2 ± 0.1°C.

is obtained in a cooler environment. This change in hypothalamic thermal sensitivity is attributed to the effect of a change in the input from peripheral receptors exposed to the environment. The effect of peripheral inputs on hypothalamic sensitivity has been of interest to temperature regulation physiologists since its demonstration in cats by Jacobson and Squires (1970).

The effect of the environment on this sensitivity in the sheep is merely a demonstration of the phenomenon in another species. What is of more interest here is the difference between intra-abdominal thermoreceptors and cutaneous thermoreceptors with respect to their influence on preoptic thermosensitivity. Unfortunately this difference in effect cannot be explained by currently available data, nor is its physiological significance apparent.

In summary, another important extra-hypothalamic thermoreceptor system has been described which appears to play a role in the body temperature regulation of sheep. The reduction in cranial temperature that follows increased panting of a sheep under heat stress probably serves to protect the brain from damaging hyperthermia. Peripheral receptors in contact with the external environment such as those in the skin, serve immediate warning of external heat stress to central integrative areas. Monitoring of deep-body temperatures on the other hand requires receptor systems other than in the hypothalamus in this animal because cranial temperature undergoes such significant reduction when panting begins. Under such circumstances, the hypothalamus hardly appears qualified to dictate thermoregulatory requirements. This, together with the significant exothermic reactions occurring in the rumen, amplify the need for extrahypothalamic thermal receptors to monitor specific deep visceral temperatures in sheep.

Where these receptors are specifically located is still undetermined. The experimental data is insufficient to clearly delineate between gut wall, great veins, or parietal peritoneum as possible loci. A study of single unit neuron activity as a function of its temperature in selected intra-abdominal tissues, and recording from splanchnic afferent axons while heating or cooling specific tissue or viscera in the abdomen would be helpful in localizing these receptors.

The input from these sensors to the central nervous system is most certainly via the splanchnic nerves, although its ultimate influence on the effector system appears to be one of simple summation, rather than one of multiplication. The autonomic responses of panting, shivering, or vasomotion that occur from exposure of sheep to hot or cold environmental temperatures can be diminished or increased by a change in intra-abdominal temperature. The sheep must be considered to be unique in this respect until other animal species possessing anatomical structures similar to the carotid rete have been studied.

REFERENCES: see p. 524.

Chapter 1, Section 4c

Temperature Regulation in Pigs

by

D. L. Ingram

See Chapter 1, Section 5d and Section 18c (2).

Temperature Regulation in Marine Mammals

by

G. C. Whittow

INTRODUCTION

Marine mammals have adapted successfully to life in the sea. In so doing they have become extensively modified, both anatomically and physiologically, from the typical mammalian pattern. In many respects, they are of great intrinsic interest. Recently, public and scientific attention has been directed increasingly toward marine mammals for a number of reasons. They are now represented in oceanaria all over the world, and there is great concern for the survival and conservation of many species. Marine mammals have considerable potential in marine agriculture, and they have been used successfully by the U.S. Navy as an extension of man's capability in the sea (Ridgway, 1972). Physiological and behavioral temperature regulation in marine mammals was reviewed recently by Irving (1973).

CLASSIFICATION

Marine mammals may be broadly classified as indicated in Figure 1. Dolphins, whales,

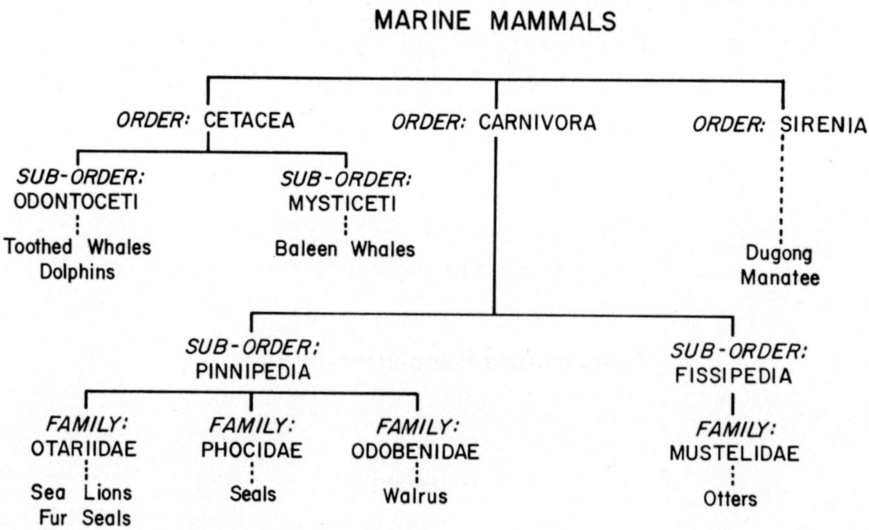

Fig. 1. A simple classification of marine mammals.

and sea-cows (dugong, manatee) are entirely aquatic. Seals, sea lions and the walrus breed on shore and, quite apart from their reproductive activity, many of them, and this includes the sea otters, elect to spend a greater or lesser amount of their time ashore.

SPECIAL PROCEDURES

Marine mammals are generally large animals, difficult and expensive to obtain and to transport, and expensive also to feed and maintain. Cetaceans require to be kept in sea water, but pinnipeds may be kept in fresh water. Three California Sea Lions were kept indoors in fresh water at the author's laboratory for over two years with no ill effects. The handling and care of marine mammals are discussed by Ridgway (1972). Marine mammals are capable of being trained to perform quite complex procedures. A technique for measuring the oxygen consumption of free-swimming dolphins, seals and sea lions has been described by Pierce (1970). Procedures for measuring the oxygen uptake and evaporative heat loss of pinnipeds in air are described by Matsuura (1972) and Matsuura and Whittow (1973a,b).

Table 1. Convective constants for water and air at 28°C. (From Webb, 1970).

	Water	Air
Convective constant*	102	0.61

* The convective constant is the product of the density (ϱ) in g/1, specific heat (C_p) in cal/g °C, and thermal conductivity (K) in cal/min.cm.°C, divided by the viscosity (η) in centipoises.

RESPONSES TO COLD

A comparison of the thermal properties of water and air (Table 1) emphasizes the main thermal problem faced by homeothermic animals that live in the ocean, namely, that the cooling power of water is enormously greater than that of air. The problem is most acute for a small animal because of its relatively greater surface area. The larger the surface area in relation to the body mass, the greater the rate of heat loss. Considerations of heat loss to water probably impose a lower limit on the size of marine mammals. The smallest adult cetacean weighs approximately 50 kg (Irving, 1973), while terrestrial mammals may weigh only a few grams. A corollary to this is the comparatively large size of newborn marine mammals (Irving, 1973). Newborn marine mammals increase their body weight at a phenomenal rate (Irving, 1973).

Heat loss to water increases as the temperature of the water decreases. Marine mammals encounter colder water when they dive, the temperature of the water decreasing from 27°C at the surface, in equatorial regions to 4°C at 600 m (Irving, 1969). The difficulty in maintaining thermal balance is compounded by the reduction in metabolic rate that marine mammals incur when they dive (Irving, 1973). Water temperature diminishes also with increasing latitude. The temperature of sea water in the arctic and antarctic is approximately −1°C. The immense numbers of marine mammals in both polar regions testifies to the successful adaptation of marine mammals to cold.

How is this adaptation accomplished? Theoretically, there are three ways in which animals may adapt to cold. They may (1) maintain a relatively low body temperature, (2) increase their rate of heat production; or (3) reduce the rate of heat loss. There is no evidence that marine mammals become hypothermic, and their size would militate against the occurrence of torpidity simply because the process of arousal would be extremely slow. The rate of cooling in air would also be very slow. It would be considerably faster in water, but the animals would either have to remain at the surface or interrupt the period of torpidity in order to breathe. The deep-body temperatures of marine mammals are within the range of temperatures of terrestrial mammals (Irving, 1973). Marine mammals do, however, incur a considerable degree of cooling of peripheral tissues, and there is evidence that their superficial tissues are adapted to function at low temperatures (Irving, 1973).

There is good evidence that marine mammals both increase heat production and reduce heat loss as an adaptation to living in the sea. Their heat production at rest, in water or in air, under thermoneutral conditions, is approximately double that of terrestrial mammals of the same size (Irving, 1973; Matsuura and Whittow, 1973a). The only marine mammals that do not appear to have an unusually high heat production are the Sirenia. The manatee and dugong are vegetarians, are relatively inactive, and are found only in the warm waters of Florida, South America and Asia. Little information is available on the effects of cold on the heat production of marine mammals. Hart and Irving (1959) determined that the heat production of Harbor Seals (*Phoca vitulina*) in water increased at a water temperature of approximately 13°C, i.e. the lower critical temperature was 13°C in winter. The lower critical temperature of animals tested during the summer was higher, approximately 20°C. In air, the lower critical temperature during the summer was −20°C. On the other hand, Harp Seals (*Pagophilus groenlandicus*), a more cold adapted species, did not increase their heat production in ice-water (Irving, 1969). The lower critical temperature of newborn Harbor Seals in air was 5°C, decreasing to −5°C in the 3-5 month old animal (Miller et al., 1973). The lower critical temperature of the Common Harbor Porpoise (*Phocoena phocaena*) appeared to be lower than 8°C (Kanwisher and Sundnes, 1966); but the Hawaiian Spinner Dolphin (Stenella longirostris hawaiiensis) – a tropical species – shivered at a water temperature of 24°C (McGinnis et al., 1972).

In addition to a higher heat production, marine mammals are endowed with a very effective insulating layer of blubber. In the Right Whale (*Balaeninae*) the blubber may be 70 cm thick and amount to 45% of the body weight of the whale (Irving, 1973). Cetaceans and Sirenians are essentially glabrous, but seals and sea lions are invested in a hair coat. In phocid seals and sea lions, the hair coat is wettable; and consequently, it provides little insulation in the water. Fur seals and sea otters, on the other hand, have a coat through which water does not penetrate; their coat affords effective insulation on land and in the water. As far as is known, pinnipeds do not have arrector pili muscles (Harrison and Kooyman, 1968). They are not therefore able to vary the thickness, and consequently the insulation, of the coat in response to cold. In the polar regions, the combination of very low temperatures and high winds results in rates of heat loss greater than that to ice-water. Under such conditions the Walrus (*Odobenus rosmarus*) has been observed to enter the water (Fay and Ray, 1968).

Many marine mammals have an anatomical arrangement of the arteries and veins in their appendages, highly conducive to the counter-current exchange of heat between arterial and venous blood (Elsner, 1969). These are especially evident in the fins, flippers, and flukes of cetaceans. Each artery is surrounded by a concentric ring of veins. These structures permit blood to flow to the extremities without the excessive loss of heat that this would entail in the absence of such vascular structures. Heat is transferred directly from the artery to the veins. Consequently, the arterial blood reaching the distal parts of the extremities is at a relatively low temperature, and heat, instead of being lost to the water, is returned to the body.

The seasonal variation in the lower critical temperature of the Harbor Seal, to which reference was made above, implies a change in the insulation. As this variation is apparent in animals tested in the water, it represents largely a seasonal alteration in the amount of subcutaneous fat, or in the peripheral vasomotor regulation, rather than in the pelage.

RESPONSES TO HEAT

Because of the facility with which heat is lost to water, it seems unlikely that any animal in the water is ever heat stressed under natural conditions. Nevertheless, it is theoretically possible that a large whale, swimming vigorously in warm water might not be able to dissipate heat rapidly enough to prevent hyperthermia. Large whales do migrate through tropical waters, but they are generally not feeding when they do so, which would lessen the heat load with which they have to deal. Many whales give birth to their young, and suckle the calves in tropical seas. Consequently, the mature females are actually consuming their complement of insulating blubber which would again alleviate the thermal problems of large mammals in warm waters. Measurements of the body temperature of large free-swimming whales in tropical waters have not been made, but such data for odontocete whales weighing up to 2,727 kg indicated that they were not hyperthermic (Whittow et al., 1974). Measurements of regional heat flow and body temperature in a large Atlantic Bottlenose Dolphin (*Tursiops truncatus*) in Hawaiian waters revealed that the animal was actively curtailing heat loss when it was at rest (Hampton et al., 1971). Moderate swimming did not result in an increase in body temperature, although the activity provoked some increase in heat flow from the extremities. The efficacy with which young Harbor Seals were able to lose heat to warm water testifies to the heat tolerance of at least some marine mammals in water (Miller et al., 1973).

Nevertheless, very few pinnipeds have penetrated the tropics and, as pinnipeds breed on land, a priori, it is the climate ashore that represents a deterrent to the colonization of hot areas. Information on the responses of pinnipeds to warm air is available only for the California Sea Lion and to a lesser extent for the Harbor Seal. The sea lion's strategy of temperature regulation, in the face of heat stress, is to store heat; consequently, it becomes hyperthermic (Whittow et al., 1972). Almost a quarter of the heat produced by sea lions at an air temperature of 30°C was stored (Matsuura 1972). There is clearly a limit to the amount of heat that may be stored. This limit is set by the lethal temperature for the animal. In early experiments on sea lions in the author's laboratory, an animal died at a rectal temperature of 42.2°C. The amount of heat that may be stored before the lethal temperature is reached is a function of the size of the animal. A large territorial bull may store a considerable amount of heat. The disadvantage of this type of response is that hyperthermia itself engenders an increased heat production – a van 't Hoff-Arrhenius, or Q_{10} effect – which, in turn, promotes a further increase in body temperature (Matsuura and Whittow, 1973a).

Sea lions were not able to reduce their heat production during exposure to heat (Matsuura and Whittow, 1973a). This is surprising in view of the reputed ability of not only sea lions, but marine mammals in general, to depress their metabolism during diving (Irving, 1973). It is also unfortunate because the high rate of heat production of marine mammals, even under thermoneutral conditions (Irving, 1973; Matsuura and Whittow, 1973a), places them at a disadvantage as far as their tolerance to heat is concerned. However, the single Harbor Seal examined was able to maintain its heat production constant during exposure to heat and even in the face of a hyperthermia (Matsuura and Whittow, 1973a). By virtue of its ability to do so, the Harbor Seal was considerably more heat tolerant than were the sea lions.

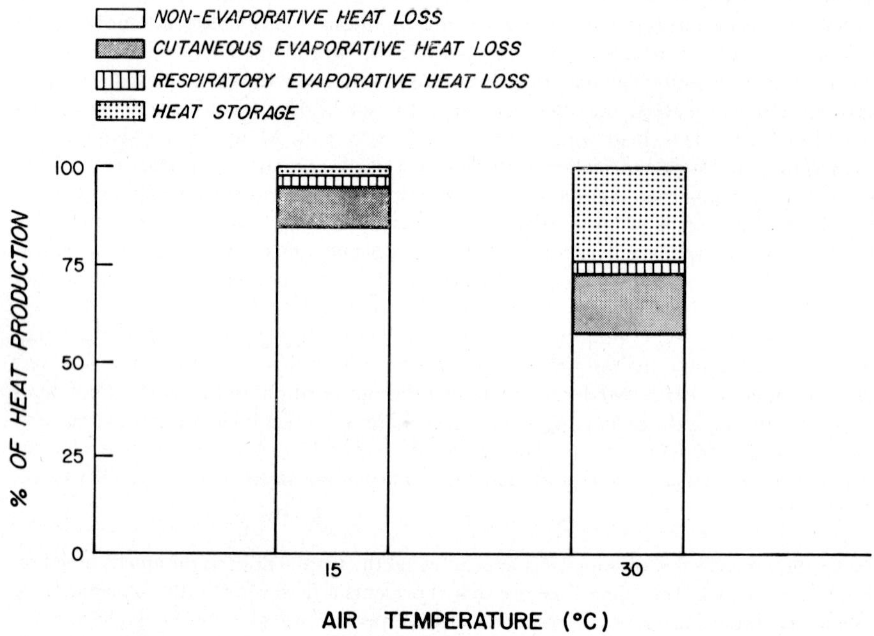

Fig. 2. Partition of heat loss in the California Sea Lion, at two air temperatures. (Based on Matsuura, 1972; Matsuura and Whittow, 1973a, b).

Evaporative heat loss mechanisms are vestigial in sea lions. There is no thermal polypnea or panting, and sweating occurs only on the bare areas of the flippers (Matsuura and Whittow, 1973b). Respiratory evaporative heat loss amounted to only 2-3% of the heat production at an air temperature of 30°C. The total amount of heat lost by evaporation of moisture from the skin surface represented approximately 16% of the heat produced by the sea lions. Consequently, less than 20% of the heat produced may be lost by evaporative means.

Most heat is lost by non-evaporative pathways, even at an air temperature of 30°C, when the animal is hyperthermic (Fig. 2). Heat may be lost to the air and by conduction to the sand or rock on which the sea lion hauls out. However, there is a limit to the amount of heat lost in this way. On dry sand it was found that the sand beneath the sea lion warmed, thus reducing the temperature gradient between the animal and its substratum, and hence, the heat loss from the animal. On wet sand, cooling of the skin in contact with the sand provoked a cutaneous vasoconstriction which increased the insulation of the peripheral tissues. Heat flow through the tissues to the sand was correspondingly reduced (Ohata and Whittow, 1973). In other words, the increased heat loss to wet sand, expected from purely physical considerations, was offset by a physiological response of the animal to peripheral cooling of tissue in contact with the wet sand.

In the absence of effective physiological responses to heat it is not surprising that sea lions, and other pinnipeds also, display a marked behavioral response to heat (Fay and Ray, 1968; White and Odell, 1971; Gentry, 1973). Under natural conditions, California Sea Lions enter the sea on warm days off the coast of California (Peterson and Bartholomew, 1967). In an experimental situation, sea lions chose to remain in water rather than seek shade out of the water at air temperatures of 26-33°C. Exposed to direct sunlight, California Sea Lions rapidly became hyperthermic. Although they executed fanning motions with their flippers and were able to secure some evaporative cooling as a result of wetting of the ventral body surface with urine, these behavior patterns attenuated the rate of rise of body temperature rather than prevented it. Sea lions were trained to press a switch in order to turn on a water shower. With this device, they could easily tolerate exposure to direct solar radiation at an air temperature of approximately 30°C (Whittow et al., 1971). The body temperature and heat production of sea lions in a warm environment were affected to a considerable extent by the behavior and activity of the animals (Matsuura and Whittow, 1973a). During sleep, the heat production and body temperature diminished by 24%. Mild activity elevated the heat production and body temperature above the level in the awake, resting animal. A sea lion that sleeps during exposure to hot conditions is clearly at an advantage, and this conforms with observations made on sea lions out of doors (Whittow et al., 1971).

CONCLUSIONS

The price that marine mammals have paid for living in the sea is a level of heat production approximately double that of terrestrial mammals. In addition, they are enveloped in an insulating layer of blubber and endowed with extensive counter-current heat exchange arrangements in the peripheral tissues in order to curtail heat loss to water.

These features, which have permitted marine mammals successfully to adapt to life in the sea, are an embarrassment to those species (pinnipeds, walrus) which have to breed on shore. In a hot environment on land, their strategy of thermoregulation is to store heat, increase non-evaporative heat loss and to invoke behavioral responses to heat. Their evaporative cooling mechanisms are relatively ineffective.

A great deal of quantitative work needs to be performed on marine mammals, particularly on cetaceans. Critical water temperatures are not known precisely for any species. Nothing is known about the role of central and peripheral temperature receptors in the control of thermoregulatory processes. This is an intriguing problem for animals that have a uniform surface temperature close to water temperature. Little information is available on the acclimatization of marine mammals to heat and to cold. Practically nothing is known about the temperature regulation of marine mammals during exercise, either on land or in the sea. The greatest challenge to the marine mammal physiologist is presented by the great whales; we simply do not know the nature of the thermal problems that they face or how they deal with them.

ACKNOWLEDGEMENTS: The preparation of this manuscript was supported by NSF Grant GB-29287X.

REFERENCES: see pp. 524-525.

Temperature Regulation in Fish

by

F. E. J. Fry

With the notable exception of the tunas and the lamnid sharks (Carey and Teal, 1966, 1969), fish attain what constancy of body temperature they achieve by behavioral regulation. That is, they are likely to be found in waters of congenial temperature whenever these are available to them. The sauger for example, in observations in Norris Reservoir (Fig. 1), was concentrated in a temperature stratum approximately 5°C wide about a median sloping gently from about 8°C in April to 20°C by the end of June and remained essentially at that level, well below the surface water temperature, for the rest of the summer with the exception noted in the legend when oxygen depletion in lower waters was severe (Dendy, 1945, 1948). All fish are found in a range of temperature characteristic for the species. Reviews on temperature selection in fish are given by

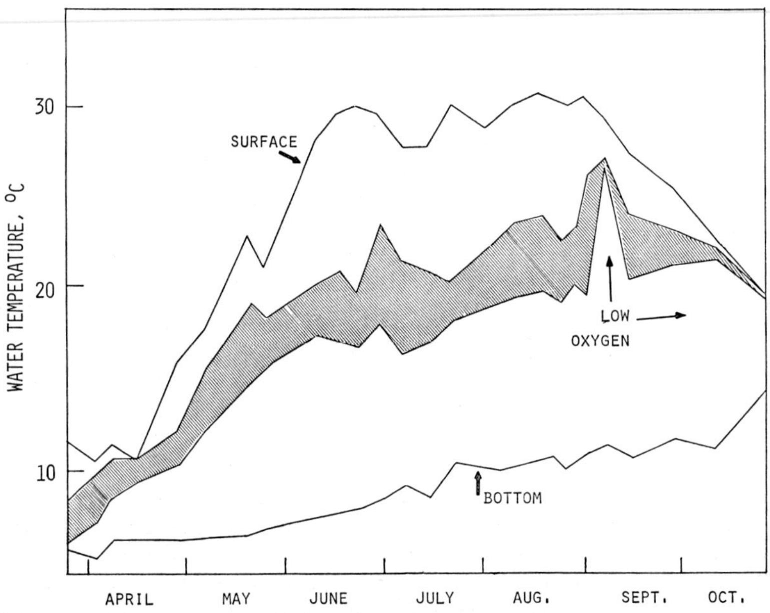

Fig. 1. Range of temperature in which the median 50% of the sauger, *Stizostedion canadense*, population was found in Norris Reservoir, Tennessee, 1943, in relation to the total range of water temperature available. Modified from Dendy (1948). Note that severe low oxygen conditions greatly restricted the downward range in late summer. For full details see also Dendy (1945).

Sullivan (1954) and Ferguson (1958). The subject is dealt with briefly in Frv and Hochachka (1970) and Fry (1971).

The classical laboratory experiment to show thermoregulatory behavior is to observe the distribution of organisms in a temperature gradient. A convenient gradient for fish can be set up which is essentially a 130 × 60 cm tank with removable double coils and connections for circulating hot and cold water but there are many other designs (e.g., Schmein-Engberding, 1953 and various of the authors cited here). Figure 2 shows

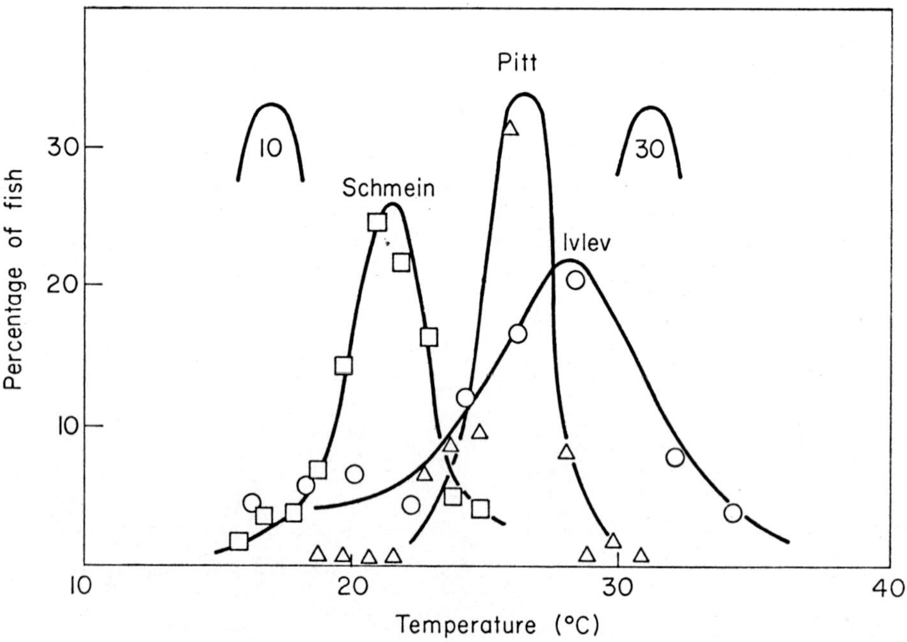

Fig. 2. Various observations on the distribution of carp in temperature gradients. From Fry (1971). The data are from Ivlev (1960); Pitt et al., (1956); and Schmein-Engberding (1953).

the results of experiments on carp in temperature gradients. The fish show a decided statistical propensity to congregate about a given temperature. That temperature measured variously as the mode, the median or the mean of the various temperatures frequented, has been ordinarily termed the PREFERRED TEMPERATURE or the TEMPERATURE PREFERENDUM. The temperature preferendum varies with the history of the fish, particularly its thermal history, as the two modes in Figure 2 labelled 10 and 30 indicate. These are modes for frequency distributions obtained for carp which had been acclimated respectively to 10 and 30°C prior to experiment. The main curves in the figure are for carp acclimated to 20°C. The effects of thermal acclimation on the temperature preferendum can be accounted for in a diagram such as Figure 3. Here a cardinal point of thermal preference is derived by adding a construction line to determine where the preferred temperature and the acclimation temperature coincide. That point is designated in the figure as the "final preferendum".

The relation between the thermal preferendum and the acclimation temperature is not always a simple one and varies both from species to species and apparently often between seasons within species as is shown below.

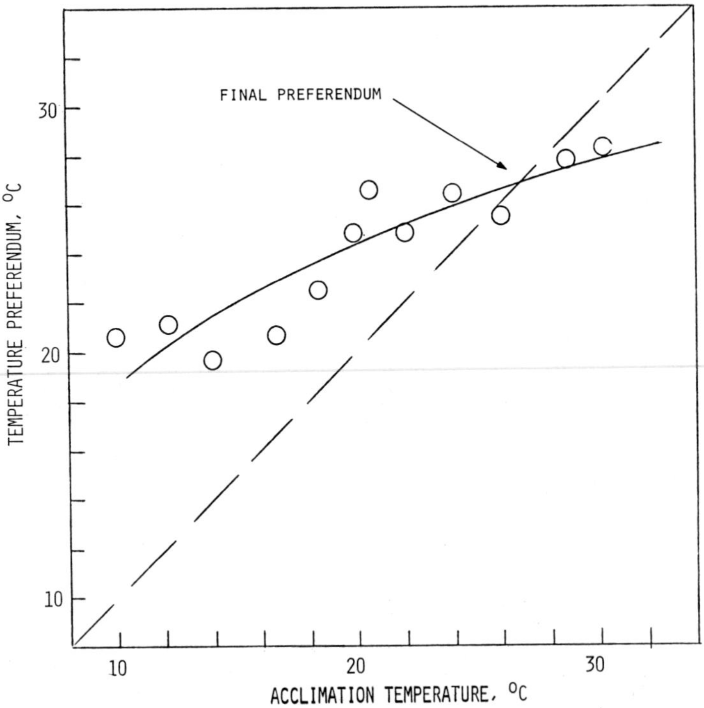

Fig. 3. The effect of thermal acclimation on the thermal preferendum of the smallmouth bass, *Micropterus dolomieui*. The points given are the modes of the selection curves. Late summer observations. The preferendum shown is the modal selected temperature.

While much work needs yet to be done, it is evident that season as well as acclimation temperature affects the level of the final preferendum (Fig. 4). It appears likely that final preferendum is adjusted downward in winter in species of the extra-tropical regions.

The final preferendum seems to be reached at a temperature which is a physiological optimum for the species concerned. Thus Brett (1971) points out for the sockeye salmon, *Oncorhynchus nerka*, that its final preferendum (15°C) is also the temperature at which growth on an unrestricted diet is fastest, as is also the speed for steady swimming. To support these activities the voluntary food intake, the effective output of the heart, the

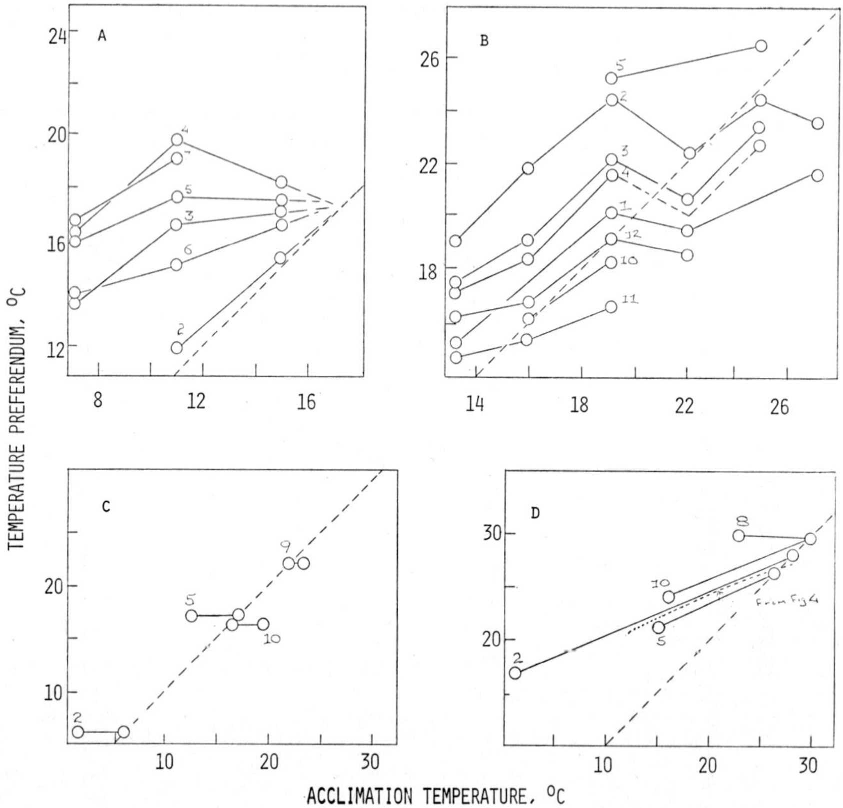

Fig. 4. Seasonal changes in temperature selection. (A) plaice, *Pleuronectes platessa*; (B) bitterling, *Rhodeus sericeus*, after Zahn (1963); (C) emerald shiner, *Notropis atherinoides*; (D) smallmouth bass, *Micropterus dolomieui*, after Barans and Tubb (1973). The acclimation temperature taken for (C) and (D) is the prevailing lakewater temperature at the time of test. The preferendum at acclimation temperature is the initial observed selection estimated from the authors' Fig. 2. The final preferendum is the temperature selected after continued exposure to the gradient. The numbers associated with the various lines indicate the months in which the observations were made. The irregularities in the order of the months in panel B may be due to two samples being used with different light histories. The second sample was added in February.

maximum uptake of oxygen available for activity and the greatest capacity for oxygen debt are also found at approximately 15°C.

The behavior by which fish reach the thermal preferendum appears to be by changing locomotor activity (kinesis), a matter discussed by Sullivan (1954). Thus Ivlev (1960) showed that movement per unit time in a continuously changing temperature was greater the greater the distance the fish was from the preferred temperature. Often such degree of movement persists only for a short time after a change of temperature has been experienced (Peterson and Anderson, 1969); but Olla and Studholme (1971) demonstrated persistent thermokinesis in the bluefish. Sullivan (Personal Comm. in Fry, 1964) emphasizes that movement of a fish at one instant in a gradient is influenced by its immediate temperature experience and thus is not a simple kinesis in the sense of Fraenkel and Gunn (1961).

It now appears that what the fish achieves by movement in a gradient is the attainment of a constant internal temperature. Observations on the body temperatures of fish in the vicinity of a thermal power plant (Table 1), where both cold lake water and the warm effluent were available to the fish, showed that they were travelling back and forth to the effect that their body temperatures were very close to their final preferenda.

Apparatus has been devised (Rozin and Mayer, 1961; Neill et al., 1972) in which fish

Table 1. Body temperatures of various fish captured in thermal effluents in relation to their temperature preferenda as found in field observations or laboratory experiments. (a) body temperatures observed by Spigarelli and Thommes (1973); (b) Bennett (1971). The preferenda are from Ferguson (1958, Tables II and III).

		Lake temp.	Discharge temp.	Body temp.	Preferred temp. Field	Lab.
	(a)	°C	°C	°C	°C	°C
Lake trout	*Salvelinus namaycush*	8.2	17.0	11.9	10-15	12
Brook trout	*Salvelinus fontinalis*	10.2	19.2	15.4	12-20	13-16
Rainbow trout	*Salmo gairdneri*	9.7	18.9	14.4	-	14
Brown trout	*Salmo fario*	11.3	20.6	16.1	-	12-17
White sucker	*Catostomus commersoni*	13.9	22.8	18.7	14-18	-
Carp	*Cyprinus carpio*	15.9	25.2	23.1	-	32
Yellow perch	*Perca flavescens*	17.2	26.6	22.5	20-21	24
	(b)					
Largemouth bass	*Micropterus salmoides*	23	33	26	27-28	28-32

can adjust the water temperature by some appropriate action. In the simpler case (Neill et al.) the action need only be to swim toward a cooler chamber to lower the temperature of the whole apparatus or to a warmer one to raise it. In Neill's apparatus the fish appear to adjust the temperature to their final preferendum. In Rozin's apparatus, where the response was more complex, the regulation was to maintain the water temperature at a level somewhat higher than the final preferendum; but the only response available to the fish in this case was to call for repeated doses of cooling water in the face of an otherwise ever increasing temperature.

There is evidence that both peripheral and central components of the nervous system are involved in behavioral temperature regulation in fish as in other vertebrates. Sullivan (1954) found that fish turned back from warm water too quickly for their internal temperature to be sensibly affected and reported also that in unpublished experiments she performed with K. C. Fisher, trout with their body surfaces insensitized by cocaine would no longer select in a temperature gradient. Trout with the forebrain destroyed but with the skin sensitivity intact would not select either. Crawshaw and Hammel (1973), using an apparatus in which fish can alter their body temperature by varying their residence time in two connecting chambers, each set at a different temperature, combined with an implanted thermode that could change the temperature of the rostral brain, found that warming the brain caused the fish to spend more time in cooler water. Conversely, when the brain was cooled, the fish ordinarily spent more time in the warm water. Cooling the brain, however, sometimes rendered the fish sluggish; and it would then remain in the cold water. The authors concluded that only if cooling were not too extreme for the fish to remain active did it give a regulatory response. Their data on changing the brain temperature together with the muscle temperature of a similar fish swimming freely in a gradient are shown in Figure 5. These data tempt one

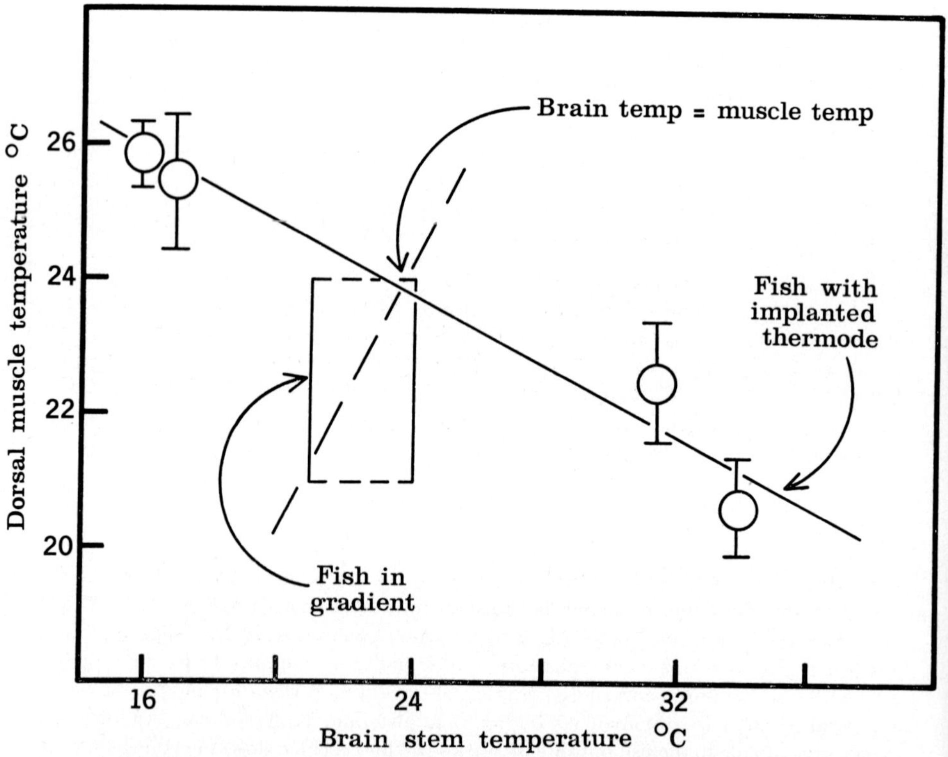

Fig. 5. The effect of changing brain stem temperature on muscle temperature of the bullhead, *Ameiurus nebulosus*, as adjusted by movement in a two-temperature apparatus. The rectangle delimits the range of muscle temperature, and by implication brain temperature, observed in a fish swimming freely in a temperature gradient with a thermistor implanted in the dorsal muscle. The range associated with the points is $\pm$ 1 SD. From Crawshaw and Hammel (1973).

to conclude that thermoregulatory behavior leads the fish to a certain set point where brain temperature and muscle temperature can be the same. Further work on this subject will be of great interest.*

While the vast majority of fish achieve whatever constancy of body temperature they do attain by thermokinetic behavior there are two groups well apart phylogenetically, the lamnid sharks and the tunas, which have a limited physiological regulation of their body temperature, chiefly of the muscles. In these fish, which are open water species and swim continuously, the heat produced by the activity of the red swimming muscles is conserved by a special arrangement of blood vessels to give a counter current heat exchanger. In all fish the blood temperature is brought almost exactly to water temperature as it passes through the gills and all blood passes through the gills after each passage through the heart before being returned to the body. Fish are drained of their metabolic heat after each circuit of the blood except for the minor heat exchange between the ordinarily parallel venules and arterioles, which in large fish can give an excess temperature of the deep muscles of a degree or so Celsius (e.g., Morrow and Mauro, 1950). In the lamnids and the tunas, however, the red muscle which supports continuous activity (Bone, 1966) is segregated into a discrete muscle running along the length of the fish and imbedded in the white muscle which supports the burst activity. In these fish, instead of being fed from the dorsal aorta, the red muscle receives its major supply from branches of the greatly developed lateral artery and is drained by tributaries to the lateral vein. These branches and tributaries are elaborated into *retia mirabilia* which could form a highly efficient heat exchanger so that the cold entering arterial blood is warmed by the leaving venous blood and the metabolic heat thus conserved in the muscle. Moreover, the red being imbedded in the white muscle, most of its heat passes to the latter thus heating it. Temperature control is by no means absolute but has a constancy somewhat of the order of the shell temperature of the mammals (Carey and Lawson, 1973). These authors also report considerable constancy in the stomach temperature of the bluefin which among others of the tunas also has visceral retia. The tuna brain, while maintaining a slight temperature differential over the environment (Stevens and Fry, 1971), has no heat conserving system like the muscle. There appear to be no records of brain temperatures for the lamnid sharks.

In very large tunas (Fig. 6a) the deep muscle temperature can be very constant over a wide range of water temperatures but in fish of moderate size e.g., the skipjack (Fig. 6b) such constancy is limited. It appears that in this species the maximum temperature differential that can be achieved is approximately 10°C. Since control is at an internal temperature of ca 35°C, then the best this species can do in water below 25°C is to maintain the maximum differential. It can be presumed that small bluefin are no better heat conservers than the skipjack, and it is a fact that it is the large fish of the species that have the greatest northerly range. For example, bluefin under 10 kilos are uncommon in Canadian waters where this species is a summer visitor. On the other hand, specimens of

* Further references which could not be examined at the time of writing are:
Crawshaw, L. I. and Hammel, H. T. (1974). Behavioral regulation of internal temperature in the brown bullhead. Comp. Biochem. Physiol., 47A: 51-60.
Crawshaw, L. I. and Hammel, H. T. (1973). Behavioral temperature regulation in the California horn shark, *Heterodontus francisci*. Brain, Behav. Evol., 7: 447-452.

175 to 300 kilos are angled there frequently and the world record taken by rod and line (444 kg) was taken in St. Ann Bay, Nova Scotia (Leim and Scott, 1966). The species does occur as far north as off the coast of Norway. Carey and Teal (1969) state that the temperatures of a sample of 36 bluefin observed by Jones and Mather all less than about 10 kg had muscle temperatures below the mean for the larger fish shown here in Figure 6.

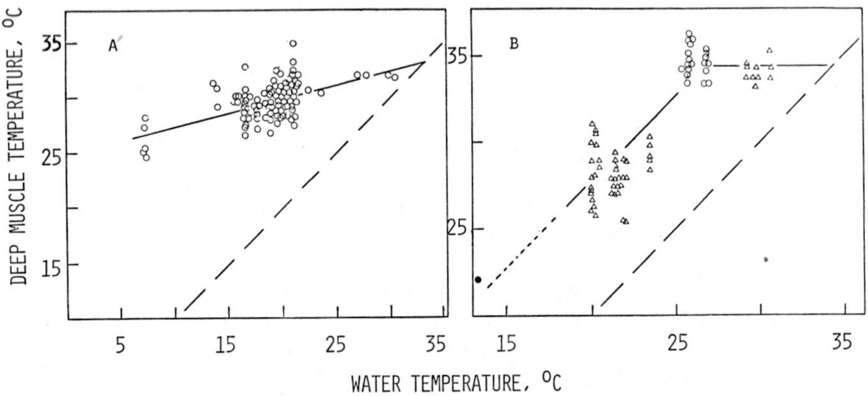

Fig. 6. Regulation of muscle temperatures in tunas. (A) large bluefin, *Thunnus thynnus*, observations on trap caught fish (Carey and Teal, 1969); (B) skipjack, *Katsuwonus pelamis*, a small species. Circles, data of Stevens and Fry (1971); the solid dot represents an individual which died in a cooling experiment. Triangles represent data figured by Carey and Teal (1969).

While the emphasis in temperature regulation above has been in the maintenance of an excess body temperature, the real element of control is the relaxation of control at the higher water temperatures so that the excess temperature becomes less and less (Carey and Teal, 1969). Whether such control is achieved by relaxation of the heat conserving system or by the reduction of metabolism is not known.

REFERENCES: see pp. 525-526.

Temperature Regulation in Birds (Effects of Cold on Metabolism and Endocrinal Reactions in Birds)

by

J. D. Rising

Birds are homeothermic, commonly are small (and consequently have relatively large surface areas from which heat is lost), and have high normal body temperatures (commonly higher than 40°C – that is, higher than those of mammals; NcNab, 1966). Thus the problems of heat loss are potentially acute for them. Certain non-physiological responses to cold, such as the development of highly effective insulative layers of feathers and fat, modifications of the circulatory system to reduce the rate of transfer of deep body heat to superficial areas, and communal nocturnal roosting have rightly received much emphasis from students of avian cold adaptation (Brooks, 1968; Irving, 1960; 1964; Steen, 1958; Veghte and Herreid, 1965). Certainly, modifications of the insulative coat play a key role in the annual adjustment that a bird makes – ultimately in response to the rigors of hibernal cold – and cold-adapted species possess better insulation than do warm-adapted ones. Whether there are comparable seasonal metabolic adjustments to cold, or whether interpopulational variation in metabolic parameters exists, has been debated, but the evidence now available suggests that these occur in at least some species (but are probably of less thermoregulatory importance than are behavioral and insulative modifications). The scope of this Section is limited to a discussion of metabolic and endocrine responses of birds to cold.

The energy metabolism and especially the endocrine reactions have not been so well studied in birds as in mammals (Chaffee and Roberts, 1971). The data available, expecially on endocrine reactions, are to some extent conflicting and inconclusive; progress has perhaps been somewhat impeded because the chicken, a convenient experimental animal, is a bird of tropical origin and may not have well-developed physiological capacities to cope with cold stress. Chickens additionally have precocial young that quickly develop adult thermoregulatory capabilities.

The metabolic enchancement associated with thermogenesis, of course, requires increased energy input (Chaffee and Roberts, 1971; Hart, 1962; Kendeigh, 1973), in many cases requiring small birds to spend much of the day foraging. In temperate regions many birds migrate and thereby avoid the most severe climatic conditions, but migration is itself an energetically expensive activity. It appears that migration has been especially selectively advantageous for those species that have difficulties finding food during the winter (most ground-foraging, aquatic, or insectivorous species). Migration appears to be an adaptation to avoid food shortage rather than cold temperature per se, and in this way is analogous to hibernation in north temperate mammals. Some species of birds, even small species [such as chickadees, *Parus* (10-20 g), and redpolls, *Acanthis* (12-16 g)], are exposed to extremely cold temperatures in nature, so the general physiological ability to withstand the cold is not beyond avian capabilities.

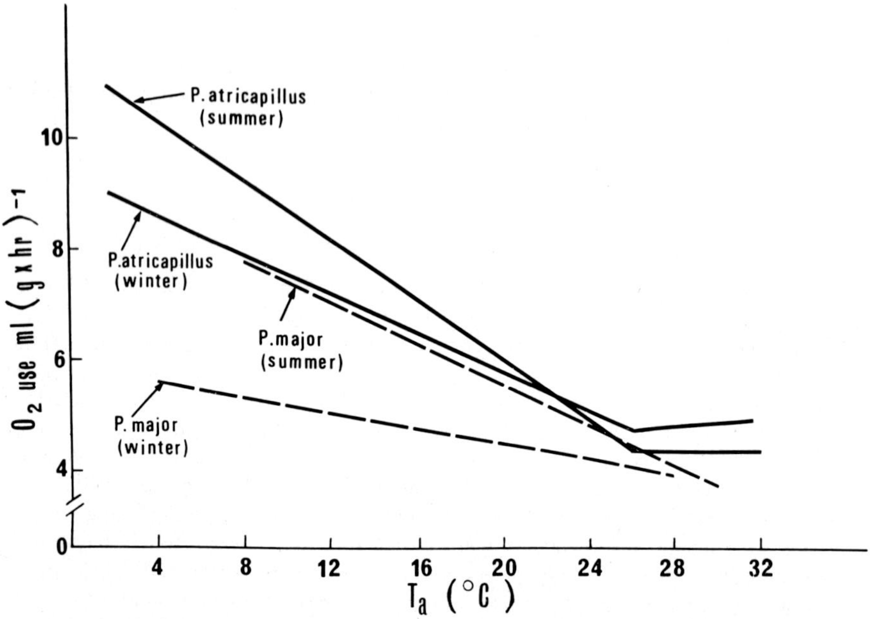

Fig. 1. Oxygen use of black-capped chickadees, *Parus atricapillus*, and great tits, *P. major*, during summer and winter, when exposed to a variety of ambient temperatures. A "thermoneutral range" of approximately 25°C-32°C is indicated for *P. atricapillus*. *P. atricapillus*, the smaller of the two species (ca. 11.0 g), has a higher SMR than does *P. major* (ca 19.0 g). Both species show a more rapid increase in oxygen consumption at progressively lower temperatures in the summer than in the winter, indicating less efficient insulation in the summer; during both seasons, the rate of increase is higher for *P. atricapillus* than for the larger *P. major*. See Hissa and Palokangas (1970) for data on *P. major* and Rising and Hudson (1974) for data on *P. atricapillus*.

There have been many studies of the oxygen consumption (reflecting metabolic activity) of quiescent birds exposed to ranges of ambient temperature conditions (Aschoff and Pohl, 1970; Lasiewski and Dawson, 1967). Oxygen use is minimal (standard metabolic rate = SMR) at a point or range (thermoneutral range, between the "upper" and "lower critical temperatures") of ambient temperatures where heat loss is presumably compensated for by the heat produced by basic maintenance metabolism, and progressively increases, usually linearly, at decreasing ambient temperatures, reflecting the increased metabolism required for thermogenesis (Fig. 1). The slope of the line that graphically represents the progressive increase in oxygen use varies from species to species, individual to individual, and season to season, and reflects not only variation in size, but doubtless also the efficacy of the insulative coat, behavior, body shape, amount of hypothermia tolerated, and the experimental milieu. Most interspecific variation in SMR appears to be explicable in terms of interspecific variation in size, smaller birds having higher SMRs (Aschoff and Pohl, 1970; Lasiewski and Dawson, 1967). However, relatively elevated SMRs, or other evidence of metabolic enhancement or acclimation, have been found in some cold-acclimated species (Fig. 2) [e.g., in chickens, *Gallus domesticus* (Chaffee et al., 1963; ElHalawani et al., 1970), pigeons, *Columba livia* (Hart, 1962), black-capped chickadees, *Parus atricapillus* (Rising and Hudson, 1974), and house sparrows, *Passer domesticus* (Hart, 1962; Miller, 1939)]. Also, depressed lower critical temperatures are characteristic of birds acclimatized to winter. Of course, seasonal variation in metabolism is influenced by many factors, such as molt, egg-laying, and lipid deposition (Kendeigh, 1973), but this variation suggests additionally that fundamental, if subtle, metabolic alterations may be associated with cold acclimation in birds as they are in some mammals (Jansky, 1973).

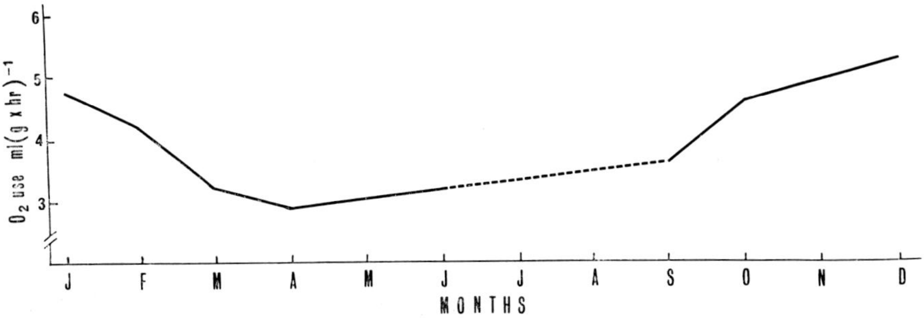

Fig. 2. Annual variation in oxygen used by house sparrows, *Passer domesticus*, at 28°C. Data from Miller (1939).

Steen and Enger (1957) and Hart (1962) have maintained that elevated SMRs in cold-acclimated birds were a reflection of a greater capacity of these birds to derive heat from shivering, and West (1965) in a detailed study was able to account for metabolic enhancements in evening grosbeaks (*Hesperiphona vespertina*) entirely by increased shivering thermogenesis (the relationship between oxygen use and shivering was linear). Nonetheless, the existence of nonshivering thermogenesis (NST) in mammals has stimulated a search for evidence of its occurrence in birds. In mammals, nonshivering thermogenesis is apparently especially pronounced in newborn individuals, and in arousal from torpor (Janský, 1973). NST is an efficient means of heat production which would be especially important for such mammals. Wekstein and Zolman (1968) and ElHalawani et al. (1970) have found evidence of NST in chickens. The study of ElHalawani et al. is especially interesting in this regard: they found the level of muscle activity (reflecting shivering thermogenesis) of Leghorn chickens decreased without corresponding decreases in oxygen consumption (which actually increased slightly) after prolonged (five and nine month) exposure to cold (5°C) (Fig. 3). This is the only study available where the same experimental individuals were exposed to progressively prolonged cold stress (in fact, unrealistically prolonged stress for chickens, but not so for some birds species) and strongly suggests that these birds possess the capability for nonshivering heat production. This capability would appear to develop gradually after prolonged exposure to cold and not as a response to short-term exposure.

In mammals, NST is primarily dependent upon catecholamines and brown adipose tissue is an important site for such thermogenesis (Jansky, 1973). This tissue has not been found in birds. The possible thermogenic role of catecholamines in birds is uncertain. Hart (1962) found that a single intramuscular injection of noradrenaline into pigeons paralyzed by curare had no thermogenic effect; Chaffee et al. (1963) and Freeman (1966) in a variety of experimental situations were unable to find any thermogenic effect of catecholamines in either adult or newly hatched chickens. Similarly, Hissa and Palokangas (1970) concluded that catecholamines had no metabolic stimulating effect on great tits (*Parus major*). However, Wekstein and Zolman (1968) found that injection of ganglionic blockers interfered with young chickens' thermoregulatory abilities which suggests that catecholamines are important to their thermoregulation; Lin and Sturkie (1968) found elevated plasma levels of both adrenaline and noradrenaline in chickens

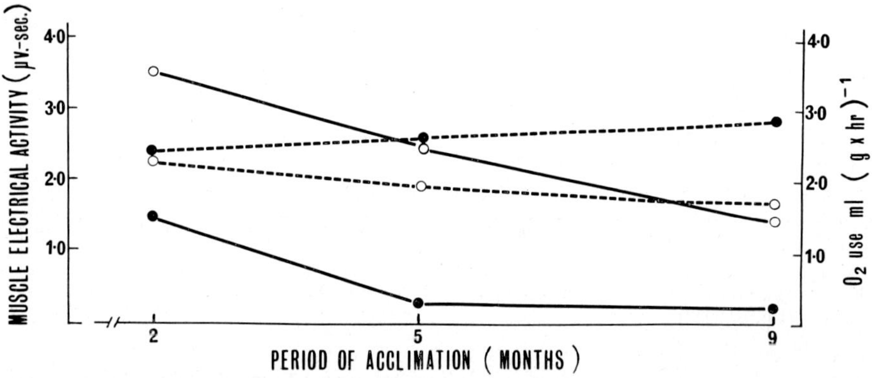

Fig. 3. Mean muscle electrical activity (solid lines) and oxygen use (dashed lines) of male Leghorn chickens subjected to 5°C temperatures, after being acclimated for two, five, and nine months to 5°C (dots) and to 23°C (circles). Variability was high, but differences in corresponding measurements after five and nine months are probably real. Data from ElHalawani et al., 1970.

exposed to prolonged cold. Thyroid hormones may be involved in thermogenesis. Freeman (1970) demonstrated that injections of both thyroxine and triiodothyronine increased metabolism in newly hatched chickens. Miller (1939) found that the thyroid of house sparrows was most active during the winter at a time when their SMR was elevated (Fig. 2) and Rising and Hudson (1974) found that rates of uptake and release of radio-iodine (probably reflecting enhanced thyroid activity) were higher in cold-acclimated (winter) than in warm-acclimated (winter and summer) black-capped chickadees. Hissa and Palokangas (1970) found that metabolism of great tits was increased by injections of corticosterone; however, this may not be of any thermoregulatory importance as high activity of the adrenal cortex is not long sustained in animals exposed to cold (Janský, 1973).

Possible sites for NST in birds, if it occurs, are obscure. The livers of cold-acclimated white-crowned sparrows (*Zonotrichia leucophrys*) were found to be greatly hypertrophied after cold exposure (Chaffee et al., 1963), indicating that the liver may be an important site for thermogenesis. Freeman (1970) suggests that newly hatched chicks may rely on

increased use of carbohydrate in muscle tissue for NST. The pectoral muscles, because of their considerable mass, would be an important site; they are certainly important for shivering thermogenesis (Hart, 1962; Steen and Enger, 1957).
Regulated hypothermia, such as frequently occurs in hummingbirds (Trochilidae) during the night (Lasiewski, 1964), may likewise be important to some other small temperate species (Steen, 1958). Chaplin (1974) found that black-capped chickadees regulated hypothermia to body temperatures as low as 30°C. Certainly, even the slight nocturnal hypothermia that is characteristic of homeotherms in general could result in significant energy savings for some species of birds, especially the small ones.

Evidence shows that behavioral adjustments and modifications of insulation are of great importance to birds exposed to prolonged cold. Regarding metabolic responses to cold, it is known that when exposed to cold stress, birds shiver, using heat generated to compensate for heat loss. Birds doubtless rely on such shivering thermogenesis during short-term exposure to cold, and it seems that at least some species rely on it in all situations. Nonshivering thermogenesis appears to occur in chickens and may occur in other taxa. Perhaps it develops slowly after prolonged exposure to cold; thyroid hormones, or perhaps catecholamines, may be involved in such thermogenic responses.

REFERENCES: see p. 527.

Thermoregulation in Wild Ungulates

by

M. K. Yousef

I. INTRODUCTION

Ungulates of the world or hoofed animals include an assemblage of orders not closely phylogenetically related. The major ungulata orders are the Perissodactyla which are odd-toed and the Artiodactyla which are even-toed. Other ungulate orders not covered in this section include the Proboscidea and the Hyracoidea. The distribution of the surviving Perissodactyla is limited to Africa; Asia; Central and South America. This order of mammals has a limited number of recent species (16) as shown in Table 1 and are restricted in their present distribution. In contrast, the Artiodactyla occur on all continents except Australia, are widely distributed in most natural areas of these continents, and have a large number (194) of recent species (Table 1). The taxonomic differences of the two orders of ungulates can be found in any mammalogy textbook. Because of their wide ecologic distribution in the arctic, desert, grasslands and tropics, ungulates display a variety of physiological patterns of temperature regulation. In general, due to their large size, ungulates can not avoid climatic stress as do small mammals but must actively cope with it. Thermoregulation studies on ungulates have been heavily oriented toward domestic animals primarily because of their agricultural and economic value. There are some excellent reviews and even books dealing with comparative thermoregulation of domestic ungulates (Bianca, 1961; Macfarlane, 1964; Schmidt-Nielsen, 1964; Johnson, 1965; Hafez, 1968; Whittow, 1971; Lee, 1972). Some investigators began to realize that studying only domestic ungulates might obscure some of the unique physiological processes acquired by wild ungulates living in their natural habitats. Attention to physiological adaptive mechanisms of wild ungulates began with a study by Dill (1938) on one of the oldest domestic animals which roams wild in the deserts of the southwestern United States i.e., the burro also known as the ass or donkey. In the early 1950's, Schmidt-Nielsen (1964) working in the Sahara confirmed Dill's finding on the donkey and extended our knowledge of another species, the camel. These classic studies have encouraged several workers to search for better insight into the essence of the mechanisms which allow wild ungulates to cope with their stressful environments.

Since this book is devoting sufficient space to several ungulates (i.e., goats, sheep,pigs, camels and domestic and wild cattle) which are primarily domesticated, this section concentrates on wild or feral species including several species of antelopes, reindeer, and donkey (burro or ass). Due to limited space, this section will contain references to only a few of the papers published in the past five years covering different areas of thermoregulation in wild ungulates.

Table 1. Summary of Classification of odd-toed and even-toed ungulates

Order	Family
I. Perissodactyla (Odd-toed) Total species = 16	Equids (Horses, Asses, Zebras) 7 species
	Tapirids (Tapirs) 4 species
	Rhinocerotids (Rhinoceroses) 5 species
II. Artiodactyla (Even-toed) Total species = 194	Suids (Pigs & Hogs) 8 species
	Tayassuids (Peccaries) 2 species
	Hippopotamids (Hippopotamuses) 2 species
	Camelids (Camels and Llamas) 6 species
	Tragulids (Chevrotains) 4 species
	Ceruids (Deer) 41 species
	Giraffids (Giraffe and Okapi) 2 species
	Antilocaprids (Pronghorn) 1 species
	Bovids (Cattle, sheep, goats, gazelles, antelopes and Duikers) 128 species

II. TEMPERATURE REGULATION

A discussion of thermoregulation must include the sources and avenues of heat production (HP) and heat loss (HL). This section deals specifically with the following questions: (1) Do wild ungulates balance HP and HL to maintain strict homeothermy? Is there

Table 2. Rectal temperature, T_{re}, and respiratory frequency, RF, of 20 species of wild ungulates at rest.

Species	Common name	RF/min	T_a, °C	T_{re}, °C	References
I. Even-toed ungulates					
Tayassu tajacu	Collared Peccary	-	Summer	37.5-40.0	Zervanos & Hadley (1973)
			Winter	37.5-40.0	
Giraffa camelopardalis	Giraffe	-	15	38.8	Bligh & Robinson (1965)
Ovibos moschatus	Muskox	-	-	40.0	Irving & Krog (1954)
Oryx beisa	Oryx	17	30	39.4	Taylor (1969)
Taurotragus oryx	Eland		30	39.8	Taylor (1969)
Gazella thomsonii	Thomson's Gazelle	25	30	39.4	Taylor (1970)
Gazella granti	Grant's Gazelle	42	30	39.8	Taylor (1970)
Cervus elaphus	Red deer	15-70	20	38.5-39.6	Johnson et al. (1972)
Connochaetes taurinus	Wild beest	34	30	39.0	Taylor et al. (1969)
Aepyceros melampus	Impala	45	30	39.2	Mailoy & Hopcraft (1971)
Alcelaphus buscelaphus	Hartebeest	28	30	38.7	Mailoy & Hopcraft (1971)
Rangifer tarandus stonei	Stone Caribou		−12to −26	39.0	Irving & Krog (1954)
Rangifer tarandus	Reindeer	50	30	38.6	Yousef & Luick (1975)
Kobus defassa	Waterbuck	25	30	39.6	Taylor et al. (1969)
Lama guanico	Guanaco	25	30	38.7	Rosenmann & Morrison (1963)
Camelus dromedaris	Camel	12	30	36.7	Maloiy (1971)
Syncerus caffer	Buffalo		18	38.5	Bligh & Robinson (1965)
Hippopotamus amphibus	Hippopotamus		32	35.4	Whittow (1971)
II. Odd-toed ungulates					
Ceratotherium	White Rhinoceros		25	35.2	Whittow (1971)
Equus assinus	Ass, Burro, Donkey	15	30	37.1	Maloiy (1971)

a relation between body temperature and climate; (2) What is the relationship of HP and body size in different species? Is there a relation between HP and climate; (3) What are the avenues and controlling mechanisms of HP and HL in wild ungulates?

A. BODY TEMPERATURE

1. Maintenance of Strict Homeothermy

Rectal temperature, T_{re}, is a simple and an easily measured index of body temperature and the balance between HP and HL. Normal or resting records of T_{re} for 20 species of wild ungulates are summarized in Table 2. With the exception of two species, white rhinoceros and hippopotamus, T_{re} ranges between 37 and 40°C. The degree of precision of thermoregulation in these species is reflected in their diurnal variations in T_{re}. The rather pronounced species differences in diurnal variations in T_{re} are illustrated in Table 3. While certain species show rather small diurnal variations in T_{re}, (1-2 C), for example the giraffe, impala, and hartebeest, others have a markedly labile T_{re} with wide variations up to 7°C as exemplified by the oryx and eland. Fluctuations in T_{re} seem to be directly related to environmental heat load as shown by Yousef and Dill (1969). In general, T_{re} rises slowly during the day reaching the maximum value in the afternoon and then decreasing gradually to the minimum in the early hours of predawn. This marked thermoinstability displayed by some ungulates represents neither a primitive thermoregulatory response nor a failure in the precise regulators. It does however,

Table 3. Diurnal changes in body temperature (T_b) of hydrated wild ungulates in the field or in a simulated desert environment in the laboratory (22-40°C)

Species	Range in T_b, °C	Type of study	ΔT_b	References
Collard Peccary	37.5-40.0	Field	2.5	Zervanos & Hadley (1973)
Giraffe	37.7-39.0	Summer & winter 28-42 5-18 10-25	1.3	Bligh and Robinson (1965)
Camel	35.2-37.8	22-35	2.6	Bligh and Robinson (1965)
Buffalo	36.9-40.2	10-25	3.3	Bligh and Robinson (1965)
Burro (Donkey, ass)	36.5-39.0	18-35	2.5	Yousef and Dill (1969)
Eland	33.9-41.2	22-40	7.3	Taylor (1969)
Oryx	35.7-42.1	22-40	6.4	Taylor (1969)
Impala	38.8-40.4	22-40	1.6	Maloiy & Hopcraft (1971)
Hartebeest	38.8-40.4	22-40	1.6	Maloiy & Hopcraft (1971)
Grant's Gazelle	39.4-40.4	22-40	1.0	Taylor (1970)
Thomson's Gazelle	39.0-40.0	22-40	1.0	Taylor (1970)

represent a useful adaptive mechanism to cope with stressful environments as discussed later.

Seasonal variations in T_{re} of wild ungulates have not been given attention. In our observations on reindeer under Alaska climate conditions, there were no measurable seasonal differences in T_{re} of resting animals. Recently, however, Zervanos and Hadley (1973) reported equal diurnal variations in T_{re} during the summer and winter seasons on the peccary under Arizona desert conditions.

Is there a taxonomic or a geographic trend in T_{re} of ungulates? Although limited and inconclusive, the available data show rather similar T_{re} of arctic, desert and tropic ungulates (Table 1) suggesting that T_{re} is not adaptive to climate.

2. Body Temperature versus Environmental Temperature

Changes in T_{re} resulting from changes in environmental temperature, T_a, point to the various patterns of thermoregulation. From the data presented on 11 different species (Fig. 1), it is obvious that: (a) the extent of changes in T_{re} over a wide range of T_a are not related to ecologic distribution or to phylogeny; (b) at T_a above T_{re} i.e., 45-50°C, all wild ungulates studied to date are able to effectively thermoregulate as shown by a stable high T_{re} that is well below T_a. The highest T_{re} recorded at 50°C was 41.5°C in the impala and hartebeest. The ability to tolerate varying degrees of hyperthermia is noteworthy since it is exhibited by both arctic and desert species, large and small. In a

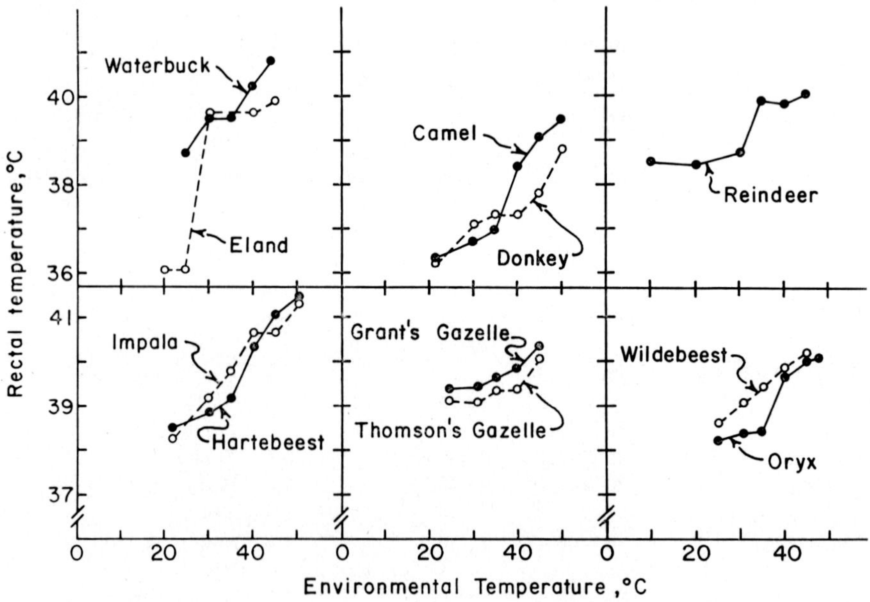

Fig. 1. Effects of environmental temperature on rectal temperature of 11 species of ungulates (Data from Taylor; Maloiy; Yousef and Luick; see references).

recent experiment on burros we found that at a T_a of 46°C they were able to maintain a T_{re} of about 38.5°C for a period of 24 hrs without any apparent distress (one of the burros did not even decrease its food consumption). In addition, various species of antelope were able to maintain a remarkable degree of regulated hyperthermia for as long as 12 hrs.

B. HEAT PRODUCTION

Measurement of basal metabolic rate (BMR) in ungulates is not practical since it is difficult to attain the state of postabsorptive and complete rest. Measuring the rate of oxygen consumption, $\dot{V}_{O_2}$, in standing, resting and under a comfortable T_a is commonly accepted as an index of minimal heat production. This is usually termed "resting" or "standing" $\dot{V}_{O_2}$. Resting heat production of mammals is expressed by an exponential relation to body weight. The following equation is modified from the original one by Kleiber (1961):

$$\dot{V}_{O_2}, \text{ liters/day} = 14.6\ W^{0.75}, \text{ where W is the body weight in kg.}$$

Data reported on $\dot{V}_{O_2}$ of 10 species of wild ungulates plotted against body weight are compared to the values predicted from the equation (Fig. 2). Values of $\dot{V}_{O_2}$ greater than predicted depend in part on the animals' not being completely inactive. However, one cannot rule out the possibility of a phylogenetic role in this observation as both desert and

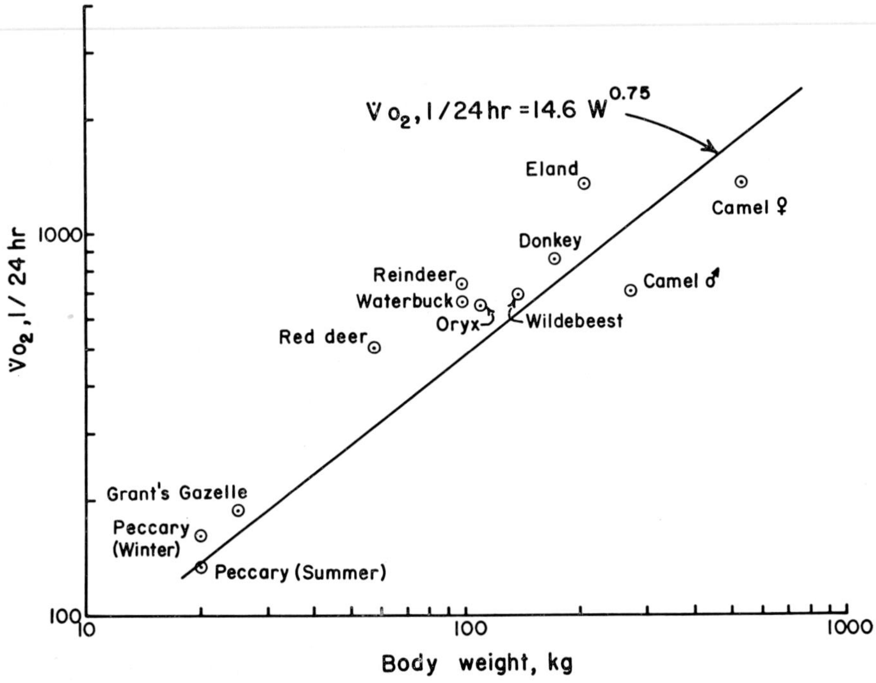

Fig. 2. Rate of oxygen consumption of 10 species of ungulates as compared to the standard mouse to elephant curve.

arctic ungulates studied have a higher $\dot{V}_{O_2}$ value than expected with the exception of the camel. This suggests that $\dot{V}_{O_2}$ in wild ungulates probably is not adaptive to climate.

Periodic variations in $\dot{V}_{O_2}$ may be seasonal and/or diurnal. Few studies have been devoted to this problem. Diurnal variations in $\dot{V}_{O_2}$ have been reported on the burro under Nevada desert conditions during the summer (Yousef and Dill, 1969). These daily changes in $\dot{V}_{O_2}$ were related to variations in environmental heat load, T_{re}, and times of feeding. In our studies on the burro under desert conditions and on reindeer under arctic conditions, there are no significant differences between $\dot{V}_{O_2}$ values measured during winter and summer. However, Zervanos and Hadley (1973) reported a 22% higher $\dot{V}_{O_2}$ in winter than summer under Arizona desert conditions on the collared peccary. The discrepancy among the burro, reindeer and peccary may be related to reproductive and nutritional status, size, and the degree of physical and physiological insulation. Hence further comparative studies are needed for a better understanding of the daily and seasonal variations in $\dot{V}_{O_2}$. Although well planned studies have dealt with $\dot{V}_{O_2}$ and body size relationship within the species of domestic ungulates, there are no systematic studies on wild ungulates to warrant any conclusions at this time.

What are the mechanisms controlling $\dot{V}_{O_2}$? Specific studies on wild ungulates cannot be found in the literature. However, one can speculate that these mechanisms are the same or similar to those discovered for domestic ungulates since the changes in $\dot{V}_{O_2}$ in response to changes in T_a are similar in the wild and domestic ungulates. A T_a below the lower critical temperature, $\dot{V}_{O_2}$ increased in the oryx, eland, donkey, and waterbuck (Taylor, 1969; Taylor et al. 1969; Maloiy, 1971). When T_a was increased above the upper critical temperature, $\dot{V}_{O_2}$ and T_{re} increased as reported on the species mentioned above and also on reindeer (Yousef and Luick, 1975). Regulatory mechanisms for changes in $\dot{V}_{O_2}$ were reviewed recently by Whittow (1971) including the peripheral and central nervous system and the neuroendocrine system.

C. HEAT LOSS

Avenues of heat loss (HL) in ungulates are the same as in all mammals and include nonevaporative and evaporative pathways. In this section brief attention will be given to the details of nonevaporative cooling since this has been adequately reviewed by Whittow (1971).

1. NONEVAPORATIVE COOLING

The insulative value of the coat or "physical insulation" in comparative thermoregulation was first emphasized by Scholander et al. (1950) in their classic study on body insulation of arctic and tropic mammals which included two wild ungulates, reindeer and Dall sheep. This study was followed by many experiments dealing with the insulative value of the coat in domestic ungulates and the factors affecting insulation (Whittow, 1971).

"Physiological" insulation, i.e. tissue indulation, specifically skin and peripheral tissues, plays an important role in HL. These phenomena are termed by Irving (1964) peripheral heterothermy. Reindeer under winter conditions in Alaska decrease the skin temperature of the extremities to near freezing thus conserving energy and preventing hypothermia in the deep body. Regulation and factors affecting physiological insulation in ungulates have been discussed in an excellent review by Whittow (1971).

2. *EVAPORATIVE COOLING*

Evaporative cooling involves respiratory and cutaneous avenues. Ungulates utilize either one or both avenues to varying degrees. The relative importance of each avenue varies greatly among different species.

a. *Respiratory Evaporative Cooling*

An increased rate of heat loss from the respiratory tract can be accomplished either by

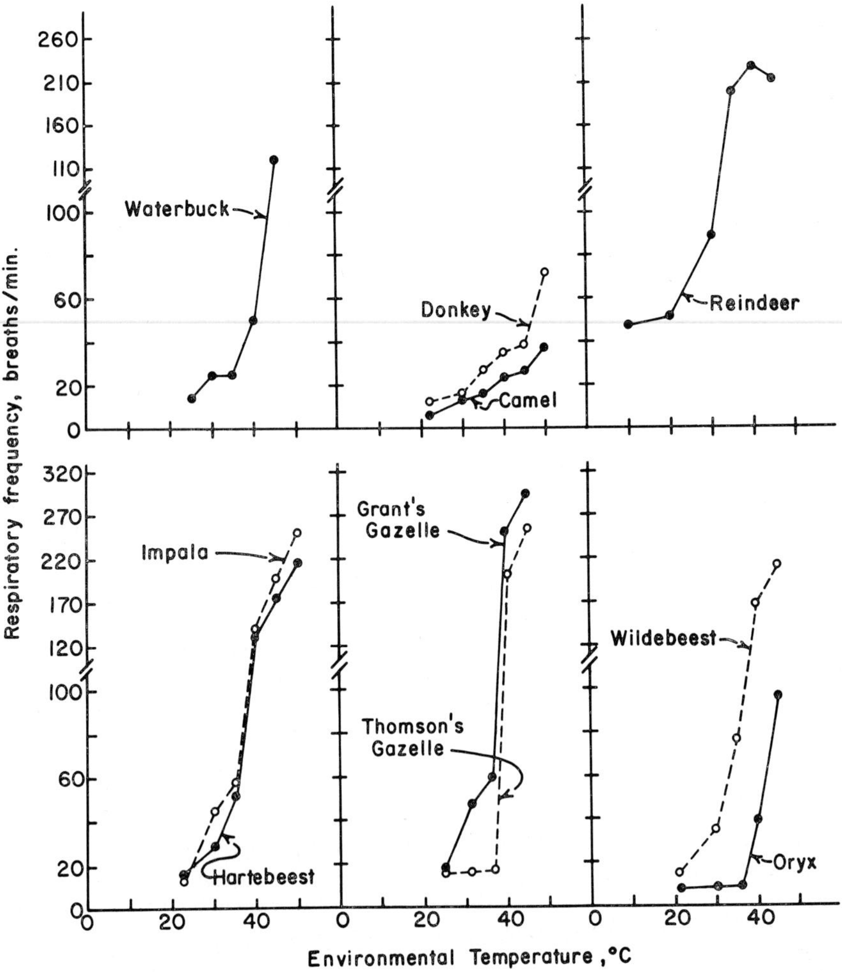

Fig. 3. Effects of environmental temperature on respiratory frequency of 10 species of ungulates (Data from Taylor; Maloiy; Yousef; see references).

an augmentation of the respiratory minute volume or by an increase in the aqueous vapor pressure gradient between the air which the animal inhales and the moist mucosa of the respiratory tract (Whittow, 1971). Data published on the changes in respiratory frequency (RF) as a result of increasing T_a on 10 species of wile ungulates are summarized in Figure 3. When T_a increased from 20-25°C to 40-50°C, RF increased ranging from 4 to 15 times pointing to the relative importance of respiratory evaporative cooling in some species over others. The lowest increase in RF is found in the camel, while the highest increase was observed in the gazelles. At high T_a, the lowest RF was about 38 in the camel and the highest was 295/min in Grant's gazelle. The amount of heat lost by wild ungulates via respiratory evaporative cooling varies in different species. For example, the eland loses between 20 and 30% of their total evaporative loss by respiratory route; but the Wildebeest depends primarily on panting for heat loss.

Diurnal variations in RF were recorded for the burro during a summer day in the desert (Yousef and Dill, 1969). These variations corresponded to variations in the respiratory minute volume and with little change in tidal volume. All of these changes were accompanied by similar changes in T_a, T_{re}, and $\dot{V}_{O_2}$.

Ungulates differ in the pattern of their panting; some pant through the nose and others pant with an open mouth and protruding tongue. Few studies have dealt with the physiology of panting and its energetic cost in wild ungulates. Data on domestic ungulates were recently reviewed (Whittow, 1971; Jenkinson, 1972).

b. Cutaneous Evaporative Cooling

Skin loses water in two ways: (a) by diffusion of vapor through the skin, or (b) by sweating. The skin of ungulates depending on the species, contains a wide range in number and size of sweat glands which occur in association with the hair follicles. This type of sweat gland is termed "epitrichial" by Bligh (1967) who opposes the use of the term "apocrine" which implies a breakdown of the integrity of the secretory cell membrane and the extrusion of cellular material into the lumen of the gland. This process of discharge is slow and cannot be used for the rapid discharge of sweat for thermoregulatory purposes.

Under cool conditions, cutaneous evaporative cooling in ungulates seems to depend primarily on simple passive diffusion of water. On the other hand, under heat stress, evaporative cooling depends on sweating. The effects of a wide range of T_a on sweet rate, SR, of 10 species of wild ungulates are shown in Figure 4. It is evident that some ungulates i.e., wildebeest and Grant's gazelle, show no change in the rate of cutaneous water loss at a T_a higher than T_{re}. In contrast, there are species which increase their SR greatly, i.e., the eland, oryx, and donkey. The data on SR and RF, shown in Figures 3 and 4, demonstrate the presence of a wide spectrum of evaporative HL as has been shown on domestic ungulates by Bianca (1968). Some species only pant (wildebeest and Grant's gazelle) while other species only sweat (camel). Intermediate between these two extremes are the species which both sweat and pant to different degrees.

The pattern and control of the sweat glands have been studied in several wild ungulates (Robertshaw and Taylor, 1969; Allen and Bligh, 1969; Bullard et al., 1970; and John-

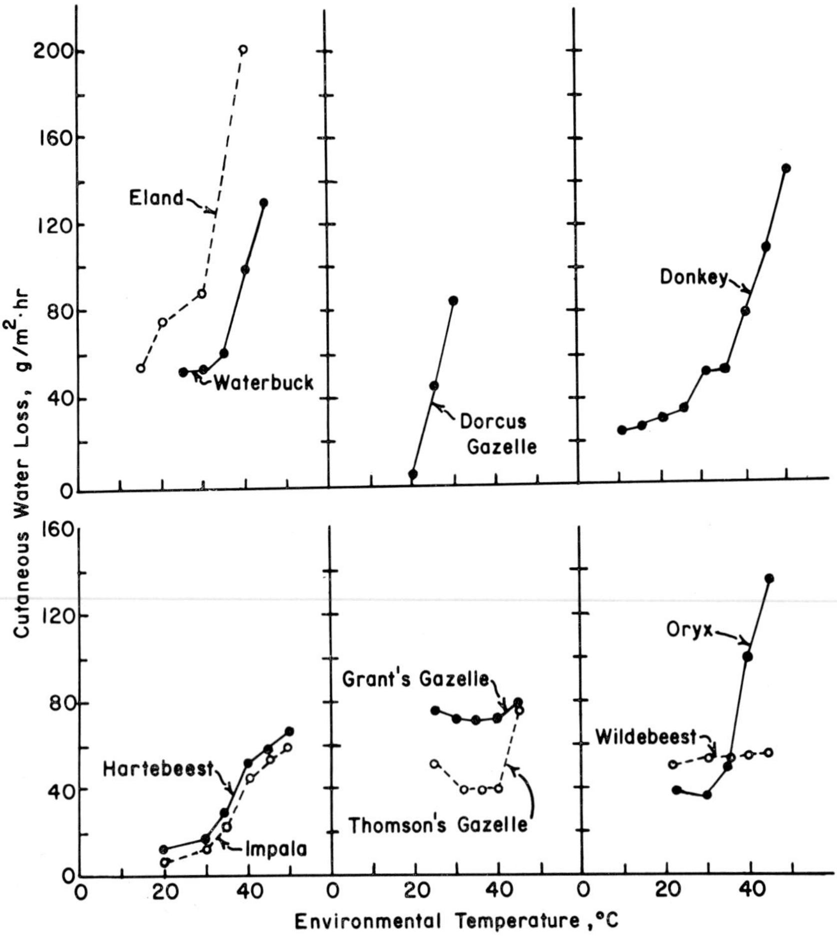

Fig. 4. Effects of environmental temperature on cutaneous water loss of 10 species of ungulates (Data from Taylor; Maloiy; see references; data on Dorcus gazelle are taken from Ghobrial, 1970, Physiol. Zoo. 43:133).

son et al., 1972). In all of the species studied, thermal sweating is under adrenergic neuro control. Sweat glands respond to adrenergic drugs but not to cholinergic drugs. The sweat glands of the various species studied release their secretion in three different patterns which are summarized in Table 4. No correlation was found between the patterns of sweating and the phylogenetic position of the species (Robertshaw and Taylor, 1969). Bligh (1967) suggested that periodic or cyclic discharge by the sweat glands is caused by the contraction of the myoepithelium surrounding the gland while the continuous discharge of sweat could be caused by a sustained overflow of fluid onto the skin surface. Macfarlane (1964) has pointed out that: "Evaporation of sweat under a thin layer or insulating hair is probably the best physical compromise for allowing the skin to heat while providing protection from heavy solar loads."

Table 4. Pattern of sweating in ungulates

Sweat pattern	Species	Reference
1. Cyclical discharge	Duiker, Grant's Gazelle, Thomson's Gazell, Oryx, donkey	Robertshaw and Taylor (1969)
		Bullard et al. (1970)
2. Continuous but stepwise rise in discharge	Waterbuck, eland	Robertshaw and Taylor (1969)
3. Continuous but immediate rise in discharge	Buffalo	Robertshaw and Taylor (1969)
	Llama, donkey	Allen & Bligh (1969)
	Red deer	Johnson et al. (1972)

III. THERMOREGULATION IN ARCTIC UNGULATES

Relative to number of studies on desert ungulates, little work has been done on arctic ungulates. The one species that has been given attention is the feral reindeer, therefore, this section deals primarily with the adaptive responses of reindeer studied under Alaskan conditions.

A. PREVENTION OF HYPOTHERMIA

The classic work of Scholander et al. (1950) has revealed a good correlation between the thickness of fur and insulation, but no correlation was found between insulation and body size among large mammals i.e., large mammals have about the same insulation per unit of surface area. Also it was shown that large arctic animals have a very wide thermoneutral zone (from +30 to −40°C). This means that total insulation would reach a maximum value only at T_a as low as −40°C, at which $\dot{V}_{O_2}$ must increase to prevent hypothermia. In 1955, Irving and Krog showed the importance of the skin in reindeer as a regulator of heat. Although $T_{re}$was 38°C at a T_a of −31°C, there was a wide range of variations in skin temperature of the extremities where temperature was recorded as low as 9°C on the lower part of the leg. The low temperatures of the appendages of reindeer serve an important role in conservation of heat during rest and also function as an effective avenue of heat loss during activity i.e., increased HP. The effectiveness of the apendages as a heat regulator is primarily due to a suitable arrangement of veins and arteries to operate as heat exchanger i.e., a counter current heat exchange system (Scholander and Krog, 1957). The low temperature of reindeer legs was found to be associated with low melting points of fats, thus the adipose tissue of the legs remain soft under cold; a condition essential for the effective operation of the legs at low T_a (Irving et al., 1956).

The importance of fur insulation and peripheral heterothermy in reindeer was confirmed by data on HP of large (9 month old) caribou (reindeer and caribou are probably the same species, *Rangifer tarandus*) by Hart et al. (1961). These authors have stressed that temperature regulation of Caribou was well established at birth and that the young calves, contrary to older calves, rely upon metabolic responses upon exposure to cold. In the older calves, overall insulation was sufficiently great so that they would remain near the thermoneutral state even under severe winter conditions, an ability that was not shown by young calves. The role of the endocrine system in winter acclimatization of reindeer was recently reported (Yousef and Luick, 1971; Yousef et al., 1971). The adrenal cortical activity was increased during winter, while the thyroid gland appeared to play only a minor role in acclimatization to arctic cold. These hormonal changes suggest that metabolic acclimatization is of less significance than insulative acclimatization in prevention of hypothermia in adult reindeer during winter in the arctic.

B. PREVENTION OF HYPERTHERMIA

Arctic ungulates may experience exposure to heat stress as a result of their effective insulation even at moderate T_a or by exposure to high T_a during the summer season. Studies on physiological responses of arctic ungulates to heat are limited. Rosenmann and Morrison (1967) concluded that reindeer have a good capacity for heat resistance when water is available but very poor resistance to water deprivation with or without heat stress. Yousef and Luick (1975) found that thermal ploypnoea in reindeer is an important mechanism permitting adequate respiratory evaporative cooling which kept T_{re} at 39.9°C when T_a was 45°C for a period of 5 hrs. Since indices were developed to measure heat tolerance in domestic ungulates (for a review see Bianca, 1963), we used two formulas to calculate an index for reindeer and compared it with other wild and domestic ungulates as shown in Table 5. It appears, regardless of which index is used, that reindeer are as heat tolerant as either wild desert or domestic ungulates.

Table 5. Comparative data on heat tolerance of domestic, wild Ezst African and Arctic ungulates.

Ungulate	Rhoad's* Coefficient, %	Benezra's** Index	Reference
Ayrshire	65	4 to 11	Bianca (1963)
Waterbuck	55	6	Taylor*** et al. (1969)
Reindeer	69	11	Yousef and Luick (1975)

* Rhoad's Coefficient $= 100—[18\ (T_{re}—38.3)]$
** Benezra's index $= T_{re}/38.3 + RF/23$.
*** The indices were calculated from data by Taylor et al. (1969).

How much do antlers contribute to heat loss? This question was studied by Krog and Wika (1974) who found that the large local blood flow to the antlers was necessary for the rapid growth of the antlers and not due to a special thermoregulatory adjustment as had been suggested. Our work on reindeer supports this finding since the animals used in the heat stress studies had no antlers but exhibited excellent thermoregulatory control.

During exercise (running while pulling a heavily loaded sled), reindeer increase their HP up to 4-5 times the resting or standing rate and are able to regulate their T_{re} at about 39°C i.e., an increase of about 0.8°C (Hammel et al., 1962). About 25% of the total HP during exercise was dissipated by panting, and the remaining portion was lost by nonevaporative cooling. This is accomplished by increasing the surface temperature of the appendages and the thickly furred parts of the body during exercise. The higher surface temperature was due to increased flow of warm blood to the skin in order to dissipate the heat of exercise.

IV. THERMOREGULATION IN DESERT UNGULATES

Several species of wild desert ungulates have been investigated. The main problem facing animals in the desert is not how to stay warm but how to keep cool with and without a source of water. Wild ungulates in the desert utilize several fascinating mechanisms to cope with heat and aridity.

A. LABILE HOMEOTHERMY

Schmidt-Nielsen (1964) was the first to demonstrate that variable diurnal T_{re} of the camel is an important mechanism for survival in the desert. Hyperthermia during the hottest part of the day would result in: (a) conservation of water as a result of heat storage in the body and (b) reduction of the amount of heat gain from the environment as a result of reducing the gradient between T_{re} and T_a. This mechanism has been found to be used to varying degrees by several other species (Table 3) even when hydrated and resting.

The degree of hyperthermia is greatly amplified by dehydration and exercise in some species to a degree that one species, the Grant's gazelle, which had a T_{re} of 46°C that was in excess of T_a Other ungulates are capable of running with a much smaller increase in T_{re} as the eland (Taylor and Lyman, 1972) and the burro can walk for more than 10 hrs on a hot day in the desert even during dehydration with a T_{re} below 40°C (Bullard et al., 1970 and Yousef et al., 1970). Taylor (1972) observed a Grant's gazelle which had a T_{re} of 46.5°C for as long as 6 hrs with no observable ill effect. Such high T_{re} would be lethal to most other mammals. The question then is how can the brain and other heat sensitive areas of the body tolerate this degree of hyperthermia? In some ungulates the internal carotid artery is absent and blood is supplied to the brain primarily by the external carotic artery. At the base of the brain, the carotic artery forms a carotid *rete mirable* consisting of hundreds of small parallel arteries and then reforms to supply the brain. This *rete* lies in a venous pool, the cavernous sinus (Daniel et al., 1953). Venous blood returning from the nasal passages is cooled by evaporation as the animal breathes and drains into the cavernous sinus (Baker and Hayward, 1968). This cool venous blood exchanges heat with the arterial blood thus cooling the blood supplying the brain (Taylor and Lyman, 1972). This counter-current cooling of the brain is of importance for tolerance of extremely high T_{re}.

B. HEAT PRODUCTION

A low HP in the summer would help an animal to better tolerate heat. Although a low HP in the summer was measured in the collared peccary, other ungulates (the burro) do not show such metabolic adaptation. In dehydrated ungulates, HP is decreased thus helping in conservation of water. Some ungulates even during the hottest parts of the day are seen seeking food and water. Do wild ungulates produce less heat during this activity? To my knowledge there are only two studies dealing with the energetic cost of walking in the burro as compared to man (Yousef and Dill, 1969; Yousef et al., 1971). In these studies, it was concluded that the burro walking (at the same rate and under the same environmental condition) with or without a load and up or down various grades expend only about one half of the energy used by man. This low HP during dehydration and walking results in a significant savings in water and allow the animal to increase its daily activity while seeking water and food. Additional studies of this type are needed on other species.

C. DEHYDRATION

The great tolerance of dehydration in ungulates was first shown by Dill (1938) and later by Schmidt-Nielsen (1964) who studied physiological responses to dehydration in the camel and donkey. The maximum ability for tolerance to dehydration is not known but some ungulates are shown to tolerate a loss of water corresponding to 30% of their original body weight even at T_a of 40°C (Maloiy, 1970). During dehydration, most ungulates studied have shown various degrees of decreasing HP and evaporative water loss. In some species, Grant's and Thomason's gazelles, dehydrated animals abandoned evaporative cooling and instead allowed T_{re} to increase to a point where T_{re} and T_a were about the same. Dehydrated gazelles began to pant when T_{re} exceeded 40°C (Taylor, 1972).

During dehydration, ungulates decrease their urine output and water loss in feces. Associated with the reduced urine volume some ungulates excrete concentrated urine (the dik-dik, camel and oryx) whereas some ungulates fall short of this ability i.e., donkey (Maloiy, 1972).

What are the mechanisms underlying tolerance to dehydration? Some ungulates seem to possess effective plasma volume maintenance during dehydration as had been reported on the camel and the burro (Schmidt-Nielsen, 1964; Yousef et al., 1970; Bullard et al., 1970). However, in other ungulates such as the guanaco and reindeer, plasma volume decreases significantly with dehydration (Rosenmann and Morrison, 1963; 1967). It seems therefore, that there is no one common mechanism to explain tolerance to dehydration and further studies are needed to reveal other mechanism(s).

The rapidity of rehydration is remarkable in wild ungulates. In about 5 min or less, the burro can drink enough water to restore its body water deficit (Bullard et al., 1970; Yousef et al., 1970; Maloiy, 1970). The nature of the signal that stops drinking when water deficit is restored is a mystery. It was suggested that a hypothalamic volumetric inhibitor receptor responds to gastric distension with a turning-off signal (Yousef et al., 1970). Investigations of this neurocontrol of drinking behavior would be well worth the effort. The ability to fill up with water rapidly to escape predatory enemies lurking at water holes has been suggested as having an important survival value (Schmidt-Nielsen, 1964).

D. SURVIVAL WITHOUT DRINKING WATER

Several species of antelope continue to inhabit inhospitable areas of the desert even during prolonged droughts (Taylor, 1972). However, when minimum water requirements of these same animals were measured, it was found that they require between 2-3 l/100 kg of body weight per day (Maloiy and Hopcraft, 1971). How then can these animals survive without drinking? Taylor studied this problem (1969; 1972) and found that the eland which seeks shade during midday beneath *Acacia* trees and eats their leaves, can obtain a sufficient amount of water from these leaves and probably can live indefinitely without drinking. The oryx and the gazelle, however, neither seek shade nor eat moist leaves. These snimals eat dry grasses and shrubs. Taylor (1968) confirmed the early observation by Buxton (1924) that some dried grasses collect moisture from the desert air at night when relative humidity is highest i.e., these plants are hydroscopic. It is then possible to see that by grazing at night, these wild antelope could meet their water requirements without drinking.

V. SUMMARY AND CONCLUSIONS

Wild ungulates present a unique fauna which is capable of thriving in the most inhospitable arctic and desert environments. This group of animals displays not only a varied, but a fascinating array of physiological mechanisms for survival under stressful climatic conditions. There are both qualitative and quantitative differences in the thermoregulatory responses among the different groups of ungulates. However, there are also some obvious similarities in their responses to stress. The mechanisms controlling the varied thermoregulatory responses are ill-defined, and further work is needed for better understanding of these responses. Although Perissodactyla other than equids are greatly restricted in both numbers and present geographic range, little effort has been given to study the tapirids and Rhinoceroses. Comparative studies on these groups are needed to help conserve these interesting and unique species in their native areas. Many groups of Artiodactyla have not received sufficient scientific attention. The husbandry of wild ungulates on game ranches in arctic and desert regions will become more prevalent in the future to meet the increasing need for animal protein for the peoples of these areas. However, a much more concentrated effort must be given to nutritional, biochemical, and physiological studies of the more suitable species for such ventures to be truly successful. Such efforts have begun at the Institute of Arctic Biology in Alaska under the leadership of Dr. Jack R. Luick (Luick et al., 1973; Cameron and Luick, 1972).

REFERENCES: see pp. 527-529.

heavily on panting for evaporative heat loss but does not show the change to slower deeper panting (Ingram and Legge, 1969). Mild heat stress of the above species, during which deep body temperature remains unchanged or is raised by less than about 0.5°C, results in respiratory activity typical of the early stages of severe heat stress.

Total respiratory ventilation has been partitioned into alveolar and dead space ventilation in the dog (Albers, 1961; Siemon et al., 1966), sheep (Hales and Webster, 1967) and ox (Fig. 2). During mild heat stress of the ox there is little change in alveolar ventilation; under these conditions and in the early stages of severe heat stress dead space ventilation increases at a rate approximately 15 times greater than does alveolar ventilation, there is a slight decrease in arterial P_{CO_2} and oxygen consumption, and a slight increase in arterial P_{O_2} and pH (above references; Findlay and Whittow, 1966; Hales, 1973). There is, therefore, a balance between the thermoregulatory drive to promote thermolysis, and respiratory control mechanisms regulating alveolar ventilation, the increased ventilation required for thermolysis being restricted to the respiratory dead space. However, during continuing severe heat stress, respiratory frequency eventually increases out of proportion to the reduction in tidal volume so that by the time the peak of rapid shallow panting is reached, alveolar ventilation has been significantly increased.

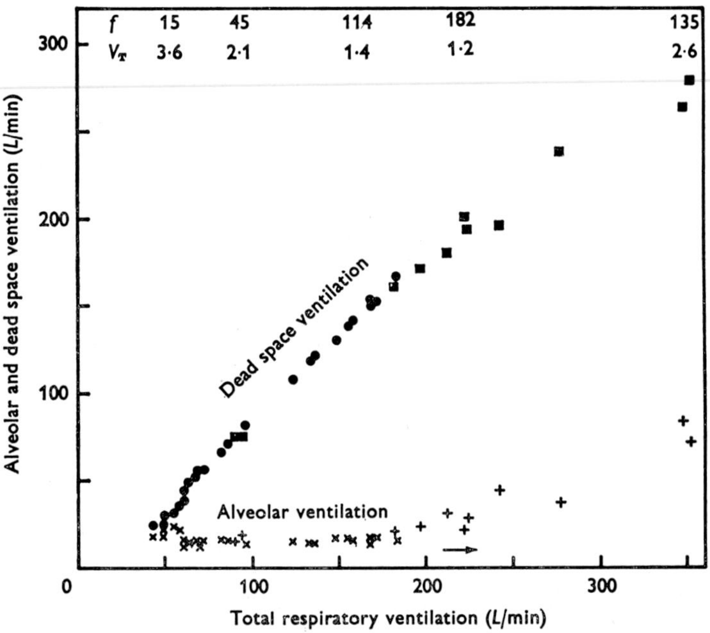

Fig. 2. The changes in alveolar and dead space ventilation with increasing total respiratory ventilation during panting. All results from 8 experiments on 2 oxen in environments of 40°C T_{db}, 28°C T_{wb} (●, ×), 40°C T_{db}, 38°C T_{wb} (■, +) and 16°C T_{db}, 14°C T_{wb} (the ten points of lowest ventilation which each represent the mean of three control measurements). The arrow indicates the transition from rapid shallow, to slower deeper panting; f, average respiratory frequency (breaths/min); V_T, average tidal volume (litres). (From Hales, 1967; courtesy of The Physiological Society).

Compared with the dog, the ox and sheep show a much smaller increase in alveolar ventilation relative to the increase in dead space ventilation and therefore must be considered as more efficient than the dog in their control of the distribution of ventilation. Alveolar ventilation in the goat increases when respiratory frequency increases even under very mild heat stress (Heisey et al., 1971); and the goat must therefore be considered the least efficient of the species studied. The ostrich is unique in showing no hypocapnia (Schmidt-Nielsen et al., 1969) and, therefore should be regarded as the most efficient of the species studied to date, whether it be that alveolar hyperventilation is avoided by a shunting of air through the air sacs or that CO_2 lost through hyperventilation is replaced by metabolic production. During the second phase of slower deeper panting, alveolar ventilation in the ox and sheep increases to up to 5 times control levels and results in a severe respiratory alkalosis. Arterial P_{CO_2} may decline to less than 10 mmHg and pH may exceed 7.7 units; changes of similar magnitude may occur in some birds (Richards, 1970b). Thus, during severe heat stress, there is ultimately a breakdown in the balance between thermoregulatory and respiratory requirements.

It is somewhat surprising that alveolar ventilation is maintained or even slightly increased during rapid shallow panting. At this time the volume of each breath may be less than the anatomical dead space of the respiratory tract in the sheep (Hales and Webster, 1967) and ox (Hales and Findlay, 1968a). This is probably explained by the cyclic raising and lowering of mid-respiratory position and intermittent periods of slower deeper panting which seem to be typical of the repiratory pattern in many species; a further contributing factor may be that physiological dead space decreases linearly with tidal volume during rapid shallow panting (Albers, 1961; Hales and Findlay, 1968a).

Blood oxygenation changes little during mild heat stress; however, as would be expected, the hyperventilation occurring during severe heat stress causes a marked increase in arterial P_{O_2} with small changes in oxyhaemoglobin saturation. Relative to many other species, the sheep normally has a low blood oxygen capacity and content which would endow it with a greater independence between oxygen transport and pH and therefore benefit it during panting. Despite this enhancement of arterial oxygenation, mixed venous P_{O_2} and oxyhaemoglobin saturation decrease, and the development of marked tissue hypoxia is indicated by increased blood levels of "excess lactate" in the dog (Frankel et al., 1963), ox (Hales et al., 1967) and sheep (Hales and Mabon, unpubl. observ.).

CONTROL OF THE RESPIRATORY PATTERN DURING PANTING

The extremely potent nature of thermoregulatory influences on respiration is illustrated by the way in which total respiratory ventilation continues to increase in the face of an extremely severe respiratory alkalosis (which would normally be expected to inhibit breathing) as described in the preceding section. Also, Schläfke and Loeschcke (1967) have found that cooling bilateral temperature sensitive areas on the ventral surface of the medulla oblongata causes a marked depression of respiration; however, if one area is cooled following ablation of the other and denervation of the peripheral respiratory chemoreceptors, respiratory activity may be completely arrested in the normothermic but not in the hyperthermic animal.

Details of specific thermoreceptor control of panting are presented in Section 4. It will suffice here to note that heating the hypothalamus, midbrain, spinal cord, abdominal viscera or skin will stimulate panting, and the magnitude of the response to a temperature change induced at one site depends largely upon the temperature of the other sites. Further, whereas the activity of some neurones in the preoptic region of the hypothalamus are sensitive to their own temperature, others are sensitive to the activity of extrahypothalamic thermoreceptors, and the hypothalamus, therefore, forms an important integrative control centre (Hellon, 1971).

The ultimate control of respiratory activity during panting must involve control phenomena which are traditionally regarded as more specifically respiratory than thermoregulatory. Control of panting may be conveniently envisaged in three phases, as illustrated in Figures 3, 4 and 5. Phase 1 is the initiation of panting and possibly the control of panting during very mild heat stress. Phase 2 is the control of rapid shallow panting commonly seen during mild heat stress, as classified in this Chapter, or the early stages of severe heat stress. Phase 3 is the control of maximal rapid shallow panting leading to slower deeper panting during severe heat stress, the panting mechanism having passed through Phases 1 and 2.

PHASE 1 (Fig. 3) covers the control of panting which appears to be entirely thermal and requires little elaboration. Information that temperature of the skin or nasal passages

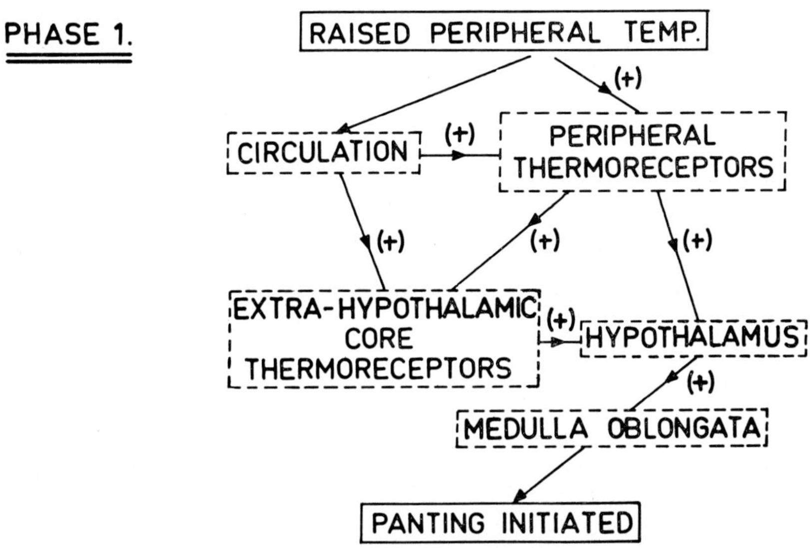

Fig. 3. A proposed scheme for control of the initiation of panting. Specific phenomena are enclosed by solid lines, and pathways by broken lines (+) = stimulation.

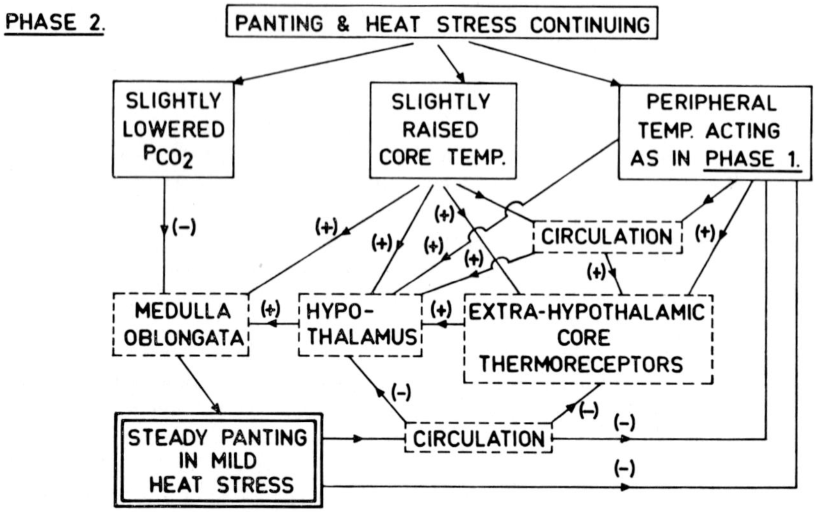

Fig. 4. A proposed scheme for the control of rapid shallow panting during mild heat stress, in which a balance between respiratory and thermoregulatory demands on the respiratory system is attained. (+) = stimulation, (—) = inhibition, and see legend to Fig. 3 for further details.

is raised is conveyed by nervous or circulatory pathways to the hypothalamus either directly or via other prime sites of body core thermoreceptors; the hypothalamus then stimulates the respiratory centres in the medulla oblongata so that changes in activity of the pneumotaxic centre and Hering-Breuer reflex cause an increase in frequency and a decrease in depth of breathing.

PHASE 2 (Fig. 4) is considered to involve the stimulating effects of warming both peripheral and core thermoreceptors, acting on the medulla oblongata to cause an increase in respiratory minute volume by rapid shallow respiratory movements as in Phase 1. A tendency for washout of CO_2 helps maintain the shallowness of respiration by exerting an inhibitory influence on the respiratory centres, and cooling resulting from panting reduces thermoreceptor stimulation via nervous and circulatory pathways: thus, a steady state is attained during mild heat stress.

PHASE 3 (Fig. 5) involves a number of less familiar phenomena, besides the continuing direct thermoreceptor effects. Thus, raised temperature is associated with a paradoxical CO_2 effect. It can have a direct influence on the medulla oblongata, and on the nervous pathways involved in pulmonary control. Respiration in normothermia is depressed by hypocapnia; however, respiratory sensitivity to CO_2 is reduced during hyperthermia (Kappey et al., 1962). Further, some fibres associated with the slowly adapting pulmonary stretch receptors in the anaesthetized rat (Schoener and Frankel, 1972), and the phrenic nerve in the anaesthetized dog (Chapot, 1967; Pleschka, 1969) exhibit an increase in discharge rate with raised temperature, and this effect is augmented by hypocapnia (i.e., the reverse to Phase 2). Tidal volume, deduced from the activity of phrenic

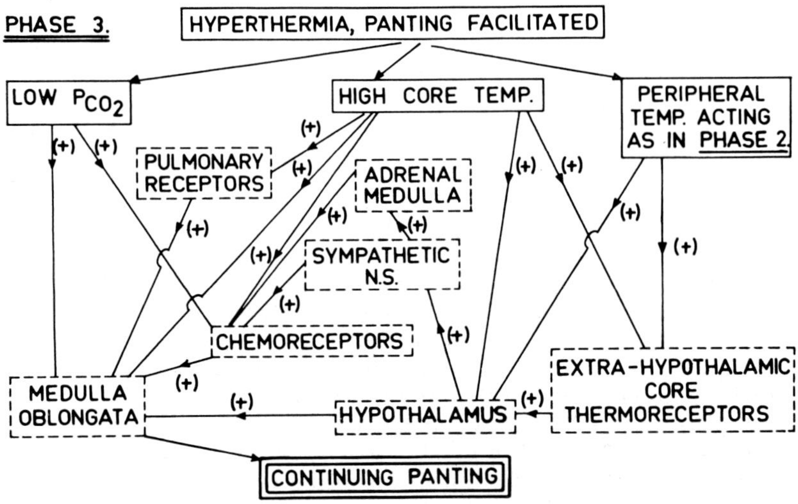

Fig. 5. A proposed scheme of regulatory factors involved in bringing about a continual increase in panting ventilation during severe heat stress. See legend to Figure 4 for further details.

nerve fibres, shows responses to temperature and CO_2 which are the converse of respiratory frequency. Also, the responses of specific neurones in the respiratory centres to changes in CO_2 vary considerably, and in fact, some increase their discharge rate when P_{CO_2} is lowered (Cohen, 1968). It therefore seems that over and above the stimulating effects of thermoreceptors on the respiratory centres via the hypothalamus, respiratory frequency is further increased and depth further decreased by the effects of temperature and hypocapnia on specific pulmonary nervous pathways. In fact, Grunstein et al. (1973) have illustrated very nicely in their experiments on anaesthetized cats that at moderate body temperatures (36-40°C), changes in respiratory frequency resulting from body temperature may be explained in terms of thermal "activation" of the vagal and bulbopontine mechanisms controlling respiratory frequency.

In the third phase there is ultimately an apparent breakdown in the balance between thermoregulatory and respiratory demands in the respiratory system; with the change from rapid shallow to slower deeper panting when deep body temperature exceeds about 40.5°C during severe heat stress, there is a strong drive to increase respiratory ventilation in the face of a worsening alkalosis. Because total respiratory ventilation continues to increase and/or because slower deeper panting rapidly reverts to the rapid shallow form when the ambient heat load is reduced, the following previously offered explanations are untenable: A decrease in thermoreceptor discharge rate when body temperature exceeds certain levels, muscle fatigue, heat damage to central mechanisms, tissue hypoxia, and the regulation of alveolar ventilation to meet oxygen requirements.

Many experiments on oxen and sheep indicated that the change in pattern of panting was due to the hypothalamus attaining a critical temperature level of about 40.5°C, above which the regulation of body temperature took complete precedence over the regulation of acid-base balance (Findlay and Ingram, 1961; Ingram and Whittow, 1962; Hales, 1969); however, experiments involving the manipulation of hypothalamic temperature with chronic brain thermodes in the ox have failed to support this hypothesis (Findlay and Hales, 1969).

One might suggest that the apparent tolerance to the alkalosis is due to the chemoreceptors of the medulla oblongata being protected against changes which occur in the blood. For example, Pleschka et al. (1965) were unable to produce classical post-hyperventilation apnoea in anaesthetized dogs when body temperature was above about 40°C, even with arterial P_{CO_2} lowered to 5 mmHg. The protection referred to might be achieved by changes in brain blood flow or metabolism. However, examination of the cerebrospinal fluid (CSF) revealed smaller and slightly slower changes in gas and acid-base status compared with those in the blood, and a decrease in CSF [HCO_3-] during the slower deeper panting (Hales et al., 1970). Thus, viewed according to conventional theories, the chemoreceptors would still be subjected to inhibition, although to not so marked a degree as would be deduced from blood analyses, and the decreasing CSF [HCO_3-] might even participate in the stimulation of ventilation. Set against this, however, are the reduction in respiratory sensitivity to CO_2 (Kappey et al., 1962) and the reversal in response to CO_2 of some respiratory neurones (Cohen, 1968) which must be borne in mind.

When body temperature is raised to a critical level, there is a decline in phrenic discharge frequency and an increase in the derived tidal volume (Pleschka, 1969). Also, in Schoener & Frankel's studies (1972), some slowly adapting pulmonary stretch receptors were found to respond to temperature only above a critical level of 40-43.5°C. In some receptors in which a generally progressive facilitation of activity occurred with increasing temperature, the effect was markedly greater with a change from 41 to 42°C than at lower levels of temperature.

Controversial results have been obtained in response to heating the medulla oblongata, but an increase in respiratory activity seems virtually beyond doubt (Holmes et al., 1961; Chai and Lin, 1972; Lipton, 1973). In fact, it is particularly noteworthy that an intact hypothalamus is required for respiratory stimulation by heating the spinal cord (Kosaka et al., 1969), but not for respiratory stimulation by heating the medulla oblongata (Lipton, 1973).

The carotid body chemoreceptors normally play an extremely important role in respiratory regulation, and there are at least four possible ways in which they may be stimulated during severe heat stress (Fig. 5). First, the most potent stimulus, a decrease in carotid body P_{O_2} could be induced by a decrease in carotid body blood flow caused by the prevailing hypocapnia (Purves, 1970a). Secondly, the increased temperature can cause stimulation (Eyzaguirre and Lewin, 1961). Thirdly, increased sympathetic activity has an excitatory effect (Purves, 1970b). Fourthly, respiration is stimulated via the carotid bodies by adrenaline and noradrenaline (Joels and White, 1968), and plasma levels of these substances increase markedly in the ox when body temperature exceeds about 40.5°C (Robertshaw and Whittow, 1966).

Thus, there are a number of possible modes of stimulation of the respiratory centres including, of course, continuing thermoreceptor drive via the hypothalamus. It seems that the increasing total respiratory ventilation should be attributed to thermoregulatory drives but that the change from rapid shallow to a slower deeper form of panting should be attributed to the requirements of respiratory mechanics. That is, although respiratory frequency at the peak of rapid shallow panting usually falls far short of the maximum possible, further stimulation to the depth of breathing necessitates a reduction in frequency (Otis and Guyatt, 1968; Clark and Von Euler, 1972). The severe heat stress as considered in this Chapter (viz., ambient dry bulb temperature above body temperature and ambient water vapour pressure equal to 100% saturated at body temperature) does not permit evaporative heat loss; and therefore, there is no negative feedback from panting to thermoreceptors as in Phase 2; rather, panting continues to increase until collapse of the animal. (A similar effect results with a non-saturated vapour pressure, if dry bulb temperature is sufficiently high e.g., 60°C).

CONCLUSIONS

A detailed picture of the pattern of changes in respiratory ventilation, blood gases and acid-base balance during panting is now available for some species.

Available evidence leads to the conclusion that panting is initiated by a specific thermoreceptor input. After initiation, the typical rapid shallow pattern of respiration is the net result of a specific thermoreceptor drive and possibly heat stimulation of the medulla oblongata to increase total respiratory ventilation; this is put into effect in the form of a rapid but shallow respiratory movement following modification of the temperature drive by factors directly concerned with metabolism. Finally, respiratory ventilation continues to increase due to heat stimulation of the respiratory centres via the hypothalamus, the carotid bodies, pulmonary receptors, or to direct heat stimulation of the medulla oblongata, and possibly also carotid body excitation by factors besides heat; further, the raised body temperature causes a reversal in the influence of hypocapnia on respiration. At this time, both depth and frequency are stimulated, but because of the interdependence between these two parameters, the already very high frequency is reduced to permit the depth increase.

Further research is required to elucidate the precise relationship between CO_2 and temperature influences on respiration in the conscious animal, as has been done for the anaesthetized animal. Also, the cause of the apparent breakdown in the balance between respiratory and thermoregulatory demands on the respiratory system during heat stress must be determined; this might involve, for example, the assessment of chemoreceptor function and manipulation of the temperature of the medulla oblongata and "prime" sites of thermoreceptors, and also careful examination of respiratory frequency/depth relationships. Finally, an examination of the influence of exercise, a common natural influence on both metabolism and temperature, on the relationships between respiratory activity and thermoregulation would be of considerable value.

REFERENCES: see pp. 529-531.

Evaporative Temperature Regulation in Ruminants

by

D. Robertshaw

INTRODUCTION

The term "ruminant" covers a widely dispersed group of herbivores that ruminate and have a highly specialized and complex stomach. They were at one time classified into a sub-order of their own, the Ruminantia, but this has little taxonomic value; and now it is generally considered that the Tylopoda, which includes the camels, have evolved the ruminating habit separately from the Bovidae, Cervidae, Tragulidae and Giraffidae. However, the term "ruminant" has some merit in that it covers the most successful, numerically, group of herbivores distributed over a variety of climatic and vegetational zones, and of great importance within the structure of an ecosystem. Of the ruminant species so far studied, their evaporative heat loss mechanisms are very similar; under hot conditions they all pant and sweat, the amount of heat lost by either of these two mechanisms varying with the species.

PHYSICAL CONSIDERATIONS

Loss of heat by the evaporation of water is made use of when so-called "sensible" methods of heat dissipation (i.e., conduction, convection and radiation) fail to prevent a rise in body temperature. When these methods of heat dissipation are adequate, the animal is said to be within its thermoneutral zone; evaporation of water is then a passive process and consists of water lost from the respiratory system by the humidification of inspired air within the upper respiratory tract and the diffusion of water across the skin surface. In cold environments when metabolic heat has to be conserved, then this passive diffusion of water represents a loss of heat that the animal could well do without. Passive evaporation is dependent on the vapour pressure gradient across the evaporating surface, the resistance to diffusion (the diffusion coefficient) and the movement of air over the surface. If the resistance to diffusion is high in relation to that of air, then an increase in air movement will have little effect on insensible loss. This is true for the skin but not for the mucosa of the respiratory tract so that increases in respiratory ventilation will lead to relatively greater increases in respiratory water loss. Since diffusion of water across either the skin or the nasal mucosa is related to the vapour pressure gradient, then an increase in ambient vapour pressure will reduce the total passive evaporation; however, the magnitude of the suppression will be dependent on the diffusion coefficient. A rise in ambient humidity would, therefore, suppress respiratory water loss to a greater extent than that from the skin. In a cool environment, inspired air cools the first part of the nasal passages both by direct heat transfer and by the process of humidification. Air leaving the alveoli in expiration is saturated with water vapour at deep body temperature. The upper respiratory passages then act as a regenerator of both sensible and latent heat so that air, although saturated, is at a lower temperature, thereby effecting an economy of water and reducing heat loss.

An active increase in evaporative water loss occurs as a result of increasing the ventilation of the nasal passage (i.e., panting) the water being provided by diffusion across the nasal mucosa and from submucosal glands which open onto the mucosal surface. In addition, there is a small gland, the lateral nasal gland, sometimes known as Steno's gland, which opens near the cutaneous mucosal junction and presumably prevents desiccation of the anterior part of the nasal passage. Its thermoregulatory function has recently been demonstrated in the dog (Blatt et al., 1972); but since it exists in other species, it presumably has a similar function.

The secretion of water onto the surface of the skin by-passes the epidermal barrier to diffusion; but the rate of evaporation will still be dependent on the ambient humidity, a rise in humidity suppressing evaporation. An increase in air velocity will increase the rate of evaporation from sweating skin to a greater extent than from non-sweating skin. If the rate of sweat secretion exceeds the rate of evaporation, there will be a build-up of moisture on the surface of the skin, which will then increase the rate of evaporation. There will, however, be a maximal evaporative capacity from the skin; and if this state is achieved, then the skin becomes visibly wet. The sweating rate of man often reaches the maximum evaporative capacity but that of ruminants is much lower, and a wet skin is usually only observed when ambient humidity is close to saturation. This has led many people, including eminent scientists, to speculate in print that ruminants do not sweat. Measurements of cutaneous moisture loss have revealed otherwise.

EVAPORATIVE HEAT LOSS IN THE COLD

As indicated above, evaporative heat loss within the thermoneutral zone or below the lower critical temperature is a passive process dependent on ambient humidity. Within the thermoneutral zone, the percentage of heat lost by evaporation is approximately 18% for cattle (Blaxter and Wainman, 1961; McLean and Calvert, 1972) and 23% for sheep (Blaxter and Wainman, 1961).

Below the critical temperature, cutaneous water loss will be virtually unchanged; but respiratory evaporative loss should increase as pulmonary ventilation increases to meet the demands of increased metabolism. Total vapour loss in cattle and sheep, however, is less at low temperatures and two mechanisms probably exist to account for this:

1. The regeneration of water in the nasal passages already referred to. Panting animals have nasal passages which are specifically constructed as a heat exchanger and, therefore, allow this mechanism to be more effective than the nasal passages of man whose large dimensions and laminar air flow do not lend themselves so readily to such exchanges (Schmidt-Nielsen, 1972);
2. In contrast to man, sheep (Joyce and Blaxter, 1964), cattle and the East Affican antelope, the eland (*Taurotragus oryx*) (Taylor, 1969), are able to increase their oxygen

extraction from inspired air instead of substantially increasing their pulmonary ventilation. This is associated with a decrease in respiration rate and increase in tidal volume. It would appear to be due to a direct effect of deep body temperature on respiration, there being a negative linear relationship between oxygen extraction and body temperature (Fig. 1). It is not known whether or not this relationship is peculiar to ruminant animals. Although this mechanism has been studied largely from its importance as a mechanism of water conservation, it will also have some advantages for heat conservation under cold conditions.

EVAPORATIVE HEAT LOSS IN THE HEAT

As air or radiant temperatures approach skin temperature or if metabolic rate increases, as in exercise, then evaporative heat loss is essential for the maintenance of homeothermy. The relative proportions of evaporative heat lost by sweating and panting varies from species to species. In a study of eight species of East African bovid, Robertshaw and Taylor (1969a) noted that sweating as a heat loss mechanism increased in significance with the increase in the adult body size of the animal. Thus the small duiker (*Sylvicapra grimmia*) weighing approximately 7 kg had a cutaneous moisture loss in the heat of approximately 30 g/m^{2}hr. whereas that of the buffalo (*Syncerus caffer*) at a body weight in excess of 240 kg was almost 250 g/m^{2}hr. Species such as the oryx (*Oryx beisa*) and waterbuck (*Kobus defassa*) weighing approximately 100 kg had a cutaneous moisture loss of 100 g/m^{2}hr. Domesticated ruminants demonstrate a similar diversity in their evaporative heat loss mechanisms. In the smaller ruminants, typified by the sheep, cutaneous moisture loss represents 40% of the total evaporative loss (Hales, 1974); whereas, the figure for cattle is given as 84% at 40°C (MacLean, 1963a). The figure of 40% for a small ruminant may seem surprisingly large when the relatively low sweat rate is considered but even in the rabbit, which does not have sweat glands, insensible water loss represents some 40% of the total evaporative heat loss (Gonzalez et al., 1971). This leads to the obvious question of why some ruminants adopt panting as their main evaporative heat loss mechanism, whereas, others have supplemented panting with sweating. Inhibition of the cutaneous evaporative heat loss of cattle by pharmacological agents (Findlay and Robertshaw, 1965) or animals with a failure of the sweat gland function (Bianca and Hales, 1970) results in an inability to prevent hyperthermia at air temperatures of 35°C to 40°C; their respiratory evaporation is unable to dissipate their heat of metabolism. The reason for this is hard to discern; some factors that have been considered do not provide an adequate explanation. From the "surface law" a smaller animal has a relatively larger surface area than a larger animal and, therefore, the proportion of evaporative heat lost by insensible means will be greater. This may explain what at first sight appears to be a rather large amount of cutaneous heat loss lost by a small non-sweating animal. However, since metabolic rate is also related to surface area, then the inability of the larger animal to dissipate its heat of metabolism is not related to surface area. Neither is it related to the efficiency of panting being less for a larger animal; it has been suggested that the energy cost of moving the respiratory apparatus will be minimal at the natural resonant frequency of the system (Crawford, 1962). By this reasoning, the larger the animal the lower the resonant frequncy. However, the oxygen cost of panting for a cow is no different than that for a sheep, in both cases it being too small for measurement (Hales, 1974).

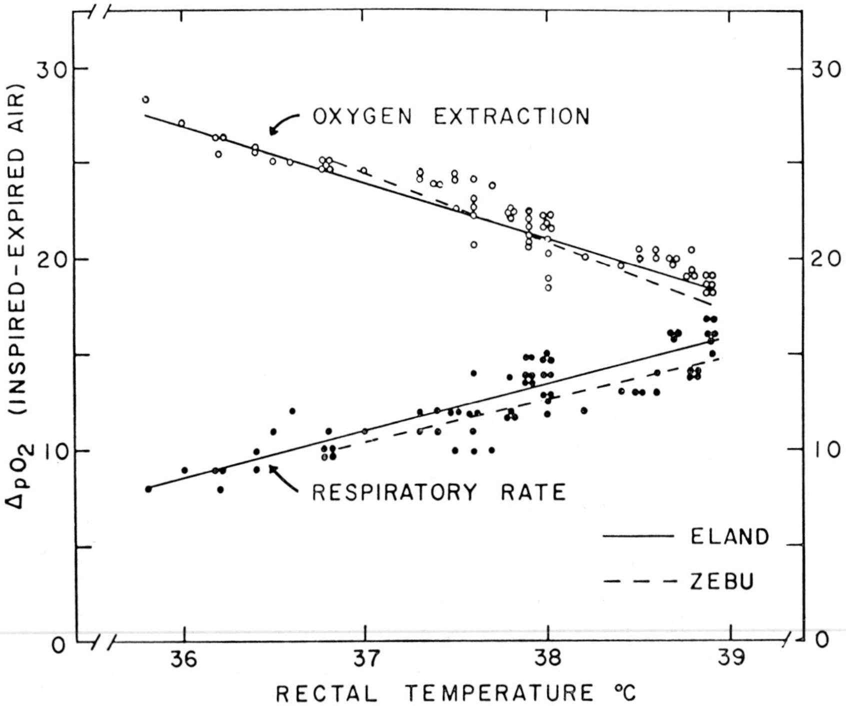

Fig. 1. Relationship between oxygen extraction ($\triangle pO_2$, mm Hg) and respiration rate and rectal temperature. Solid lines and points are for data from eland. The dashed lines are for data from zebu cattle (From Taylor (1972), with permission).

The development of sweating in ruminants in the course of evolution may be related to the need to dissipate the heat of exercise. Under these circumstances, a relatively large surface area would be an advantage in dissipating the heat of exercise by convective means. A large animal with a relatively small surface area would have a proportionately smaller convective heat loss. Thus, depending on the magnitude of the increase in heat production associated with exercise a supplementary mode of evaporative heat loss is required to prevent hyperthermia. It is, therefore, as a method for the dissipation of the heat of exercise that the sweating of the larger ruminant species may have evolved. (Taylor, 1974).

PANTING

All ruminants pant on heat exposure. The increase in ventilation is confined to the dead space so that although ventilation increases, there is a concurrent fall in tidal volume. This is shown strikingly in Figure 2 which demonstrates the partitioning of alveolar and dead space ventilation; alveolar ventilation remains constant as dead space ventilation increases. Gaseous exchange within the lungs, therefore, remains constant and arterial pO_2, pCO_2 and pH are little affected. During severe heat stress associated with hyperthermia respiratory frequency increases to a maximum value; thereafter there is a

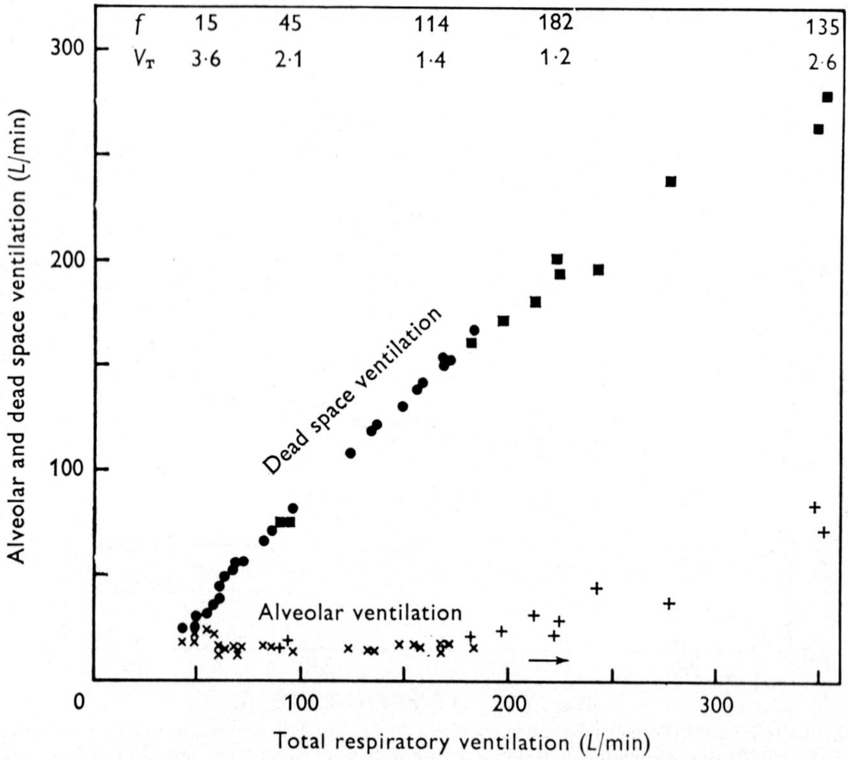

Fig. 2. The changes in alveolar and dead space ventilation with increasing total respiratory ventilation of the ox. The arrow indicates the transition from rapid shallow to slower, deeper breathing; f, average respiration rate (breaths/min); VT, tidal volume, (From Hales (1967). J. Physiol. Lond. 188: 45-46P, with permission.)

change in respiratory activity in that tidal volume increases with a fall in respiratory frequency. This brings about increased alveolar ventilation and leads to respiratory alkalosis; increased respiratory evaporation is, therefore, achieved at the expense of acid-base homeostasis. The transition from rapid shallow to slower, deeper breathing is also shown in Figure 2.

Ruminants pant with the mouth closed and heat exchange must, therefore, take place at the nasal mucosa. Indirect evidence for this is obtained from the work of Bligh (1957) who demonstrated that there is no change in the temperature of the blood as it passes through the lungs during panting, but there is considerable cooling of blood draining the head. (Ingram and Whittow, 1962).

The cool blood from the nasal passages enters the venous sinuses at the base of the skull. The arterial supply to the head also passes through these sinuses and breaks up into a rete of vessels. This arrangement allows the exchange of heat between the two blood streams which is enhanced by the fact that they are flowing in opposite directions and thus are in effect a counter-current heat exchanger. Consequently, the panting mechanism allows for the localised cooling of the most heat sensitive tissue in the body, the brain.

As already stated, the oxygen cost of panting in ruminants appears to be immeasurably small in that there is only a small and usually insignificant increase in oxygen consumption associated with the onset of panting, (Hales and Brown, 1974; Hales and Findlay, 1968; Whittow and Findlay, 1968). This has proved puzzling since there must be an increase in oxygen consumption of the muscles involved in panting. Hales (1973) realised that there need be no net increase in metabolic rate if the increase in oxygen consumption of the respiratory muscles is offset by a decrease in non-respiratory muscles. He, in fact, demonstrated that a reduction of blood flow to non-respiratory muscles counter-balanced the increase in blood flow to respiratory muscles. In addition, he showed that normal panting is achieved largely by movement of the diaphragm and the slower deeper phase of panting involves in addition the intercostal muscles. This increased muscle activity in the later stages of panting is not counter-balanced by a reduced metabolism in other tissues and so there is a net increase in heat production which further compounds the heat stress (Hales and Findlay, 1968; Whittow and Findlay, 1968; Hales and Brown, 1973). Although the hyperventilation of heat stress is less efficient energetically than normal panting, no simultaneous measurements have been made of respiratory heat loss and total heat production to establish whether or not there is a net gain in heat loss. Calculations from the data of Hales and Findlay (1968) on the ox using derived estimates of respiratory heat loss from total ventilation suggest that the increase in heat loss is more or less completely matched by the increase in heat production, thereby, providing the animal with little or no additional advantage.

SWEATING

All ruminant species possess sweat glands and of those so far studied, with the exception of the red deer (*Cervus elaphus*) which belongs to the Cervidae, are stimulated by heat exposure (Johnson et al., 1972). As previously discussed, however, its importance as a heat loss mechanism varies with adult body size. The glands of ruminants are variously described as apocrine or "epitrichial" (Bligh, 1967), the latter term referring to their association with hair follicles. This is in contrast to the sweat glands of man and those of the friction surfaces of many animals e.g., the dog, which are described as eccrine or "atrichial". Each sweat gland of the ox is associated with a hair follicle so that the hair density is the same as the sweat gland density. In the sweat glands of the sheep, goat and camel there are two types of hair follicle, the primary and secondary hairs; sweat glands being only associated with the primary hair follicles.

Three distinct patterns of discharge of sweat are apparent:

1. Sudden and intermittent discharges of moisture onto the surface of the skin. The subsequent discharges are usually smaller than the earlier ones, giving a fatigue effect. This is seen in sheep (Bligh, 1961; Robertshaw, 1968).
2. A stepwise increment in cutaneous moisture loss as seen in the ox (MacLean, 1963b).
3. A smooth increase in water vapour loss. This has been described in the llama (*Lama*

glama) by Allen and Bligh (1969) and the camel by Robertshaw and Taylor (Unpubl. Observ.).

The sudden fluctuations in water loss have been ascribed to myoepithelial contractions which occur synchronously all over the body (Bligh, 1967). The comparative study of Robertshaw and Taylor (1969a) also demonstrated the two types of discharge among the eight species of wild East African ruminants examined. Jenkinson and Robertshaw (1971) in a study of the rate of sweat gland secretion in the goat concluded that the large discharges of moisture were due to strong myoepithelial contractions which emptied the glands; and although there is an enhanced secretion of sweat, it was unable to supply sufficient fluid to maintain each discharge at that of the first one. The actual rate of moisture production per sweat gland was comparable to that of the ox; the difference in total sweat production simply being a reflection of the difference in the number of sweat glands, there being four to five times more glands in the skin of the ox than in that of the sheep.

The sweat glands of all ruminants appear to be under the control of the sympathetic nervous system. However, in contrast to the sweat glands on the general body surface of man which are under cholinergic control, those of ruminants are innervated by adrenergic nerves. Thus although atropine inhibits sweating in man, it has no effect on ruminant sweating. On the other hand, adrenergic neurone blocking drugs will block sweating of the cow, sheep and goat; and the sweat glands can be stimulated by intravenous adrenaline administration (Findlay and Robertshaw, 1965; Robertshaw, 1968). Neural adrenergic activity is supplemented under conditions of stress by adrenaline secreted from the adrenal medulla. Although intravenous adrenaline administration is a potent stimulus of sweating in ruminants, circulating adrenaline does not appear to be important in heat-induced sweating. The evidence for this is two-fold:

1. Denervation of the adrenal medulla has no effect on the magnitude of sweat gland activity resulting from heat exposure (Findlay and Robertshaw, 1965: Robertshaw, 1968).
2. The circulating levels of plasma adrenaline do not increase on exposure to a dry environment when sweating is almost maximal (Robertshaw and Whittow, 1966).

Circulating levels of adrenaline are, however, increased during exercise (Vendsalu, 1960) and it has been shown in the donkey that the sweating of exercise may have a humoral component of adreno-medullary origin (Robertshaw and Taylor, 1969b). Similar studies have not yet been carried out on ruminants.

The receptor sites where the adrenergic transmitter acts were originally classified by Ahlquist (1948) into alpha and beta types. This was an attempt to explain the opposite actions of adrenaline at various sites in the body. This classification received support with the discovery of drugs which would specifically block these receptors, and the use of these drugs on the sweat gland system of ruminants reveals that they are predominantly of the alpha variety (Findlay and Robertshaw, 1965; Robertshaw, 1968; Johnson, Maloiy and Bligh, 1972). However, those of the camel are beta-receptors since sweating can be blocked by beta-receptor blocking drugs (Robertshaw, 1974). This property of the camel sweat glands is not unique in that equine glands can also be blocked by beta-receptor blocking drugs. This difference in the types of receptor involved in sudomotor transmission is difficult to interpret but presumably must be an example of convergent evolution of the sweat gland mechanism and supportive evidence for the taxonomic separation of the Tylopoda from the other ruminants.

CONCLUSIONS

It will be apparent from the above brief review of the subject that certain aspects of evaporative heat loss in ruminants still require investigation. Thus, for example, no firm evidence exists for a decrease in respiratory evaporation at temperatures below the critical temperature. The mechanism of the slower deeper type of panting still remains to be elucidated. Little is known of the role and control of the mucosal glands of the nasal passages and their importance in providing moisture for respiratory evaporation. Transmission to the sweat glands of ruminants requires further study; the transmitter substance itself is not known, and the anatomical location of the nerve endings has not been revealed. A possible role of the adreno-medullary hormones in stimulating the sweating of exercise requires examination. The interesting problem of the relationship between body size and route of moisture vapourization, particularly the reason for the fact that larger animals cannot use panting as their sole means of evaporative heat loss, requires explanation.

REFERENCES: see pp. 531-532.

Evaporative Temperature Regulation in Sheep

by

K. G. Johnson

INTRODUCTION

Sheep lose water continuously by evaporation, as an unavoidable concomitant of respiration and due to passive diffusion and active expulsion through the skin. Because sheep carry a fleece of continuously changing length, the control and importance of the contributions to total water loss and to heat loss are in many ways different to those in other genera.

The rate of water vapour loss of an animal depends on the water vapour pressure (WVP) gradient from its surface to the adjacent air. The WVP gradient in turn depends on the absolute humidity of the air and on the availability and temperature of aqueous solutions at the interface between the animal and its environment. The concentration of the aqueous solutions also affects the WVP slightly (Kerslake, 1972). On the skin where salts accumulate as sweat evaporates, surface fluids may become concentrated; this reduces the WVP and the rate of sweat evaporation. So long as sweat does not accumulate and drip off the body as a result, the total heat loss remains unchanged since the heat of vaporization of a solution is virtually independent of its concentration. In the respiratory tract where the evaporating solution is dilute, this effect is negligible.

Sweating and panting lead to loss of body heat only so long as the heat of vaporization of water from body surfaces is drawn from the body and not from the environment. If sweat evaporates at the tips of the wool away from the body surface and draws its heat of evaporation from the air, no cooling of the body occurs. Similarly saliva which drools from the mouth confers no thermal benefit on the animal. Evaporation from the respiratory tract can be measured either as heat loss, or as water loss when the latent heat of vaporization will represent the heat loss by the body. But for sweat, the site of evaporation is difficult to identify. When calculating rates of cutaneous evaporative heat loss from rates of water loss, allowance should be made for the possibility that some fluid did not evaporate at the body surface.

EVAPORATIVE AND NON-EVAPORATIVE HEAT LOSSES

Unshorn sheep exposed to a thermoneutral environment which does not increase the rates of metabolism or evaporation (0-10°C T_a) lose approximately 25% of their heat by evaporation, about half being from the respiratory tract and half from the skin (Fig. 1). Shorn sheep have thermoneutral temperatures of approximately 25-30°C (Blaxter et al., 1959) and at this T_a the proportions of total heat loss attributable to respiratory and

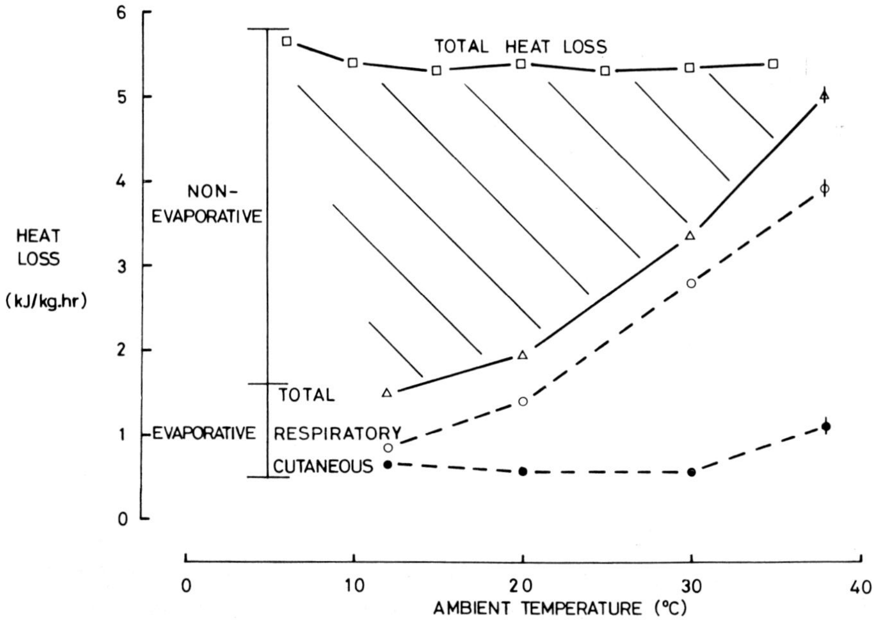

Fig. 1. The evaporative and non-evaporative components of total heat loss of sheep. Values at 12-35°C T_a are for animals carrying 3-8 cm wool (from Brockway et al., 1965) and at 38°C T_a for shorn animals (from Hofmeyr et al., 1969).

cutaneous evaporation and to non-evaporative processes are similar to those of unshorn animals at 10°C T_a (Brockway et al., 1965).

Below thermoneutrality, both heat production and heat loss increase. In shorn and unshorn sheep the increase in heat loss is almost entirely non-evaporative, although exposure to very cold conditions does cause a small increase in pulmonary ventilation and, presumably, in respiratory evaporation (Fig. 2a). Above thermoneutrality, evaporation accounts for an increasing proportion of heat loss; when T_a is equal to, or greater, than, T_b heat loss is entirely evaporative (Fig. 1).

In unshorn sheep the increase in evaporation results largely from panting and only slightly from an increased cutaneous water loss (Fig. 1). The results of Brown and Shanklin (1970) on unshorn sheep and of Alexander and Williams (1962) on lambs are consistent with these results, although Brown and Shanklin suggest that in shorn sheep cutaneous water losses can account for 50% of total evaporation. This may only be at an early stage of heat exposure: the high initial rate of cutaneous water loss upon exposure of sheep to heat is not maintained (Hofmeyr et al., 1969). Ames et al. (1971) have indicated that cutaneous water losses may amount to 75% of total evaporation, but these high values were recorded with a capsule on a small skin area which may not be representative of the entire skin surface.

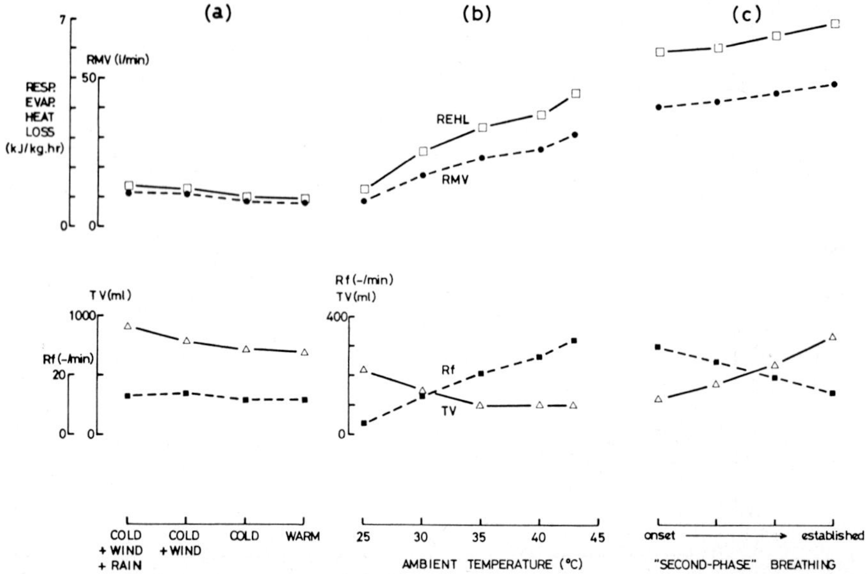

Fig. 2. Respiratory function and respiratory evaporative heat loss of sheep exposed to cold (a, from Joyce and Blaxter, 1964) or heat (b and c, from Hales and Webster, 1967) during normal panting (b) and 2nd-phase breathing (c).

RESPIRATORY EVAPORATION

Apart from a probable small increase in respiratory evaporation at low T_a (Fig. 2a) the respiratory pattern remains virtually constant within and below the thermoneutral zone. As T_a rises above thermoneutrality, respiratory frequency (R.f.) and respiratory minute volume (RMV) increase while tidal volume (T.V.) decreases (Fig. 2b). This respiratory response increases the rate of evaporation with little increase in alveolar ventilation. Reports by Lee (1950), Knapp and Robinson (1954), Alexander and Williams (1962) and Hofman et al. (1971) have indicated that the fall in T.V. does not always occur when respiratory frequency increases during heat exposure. However, such responses must be of brief duration or of rare occurrence amongst animals living in hot environments for, without a decrease in T.V., a progressive alkalosis would seem inevitable.

If heating continues, the pattern of rapid shallow nasal breathing changes to become slower and deeper with the mouth open, described by Bianca (1962) as "2nd-phase" breathing (Fig. 2c). This change in respiratory pattern permits a further rise in RMV, but alveolar ventilation then increases rapidly leading to severe alkalosis (Hales and Webster, 1967). The change to 2nd-phase breathing possibly results from an increasing demand for extra ventilation either of the upper respiratory tract to promote evaporation or of the lungs to relieve hypoxia (Hales and Webster, 1967). Another possibility is that breathing with the mouth open allows inspiration and expiration to be directed preferentially through the nose and mouth respectively. Schmidt-Nielsen et al. (1970) have observed such unidirectional respiratory flow in panting dogs and demonstrated its effectiveness in maximising evaporative heat loss. Whether this pattern of respiratory air flow occurs in sheep is unknown.

Expired air has often been assumed to be at T_b and saturated with water vapour, but in some mammals and birds, air is expired at a temperature well below that of the body owing to heat exchange with the respiratory passages cooled by evaporation during the previous inspiration. The expired air, although still saturated, therefore contains considerably less water than when at T_b (Schmidt-Nielsen et al., 1970). Such counter-current heat exchange promotes water conservation at the expense of heat loss; to what extent it occurs in sheep is unknown. The calculations of respiratory water loss in Figure 2 are based on the assumption that expired air is at T_b and saturated; if this should prove to be unjustified, the calculated values will be inaccurate and high.

An increase in R.f. can be induced in sheep by heating the whole animal, or by heating only the upper respiratory tract (Bligh, 1959), scrotum (Waites, 1962), rumen (Webster and Johnson, 1968), abdominal cavity (Rawson and Quick, 1971), hypothalamus (Maskrey and Bligh, 1971) or spinal cord (Bligh, pers.comm.). Heat sensitive neurons must exist in all these structures and probably at other sites as well, but the roles they play when the whole animal is heated are not clear. Hypothalamic temperature is apparently not a major factor in the control of respiratory frequency since panting can begin in the absence of a rise in central arterial blood temperature (Bligh, 1959). Sensors in the respiratory tract usually exert no disproportionate influence on respiration (Bligh, 1959) but they appear to play an important role in eliciting a rise in R.f. when ambient humidity increases (Bligh, 1963a). Some exposed skin areas such as the inguinal region (Waites, 1962; Linzell and Bligh, 1961) contain warm sensors of greater sensitivity, or in greater density, than trunk skin since heating such areas leads to greater tachypnea than seen after heating equivalent areas of the trunk. In contrast, trunk skin apparently contains cold sensors which, when exposed by close shearing to the influence of a low ambient temperature, cause a depression of R.f. which persists for some while after the shorn sheep is exposed to heat (Bligh, 1963b). Cold sensors, stimulation of which leads to respiratory inhibition, also occur in or near the vena cava (Bligh, 1961b). The thermoregulatory control of R.f. seems to be determined by the excitation of temperature sensors in many parts of the body and to be affected by many body temperatures, with greatest influence being derived from peripheral sensors, in particular those in skin areas not normally covered by wool, and with relatively less influence being derived from core tissues including those of the brain.

CUTANEOUS EVAPORATION

Thermoregulatory sweating in sheep was demonstrated by Brook and Short (1960) who showed that in hot conditions sheep with sweat glands lost water from the skin at a greater rate than could be accounted for by passive diffusion of water vapour and at a greater rate than from sheep lacking sweat glands. Muto (1916) had shown that sweating could be elicited in sheep by I.V. or S.C. injections of adrenaline, but he observed no

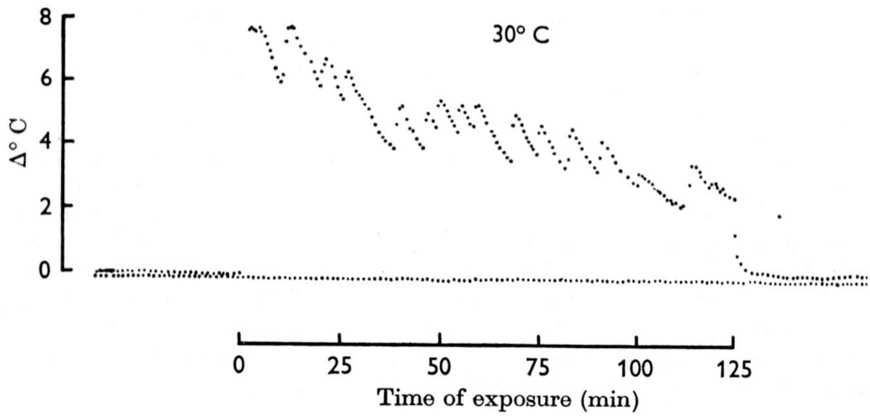

Fig. 3. Water vapour loss (△°C) from the trunk skin of a Suffolk sheep exposed to 30°C T_a (from Robertshaw, 1968).

thermally induced sweating. A possible explanation of the early confusion about the existence of sweating in sheep and a clear demonstration of the basic pattern of discharge was given when Bligh (1961a) found that Welsh Mountain sheep discharge sweat intermittently in pulsatile bursts occurring synchronously over the whole body surface. Bligh observed that the frequency of sweat discharge depended on T_a and that no sweat appeared to reach the skin surface other than during pulsatile discharges. Although in some animals sweat discharges occur so frequently that sweat output appears to be almost continuous (Fig. 3), a basic pattern of intermittent discharging has been found in all breeds of sheep studied (Robertshaw, 1968; Allen and Bligh, 1969) and the pattern is especially obvious in the heated scrotal skin of rams (Waites and Voglmayr, 1963).

The discharges from both scrotal and flank skin decrease in size progressively during long exposures to heat, although the discharging, even if only of small amounts of sweat, continues for many hours and perhaps indefinitely (Hofmeyr et al., 1969; Johnson, 1973). The decline in output has been described as "fatigue" (Jenkinson and Robertshaw, 1971) and is apparently due to a depletion of fluid stored in the gland and an inadequate rate of secretion to maintain the initial rate of output (Johnson, 1973).

Surgical denervation or α-, but not β-, adrenergic blockade prevents sweat discharge during heat exposure, so normal function of the sweating apparatus apparently requires an intact sympathetic nervous system (Waites and Voglmayr, 1963; Robertshaw, 1968). Histological studies indicate that nerves do not extend to the sweat glands (Jenkinson and Blackburn, 1967), so the nature of the final pathway from sympathetic nerves to the glands is unknown. Adrenaline is a more effective stimulus to sweat discharge than noradrenaline (Waites and Voglmayr, 1963; Johnson, 1971a) and this may indicate that adrenaline is the transmitter from the nerves to the glands, or only that noradrenaline-uptake mechanisms remove adrenaline less rapidly than noradrenaline from sensitive areas. The pattern of sweat discharge in sheep has been attributed to a combination of two processes; secretion at a slow rate insufficient to cause overflow from the

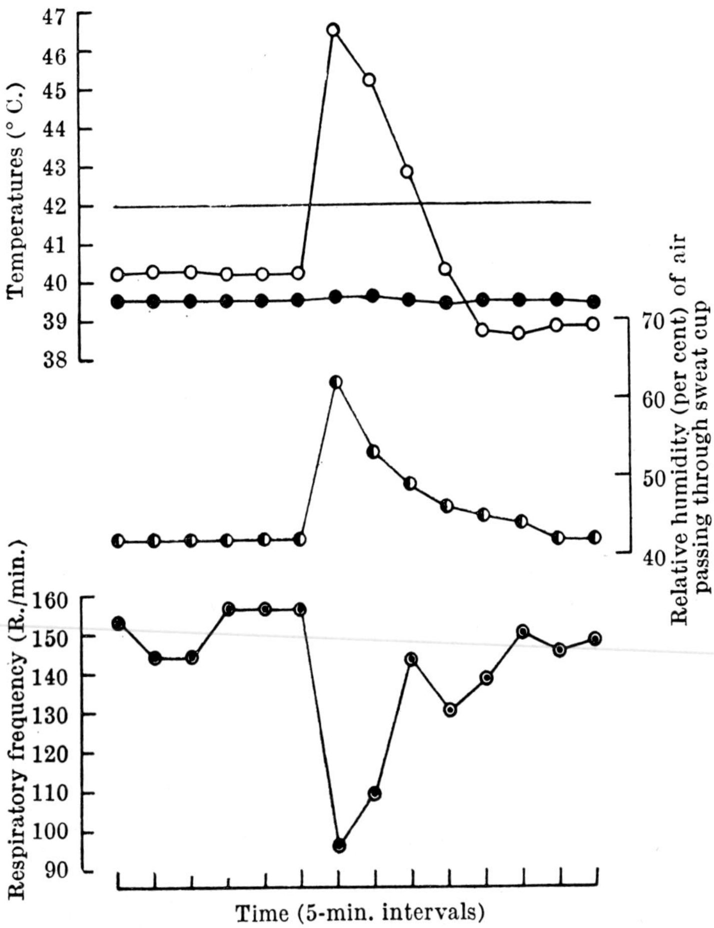

Fig. 4. The effect of a spontaneous discharge of sweat from the skin of a Welsh Mountain sheep on T_{sk} (○), the r.h. of air passing through a sweat capsule (◐), respiratory frequency (⊙), and the temperature of the blood supply to the brain (●) at 42°C T_a (from Bligh, (1961a).

gland, and expulsion onto the skin surface due to contraction of the myoepithelium which surrounds each gland (Allen and Bligh, 1969). Control of these mechanisms may be independent but attempts to dissociate them have been unsuccessful (Johnson, 1971a).

Water can move through fleece as vapour or liquid between fibres, and as molecules adsorbed on or absorbed within fibres (Wool Sci. Rev., 22, 1963). Furthermore, both liquid water and water vapour adsorb to wool fibres with the generation of heat (Wool Sci.Rev. 16, 1956). Therefore, when an aqueous solution is discharged at the inner boundary of a thermally insulative fleece, temperature and WVP gradients immediately change due to simultaneous processes of vaporization, condensation, absorption and desorption at different levels of the fleece. One effect is that T_{sk} rises rapidly, sometimes by as much as 9°C (Fig. 4). This large rise in deep fleece temperature inverts the temperature gradient at the skin surface, at least temporarily, and heat presumably flows from the fleece to the animal. The period during which T_{sk} exceeds T_b is short, and T_{sk} then falls below the level observed before discharge for perhaps 10 min or more, so it would appear that the rise in T_{sk} immediately after sweating may be balanced thermally by the subsequent fall in T_{sk}.

Any analysis of the transfer of heat and water vapour through animal coats is bound to be complex. Cena and Monteith (1975) have made such an analysis, but only for steady-state conditions, and considering the large moment to moment variation in intra-fleece temperatures and water relations, the final benefit to the animal cannot be deduced from their results. The crucial question is what proportion of the heat of vaporization of sweat is drawn from the animal and what proportion from the environment. Even in a gradient-layer calorimeter (Brockway et al., 1965) the calculated non-respiratory evaporative loss cannot with certainty be attributed to evaporation at the skin surface. Until some technique is developed to permit measurement of cutaneous heat loss during sweating in sheep, calculations of the rate of cutaneous evaporation based on measurements of water loss will have to be accepted, and with caution. In practice, cutaneous evaporation probably contributes only slightly to total evaporation (Fig. 1), so the efficiency of sweat evaporation makes little difference to its contribution to total heat loss.

INTERACTION BETWEEN RESPIRATORY AND CUTANEOUS EVAPORATION

Discharge of the sweat glands of sheep is frequently, and perhaps invariably, associated with a large transient fall in R.f. (Fig. 4). The respiratory inhibition is not a consequence of skin cooling since it can occur when T_{sk} is rising (Bligh, 1961a; Waites and Voglmayr, 1963). While it is not certain that this respiratory inhibition results in a fall in respiratory heat loss, the decrease in R.f. is so great that this seems likely. When the effect of possible inefficient sweat evaporation is added to the disadvantageous effect of a simultaneous fall in R.f. and presumably in respiratory evaporation, the thermoregulatory benefit derived from and during a sweat discharge in sheep must be doubted.

The relative utilization of sweating and panting for heat loss by sheep appears to be determined not only by the changing insulation of the fleece but also by changes in central control of effector function. Unshorn sheep have lower frequencies of sweat discharge and higher respiratory rates than shorn sheep (Johnson, 1971b), an alteration in the pattern of heat loss clearly indicating a change in central control. Since sweating and panting differ in their sensitivity to changes in central and peripheral temperatures, such an alteration might arise because panting lowers hypothalamic temperature as a consequence of heat exchange in the carotid rete (Baker and Hayward, 1968).

CONCLUSIONS

Sheep incur evaporative heat loss in all environments. At thermoneutrality or below, respiration and cutaneous evaporation contribute approximately equally to probably 25% or less of total heat loss. Above thermoneutrality, evaporative heat loss increases progressively to equal heat production when T_a equals T_b. The present best estimates indicate that effective evaporation occurs principally from the respiratory tract and that cutaneous evaporation is small and of doubtful importance in thermoregulation.

Problems remain concerning the effectiveness of evaporation from both respiratory and cutaneous surfaces. Measurements need to be made of the temperature of expired air to determine the importance of nasal counter-current heat exchange, of the fluid for respiratory evaporation contributed by Steno's gland (Blatt et al., 1972), of the nasal and oral respiratory flows during "2nd-phase" breathing, and of the effect of A-V heat exchange in the carotid rete on patterns of respiration and sweating. When knowledge of the interrelation between sweating and panting is added to knowledge of the efficiency of sweating from a fleeced surface, it should be possible to determine whether the intermittent discharging of sweat is an integral part of body temperature control, and to estimate accurately the evaporative heat losses of sheep in hot and cold environments.

REFERENCES: see pp. 532-534.

Evaporative Temperature Regulation in Pigs

by

D. L. Ingram

INTRODUCTION

The evaporation of water derived from within the body occurs both from the respiratory tract and from the skin surface under almost all climatic conditions. Water which has fallen as rain or has come into contact with the body as a result of wallowing may also be vaporized by heat from the body, and to a limited extent water may evaporate at the tips of the bristle coat where a large part of the latent heat can be drawn from the ambient air. In this section the magnitude of the heat loss by these channels will be considered, as well as the extent to which the heat loss can be controlled by the pig.

METHODS

The vaporization of water from the whole body, or from groups of pigs can be found by reference to the humidity of the air entering and leaving a chamber containing the animals (Mount, 1968). This method has the advantage of causing the least disturbance to the animal. As a development of this approach the chamber containing the pig can be divided so that the head region (chiefly respiratory loss) is ventilated by a separate system to the body (cutaneous vaporization). The measurement of the actual amount of air taken in at each breath (tidal volume), however, involves the use of a mask which can only be worn by highly trained animals.

Cutaneous water loss can be detected by first painting the skin with a solution of iodine in alcohol and then with a solution of starch in glycerine. Water passing through the skin then causes the starch iodine reaction to occur and the area turns black. The quantitative measurement of water loss through the skin can be measured using a ventilated sweat cup of the type described by McLean (1963a). These consist of a small cup 3.5 cm in diameter held onto the surface of the skin by means of elastic straps. The cup is ventilated by sucking a stream of air through it so that air movement inside the cup is similar to that outside. Water loss can then be determined by reference to the wet bulb temperatures of air entering and leaving the cup and the rate of ventilation.

EVAPORATION OF WATER DERIVED FROM WITHIN THE SKIN

Histological examination of the pig's skin reveals that there are glands associated with each hair follicle (epitrichial) and that these glands are relatively large in size. They do not, however, appear to be associated with thermoregulation even though they respond

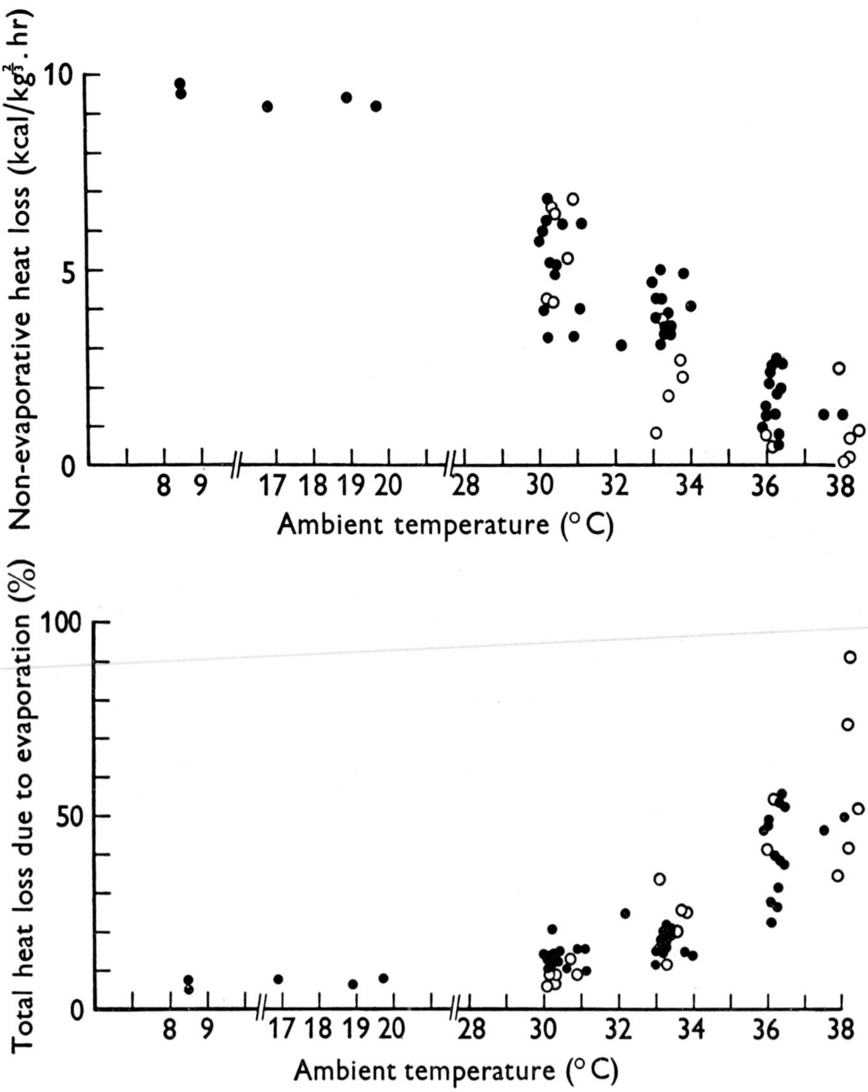

Fig. 1. (a) The non-evaporative heat loss, and (b) the percentage of the total heat loss due to evaporation from young pigs (From Mount, 1962).

to adrenaline and noradrenaline as do the epitrichial glands of other species which do have a thermoregulatory function (Ingram, 1967). Mount (1962) exposed new-born pigs to a series of environmental temperatures in a calorimeter and measured the total amount of water lost, taking precautions to exclude that from the urine and faeces. From the results of this study it was possible to relate both the non-evaporative heat loss and the percentage of the heat loss due to evaporation, to ambient temperature (Fig. 1).

Above the critical temperature the percentage of the total heat loss due to the vaporization of water soon increased to as much as 60% and similar results were obtained by Butchbaker and Shanklin (1964). Nevertheless, Mount (1962) concluded that this water loss could readily be accounted for by evaporation from the respiratory tract, and it therefore appeared that the pig does not sweat, a conclusion also drawn by Brody (1948) and Heitman and Hughes (1949). By contrast, Robinson and Lee (1941) regarded the total water loss of pigs to be greater than that which could be accounted for by panting.

Direct measurement of the water loss from the pig's skin has been made using a ventilated capsule of the type described by McLean (1963a). The results obtained by this method were displayed as a continuous line on a chart recorder as the difference between the two wet bulb temperatures and followed over several hours (Ingram, 1967). During such studies at high ambient temperatures the water loss was low and remained constant with no brief increases of the type observed in the sheep and some other species by Allen and Bligh (1968). During a series of observations at environmental temperatures between −5° and 35°C (Ingram, 1964) it was found that below the critical temperature for animals of the age used water loss through the skin was between 7 and 16 $g.m^{-2}.h^{-1}$ and above the critical temperature the loss was 24-32 $g.m^{-2}.h^{-1}$. The value for water loss in a cool environment is similar to that found for cattle, sheep and man and can be accounted for by the passive diffusion of water through the skin along the vapour pressure gradient. In a warm environment the rate of blood flow through the skin increases and the skin temperature rises. This in itself would be expected to influence the rate of passive diffusion of water through the skin and would account for the increased water loss in the heat. For the passive diffusion of water through the skin the ratio of the rate of water loss at one environmental temperature to that at another is equal to the ratio of the vapour pressure (V.P.) difference. Thus:

$$\frac{\text{Rate of water loss at } T_1\text{°DB}}{\text{Rate of water loss at } T_2\text{°DB}} = \frac{\text{Saturation VP at skin temperature-VP at } T_1\text{°DB}}{\text{Saturation VP at skin temperature-VP at } T_2\text{°DB}}$$

Table 1. Change in moisture loss from the skin at various temperatures compared with the vapour pressure difference between skin and air (From Ingram, 1964)

Change in DB °C	Change in V.P. in mmHg	Change in water loss $g.m^{-2}.h^{-1}$
− 5 to 10	19 to 21	7 to 10
10 to 25	21 to 31	10 to 16
− 5 to 25	19 to 31	7 to 16
25 to 30	31 to 33	16 to 32
30 to 35	33 to 38	32 to 24
25 to 35	31 to 38	15 to 24

As can be seen from Table 1, the figures obtained for the pig are close to those expected on this basis. On some few occasions higher values of 70 $g.m^{-2}.h^{-1}$ were recorded from pigs and small patches of skin with a high water loss were also detected using a starch

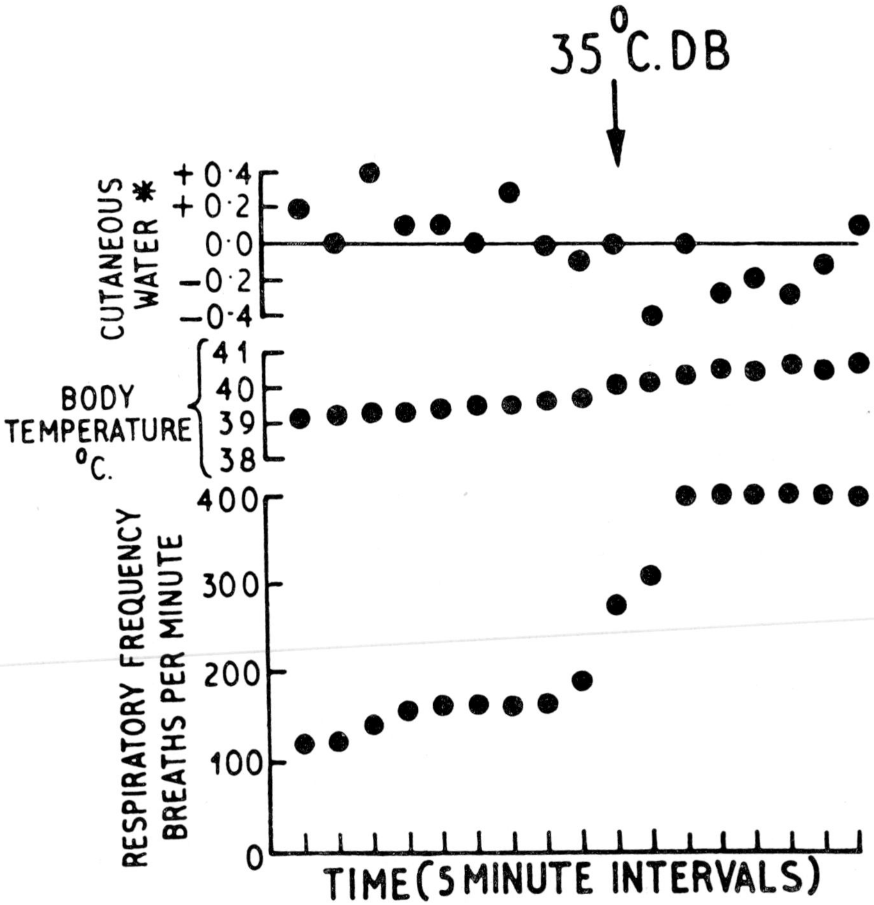

Fig. 2. Effect of an increase in humidity (at arrow) on respiratory frequency, body temperature and cutaneous water loss (*indicated as the difference in WB temperatures of air entering and leaving a sweat cup) at a DB temperature of 35°C. When the humidity increases water is gained through the skin (From Ingram, 1965a).

iodine technique. It was found that the patches of skin which darkened rapidly corresponded to areas which began by being reddish in colour probably as the result of the pig rubbing itself and removing the upper layers of the epidermis. The removal of a layer of skin would affect water transfer, and there is thus no evidence to support the idea that the pig can sweat. In fact under conditions of high humidity with the vapour pressure above 22 mm Hg water is actually gained through the pig's skin from the ambient air (Fig. 2) (Ingram, 1965a). A similar passive gain of water also occurs in man, but since true sweating occurs in man at high ambient temperatures there is of course still a net loss of water.

EVAPORATION OF WATER AFTER WETTING OF THE SKIN

Although the pig does not sweat it has the habit of wallowing in mud or water during hot weather and in the absence of water the pig may wallow in its own urine and faeces. In order to examine the possible effectiveness of this type of evaporative cooling studies were made of the evaporative loss from pigs using a ventilated capsule. The pig was placed in a hot environment of 35°C and the skin and bristle soaked in water or smeared with mud (Ingram, 1965b). The water loss was then followed until it returned to near the control value. The rate of water loss increased in both instances to between 700 and 800 $g.m^{-2}.h^{-1}$, a figure similar to that for man working in a hot climate. The application of mud, however, prolonged the time for which this high rate of loss continued to 2 hrs so that a pig would need to make only infrequent visits to a mud wallow in order to maintain a high rate of evaporative heat loss. Moreover, because the pig is a relatively naked species the mud or water would be in contact with the skin rather than lying on the surface of the hairs. This is important because it means that the latent heat needed to vaporize the water is likely to be drawn in a large part from the pig's skin and so cool the body. Water which is evaporated from the surface of the hair on the other hand may be vaporized with heat drawn mostly from the ambient air and so produce little cooling of the body. In addition evaporative heat loss achieved as a result of wallowing avoids the problems of water and salt balance which can be encountered in sweating and panting animals.

Under cold conditions after rain the absence of thick hair coat would of course prove to be a disadvantage because the cooling of the body by evaporation would be so much greater than in an animal like a sheep which is protected by a thick fleece and where the evaporation of water takes place from the tips of the hair rather than on the skin.

EVAPORATION FROM THE RESPIRATORY TRACT IN THE COLD

When air is inspired it is warmed to body temperature and saturated with water. At low ambient temperatures even if the air is saturated with water, the saturation deficit is very large when the dry bulb temperature is increased to body temperature and consequently the potential for evaporating water is great. The amount of water which is actually lost from the respiratory tract in the cold depends on the degree to which heat exchange mechanisms operate in the nasal cavity. These heat exchanges have been investigated in several species (Schmidt-Nielsen et al., 1970a; Schmidt-Nielsen et al., 1970b). Air is warmed and saturated as it passes through the nasal cavity and as a consequence the turbinate region is cooled. On expiration the air temperature is reduced by contact with the cool surfaces and water is condensed ready to be used again at the next inspiration. The extent to which this mechanism saves water loss depends on the anatomy of the respiratory tract but in some desert species it is very efficient. The pig has not been investigated in this respect and for this reason calculations about evaporative losses must be made with some caution.

Respiratory evaporative loss has been investigated in the fattening pig by Morrison et al. (1967) and in pigs just after weaning by Ingram and Legge (1970). Below the critical temperature the minute volume increases as the ambient temperature falls and oxygen consumption increases. The increase in minute volume which is related chiefly to an increase in tidal volume as ambient temperature falls from 25°C to 0°C is about 40%, but oxygen consumption doubles over the same range. It was found (Ingram and Legge, 1970) that the quantity of oxygen in expired air decreased and the amount of carbon dioxide increased over the same range of temperature, as has been found in the sheep (Joyce and Blaxter, 1964). If the efficiency of the water and heat exchange system remained constant when the temperature fell this increase in the efficiency with which oxygen is extracted from the inspired air could represent a saving in heat production of about 15% simply because less air needs to be humidified in a given period.

EVAPORATION FROM THE RESPIRATORY TRACT IN THE HEAT

On exposure to high ambient temperatures pigs increase their respiratory frequency and decrease their tidal volume in such a way that the minute volume increases (Fig. 3). At the same time the heat exchange system is by-passed because panting occurs through the open mouth. Unlike most other panting animals (e.g., dog (Hemmingway, 1938); sheep (Hales and Webster, 1967); cattle (Findlay, 1957); chicken (Randall and Hiestand, 1939)) there is no obvious secondary change in the pattern of respiration at high body temperatures involving an increase in tidal volume and decrease in frequency. When the body temperature was between 42 and 43°C in the 10 week old pig, however, Ingram and Legge (1970) found some evidence of a very slight increase in tidal volume, suggesting perhaps that the difference between pigs and other panting animals is one of degree rather than one of kind.

The minute volume (ventilation per unit time) increased by a factor of about 3 in pigs with high body temperatures, which is less than that possible in sheep (Hales and Webster, 1967) but similar to that of cattle (Findlay, 1957). The ox, however, can also lose a significant amount of heat by sweating (McLean, 1963b).

EFFECT OF HUMIDITY ON TOLERANCE OF HIGH AMBIENT TEMPERATURES

Because the pig cannot sweat and cannot lose a large proportion of its heat production by panting it would be expected that the species would be relatively unaffected by changes of humidity in a hot environment. Estimates of the weighting which should be given to the WB temperature in the pig with respect to heat stress confirms this conclusion (Ingram, 1965a; Morrison et al., 1967). The reaction of pigs to a series of hot environments at different humidities was assessed from the rate of rise of rectal temperature and this figure was plotted against a figure obtained by adding a percentage of the DB temperature to a percentage of the WB temperature. The weighting obtained by Ingram (1965a) was WB × 35% + DB × 65%, while Morrison et al. (1967) working with older pigs suggested a slightly lower weighting for the WB. This compares with a weighting of 65% for the WB in cattle (Bianca, 1962) and 85% for man (Provins et al., 1962).

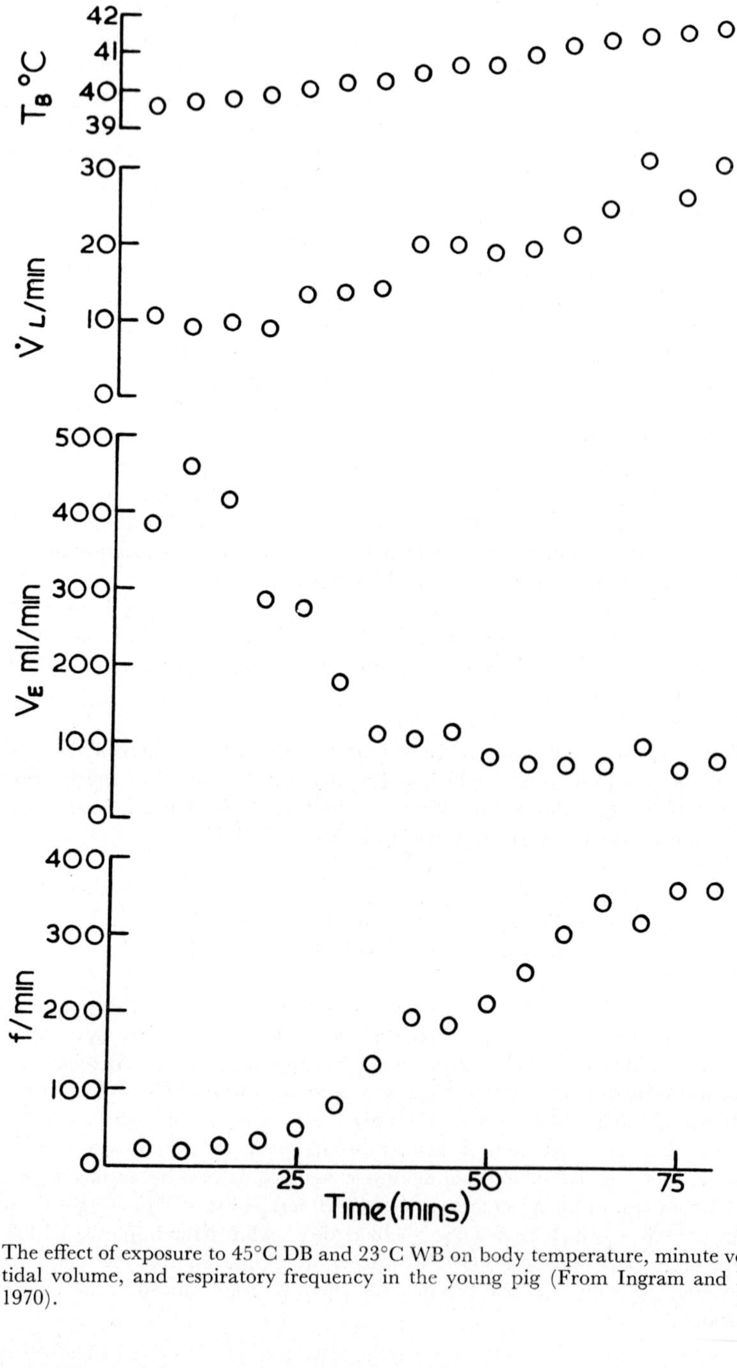

Fig. 3. The effect of exposure to 45°C DB and 23°C WB on body temperature, minute volume, tidal volume, and respiratory frequency in the young pig (From Ingram and Legge, 1970).

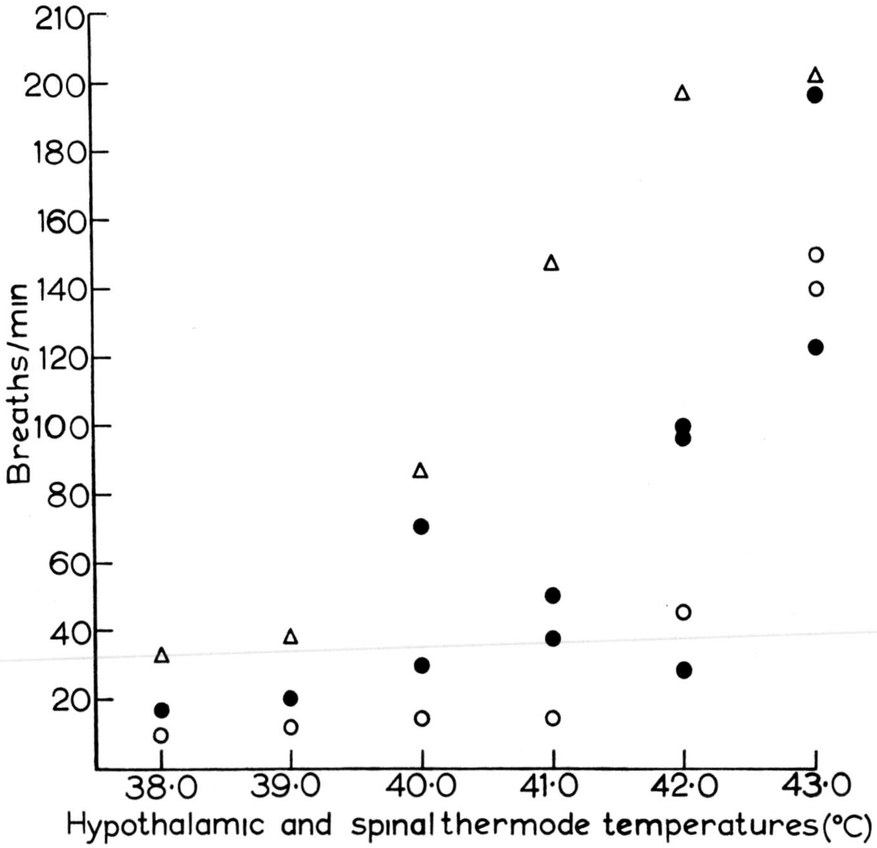

Fig. 4. The effect of heating both the hypothalamus, and the spinal cord to various temperatures while the pig was in a room at an ambient temperature of 30°C. The skin on the pig's trunk was controlled at 37°C (○), 39°C (●) or 40°C (△) by means of a temperature controlled coat (From Ingram and Legge, 1972a).

PROPORTION OF TOTAL HEAT PRODUCTION LOST BY EVAPORATION

When the ambient temperature is greater than the body temperature evaporation represents the only means of heat loss and hence the proportion of the total heat loss accounted for by evaporation rises towards 100% as the ambient temperature rises (Fig. 1). In the pig the maximum evaporative heat loss does not amount to more than about 80% of the heat production, unless the pig is allowed to wallow, and hence adult pigs frequently die of heat stroke in hot climates. The partition of heat loss from the pig into the various channels of non-evaporative and evaporative heat loss has been discussed by Mount (1968); below the critical temperature evaporation accounts for 8-15% of the total heat loss in both small and large animals (Bond et al., 1952; Butchbaker and Shanklin, 1964; Holmes, 1968; Ingram and Legge, 1970). In groups of pigs in a calori-

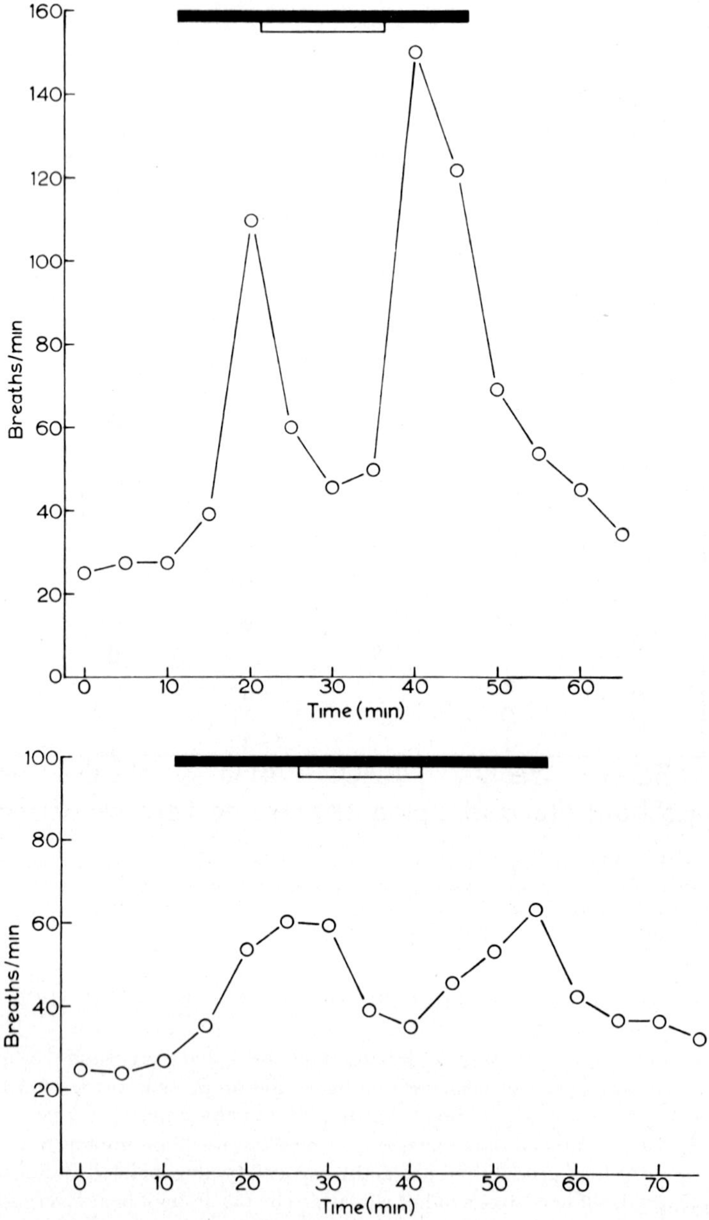

Fig. 5. The effect of warming the scrotum (black bar) on respiratory frequency at an ambient air temperature of 33°C (top) and 31°C (bottom). During the period the scrotum was heated the hypothalamus was cooled (white bar) (From Ingram and Legge, 1972b).

meter the total evaporative heat loss occurs from the whole pen including the floor and bedding, if any. The proportion of the total heat loss which occurs by evaporation therefore depends partly on the design of the pen. In the studies of Holmes and Mount (1967) 30% of the total heat loss from the pen was by evaporation at 20°C D.B. and at 30°C it rose to 60%.

CONTROL OF EVAPORATIVE HEAT LOSS

The mechanisms by which evaporative heat loss is controlled by panting appear to depend on the interaction of thermal signals originating from various places on the periphery and from centres in the body core. Ingram and Legge (1972a) implanted thermodes into the hypothalamus and along the spinal cord of the pig and also controlled the temperature of the skin on the trunk by means of a coat through which water could be circulated. At an ambient temperature of 30°C the skin on the trunk was kept at 37°, 39° and 40°C while the hypothalamic and spinal cord temperatures were changed (Fig. 4). It was found that the respiratory frequency depended on both the skin temperature and the core temperature so that the respiratory frequency associated with a given hypothalamic and spinal temperature, was higher at greater skin temperatures. There was also evidence from the same study that there are other regions deep in the body which are sensitive to heat, since even when the hypothalamic temperature, spinal cord temperature and trunk skin were controlled at constant ambient temperature it was found that respiratory frequency depended on rectal temperature. In contrast to other panting animals the pig does not pant in response to heating either the hypothalamus or the spinal cord unless the ambient temperature is above the critical temperature, indicating a strong inhibition on the part of cold receptors in the skin.

As in some other species (Waites, 1961) heating the scrotum of the pig caused an increase in respiratory frequency (Ingram and Legge, 1972b) while heating a similar region of the skin elsewhere had no effect. The signals from warm receptors in the scrotum are, however, integrated into the control system since the effect of warming the scrotum can be offset, by cooling the cord or hypothalamus; or enhanced by heating these regions (Fig. 5). Moreover, the effect on respiratory frequency of heating the scrotum depends on the temperature of the ambient air.

CONCLUSIONS

The pig does not sweat and has only a limited ability to evaporate water from the respiratory tract. Survival in a hot climate, therefore, depends on the availability of a wallow. The extent to which heat and water exchange occurs in the respiratory tract in a cold environment has not been investigated but may be important in the limitation of water loss. The mechanisms controlling thermoregulatory behavior in the heat should also receive more attention.

REFERENCES: see pp. 534-535.

Evaporative Temperature Regulation in Birds

by

M. van Kampen

INTRODUCTION

In recent years evaporative water loss in birds has become a subject of extensive study. It has been commonly assumed that most birds are unable to dissipate more than approximately half of their metabolic heat production by panting and that evaporation from the skin is a negligible part of the total water loss of the bird because the bird lacks sweat glands. However, Lasiewski et al. (1966) found that a wide variety of birds are capable of maintaining their body temperature below environmental temperatures of 44° to 46°C for periods of from one to four hours by evaporative cooling, provided that the water vapour tension is low.

Indeed Schmidt-Nielsen et al. (1969) have shown cutaneous evaporation to account for less than two per cent of total evaporation at 40°C, whilst recent data of Lasiewski et al. (1971) indicate that the skin provides an important avenue of evaporative water loss in several species of birds ranging in weight from 12.5 to 300 g.

INFLUENCE OF AMBIENT TEMPERATURE AND HUMIDITY ON RESPIRATORY EVAPORATION

In order to compare the evaporation from animals of widely different sizes Schmidt-Nielsen (1972) suggests the amount of water evaporated should be related to the volume of oxygen consumed, since lung ventilation and thus evaporation, varies with the oxygen demand. As recent research points to a significant contribution of the skin to total evaporation this will only be true for respiratory evaporation. However, there is an objection to this proposal. In Table 1 the respiratory evaporation of fully-fed White Leghord chickens (*Gallus domesticus*) at different ambient temperatures is shown. It will be seen that as the temperature falls, oxygen consumption increases, but without a concomitant increase in respiratory evaporation. Possible explanations for this phenomenon include (a) a more efficient removal of oxygen from tidal air, (b) the temperature of the exhalad air is reduced, or a combination of the two. Indeed evidence for both explanations is available. Van Kampen (1973), working with North Holland Blue cocks (*Gallus domesticus*) and White Leghorn hens, found at an ambient temperature of 0°C a significant lowering of the pH and an increase of pCO_2; whereas Schmidt-Nielsen et al. (1970) found in six species of birds, over a range of ambient air temperatures of 12° to 30°C, the temperature of exhaled air to be within a few degrees (1° to 7°C) of the inhaled air temperature.

Table 1. Influence of ambient temperature on heat production and respiratory evaporation.

Ambient Temp. °C	Body Weight kg	pH_2O[a] mm Hg	Heat Prod. kJ/24 hrs	Resp. Evap. g/24 hrs	Resp. Evap. g/l O_2	ER/H × 100[b] %
− 5	1.688	3.7	1323	25.0	0.40	4.6
0	1.752	5.3	1222	28.6	0.49	5.6
5	1.785	7.2	1161	35.6	0.64	7.4
10	1.784	7.1	998	30.9	0.63	7.4
15	1.898	8.1	891	41.5	0.94	11.2
20	1.685	9.0	729	52.0	1.46	17.2
25	1.699	10.5	651	70.0	2.20	25.9
30	1.642	10.9	753	74.0	2.03	23.6
35	1.554	17.4	679	133.0	3.96	47.1
38	1.480	17.2	602	136.4	4.59	54.5

[a] Water vapour tension of the inspired air
[b] The percentage metabolically produced heat (H) dissipated by respiratory evaporation (Er) assuming 1 g water = 2.406 kJ.

At temperatures over 30°C the amount of water evaporated from the lungs, expressed in grams per litre oxygen consumed, increases considerably. This is a result of (a) the greater capacity of the warmer air for carrying water vapour and (b) the increased respiratory rate which occurs in spite of a falling oxygen requirement. It should be noted, however, that oxygen consumption increases again at very high temperatures. It is also evident from Table 1 that the ambient water vapour pressure will increase markedly when the ambient temperature exceeds 35°C if the ventilation rate is low and this will lead in turn to a depressed rate of respiratory evaporation. This is illustrated further in Table 2 in which the influence of temperature and water vapour tension on respiratory evaporation is given. Increasing water vapour causes the percentage of metabolically produced heat that is dissipated by respiratory evaporation to diminish.

Table 2. Influence of humidity on heat production and respiratory evaporation.

Ambient Temp. °C	Body Weight kg	pH_2O[a] mm Hg	Heat Prod kJ/24 hrs	Resp. Evap. g/24 hrs	Resp. Evap. g/l O_2	Er/H × 100[b] %
20	1.665	10.3	861	47.4	1.14	13.2
20	1.671	15.2	891	36.2	0.85	9.8
30	1.620	13.0	687	47.0	1.41	16.5
30	1.599	26.9	638	8.7	0.28	3.3
35	1.473	16.8	609	93.7	3.11	37.0
35	1.384	36.2	516	20.5	0.79	9.6

[a] and [b] see Table 1

At high ambient temperatures the temperature gradient between the animal and its surroundings, and therefore the sensible heat loss (by radiation, convection and conduction) is small. Even so many birds succeed in maintaining an equilibrium between heat production and heat loss by means of evaporation, provided the ambient water vapour tension is low. At a temperature of 44°C and a water vapour pressure of 4 mm Hg the White Leghorn hen is able to maintain body temperature at least 0.9°C below ambient temperature for more than 24 hrs (Van Kampen, 1971).

The heat loss by evaporation of the brown-necked raven (*Corvus corax ruficollis*) at 50°C is equal to 167 per cent of the metabolically produced heat (Marder, 1972). The ostrich (*Struthio camelus*) shows no change in cloacal temperature as ambient temperature increases from 25° to 51°C, even at the end of an exposure of 7.5 hrs (Crawford and Schmidt-Nielsen, 1967). At air temperatures higher than body temperature, evaporative heat loss will account for metabolically produced heat and heat gain from the environment. This means that the birds are in fact cooling the environment.

Table 3. Acclimation of the respiratory evaporation to an ambient temperature change.

Ambient Temp. °C	Body* Weight kg	pH_2O[a] mm Hg	Heat Prod. kJ/24 hrs	Resp. Evap. g/24 hrs	Resp. Evap. g/l O_2	Er/H × 100[b] %
25	1.900	14.7	811	58.2	1.51	17.3
	1.957	13.7	852	48.0	1.20	13.6
	2.025	14.2	838	52.8	1.34	15.1
	2.042	13.8	829	48.6	1.25	14.1
	2.018	14.3	836	53.1	1.34	15.3
	2.054	14.1	830	51.2	1.30	14.9
	2.051	14.2	829	52.5	1.33	15.2
35	2.002	21.1	661	112.1	3.44	40.8
	1.913	21.1	664	111.1	3.44	40.3
	1.904	20.0	638	100.1	3.23	37.7
	1.854	15.8	501	59.4	2.39	28.5
	1.763	19.7	588	97.3	3.42	39.8
	1.729	18.4	575	85.0	3.04	35.5
	1.685	17.7	568	78.0	2.84	33.1
	1.650	14.6	441	49.6	2.30	27.1
	1.626	17.3	543	75.5	2.89	33.4
25	1.597	10.9	544	21.3	0.82	9.4
	1.609	10.7	545	18.9	0.73	8.3
	1.682	12.0	688	31.6	0.98	11.0
	1.699	11.3	615	24.7	0.87	9.7
	1.699	12.6	743	37.2	1.07	12.0
	1.779	13.1	760	42.1	1.20	13.3
	1.795	12.8	773	39.5	1.09	12.3

* Results of 1 hen on consecutive days [a] and [b] see Table 1.

INFLUENCE OF ACCLIMATION ON RESPIRATORY EVAPORATION

Acclimation (non-inherited adaptive adjustments to specifically controlled factors of the physical environment; IUPS-definition) of the respiratory evaporation system to a change in ambient temperature in the White Leghorn hen was determined by measuring respiratory evaporation during seven days at an ambient temperature of 25°C, and continuing this thereafter at a temperature of 35°C.

Immediately following the rise in temperature respiratory evaporation increased significantly (Table 3) and was accompanied by a substantial lowering in food consumption (Van Kampen, 1974). Both respiratory evaporation and body weight decreased during the whole period at 35°C particularly on the 4th and 8th days, pointing to an incomplete acclimation. By reducing food consumption, heat production and heat loss will decrease. The reduction in heat loss is attained by decreasing respiratory evaporation. At higher temperatures it is respiratory evaporation which is the first to affect the equilibrium between heat production and heat loss. The contribution of respiratory evaporation to the total heat loss is therefore at a maximum in the beginning of the period at 35°C, decreasing thereafter for six days before coming to a new equilibrium. This means that heat loss via other routes has to increase. This may be especially accomplished by a rise in sensible heat loss by radiation and convection but the possibility of an increase in cutaneous evaporation cannot be excluded. It may be concluded from the data of respiratory evaporation, expressed in grams per litre O_2-consumption, that a new balance is attained six or seven days after the change in temperature. When the air temperature is lowered again to 25°C, respiratory evaporation appears at first to be less than during the first period at 25°C. This can be ascribed, on the one hand, to the lower body temperature, by which the temperature of the exhaled air will be lower and thus can contain less water vapour, and on the other, to a decreased rate of lung ventilation as a result of the decreased oxygen consumption. After five to six days the evaporative rate had virtually returned to its original level.

Table 4. Respiratory and cutaneous evaporation at different ambient temperatures. From Van Kampen, Brit. Poultry Sci., April 1974.

Ambient Temp. °C	Body Weight kg	pH_2O[a] mm Hg	Heat Prod. kJ/h	Cut. Evap. g/h	Resp. Evap. g/h	O_2-Cons. l/h	Et/H × 100[c] %	Er/Et × 100[d] %
10	1.697	4.1	27.38	0.289	0.436	1.36	6.4	60.1
20	1.656	4.1	21.30	0.350	0.481	1.05	9.4	57.9
30	1.689	4.8	18.51	0.562	0.839	0.93	18.2	59.9
35	1.665	5.9	22.06	1.039	2.714	1.10	40.9	72.3
40	1.645	8.8	22.27	0.861	5.987	1.09	74.0	87.4

[a] water vapour tension in the "head" compartment (in the "body" compartment 4-5 mm Hg)
[c] the % metabolically produced heat (H) dissipated by total evaporation (Et), assuming 1 g water = 2.406 kJ.
[d] respiratory evaporation (Er) in % of total evaporation (Et).

INFLUENCE OF AMBIENT TEMPERATURE ON RESPIRATORY AND CUTANEOUS EVAPORATION

From Table 4 it can be seen that respiratory evaporation increases slowly at first with increasing environmental temperature but then more rapidly when the temperature exceeds 30°C. The same is true for cutaneous evaporation, with the exception of the results obtained at 40°C. Cutaneous evaporation even at lower temperatures can represent up to 40 per cent of the total evaporation. Seven avian species, varying in weight from 12.5 to 300 g, lost 45% or more of their total water loss by evaporation via the skin at 30° and 35°C (Lasiewski et al., 1971). These data suggest therefore that cutaneous evaporation occurs in many species of birds. Cutaneous evaporation increases as the temperature increases from 10° to 35°C, but falls again at 40°C. Possibly at this high temperature diffusion through the skin is diminished, caused by a decrease in blood flow, as heat transport by convection in the body (via the blood) now mainly takes place in the direction of the respiratory system.

At 40°C 74.1% of total heat loss is achieved by evaporation. Respiratory evaporation accounts for 64.5% of this, leaving 9.6% for cutaneous evaporation. The principal contribution to heat regulation in the White Leghorn hen at high temperatures is made by respiratory evaporation but cutaneous evaporation also plays an important part.

The fact that some birds evaporate more water via the skin than was originally supposed has important consequences for different fields in the physiology of birds, as for example, the calculation of respiratory volume and heat loss.

The great amount of water lost at high ambient temperatures has to be balanced by a large water intake because at temperatures ranging from 38° to −5°C the amount of metabolic water produced is 0.6-0.7 g per litre O_2 consumed; and only at −5°C will it suffice as a replenishment of water, lost by the fowl by respiratory and cutaneous evaporation. Generally the bird succeeds well in balancing water intake and water loss over a wide range of environmental temperatures (Table 5). It is noteworthy that water loss via the excreta and cutaneous evaporation at high ambient temperatures is very high. Considering the need for economizing water under these conditions one would expect drier excreta, but instead these appear to be very wet. This can be explained partly by the lowered food intake and consequent small amounts of faecal material, whereas the capacity of the kidney to concentrate the urine is very limited. This will result in an increase of the ratio of urine to feces, though this does not explaining the increase in the absolute amount of water in the excreta. It is possible that as a result of the intake of much drinking water and little food, a rather great amount of water passes the alimentary tract very quickly. As long as the temperature of the drinking water is below body temperature, the transport of this water will contribute to heat loss.

Cade (1964) describes an inverse relationship between evaporative water loss and body weight for various species at ambient temperatures near 25°C with a sharp inflexion at a body weight of about 50 g. Thus the evaporative water loss from the house wren with a body weight of 9 g amounts to 36% of the body weight per day, whereas the ostrich

Table 5. Water balance at different ambient temperatures.

Ambient Temp.	pH_2O[a]	WATER INTAKE				WATER LOSS				Balance
		Food	Drinking water	Metab. water	Total	Resp. evap.	Eggs	Excreta + cut. evap.	Total	
°C	mmHg	$g/kg^{\frac{3}{4}} \cdot 24$ hrs					$g/kg^{\frac{3}{4}} \cdot 24$ hrs			$g/kg^{\frac{3}{4}} \cdot 24$ hrs
− 5	3.7	12.9	133.9	29.6	176.4	16.9	–	140.8	157.7	+18.7
0	5.3	9.9	123.4	26.3	159.6	18.8	13.2	103.6	135.6	+24.0
5	7.2	9.4	124.0	24.7	158.1	23.1	19.8	99.3	142.2	+15.9
10	7.1	6.7	101.3	20.2	128.2	20.1	12.1	90.6	122.8	+ 5.4
15	8.1	3.2	74.3	16.5	94.0	25.7	12.1	74.6	112.4	−18.4
20	9.0	5.6	107.2	15.3	128.1	35.2	22.3	77.7	135.2	− 7.1
25	10.5	5.6	112.2	13.7	131.5	47.0	22.2	76.9	146.1	−14.6
30	10.9	7.0	172.8	16.6	196.4	51.0	22.8	117.1	190.9	+ 5.5
35	17.4	4.9	296.0	14.7	315.6	95.5	23.1	180.1	298.7	+16.9
38	17.2	3.6	250.0	13.5	267.1	101.6	23.1	204.9	286.5	−19.4

[a] see Table 1

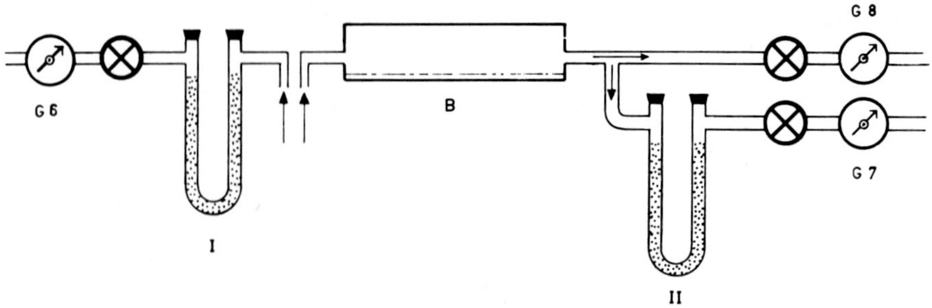

Fig. 1. The open circuit method for the measurement of evaporation (B = chamber; I and II = drying columns; G6, 7, 8 = gasmeter).

(body weight 100 kg), loses only 0.6 %per day (Crawford and Schmidt-Nielsen, 1967). This clearly indicates that small birds have a greater need for access to water.

DETERMINATION OF EVAPORATION

For the determination of evaporation the open circuit method, schematically reproduced in Figure 1, is generally used. Air passes through a chamber (B) and a drying column (Drierite: anhydrous $CaSO_4$). From the rate of flow and the increase of weight of the drying agent the water vapour content of the air can be calculated. In fact the water vapour comes not only from B, but also from the incoming air. The water content of this air can be determined also by drawing a known amount of it through the absorption columns and weighing these columns.

The vapour pressure in B depends on (a) the water vapour pressure of the incoming air; (b) the amount of water, added in B; (c) the degree of mixing and (d) the rate of ventilation.

DETERMINATION OF TOTAL EVAPORATION

A. Directly

The bird is placed in the chamber (B) on a hardware cloth platform above a pool of mineral oil. Voided excreta will be covered by this oil, thus preventing the evaporation of water from the excreta. The most convenient way is to use a bird in the postabsorptive state, giving it neither food nor water during the experiment in order to eliminate extra sources of evaporation.

B. Indirectly

The weight of the bird decreases with time as a result of the production of eggs, the loss of feathers, excreta, CO_2 and water vapour, whilst by taking in oxygen, water and food, weight will increase. With the exception of CO_2 and O_2, which can be measured by gas analysis, the above mentioned parameters can be determined by weighing. The only unknown factor remaining is the amount of water vapour lost; this can be calculated from the other parameters.

DETERMINATION OF RESPIRATORY AND CUTANEOUS EVAPORATION

For this the chamber B is divided into a "body" compartment and a "head" compartment. The head protrudes through an opening in the partition into the "head" compartment, a rubber sheet stretching around the neck across the opening to prevent air flow between the compartments. By duplicating the ventilation and absorption systems cutaneous and respiratory evaporation can be determined simultaneously. Respiratory evaporation will be overestimated because of evaporation from the skin of neck, head and eyes, the cutaneous evaporation may correspondingly be estimated too low.

CONCLUSION

Dependent on such factors as food intake, the state of acclimation and the relative humidity, the parameters given in Figure 2 will shift with regard to the ambient temperature.

In order to compare the efficiency of heat loss by respiratory evaporation in diverse animals, the amount of water evaporated is expressed per unit volume of oxygen consumed. To attain a just comparison not only the temperature of the inspired air but also

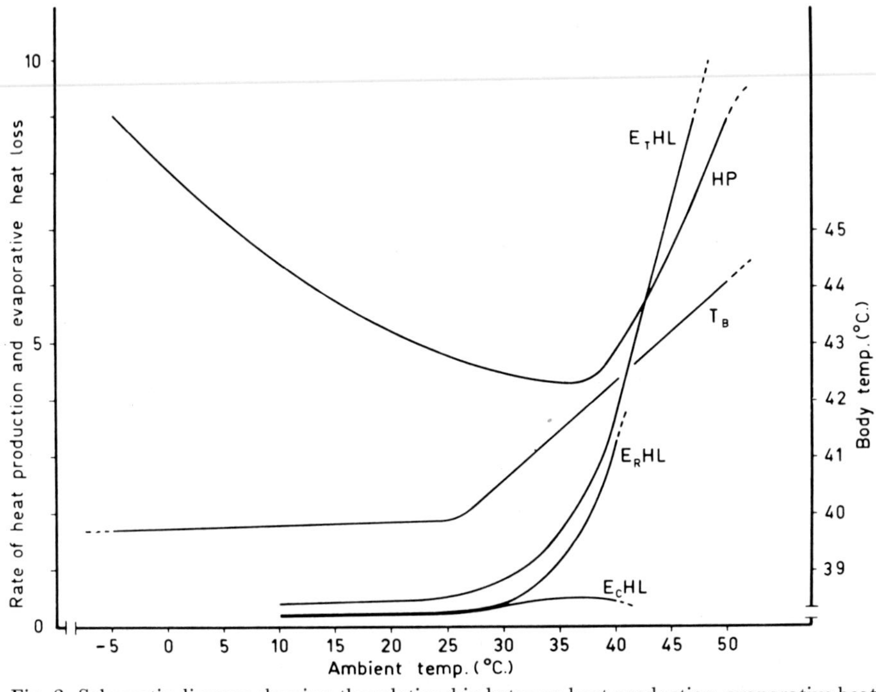

Fig. 2. Schematic diagram showing the relationship between heat production, evaporative heat loss and body temperature and ambient temperature. (Tb = body temp.; HP = heat production; Et-, Er- and EcHL = total, respiratory and cutaneous evaporative heat loss).

its water vapour pressure should be the same, while food intake and the degree in which the animal is acclimated should be similar.

A wide variety of birds is capable of dissipating more heat by evaporation than is produced metabolically and can therefore cool the environment.
Several species of birds have a significant evaporative water loss via the skin.

At high temperatures the most important method of losing heat is by evaporation. This can work well only when the ambient water vapour pressure is low.

Small birds have a greater need for access to water than the larger ones. In order to advance understanding of this subject we need more information about (a) the effect of acclimation; (b) the influence of ambient temperature and humiditty; (c) the temperature at which the water evaporates and (d) evaporation under normal conditions, by experimenting with acclimated animals, disposing of food and drinking water ad libitum.

REFERENCES: see p. 536.

Possible Mechanisms Involved in Death in Laboratory Animals Due to Thermal Stresses (Heat and Cold)

by

G. L. Wright

INTRODUCTION

Every living organism has a definite range of body temperatures within which it must remain in order to maintain its functional integrity. Thermal tolerance varies considerably among species and may differ among individuals of a given species with changes in environmental conditions and the physiological state of the animal. The thermal ranges of a typical poikilotherm and homiotherm are presented in Figure 1. In each case, the animal exhibits a voluntary range of temperatures for normal activity which it will attempt to maintain by physiological or behavioral means. The temperatures ranging immediately above and below the voluntary range are generally associated with short term beneficial effects on activity or caloric conservation as well as providing a buffer zone in which the organism remains operable during periods in which the thermoregulatory mechanisms are being activated. Further increases in the severity of heating or cooling lead to the critical thermal maximum (CTMax) and critical thermal minimum (CTMin), respectively. Cowles and Bogert (1944) first developed the concept of the critical temperature as the potentially lethal temperature which renders the animal incapable of escaping its imminent death. Lowe and Vance (1955) modified the concept of the CTMax to be the mean of the thermal points of a sample of animals at which locomotory activity becomes disorganized and Hutchison (1961) has discussed the importance of the rate of temperature change in the critical temperature determination. Theoretically, an animal exposed to the critical thermal maximum or minimum may show complete recovery if it is immediately removed from the adverse environment. This indicates that the damage incurred is largely reversible or possibly represents the loss of only a minor portion of the total functional capacity so that the animal appears to be normal upon the return to the less demanding voluntary temperature range. In contrast, exposure to the lethal maximum or minimum temperature results in the accruement of damage of a quantitative or qualitative nature that precludes an overwhelming of the recovery mechanisms and the inevitable death of the animal.

The CTMax or CTMin determination provides a reproducible, easily performed technique for measuring the relative thermal sensitivity of the intact animal. These thermal points are generally determined as the core temperature at which the animal exhibits evidence of neuromuscular malfunction. The internal temperature may be obtained from water bath temperatures for small amphibious poikilotherms (Mahoney and Hutchison, 1969) or from intra-intestinal probes inserted into the cloaca and rectum of larger animals when the study is conducted in air. Additional information concerning

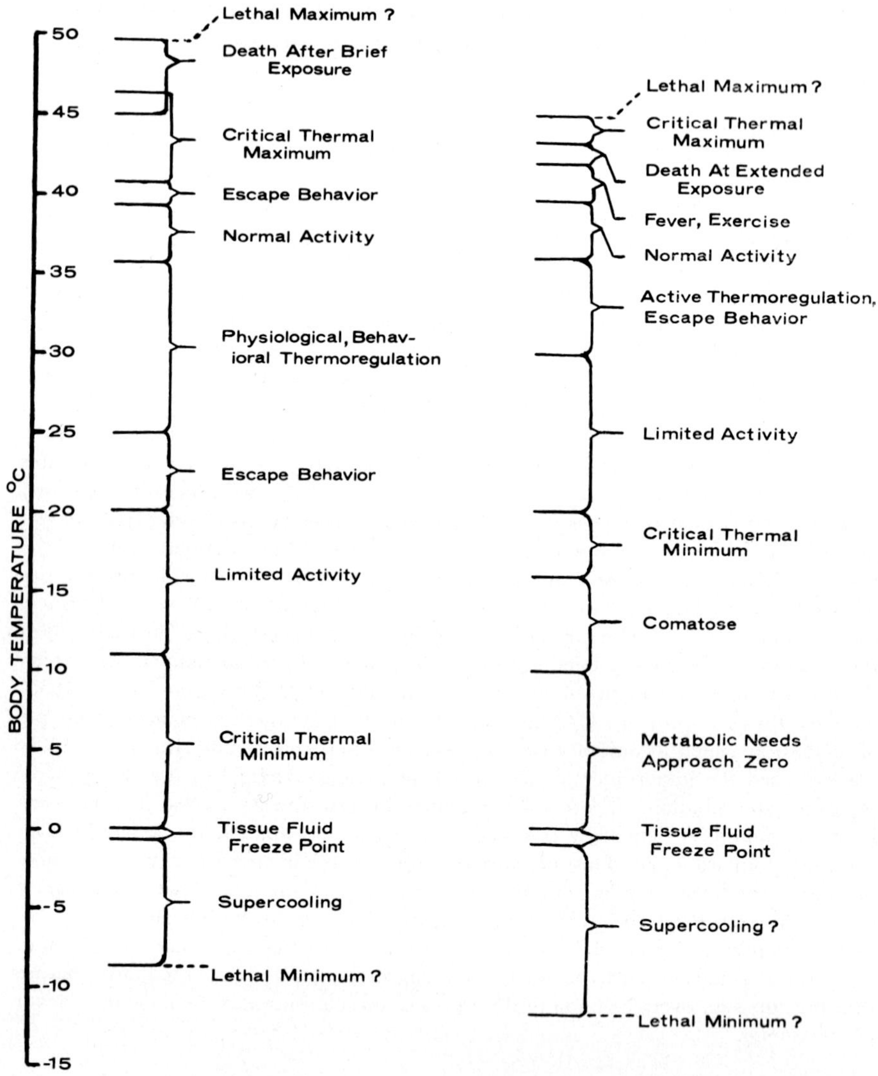

Fig. 1. Typical reptilian and mammalian thermal ranges. (Constructed from Bogert, 1959: Kour and Hutchison, 1970: Lowe et al., 1971: Wright, unpublished data).

the thermal state of the animal may be obtained from intra-esophageal, tympanic and tissue implanted probes but these have the drawback of necessitating the imposition of undue restraint on the animal (Frankel, 1959) as well as compromising the physiological condition of the experiment. The end points that have been utilized include the initial onset of spasms (Zweigel, 1957; Heatwole et al., 1965), loss of the righting reflex (Brattstrom and Lawrence, 1962), and cessation of body movements (Dunlap, 1968) for the CTMax determination while the disappearance of the righting response is the criteria generally employed in the CTMin determination (Kour and Hutchison, 1970). The thermal range limits of an animal are influenced by a number of factors including season (Hoar, 1955) the diurnal cycle (Kosh and Hutchison, 1968; Mahoney and Hutchison, 1969) previous thermal history, and the physical characteristics of the method used for heating and cooling (Hutchison, 1961). A comparison of the 24-hr cyclic fluctuation in the CTMax of a homiothermic and a poikilothermic animal is provided in Figure 2. The temporal variation in the CTMax that is exhibited here

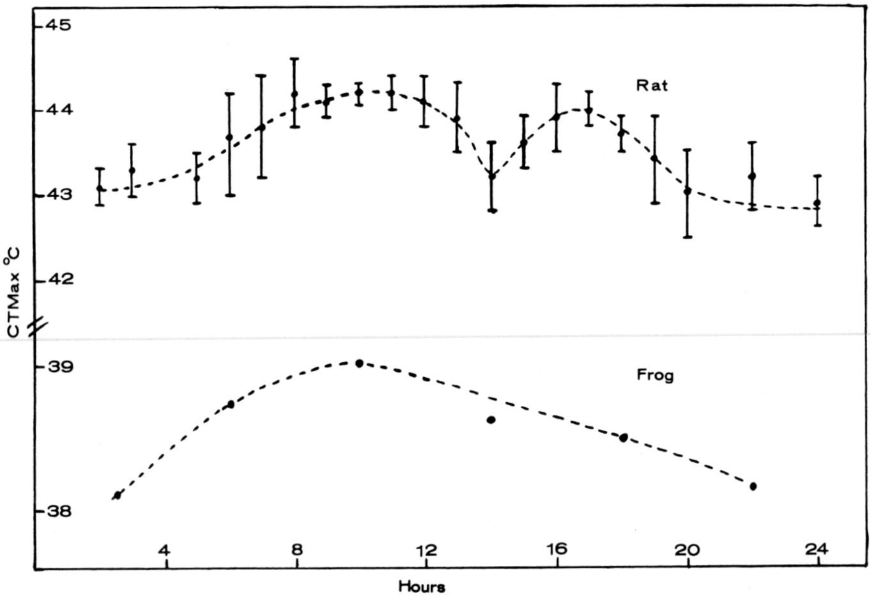

Fig. 2. A comparison of the 24 h variation in the critical thermal maximum of the albino rat (Wright unpublished) and the frog. *Hyla labialis* (Maloney and Hutchison). Each group of animals was maintained at an ambient temperature of 25°C with a 12/12 L/D photoperiod. A total of 212 rats and 149 frogs was utilized. The dots represent the mean value while the bars above and below the mean equal one standard deviation.

provides a remarkable demonstration of the inherent ability of many animals to undergo short term systemic or molecular adjustments in meeting the changing probabilities of high temperature exposure and illustrates the importance of careful experimental standardization and scheduling in studies of this type.

THE HEAT DEATH MECHANISM

The marked loss of coordination and the convulsive movements of animals exposed to critically high temperatures tends to implicate the neuromuscular system as the primary site of heat damage. This does not necessarily indicate a unique thermal sensitivity of nervous or muscle tissue but merely demonstrates the immediate importance of this system in the control of vital functions. Thermal resistance is variable among different cell types (Burger and Fuhrman, 1964). However, it is doubtful that any tissue escapes damage at temperatures producing convulsions or other obvious impairment of the animal, and the deleterious effects that are often observed to occur some time after heating has taken place are probably the result of the cumulative effect of cellular damage throughout the body.

Heat damage increases progressively with temperature and encompasses a range of effects extending from subtle biochemical changes (Burger and Fuhrman, 1964) to disruption of mitochondrial structure (Davison, 1971) and disintegration of the cell (Buchsbaum, 1970). There has been considerable speculation as to the mechanism underlying these events and several agents have been described as the principal causal factor. Perhaps, the most widely accepted theory of heat induced damage is protein denaturation resulting in deleterious structural and enzyme activity modifications. The most convincing evidence supporting this hypothesis is the high energy of activation for cellular heat damage (50 cal/mol) which is similar to values obtained for heat denaturation of proteins (Ushakov, 1964). This type of analysis has also been applied to thermal death curves of multicellular organisms. The heat death rate in *Drosophilia melanogaster* has been shown to follow a compensation law describing a linear relationship between the activation entropy and enthalpy, which was in excellent agreement with data from unicellular killing curves, and which gave constants very similar to those obtained for thermal protein denaturation (Rosenberg et al., 1971). The major criticism of the protein theory is that isolated proteins generally show a greater degree of thermal resistance than their tissue of origin. However, the discrepancies between the thermosensitivities of the intact cell and its component proteins probably reflect the alteration of intracellular binding conditions that affect the thermostability of the protein during in vivo heating. Stearn (1949) has suggested that thermally induced protein denaturation involves the progressive breakage of hydrogen bonds with increasing temperatures until a point is reached at which there is a precipitous change in the tertiary structure of the molecule. Presumably, the low incidence of bond breaking that occurs during the early stages of heating produces the readily reversible damage observed in subcritical heating. The point of tertiary structure modification is associated with an increase in entropy and the irreversible loss of the enzyme or structural protein, and is responsible for the permanent damage incurred during hyperthermia.

The lipoid liberation has theory been proposed as a second mechanism of heat death (Belehradek, 1931). It states that heat death is due to the melting of cellular lipids, and the thermostability of a system is dependent on the melting point of a key lipoprotein component rather than the thermosensitivity of a specific protein. Evidence supporting this hypothesis stems mainly from studies showing a correlation between diet induced changes or species differences in the lipid composition of animals and their upper thermal limit (House et al., 1958; Chan et al., 1971). However, numerous studies are available which dispute the reports of a well defined relationship between body lipid melting points and thermal resistance (Frankel and Hopf, 1940; Cherry, 1959).

Ushakov (1964) has suggested that the lipid effects on thermostability are mediated through the complexes they form with proteins. The observed decrease in the thermal sensitivity of the cell that occurs with an increase in lipid melting points may actually be indicative of the decreased availability of unsaturated double bonds and thus the diminished susceptibility of the system to disruptive reactions such as peroxide formation. In any case it is now generally recognized that lipids play an important role in determining cellular thermostability, and it is thought that their influence is exerted through the interactions they maintain with proteins.

Disruption of cell protein and lipid complexes would promote the death of the organism through (1) an inactivation of enzymes involved in energy yielding reactions and synthesis processes, (2) loss of selective permeability and functional structuring of the membranes and (3) the release of toxic materials or other indirect effects which would act through the first two pathways. The well documented decrease in blood pH and associated high lactate concentrations in heat stressed animals (Ahearn, 1970; Frankel et al., 1963) indicate a malfunctioning or inadequate aerobic metabolism. The proposals of heat death due to general hypokinetic and overutilization hypoxia (Mayer, 1914; Frankel et al., 1963) have been rejected by Wright et al. (1972) who showed that tissue hypoxia in heat killed rats was less severe than that of animals dying from a known anoxia. This does not eliminate the possibility of deficient ATP synthesis and recent work with muscle sarcomeres from heat killed flies indicates that this may be a contributing factor in heat death (Davison and Bowler, 1971). Granger's (1969) findings of decreased oxygen utilization and increased sodium-potassium permeability at temperatures near the death point provide the basis for his proposal of heat death due to the failure of neuromuscular transmission. Granger's hypothesis is made especially attractive by its capacity for the expression of the heat death mechanism as the result of the interaction of the several factors which have been shown to affect high temperature tolerance while it remains consistent with the behavioral implications of nervous dysfunction in the animal during critical hyperthermia.

THE COLD DEATH MECHANISM

Cold injury and death can occur through the thermo-chemical suppression of various vital activities at non-freezing temperatures or to the disruption of cellular structure and biochemical processes by ice formation at freezing temperatures. The mechanism of non-freezing cold injury is not known although Salt (1961) has discussed an imbalance of various segments of metabolism due to unequal rate reduction in all reactions as a probable mechanism in poikilotherms. Sullivan and LeBlanc (1957) showed that the rate of rewarming was a principal determinant of the degree of cold injury and suggested that a rapid restoration of blood flow was an important aspect of the recovery from deep hypothermia. A similar view was held by Fairfield (1948) who concluded that non-freezing cold death was due to rewarming hypoxia which developed as the result of a lag between the increase in tissue metabolism and the respiration and heart rates.

The anesthetic effect of cold on excitable tissue is well known and is probably produced by either an alteration in the membrane ionic permeabilities or a suppression of the sodium-potassium pump activity. Squid nervous tissue exhibits an increase in the resting K^+ loss and does not appear to reabsorb K^+ during interim periods between stimulation when cooled to 6.1° C (Shanes, 1954). The evoked cortical potential disappears at 13°–15° C in the rat (Battista, 1957), and the heart ceases to beat at about 10° C. Each of these processes appears to be reversible and produces no apparent permanent damage with controlled rewarming. The supressant thermo-chemical effect of non-freezing cold on cellular process produces a reversible type of damage which, if sustained, may lead to systemic changes in oxygen availability, membrane permeability, metabolism, viscosity etc., which can cause permanent damage and death. For example, it is now fairly certain that the lethal effect of long term exposure to non-freezing cold in marine fishes is the result of the failure of osmoregulation (Umminger, 1971; Woodhead, 1964). If the imbalances created by cooling, maintained hypothermia or the rewarming process exceed the capacity of the homeostatic mechanisms for the progressive readjustment to the normal condition the animal will exhibit damage.

The mechanism underlying the irreversible functional and structural alterations that occur in frozen tissues is unresolved. Lovelock (1953) and Meryman (1968) concluded that the deleterious action of ice crystal formation was attributable to the concentration of cellular electrolytes or dehydration produced by water freezing out rather than to a mechanical effect as proposed by Luyet (1949) and Asahina (1966). It is thought that the principal molecular events occurring during freeze injury are the conversion of sulfhydryl groups to the disulfide bond (Levitt, 1967) or the disruption of intramolecular hydrophobic bonding (Scheraga, 1963) which gives rise to denaturation of proteins.

There is considerable evidence that animals may survive subfreezing temperatures either by supercooling or the development of a resistance to the damaging effects of freezing and thawing. Supercooling refers to the metastable condition in which tissue temperatures may be several degrees below their freezing point without ice formation. Lowe et al. (1971) found that nearly all poikilotherms examined were capable of utilizing supercooling in the environment. Mammalian species may not generally supercool in the environment but possess the ability to survice sub-freezing temperatures under special laboratory conditions (Smith, 1961).

Freezing tolerance in intertidal mollusks has been associated with the maintainence of a constant percentage of total water in an osmotically inactive state. The chemical or physical nature of the influence exerted by a decrease in water activity on freezing tolerance is unknown although it may provide for the stabilization of a fixed quantity of cellular water that in turn would affect a freezing tolerance limit defined by a maximum level of cellular dehydration or by a minimum cell volume (Williams, 1970).

SUMMARY

It has become apparent that many factors are involved in heat and cold death, and an investigator attempting to elucidate the mechanism of these phenomena must decide the level of severity of thermal stress that is pertinent to the environmental condition he wishes to recreate for examination in the laboratory. Hutchison's (1961) original conclusions concerning the unique evolutionary and ecological significance of the CTMax and CTMin must be revised with the observations of supercooling and sub-CTMax lethal temperatures in the environment. Thermal damage is a function of the tissue temperature and the period of exposure and variations in either of these parameters may alter the hierarchy of factors defining the thermal limits tolerated.

The frustration that has plagued attempts to conclusively describe a heat or cold death mechanism demonstrates the necessary naivete that has influenced our experimental design in this area. Cellular structure is based in complexity and the occurrence of unbound components acting alone is the rare exception. In view of the critical interdependence existing between biological systems from the molecular to the intact animal level it is not surprising that investigations of isolated entities have failed to produce satisfactory evidence for the relatively simple and clear-cut mechanism for which we have been searching.

It is highly probable that a thorough comprehensive evaluation of the molecular mechanism of thermal damage to a tissue will involve the integration of the physical chemistry of temperature induced bonding and configurational changes of membrane and enzyme molecules with the thermal effect on the enzyme molecule's relationship with its immediate complexing materials and the subsequent effect on the sequential relationship between enzyme controlled points in the reaction sequence. Similarly, an overview of the interrelated effects of altered metabolism, humoral activity, cardiovasculature function, CNS function, etc. may ultimately provide a more realistic assessment of thermal lethality at higher levels of organization than a "weak link" approach aimed at defining the first system to demonstrate apparent critical loss of functional capacity.

REFERENCES: see pp. 536-537.

Environmental Temperature and Feeding Behaviour

by

C. L. Hamilton

INTRODUCTION

During the past decade, factors related to the control of food intake have received intensive study. Most investigators have attempted to place feeding behaviour within the context of control and regulatory systems involved in the maintenance of homeostatic set points. Among these have been included the regulation of (1) blood glucose, (2) size of adipose tissue stores, and (3) body temperature. It is the latter regulatory mechanism which will be discussed in this Section.

In the normal life style of lower animals, few environmental influences affect food intake more than that of environmental temperature. In homeotherms, food intake (within limits) is inversely related to environmental temperature. In the laboratory rat, for example, daily food intake may range from approximately 130 kcal at an ambient temperature of 7° C to 20 kcal at 35° C.

ENVIRONMENTAL TEMPERATURE AND FEEDING

One of the earliest laboratory studies of the relationship between environmental temperature and food intake was that of Gasnier and Mayer (1939). They exposed shaved and unshaved rabbits to ambient temperatures of from −3° to 30° C. The shaved animals ate more than the unshaved at all environmental temperatures and both groups ate more in the cold than in the heat.

More recently, time spent in feeding by goats has been studied as a function of ambient temperature over the range of 0° to 40° C. (Appleman and Delauche, 1958). In this study body temperatures were also recorded and it was found that at ambient 41° C rectal temperatures averaged 40° C and feeding was suppressed. In a non ruminant, the pig, Heitman and Hughes (1949) observed that when the animals were kept chronically at 38° C ambient, food intake fell from 7 to less than 2 pounds/day. Results similar to the above have been reported for cows and chicks by Kleiber (1961) and by Brody and his colleagues (1964).

In many terrestrial animals of the Artic, the onset of cold weather is marked by an increase in the insulative capacity of the fur. Another adaptation to cold is an increase in food intake. Durrer and Hannon (1962) have observed that the Alaskan Husky increases caloric consumption from a low in summer of 49-kcal/kg/day to a high of 87 kcal/kg/day in winter. In addition they noticed that body weight gains occurred during the warm weather indicating that much of the food consumed in winter was used for heat production.

Most of the work on temperature and feeding behavior has been conducted using the laboratory rat as the subject. Figure 1 shows the relationship between food intake and body temperatures in these animals after 24-hr exposure to test temperatures. It should be noted that the lowest body temperature was recorded at ambient 32° C. This is approximately thermoneutrality for the rat, that environmental temperature at which

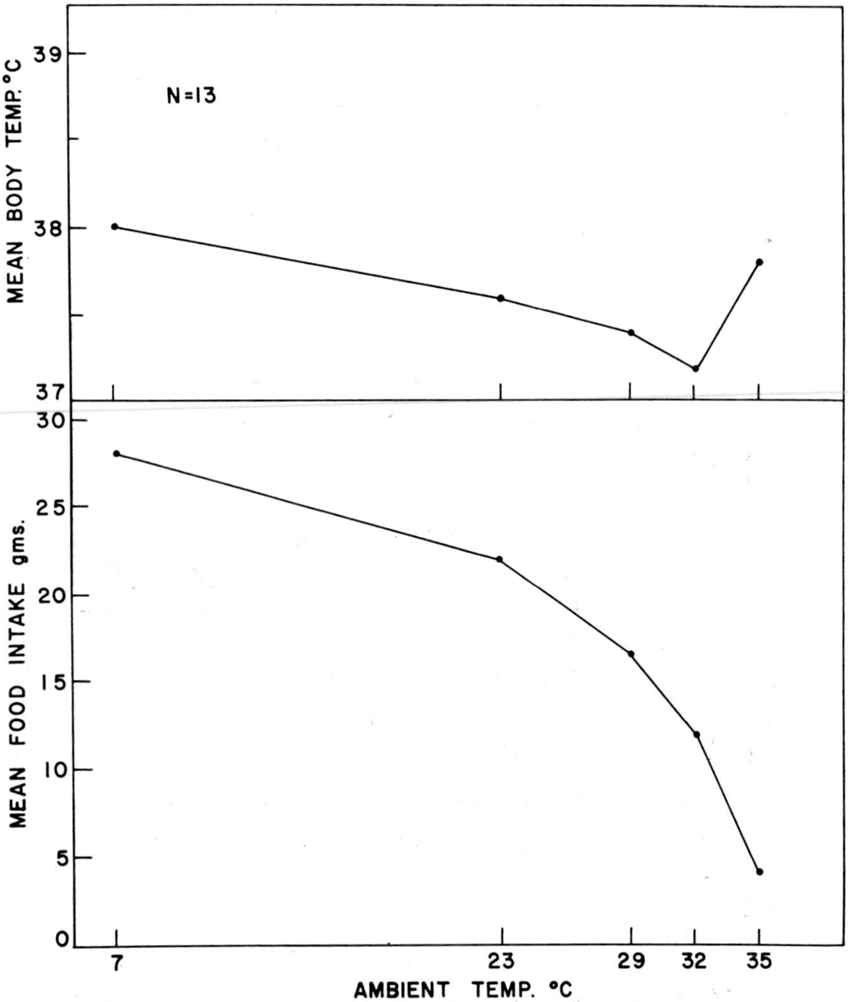

Fig. 1. Food intake and body temperature of rats as a function of environmental temperature. (Hamilton, 1963).

one observes the lowest basal metabolic rate (BMR). At 35° C ambient, body temperature rises again and food intake is near zero. Thus, body temperature is high both in the cold and extreme heat and it is obvious that food intake cannot be predicted solely from body temperature.

Data from the study of chronic exposure to high ambient temperature are presented in Figure 2. In that study separate groups of rats were exposed to environmental temperatures of 32° and 35° C for 21 days. The data reveal many of the adaptations the rat makes to heat stress. Body temperatures at ambient 32° C were regulated lower than at 24° C (Fig. 2a) while at ambient 35° C they were regulated at higher than

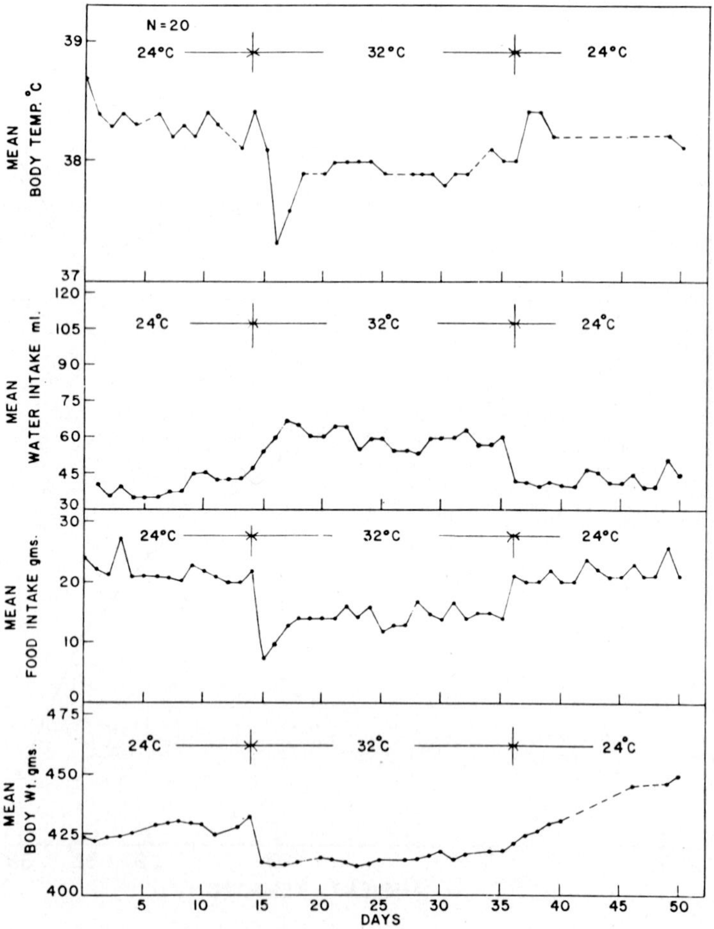

(Fig. 2a)

"normal" levels (Fig. 2b). In both cases food intake was depressed the first several days in the heat but, after adaptation, food intake was stabilized at approximately 2/3 the amount consumed at 25° C. Water intake at 35° C was double that of the group at 32° C. These data reflect the evaporative water loss of the animals at the higher temperature. Since the rat does not perspire it uses saliva spreading as a means of enhancing evaporative cooling. The sequence of events is first a profuse salivation then, within 24 hrs, the animals appear to learn to spread the saliva over the body. Heat stress at 32° C ambient is not severe enough to result in saliva spreading. The gradual increase of food intake in both groups after exposure to heat also accounts partially for the rise in water consumption.

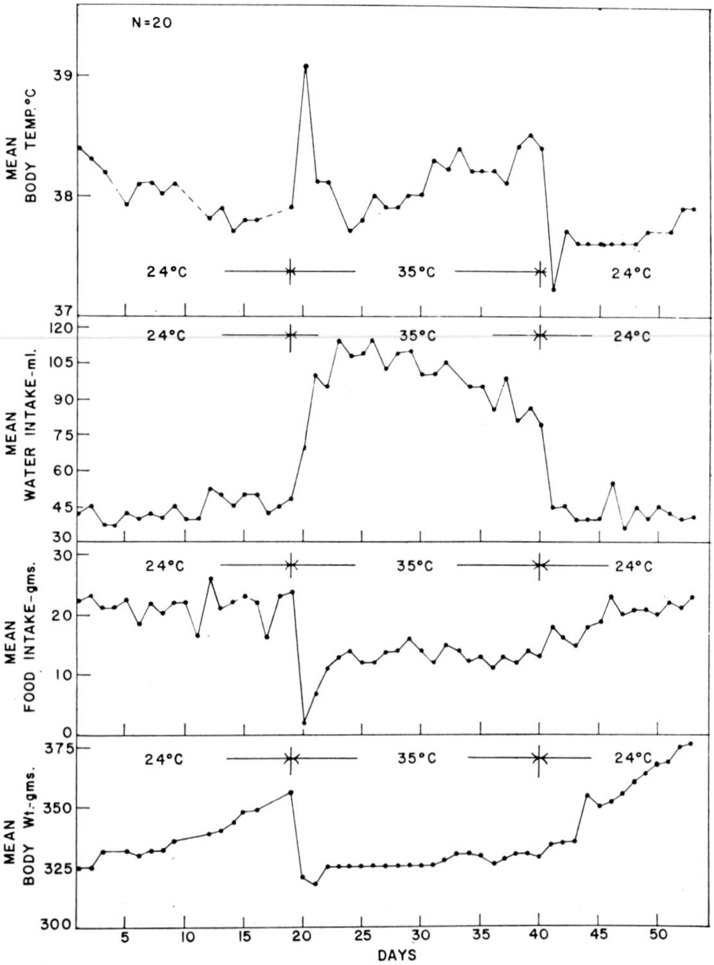

Fig. 2. Response of rats to ambient temperatures of 32°C and 35°C over 21 days exposure. (Hamilton, 1963).

Adaptation of rats to cold stress (8° C) over 21 days results in body temperature regulation at approximately 1° C above normal along with a 70% increase in daily food consumption (Hamilton and Brobeck, 1964).

Of course there are lethal limits of acute heat and cold stress for mammals. However, between these extremes there are temperatures at which the homeotherm has great difficulty in making the appropriate feeding adjustments. If the temperature is too cold, the body's need for heat may exceed its feeding capacity. On the contrary in hot environments body temperature may be so high that food intake sufficient to maintain life may further embarrass the temperature regulating mechanisms of the body because of the specific dynamic effect of food. This latter condition was dramatically demonstrated in a study in which rats were tube fed various caloric loads at an ambient temperature of 38° C. (Hamilton and Brobeck, 1962). Figure 3 shows that the mortality

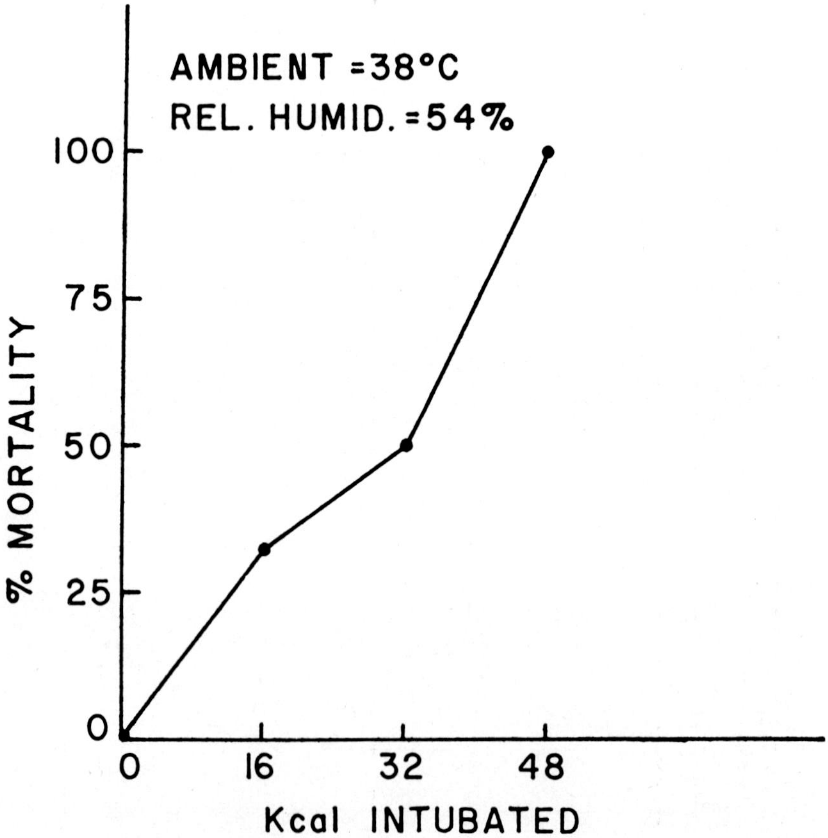

Fig. 3. Mortality rate of rats force fed various caloric loads. (N = 6 each group). Data replotted from Hamilton and Brobeck, 1962).

rate of the rats (from lethal hyperthemia) was a function of the amount of calories intubated.

It is obvious that in hot environments homeotherms must schedule their feeding so as not to overtax the heat dissipating mechanisms of the body. We do not have carefully collected data from the laboratory on this point, but it is common knowledge that cattle tend to schedule their feeding periods to coincide with the cooler parts of the day during hot weather.

A question arising from the above series of observations is whether the alterations of food intake during thermal stress are directly related to ambient temperature changes or only indirectly through other physiological adaptations. In the cold one might postulate that the increase in food intake is the result of an increase in metabolic rate during cold stress. The fact that thyroidectomized rats also increase food intake in the cold is only a partial answer since those animals also increase activity levels and therefore "non-chemical" heat production. A better approach to the problem is to determine whether the rat responds to short term thermal stress. Spector et al. (1972) have shown that rats housed at room temperature 24° C and trained to feed for only 2 hrs/day reduced food intake significantly during the first 2 hour feeding session conducted at 32° C. When tested at 7° C ambient, however, the rats did not exhibit an increase in food intake during the first 2-hr session, indicating no food response to the cold stress. In the same work, Spector et al. (1972) repeated the study using rats with shaved coats and found that they did increase food intake significantly on the first 2-hr exposure to 7° C .(c.f., Gasnier and Mayer, 1939). These data indicate that the initial response of the rat to cold stress is the use of the fur as insulation through piloerection. Such a response is under control of the autonomic nervous system and occurs very quickly. Feeding, a voluntary response, is a second line of defense. In unpublished work from our laboratory we have seen increases in food intake on a one hour test schedule when rats were shifted from a room temperature of 25° C to 18° C. Such mild cold stress is not sufficient to result in piloerection and the feeding behaviour is not obscured by the autonomic adjustment. Presently more data are needed before we can accept the direct relationship between ambient temperature and feeding behaviour especially with respect to cold stress.

The importance of food for animals subjected to environmental thermal stress has been emphasized by Kleiber (1961). He showed that survival time for starving rats could be increased from 10 to 15 days by altering room temperature from 20° to 30° C, indicating that in the warm environment the rat "feeds less rapidly on itself."

In poikilothermic animals little experimental work has been accomplished relating feeding to ambient temperature. However, representative studies by Harris (1959) in the toad and McMahon (1965) in *Daphnia magna* showed a positive relationship between feeding and ambient temperature. Since body temperature regulation in the poikilotherm is poor, it probably precludes any more than a coincidental relationship between feeding and environmental temperature.

DIET AND ENVIRONMENTAL TEMPERATURE

The increase in caloric intake by rats when cold stressed has interesting overtones. It has been shown that rats fed a protein deficient diet (5% instead of 22% protein by weight) fail to gain weight at room temperature 21° C and die within 6 weeks. In contrast, rats on the same deficient diet but maintained at a room temperature of 5° C survived indefinitely (Andik et al., 1963). Similar results have been observed by Meyer and Hargus (1959). These data not only demonstrate the appetite suppressive effects of protein deficient diets (at 21° C) but also show that when conditions are established that stimulate animals to increase caloric intake the protein deficiency is covered and the animals survive.

The anorexia of protein deficiency is unexplained but it also occurs in rats maintained on diets lacking a single essential amino acid. However, as with protein deficiency, rats placed in the cold double their food intake and survive (Klain and Winders, 1964). In essence, in warm environments rats do not increase caloric intake to "get" more protein. Cold stress appears, however, to be a sufficiently potent stimulus to overcome protein deficiency anorexia.

In cafetaria type of experiments in which rats were given a choice between a high carbohydrate and high fat diet with protein content kept constant, contradictory results have been reported. Donhoffer and Vonotzky (1967) reported that their animals shifted from a preference for a high fat diet in the warm to a high carbohydrate diet in the cold (10 °C). Dugal et al. (1945) reported opposite results but pursued the issue farther and found high fat diets to be superior to high carbohydrate diets as measured by survival time of the rat kept at −2° C.

We have little data on food selection and ambient temperature on animals other than the laboratory rat. Hale and Schein (1962) have summarized some unpublished studies with turkeys by Almquist and Margolf. They found that when given the choice of oats, wheat and corn ad libitum, the birds took only 18% of their total caloric intake from the high energy wheat and corn during summer and 48% of their calories from those food stuffs in winter.

In general, we may conclude that in mammals food intake is enhanced in cold and suppressed in warm environments. The exact nature of the stimulus for these alterations in feeding is yet to be elucidated but there is suggestive evidence that it is the ambient temperature per se. We need more definitive work regarding the subtle aspects of diet and heat and cold stress. Food utilization efficiency as a function of diet and ambient temperature needs to be studied in a wide range of animals. In addition, the need for minerals and vitamins under conditions of thermal stress should be investigated.

REGULATION OF BODY TEMPERATURE AND FEEDING

If one assumes that food intake is controlled as part of the mechanism regulating body temperature, then it appears logical that a search for the central nervous system components of feeding behavior should be centered around the area of the brain directly involved in temperature regulation. Brobeck (1945) has suggested that since experimental lesions of the hypothalamus result in abnormalities of body temperature, food intake, activity and body weight, this area of the brain may be the integrator of energy balance for the organism. In particular, the area labeled the pre-optic anterior hypothalamus (POAH) has been shown to be intimately involved in the regulation of body temperature. In all mammals studied so far, lesions of this area are followed by labile temperature regulation. The animals are hypothermic in the cold and hyperthermic in the heat. In the POAH, which lies dorsal to the optic chiasma and ventral to the anterior commissure, are cells sensitive to the temperature of the blood profusing them (Nakayama et al., 1963). Some of these cells increase their firing rate when warmed above normal body temperature, others increase firing rate when cooled. A third type of cell found in this area does not respond to heating or cooling. These cells are postulated to be neurones that modulate the activity of the heat and cold sensitive cells. It is believed that their activity may be influenced by information reaching them via pathways from the peripheral temperature receptors. According to the present scheme of organization, the brain "knows" not only its own temperature, but also temperature conditions at the periphery.

During experiments in which the cells of the POAH have been artificially heated or cooled, animals respond as though they were environmentally heat or cold stressed respectively (Magoun et al., 1938) (Fusco et al., 1961). There is therefore considerable evidence that the POAH is intimately involved in the regulation of body temperature although other parts of the nervous system may contribute some input. It was logical to assume that the neurological interaction of feeding and body temperature regulation might stem from effector mechanisms controlled by the POAH.

In studies using rats with lesions of the POAH it has been determined that the animals show not only abnormalities in the regulation of body temperature but also fail to make appropriate feeding adjustments to heat and cold stress. Figure 4 shows the effects of heat stress on food intake and temperature regulation in POAH lesioned rats. The lesioned animals were hyperthermic at ambient 32° C and failed to reduce feeding to control levels during the period of the stress. In more extensive testing, it was shown that at all temperatures below ambient 29° C the lesioned animals ate less than their controls, and were relatively hypothermic. (Hamilton and Brobeck, 1964).

Further exploration of this problem has resulted in paradoxical results. Spector et al. (1968), using thermodal heating and cooling of the POAH in the rat demonstrated increased food intake with artificial heating and decreased feeding with cooling. Since heating the POAH of the rat results in physiological adjustments seen when the animal is exposed to environmental heat, and cooling the POAH in adjustments compatible to cold stress, one would have expected the feeding adjustments to be of a similar nature. Presently we have no explanation for these data, and it is this dilemma that poses problems for future research.

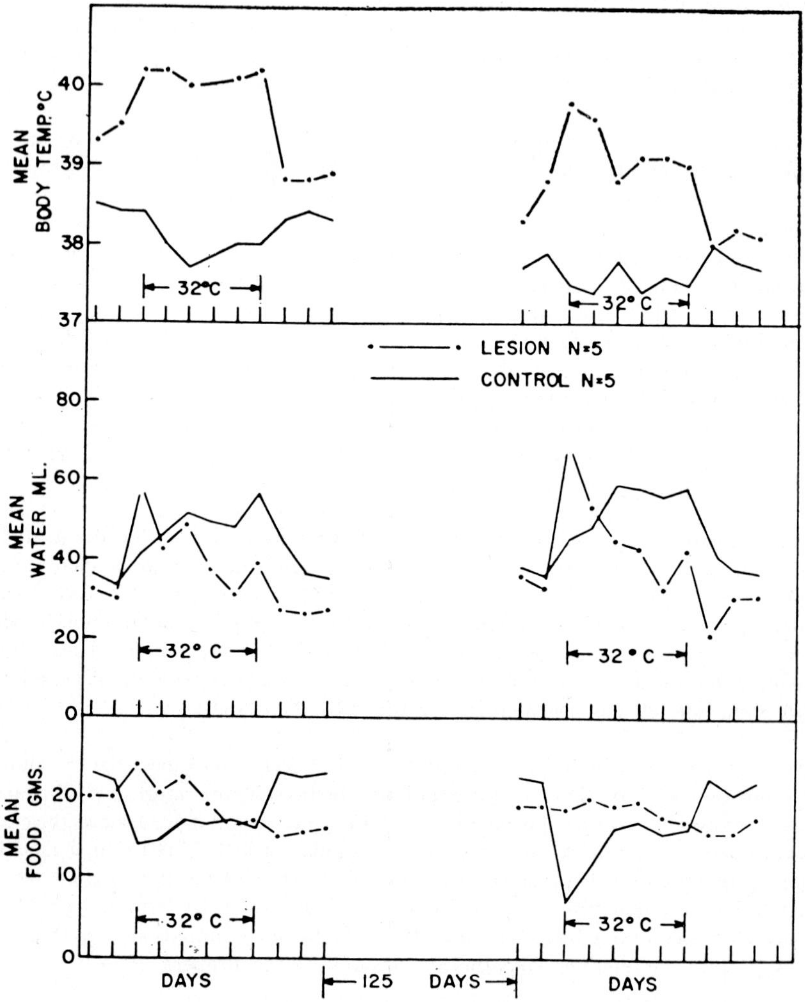

Fig. 4. Response of rats with an without POAH lesions to mild heat stress. (Hamilton and Brobeck, 1964).

CONCLUSION

Physiology is the study of control and regulation as first suggested by Bernard (1949) and later elaborated by Cannon (1939) through his homeostatic concepts. Considering body temperature to be REGULATED, then certain physiological systems are CONTROLLED in appreciation of that regulation. There is strong evidence that in mammals, and under certain conditions, food intake is controlled as part of the mechanism regulating body temperature. The evidence began with gross observations of the interaction between feeding and ambient temperature and has proceeded to the more basic reports showing that rats with lesions of the POAH fail to adjust food intake to environmental temperature. Presently, it is the interaction between the environment and the central nervous system that attracts our attention. The dilemma of the data presented by Spector et al. (1968) must be resolved along with the discovery of the error signal in the temperature system relating feeding to thermal stress.

REFERENCES: see pp. 537-538.

Effects of Heat and Exercise on Energy Metabolism of Cattle

by

J. A. McLean

INTRODUCTION

Energy metabolism is the conversion within the body of chemical energy, initially in the form of food into other forms of energy, mostly heat. Some of the chemical energy may be finally synthesised into another chemical form as in growth or used to do external mechanical work, but in many instances all or most is given off as heat. The study of energy metabolism has been comprehensively treated by Brody (1945) and its application to ruminant animals by Blaxter (1962). More recent developments in relation to climate have been reviewed by Bianca (1965) and Thompson (1973).

The level of heat production of a ruminant animal is determined primarily by its nutritional status and its thermal environment. Under cold conditions the environment is the more dominant factor and this is dealt with in Chapter 1, Section 8b (1) by Thompson. In warm conditions nutritional status is of over-riding importance. A fasting metabolic level is achieved in cattle only after several days without food, and a true basal level (a state defined for humans as the metabolism after several hours without food of a recumbent conscious subject in a thermoneutral environment) is virtually never attained even experimentally with cattle. Even the fasting level of heat production in a thermoneutral environment depends on the general level of food intake prior to the fast. Cattle have an unusually high level of metabolism in comparison with mammalia in general for which Brody (1945, p. 370) proposed the equation

$$M = 3.416\ W^{0.73}\ \text{Watts}$$

where M is basal metabolism and W is body weight in kg. According to Blaxter (1962, p. 98) cattle have a fasting metabolic level 100% higher at birth, and falling to 30% higher at maturity than that predicted by Brody's equation.

EFFECTS OF ENVIRONMENT AND EXERCISE

Two distinct types of metabolic response are observed in normal cattle when exposed to environments warmer than thermoneutral.

When animals LIVE in a hot environment their metabolic level is depressed and this is accompanied by a decrease in food intake and usually a slight increase in rectal temperature. The environmental temperatures at which the effects of sustained heat on metabolism become evident varies from 21° C in lactating cows to 38° C in some heifers.

When cattle which are normally housed in a temperate environment, are exposed FOR A FEW HOURS to hot conditions a slight increase in heat production and a considerable rise in rectal temperature are observed. The food intake of such cattle is not affected by short heat exposures provided that they are returned to thermoneutral conditions between exposures. Examples of each of these types of response have been frequently reported and Kibler (1960) demonstrated both types of response in one group of animals. A third type of response to heat has been demonstrated by Kibler et al. (1970) who maintained the thermoneutral level of food intake of cattle exposed to sustained heat by force-feeding them through rumen cannulae. The heat production of these animals was 21% above the thermoneutral level in moderate heat (30° C) and 5% below it in more severe heat (35° C). Rectal temperatures of these animals increased with increasing air temperature. The humidity of the environment has no effect on the rectal temperature or on metabolism of cattle at low air temperatures. But at high air temperatures, where the animals are relying on evaporative cooling, increasing humidity results in increased rectal temperature and a small increase in heat production (McLean, 1963; McLean and Calvert, 1972). The effect of humidity at high air temperature may thus be regarded as an additional heat load imposed on the animals. Bianca (1962) has shown that over the air temperature range from 30 to 40° C, a rise of 1° C in wet bulb temperature of the air is equivalent in its stress effect on cattle to a rise of approximately 2° C in dry bulb temperature.

Solar radiation may also be regarded as an additional heat load imposed on the animal but no quantitative measurements appear to have been made on the effect of solar radiation on heat production of cattle. The effect of environmental warmth on the level of heat production, whether due to high air temperature alone or whether allied with solar radiation or with high air humidity, is quantitatively small compared to the effect of severe cold.

Exercise has a marked effect on the level of metabolism. Walking about whilst foraging for food, especially if walking uphill, the acts of getting up and lying down, even remaining in the standing position and, of course, doing draught work all raise the metabolism above the resting level. Despite the high metabolic cost of some of these activities which were extensively studied many years ago and also by Vercoe (1973), their overall additional net cost per day is not as great as might be expected because cows spend most of their time lying down. Blaxter (1962, p 110) calculates that the total energy expenditure of cattle grazing out-of-doors is only about 15% higher than that observed for animals confined in a calorimeter. Cattle tend to adapt their behaviour in such a way that the time pattern of their metabolism is best suited to diurnal climatic changes. Thus they seek shade from the sun during the hottest part of the day and do much of their grazing by night.

MEASUREMENT OF HEAT PRODUCTION

The rate of heat production may be measured by means of indirect calorimetry. The principal measured quantity is the rate of oxygen consumption and further accuracy can be obtained by additionally measuring the rates of production of respired carbon dioxide and methane and of urinary nitrogen. The respiratory quantities are measured by making the animal wear a ventilated face-mask or head-cage or by confining the entire animal in a ventilated chamber. In CLOSED CIRCUIT CALORIMETRY the ventilated system is sealed and gaseous products are absorbed and measured gravimetrically. In OPEN CIRCUIT CALORIMETRY air is washed through the ventilated system continuously and the volume and composition of air entering and air leaving is measured. Heat production (M) is calculated from equations based on known values of the quantities of oxygen consumed and of heat evolved during the combustion of carbohydrate, protein and fat (Blaxter, 1962, Ch. 3). For closed circuit calorimetry of ruminant animals the equation used is

$$M = (16.18 \times O_2) + (5.02 \times CO_2) - (2.17 \times CH_4) - (5.99 \times N) \text{ KJ}$$

where O_2, CO_2 and CH_4 represent volumes (dm^3 at s.t.p.) of oxygen consumed and of carbon dioxide and methane produced, and N is the quantity of urinary nitrogen excreted (g) (Brouwer, 1965). For open circuit calorimetry a high degree of accuracy may be obtained from the simplified equation

$$M = 0.2047\ V_0X$$

where V_0 is the volume (dm^3 at s.t.p.) of air exhausted from the ventilated system and X is the difference in percentage oxygen concentration between air entering and air leaving the system (McLean, 1972).

Since all heat produced must eventually be given off from the body by radiation, conduction, convection or by evaporation of water, measurement of these quantities by direct calorimetry provides another means of measuring heat production. This method was employed over 70 years ago for cattle but was cumbersome to use and gave way to indirect methods many of which have been described in detail at a symposium of the European Association of Energy Metabolism (EAAP, 1958).

In recent years the advantages of the gradient layer system have been applied to construct a fast acting direct calorimeter for adult cattle (McLean, 1971). Many of the calorimetric chambers can be controlled to operate at different temperatures and humidities so allowing their use in environmental studies.

DISCUSSION

The effects of warm environments on heat production may be considered in terms of energy balance. In thermoneutral or warm conditions the total amount of food digested by cattle supplies three energy requirements.

1. The metabolic cost of maintaining the physiological functions of the resting animal;
2. The metabolic cost of physical activity;
3. The metabolic cost of growth or milk production.

These three requirements are met in the order listed from the energy source available, and except for some of the third, all of the energy is eventually dissipated from the animal in the form of heat. When the animal eats less in a warm environment, growth or productivity is the first form of energy turnover to be reduced; this is apparent in the reduced heat production and live weight gains or milk production of animals exposed to sustained heat. Conversely the metabolic cost of maintaining physiological function tends to be increased in a hot environment due to the van 't Hoff-Arrhenius effect and to a lesser extent because of the additional energy expended in panting. Rapid shallow panting involving a 6-fold increase in respiratory rate requires little energy (Hales and Findlay, 1968); but in the laboured form of panting, associated with severe hyperthermia, the increase in O_2 consumption that can be attributed to panting has been estimated as 11% (Whittow and Findlay, 1968). The van 't Hoff-Arrhenius effect is usually expressed in terms of Q_{10}, the factor by which the speed of a chemical reaction increases for each 10° C increase in temperature. Metabolism being the sum of very many chemical processes would be expected to obey the law

$$M_2/M_1 = Q_{10}^{(T_2-T_1)/10}$$

where M_2 and M_1 are metabolic rates at temperatures T_2 and T_1° C. In passing it is pointed out that the alternative form of equation

$$Q_{10} = 1 + \{(S_2 - S_1)/\tfrac{1}{2}(S_2 + S_1)\} \times \{10/(T_2 - T_1)\}$$

given by Brody (1945, p 269), though still often used, is very inaccurate and should be avoided. The appropriate temperature to insert into the equation is that of the body temperature at which the chemical reactions take place. The deep body temperature of cattle frequently increases by more than 1° C as a normal response to environmental warmth and the temperature of the musculature is subject to very much wider variation. A Q_{10} of 2 (Macfarlane, 1964) would be enough to account for the increase in heat production observed during short term heat exposures as well as for the increase observed at 30° C air temperature in the forced-fed cattle of Kibler et al. (1970). The reason why these animals had a reduced metabolism at 35° C despite forced feeding however requires alternative explanation and may be connected with thyroid activity.

It is well known that thyroxine stimulates metabolism and that thyroid activity is reduced during sustained exposure to heat; this effect is not apparent for the first two days of heat exposure of cattle (Yousef et al., 1967). It has also been shown that thyroid activity in cattle is reduced by starvation (Blincoe and Brody, 1955). Loss of appetite due to heat exposure could therefore be an initiating factor which through depressed thyroid activity causes a reduction in heat production after prolonged, but not after short term, heat exposure. However, reduced food intake is not an essential part of this process because thyroid activity and heat production can be still depressed by sustained heat even if the animal's food intake is kept at the thermoneutral level by forced feeding (Yousef and Johnson, 1966). Adrenal cortex activity and the level of corticoids in the blood are also reduced during sustained heat exposure (Christison and Johnson, 1972) and appear to be related to heat production (Yousef and Johnson,

1967). It would, therefore, seem that a number of factors including nutritional status, body temperature and endocrine activity are all involved in the effects of heat exposure on metabolic rate and the relative importance of these various factors is still incompletely understood. A detailed treatment of the effects of heat exposure on endocrine activity is the subject of another Section (Chapter 1, Section 2a (1)).

REFERENCES: see pp. 538-539.

Effects of Heat on Energy Metabolism of Sheep

by

K. L. Knox

CLIMATE, AREA AND CONDITIONS OF SUMMER SHEEP PRODUCTION

Sheep production is concentrated in the Western one-half of the United States. This area (desert, mountain valleys, high plateaus, prairie and mountains) is characterized by diverse vegetation and hot, dry summers. The vegetation fluctuates in palatability, availability and nutrient content in accordance with local environmental conditions (Wallace, 1969). Sheep are selective; for the most part they eat the more digestible feeds and those having the higher protein content during the summer (Wallace and Denham, 1970). Range vegetation may vary drastically from spring through late summer, e.g., coefficients of protein digestion of 69% in June and 34% in September have been noted (Wallace, 1969). Daytime temperatures frequently exceed 35° C. However, relative humidity is in general low (e.g., less than 60%) as is the rainfall, which averages less than 10 cm/month in much of the sheep producing regions during June through September. Nights are cool, frequently 10–15° C lower than daytime peak temperatures, and clear, making night-radiation possible. Heat stress does occur during the summer under practical conditions of production.

Mature ewes enter the summer in the shorn state and lambs have 1–3 months wool growth as the hottest part of the summer approaches. Water is not abundant as most sheep are on range during the summer. Under such conditions sheep travel 1 to 2 miles/day, some of which is frequently at a gradient (Cook, 1970). Heat production necessitated by activity can add a significant thermal load, particularly when environmental temperatures approach or exceed the thermoneutral zone, e.g., the daily energy expenditure associated with activity (travel in search of feed) has been estimated to be in excess of 200 kcal/day (Blaxter, 1962; Cook, 1970).

PHYSIOLOGICAL RESPONSE TO HEAT

Although all sheep respond physiologically to heat stress in a similar manner, different breeds are more suited to high temperature environments than are others. Sheep breeders have long recognized that specific breeds are more tolerant to extreme climatic conditions and through constructive breeding practices they have introduced these characteristics into their flocks in an effort to increase productivity. For example, Merino sheep have been found to be particularly suited to regions where high environmental temperatures challenge the ability of sheep to regulate body temperatures and metabolic processes (Blaxter, 1962).

In order to cope with an environmental temperature near or above the thermoneutral zone, sheep may make several physiological adjustments:

1. Reduction in voluntary feed consumption (Knox and Soderquist, 1969; Kotb and Pfander, 1964; Bhattacharya and Hussain, 1974).
2. Heat storage may occur (Brockway, 1965; Soderquist, 1967).
3. Sweating from both poorly insulated and well insulated areas of the body (Hofmeyr et al., 1969).
4. Increase frequency of respiration (Bligh, 1959).
5. Reduction in metabolic rate (Graham et al., 1959; Blaxter et al., 1959).

Physiological reactions either reduce heat production or increase its dissipation.

Partitioning of heat loss is difficult to accomplish and can be done experimentally only under rigorous conditions, including confinement. Whether one can extrapolate experimental findings under such rigid conditions to animals under practical production conditions has not been tested. However, data from studies using beef cattle in practical feed-lot conditions, suggest that there is sufficient similarity to laboratory results to be of value in estimating the environmental effect on energy utilization (Knox and Handley, 1973).

Although heat stress has been studied in sheep, considerably more data is available on the effects of cold stress on energy metabolism in sheep than is available on similar effects of heat stress. Animals forced to acclimate to high environmental temperatures, such as those existing in the sheep producing areas of the U.S. during the summer, can either reduce the rate of heat production or increase the rate of heat loss. Sensible heat loss (conduction, convection and radiation) is a function of the animal's surface area, conductance of the insulating layers and the difference in temperature between the animal and the environment as noted by Kleiber (1961):

$$\frac{dQ}{dt} = k_L A\ (Ts - Te) \text{ where:}$$

dQ/dt = rate of heat loss
A = surface area, which is proportional to the 2/3 power of the body weight
k_L = constant for heat loss, which is a function of resistance to heat flow
Ts = surface temperature
Te = temperature of the environment

The sensible heat loss through fleece obviously decreases with an increased thickness, which does vary with time after sheering, as the fleece continues to grow. At environmental temperatures near or above the thermoneutral zone, it follows that sensible heat loss would be zero (or heat could accumulate) when Ts = Te, and the animal would gain heat when Te is greater than Ts. This would lead to a rise in body temperature. Although heat storage may occur for short periods of time, this is not likely to be a significant factor in the response of sheep to high environmental temperatures (Brockway, 1965). What occurs, in order to prevent such an event, is that the rate of vaporization increases. Evaporative heat losses are particularly effective under practical conditions of low humidity and high temperatures. In the sheep, respiratory and skin evaporation may be in excess of 100 kcal/m²/hr. Vasomotor control of blood circulation to poorly insulated areas such as the face and legs may also improve evaporative losses. Under such conditions sweating can be an effective vehicle for heat elimination (Hofmeyr et al., 1969).

At environmental temperatures where Te exceeds Ts and where evaporative heat loss has reached a maximum, the body temperature must increase; metabolic rate thus increases in accordance with the Van't Hoff effect. The ultimate effect is death if control is not instituted.

Two other factors which can add to the heat production are activity and food intake. Sheep reduce muscular activity during periods of high environmental temperatures. Movement is reduced during the hottest part of summer days when environmental temperatures are 30° C and higher. Under experimental conditions sheep also reduce the voluntary food intake, thus reducing heat produced from the calorigenic effect.

ENERGY METABOLISM DURING HEAT STRESS

Food energy may be retained or eliminated from the body. Energy may be lost via the feces, urine or emitted as heat, and it may be retained as body substance or stored. The latter would result in a rise in body temperature.

FECAL LOSSES

At high environmental temperatures fecal energy losses tend to decrease (Graham et al., 1959; Kotb and Pfander, 1964; Bhattacharya and Hussain, 1974). This effect may be due to a decreased transit time for food in the digestive tract; a possible result from a reduced food intake.

URINARY LOSSES

In sheep exposed to environmental temperatures above 30° C, the urinary energy losses are inconsistant. Graham et al. (1959) exposed closely clipped sheep, fed at 3 separate levels, to environmental temperatures from 8° C, to 38° C at 5° C intervals. The urine energy content varied with the level of feeding, but at each feed level, it decreased with increasing environmental temperature. On the other hand Knox and Soderquist (1969) reported that urinary energy losses of sheep fed at 31–36° C were significantly greater than for control sheep fed at 20–25° C. This variation may be due to breed differences in the experimental animals (Blaxter, 1962).

METHANE LOSSES

Energy losses via methane are not drastically effected by increasing environmental temperatures (Graham et al., 1959) or they may increase (Knox and Soderquist, 1969). Graham et al. (1959) suggested that methane production reached a maximum at temperatures of 23–28° C. Both of the above reports noted that methane production was higher than noted in other studies. Increased methane production, which is a function of the activity of specific rumen microbes, may result from a higher deep body temperature, or increased water intake (Blaxter et al., 1959). Such changes in the environment of the digestive tract might effect the bacteria flora.

HEAT PRODUCTION

Heat production of sheep exposed to a wide range in environmental temperature fluctuates in a predictable manner as illustrated in Figure 1. The rate of decline in heat production, as the environmental temperature increases from below the critical temperature to that of the thermoneutral zone is a function of the insulation capacity (Kleiber, 1961). Heat production in the thermoneutral zone is dependent upon the quality and quantity of the diet; the heat production reflecting differing calorigenicity of the particular diet. As the environmental temperature continues to increase above the thermoneutral zone the heat product continues to decrease, but at a lower rate, and then increase. This increase in heat production occurs once the animal is unable physiologically, to dissipate the heat and death from heat exhaustion will occur if not relieved. This upper critical temperature has not been extensively studied but is probably around 37–40° C.

As the animal metabolically responds to a changing environmental temperature, Graham et al. (1959) have reported that the source of energy also changes. Under

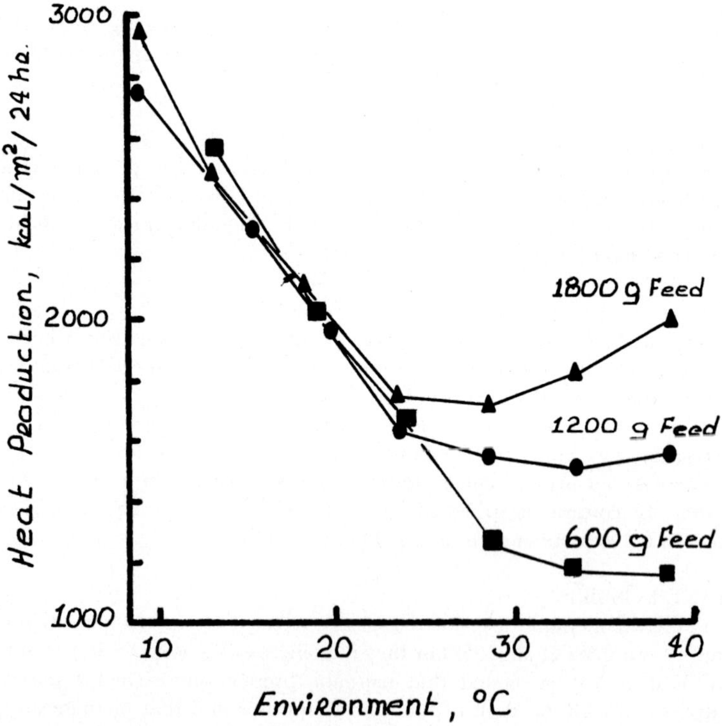

Fig. 1. The heat production of a closely clipped sheep at three feeding levels in relation to environmental temperatures. (From Graham et al., 1959: J. agric. Sci., 52:13).

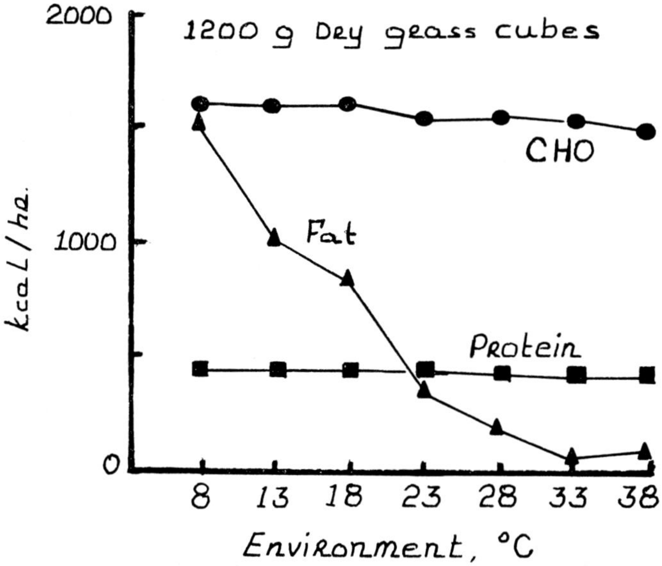

Fig. 2. The mean number of calories arising from the metabolism of carbohydrate, fat and protein by two sheep in relation to environmental temperature. (From Graham et al., 1959: J. agric. Sci., 52:13).

conditions of cold stress, sheep mobilize and oxidize fat as a primary source of energy. As the environmental temperature approaches the thermoneutral zone the utilization of fat as a source of fuel is practically diminished and the animal utilized increasing amounts of carbohydrate and protein as a source of fuel (Fig. 2).

ENERGY RETENTION

The consequence of the metabolic events which occur relative to energy metabolism and environmental temperatures above the thermoneutral zone is that the animal goes into a negative energy balance (Graham et al., 1959; Knox and Soderquist, 1969). At high environmental temperatures the dietary energy intake generally is not sufficient to allow for energy retention.

In Figure 3, it can be seen that energy retention as related to metabolizable energy intake was not altered by increasing the environmental temperature. In other words, the efficiency of utilization of energy for productive processes was not grossly affected by the environmental temperature. What is, in fact affected, is energy (feed) intake. Although some changes in the partitioning of energy occur relative to changes in the environmental temperature, they tend to cancel each other, such that the efficiency appears to remain approximately constant. It is, therefore important to recognize that one of the major effects of high environmental temperatures, is its effect on intake. If intake is indeed a major determinant in energy retention, it is of practical concern to understand this aspect relative to temperature stress.

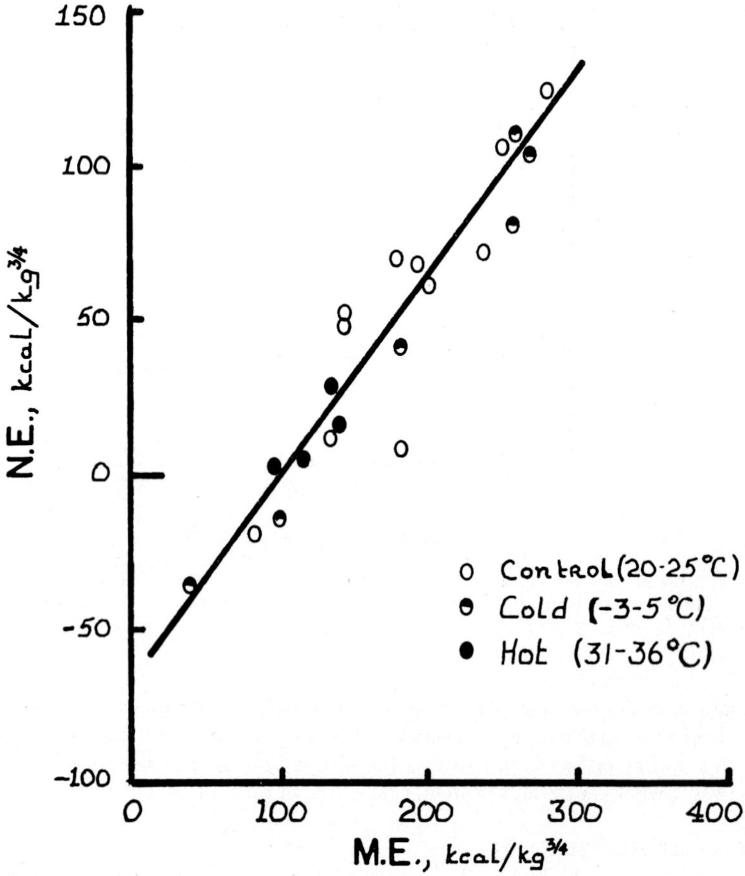

Fig. 3. Linear relationship of net energy retention on metabolizable energy intake.

Factors affecting food intake in sheep and other animals have been studied but all of the component controls and their interrelationships are not well understood. Hormones certainly play a role in controlling food intake (Bray, 1974). Levels of metabolites and environmental temperatures are also factors influencing food intake but, the precise mechanisms of their action have not been delineated to date. The thyroid and the hypothalamus have been the two organs most studied regarding regulation ot food intake. Each has shown to be involved in the process but the interrelationships, the control mechanism(s) etc., have not been described to date. There is indeed, a paucity of information on the control of food intake at high environmental temperatures.

REFERENCES: see pp. 539-540.

Effects of Heat on Energy Metabolism and Related Endocrine Functions in Rodents

by

M. K. Yousef

INTRODUCTION

For simplicity, the thermoregulatory processes in rodents are phrased in a control system terminology i.e., sensor, controller and controlled elements. The elements involved in sensing include: peripheral and internal temperature receptors. The elements composing the other two subsystems (for the purpose of this presentation) are summarized in Figure 1. For many years, enormous effort has been directed toward the understanding of the right side of this model. Many of these elements are identified and quantified since such data are essential for the study of the other half of the model i.e., physiological thermoregulatory responses and associated endocrine adjustments.

Studies on effects of heat on homeostatic adjustments in rodents are limited as compared to studies concerned with the effects of cold. Moreover, data on responses of wild rodents to heat are scarce. Recently, Chaffee and Roberts (1971) and Hart (1971) have published comprehensive reviews on thermoregulation in rodents in general, whereas, Bradley and Yousef (1972) have surveyed the literature on physiological and

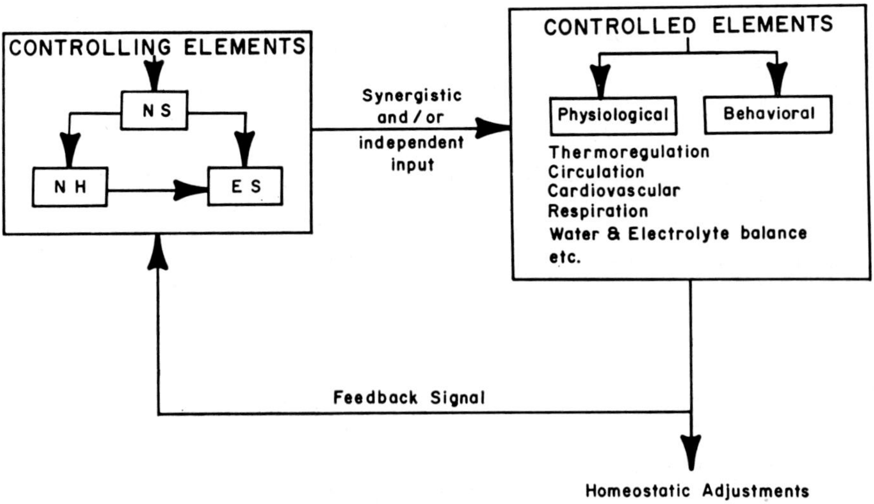

Fig. 1. A control system of the thermoregulatory processes in rodents with emphasis on the controller and controlled elements. (NS = nervous system; NH = neurohumoral factors; ES = endocrine system).

behavioral adaptations of desert rodents. Earlier, excellent reviews should be consulted for more background (Schmidt-Nielsen, 1964; Hudson and Bartholomew, 1964; Bartholomew and Dawson, 1968). Why are rodents used extensively in research on thermoregulation? Experimentation on thermoregulation and its physiological and biochemical mechanisms has been concentrated on rodents because of their size, economy, and convenience of obtaining and/or raising these animals in the laboratory. Some studies on rodents have been made with the hope that the findings can be applied to man and other mammalian species. The question arises whether or not such a hope is true. Chaffee and Roberts (1971) have emphasized that not all of the adaptive responses, such as biochemical changes, seen in temperature-acclimated rodents can be extrapolated to other species, such as primates. Therefore, generalizations derived exclusively from studies on rodents should be made with care and caution.

In general, this Section gives emphasis on data related to changes in heat production, heat loss and associated hormonal adjustments of rodents during exposure to heat. Recent findings on wild rodents and thermoregulation during exercise are given special attention. Because of restricted space, references are made to recent reviews, symposia and some major research papers.

TEMPERATURE REGULATION IN HEAT

In this section, the basic principles of thermoregulation in heat are presented. Homeothermy, in general, is maintained as a result of a sensitive balance between heat production and heat loss.

A. HEAT PRODUCTION

Measurements of basal metabolic rate (BMR) in rodents are not practical due to the difficulty in applying the strict requirements of fasting (postabsorptive state) and complete rest (minimal activity). Available data on heat production of rodents are usually termed "resting" and/or "standard" metabolic rate.

In general, measurements of resting metabolic rate are made indirectly using the volume of oxygen consumed per unit of time as an index. Measurements of the rate of oxygen consumption, $\dot{V}_{O_2}$, are made using closed or open indirect calorimetry methods. The principles of these techniques are available to the reader (Brody, 1945). Resting $\dot{V}_{O_2}$ of rodents as well as other mammals is an exponential function of body weight. This relationship is expressed by the following equation derived originally by Brody (1945).

$$\dot{V}_{O_2}, \text{ml/g}\cdot\text{hr} = 3.8\ \text{Wg}^{-0.27}, \text{ where Wg is body wt in grams.}$$

However, there are a number of exceptions to this generalization. Acclimation and/or acclimatization to heat and cold results in modification of $\dot{V}_{O_2}$. Additionally, there is a number of species of desert rodents which have a lower $\dot{V}_{O_2}$ than the expected values calculated from the equation. The available evidence is convincing that there is metabolic adaptation in rodents. This lowered $\dot{V}_{O_2}$ found in desert rodents is a useful

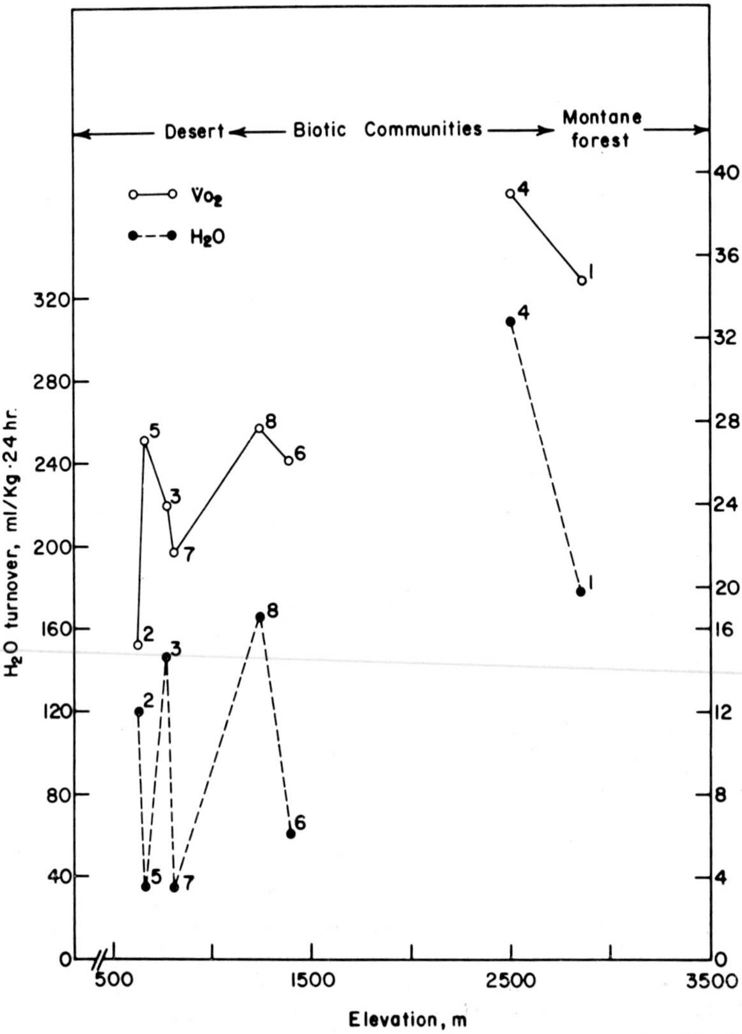

Fig. 2. Water turnover rate in ml/kg 24 hrs and rate of oxygen consumption, V_{O_2}, in l/kg 24 hrs in 8 species of desert and mountain rodents (From Yousef et al., 1974).

adaptive mechanism since it minimizes the need for food and water which are at a premium in the desert. Recent work on desert rodents in our laboratory has pointed to a positive correlation between water turnover rate and $\dot{V}_{O_2}$ as shown in Figure 2 (Yousef et al., 1974). Also, it has been shown that some desert rodents are capable of tailoring their energy budget with availability of food to a degree that these rodents deviate from strict homeothermy under food deprivation (Yousef and Dill, 1971[a]). Labile body temperature in some desert rodents is an effective mechanism of adaptation to heat. This mechanism is employed more effectively by estivators.

What are the mechanisms controlling heat production? Available data on rodents suggest that the neuroendocrine system seems to be an important controlling factor of V_{O_2}. The role of the endocrine system in controlling V_{O_2} in the thermoneutral zone was well demonstrated by the work of Evans and his workers (1957). Their data proved that the thyroid hormones, adrenal cortical hormones, and growth hormone are calorigenic. Recently, Cassuto and Amit (1968) have shown that thyroxine and norepinephrine are calorigenic whether acting independently or synergistically in normothermic and heat acclimated hamsters. Further experimental work is needed to elucidate the role of the pituitary trophic hormones and the hypothalmic releasing factors in controlling heat production.

B. HEAT LOSS

Heat loss avenues in rodents are the same as in other mammals: sensible (conduction, convection, and radiation) and evaporation. Very little information is available on partition of sensible heat loss in rodents (Hart, 1971). Evaporative heat loss includes pulmocutaneous water loss and saliva. The contribution of the skin and respiratory tract to evaporative heat loss differs among species. In some laboratory and wild rodents, cutaneous and respiratory water loss are about equal (Shkolnik and Brout, 1974). However, in the kangaroo rat *Dipodomys merriami*, the evaporative water loss by the skin was found to be insignificant (Schmidt-Nielsen, 1972). In general, some desert rodents seem to have lower evaporative water loss than non-desert species.

Salivation has been considered as an emergency mechanism for temperature regulation in rodents. However, the importance of saliva as a major and effective avenue for evaporative heat loss in rodents has been well documented by Hainsworth and Stricker (1970). They found that the salivary glands do not contribute equally to evaporative cooling: the submaxillary-sublingual glands could continue to evaporate saliva, whereas rats with only parotid glands could not. What are the mechanisms controlling heat loss in rodents? Knowledge of these mechanisms is incomplete. Few studies have demonstrated that evaporative water loss is related to changes in peripheral blood flow i.e., the tail (Thorington, 1966). Vasodilation of the peripheral blood vessels i.e., the tail, is under neurocontrol. Also Hainsworth and Stricker (1970) have proven that salivation of rodents is under neurocontrol. This explains at least in part the control of cutaneous and salivary water loss. What then is controlling respiratory water loss? If respiratory frequency increases to account for the increased respiratory water loss, this would have been a logical explanation. Rodents do not pant, but recent studies have revealed that some rodents significantly increase their respiratory frequency (Hudson and Deavers, 1973). Hart (1971) suggested that the increase in respiratory water loss at high temperatures without salivation is probably due to an increase in tidal volume rather than in respiratory frequency. Further studies are needed to test this possibility and to discover the controlling mechanisms of evaporative water loss.

ACCLIMATION TO HEAT

The Glossary Committee of the International Union of Physiological Sciences has defined acclimation as "a physiological change, occurring within the lifetime of an

organism, which reduces the strain caused by experimentally induced stressful changes in particular climatic factors" (Bligh and Johnson, 1973).
In this section, emphasis is given to the effects of short-term and acclimation to environmental heat on heat production and endocrine activity.

A. HEAT PRODUCTION

Short-term exposure to heat results in an increase of rate of oxygen consumption ($\dot{V}_{O_2}$). This finding is seen in laboratory and wild rodents in studies dealing with determination of the thermoneutral zone. The increase in $\dot{V}_{O_2}$ was suggested to be related to: hypersecretion of the adrenal cortex, relatively high thyroid activity and increase in body temperature as would be expected from the Van 't Hoff-Arrhenius law (Yousef and Johnson, 1967).

As the duration of exposure increases and the animals become acclimated, $\dot{V}_{O_2}$ decreases although hyperthermia persists. Exposure of young rats to 34° C for 205 days caused a decrease in $\dot{V}_{O_2}$ expressed in ml/hr throughout the experimental period. However, when $\dot{V}_{O_2}$ was related to body weight i.e., ml/g-hr, the decrease in $\dot{V}_{O_2}$ was not evident after the first 50–100 days although rectal temperature was more than 1.0° C higher in the heat acclimated group (Kibler and Johnson, 1967).

B. ENDOCRINE ACTIVITY

Most of the available data are related to thyroid and adrenal cortical hormones and little has been reported on growth hormone and antidiuretic hormone. Endocrinological aspects of exposure to heat have been reviewed by Collins and Weiner (1968).

1. *Thyroid Activity*

Little is known about thyroid function of rodents after short-term exposure to heat. Data on large mammals have shown that short-term exposure to heat caused no change in thyroid function. Long-term exposure to heat or acclimation causes a decrease in thyroid activity. However, exposing rats to heat (34° C) continuously for over two years resulted in higher thyroidal release rate after the first two months of exposure (Johnson et al., 1966). A further study on long-term (220 days) exposure to heat has revealed that thyroxine secretion rate, TSR, was decreased during the first 200 days of exposure and then TSR was similar to that of the control group. Although TSR of heat exposed rats was the same as that of their controls kept at 28° C, the heat exposed rats utilized less thyroxine than their control (Yousef and Johnson, 1968[a]). Also, long-term exposure of rats to 34° C caused a shift in the relative amounts of iodine compounds in the plasma: the amount of thyroxine decreases, while the concentration of free I^- and monoiodotyrosine increases (Yousef and Johnson, 1968[b]). The decreased thyroid activity in heat-acclimated rodents was thought to be related, at least in part, to the voluntary feed restriction induced by heat (Yousef and Johnson, 1968[a]). Reduced excretion of thyroid hormones in heat-acclimated rats was thought to be due to a reduced biliary clearance of circulating hormones and not to the observed reduction in bile flow (Chaffee and Roberts, 1971).

2. *Adrenal Cortical Activity*

Short-term exposure to heat leads to increased plasma glucocorticoids, while prolonged exposure lowers plasma levels and decreases urinary excretion of their metabolites suggesting a decrease in secretory activity of the adrenal cortex (Collins and Weiner, 1968). The time course of changes in plasma levels of glucocorticoids post exposure to heat was studied by Kotby and Johnson (1967). They found that the highest levels occurred after 24 hrs of exposure to 34° C and this level remained high for about 12 days and then returned to normal or below normal between 13–15 days following exposure. It should be emphasized here that the time course of changes in plasma corticosterone would depend upon the degree of heat stress, age of the animals and the previous state of acclimation and/or acclimatization.

3. *Other Hormonal Activity*

Few studies have been reported on the effects of heat on growth hormone releasing factor (GHRF) and antidiuretic hormone (ADH). Exposure to heat (35.5° C) for 1, 4 and 96 hrs increased GHRF activity which was returned to normal at 240 hrs post exposure (Parkhie and Johnson, 1973). These authors have shown that not only GHRF activity increases in response to short-term exposure to heat but also GHRF synthesis continue during these periods.

Studies on ADH have revealed that desert rodents, specifically Merriam's kangaroo rats, excrete a high concentration of ADH in urine and the concentration of ADH in blood is significantly high (Ames and Van Dyke, 1950). An increase in environmental temperature results in a marked increase in plasma ADH of rats but it was concluded that the rise in plasma ADH levels is in response to dehydration caused by heat and not from the response of receptors to heat as such (Robinson and Macfarlane, 1956).

ACCLIMATIZATION TO HEAT

Acclimatization is defined (Bligh and Johnson, 1973) as "a physiological change occurring within the lifetime of an organism which reduces the strain caused by stressful changes in the natural climate (e.g., seasonal or geographical)." Studies on seasonal acclimatization in rodents have been recently reviewed by Hart (1971).

A. HEAT PRODUCTION

Some species have higher heat production in winter than in summer and other species show the opposite pattern. This difference may be related to several factors including: ecologic distribution, changes in physical and physiological insulation and types of behavioral adaptations displayed by the various species.

Little attention has been given to responses of winter and summer acclimatized rodents to heat. Winter kangaroo rats, when exposed to 34° C, had a higher $\dot{V}_{O_2}$ and body temperature and lost weight at a faster rate as compared to summer animals (Yousef and Dill, 1971[b]). The difference in $\dot{V}_{O_2}$ between summer and winter kangaroo rats is not associated with a reduction in the relative mass of heat-producing organs (liver, kidney, heart... etc.) since relative organ weights were similar in both groups. This finding is contrary to data reported on heat-acclimated laboratory rodents (Chaffee and Roberts, 1971).

B. ENDOCRINE ACTIVITY

Seasonal changes in the endocrine system of rodents were well revealed by Heroux (1960). In general thyroid function is decreased during the winter and increased during the summer. This finding is contrary to data reported in heat and cold-acclimated rodents. Adrenal cortical activity was shown to be greater in winter than in summer (Heroux, 1960). The enhanced adrenal cortical activity during winter acclimatization may be related to the daily changes in environmental temperature. However, in a study of two desert rodents, it was found that adrenal cortical activity was higher in summer than in winter (Haggag and El-Husseini, 1974).
Seasonal changes in ADH activity of two desert rodents have been reported by El-Husseini and Haggag (1974): summer values were much higher than winter values.

THERMOREGULATION DURING EXERCISE IN HEAT

Most observations on thermoregulatory responses during exposure to heat have been made on resting animals. This experimental approach is an oversimplification of what wild rodents do in their natural habitats. Wild rodents are exposed to the environmental heat load in addition to an endogenous heat load resulting from the activity required for food collection, reproduction and other functions essential for the maintenance and survival of the population. Although several studies have dealt with thermoregulation during exercise under cold conditions, very little attention has been given to this aspect at high environmental temperatures. In general, $\dot{V}_{O_2}$ increases by exercise linearly as the speed of running increases (Fig. 3). The energetic cost of running in rodents is higher than that of large mammals.

In two species of desert rodents running on a motor-driven treadmill, Yousef et al. (1970, 1973) reported that the actual energetic cost of running (net $\dot{V}_{O_2} = \dot{V}_{O_2}$ running $- \dot{V}_{O_2}$ resting) was the same in heat as in the thermoneutral zone but was higher at low temperature (Fig. 3). Rectal temperature after running increased in heat and in the thermoneutral zone and showed no change in cold. Given a constant load of work the increase in rectal temperature at the thermoneutral zone was greater than that at high temperature. This suggests that activity under heat stress seems to facilitate a high level of heat loss thus preventing an explosive rise in body temperature. The ability to maintain a regulated degree of hyperthermia under heat stress is of significance since these ground squirrels are diurnal and inhabit low deserts.

Does the increase in $\dot{V}_{O_2}$ during running influence evaporation? Does a running rodent at high environmental temperature maintain a positive water balance and if so, for how long? These questions have not yet been answered. When kangaroo rats were running at moderate air temperature, the ratio of evaporative water loss to $\dot{V}_{O_2}$ did not change significantly with running, suggesting that the animals remain in water balance. However, with increasing air temperature the ratio of evaporative water loss to $\dot{V}_{O_2}$ increases and the water balance becomes increasingly precarious (Schmidt-Nielsen, 1972).

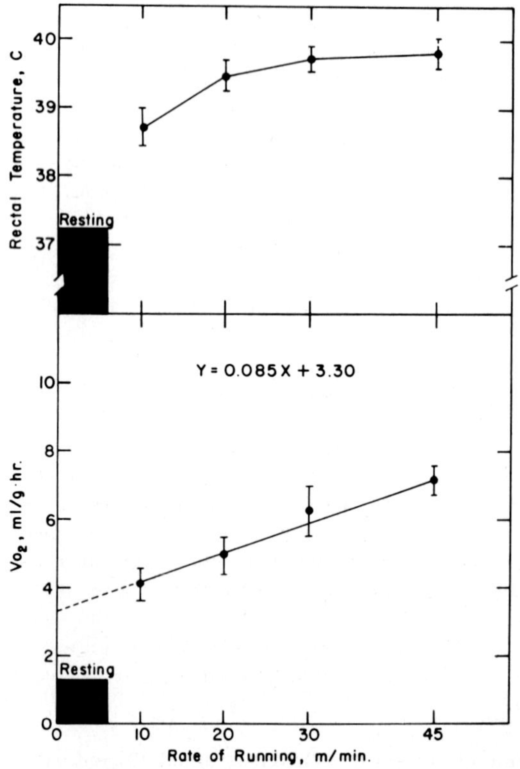

Fig. 3. The right side of the figure shows the effects of rate of running in m/min on rate of oxygen consumption, V_{O_2}, and rectal temperature. The left side shows the effects of environmental temperature on resting and running (30 m/min) $\dot{V}_{O_2}$ and rectal temperature. (From Yousef et al., 1973).

SUMMARY AND CONCLUSIONS

Physiological responses of rodents to heat are much better understood than the controlling mechanisms of their various responses. Heat production of rodents similar to that of other mammals is an exponential function of body weight, however, this concept is modified by acclimation or acclimatization to heat and geographic distribution. Many deserts rodents have a lower rate of oxygen consumption, $\dot{V}_{O_2}$, than expected from the $\dot{V}_{O_2}$-body weight established equation (mouse to elephant curve; Brody, 1945). The controlling mechanisms of heat production are not completely understood, but it is known that the nervous and endocrine systems play an important role. Heat loss avenues in rodents are also similar to other mammals, where they include: sensible and evaporative avenues. The partition of the sensible avenues requires investigation. The avenues of evaporative water loss seem to follow two main patterns depending on the species: (a) respiratory and cutaneous water loss seems to be equal in some species

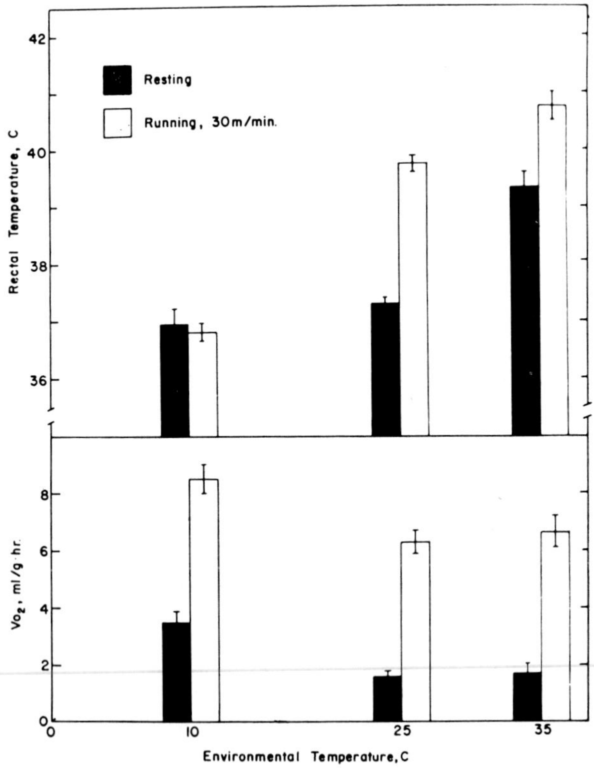

and (b) respiratory is much greater than cutaneous water loss in other species. Spreading saliva is a normal and effective avenue for evaporative water loss in some rodents.

The mechanisms controlling heat loss in rodents are partially understood to be: peripheral vasomotor and neural factors.
Acclimation to heat reduces $\dot{V}_{O_2}$, thyroid function and adrenal cortical activity. Acclimatization to heat (summer), causes a decrease in $\dot{V}_{O_2}$ in some species and an increase in others. Acclimatization to summer is shown to increase thyroid activity and has little effect on adrenal cortical function. Summer acclimatized rodents seem to tolerate heat better than winter acclimatized rats.

Exercise i.e., activity, increases $\dot{V}_{O_2}$ linearly with the speed of running with a different slope for the various rodents studied to date. Rodents in general are less economical in doing a given work load as compared to large mammals. The energy cost of running during heat stress is about the same as in the thermoneutral zone. Mechanisms for controlling heat loss in exercising rodents at high air temperature require further studies.

REFERENCES: see pp. 540-541.

Effects of Heat on Energy Metabolism in Birds

by

H. S. Siegel

INTRODUCTION

The large numbers and the diversity of species that live in hot climates are indicative of the adaptability of birds to high temperature environments, moreover, because most birds are diurnal and generally do not utilize underground burrows for shelter, their chances to avoid the midday heat are further reduced.

In ambient temperatures below the upper limit of thermoneutrality, the bird loses heat to the environment by physical means (radiation, conduction, convection, and evaporation). But as temperatures approach the upper critical limit, sensible heat loss becomes negligible and insensible heat loss (i.e., evaporative) becomes the only method for heat dissipation. However, in most species, with the notable exception of the Caprimulgids (Lasiewski and Bartholomew, 1966; Dawson and Hudson, 1970), it is rare when the upper limit of thermoneutrality is reached that more than one-half the metabolic heat can be dissipated by evaporative means (Benedict and Lee, 1937; Dawson, 1954; Dawson and Tordoff, 1959; Kayser, 1930). It is evident that this excess heat must be stored in the body; this is reflected in elevations in body temperature of 2 to 4° C (Dawson, 1954; Dawson, 1958; Bartholomew and Cade, 1957; Bartholomew and Dawson, 1958).

BODY TEMPERATURES IN HOT ENVIRONMENTS AND THE VAN'T HOFF-EFFECT

When body temperature increases, the rate of heat production increases (Whittow, 1965). Initially this is the result of increased cardiac activity and the increased muscular work associated with panting by which the bird facilitates its rate of evaporative cooling. However, as body temperature rises, metabolic intensity rises exponentially according to the van't Hoff equation. This equation predicts that the velocity of a chemical reaction will increase two or three times for each increase of 10° C. Thus a 1° C rise in body temperature causes a 10 to 20% increase in basal metabolism. This relationship is often expressed numerically as the Q_{10} or the heat production at a given temperature divided into the heat production at a temperature 10° C higher. Heart rates and other physiological activities of birds seem to be related to body temperature in a similar fashion (Whittow et al., 1964).

Many species of birds are able to tolerate a temporary hyperthermia of 2° to 4° C (Bartholomew and Dawson, 1954, 1958; Baldwin and Kendiegh, 1932; Yeates et al.,

1941; Hutchinson, 1954; Von Saalfeld, 1936). Initially this temporary hyperthermia aids heat dissipation because of the increased thermal gradient between the bird and the environment and the increased vapor pressure at the evaporative surfaces of the respiratory tract. It is thus of adaptive advantage in extending the range of heat tolerance, but its usefulness is limited by the specific tolerance of a species to high core temperatures.

A model of how temporary hyperthermia may be used as an adaptive device in heat stress and its limitations based on data from Ortalan buntings (*Emberiza hortulana*) has been elegantly presented by King and Farner (1965). The interaction of changes in heat loss, heat production, and heat content of the body during hyperthermia expressed as multiples of resting metabolic rate are shown in Figure 1. The model assumes that at each level of activity, hyperthermia potentiates heat production in accordance with the van't Hoff equation (where $Q_{10} = 2$). In this model, hyperthermia would be an advantage at levels of metabolism up to about three times the resting rate. Above this level, the increase in body temperature produced by the metabolic effort would augment heat production faster than heat could be lost and would thus be detrimental to thermoregulation.

Despite the fact that many birds tolerate temporary hyperthermia and can use this for thermoregulation, there is little evidence that given species inhabiting areas of extreme heat have higher resting body temperatures than the same species living in milder climates (Dawson and Hudson, 1970). Some species, such as ostriches (*Struthio camelus*) (Crawford and Schmidt-Nielsen, 1967) or the Caprimulgids (Lasiewski and Bartholomew, 1966) possess high capacities for temperature regulation in hot climates, however in most birds, body temperature rises during heat stress (Wilson, 1948; Calder, 1964; Lasiewski et al., 1966; Calder and Schmidt-Nielsen, 1967). As shown above, a rise of 2 to 4° C may be tolerated, but body temperatures of 46 to 47° C can be maintained for only short periods before death ensues (Calder, 1964).

The exact body temperature reached by birds subjected to a heat load may be influenced by several physical factors e.g., ambient temperature or air movement. However, the body temperatures of small passerines often closely approach the ambient, whereas the body temperature of non-passerines of similar size do not increase as rapidly under the same physical conditions. Lasiewski et al. (1966) suggest that these small passerines have a relatively higher weight-specific level of standard metabolism than non-passerines of comparable weight as indicated by the coefficients in the following equations for heat production (Lasiewski and Dawson, 1967):

$$\text{passerines } M = 129\ W^{0.72}$$
$$\text{non-passerines } M = 78\ W^{0.72}$$

where: M = kcal/24 hrs and W = wt. in kg. Thus temporary hyperthermia should occur more rapidly in passerines which may be of adaptive significance in view of their somewhat reduced ability to dissipate heat by evaporative means (King and Farner, 1964; Dawson and Hudson, 1970).

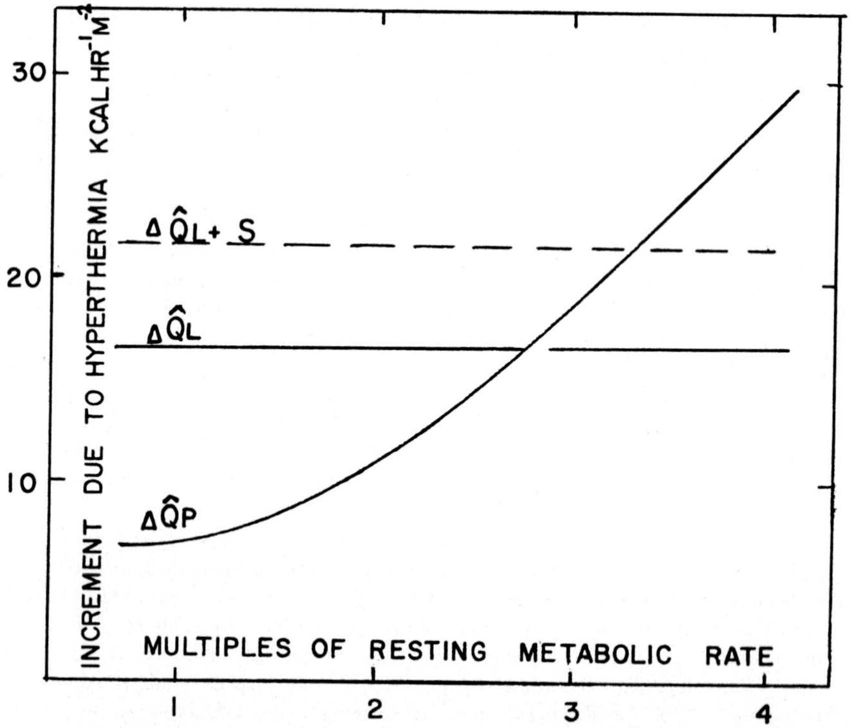

Fig. 1. Effect of hyperthermia on heat production ($\hat{Q}_p$), heat loss ($\hat{Q}_2$) and heat storage (S) in the ortolan bunting, *Emberiza hortulana*. (From King and Farner, 1964).

ACCLIMATIZATION BY REDUCTION IN METABOLIC RATE

Experimentally continued exposure in the laboratory to warm ambient temperature 29° to 32.5° C produces lower resting metabolic rates in several passerine species (Gelineo, 1964). In the canary (*Serimus canaria*) and siskin (*Chrysomitris spinus*), previously acclimatized to 6° C, heat production at 30.6° C decreased about 40% (Gelineo, 1964). Similar results have been reported for domestic fowl (*Gallus domesticus*) (Hoffman and Shaffner, 1950, Huston et al., 1962; Hutchinson and Sykes, 1953; Berman and Snapir, 1965). Reduction in thyroid activity is indirectly indicative of lower metabolic intensity after heat exposure in the fowl (Reineke and Turner, 1945; Hoffman and Shaffner, 1950; Huston and Carmon, 1962).

Theoretically it would seem that a lower resting metabolic rate would be of adaptive value in hot environments because the heat load would be reduced. However, King and Farner (1965) feel that there is little evidence for acclimative changes in resting metabolic rates in natural environments.

First, the evidence indicates that the resting metabolic rate is basically determined by body size (Brody and Proctor, 1932; King and Farner, 1961) as described by the equation: $\text{Log } Q_r = \log 74.3 + 0.744 \log W \pm 0.074$, where: Q_r resting metabolic rate in kcal/bird/day; W. body wt. in kg.

Second, in species where resting rates are lower than those predicted by the equation when the birds are in high temperature environments, the adaptation appears to be largely genetic. Examples of genetic adaptation in lowered metabolic rate are found in the tropical species, *Fregata magnificous* and *Cathartes aura* (Enger, 1957) and among the goatsuckers such as the tropical nighthawk (*Nyctidromus albicollis*) (Scholander et al., 1950).

The apparent difference between laboratory experiments and natural environments may lie in the difference in the degree of diurnal fluctuation. Resting metabolic rates of yellow buntings (*Emberiza citrinella*) did not vary seasonally when the birds were kept in natural temperature environments, but metabolic resting rates of both yellow buntings and ortolan buntings did decline as the birds became acclimatized to constant temperatures of 27.5° or 32.5° C (Wallgren, 1954). This decline did not occur when individuals were placed in 10 to 14° C environments for 8 hrs of the day. Interestingly, suppression of thyroidal function by feeding thiouracil reduced the amplitude of the diurnal body temperature cycles in domestic chickens (Washburn et al., 1962). Thus metabolic acclimatization to high temperature may be delayed by periods of low temperature by modifying thyroidal stimulation of metabolism.

CONTROL OF CHEMICAL THERMOREGULATION DURING HIGH TEMPERATURE

Although neural-humeral control of the physical means of heat dissipation in birds has been described by the classical work of Rogers (1923), Hiestand and Randall (1942), and Randall (1943a, b), relatively little is known of the mechanism of control of metabolic function in response to hyperthermia. The avian hypothalamus has been found to be thermosensitive (Rogers, 1928) but most of the evidence that it influences the metabolic responses to high temperature is indirect.

For example, reduced thyroid activity is related to reduced metabolic intensity during periods of high environmental temperature (Reineke and Turner, 1945); thus it may be of significance that the rate of production of thyroid stimulating hormone by the pituitary is controlled by the hypothalamus (Egge and Chaisson, 1963). Furthermore, the fact that appetite and water intake are also controlled by the hypothalamus (Feldman et al., 1957; Lepkovsky et al., 1968) suggests that if the hypothalamus is sensitive to thermal stimulation, changes in metabolic rate should occur.

Both mammalian and avian hypothalami contain relatively high concentrations of the biogenic monoamines, epinephrine (E), norepinephrine (NE) and 5-hydroxytrpytamine (5-HT) (Juorio and Vogt, 1967; Vogt, 1954; Amin et al., 1954) and this has been thought to be related to temperature regulation (Euler, 1961; Feldberg and Myers, 1963). NE caused hypothermia in young and adult chickens when microinfusions were made into the third ventricle of the brain (Marley and Stephenson, 1969). By contrast, 5-HT given to chicks that were pretreated with a monoamine oxidase inhibitor caused an increase in body temperature (Allen and Marley, 1967).

However, the hypothalamus apparently is not the only thermoregulator that influences metabolic rate. Mills and Heath (1970) found localized heating of the preoptic area of house sparrows (*Passer domesticus*) caused decreased metabolic rates as measured by oxygen consumption. Localized thermostimulation of the spinal cord has also been shown to elicit thermoregulatory responses in birds (Rautenburg, 1969).

METABOLIC EFFECTS OF EXERCISE

Using muscle energy alone, some birds can perform flights of extraordinary length. Golden Plovers (*Gambianus pacificus*) fly over water for 3900 km without opportunities to feed (Johnston and McFarlane, 1967) and Blackpoll Warblers (*Dendroica striata*) travel 1300 km over water without stopping to feed (Nisbet et al., 1963). Ducks have been known to maintain speeds up to 20 m/sec for many hours (Tucker, 1971).

Such expenditures of energy produce hyperthermia even when high environmental temperatures are not present. Farner (1956) found that the body temperatures of Fairy Prions (*Pachyptila turtur*) rose an average of 2.6° C when flight was completed, and Hart and Roy (1967) reported that body temperatures of rock doves stabilized between 1.5 and 2.0° C above resting temperatures during flight. In wind tunnel experiments with temperatures of 36° to 37° C, budgerigars (*Melopsittacus undulatus*) tend to become overheated with 20 min of level flight (Tucker, 1968). Pigeons in level flight with ambient temperatures of 25° to 29° C were unable to fly more than 10 min at 55 km/hr before becoming overheated (Aulie, 1971).

Original estimates of the increase in metabolic rate required by flight were made on 300 to 350 gm rock doves by Zeuthen (1942). Using indirect estimates based on 60% of the heat loss being by evaporative means, he concluded that these doves should have metabolic rates nearly 27 times the resting rate when flying at 70 km/hr.

More recently however, direct measurements of O_2 consumption by hovering hummingbirds (*Calypte costae*) indicated that these 3.2 gm birds used an average of 42.4 ml O_2 (gm hr)$^{-1}$ which is about 7 times the standard rate at the ambient temperature of measurement 24° C (Lasiewski, 1963). Studies on pigeons in flight (Pearson, 1964; Lefebvre, 1964; Aulie, 1971) also indicate that the estimates by Zeuthen (1942) were probably too high. Apparently Zeuthen's estimate that 50 to 60% of the heat of rock doves in flight would be evaporative was excessive. The data of Hart and Roy (1966) indicates that about 18% of the heat is lost by evaporation; the rest apparently is lost by convection into the air stream and by radiation (Eliassen, 1963; Hart and Roy, 1967).

The problems of birds in flight during periods of high environmental temperature have been studied in budgerigars (Tucker, 1968). Metabolic rates in level flight at 35 km/hr were maintained at about 3.7 kcal/hr at temperatures from 20° and 37° C. From 20° to 30° C nearly 85% of heat production was sensible (convective and radiative) with the remainder non-sensible (evaporative). However, at 37° C non-evaporative cooling only accounted for 40% of heat production; evaporative cooling made up another 47% and the remainder was stored. The thermal conductance of these birds flying at 37° C reached 5 cal (gm hr °C)$^{-1}$, about five times the maximum observed in resting birds. Similar data have been reported for flying rock doves (Hart and Roy, 1967).

As shown above, it is at high temperature that evaporative cooling becomes important for heat dissipation during flight. This would be important when considering distances covered during migration. Tucker (1968) has calculated from data collected on budgerigars that the birds would have a net water loss of 11 mg (gm hr)$^{-1}$ which would correspond to 1.1% of the body weight/hr. If the bird could tolerate a water loss equivalent to 15% of body weight, it could fly 14 hrs at 35 km/hr for a distance of 450 km. However, some birds fly as long as 50 hrs without drinking (Nisbet, Drury and Baird, 1963). Fuıther studies are required to determine why some species do not become dehydrated during migration.

The metabolic requirements for long distance flight have been reviewed by Tucker (1971). The primary energy source is necessarily fat because of its high energy content per unit mass (Weis-Fogh, 1952). This is confirmed by studies that show that up to 50% of total body weight is fat prior to migration (King and Farner, 1965; George and Berger, 1966; Drummond, 1967) and that carcass fat drops to as low as 5% after a long migration flight (King and Farner, 1965). O_2-CO_2 measurements during flight indicate RQ's between 0.7 and 0.8, an indication of primarily fat metabolism (Tucker, 1970).

The distance traveled through the air can be calculated as a function of initial mass and the percentage of the initial mass lost during flight. Fat is the fuel (Tucker, 1971).

$$X = 100\% \frac{(M_o - M)}{M_o} 7.46 \times 10^4 M_o^{0.227}$$

where: X = distance through the air in km
M_0 = initial mass
M = final mass.

The actual flight distance varies with the direction and velocity of the wind; however, hummingbirds with a mass of 5×10^{-3} fly over 800 km (Lasiewski, 1962) and Golden Plovers with a mass of 0.2 kg fly 3900 km in a single flight (Johnston and McFarlane, 1967). This would indicate that these birds must lose nearly 50% of the body mass during the flight, and one must assume that such flight performance must fall just at the border of the margin of safety.

REFERENCES: see pp. 541-544.

Effects of Cold on Energy Metabolism of Cattle

by

G. E. Thompson

THE RATE OF HEAT PRODUCTION IN THE COLD

The amount of heat produced by cattle can be estimated using the Brouwer equation:

$$H = (16.18 \times O_2) + (5.02 \times CO_2) - (2.17 \times CH_4) - (5.99 \times N)$$

where H = heat production (kJ), O_2, CO_2, and CH_4 are respectively volumes of O_2 consumed and of CO_2 and CH_4 produced (dm^3) and N is the quantity of urinary N excreted (g) (Brouwer, 1965). Measuring O_2 consumption alone, with an open circuit, gives a close approximation to heat production if the animal is fed normally, when

$$H = 0.2046\ V_0X$$

where H = heat production (kJ), V_O is the volume of outlet air from the open circuit (dm^3) and X is the difference in O_2 concentration between inlet and outlet air (%) (McLean, 1972).

Heat production normally equals heat loss over a period of time but temporary imbalances result in a rise or fall of average body temperature. The changes in heat production and heat loss due to change in environmental temperature are diagrammatically represented in Figure 1. Over a limited range of environmental temperatures, the thermoneutral zone, heat production is minimal and is independent of environmental temperature. The lowest temperature in the thermoneutral zone is called the "critical" temperature. At lower temperatures than critical, the animal increases its heat production to equal the basal evaporative heat loss plus the sensible heat loss. The sensible heat loss is proportional to the difference in temperature between the animal and its surroundings and so is zero at an environmental temperature equal to body temperature. The slope of the line representing sensible heat loss depends on the thermal insulation of the body which is maximal in a cold environment and minimal in a hot one. As the environment becomes colder, extra heat is produced to offset the extra sensible heat loss, until the maximum heat production, or summit metabolism, is inadequate. At this extreme the body temperature falls.

The amount of heat produced by the animal at thermoneutrality increases with increase in body weight, and for comparative purposes heat production is usually expressed per unit surface area (m^2) of the animal, or alternatively per unit body weight (kg) raised to the 3/4 power (Kleiber, 1965). The basal heat production of cattle, starved

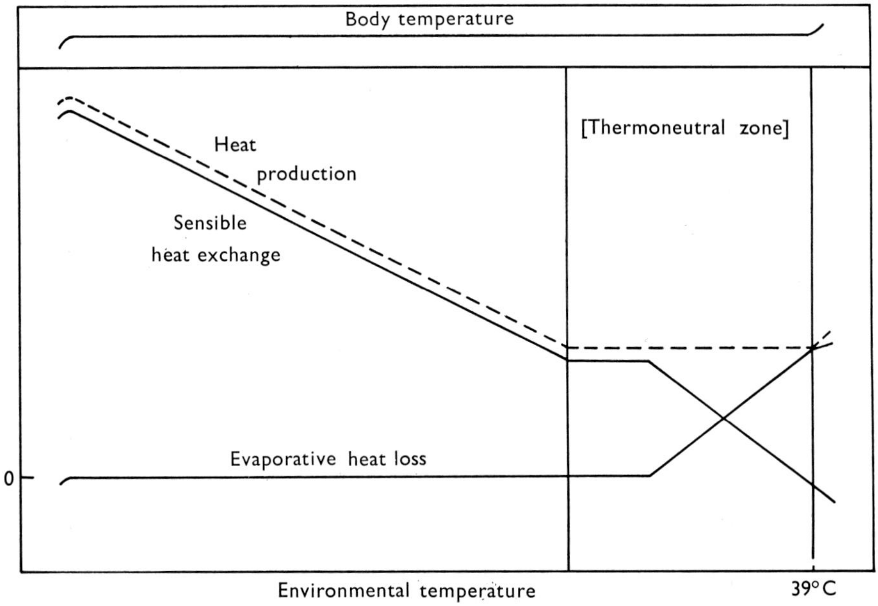

Fig. 1. Diagram showing heat production and heat loss at different environmental temperatures.

and at thermoneutrality, is about 70 W/m². The resting metabolism of the animal fed a diet to maintain body weight, is slightly less than 100 W/m². These values have been published by a number of people. The amount of heat produced at thermoneutrality increases with the level of feeding and productivity of the animal, and can be as high as 170 W/m². The resting heat production of the young calf fed 4 dm³ milk daily is about 100 W/m² (Gonzalez-Jimenez and Blaxter, 1962).

It can be seen from Figure 1 that an increase in the heat production of the animal at thermoneutrality will reduce the critical temperature. The effect of this factor, and others which influence critical temperature, has been incorporated by Blaxter (1967) into the formula

$$T_c = T_B - \frac{H_P - H_W - H_E}{A_B K_C}$$

where T_c = critical temperature (°C), T_B = body temperature (°C), H_P = heat production in a thermoneutral environment (W), H_W = average heat of warming food (W), H_E = basal evaporative heat loss (W), A_B = surface area of the body (m²) and K_C = cooling constant. H_E/A_B is about 16.9 W/m² in the adult (Blaxter and Wainman, 1961) and the young calf (Gonzalez-Jimenez and Blaxter, 1962). It can be calculated that the critical temperature of fasted, maintenance-fed and high-yielding cattle is about 18, 7 and −1° C respectively. The critical temperature of the young calf given 4 dm³ of milk daily is 13° C (Gonzelez-Jimenez and Blaxter, 1962). The heat required to warm food or water to body temperature can be significant in the total heat production of young calves fed cold milk (Holmes, 1971a), but this factor has been omitted in a simplified equation for estimating the critical temperature of dry and lactating cows (Hamada, 1971).

Below the critical environmental temperature, cattle produce approximately 2.4 W/m^2 more heat for every 1° C fall in environmental temperature (Blaxter and Wainman, 1961) and new-born calves approximately 3.0 $W/m^2 . °C$ (Gonzalez-Jimenez and Blaxter, 1962). This equals the extra sensible heat loss when the insulation of the animal's skin and coat is maximal. At a given cold environmental temperature, sensible heat loss is further increased by wind but basal evaporative moisture loss is not increased (Blaxter and Wainman, 1964). Rain, in a cold environment, presumably increases the evaporative moisture loss above basal, but while it is raining and the air is saturated with water, rain probably cools the animal by reducing its external insulation. Evaporative heat loss is also high during the first hour of life when the calf is covered with amniotic fluid (Thompson and Clough, 1970) and this is probably an important cause of the mild hypothermia which is found in calves during the first 3 days of life (Markovič, 1963; Szekely, 1967).

The effect of a long period of exposure to cold on the resting heat production of cattle has been studied by Young (1972), who found higher values in beef cows exposed to a Canadian winter than control animals kept indoors, although earlier experiments on heifers in Canada (Webster et al., 1970) and cold-reared calves in Russia (Rogal, 1955) failed to show this effect.

The maximum ability of cattle to produce heat in the cold, i.e., their summit metabolism, is not known.

THE SOURCE OF THE EXTRA HEAT PRODUCED IN THE COLD

The increased heat production of cattle in the cold must use up extra substrate for oxidation. In the long term, cows fed ad libitum increase their food intake when exposed to cold, but in the short term, this extra heat production uses up energy stores in the body. Blaxter and Wainman (1961) studied the effect of exposure to cold on the metabolism of 2 steers fed a diet sufficient for maintenance in thermoneutral temperatures, a submaintenance ration, or fasted for 4 days. The animals were exposed to a given temperature for 4 days, during which time O_2 consumption, CO_2 and CH_4 production, and urinary N were measured. They concluded that the metabolism of carbohydrate and protein is independent of environmental temperature and the increase in heat production in the cold is due to an increase in the catabolism of fat.

The changes in blood lipids that occur during exposure to cold support the idea that fat is a fuel for heat production. In new-born calves there is a progressive increase of plasma free-fatty-acid concentration in the first 2 hrs after birth if they are kept at 25° C, and an even bigger increase if they are kept at 5° C. Similarly, the steer has a higher plasma free-fatty-acid concentration after 2 hrs at −20° C than at 15° C.

Short periods of exposure to cold have no effect on plasma triglycerides, cholesteryl esters or phospholipids in the new-born or the adult (Thompson and Clough, 1972). The circulating free fatty acids probably originate from triglycerides in adipose tissue and perhaps other tissues. At birth, before suckling takes place, the adipose tissue of the calf contains triglyceride stores, and immediately after parturition and before suckling the calf can increase its heat production in response to a cold environment (Thompson and Clough, 1970). The high levels of free fatty acids in plasma persist during 3 days of exposure to cold (−19 to −26° C) (Olsen and Trenkle, 1973) and through the winter months in Canada (Dietz and Young, 1972). Cows kept in an unheated stable in Romania, where ambient temperatures varied from 15.9 to −6.6° C, had lower serum cholesterol concentrations when the weather was cold (Lungu et al., 1972).

The effect of exposure to cold on plasma glucose concentration has also been studied. Khan et al. (1970) exposed new-born calves to cold for 72 hrs, but found no significant effect on plasma glucose concentration. Olsen and Trenkle (1973) found an increase of plasma glucose in non-lactating cows exposed to cold for 3 days, and Lister et al. (1973) observed the same effect in pregnant beef cows wintered out-of-doors with only a wind-break shelter in Canada. Dietz and Young (1972) also measured the plasma glucose concentration of pregnant beef cows during winter in Canada and noted an increase only during the first severe cold spell. Subsequent lower levels were, however, high enough to prevent any increase in blood ketone concentration.

A fall in plasma creatinine concentration has been observed in pregnant beef cows kept out-of-doors in the winter and spring in Canada. These animals had a greater intake of dry matter but a smaller body weight gain than control animals kept in a barn (Lister et al., 1973).

It is not known precisely how many organs of the body produce the extra heat needed during cold exposure, but there is some evidence on this subject. The new-born of several animal species, including lambs, have brown adipose tissue which is capable of producing heat in response to cold stimulation: calves also have brown adipose tissue (Alexander et al., 1975). Calves shiver in the first few days of life (Gonzalez-Jimenez and Blaxter, 1962) and in older animals shivering in the muscles of the hind leg is especially obvious. Bell et al., (1974) have measured the blood flow and oxygen consumption in the hind leg of young steers in thermoneutral and cold conditions. When the total oxygen consumption of the animals was doubled, due to cold exposure, leg blood flow increased to about 4 times, and leg oxygen consumption increased to about 10 times, the thermoneutral values. It is hoped to measure the conversion of [^{14}C]-labeled free fatty acids to $^{14}CO_2$ in this preparation and find how much their metabolism contributes to heat production in the cold. A greater fraction of the cardiac output passes through the hind leg, during exposure to cold, because its resistance to flow is reduced (Bell and Thompson, 1974). Consequently, these tissues are presented with more of the products of digestion, and the uptake of volatile fatty acids by the leg is increased during exposure to cold. However, assuming their complete oxidation, this would account for less than 30% of the increased oxygen consumption of the leg in the cold and some of these circulating volatile fatty acids probably originate from incomplete oxidation of free fatty acids (Bell et al., 1974). Kelley et al. (1967) found no difference in the concentration of volatile fatty acids in the rumen fluid of non-lactating cows in a neutral environment (18° C) and a mildly cold environment (1.6° C) with the same food intake. The direct effect of coolness, if it reached the rumen, would be to reduce the rate of fermentation.

THE CONTROL OF HEAT PRODUCTION DURING EXPOSURE TO COLD

It is assumed that in cattle, as in other animals, the skin is sensitive to cold and that receptors there signal a fall in environmental temperature to the animal's brain, since the thermoregulatory response to cold sometimes correlates better with skin temperature than with central temperature. For example, when the calf is exposed to a cold environment shivering occurs while skin temperature is reduced and rectal temperature is increased (Gonzalez-Jimenez and Blaxter, 1962). The hypothalamus of cattle is also sensitive to cold, and hypothalamic cooling produces an immediate increase in plasma free-fatty-acid concentration, a fall in respiratory quotient and an increase in the animal's heat production which is followed by an increase in body temperature (Calvert et al., 1972). Cooling the spinal cord of cattle has very little effect (McLean et al., 1970; Jessen et al., 1972).

In the brain, the input nerves from the cold-sensitive areas of the body synapse with output nerves to the various organs of the body, which influence heat production. The neurotransmitter substance noradrenaline, injected into the lateral ventricle of the brain of the ox, has no effect unless the animal is in a cold environment when it stops the animal shivering, inhibits heat production (Findlay and Thompson, 1968) and prevents fat mobilization (Thompson and Clough, 1971). Noradrenaline may, therefore, be an inhibitory neurotransmitter at this synapse.

The output nerves from the brain which transmit the signals that promote shivering in the muscle are probably the somatic nerves, as in other animals. The contraction of muscle during shivering presumably results in adenosine triphosphate utilization and adenosine diphosphate production, and the supply of adenosine diphosphate to the tissue will stimulate O_2 consumption and substrate utilization by the tissue.

The output nerves from the brain which transmit the signals that mobilize free fatty acids into the circulation are the sympathetic nerves, and in steers blockage of these nerves with bethanidine prevents any increase in plasma free-fatty-acid concentration during acute cold exposure. Denervating the adrenal medullae does not alter the normal response (Thompson and Clough, 1972). In addition to catecholamine-containing sympathetic nerves, there are also a large number of catecholamine-containing mast cells in the white adipose tissue of the steer. The main catecholamine in the nerves is probably noradrenaline and in the mast cells dopamine, and both of these substances when infused intravenously into steers produce an increase in the plasma concentration of free fatty acids (Clough et al., 1971). Alvarez and Johnson (1970) compared the urinary excretion of catecholamines by non-lactating cows in a neutral environment (18° C) and at 2° C, and found an increase in urinary noradrenaline, but not adrenaline, at 2° C which indicates a release of catecholamines by sympathetic nerves rather than the adrenal medullae.

To sum up, when an animal is exposed to cold there is an increased demand for substrate by shivering muscle, which is probably stimulated through the somatic nerves. This coincides with an increased availability of free-fatty-acid substrate in the circulation, which is stimulated through the sympathetic nervous system.

It is not known whether cattle continue to shiver when exposure to cold is prolonged for weeks or months; shivering continues for at least 3 days (Olsen and Trenkle, 1973), but it is possible that different mechanisms for producing heat may eventually substitute for shivering. The possibility that an increased activity of the thyroid gland is at least partly responsible for the increased heat production of cattle in the cold, is being investigated. It has been confirmed that injections of thyroxine increase the heat production of lactating cows in an environment of 18° C, after a latent period of 3-4 days (Yousef and Johnson, 1966a). These injections increase the [^{131}I]-thyroxine disappearance rate, and also increase the concentration of acetate in the plasma (Yousef and Johnson, 1965a). Studies on the effect of cold exposure on thyroid secretory activity have been complicated by the fact that an increase in food intake increases thyroid activity. Fasted cows exposed to a temperature of 1° C have a higher plasma protein-bound iodine concentration and faster [^{131}I]-thyroxine disappearance rate, than when they are similarly fasted at 18° C (Johnson and Yousef, 1966). The same difference between environments of 1 and 18° C is found if the animals are fed at 1° C the amount of food eaten ad libitum at 18° C. Ad libitum feeding at 1° C results in a further increase in thyroid activity (Yousef and Johnson, 1966b). The effect of an environment at 1° C, on thyroid activity, takes about 36 hrs to develop (Yousef and Johnson, 1965b). These experiments show that, in cattle, exposure to 1° C results in an increase in thyroid activity, but this was a very mild exposure to cold and did not always stimulate an increase in heat production (Yousef et al., 1967). Dietz and Young (1972) measured the plasma protein-bound iodine concentration of pregnant beef cows over winter in Canada, and found an increase during the first 4 weeks of exposure to cold which levelled off thereafter.

Khan et al. (1970) could find no significant effect of 72 hrs of exposure to cold (−4° C) on the plasma concentration of cortisol and corticosterone in new-born calves. Olsen and Trenkle (1973) exposed non-lactating cows to cold (−19 to −26° C) for 3 days, and their plasma growth hormone concentration increased over the 3 days with the highest value occurring on day 3. The high levels persisted after the animals were returned to the control (23° C) temperature. An injection of growth hormone (200-300 mg/cow, i.v.) increases the heat production of cows and simultaneously increases the [^{131}I]-thyroxine disappearance rate. It is suggested that growth hormone and thyroxine have a synergistic effect on metabolic rate, although growth hormone has a shorter latent time before having its effect (Yousef and Johnson, 1966c).

These long-acting, hormonal, mechanisms for increasing heat production in cattle which have been exposed to cold for long periods would increase their resting heat production at thermoneutrality (see above).

NUTRITION AND EXPOSURE TO COLD

The increased heat loss in the cold is normally associated with an increase in heat production and an increased intake of substrate. Ragsdale et al. (1949) found that if lactating cows were gradually or rapidly taken from 10 to −15.5° C they increased their consumption of hay and total digestible nutrients. MacDonald and Bell (1958) confirmed this and found that the stimulation carried over for about 24 hrs after the exposure to cold had ended. It has been suggested that the sensation of cold is a direct stimulant of appetite (Bhattacharya and Warner, 1968) although muddy pen conditions can seriously reduce the voluntary food intake of steers (Bond et al., 1970). In very cold conditions, the increased voluntary food intake does not necessarily match the increased energy loss and ad libitum fed animals exposed to Canadian winters still have a reduced growth rate (Webster et al., 1970; Hidiroglou and Lessard, 1971; Milligan and Christison, 1973).

It has proved difficult to recommend a standard winter energy intake for cattle in North America because of the large geographical variations in climate. The U.S. National Research Council (1963) standard for maintenance or slight gain of body weight for 454 kg (1000 lb) wintering, mature, pregnant beef cows was 75.3 MJ digestible energy/day (16.6 MJ/100 kg/day). However, pregnant cows wintered in Eastern Canada with access to barn shelter tended to become fat on this standard, and a maintenance requirement of 14.2 MJ digestible energy/100 kg . day was found (Jordan et al., 1968b). In another feeding trial, beginning with lighter animals which were wintered without shelter in range conditions in Alberta, it was found that energy intake levels below the National Research Council (1963) recommended level were inadequate for periods of high winds and low temperatures (Hironaka and Peters, 1969). Young and Berg (1970), also in Alberta, measured the energy requirement for maintenance of pregnant beef cows weighing about 470 kg, during cold weather, and found values averaging 69.5 MJ metabolizable energy/day. Lister et al. (1972) estimated the energy requirement for maintenance of 3- to 10-yr old pregnant beef cows during the winter in Ontario with mean mid-winter daily temperatures approaching −20° C, and obtained a value of 17.2 MJ digestible energy/100 kg . day. To predict the energy required to maintain body weight of cattle in a given set of meteorological conditions, Webster (1971) has used a "wind-chill index".

To make farming more profitable in areas with a severe winter, it is economically important to minimise expensive winter feeding and make the best use of cheap summer pastures. It has been suggested that pregnant beef cows can be safely fed to lose body weight in the winter and then regain it in the summer (Jordan et al., 1968a) without adversely affecting calf survival, birth weight, or daily weight gain up to the time of weaning (Jordan et al., 1968b). Measurements of blood ketone concentration in thin and fat pregnant cows which were respectively fed to gain and lose weight over winter has shown no ketonaemia due to either treatment (Dietz and Young, 1972). When rearing beef calves, underfeeding is not advisable because bull calves fed a low plane of nutrition during the winter do not compensate fully in the following summer and their weight gain at the age of 18 months is less than that of calves wintered on a high nutritional plane (Lawson, 1967).

To produce cold-tolerant cattle, private breeders have crossed domestic cattle with American bison and produced hybrid cows. These were tested at the Range Experimental Station, Manyberries, Alberta, in the 1950's, and found to have outstanding ability to forage in winter range conditions (Smoliak and Peters, 1958) and wean heavy calves (Peters and Slen, 1966), but perinatal problems in these crosses were not solved.

The blood vitamin A and blood carotene levels of calves may be lowered during cold weather, suggesting that there may be a greater carotene requirement in cold weather (Keener et al., 1942). Measurements of the concentration of vitamin A and carotene in the liver of beef cattle wintered outside in Canada show a similar trend (Hidiroglou and Lessard, 1971). The effect could result from an increased requirement for vitamin A, a reduced content in the hay, or both. Supplementing the diet of pregnant cows with vitamin A during the winter in Canada had no effect on their body weight (Jordan et al., 1968a) and they produced calves with a poorer survival rate than the calves of control dams not given this supplement (Jordan et al., 1968b).

Fisher et al. (1972) found an increase in the concentration of copper in the plasma of cows kept outside during the winter in Ontario, whereas Hidiroglou et al. (1970), studying similar animals, found no effect.

LACTATION AND EXPOSURE TO COLD

The effect of exposure to cold on lactation was studied by Ragsdale et al. (1949) who found that a drop in the environmental temperature of lactating cows, from 10 to −15.5° C, resulted in a fall in milk yield and an increase in fat percentage. There was a greater effect if the change in temperature was rapid instead of gradual, and a greater effect in small Jersey cows than in big Holstein cows. The increased intake of total digestible nutrients and the reduced milk yield at −15.5° C meant that there was a reduced efficiency of milk production. A more detailed examination of the composition of the milk from cows exposed to −15° C showed an increase in butter fat percentage and increases in the yield of milk total solids, solids-not-fat and total N. The ratio of butter fat to solids-not-fat was low (Cobble and Herman, 1951). Broadly speaking there are 3 different levels at which exposure to cold could have its depressing effect on lactation. First, it could result from the fact that there is a greater energy requirement of the whole body because of its increased heat production. Second, there may be a reduced supply of nutrients to the mammary gland by the circulation because in species other than bovines it has been shown that the blood flow of the mammary gland is reduced by lowering its temperature. Third, there may also be a direct effect of cooling the udder on milk synthesis and secretion. Holmes (1971b) clipped the hair off one half of the udder of lactating cows and covered the other half. The air temperature was 6 to 10° C and air movement over the clipped udder was artificially raised to 3 to 5 m/s. After 22 hrs of cooling, the milk yield of the exposed half of the udder was lower than that of the covered half and fat, protein and lactose yields were reduced. Further studies are required to explain the deeper mechanisms involved in the various processes described in this section.

REFERENCES: see pp. 544-546.

Effects of Cold on Energy Metabolism of Sheep

by

A. J. F. Webster

INTRODUCTION

In the temperate and cold regions of the earth the domestic sheep has been bred and selected over the centuries primarily for its fleece which has provided an excellent form of thermal insulation for man. This rather unnatural form of selection has produced an animal which is very resistant indeed to cold stress. There are, however, circumstances in which the sheep may be vulnerable to cold. The cold resistance of the sheep is principally dependent on the thermal insulation provided by the fleece and can be markedly reduced by wind and rain or snow. Once or twice a year the fleece is shorn mechanically, or more recently by treatment with cytotoxic drugs. The former approach leaves the sheep about 1 cm of fleece; the latter leaves it bald. The lamb that is born saturated with body fluids in severe winter conditions is in a perilous situation immediately. There is a fourth and only slightly cynical point, which is that to a casual observer a very thin sheep in full fleece looks much like a fat sheep. Ewes can therefore become severely stressed by the combined effects of cold and undernutrition (see Chapter 1, Section 6) without the owner being aware of it.

DEFINITIONS AND EQUATIONS OF HEAT EXCHANGE

UNITS. Biologists who use calories as a unit of energy have become increasingly isolated from the main stream of the physical and chemical sciences and from the teaching of science in the schools and universities. It has been proposed that the standard unit of energy in the nutritional sciences should be the joule (J). Requirements of animals for food energy are therefore best expressed in kJ or MJ/24h, but for biometeorological purposed rate of energy expenditure is probably best expressed in watts (J/s). Heat losses are presented per unit of surface area (m^2) which is itself taken as a function of body weight (kg) ($0.09\ W^{0.67}$).

THERMONEUTRAL ZONE is the thermal zone in which metabolic rate is independent of air temperature and homeothermy is maintained by fine adjustments to evaporative and sensible heat loss. Metabolic rate at thermoneutrality is given by M^* (W/m^2).

THERMAL INSULATION describes the resistance to heat loss conferred first by the skin and superficial tissues of the body shell, TISSUE INSULATION, I_t, and secondly by the coat of wool and the air trapped therein and thereon, EXTERNAL INSULATION, I_e.

If sheep are exposed to cold and measurements are made of metabolic rate $M(W/m^2\cdot$ of body surface area) and deep body temperature T_b, air temperature T_a, and the mean temperature of the skin surface T_s, then

$$I_t = (T_b - \bar{T}_s)/M \quad (1)$$

This is the equation that has been used in most descriptions of the tissue insulation of sheep (Blaxter, 1967; Burton and Edholm, 1955). It makes the pedantically imprecise but practically convenient assumption that all the heat produced in metabolism, including that proportion lost ultimately by evaporation from the respiratory tract, is conveyed to the skin surface by the sensible means of convection and conduction. To render equation 1 absolutely precise one would need to know the mean temperature and area of the evaporating surface of the respiratory tract.

Heat is lost from the surface of the skin and respiratory tract by sensible means (convection, conduction, radiation) and by evaporation. In cold conditions total evaporative heat loss from sheep is low and relatively constant at about 15 W/m^2 of surface area (Joyce and Blaxter, 1964). Thus

$$I_e = (\bar{T}_s - T_a)/(M - 15) \quad (2)$$

The units of both I_t and I_e are $°C\cdot m^2\cdot W^{-1}$.

LOWER CRITICAL TEMPERATURE, T_{ac} describes the lower limit of the zone of thermal neutrality i.e., that air temperature below which an animal must elevate metabolic rate in order to maintain homeothermy. T_{ac} may be derived by solving equations 1 and 2 when $M = M^*$, thermoneutral metabolic rate.

$$T_{ac} = (T_b + 15I_e) - M^*(I_t + I_e) \quad (3)$$

Out of doors, the effects of wind and precipitation are to increase M at any stated $(\bar{T}_s - T_a)$. They are therefore properly described as factors influencing I_e. Heat losses by radiation are not related to $(\bar{T}_s - T_a)$ but to the fourth power of the difference between the radiant surface of the animal and the effective radiant temperature of the outdoor environment, an expression that almost defies measurement.

In fact, indoors when wall temperature is the same as air temperature, or outdoors when the sky is completely overcast, I_e and thus heat loss in the cold are adequately described by equation 2 (Burton and Edholm, 1955). When there is direct or indirect radiation heat gain from the sun or radiant heat loss to cloudless night skies the effective radiant temperature of the environment may differ markedly from T_a. If net radiation exchanges $R(W/m^2)$ between the animal surface and the environment can be measured, sensible heat loss to the outdoor environment can be described by

$$(M - 15) = (T_b - T_a)/I_t + I_e - \beta R \quad (4)$$

MEASUREMENT OF ENERGY EXCHANGES

DIRECT CALORIMETRY. Direct measurement of heat losses (H_1) from sheep can be made using heat sink or thermal gradient calorimeters. In the heat sink calor meter heat loss from an animal is measured as a rise in temperature in an absorbing medium which may be the air stream ventilating the chamber or water circulating outside its walls. Evaporative heat loss (H_e) can be measured by the increase in relative humidity of the ventilating air stream so long as there is no condensation of water vapour within the animal chamber. Sensible heat loss (H_n) is obtained from $H_1 - H_e$. Heat sink calorimeters are robust, relatively simple to construct and reliable (Mount, 1968), although their response to a sudden change in the pattern of heat loss from an animal is rather slow.

The thermal gradient calorimeter measures the temperature difference across a conducting layer interposed between an animal and a constant temperature source. H_n is obtained rapidly and precisely from the e.m.f. created as the result of the temperature difference across a thermopile between the inner walls of the animal chamber and the outer walls surrounded by a constant temperature water jacket. H_e can be measured with equal rapidity and precision by including gradient layer thermopiles into the air conditioning system for the chamber (Pullar, 1969). The "gradient layer" calorimeter is undoubtedly the most precise and elegant tool for measuring heat losses from animals. It is, however, both expensive and laborious to construct. Moreover it can only be used to study the effects of constant air temperatures and existing models do not operate effectively at air temperatures below about +3° C.

INDIRECT CALORIMETRY. The heat production of an animal metabolizing a mixture of carbohydrates and fats can be estimated from measurements made of its oxygen consumption and carbon dioxide production. Not all substrates are completely oxidized to CO_2 and H_2O, however. Incomplete combustion of proteins leads to the production of urea and other organic nitrogenous compounds. In the sheep, moreover, anaerobic fermentation of carbohydrates in the rumen and large bowel yields copious amounts of methane. The most widely used equation used for estimating M from measurements of chemical exchanges in metabolism is (Royal Society, 1972)

$$M(kJ) = 16.17 \times 1O_2 + 5.021 \times 1CO_2 - 2.167 \times 1CH_4 - 5.987 \times g\,N \quad (5)$$

The correction factors for CH_4 production and for urinary N excretion are small, and if they are neglected altogether the error of estimate will only exceed 2% in exceptional circumstances. For short-term measurements in ruminants it has been shown that M can in most circumstances be estimated simply from oxygen consumption (MacLean, 1972).

$$M(kJ) = 20.46 \times 1O_2 \quad (6)$$

Many of the definitive measurements of the energy exchanges of adult sheep (Blaxter, 1967) and lambs (Alexander, 1961) were made using closed circuit respiration calorimeters (Fig. 1). In this system the sheep is confined in a temperature controlled

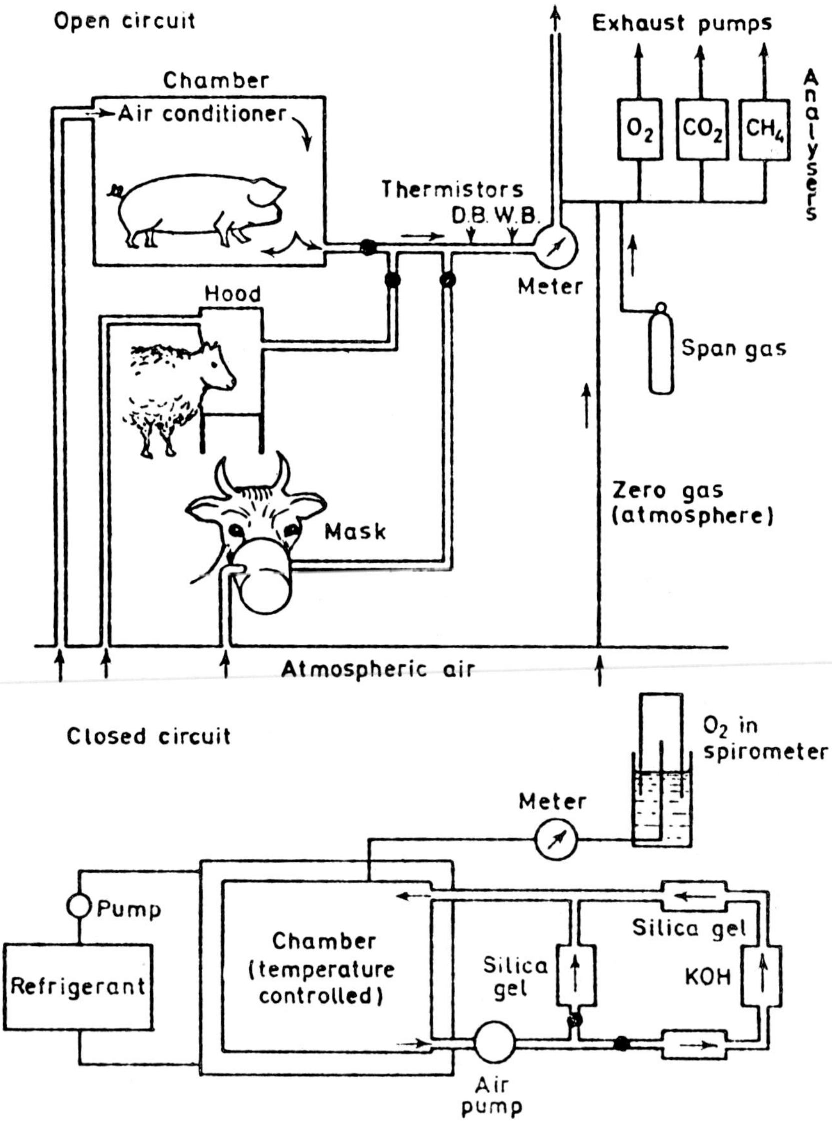

Fig. 1. Principles of open and closed circuit measurement of respiratory exchange.

chamber. Water vapour and CO_2 produced by the animal are absorbed continually and the pressure inside the system is maintained by a supply of pure oxygen. Since the method relies on gravimetric measurement of CO_2 production and volumetric measurement of O_2 consumption it is precise and requires no calibration. Moreover, since the method gives not only M, but also H_e, from loss of water vapour (approx. 2.5 kJ/g

H_2O), H_n can be obtained by difference when the period of measurement is sufficiently long to ensure that the heat storage term is small. In practice, measurements are usually made for periods of 24 hrs and air temperature in the chamber must be kept constant throughout.

Open circuit respiration apparatus depends on the continuous analysis of the concentrations of O_2, CO_2 and CH_4 in a stream of air moving at a precisely measured rate which includes the air expired by the animal. This may involve the use of a mask, a hood enclosing the head, or a chamber accommodating the entire animal (Fig. 1). The system is attractively flexible since it can be applied indoors or out and the animals can be exposed to a wide range of thermal environments. It is also relatively simple to construct from off the shelf components. The method is, however, dangerously susceptible to error. In order to estimate oxygen consumption to an accuracy of $\pm 1\%$ when the concentration of oxygen in sampled air is 0.5% below atmospheric, that concentration of oxygen must be measured to an accuracy of 1 part in 20,000. If both analysers and meters used in open-circuit respiration apparatus are calibrated using reliable volumetric or gravimetric procedures (Brockway et al., 1971) then one can approach this degree of precision. The accuracy of calibration of most commercial cylinders of mixed gases used with respiration equipment is quite unacceptable.

FIELD METHODS. Biometeorological studies are best served by measurements made of the metabolic rate of animals ranging free out of doors. Corbett and his colleagues built a small apparatus for collection of expired air that was drawn by sheep on a small cart (Corbett et al., 1969). More recently they have estimated M from CO_2 production obtained from the specific radioactivity of labelled bicarbonate infused subcutaneously or intravenously (Corbett et al., 1971). Samples are taken of blood or urine and the whole apparatus is light enough to be carried by the sheep. The method has proved most successful in Australia, although below 0° C there is the added complication that the liquids being infused or withdrawn must be prevented from freezing. In both these techniques the sheep must be sufficiently close to hand to permit samples to be collected.

The alternative approach is to telemeter continuously a function such as heart rate (Webster, 1967) or preferably cardiac output, that is closely correlated with M. The advantages of this technique are that the equipment is very light and the sheep can roam free. The disadvantages are that the correlations are imperfect and heart rate or cardiac output must be related to M for each individual using some form of calorimetry.

METEOROLOGICAL MEASUREMENTS

Most of the equipment that has been used in studies with sheep is standard. Briefly, skin surface temperatures are better measured with thermocouples than with thermistors, particularly in the cold. Thermocouples are cheap and expendable. Moreover their tips can also be made of very fine wire so that (1) they record a temperature that is very close to the skin surface and (2) conduction of heat from the skin down the wire is minimal. The environmental factors that contribute to heat loss must be measured continuously and integrated or averaged over the period of interest. These are, air temperature, wind run (m/s), net radiation (W/m^2) and precipitation. The overall thermal demand of the external environment has been assessed in terms of these environmental factors using artificial sheep or cattle (Webster, 1971). These models had the approximate shape and thermal insulation of the animals in question and integrated the thermal demand of the environment as the electrical power consumption W/m^2) required to maintain the core temperature at about 39° C.

ENERGY EXCHANGES OF SHEEP IN THE COLD

MATURE SHEEP. Cold tolerance is best described by the lower critical temperature which is a function of thermal insulation and thermoneutral metabolic rate, M* (equation 3). The value of M* in a mature sheep can vary a great deal according to the quantity and quality of food that it receives and the amount of exercise that it takes. The basal metabolic rate of a starved, confined sheep is about 45-50 W/m^2. The ewe that is confined but is getting enough food to maintain energy equilibrium has an M* of about 65-70. Outside on range, it may range from 80-110. The growing lamb receiving food ad libitum has an M* of about 155 W/m^2, although in individuals with very high appetites it may be as high as 175 (Blaxter, 1967; Alexander, 1973).

Tissue insulation in sheep depends upon the physical properties of the skin and subcutaneous tissues and on the rate of blood flow through this body shell. It is greatest, therefore, when vasoconstriction is maximal. Individual values for I_t in these circumstances range from about 0.07 to 0.18° $C \cdot m^2 \cdot W^{-1}$. At air temperatures below 0° C some degree of cold induced vasodilatation in the vessels traversing the tissues of, for example, the ears and shanks, is essential to prevent these areas from freezing. In these circumstances I_t falls by about 0.002 units per °C (Webster and Blaxter, 1966).

External insulation at an air speed v, (m/sec) is linearly related to the mean depth of the fleece, F (mm) (not to the length of the staple) according to the equation

$$I_{e,v} = a_v + b_v F \qquad (7)$$

The intercept term a_v is the insulation of the boundary layer of air around the fleece and $b_v F$ is the insulation of the fleece itself. The effect of increasing air movement on I_e was summarized by Joyce et al. (1966) in the form of the following equation, where I_e is here re-expressed in $°C \cdot m^2 W^{-1}$, and r_s is the radius of the trunk of the sheep (mm).

$$I^a = \frac{r_s}{r_s + F}\left(\frac{1}{5.57 + 13.08\ V}\ 0.5\right) + r_s \log_e\left(\frac{r_s + F}{r_s}\right)(0.0116 - 0.0052\ V^{\cdot 05}) \qquad (8)$$

(boundary layer insulation) (fleece insulation)

Joyce et al., (1966) added an expression to this equation to include the effect of solar radiation. The term, however, relates only to solar radiation; it does not describe all the pathways of net radiation exchange between the surface of the animal and the effective radiant temperature of the environment. A complete description of the effects of radiation exchanges on heat losses of sheep out of doors has not been made. Some data are available for cattle (Webster, 1971).

The effects of rain or snow on I_e are difficult to quantify because it is never certain to what extent moisture has penetrated the fleece. Snow in particular either blows off the fleece altogether or cakes on the surface leaving the area underneath relatively dry. When sheep in full fleece were exposed either to wind (4 m/s) or the combined stresses of wind and artificial rain (1 cm/h) I_e values were 0.60 and 0.31 °C.m²/W respectively, a decrease of about 50% due to wetting of the fleece (Webster and Park, 1967).

Examples of some effects of variations in M*, I_t and I_e on the lower critical temperature of mature sheep are presented in Table 1. This shows that the most important single factor determining T_{ac} is M*. The confined sheep on full feed, for example, has a T_{ac} nearly 50° C lower than that of an animal being fasted while in full fleece. By comparison the maximum effect of the fleece in these examples is to lower T_{ac} by 16° C, a benefit that disappears almost entirely when the fleece is saturated.

Summit metabolism in the mature sheep is about 400 $W.m^{-2}$ (Bennett, 1972). The environmental temperature that would induce a maximum metabolic response from a shorn adult sheep in still air would be −94. For a sheep in full fleece this value would, in theory, fall to −271. Clearly the mature sheep will not succumb to acute cold stress due to an inability to sustain homeothermy. Cold stress is a chronic problem in that it further depletes the energy reserves of sheep expected to winter on little food of poor quality. The ewe that succumbs to cold does so only because she has already been starved to exhaustion.

Table 1. Approximate lower critical temperatures of mature sheep under different conditions of management and environment.

Management	Fleece depth (mm)	Wind speed (m/s)	M* (W/m²)	I_t (°C.m².W⁻¹)	I_e	Lower critical temperature (°C)
Fasting, confined	10	0.2	50	0.14	0.20	+25
	60	0.2	50	0.14	0.67	+9
Maintenance						
confined	60	0.2	70	0.13	0.67	−7
outside, shorn	10	0.9	90	0.14	0.17	+13
windy	10	4.3	90	0.14	0.10	+19
windy, dry fleece	60	4.3	90	0.14	0.39	−3
windy, wet fleece	60	4.3	90	0.14	0.20	+12
Full feed, confined	40	4.3	150	0.08	0.50	−40

NEW BORN LAMBS. The cold tolerance of lambs at birth varies considerably according to their size and the length and type of their coat (Alexander, 1973). New born, and still wet, their lower critical temperature in still air is about 26° C (22-32° C). By the time they are dry this falls to about 16° C (12-22° C). Exposed to an air temperature of −10° C and a wind speed of 5.5 m/s they would have to elevate M to 350 $W.m^{-2}$. Summit metabolism in a 5 kg lamb is about 360 $W.m^{-2}$. Lambs therefore can be overwhelmed by acute cold stress. Moreover their reserves of energy to sustain cold thermogenesis are very limited and rapidly exhausted.

METABOLIC ASPECTS OF COLD THERMOGENESIS

The increase that occurs in the metabolic rate of homeotherms when they are exposed to cold environments has frequently been considered under the two headings, shivering and nonshivering thermogenesis. These terms are however less simple and distinct than they appear. Shivering describes a rhythmic tremor in muscle recognised by visual inspection or by electromyography. It does not describe cold thermogenesis in muscle and the absence of shivering cannot be taken to indicate that muscle tissue is not participating in cold thermogenesis. Nonshivering thermogenesis (NST) is a convenient abbreviation for regulatory nonshivering cold thermogenesis activated by noradrenaline. It occurs only in those species which possess brown adipose tissue although it is most unlikely that all the heat is produced in the brown adipose tissue itself (Webster, 1974).

It would be more precise to divide cold thermogenesis into

1. SOMATIC, muscle thermogenesis controlled by the somatic nervous system, and abolished by neuromuscular inhibitors such as curare,
2. AUTONOMIC, thermogenesis in brown adipose tissue, and probably in muscle, controlled by the autonomic nervous system through adrenergic BETA-receptors and inhibited by propranolol. The terms SOMATIC and AUTONOMIC thermogenesis are used instead of shivering and nonshivering thermogenesis respectively.

The lamb at birth possesses brown adipose tissue. Noradrenaline exerts a direct thermogenic effect and cold thermogenesis is markedly inhibited by propranolol (Alexander and Williams, 1968). In the normal course of development, brown adipose tissue and the capacity of the lamb for autonomic thermogenesis (NST) disappear in the first 7-10 days of life, although they persist longer if the lambs are kept chronically cold and undernourished. The somatic nervous system of lambs is mature at birth and they can elevate heat production in muscles by shivering and other motor activities. Alexander and Williams (1968) calculated that the relative contributions of the SOMATIC and AUTONOMIC components to cold thermogenesis in lambs at summit metabolism were 60:40 at birth rising to 95:5 at about one month. Under less severe conditions of cold the contribution of autonomic cold thermogenesis in the new born lamb is probably more important.

The adult sheep has no brown adipose tissue. Noradrenaline has no significant thermogenic effect and cold thermogenesis is not inhibited by propranolol unless the stress of cold is very severe. Acclimation to prolonged moderate cold has no effect on the responses to noradrenaline and propranolol. One may conclude from this that cold thermogenesis in the adult sheep is entirely a SOMATIC phenomenon (Webster et al., 1969).

There is, however, some evidence to suggest that the nature of this somatic phenomenon does change during acclimation to cold in the sheep. Sykes and Slee (1968) reported shivering as indicated by electromyography was reduced after acclimation. Webster et al. (1969) showed that shorn sheep exposed to an air temperature of −25° and treated with the ALPHA-adrenergic inhibitor, phenoxybenzamine, were only able to sustain homeothermy if they had previously been acclimated by continuous exposure to moderate cold. This is some recent evidence to suggest that phenoxybenzamine may inhibit the fusimotor control of the muscle spindles concerned in establishing the rhythm of shivering (Webster, 1974). It seems likely at the moment therefore that acclimation to cold in the adult sheep involves changes in the pattern of somatic thermogenesis in striated muscle, although how these are brought about and what may be the significance of these changes are not known.

CONCLUSION

The energy exchanges between sheep and the environment have been described thoroughly and sufficient information is now available to permit a rational and precise approach to the design of buildings and management practices for the intensive rearing of sheep. Uncertainties still exist, the effects of radiation and precipitation for example, but these relate to situations over which man has little or no control, and the practical benefits resulting from a better description of these phenomena would be small.

The metabolic aspects of cold thermogenesis in the sheep are poorly understood. Since sheep are so tolerant to cold this is probably of no great importance in itself. However, the attention that comparative physiologists have concentrated upon autonomic thermogenesis (NST) in the last 20 years has clouded the fact that it plays no part at all in the metabolic response to cold in adult man and all mammals weighing more than about 2 kg. The sheep is a tractable animal, similar in size and metabolic rate to man and it can simply be exposed naked or fully clothed to a wide variety of environmental stimuli.
It is therefore a useful model for the study of cold thermogenesis in man and most animals of economic importance.

REFERENCES: see pp. 546-547.

Effects of Heat and Cold on Energy Metabolism of the Pig

by

L. E. Mount

INTRODUCTION

The pig was one of the earliest animals to be domesticated in primitive human settlements, obtaining food by scavenging and foraging in forests. Nowadays in industrialized countries the pig is raised intensively under conditions of confinement, so that for the animal the environment imposed by housing, rather than the out-door environment, has become dominant in many areas. Increasing study of the pig reflects the animal's great economic importance in meat production. The animal is also used increasingly in biological research, and miniature pigs have been bred for this purpose (Mount and Ingram, 1971).

The outstanding characteristics of the animal in relation to the thermal environment are that the new-born pig is susceptible to cold and the mature pig is susceptible to heat. Several major adaptations stem from these basic facts. In the case of the new-born, continued exposure of the single animal to cold conditions can lead to death in hypothermia. However, if the young animal is with the other members of the litter and the sow, the behavioral adaptation of huddling means that cold can be withstood more easily and the range of temperature conditions in which the animal can live is considerably expanded; the presence of bedding increases this range. In the case of the large mature animal, the sow or boar, exposure to hot dry conditions can lead to death in hyperthermia because the pig cannot sweat effectively and its rate of respiratory evaporative heat loss is limited. Large pigs in hot climates must therefore be supplied with sprays or wallow to provide surface evaporative cooling; the pig is basically native to swampy woodland in which it has available this external wetting facility. Pigs of intermediate size, that is the rapidly growing animals which provide pork and bacon, show some of the reactions of the two extremes, but in modern pig farming the greater concern in respect of environment in these animals lies in the effects on feed utilization and growth, and consequently on energy metabolism, which plays a central role in adaptation to environmental change (Mount, 1972).

Adaptations may be considered in the three categories of physiological, behavioral and morphological. Physiological adaptations are discussed in the remainder of this chapter and behavioral adaptations in Chapter 1, Section 18. Morphological adaptations to temperature can be described briefly. Pigs kept under cold conditions have shorter ears and denser hair coats than those kept in the warm (Fuller, 1965). Littermates raised at 35° C and 5° C and maintained at equal weights show marked differences in body conformation as well as in hair coat. The pig kept in a warm environment has longer limbs and a larger number of blood vessels in the skin (Ingram and Weaver, 1969).

PHYSIOLOGICAL ADAPTATIONS

MEASUREMENT OF ENERGY METABOLISM

The energy metabolism of the pig has been investigated using both direct and indirect calorimetry. In the direct category, the gradient layer calorimeter has been used for young pigs (Cairnie and Pullar, 1959). The heat sink type of direct calorimeter was used at an early stage (Capstick, 1921). More recently, in a modified automatic form, it has been used both for single pigs and for groups of pigs; for the work with groups, the interior of the calorimeter was designed as a pig pen (Mount, 1968). Indirect calorimetry has been used widely, with measurement of oxygen consumption and carbon dioxide production, or with carbon and nitrogen balances (Kleiber, 1961; Brody, 1945).

Use has been made of telemetry to transmit information such as body temperature and ambient air movement and air temperature from freely moving animals both in the laboratory and out-of-doors.

HEAT PRODUCTION AND ENVIRONMENTAL TEMPERATURE

In discussing the relation of the rate of heat production to environment, environmental temperature can be assumed to be standardized in terms of equal air and mean radiant temperatures with low air movement. The effects of deviations of the components of environment will be considered subsequently in relation to modes of heat loss. The rate of heat production is usually estimated from the measured rate of oxygen consumption and in the aerobic mammal is taken as synonymous with metabolic rate.

New-born pig

The mean weight of the new-born pig lies between one and 1.5 kg; the animal is born as a member of a litter of usually 6 to 12 in number and grows rapidly to double its birth weight in one week. The pig feeds shortly after birth and then at approximately hourly intervals. The animal can shiver from birth, and measurement of the thermal circulation index suggests that vasoconstriction occurs in cool conditions. These features suggest functionally established thermoregulation (Mount, 1968; Curtis, 1970).

When the single new-born pig is exposed in a metabolic chamber for periods at each of a number of standardized environmental temperatures, the metabolic rate varies as in Figure 1. This is the relation between metabolic rate and environmental temperature which is characteristic of the homeotherm (Giaja, 1938). The maximum cold-induced metabolic rate is approximately three times the minimum rate. The low level of thermal insulation in the new-born pig leads to the whole range of metabolic adaptation to cold taking place between the critical temperature of 34-35° C and the cold limit of −5° C; the animal has no subcutaneous fat and very little hair coat.

In the single new-born pig it is unlikely that continued existence, as opposed to short periods for measurement, is feasible at temperatures below a mean of 0 to 5° C. The

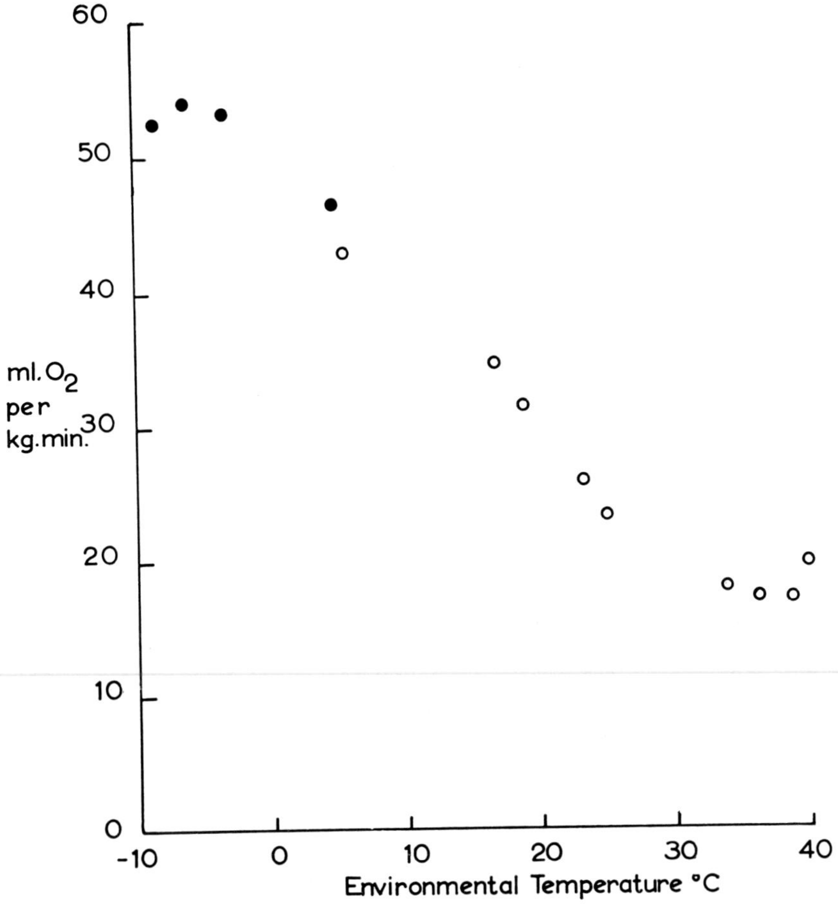

Fig. 1. Maximum and minimum metabolic rates in the new-born pig. One pig, 52 hrs of age and weighing 1.70 kg (filled circles), was exposed to a decreasing environmental temperature for the detection of the maximum rate. Another pig, 60 hrs of age and weighing 1.65 kg (open circles), was exposed to a rising environmental temperature for the measurement of minimum metabolic rate (From Mount and Stephens, by permission of J. Physiol. (Lond.), (1970), 207: 420).

slope in Figure 1 indicates the high thermal conductance, which is comparable with that of the new-born human infant and about double that of the lamb (Barnett and Mount, 1967). These observations refer to the pig more than one day old. Metabolic rate during the first day following birth, particularly, is lower than it is later. During the first one or two days after birth the metabolic rate increases with age in addition to the clearly defined increase with body weight. During the first few hours the minimum rate is 8-10 ml O_2/kg.min; on the second day this rises to a mean value of about 15 ml/kg.min.

The new-born pig's deep body temperature drops to 37-38° C within about half an hour after birth, and to lower levels in the cold, rising to the mature deep body temperature of 39° C within one to two days, the rise being associated with the rise in metabolic rate (Mount, 1968). These increases do not take place if the animal is fasted from birth.

As the body weight increases, subcutaneous fat and thermal insulation increase, the critical temperature falls, and oxygen consumption per unit body weight falls, to reach about 5 ml/kg. min at a body weight of 100 kg. These effects in the young animal are

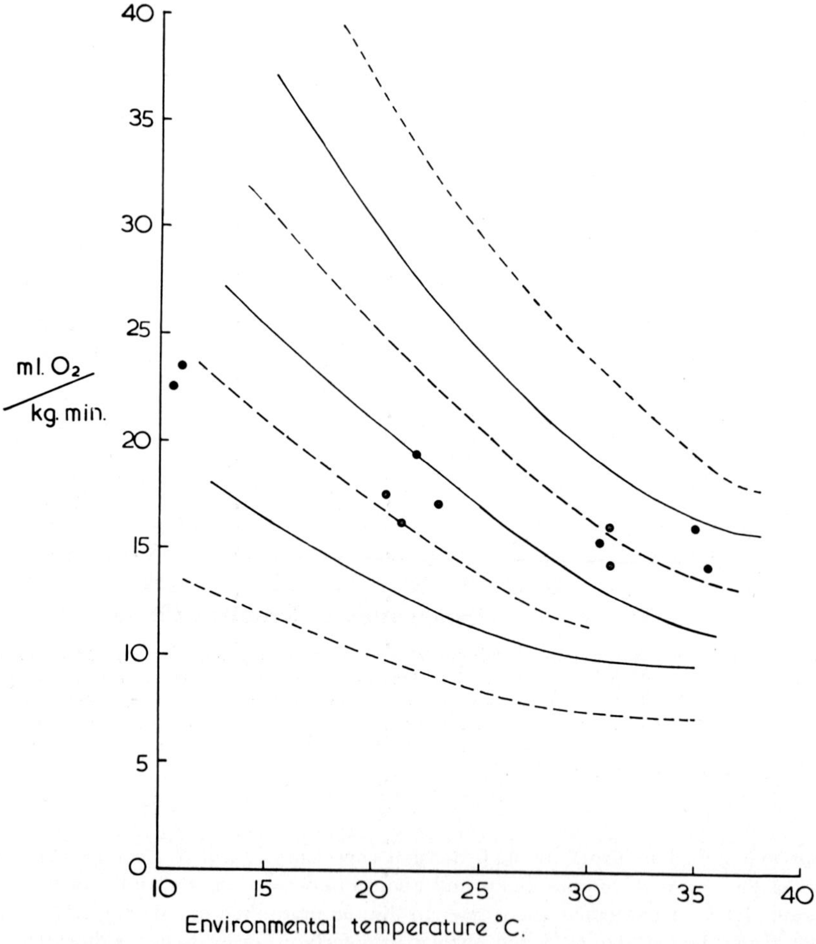

Fig. 2. Metabolic rates for three weight groups of young pigs over a range of environmental temperature. Curves refer to results on single pigs, with interrupted lines enclosing approximately 85% of results in each weight-class; points refer to results on groups consisting of either four or six animals, each weighing 1-2 kg, together in the metabolic chamber (From Mount, by permission of Fed. Proc., (1963), 22: 818).

illustrated in Figure 2, which includes values obtained from groups of four or six 1-2 kg pigs together in the metabolic chamber. At the higher temperatures the animals are spread out in the chamber and the recorded mean metabolic rate per unit weight is similar to that obtained from an individual 1-2 kg pig. As the temperature falls, however, so the animals huddle together increasingly, and the metabolic rate per unit weight approximates more to that found for a single larger pig. At low temperatures the group of pigs behaves like a single larger pig, with consequent economy of heat production, whereas in the warm the group breaks up into individual animals and heat loss is increased.

The increased facility for animal-environment interaction which is provided by the group as opposed to the individual is illustrated in Figure 3. Measurements were made at 20° C, which is cool for the baby pig, and the number of pigs together in the chamber was first progressively increased and then decreased. Metabolic rate per unit weight

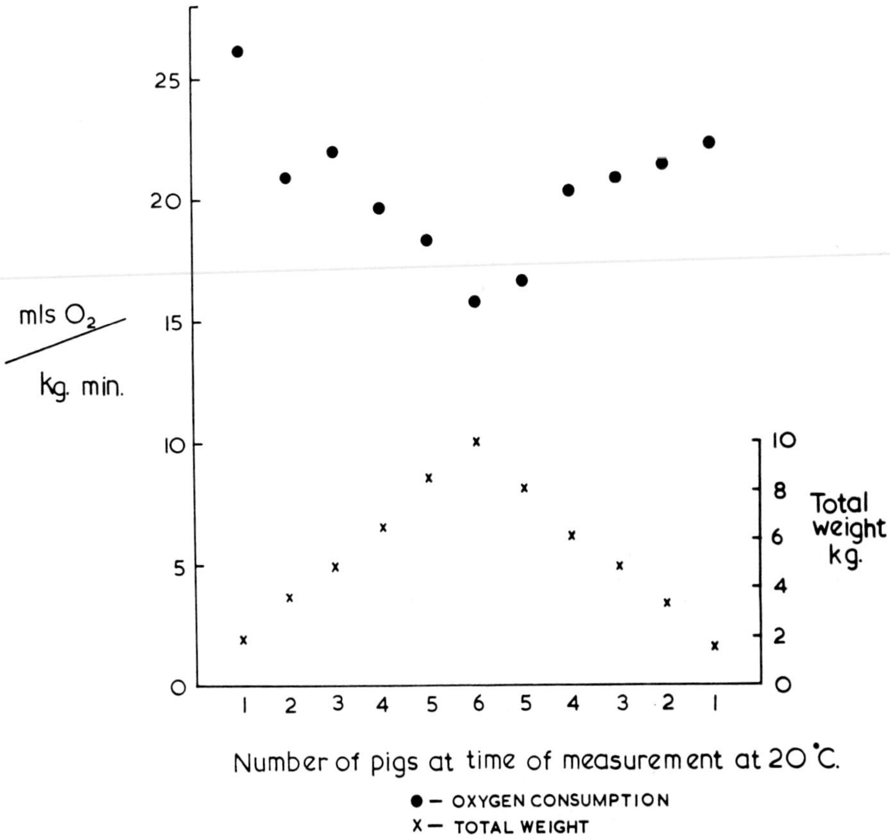

Fig. 3. The relation between metabolic rate per unit of body weight and the number of new-born pigs together in a chamber at an environmental temperature of 20°C (From Mount, by permission of J. agric. Sci., (1960), 55: 101).

varied inversely with group size, indicating the saving in energy expenditure conferred by the group at a given environmental temperature.

The importance of the group in the new-born pig's response to cold is made evident by comparison with a single bacon pig in Figure 4, in which metabolic rate is expressed per unit of surface area as a means of normalizing rates between animals of different sizes. At an environmental temperature of 20° C the intact group has a metabolic rate much closer to that of the larger bacon pig than to that of the single new-born pig. As the temperature declines, the single new-born animal reaches its limit of heat production at a point at which there is still reserve metabolic capacity for the members of the group, indicating their ability to tolerate and survive lower ambient temperatures than is the case for the single pig.

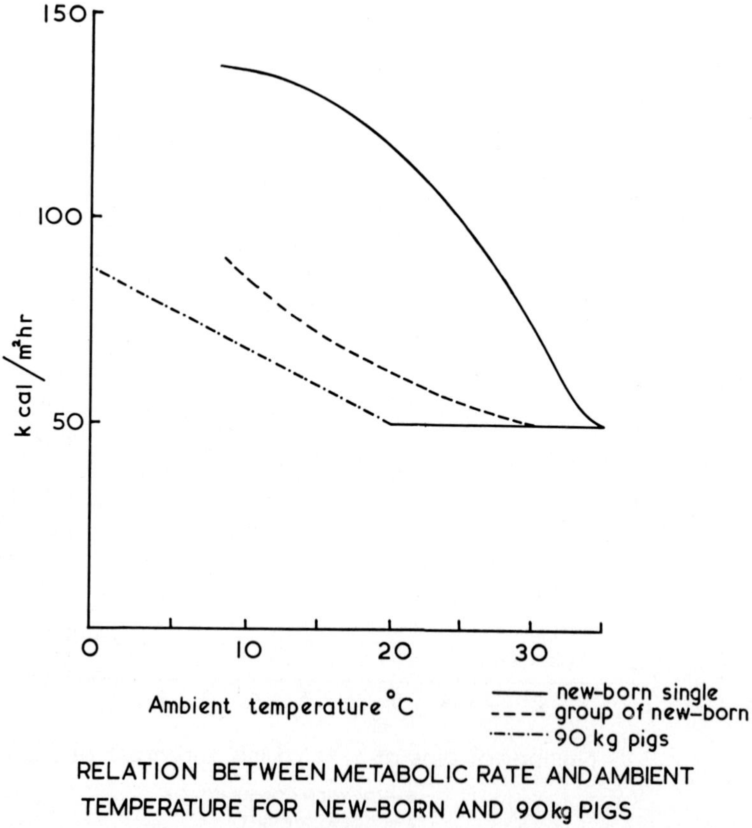

Fig. 4. The relation between metabolic rate and ambient temperature for single new-born pigs, a group of new-born pigs, and single 90 kg pigs (From Mount, by permission of Eur. Ass. Animal Prod., Energy Metab., (1965), Academic Press: 380).

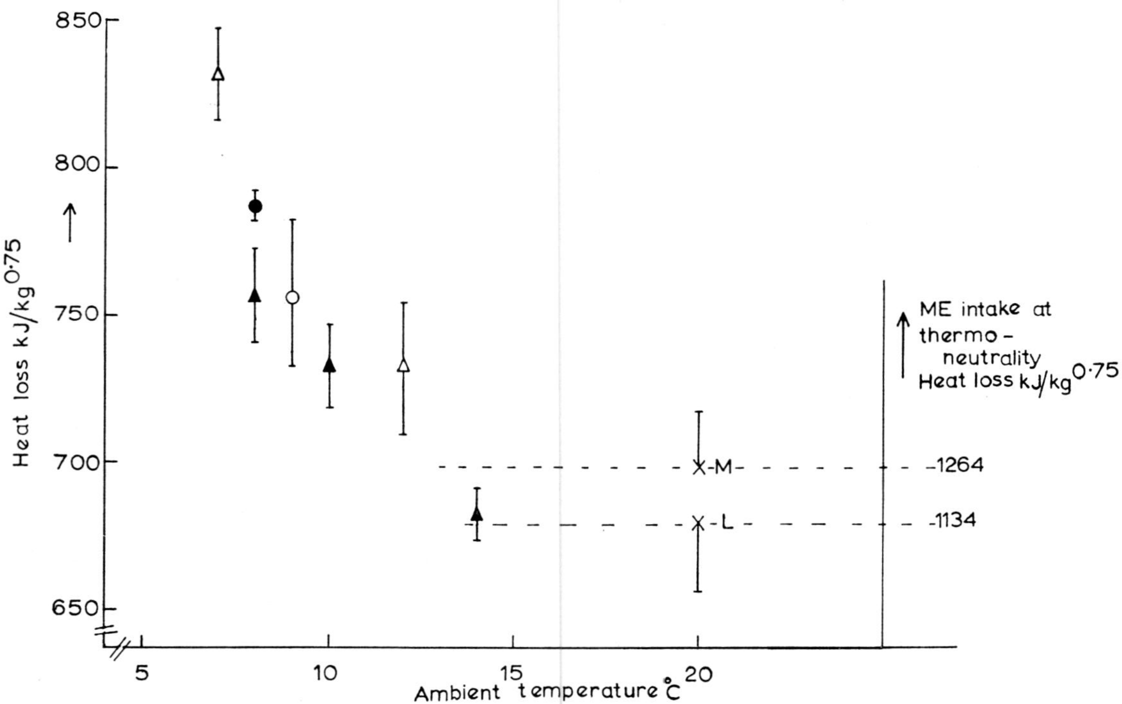

Fig. 5. Rates of heat loss from groups of pigs at several feeding levels and ambient temperatures, taken from several authors. M = Medium level intake, L = low level intake (From Verstegen et al., (1973), by permission of Br. J.Nutr.).

Growing pig

For the pig of several months of age the critical temperature might be in the region of 20° C, and the cold limit approximately −40° C. These figures are only rough indications since they depend on the level of feeding and thermal adaptation history, but they indicate the decreased tendency to cold-susceptibility and the potentially increased heat-susceptibility. Pigs adapted to the cold can live out of doors with only the protection of a roof and dry bedding; at an air temperature of −30° C, they have skin temperatures which may fall below 5° C (Irving, 1964).

As with the new-born pig, the huddling together of larger pigs in the cold reduces the thermal demand of the environment for heat production, although the temperatures at which the effects occur are lower than those for the new-born. Investigations into heat loss and energy retention in groups of four pigs at a time (Verstegen et al., 1973) have indicated effective critical temperatures of 11° C for a feeding level of 52 g food/kg. day and 17° C for a feeding level of 45 g/kg. day. Figure 5 shows the increased rates of heat loss from such groups when the environmental temperatures fall below these values. Below the critical temperature, in animals of a given body weight, the environment determines the rate of heat loss; above the critical temperature, the rate of heat loss varies with the feeding level. In addition to body weight, group size also influences the increased rate of heat loss below the critical temperature, an effect shown in Table 1.

Table 1. Approximate increases in heat loss ($kJ/kg^{0.75}$ per day) for each 1° C below the effective critical temperature for different body weights and groupings of pigs

Individual body weight, kg	Single or group	$kJ/kg^{0.75}$ per day per °C below critical temperature
1	single	40
1	group (5 pigs)	20
5	single	30
35	single	17
35	group (4 pigs)	12
35	group (9 pigs)	9

Thus the chief factors in the resting animal which influence metabolic rate and consequently heat loss are body size, grouping, environmental temperature and level of feeding. Increased activity also increases metabolic rate.

The effects of these factors are superimposed on the 24-hourly variation in metabolic rate which is shown for a group of pigs in Figure 6. At an environmental temperature of 20° C the amplitude of variation is approximately 30% of the mean 24-hourly value, that is ±15% of the mean. One of the consequences of this variation is that measurements made at different times during the 24 hrs are not necessarily directly comparable. For comparison of the effects of factors such as different physical environments and levels of feeding it is preferable to make continuous measurements during a succession of full 24-hr periods.

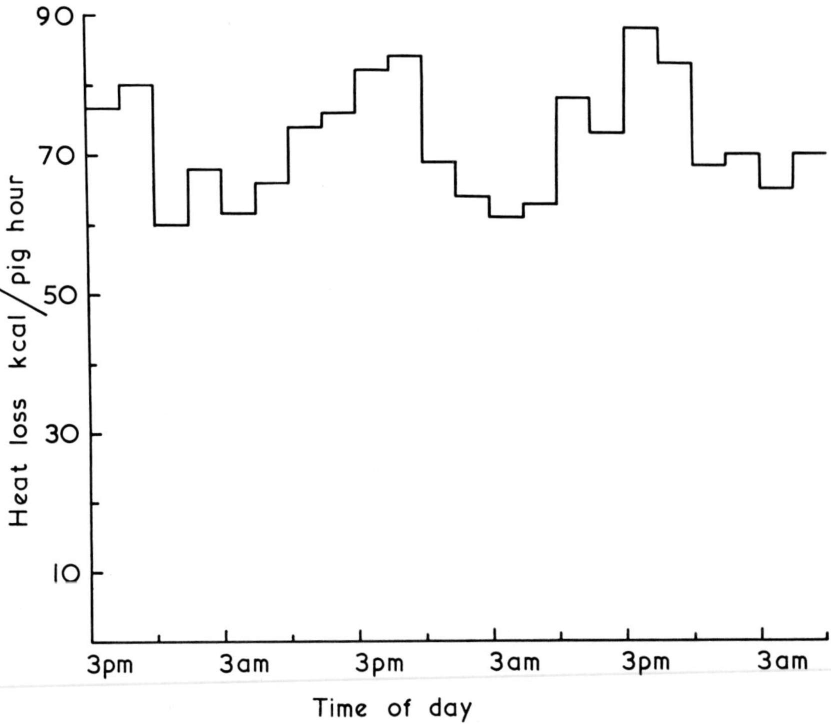

Fig. 6. 24-hr variation in heat loss from a group of six pigs, each weighing approximately 20 kg, at 20°C in the Babraham pen calorimeter (From Mount, by permission of Lea and Febiger, In: Adaptation of Domestic Animals, (1967), 286).

CHANNELS OF HEAT LOSS

The standardized environmental temperature (equal air and mean radiant temperatures with low air movement) which has been used in the discussion of heat production is not often encountered in practice. In addition to the air and mean radiant temperatures differing one from the other, they also vary with time and so does the rate of air movement. Other factors of importance to the pig, which spends much time lying down, include the temperature and thermal conductivity and capacity of the floor, and the presence or absence of bedding. At high temperatures humidity also affects evaporative heat loss.

Heat transfer between animal and environment takes place through the four channels of radiation, convection, conduction and evaporation and for a given environment it is possible to combine the quantities of heat transferred in the different channels so that an equivalent standardized temperature can be estimated. This requires estimates of the relative magnitudes of transfer in each channel and knowledge of the environmental components. A very brief account of these factors is given here; a more extensive account has been published elsewhere (Mount, 1968) to which reference may be made for more detail.

The environmental components which are of thermal significance are the mean radiant temperature, air temperature, air movement, thermal characteristics of the floor, and humidity. These are associated with the corresponding channels of heat exchange and are shown in Table 2 together with other relevant factors.

Table 2. Factors governing heat transfers between organism and environment

Mode of Transfer	Animal Characteristics	Environment Characteristics
Radiant	Mean radiant temperature of surface; effective radiating area; emissivity (variable for short wave, approaching unity for long wave)	Mean radiant temperature; solar radiation
Convective	Surface temperature; effective convective area; radius of curvature and surface type	Air temperature; air velocity
Conductive	Surface temperature; effective contact area	Floor temperature; thermal conductivity and thermal capacity of solid material
Evaporative	Surface temperature; percentage wetted area; site of evaporation relative to skin surface	Ambient humidity; air velocity

The net radiant heat exchange is proportional to the difference between the fourth powers of the absolute surface temperature of the animal and the surrounding mean radiant temperature. It also depends on the emissivity of the surface which can range from zero to unity. A perfectly reflecting surface has an emissivity of zero, and a perfectly absorbing surface an emissivity of unity. For short-wave solar radiation many surfaces have high reflectance, including pale-coloured animal coats. For long-wave radiation, however, of wavelength greater than 3 μm, the majority of surfaces, including white coats and pale skin, have an emissivity which approaches unity. For short-wave radiation the reflectance is dependent on coat colour, whereas for long-wave it is independent. The radiant heat exchange of pigs out of doors is therefore influenced by coat colour but that of pigs indoors is not.

Convective heat exchange is proportional to the difference between the mean skin temperature and the air temperature, to a function of the air velocity which is usually given as the square root, and to the effective surface area for convection.

Heat exchange by conduction between the pig and the floor is influenced considerably by the animal's posture when lying, whether relaxed or supported on its limbs, as well as by the thermal characteristics of the floor. As a mean value, approximately 20% of the pig's surface is in contact with the floor during lying. In experiments with growing pigs it was found that the effective critical temperature of animals weighing 40 kg was 11.5-13° C on straw bedding, 14-15° C on asphalt, and 19-20° C on concrete slats (Verstegen and Van der Hel, 1974). These figures indicate the higher air temperature required to counter-balance the effects of floors with high thermal conductance; the effects of straw bedding extend to providing increased thermal insulation around tne animal as well as decreasing direct heat loss through the floor. In the case of the new-born pig, moving the animal at an ambient temperature of 10° C from a bare concrete floor on to straw has the same thermal effect in reducing metabolic rate as raising the ambient temperature to 18° C (Stephens, 1971). A small component of heat transfer which may be termed conductive lies in warming ingested food and drink; usually this does not exceed 3% of the total heat loss.

Heat loss by evaporation in the pig is limited by the animal's inability to sweat and is discussed further in Chapter 1, Section 5.
Figure 7 indicates the partition of heat transfer through the different channels for an average pig in a pen where some evaporation takes place from drinking water, feces

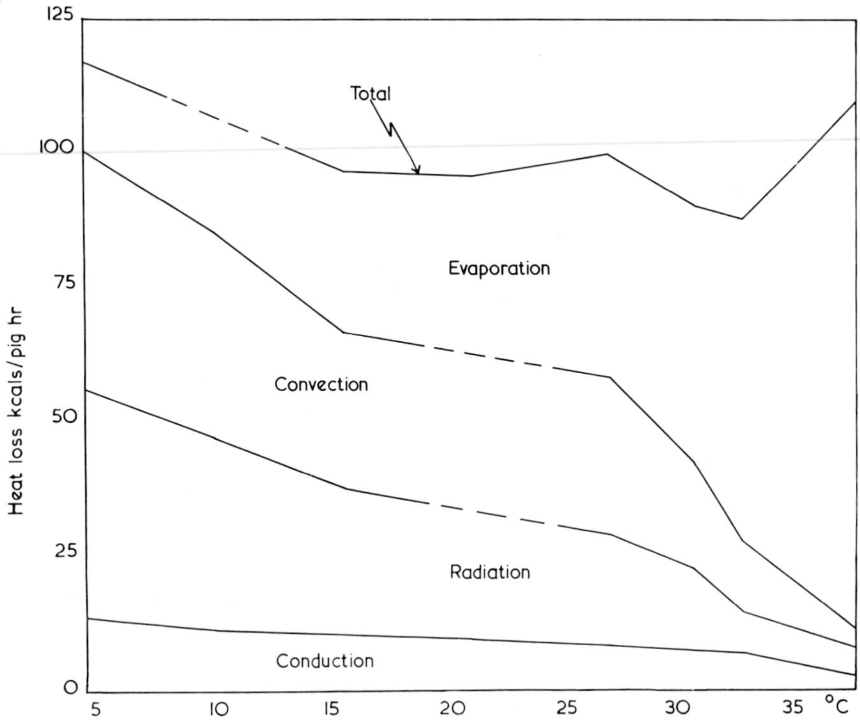

Fig. 7. The effect of environmental temperature on heat loss and its partition for an average pig of a group, with body weight ranging from 26 to 37 kg (From Bond et al., by permission of Agric. Eng., (1952), 33: 148).

and urine. Under cooler conditions convection and radiation together account for the largest part; when air movement increases, heat loss increases and the proportion of convective transfer increases and that of radiative transfer decreases. The use of infra-red lamps for new-born pigs leads to radiative heating and a consequent rise in the equivalent standardized environmental temperature.

In well-designed pens air and wall temperatures tend to be similar, air movement is low, and the floor is insulated. Under these conditions it is easier to assess a mean temperature which represents the thermal environment of the pen, then to adjust this for huddling and any bedding material, and so to derive a standardized temperature of some predictive value. It is in environments which contain widely diverse thermal factors that it is difficult to make a useful assessment.

SUMMARY AND CONCLUSIONS

The new-born pig has a homeothermic type of metabolic response to change in environmental temperature, but since it has so little thermal insulation it is susceptible to cold. The large mature pig, on the other hand, is susceptible to heat and can withstand cold. The development of standardized environmental temperature equivalents for different sizes and groupings of pigs in different ambient conditions remains to be put on a firm basis.

REFERENCES: see pp. 547-548.

Effects of Cold and Exercise on Energy Metabolism of Small Mammals

bv

L. Janský

INTRODUCTION

Small mammals are thermally sensitive animals inhabiting all regions of the earth. Seasonal and regional variations of temperatures in their habitats are great. In the Arctic, temperature variations in rodent burrows and runways range from $-15°$ C to $+15°$ C, in cool temperate zone from 0° C to $+18°$ C; and even in deserts, the summer temperature in burrows does not reach 30° C (Hart, 1971) (Fig. 1). Arctic and subarctic climates thus impose a considerable burden on small mammals which can only be partially avoided by behavioral tendencies such as huddling, nest building and seeking warmer microclimates. Under these circumstances the protection by behavioral processes may be insufficient and various physiological mechanisms must also take place during exposure and adaptation of small mammals to cold.

It is evident, that mammals during their life course, are mostly exposed to temperatures below thermoneutral zone and that in some regions temperatures in their habitats may be extremely low. With the ambient temperature decreasing below the thermoneutral zone, the heat loss from animal bodies tends to increase and in order to maintain a constant body temperature, it must be compensated by active intervention of thermoregulatory mechanisms. The amount of heat, lost under these conditions, depends on the temperature gradient between body and environment, on insulation of the body and on the body surface area, according to the formula:

$$H = K \frac{T_B - T_A}{I} \qquad \text{(Hart, 1963b)} \quad (1)$$

(H = heat loss, K = a constant representing the body surface area, T_B = body temperature, T_A = air temperature, I = insulation qualities of the body surface)

If the body surface area and body insulation remain constant, the heat loss increases proportionally to the temperature gradient. In order to maintain a heat balance, the heat loss must be compensated by heat production mechanisms, such as SHIVERING, NONSHIVERING THERMOGENESIS and MUSCULAR WORK. The intensity of heat production thus increases to a certain point which we call the LOWER SURVIVAL LIMIT (Fig. 2). Below this point an organism is no longer capable of maintaining a constant body temperature permanently and becomes gradually hypothermic. The intensity of heat production at the lower survival limit (the SUMMIT METABOLIC RATE) amounts to a quadruple of the basal metabolic rate (the METABOLIC

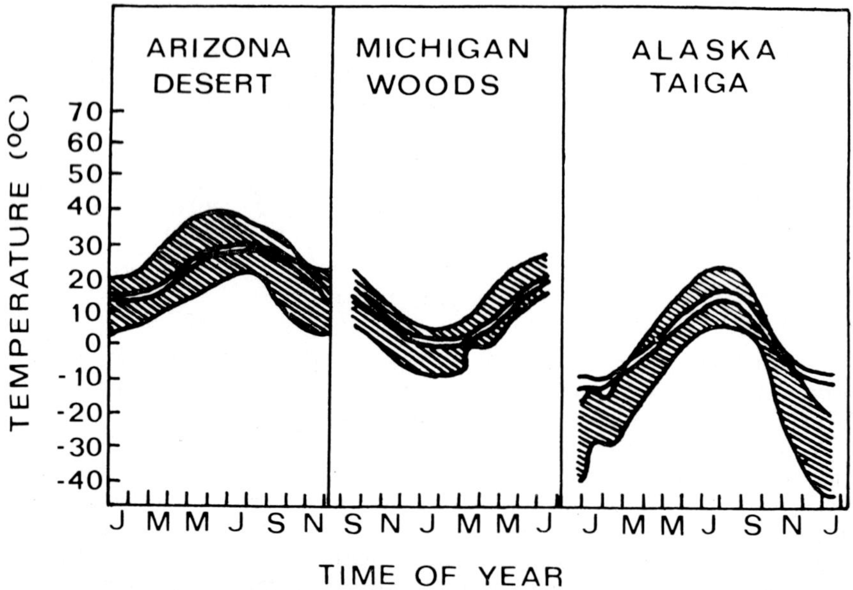

Fig. 1. Seasonal variations in habitat temperatures in different climatic regions. Air temperature range given by shaded area. Soil temperature in rodent burrows or runways given by double lines bounding unshaded area. All temperature variations are grossly smoothed to show only general trends (Hart, 1971).

QUOTIENT) in most of the species studied (Giaja, 1938). With further lowering of ambient temperature, the intensity of metabolism may be further increased temporarly reaching the level of MAXIMAL METABOLISM. The same level of maximal metabolism can be also achieved during long-term maximal work (Hart, Janský, 1963). The intensity of maximal metabolism in different species increases with the exponent of body weight similar to that of basal metabolism (Janský, 1963, 1966) so that the metabolic increase from basal to maximal level is about the same in most of the species and in rodents it represents a 7 fold of basal metabolism (Hart, 1971). The maximal metabolism as measured in the cold like those measured during exercise can only be obtained for very short periods of time (3-20 min). During the measurement, the heat loss exceeds heat production and body temperature falls rapidly. Thus, it is difficult to measure the maximal metabolic values, but it is felt that values so far obtained do not represent a large departure from the true values.

The slope of the metabolic increase in the cold depends on factors related to body surface area. Bigger species having a relatively smaller body surface increase their metabolic rate in the cold slower than smaller species (Thauer, 1939). With respect to the fact that summit metabolic rates are about the same in all species, larger animals preserve the ability to produce heat in lower temperatures and seem to be able to live within larger limits of environmental temperatures.

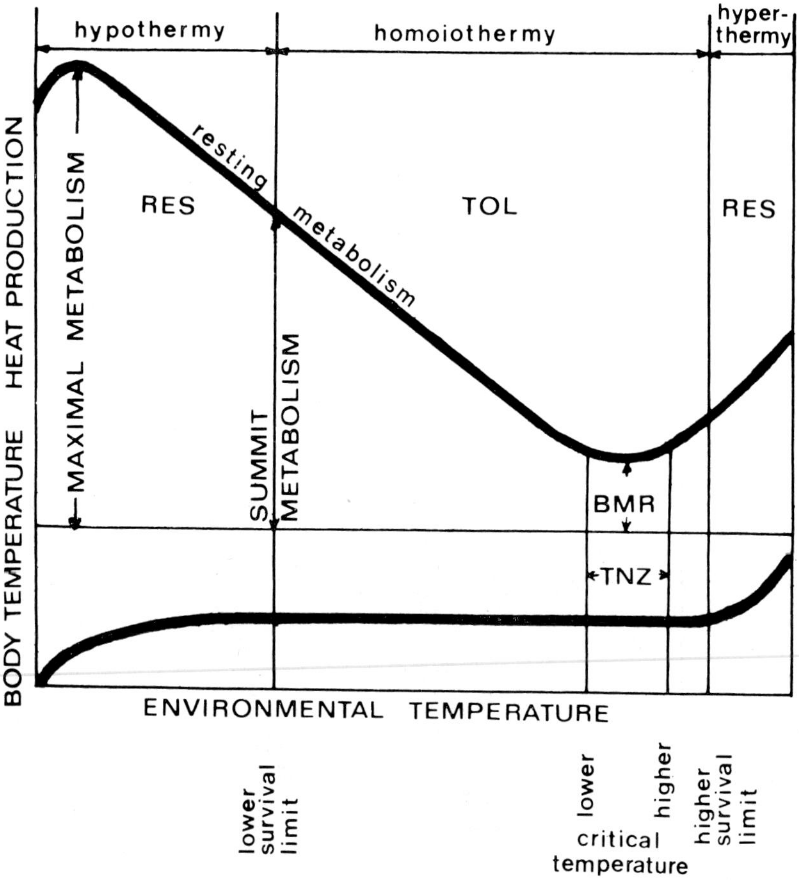

Fig. 2. Effect of ambient temperatures on heat production and body temperature of homeotherms (scheme). The zone of tolerance (TOL) is a temperature range in which no lethal effect take place, and the zones of resistance (RES) are ranges in which death results after varying periods of time depending on temperature of exposure. TNZ = thermoneutral zone, BMR = basal metabolic rate.

The resistance of small mammals to cold is surprisingly great. Even small mammals as shrews (*Sorex araneus*) are able to withstand temperatures around −20° C for 14 days (Vlasák, unpublished), ground squirrel adapted to 30° C live at −28° C for 5 hrs, while cold-adapted individuals of this species survive for about 41 hrs at this temperature (Twente et al., 1971). The best expression of survival limits is given by a correlation between exposure temperatures and logarithm of the time to death. Hart (1957) showed that winter populations of deer mice survive about 20° C lower temperatures than individuals from the summer season (Fig. 3). Temperatures at which warm-adapted rodents survive for 200 minutes are given as −7° C for white mice, −10° C for deer mice and −18° C for white rats (Hart, 1953).

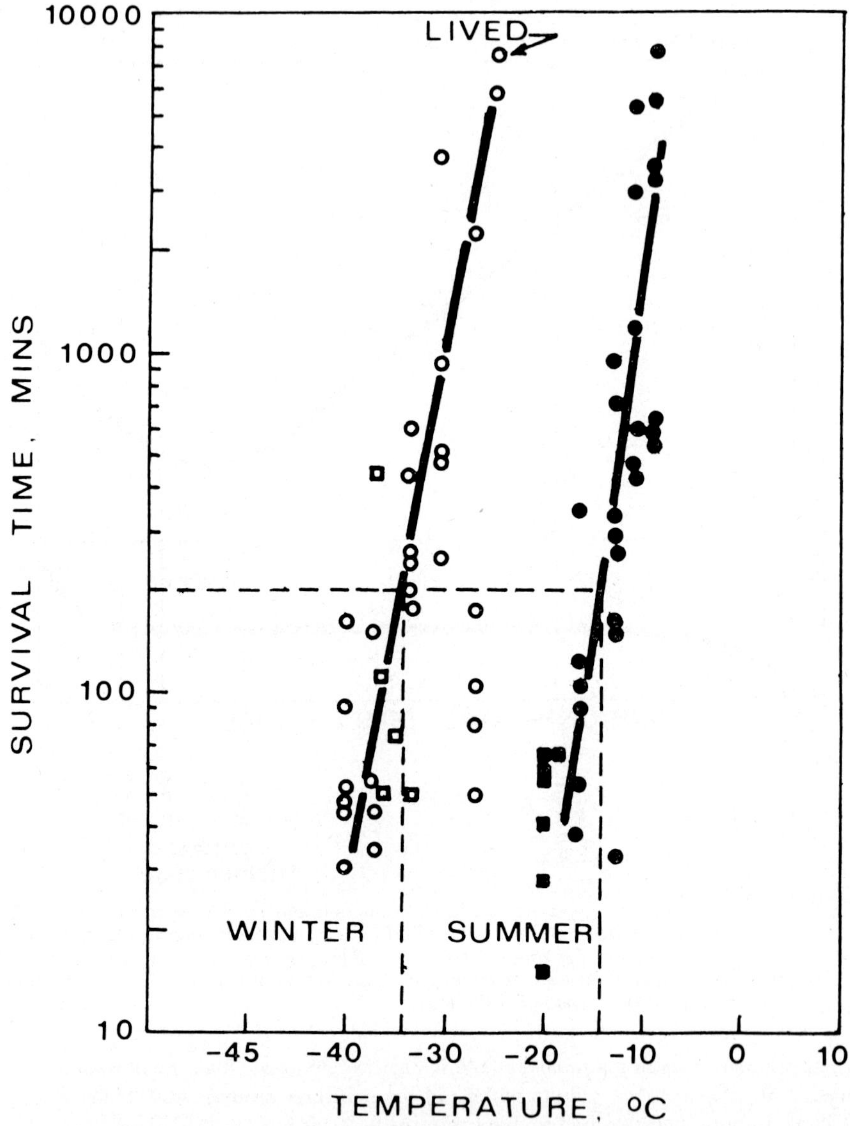

Fig. 3. Survival time in relation to temperature for winter (○ □) and summer (● ■) deer mice Winter and summer temperature limits for survival time of 200 min are shown by broken lines. (Hart, 1957

Insulation qualities of the body surface are seldom held constant during thermoregulation. If the insulation is improved then, according to formula (1), a smaller amount of heat is necessary to be produced within the body to maintain homeothermy and vice versa. This fact is very important for estimating the thermoregulatory efficiency of different heat production mechanisms. The most efficient are those processes which do not stimulate heat loss mechanisms. In order words, with the aid of mechanisms which do not increase heat loss, a constant body temperature can be maintained with lower energy expenditure.

In homeotherms adaptation to cold is manifested by the increased resistance to cold so that the lower survival limit shifts toward lower temperatures (from L_1 to L_2) (Fig. 4). Theoretically, adaptation to cold can be realized by two mechanisms: either by increasing the capacity of heat production (from C to D) or by enlarging thermoneutral zone (from B to B′) due to improved insulation (lowering of conductance), heat production capacity being unchanged (Hart, 1963b).

In this Section the thermoregulatory significance and energetic efficiency of various heat production mechanisms will be considered and their adaptability to cold will be discussed.

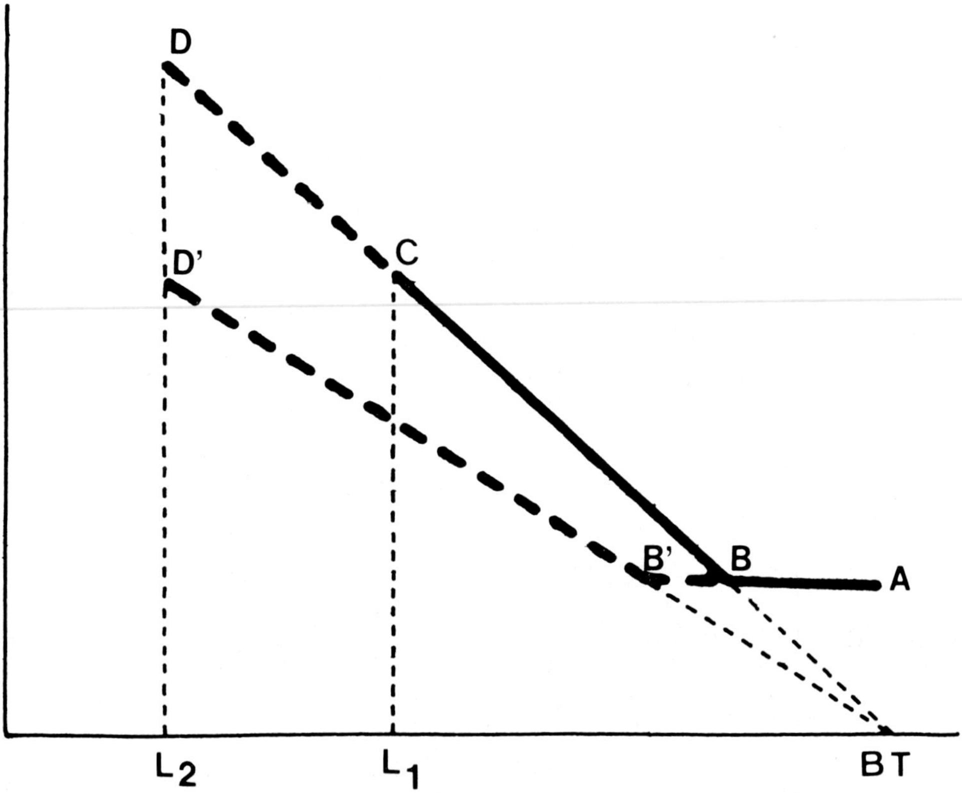

4. Scheme of two possible mechanisms of cold adaptation - metabolic (from C to D) and insulative (from B to B_1). Both types of adaptation enlarge the range of temperatures animals can survive (from L_1 to L_2). BT = body temperature: A—B, resp. A—B_1 = basal metabolism; D and D_1 = maximal capability of metabolism. In metabolic adaptation there is an increase in heat production capability; insulative adaptation enlarges the thermoneutral zone, the metabolic capability being unchanged (Hart, 1963).

SHIVERING THERMOGENESIS

The most important source of heat during thermoregulation is muscle tone and shivering. Shivering is very sensibly regulated and animals showing imperfectly developed shivering thermogenesis can be considered as primitive from the evolutionary point of view.

EMG recordings are usually used for identifying shivering. This technique is far from being adequate, however, since it records shivering in a small area of the body only and indicates little about interaction of individual muscles during this process. Because of this, it is very difficult to estimate exactly the thermal equivalent of shivering e.g., the amount of heat corresponding to a certain intensity of shivering.

Shivering takes place in almost all muscles of the body with exception of facial, perineal, extraocular and middle ear muscles (Kawamura, 1961). During exposure to cold it appears first in masseter muscle and then it extends to other muscular groups. There is still a controversy whether it occurs first on the body periphery and then extends

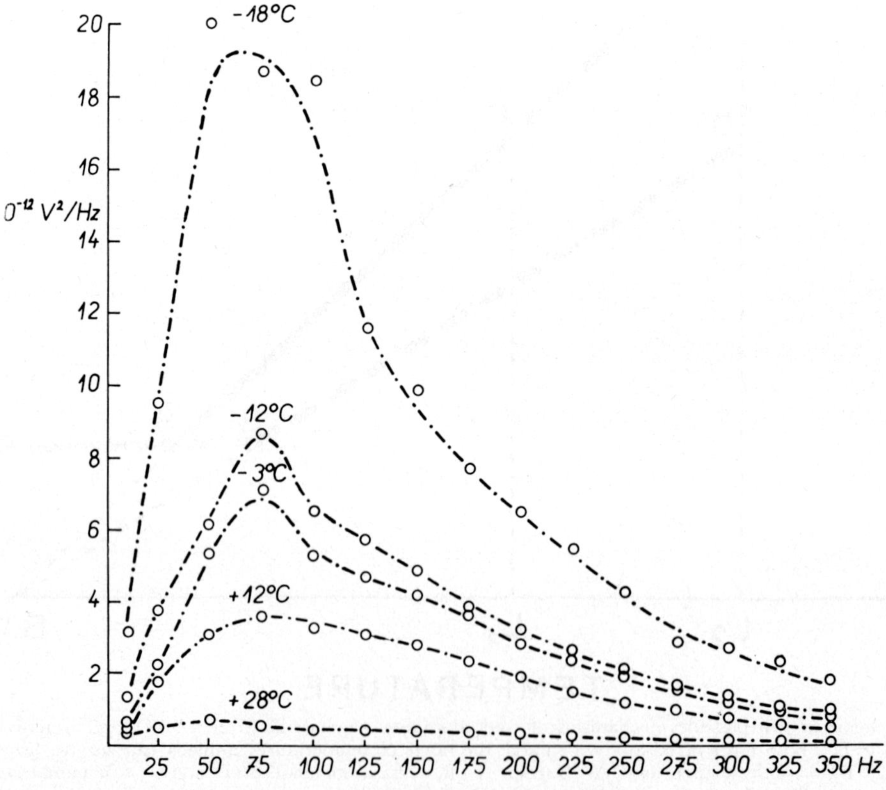

Fig. 5. Power spectral density analysis of shivering electrical activity of neck muscles of the rabbit at different environmental temperatures (Janský et al., 1970).

to muscles of the trunk (Golenhofen, 1965) or whether it starts in the centre of the body and extends to legs (Horvath et al., 1956). After exposure to cold the amplitude of EMG recordings increases (in rats from about 20 μV to 70 μV peak to peak – Hart et al., 1956), while the spectrum of frequencies is not changed (Janský et al., 1970) (Fig. 5). It is known that in bigger species shivering is a discontinued process. The total intensity of shivering expressed in terms of O_2 consumption fluctuates, similarly as does shivering in different areas of the body (Janský et al., 1970). The greatest intensity of shivering can be measured in the neck and smallest in the legs. The quality and intensity of shivering in rabbits is not affected by cold adaptation, as judged from frequency spectrum analysis (Janský, unpubl.). Some authors, however, suggest that the thermal equivalent of shivering is increased in cold-adapted rats and rabbits (Tkachenko and Ivanov, 1971).

During shivering the blood flow increases substantially in muscular organs, while in internal organs and in the skin it is practically unchanged, so that the relative distribution of cardiac output to internal organs and to the skin decreases (Janský and Hart, 1968) (Fig. 6). The heat loss from the body is thus reduced and therefore the heat produced during shivering may be efficiently utilized for maintaining a constant body temperature.

Shivering is regulated by nervous mechanisms from the posterior area of the hypothalamus. In guinea pig, afferent signals for shivering may come from receptors in the

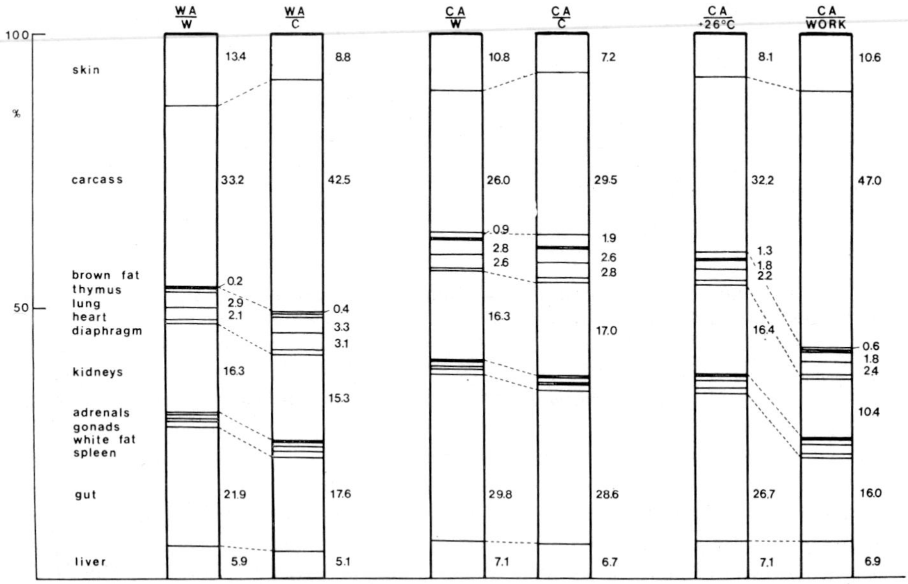

Fig. 6. Fractional distribution of the cardiac output of cold-(Ca) and warm-(WA) adapted rats exposed to cold (C), thermoneutral zone (W) and during work (work). (From Janský and Hart (1968); Bartůňková and Janský (1970)). Numbers denote the percentage.

body periphery and from temperature sensitive neurons located in the spinal cord (Brück and Wünnenberg, 1967, 1970). The induction of shivering is probably dependent on temperature changes both on the periphery and in the centre of the body; the slope of the metabolic increase during shivering being dependent on the difference between these temperatures. The mutual relation of these variables is best expressed by a hyperbola (Fig. 7). It follows from this relation that the lower is the temperature of the

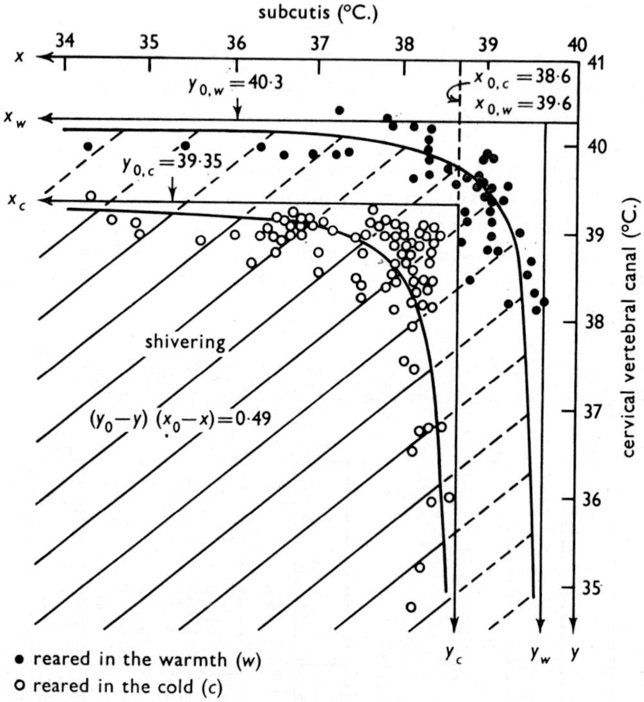

Fig. 7. Values of threshold temperatures of the spinal cord and of the body surface necessary to induce shivering in cold-adapted and warm-adapted guinea-pigs, 4-8 weeks old (Brück and Wünnenberg, 1967).

skin the higher must be the temperature of the spinal cord in order to induce shivering and vice versa. Cold-adapted animals have the threshold temperatures for shivering about 1° C lower than that in warm-acclimated animals and it was suggested that shivering is suppressed in newborn and cold-acclimated adult animals by maintenance of temperature-sensitive areas of the cervival cerebral canal above the shivering threshold by heat supplied from the brown adipose tissue. Only after greater cooling of animals, which also reaches the area of the spinal cord, shivering can be induced in addition to non-shivering thermogenesis. Suppression of shivering contrary to its induction seems to be dependent solely on the increase of temperature of the skin (Janský et al., 1973).

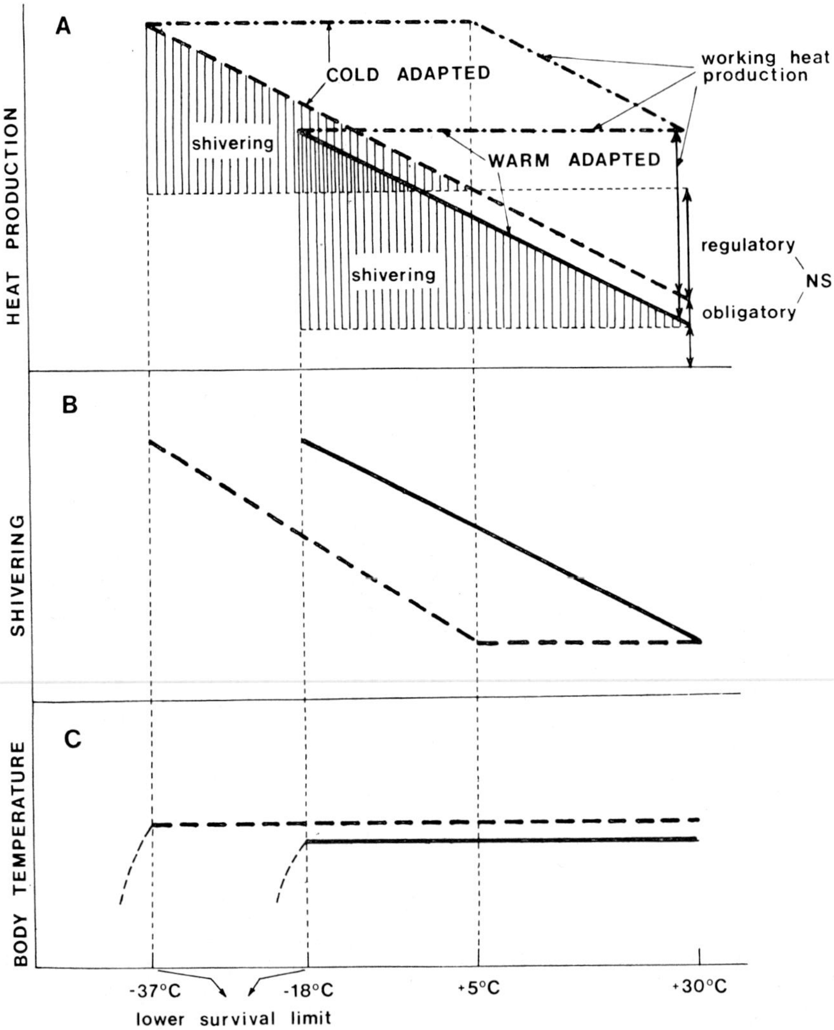

Fig. 8. Scheme of heat production mechanisms of cold- and warm- adapted rats. In warm-adapted individuals shivering is the only source of heat and maximal working heat production can substitute for shivering. In cold-adapted individuals NST takes place at temperatures close to thermoneutral zone. At lowered temperatures when the capacity of non-shivering thermogenesis is exceeded, shivering comes also into play. Working heat production is additional to the resting metabolism in the cold in that range of ambient temperatures at which NST occurs. Cold adaptation increases heat production capability and enlarges the range of temperatures animals can survive by 19° (from —18 to —37°C). Scheme according to Hart and Janský (1963).

Efferent pathways for shivering are similar to those during muscular work (Hemingway and Stuart, 1963), similarly as are the mechanisms regulating intensity of shivering at the cellular level. During muscular contractions, the frequency of which depends on the character of nervous stimulation, the ATP is split by actomyosin adenosinetriphosphatase and the higher level of ADP produced in this process induces an increase in metabolism of sarcosomes by mechanisms of respiratory control (Lardy and Wellmann, 1952). The increased oxygen consumption is used for resynthesis of ATP from ADP. Consequently, the energetic metabolism of the muscular cell is controlled by the ADP level and it increases with the quantity of ADP available.

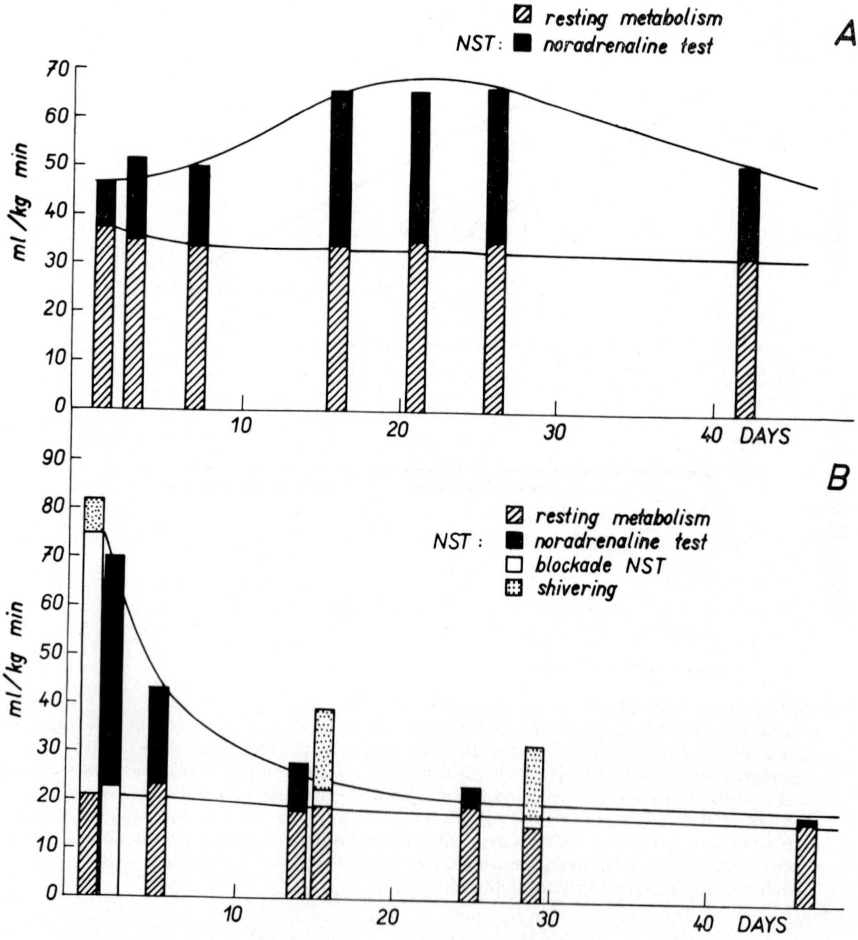

Fig. 9. Postnatal development of nonshivering thermogenesis in rats (A) and guinea-pigs (B). Arranged according to Janský et al. (1969) and Zeisberger et al. (1967).

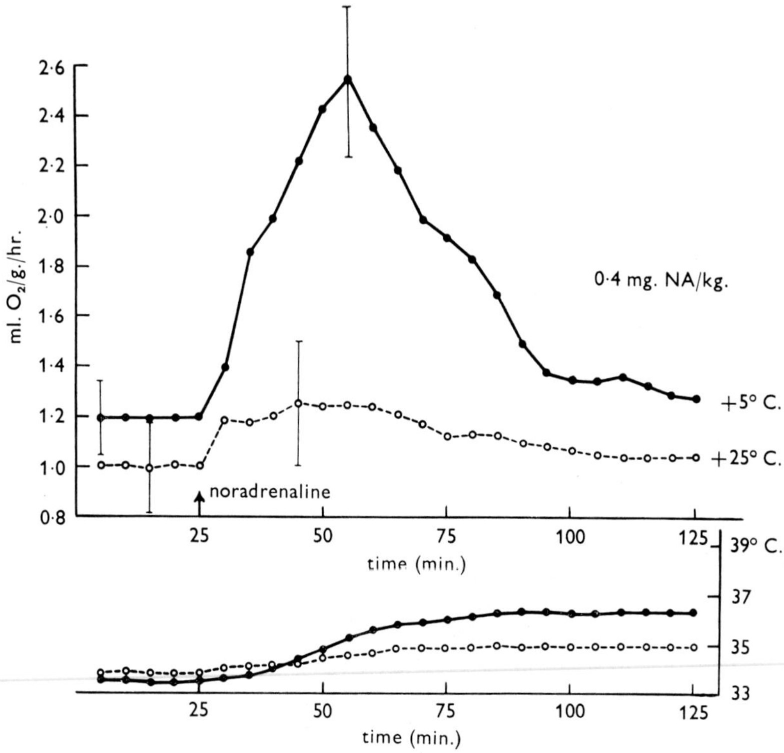

Fig. 10. Influence of intramuscular injections of noradrenaline on the magnitude of the calorigenic response in cold-adapted (5°C) and warm-adapted (25°C) rats. Lower panel shows changes in rectal temperatures (Janský et al., 1967).

NONSHIVERING THERMOGENESIS

Nonshivering thermogenesis (NST) is a heat production mechanism liberating chemical energy due to processes which do not involve muscular contractions. The heat released under basal conditions is called OBLIGATORY NST, while the heat produced during cold exposure is called REGULATORY NST (Janský, 1973) (Fig. 8).

Different methods are currently used for measuring nonshivering heat production. The magnitude of NST can be assessed from the metabolic increase in the cold which is not yet accompanied by shivering, or from the calorigenic effect of noradrenaline or from the metabolic decrease occurring in cold exposed animals after application of drugs, which inhibit the effect of calorigenic hormones (adrenolytics) or suppress the activity of the sympathetic nervous system (sympatholytics) (Mejsnar and Janský, 1971a).

Nonshivering thermogenesis is of great significance for newborn mammals. It reaches a 2-3 fold of basal metabolic rate in all species so far studied regardless of their body weight (Janský, 1973). The significance of NST changes during the postnatal period. In species which are born relatively mature (guinea pigs, kittens, lambs) it reaches its greatest significance just after birth and during the further postnatal development its participation in the total thermogenesis greatly decreases, so that in 3-week-old guinea pigs its significance is minimal. In species that are born immature (golden hamster, lemming, rat), the magnitude of NST first increases during the postnatal period reaching its maximum in the second or third week of life. In older individuals the magnitude of NST again decreases (Fig. 9).

In adult mammals the NST is the main mechanism of cold-adaptation. Nonshivering thermogenesis is associated with calorigenic action of noradrenaline. Injection of this substance in optimal concentration induces a metabolic response which is comparable in magnitude to that of NST (Janský et al., 1967) (Fig. 10). In cold-adapted individuals the NST is additional to shivering and takes place at temperatures close to thermoneutral zone shifting shivering to lower temperatures (Hart and Janský, 1963) (Fig. 8).

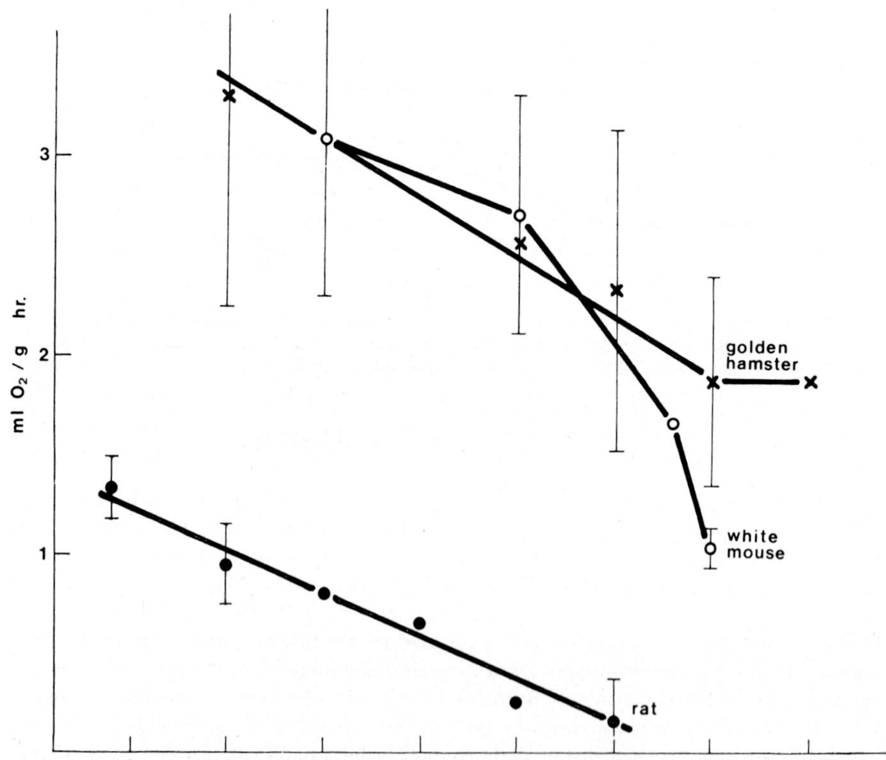

Fig. 11. Magnitude of nonshivering thermogenesis measured as maximal metabolic increase after injection of noradrenaline in white rats, golden hamsters and white mice adapted to various temperatures of the environment. (From Janský et al. (1967); Mejsnar and Janský (1971); Vybíral and Janský (1974)).

In the rat, the magnitude of NST is inversely related to the acclimation temperature. The lower the acclimation temperature, the greater is the magnitude of NST (Fig. 11) (Janský et al., 1967). In the white mouse, the maximum of NST is already reached at acclimation temperature of 20° C (Majsnan and Janský, 1971b); while in the rat the adaptational temperature must reach more than 0° C in order to obtain maximal intensity of NST. For maximal development of NST a cold exposure lasting at least 3 weeks is necessary. The rate of development of NST is the same at all acclimation temperatures. Intermittent exposure to cold also induces the development of NST (Janský et al., 1967).

The magnitude of NST in different species decreases with the increasing body weight of the animals. In the smallest mammals (bats) it reaches more than 10 times the basal metabolic rate while in mammals weighing more that 10 kg the occurrence of NST is minimal (Heldmaier, 1971) (Fig. 12). Thus, in small species the NST may be more important than shivering. Nonshivering thermogenesis is of great importance for hibernators, where it occurs not only during the exposure of animals to cold, but also during awakening from hibernation. It seems, that in hibernators the magnitude of NST is greater than in euthermics of the same size and that it is already present in a great extent in warm-adapted individuals (Vybíral and Janský, 1974). Nonshivering

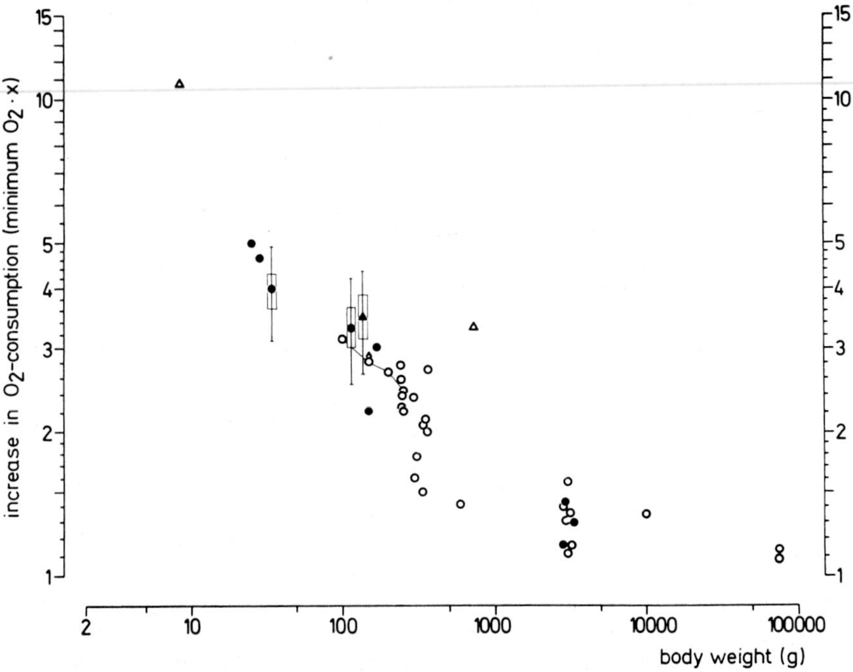

Fig. 12. Relation between the weight of various mammals and their metabolic response to the administered noradrenaline (Heldmaier, 1971). ○ ● non-hibernators; △ ▲ hibernators.

thermogenesis can supply about 40% of the necessary heat during arousal of hamsters (Hayward and Lyman, 1967) and almost as much as 80% of heat during arousal of a small bat Myotis lucifugus (Hayward, 1968).

During NST the heart frequency and the cardiac output increase by about 50%, as well as the blood flow in various organs with the highest increase in brown adipose tissue, heart and muscles. On the other hand, no significant changes in the blood flow of the skin were observed. The fraction of the cardiac output distributed to the individual organs remains practically unchanged with exception of the skin. Its blood supply relatively decreases, indicating that the heat loss from the skin is impaired during NST (Janský and Hart, 1968) (Fig. 6).

During NST the heat is being produced in various organs (Janský, 1971). The most important organ is the skeletal muscle which in the rat can generate about 50% of total heat produced during NST (Janský and Hart, 1963). Another very important thermogenetic organ is the brown adipose tissue. In newborn rabbits about two-thirds of NST can have its origin in this organ (Heim and Hull, 1966). In the initial phase of awakening from hibernation, this organ may cover about 80% of total heat, the total contribution over the whole period of awakening being less that 25% (Mejsnar and Janský, 1970). In adults, the thermogenetic significance of the brown adipose tissue may be substantially smaller (about 10%) (Janský and Hart, 1968) and this organ may have a special role in heating thermosensitive elements in the spinal cord (Brück and Wünnenberg, 1970). In adult rats a small amount of NST may be generated in the liver and in the intestine (Janský, 1971).

In guinea pigs the induction of NST depends on mutual relation between afferent signals from preoptic and supraotic areas of the anterior hypothalamus and from there on the body periphery (Brück and Wünnenberg, 1970). The mutual relation between threshold temperatures necessary for induction of NST is again hyperbolic. The signals to effectors are generated in posterior hypothalamus which at higher temperatures inhibits the sympathetic nervous system and thus the NST.

The dominant role of the sympathetic nervous system in regulation of NST was documented by depression of NST due to action of sympatholytics (Hsieh et al., 1957) (Fig. 13), by an increased amount of noradrenaline in the urine during long-term cold exposure (Leduc, 1961) (Fig. 14) and by the increased calorigenic action of noradrenaline (Hsieh and Carlson, 1957b, Janský et al., 1967). The increased concentration of noradrenaline in the blood activates β-adrenergic receptors at membranes of target cells and stimulates heat production by mechanism which has not been yet entirely clarified. β-adrenergic receptors were identified as the adenylcyclase system (Himms-Hagen, 1972). Activation of this system leads to an increased production of cAMP within the cell and to a consequent activation of lipolytic enzymes. The free fatty acids liberated by the lipolytic action of noradrenaline induce the loosening and thus the decreased effect of oxidative phosphorylation and then serve as substrate for increased mitochondrial oxidations (Lindberg et al., 1967).
Among other hormones that have the calorigenic effect, mainly adrenaline, secreted from the adrenal gland, should be mentioned. Adrenaline thermogenesis is present in a

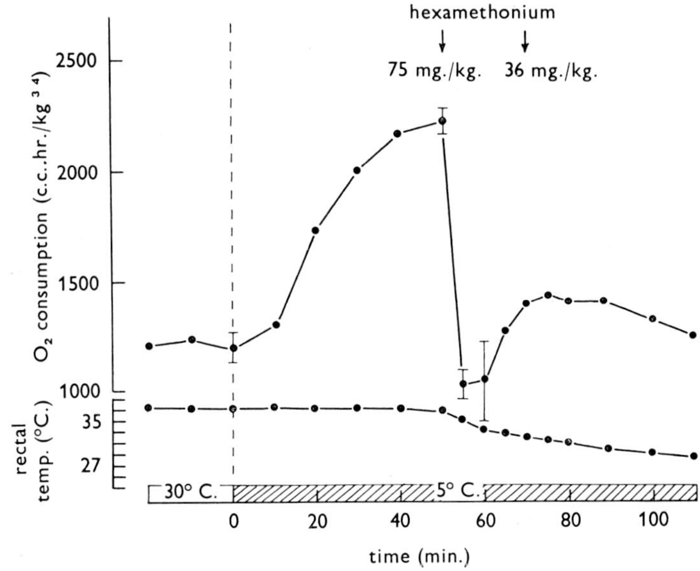

Fig. 13. Changes in the oxygen consumption of curarized rats before and after application of cold. At the arrows non-shivering thermogenesis was blocked by hexamethonium (Hsieh et al., 1957).

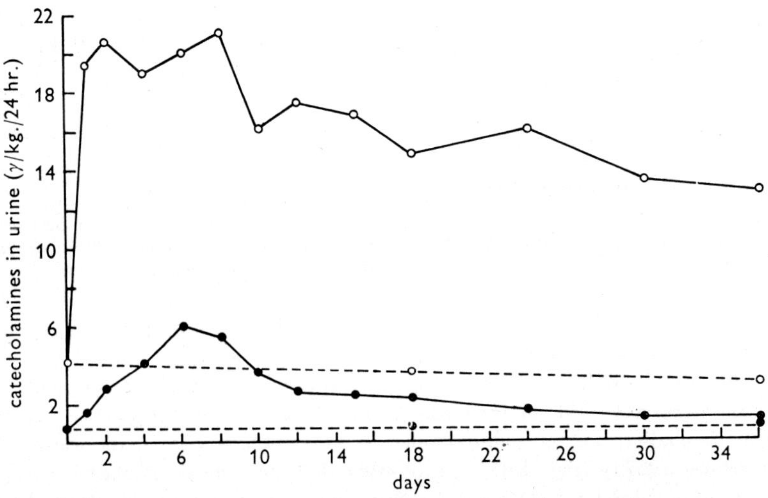

Fig. 14. Changes in the catecholamine contents in the urine during the development of cold-adaptation (Leduc, 1961). Noradrenaline (○——○) and adrenaline (●——●) after exposure to cold. Control levels of noradrenaline (○——○) and adrenaline (●——●).

great extent in warm-acclimated individuals, being almost as great as that of noradrenaline; and it could be further increased due to cold-adaptation. (Himms-Hagen, 1969). Adrenaline thermogenesis seems to be of secondary significance for thermoregulation as the extirpation of adrenal gland leads to a small decrease of NST (Cottle, Carlson, 1956) (Fig. 15). It is presumed therefore that this hormone comes into play

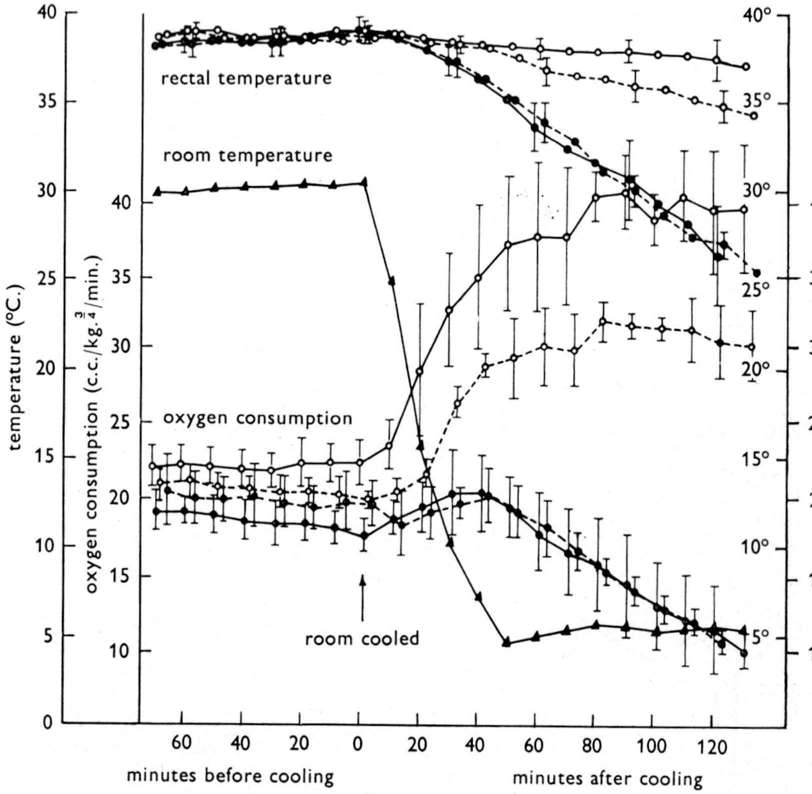

Fig. 15. Effect of adrenalectomy on the metabolism of curarized cold- and warm-adapted rats before and after cold-exposure (Cottle and Carlson, 1956). ○ cold- (5°C) adapted rats; ● non-adapted (28°C) rats; — intact rats; --- adrenal demedullated rats.

under conditions of a great stress, when the noradrenaline thermogenesis is not able to meet all demands of the organism (Janský, 1973). The role of thyroxine thermogenesis in the thermoregulation is not precisely known. Nonshivering thermogenesis still persists in thyroidectomized cold adapted rats, although the metabolic increase after cold exposure is smaller and the maintainance of the heat balance is deteriorated (Hsieh and Carlson, 1967a) (Fig. 16). It is possible that the role of thyroxine consists in potentiation of calorigenic effects of noradrenaline and adrenaline (Swanson, 1956).

ACTH, growth hormone, corticosterone, glucagon and insulin are also calorigenic. It

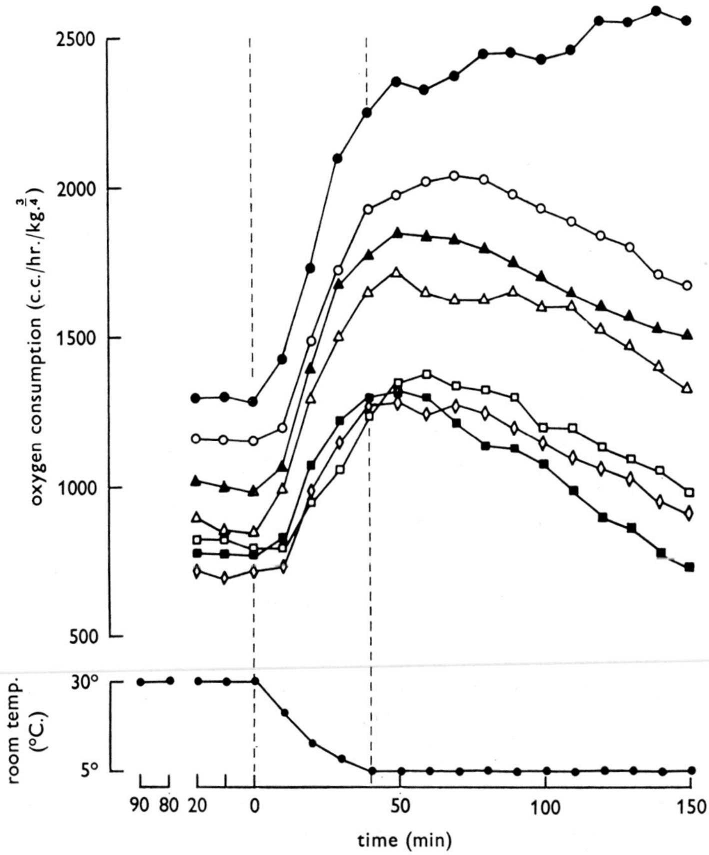

Fig. 16. Effect of a thyroidectomy on the metabolism of curarized cold-adapted rats before and after cold-exposure. ● controls; ○ 2 days; ▲ 4 days; △ 6 days; □ 8 days; ■ 10 days; ▽ 12 days. All after thyroidectomy (Hsieh and Carlson, 1957).

is not clear, however, whether thermogenesis of these hormones is of any thermoregulatory significance (Janský, 1973).

THERMOGENESIS DUE TO MUSCULAR WORK

A great amount of heat can be also generated during work and voluntary activity. Small mammals running the treadmill at their maximal performance can increase metabolism by 4 to 8 times (Hart, 1971). The question to be clarified is whether or not the heat liberated by various kinds of muscular activity can be utilized for maintaining the body temperature.

In warm-adapted rats, it was observed that the amount of heat produced during maximal work at different environmental temperatures is the same as that produced during maximal work at thermoneutral zone. In these individuals the lowered temperatures do not change the intensity of metabolism during work, so that the difference between the intensity of metabolism during maximal work and at rest becomes smaller (Hart, Janský, 1963) (Fig. 8). Since the body temperature does not change substantially, it can be concluded that the heat produced during maximal work substitutes for cold thermogenesis and thus may have a thermoregulatory significance. A considerable greater heat production at work than at rest taking place at thermoneutral zone indicates a marked reduction of insulation and thus increased conductance of the body periphery mostly due to vasodilatation. This can be demonstrated by a greater blood flow in muscles and in the skin during work at thermoneutral zone. (Fig. 6). With lowering of temperatures, the insulation of the body during work becomes close to that at rest and thus, finally, the same intensity of heat production is sufficient to maintain the heat balance.

In cold-adapted rats (Hart and Janský, 1963) and in small wild-living mammals (Hart, 1950; Hart and Heroux, 1955; Janský, 1959a, b, c) the heat production during maximal work at moderately lowered temperatures first increases proportionally to the temperature gradient, so that for the same effect – e.g., for maintaining the heat balance it is necessary to produce more heat (Janský, 1965). The heat produced during maximal work is additive to that in the cold and it is evidently lost without any thermoregulatory effect. With further lowering of environmental temperature, however, the working metabolism substitutes again for resting metabolism, as in warm-adapted individuals (Fig. 8).

What is the cause of differences in heat economy between warm- and cold-adapted animals? The substitutive relation between working and resting metabolism in the cold evidently occurs in those individuals in which both components of heat production are of the same nature. Since shivering and work produce heat by the same mechanism e.g., by muscular contractions, it is tempting to conclude that these processes can substitute each other during thermoregulation. Substitution of both components of heat production occurs in warm-adapted individuals at all temperatures below the thermoneutral zone, and it is more effective the more the environmental temperature is lowered. On the other hand, in cold-adapted animals the additive relation between resting and working heat production indicates that the heat is produced by different mechanisms and points out the existence of NST. At very low temperatures when shivering appears along with NST, addition of working and resting metabolism is again possible as in warm-adapted individuals.

It follows that the working heat production is not very economical from the thermoregulatory point of view. So far, there is no evidence that it would be used purposely as a source of heat in animals exposed to cold.

ADAPTATION TO COLD

Data presented above have shown that in rodents and in other small mammals adaptation to cold is realized mainly by increasing the heat production capability due to NST.

This adaptation is undoubtedly very efficient – in rats it enlarges the range of thermal tolerance by about 18° C. On the other hand, this adaptation has also its negative side. The increased heat production results in higher demands for energy restitution in the body, which is attained by an increased food consumption. As a result, individuals adjusted this way become more dependent on accessibility of food and they are forced to use more effort to provide it. The reduced dependence of animals on thermal factors is thus substituted by increased dependence on food factors.

Therefore, mechanisms reducing heat loss from the body by improving insulation (lowering the conductance) also commonly occur in mammals. It is well know that certain species improve body insulation in winter season (Hart, 1956) and that animals living permanently in Arctic reduce heat loss so efficiently, that even considerable reduced ambient temperatures (down to −30° C) do not result in an increase of heat production (Scholander et al., 1950). The enlarged thermoneutral zone indicating an improved body insulation was also observed in rabbits adapted to cold under laboratory conditions (Kočková and Janský, 1968).

In these animals the improved insulation was achieved by a thick fur. The same insulative qualities, however, can have a thick layer of subcutaneous fat, which appears in bare mammals, such as seal and swine (Hart, 1963a). In these animals a vasoconstriction restricts the flow of the warm blood to the body periphery. This results in the peripheral hypothermy which also efficiently reduces the heat loss.

Adaptation to cold due to improved insulation appears to be biologically very important, because it makes possible the existence of mammals under extreme cold conditions with minimal requirements as to the food supply. However, it becomes functionally justified only in species of greater size (size of fox and larger) (Hart, 1956), because in small mammals the relative heat loss is so great that the fur cannot protect them sufficiently. Smaller mammals must therefore rely on an additional mechanism – on an increased heat production capability due to NST.

CONCLUSIONS

Rodents' ability to withstand cold has been intensively studied during the last 20 years. This interest has stemmed partially from the convenience in methodology for study of animals of small body size (10^1-10^4 gm) and for the great diversity of thermoregulatory responses within the group. Much of the work in this area has been done on laboratory animals such as the white rat in which a great effort has been devoted to the study of adaptation to cold.

In small mammals there exist three basic mechanisms of heat production which can be utilized in temperature regulation: shivering, nonshivering thermogenesis and muscular work. The quantitative significance of all these mechanisms seems to be about the same, but their energetic efficiency is different. Muscular work appears to be a less efficient mechanism of heat production because at most temperatures it increases substantially the heat loss from the body, so that for the same thermoregulatory effect e.g., for maintenance the same body temperature, it is necessary to produce more heat. Heat production during muscular work can substitute for shivering, while it is additional to NST.

It was shown clearly that in small mammals the main mechanism of cold adaptation is the increased heat production capability due to development of a new thermogenetic mechanism called nonshivering thermogenesis. Nonshivering thermogenesis takes place at temperatures below thermoneutral zone, shifting shivering to lower temperatures. In larger mammals this adaptation becomes uneconomical, therefore energy conserving changes also develop and become the principal mechanism of cold adaptation in these species.

REFERENCES: see pp. 548-550.

Effects of Cold on Energy Metabolism in Birds

by

H. S. Siegel

1. INTRODUCTION

In contrast to the poikilothermic reptiles from which they were derived, birds, like mammals, are homeothermic. Thus to maintain a relatively constant temperature at the vital organs, birds must lose or conserve heat in response to changes in the environment. Although some reptiles can maintain body temperatures within a narrow range (Cowles and Bogert, 1944), birds and mammals, in addition, can also derive heat by increasing their own metabolic processes. We assume from the small but significant differences in levels at which body temperatures are maintained (Table 1) and the separateness of their evolutionary pathways that the homeothermy of birds has arisen separately from that of mammals (Freeman, 1971; Dawson and Hudson, 1970).

With exceptions among certain avian orders, body temperatures of birds in an inactive state in moderate temperature environments range from 39-44° C during the daily waking stage (Table 1). This range contrasts to the range of 36-40° C under similar conditions for higher mammals (Eisentraut, 1961). Although many factors may influence body temperature (age, sex, breed, size, activity, diurnal variations, and the insulative value of the outer integument), the above comparison indicates that in periods of low environmental temperature birds may require a higher metabolic intensity to maintain relatively constant core temperatures by endothermic means.

2. ACCLIMATIZATION TO COLD

Birds living in polar regions are remarkably tolerent to cold. Although larger species (e.g., arctic glaucous gulls, *Larus hyperboreus barrovianus*; willow ptarmigan, *Lagopus lagopus*; or ravens, *Corvus corax*), may have sufficient feather insulation to live with no special metabolic effort (Scholander et al., 1950), smaller species are not so well protected. Redpolls (*Acanthas*), chickadees (*Parus atricapillus*), and gray jays (*Perisoreus canadensis*), ranging in weight from 12 to 60 gm, live at temperatures of −20 to −50° C although their insulation appears to be inadequate. Steen (1958) found that the daytime critical temperature of redpolls in Sweden was 24° C and at −20° C their metabolic rate was some 300% greater than their resting rate. Similarly, pigeons (*Columbia livia*) with critical temperatures of 20° C had metabolic rates about 220% above resting rates at −30° C (Steen and Enger, 1957). The increases in food intake in subsequent days indicated that acclimatization was related to the ability to sustain elevated heat production by metabolic means.

Table 1. Body temperatures for birds of various orders*

Order	No. of species sampled	Range (°C)
Sphenisciformes (penguins)	6	37.0–38.9
Struthioniformes (ostrich)	1	39.2
Casuariiformes (casuaries and emu)	4	38.8–39.2
Apterygiformes (kiwis)	3	37.8–39.0
Tinamiformes (tinamous)	1	40.5
Gaviiformes (loons)	1	39.0
Podicipediformes (grebes)	4	38.5–40.2
Procellariiformes (albatrosses, shearwaters, petrels, and allies)	13	37.5–41.0
Pelicaniformes (tropic birds, pelicans, frigate-birds, and allies)	9	39.0–41.3
Ciconiiformes (herons, storks, ibises, flamingos, and allies)	12	39.5–42.3
Anseriformes (screamers, swans, geese, and ducks)	28	40.1–43.0
Falconiformes (vultures, hawks, and falcons)	12	39.7–42.8
Galliformes (megapodes, curassows, pheasants, and hoatzins)	22	40.0–42.4
Gruiformes (cranes, rails, and allies)	7	40.1–41.5
Charadriiformes (shorebirds, gulls, auks, and allies)	39	38.3–42.4
Columbiformes (sand-grouse, pigeons, and doves)	5	40.0–43.3
Cuculiformes (cukoos and plantain eaters)	2	41.9–42.3
Strigiformes (owls)	9	39.2–41.2
Caprimulgiformes (goatsuckers, oilbirds, and allies)	5	37.6–42.4
Apodiformes (swifts and hummingbirds)	25	35.6–44.6
Coraciiformes (kingfishers, motmots, rollers, bee-eaters, and hornbills)	1	40.0
Piciformes (woodpeckers, jacamars, toucans, and barbets)	10	39.0–43.0
Passeriformes (perching birds)	101	39.2–43.8

* From Dawson and Hudson (1970)

Although acclimatization to cold is not entirely metabolic (West and Hart, 1966), Irving (1964) believes that the data do not support the contention that thick feather coats alone suit birds to arctic environments. By his calculations, birds of the tropics would have to increase feather insulation eight times to adapt to arctic environments and no such difference exists. Flight aerodynamics and swimming hydrodynamics also impose restrictions on the feather cover of birds not found in mammalian coats (Irving, 1960).

The degree of metabolic adjustment is related to the temperature at which the bird is acclimatized and its genetic background. For instance, house sparrows (*Passer domesticus*) increase their metabolism sufficiently to maintain energy balance in ambient temperatures of −30° C if they have been acclimatized to winter conditions but cannot maintain such a balance even at 0° C when tested in the summer (Kendiegh, 1949; Davis, 1955). West (1960) shows, in contrast, that tree sparrows (*Spizella arborea*) do not seasonally change their capacity for maintenance of energy balance in low temperatures despite their residency in winter cold. Hart (1962) and West (1965) found that evening grosbeaks (*Hesperiphana vespertina*) increase their oxygen consumption in winter.

The compensatory effect of increased metabolic effort in cold environments has been demonstrated experimentally in house sparrows (Gelineo, 1955; Miller, 1939) and domestic fowl (*Gallus domesticus*) (Barott and Pringle, 1947, 1949). On the average, metabolic rates of birds acclimatized to low temperature (−14 to +10° C) were 38% higher than those of birds acclimatized to high temperatures (29 to 32° C). The greatest difference was the 85% cited by Gelineo (1964) for siskins (*Chrysomitris spinus*). In laboratory experiments, the critical temperatures of cold-acclimatized birds usually declines about 5° C and, according to Hart (1957), is accompanied by a DECREASE in insulation resulting from improved peripheral heating. Increased heating of the periphery would protect exposed areas from cold injury (Dawson and Hudson, 1970) but presumably would increase energy requirements.

3. THERMOGENESIS

The preceding section clearly shows that birds live in polar regions often without sufficient feather insulation to prevent significant heat loss to the environment. Thus, chemothermogenesis must increase to ensure that heat production does not fall behind heat loss.

NONSHIVERING THERMOGENESIS

a. *Neonatal.* The bird's ability to produce heat for thermoregulation without shivering (nonshivering thermogenesis) has been disputed for several years. This dispute has come from the observation that the electromyographic (EMG) activity associated with shivering increases linearly (or almost) with decreasing ambient temperatures below 30° C (Hart, 1962; West, 1965). In addition, birds lack the mitochondria-rich brown adipose tissue found in mammals which has the capacity to break down triglycerides within the cells (Dawkins and Hull, 1964; Hull and Segall, 1965; Ball and Jungas, 1961). In neonate chicks, fat stores are in the form of pads of subcutaneous white adipose tissue (Freeman, 1967) and in the unabsorbed egg yolk. The assumption therefore has been that nonshivering thermogenesis is a relatively unimportant source of heat in birds.

Research by Freeman (1967, 1969) indicates, however, that there are similarities between the metabolic responses to cold by neonatal fowl and mammals. The rise in plasma free fatty acids (FFA) and the fall in respiratory quotient (RQ) in the neonatal chick indicate that lipids are being used despite the absence of brown fat. Furthermore, the thermoregulatory response of neonatal fowl (and mammals) is impaired by pharmacological blockade of the β-adrenergic receptors (Wekstein and Zolman, 1969); this impairment suggests that the release of β-activating catecholamines is involved in temperature regulation in these young birds. In mammals the metabolic breakdown of triglycerides in the brown fat is stimulated by the release of norepinephrine from nerve endings (Joel et al., 1968).

However, several questions remain unanswered. Although treatment with norepinephrine does increase plasma FFA in domestic chicks up to 8 weeks of age (Freeman, 1969), it may be that FFA are not mediators of the thermoregulatory response in birds. The catecholamine acts via the β-receptors and although the β-receptors are blocked and the thermoregulatory response is impaired (Wekstein and Zolman, 1969), mobilization of lipids seems to be unaffected (Freeman, 1969).

Fat mobilization possibly may be dependent on other factors. For example, glycogen may be the mobilizing agent because avian fatty tissue is very sensitive to glycogen (Langslow and Hales, 1969) and treatment with β-adrenergic blockers does not appear to interfere with its lipolytic activity in vivo (Heim and Hull, 1966). A recent report (Thaxton, Wyatt, and Hamilton, 1974) suggests that carbohydrate reserves may be utilized preferentially to lipids to produce metabolic heat in the neonatal chick. Oral administration of glucose during the 12 hrs after hatch raises body temperature in chicks whose body temperatures were depressed by a hypothermic environment (2° C).

The thyroid may also be implicated in neonatal thermoregulation. Rectal temperatures of chicks in the thermoneutral or hypothermic environments are increased by injections of thyroxine or triiodothyronine (Freeman, 1971). Significantly, an increase in plasma FFA releases thyroxine from its binding sites (Itoh et al., 1969) and this release increases metabolic heat production.

b. *Adult.* Most authors assume that nonshivering thermogenesis is lost within a week or so after hatching (Freeman, 1970; Dawson and Hudson, 1970). This assumption is based on the work of Allen et al. (1969) with domestic fowl and that of West (1965) with wild birds that indicated that the EMG activity of muscles was linearly related to decreasing ambient temperatures below 30° C and that this activity in turn was related to oxygen consumption. Although Kayser (1929) had observed that cold induced an increase in the metabolic rate of rock doves (*Columbia livia*) that had been cervically transected to prevent shivering, Hart (1962) found that curarized, cold-adapted rock doves did not show metabolic responses to norepinephrine.

However, El-Halawani et al., (1970) have shown that if cold-acclimatization continues for a long enough period (5 to 9 months), nonshivering thermogenesis may occur in adult White Leghorn males (El-Halawani et al., 1970). This nonshivering thermogenesis was evidenced by the essential disappearance of EMG activity and a marked rise in oxygen consumption.

SHIVERING THERMOGENESIS

The shivering reflex is present in newly hatched prococial chicks (Randall, 1943; Allen et al., 1969; Marley and Stephenson, 1969; Freeman, 1970), and nestlings of altricial species (Odum, 1941, 1942). Freeman (1971) feels that in neonatal chicks, only when nonshivering thermogenesis fails to maintain sufficient body temperature is the shivering mechanism stimulated.

Odum (1942), recording muscle tremors with a piezoelectric device in both nestling and adult birds of several species, found that intensity of shivering fluctuated in short-term cycles. EMG studies have shown that the relationship of temperature with shivering intensity and oxygen consumption is linear (Steen and Enger, 1957; Hart, 1962; West, 1965). These data also show that there is a broad temperature range in which chemical and physical thermoregulation overlap.

SPECIFIC DYNAMIC ACTION AND EXERCISE

As in mammals, the specific dynamic action (SDA) of food contributes to overall thermogenesis in birds. The calorigenic effects, originally applied to the heat produced during deamination of proteins, has been extended to the less pronounced "heat of digestion" and metabolism of carbohydrates (Dukes, 1955). The thermal increment over the basal expenditure produced by SDA in domestic fowl is 5-18% (Barott et al., 1938); in the pigeon it is in the order of 30% (Groebbels, 1928). Although in thermoneutral environments SDA may represent waste heat, it probably aids in the maintenance of body temperature when the ambient drops below the lower critical temperature (King and Farner, 1961). The maintenance of body temperature should spare the need for shivering and may have ecological benefits in some species during periods of low temperatures (Hart, 1957; Dawson and Hudson, 1970).

Knowledge of the degree to which heat produced by exercise could be used to heat the bird during cold weather had to wait for adequate techniques for measuring heat production of birds in flight. In early work, Hutchinson (1954) showed that the metabolic rate of domestic fowl was 40 to 45% higher while standing than while sitting, and Pearson (1950) estimated the oxygen consumption of hovering hummingbirds (*Calypte anna*) to be six times that of the birds at rest. In pigeons and budgerigars (*Melopsittacus undulatus*), the metabolic rates during flight may increase by as much as eight times the resting rate (Lefebvre, 1964; Tucker, 1968). Several authors have confirmed that in diurnal birds, the metabolic rates are higher during daylight hours than during the relatively inactive nocturnal hours (Hart, 1962; Pohl, 1969).

Tucker (1971), reviewing the various methods for estimating the power input required for flight points out that literature estimates vary 100-fold for birds of a given age but that techniques based on indirect calorimetry appear to give the most accurate results (Lasiewski, 1963; Tucker, 1966, 1970; Berger et al., 1970; Lefebvre, 1964). Research by Hart and Roy (1967) and Aulie (1971) indicates that the problem for species normally living in temperate or arctic regions may be that of overheating during flight.

4. HYPOTHERMIA AND TORPOR

As previously stated, birds acclimatized to low temperatures are able to maintain normal body temperatures by an elevation of overall metabolism. However, if availability of food is reduced during cold or inclement weather, thermic regulation may be jeopardized. Depletion of energy sources, as indicated by weight loss, causes hypothermia in young birds (King and Farner, 1961). In some birds such as the turkey vulture (*Cathartes aura*) (Heath, 1962) or the Inca dove (*Scardafella inca*) (MacMillan and Trost, 1967), the adult also undergoes moderate hypothermia.

If there is a sufficient reduction in body temperature, torpor is induced. This induction has been reported in such orders as: *Apodiformes* (hummingbirds and swifts) (Lasiewski, 1963; Bartholomew et al., 1957); *Caprimulgiformes* (nightjars, poorwills, and night hawks) (Dawson and Hudson, 1970); and *Colliiformes* (Colys) (Bartholomew and Trost, 1970). Most of these birds can enter torpor and later spontaneously arouse to feed. The extensive reviews on hypothermia by King and Farner (1961) and Dawson and Hudson (1970) indicate that with one exception, the goatsuckers (*Capriimulgidae*), there is little evidence of the type of seasonal dormancy observed in some mammalian orders.

ENTRANCE INTO TORPOR

Until recently, information on the stimulus for entrance into torpor and the physiological processes involved were gained from studies on restrained hummingbirds by Lasiewski and associates (Lasiewski, 1963, 1964; Lasiewski et al., 1967). These studies suggested that entrance into torpor is largely passive because there is little or no shivering, and so the rate of entry was thought to be primarily determined by such physical considerations as body weight and surface area. Oxygen consumption data indicated that passage from the active to inactive state can be represented by a smooth curve (Howell and Bartholomew, 1959).

Continuous temperature recordings of nesting broad-tailed hummingbirds (*Selasphorus platycerous*) with sensors embedded in synthetic eggs show that entrance into torpor is correlated to weather conditions that place demands on marginal energy reserves (Calder and Booser, 1973). However, nocturnal torpor in Andian hillstar hummingbirds (*Oreotrochellus estella*) occurs more frequently and is of longer duration in winter than in summer despite similar overnight ambient temperatures and the presence of adequate food supplies (Carpenter, 1974). The proximate cue for entrance into the inactive state was suggested to be the length of the photoperiod. Both the Calder and Booser (1973) and Carpenter (1974) studies show that nesting females can undergo torpor but with less frequency than males or non-nesting females.

METABOLISM DURING TORPOR

Once the bird is in the inactive state, metabolic rates may fall to as low as one-fiftieth of basal levels with an approximate Q_{10} of 4.1 (Lasiewski and Lasiewski, 1967). This reduced metabolic rate is accompanied by a decline in the heart rate of hummingbirds from 180 beats per min at 30° C to 48 beats per min at 7° C (Lasiewski, 1964) and from 500 beats per min to less than 10 beats per min in poorwills with similar temperature change (Bartholomew et al., 1962). Respiration is slowed and irregular with periods of apnea (Bartholomew et al., 1957); these periods of apnea result in a reduction in water loss of one third to one tenth that of homeothermic individuals (Lasiewski and Lasiewski, 1967).

Body temperatures that are reached during deep torpor vary with species. Poorwills and nightjars (*Copromulgus europaeus*) tolerate body temperatures of 5 to 8° C for several hours (Howell and Bartholomew, 1959). Lasiewski (1963) reports that body temperatures below 8° C may kill hummingbirds. However, Carpenter's (1974) field study showed that Andean hummingbirds endured body temperatures of 7° C; at this temperature, periods of slow breathing were alternated with periods of apparent apnea.

Below 7° C, however, the periods of apnea ceased, indicating that increased metabolic output was required to maintain life.

AROUSAL FROM TORPOR

Spontaneous arousal from torpor is accompanied by a rapid increase in metabolic activity; the increase in body temperature depends primarily on shivering (Bartholomew et al., 1962). Heart rates may increase up to 1000 beats per min in hummingbirds (Lasiewski et al., 1967). Studies on restrained hummingbirds (Lasiewski, 1963, 1964) show that arousal is not initiated unless body temperatures are moderately warmed by ambient temperature near 13° C. Studies on poorwills (Howell and Bartholomew, 1959), night hawks (*Chordilles*) (Marshall, 1955; Lasiewski and Dawson, 1964), and swifts (*Aeronautes sexatales*) (Bartholomew et al., 1957) also indicated that some increase in ambient temperature occurs before arousal. More recent field studies indicate however that arousal can occur before the rise in ambient temperature and that it is timed by circadian rhythms synchronized by photoperiod length (Calder and Booser, 1973; Carpenter, 1974).

REFERENCES: see pp. 550-552.

Effects of Cold on Electrolyte Metabolism in Rodents and Laboratory Animals

by

W. S. Platner

INTRODUCTION

A statement made 14 years ago regarding the lack of research on electrolytes in cold exposed animals is still true (Platner, 1959). There has been very little significant animal research even on the common ions as Ca^{++}, Mg^{++}, Na^{+}, K^{+}, PO^{--}_4 SO_4 and HCO^{-}_3 and less research on the so called trace elements as Cu^{+++}, Fe^{+++}, Co^{++}, Mn^{++} etc. This probably stems from the fact that such data collected on animals was judged to be of little value and that it would be more meaningful and easier to collect such data on man himself. Furthermore caloric and vitamin needs of cold exposed animals seem to be more important thus relagating ion fluxes and dietary mineral requirements to a secondary role.

Many of the minerals are known to be important in various biochemical processes. The elevated metabolic rate of cold exposed animals would certainly increase the turnover rate and requirement for the various ions, yet for some reason very few studies have taken this direction. In the past, detection and quantitation methods have been tedious and limited. This may explain why such studies have been neglected. However, today this is not true, new detection and quantative techniques should make the task relatively easy. Atomic absorption, neutron activation, electroanalytical chemistry and flame photometry are available techniques on most university campuses.

The use of cold exposed, cold acclimated, hibernating or hypothernic animals as models presents a unique and natural means of altering metabolic rates for the study of ion movements and dietary mineral requirements. Except for experiments on hypothermia this can be done without the use of anesthetics which frequently alter metabolic pathways.

In the past most experiments were done to simply accumulate data; very few were designed to study mechanisms. The present review of the literature makes reference to the older literature; and whenever possible, an attempt is made to define the physiological significance of the ion movements.

REQUIREMENT FOR IODINE BY COLD EXPOSED ANIMALS

Many homeotherms show a significant increase in their basal metabolic rate when exposed to cold. The rabbit (Kockova et al., 1968), dog (Nagasaka and Carlson, 1968), cat (Hemingway et al., 1964), ground squirrel (Pohl and Hart, 1965) and rat (Janský et al., 1967; LeBlanc and Cote, 1967) have increased metabolic rates on cold exposure which have been related to increased thyroid activity (Locker and Weish, 1966; Bauman and Turner, 1967). The requirement for iodine increases during cold acclimation (Heroux and Petrovic, 1969). Placing rats in 5° C shows a significantly elevated urinary excretion of iodine after the first day (Cadot et al., 1969). However, rats cold acclimated for 3 months and fed a low iodine diet excrete less iodine in the urine than controls even when they receive higher doses of thyroid hormone (Galton and Visula, 1969).

The requirement for thyroxine increases many fold when rats are transferred from 28° to 6° C (Hsieh, 1966). An indication of the large requirement of cold exposed (6° C) rats can be appreciated from the work of Hsieh (1966) who showed that a dose of thyroxine lethal at 28° C may actually be insufficient when rats are exposed to 6° C. The increased requirement for iodine (thus thyroxine) in rats appears to be in the early stages of cold acclimation (Heroux and Braver, 1965). Woods and Carlson (1956) estimated the amount of thyroxine required to restore resting metabolism of thyroidectomized rats to the level of that of intact rats. They concluded that secretion at 5° C was equivalent to 7.5 μg of L-thyroxine per day and at 25° C it was 3.5 μg per day. The requirement was still more than double after 60 days at 5° C.

A similar increase in I^{131} values was also observed in the cold exposed hamster (Knigge, 1963). However, the increase in total I^{131} activity revealed a peak at 30 days of cold exposure (5-7° C) and subsequently declined at 60 days (Knigge, 1963).
Cadot et al. (1969) support this view with studies on rats after using the isotope equilibrium method and controlled dietary iodine intake. These workers reported that thyroid activity was near control levels after the cold-acclimated state was reached. Thus the increased fecal excretion of iodine in the early stages of cold acclimation is diet dependent and probably unnecessary once cold acclimation is complete (Heroux and Petrovic, 1969). This study has received additional support from Patton and Platner (1970, 1971) and Masoro (1966) who postulated that thyroxine was responsible for the phenomenon of non-shivering thermogenesis, one of the most important of the many metabolic adjustments occurring early in cold acclimation. Continued exposure reduces thyroxine requirement. In other animals such as the ground squirrel (*Citellus Citellus*), the thyroid involutes in the fall and subsequent cold exposure cannot increase I^{131} release (Lachiver and Petrovic, 1963). These differences may arise from variations in blood flow to the thyroid gland or differences in secretion rates of thyrotropin or releasing factors.

METABOLISM AND REQUIREMENT OF SODIUM, POTASSIUM AND CHLORIDE IN COLD EXPOSED ANIMALS

These three ions, Na^+, K^+ ,and Cl^-, are almost invariably considered together whenever studies are conducted on electrolytes regardless of the experimental variable. Reasons for this rest with the ease with which these ions can be determined. Sodium and potassium by flame photometry and chloride by electrotitration.

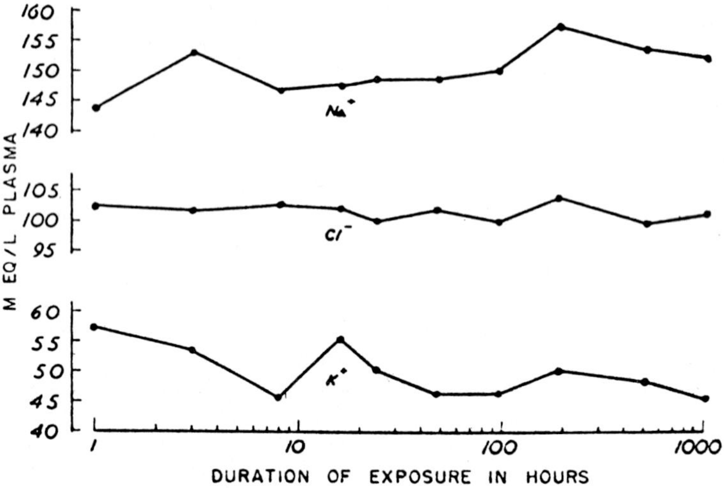

Fig. 1. Plasma electrolytes during exposure to cold of 2°C. (From Baker and Sellers, 1957).

Bass and Henschel (1956) pointed out how many organisms are able to tolerate adverse temperatures by altering their water and electrolyte compositions. Experiments by Baker and Sellers (1957) where they exposed the rat to a low environmental temperature of 2.0° C found a loss of chloride on the first day of exposure. Continued exposure to cold resulted in retention of sodium and to a lesser extent potassium (Fig. 1). Exposure to cold for 45 days (at 2.0° C) caused a sustained elevation of sodium in the plasma, with transient increase in potassium and no change in the chloride concentration. Prolonged exposure to cold (45 days) resulted in an increased blood volume and total body water content.

Baker and Sellers (1957) also used balance studies and short exposure periods of 15 days (Fig. 2). Their data show "peaks" of positive balance in potassium. The significance of these peaks is unclear but may be related to the increased metabolism of the animals exposed to cold. The decrease in chloride on the first day of cold exposure may relate to the increased gastric secretion of chloride, some of which may have been lost with increased fecal excretion.

Similar water and electolyte balance studies on rats have been performed by Sather and Rogers (1969). In these and other studies on mammals (Nungesser, 1955; Bass and Henschel, 1956; Ferrand, 1959; Bader and Eliot, 1952) exposed to cold, show that the early response is characterized by a hemoconcentration of brief duration with some diuresis. The plasma sodium and potassium increase slightly, and these increases are transitory. Prolonged cold exposure, on the other hand, shows elevations in plasma, blood extracellular and total body water volumes. The total water content of muscle is decreased while chloride space is increased. Potassium and chloride in the plasma remain in the normal range while sodium is elevated. These fluid and electrolyte

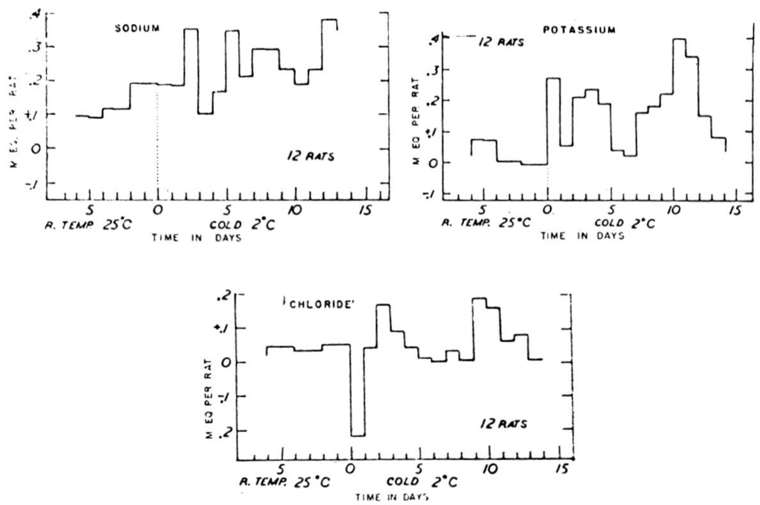

Fig. 2. Sodium balance, potassium balance and chloride balance expressed as average number of mEq lost (—) or gained (+) per day per rat. Exposure to cold started on day o and the periods of time before or after exposure are referred to the zero time. Each horizontal segment gives the time period for which the indicated balance holds. (From Baker and Sellers, 1957).

changes produced by cold exposure may have resulted from ADH release which was reflected in the production of hyposmotic urine. The retention of sodium probably results from increased aldosterone secretion. Other mechanisms may also be involved.

HIBERNATORS. They appear to respond differently to cooling than non-hibernators. Willis (1972) points out that tissues of mammalian hibernators retain relatively normal K and Na concentrations at low temperatures and in so doing avoid loss of excitability which would otherwise be fatal. Contrary to this view were the experiments of Fisher and Mrosovsky (1970). Their results support the idea that accumulation of K^+ in ground squirrels triggers arousal within 24 hours after injecting KCl solutions. The theory is that since the hypothermic kidney is non-functional in producing urine, animals have to arouse to eliminate metabolic end products from their blood stream. In hibernation such substances might gradually build up and then at a given concentration, trigger arousal of the animal. Since K ions from muscle breakdown appear as metabolic end products, potassium was a possible suspect. Potassium was tried with the results as stated above. Fisher and Manery (1967) have developed this theory in detail. However, many other chemicals have also been implicated in arousal. (See Fisher and Mrosovsky, 1970 for discussion).

POIKILOTHERMS. They appear to present a different electrolyte picture when exposed to cold. The few studies available on goldfish (Meyer, Westfall and Platner, 1956) indicate no significant change in total body sodium or potassium despite an increase water content of 2.15% after 48 hrs at 4.0° C. The question of whether teleost

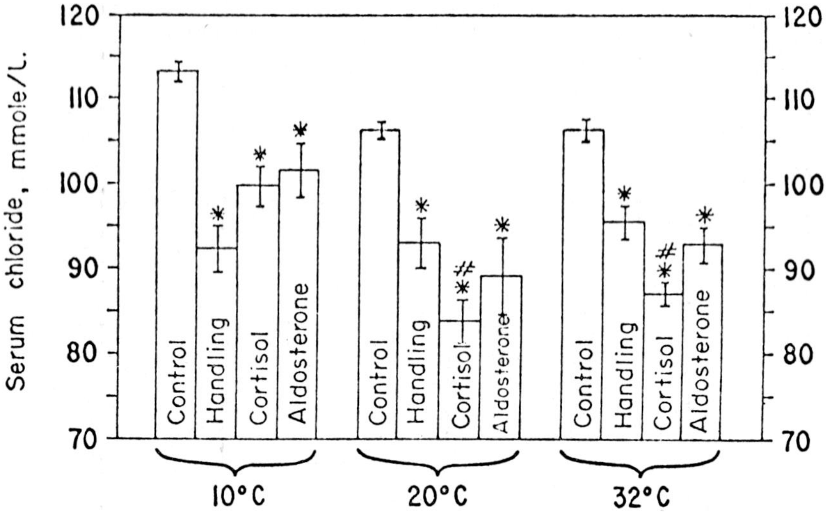

Fig. 3. Serum chloride levels in uninjected controls, goldfish subjected to handling stress, fish injected with cortisol and fish injected with aldosterone at three temperatures of acclimation. Data expressed as mean ± S.E. An * indicates values significantly different from uninjected controls whereas a # indicates values significantly different from handled fish.

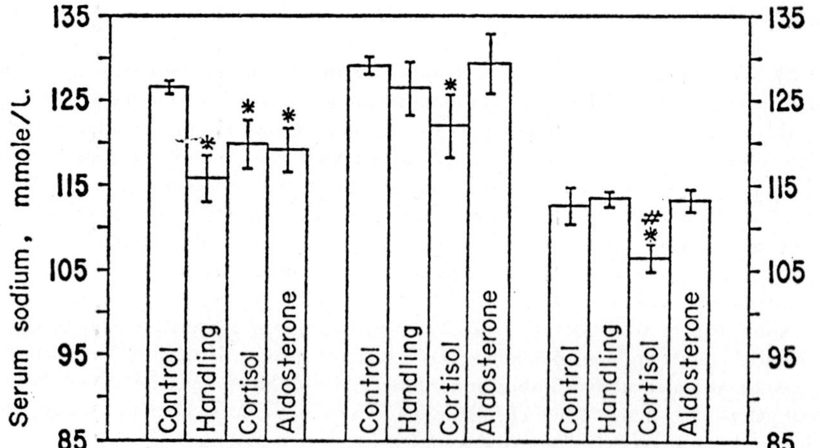

Fig. 4. Serum sodium levels in uninjected controls, goldfish subjected to handing stress, fish injected with cortisol and fish injected with aldosterone at three temperatures of acclimation. Symbols same as in Fig. 3. (From Umminger and Iist, 1972).

fish have aldosterone has stimulated considerable discussion. The pros and cons have been recently reviewed by Umminger and Gist (1972). The results of their studies suggest that aldosterone has no obvious physiological role in the control of osmoregulation in goldfish subjected to 10° C from an acclimation temperature of 32° C. Handling stress appears to have a greater effect than 10° C cold exposure on serum chloride or sodium (Fig. 3 and 4.)

FRESH-WATER MUSSEL AND EARTHWORMS. Interesting changes relating Na, K, Cl and Ca in body fluids of cold acclimated fresh-water mussels *L. Marginalis* (Fig. 5) and earthworms *L. mauritii* (Fig. 6) were described by Rao (1967). The ion changes in body fluids of cold (20°) earthworms are given in Table 1. These changes are thought to be triggered by steroidal, hormonal type secretions of neurosecretory cells in the subesophageal ganglion of the earthworm. Tissue extracts from cold acclimated worm's neurosecretory cells when injected into warm worms increased oxygen consumption 75 percent over normal. These results indicate the possibility that the contents of the nerve-tissue extracts from cold- or warm-acclimated worms, when injected into normal worms, can initiate change in a variety of biochemical processes. It was proposed that the sequence of these complex biochemical changes leads to metabolic compensations which are triggered by cold. Similar changes under hormonal control have been reported in insects.

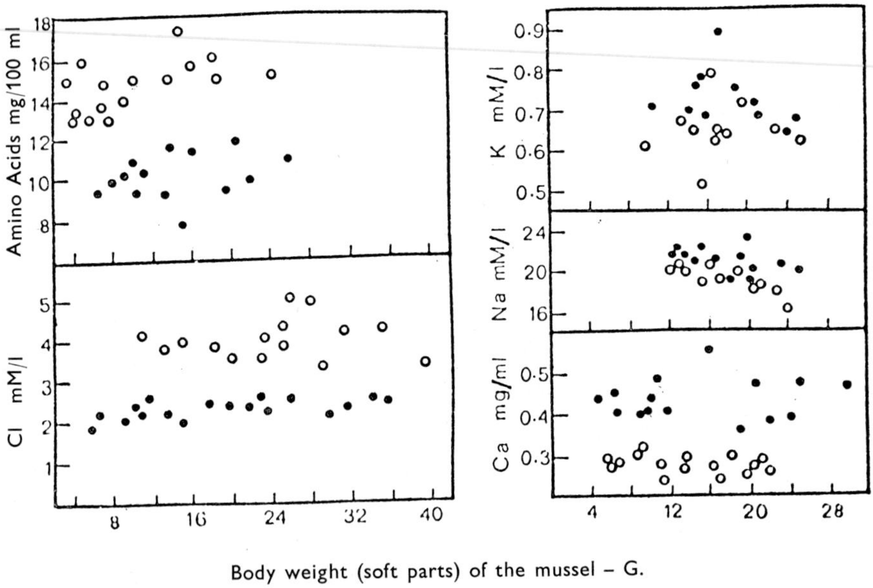

Fig. 5. Changes in the level of certain ions and of total amino acids in the blood of the fresh-water mussel, *Lamellidens marginalis*, after acclimation to low temperature. Solid circles represent cold (20°C)-acclimated: open circles, warm (35°C)-acclimated mussels. (From Rao, 1967).

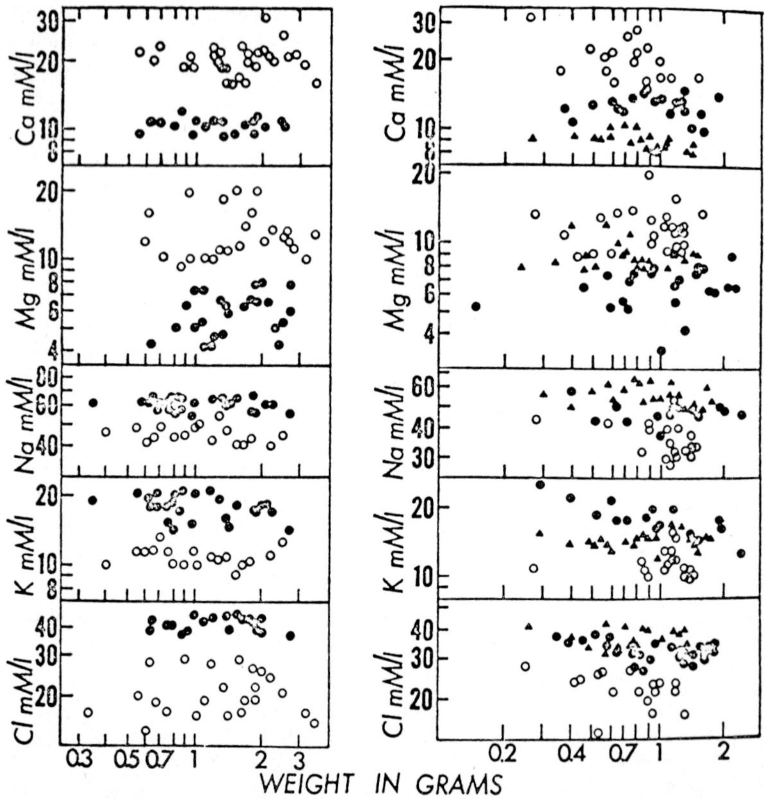

Fig. 6. Concentration of various inorganic ions of the body fluids of the earthworm, *L. mauritii*, as a function of season (left) and of temperature of acclimation (right). In the left column, solid circles represent winter worms; open circles, summer worms. In the right column, solid circles represent cold (20°C) acclimated worms; open circles, warm (35°C)-acclimated worms; triangles, normal (28°C) worms. (From Rao, 1967).

METABOLISM AND REQUIREMENT OF MAGNESIUM, CALCIUM AND PHOSPHATE BY COLD EXPOSED ANIMALS

The classic studies of Lustig, Ernst and Reuss (1937) on the snail (*Helix pomatia*) and of Suomalainen (1938) on the hedgehog have demonstrated that hibernation was accompanied by a rise in Mg and to a lesser extent Ca. Since these early reports, many publications have appeared showing a rise in plasma Mg and Ca following cold acclimation in the rat (Hannon et al., 1958; Neubeiser, 1961), hamster (Williams, 1962), dog, cat (Platner, 1953), goldfish and turtle (Platner, 1953, 1950).

The most definitive work investigating the effects of cold acclimatization on plasma electrolytes of a mammal was that of Hannon et al. (1958). Plasma from control (26° C)

Table 1. Average values and percentage increase (+) or decrease (−) of the content of various ions and free amino acids in the body fluids of cold (20° C), warm (35° C), and normal (28° C) earthworms

Constituent	Average values			Percentage changes in relation to normal	
	20° C	28° C	35° C	20° C	35° C
Calcium (m*M*/liter)	12.65	9.16	20.52	+31.7	+113.7
Magnesium (m*M*/liter)	6.55	8.88	11.45	−26.25	+ 28.6
Sodium (m*M*/liter)	47.0	58.0	36.5	−17.3	− 37.1
Potassium (m*M*/liter)	17.6	14.7	11.9	+20.0	− 19.0
Chloride (m*M*/liter)	31.73	36.7	21.55	−13.54	− 37.04
Sulfate (m*M*/liter)	2.88	4.7	4.67	−38.72	− 0.62
Free amino acids (mg/100 ml)	56.2	10.4	2.52	−45.96	+142.3

From Saroja and Rao, 1965, Table 1, p. 40.

and cold acclimatized (5° C) rats was assayed for Na^+, K^+, Ca^{++}, Mg^{++}, Cl^-, and PO^{--}_4. Significant increases were found in the levels of Na^+ and Mg^{++} following cold exposure. The sodium increase was explained as probably resulting from hemoconcentration. However, another report cited by Hannon (that of Barlow) where a 23 day exposure of troops to winter maneuvers resulted in plasma sodium increasing 8 meq/l. In these experiments the hematocrit decreased by 2%.

The rate of increase of serum Mg in the experiments cited appears to depend solely on the rate of fall in rectal temperature (Platner, 1953) or in the amount of peripheral tissue that is cooled (Neubeiser et al., 1961). Presumably cellular cooling is responsible for the release of the Mg ion. Such a movement might result from a reduction in the metabolic energy normally utilized to maintain a concentration gradient (Hannon et al., 1958).

The effect of low temperature and dietary calcium upon the magnesium requirement was studied by Hegsted et al. (1956). Growth studies with weanling albino rats exposed 24 days showed that Mg required per gram of diet to promote equal gains in weight appears to be about 4 times higher at 13° C than at 26° C. They also reported that high Ca diets and low Mg intakes apparently accentuated cold injury.

An interesting role for Ca has been described recently (Hanegan, 1973). Perfusion of the preoptic-anterior hypothalamus with excess Ca ion in ground squirrels produced a drop in core temperature. The magnitude of the drop was directly dependent on the ambient temperature. Respiration, heart rate and oxygen consumption were also reduced during Ca ion perfusion presumably due to depression of the preoptic-anterior hypothalamus. Similar experiments were performed by Heagy and Burton (1948) who showed that magnesium chloride which was injected into dogs inhibited shivering and induced panting accompanied by vasodilitation and a fall in body temperature. Many experiments using magnesium were performed many years ago (Hall and Whalen, 1951; Lecoq et al., 1960) with similar results.

One of the few studies on phosphorous was done on winter and summer bats (Myotis lucifugus) by Bruce et al. (1970). The winter bat's plasma showed an increased calcium and low phosphate presumably due to an overactive parathyroid gland when these mammals were in hibernation. The bones demeneratized during hibernation because of diminished muscular forces which normally act on bone when the animal is active. The loss of Ca and PO_4 from bone was reflected in higher PO_4 and Ca in the urine and feces, respectively.

SUMMARY

Exposure of an animal to cold (5° C) raises its metabolic rate approximately four fold. Accompanying this increase in metabolism, there is initially a water diuresis, hemoconcentration and an increased requirement for iodine. At 5° C the I^{131} requirement was found to be equivalent to 7.5 μg of thyroxine per day. This requirement gradually declines after 30 days of cold exposure and at 60 days returns to near control levels of 3.5 μg per day (Knigge, 1963).

During the early period of cold exposure, plasma sodium and to a lesser extent potassium increase. There is an increased sodium and chloride excretion. The reason for the reduced plasma volume is obscure since the reduction in plasma volume can occur without diuresis (Bass and Henschel, 1956). In this early period, plasma magnesium rises slowly so that after several days it increases to about 14% above control values (Hannon et al., 1958). However Neubeiser et al. (1961) reported a serum Mg increase of 48% after 5 days exposure at 4.0° C. Calcium also increases but to a lesser extent. While increased oxidation is necessary for increased heat production, this does not imply a concurrent increased requirement for high energy phosphates, for plasma phosphate remains unchanged.

Excluding the phenemona of hibernation and hypothermia, prolonged exposure to cold may be regarded as acclimatization to cold. The metabolic time table that takes place depends upon the parameter studied. Ascorbic acid may reach its peak concentration in tissues after 2-3 months whereas plasma magnesium reaches a peak in about 5-10 days. Plasma sodium on the other hand remains elevated throughout the cold exposure period. After the initial cold diuresis the plasma volume returns to control values as does K, Cl, PO_4, and Ca.

Poikilotherms present a different electrolyte pattern than homeotherms. Goldfish when exposed to temperatures of 4.0° C for two days show no significant change in total body sodium or potassium. Plasma magnesium increases in goldfish, and turtles exposed to cold but decreases in body fluids of winter or cold exposed earthworms. Sodium, potassium and chloride increase, in cold (20° C) worms compared to warm worms (35° C). When possible the physiological significance or mechanisms involved in the electrolyte shifts in the various animals are discussed.

REFERENCES: see pp. 553-554.

Effects of Heat on Vitamin Metabolism

by

M. L. Scott

INTRODUCTION

Animals do not grow and produce as well in very hot climates as they do in temperate or cold climates. With a sudden onset of hot weather, broilers and other meat animals grow more slowly, hens produce fewer eggs, egg size decreases, and the eggshells become thinner and more subject to breakage.

Environmental temperature has been shown to influence vitamin requirements. The effect of high environmental temperature in decreasing food consumption is generally recognized. Hertman and Hughes (1949) and Mangold et al. (1967) showed that high temperatures and humidity caused marked reductions in feed intake by swine. Thus, when the environmental temperature increases, the levels of those nutrients required in definite daily amounts, such as the vitamins, must be increased in terms of percentages of the diet. In an experiment conducted with broiler chickens at the College of Agriculture in the Philippines, a ten per cent increase in growth was achieved by feeding the broilers during the cooler nighttime hours as compared to the growth of broilers fed conventionally during the hot daytime hours.

Environmental heat, especially temperatures above the upper critical temperature of the animal, represents a serious stress which brings into play many physiological processes of adaptation required for homeostasis and survival.

Under conditions of heat stress, the animal reduces its activity to a minimum; it finds the coolest place available and tends to remain there almost motionless, panting (and sweating where possible), releasing moisture to help cool the blood stream. Animals exposed to high environmental temperatures must reduce feed intake for several reasons:

1. Less dietary energy is required because no energy is needed to keep the body warm and because the animal has reduced physical activities to a minimum. There is a slight increased need for energy for panting, but this is minute compared to the above decreases in energy requirements.
2. Ingestion of food causes an increased heat production by the animal due to the "specific dynamic action" of the food and its metabolites. When an animal is having trouble surviving because of an environmental temperature that is above its "critical temperature", feed consumption, especially of nutritionally-unbalanced diets, will cause a production of internal heat which is intolerable. Therefore, the animal must reduce feed consumption to a minimum. This minimum level of feed intake depends to a great extent upon nutrient composition of the diet. Diets imbalanced in protein,

amino acids or vitamins produce a larger "heat increment" than do perfectly balanced diets.

EFFECTS OF TEMPERATURE ON VITAMIN REQUIREMENTS

Each vitamin plays a specific role in the metabolism of the animal. Some vitamins are needed for the proper functioning of every cell of the body, others are needed only under specific conditions. For example, vitamin E is present in the lipid membrane of every cell and is necessary to neutralize free radicals which are formed as a result of oxidative reactants attacking the linoleic and arachidonic acids of the membrane phospholipids (Noguchi et al., 1973). Vitamin D_3, on the other hand, has specific functions dealing with the formation of calcium-binding proteins and calcium transport (Wasserman and Taylor, 1969).

High environmental temperatures, therefore, may bring about a general increase in requirements in terms of milligrams per kilogram of diet for those vitamins required for maintenance of body cellular metabolism simply by decreasing the total feed consumption.

Other vitamin requirements may be altered indirectly by the effects that changes in environmental temperatures may have on the endocrine system of the animal.

TEMPERATURE EFFECTS ON VITAMIN A AND β-CAROTENE

Smith and Borchers (1972) reasoned that since thyroxine influences the conversion of β-carotene to vitamin A (Johnson and Baumann, 1947; Cama et al., 1957), and since environmental temperatures influence thyroxine secretion rate (Woods and Carlson, 1956), there should be a relationship between environmental temperatures and carotene utilization. The results of three experiments with rats showed that the per cent molar biopotency of β-carotene, relative to that of pure retinyl acetate, averaged 74% at 5° C, 55% at 25° C, and 53% at 35° C.

The elevated storage of vitamin A in the liver of the rats under cold weather conditions appeared to be mainly a result of changes in vitamin A metabolism rather than an increase in the conversion of β-carotene to vitamin A. However, administration of triiodothyronine (T_3) to rats at 25° C produced an improvement in utilization of β-carotene, which was almost equivalent to that obtained in the rats at 5° C.

Smith and Borchers (1972) suggested the possibility that the increased protein intake of the rats at 5° C as compared to that of rats at the higher temperatures may have been partially responsible for the enhanced utilization of β-carotene, since Stoecker and Arnich (1971) reported that biosynthesis of retinol from β-carotene is affected by the pattern of protein feeding in rats.

Anderson et al. (1964), in studies on the effects of thyroxine, thiouracil and high ambient temperatures on utilization of vitamin A, found that hot weather and thiouracil caused enlargements of the rat livers while cold weather and thyroxine caused decreases in liver weight. Thus, while the cool weather rats and those given thyroxine (7.5 μg/day) had higher levels of vitamin A per gram of liver, the total vitamin A per liver was not different from that of the rats housed in a warm environment or of those given thiouracil.

Table 1. Influence of environmental temperature upon hepatic vitamin A and serum protein bound iodine levels in rats

Substance	Dietary vitamin A, IU/rat/day			
	150		50	
	21° C	35° C	21° C	35° C
Liver weights, g	5.05	7.41	—	—
Hepatic vitamin A, μg/g	97.04	70.71	—	—
Total vitamin A/liver, μg	484	500	—	—
Protein-bound-iodine in serum, μg/100 ml	1.67	3.56	4.52	4.12

From Anderson et al., 1964

The results of these experiments, briefly summarized in Table 1, support the hypothesis that vitamin A influences thyroid activity by acting on the hypophysis to reduce the release of thyrotropic hormone.

Nir and Ascarelli (1966a, 1966b) found that hepatic vitamin A in chicks was present largely in esterified form at temperatures below 30° C (10-27% free alcohol; 83-90% esterified), while at temperatures of 31-32° C 41 to 59% of the hepatic vitamin A was in alcohol form. Since high temperature, thyroxine and high-protein diets all increase the percentage of the total hepatic vitamin A found in the alcohol form, it is assumed by these investigators that these factors increase the metabolic need for vitamin A.

Braekken et al. (1969) showed that the feed consumption and vitamin A uptake of rainbow trout were higher in fish kept at 14° C than in water of 9° C. However, the higher water temperature did not appear to enhance the conversion by the trout of vitamin A_1 to vitamin A_2.

EFFECTS OF HIGH ENVIRONMENTAL TEMPERATURE ON VITAMIN A REQUIREMENT

In studies on the influence of high environmental temperatures upon appearance of A avitaminosis, Kurokawa (1941) reported that vitamin A deficiency signs appeared much sooner in rats at high temperatures than in similar rats receiving similar diets at moderate temperatures. Studies with rabbits showed that vitamin A deficiency markedly shortens the time required for rabbits to become heat-exhausted at high temperatures.

Heywang, in 1952, found that hot weather caused a rather marked increase in vitamin A requirements of laying and breeding chickens. A level of 2730 IU vitamin A per kg of diet was inadequate both during hot and during moderate weather; vitamin A at 5460 IU/kg of diet was adequate during moderate weather but insufficient for hot weather periods. The vitamin A level found adequate for hot weather (average maximum temperature of 38.5° C) was approximately 8200 IU per kg of diet. This increased requirement apparently is not due to differences in vitamin A absorption since Squibb, Braham and Arroyave (1958) found no differences in the rate of vitamin A absorption in New Hampshire laying hens subjected to environmental temperatures of 20° and 36° C.

Ascarelli and Bartov (1963) found no differences in vitamin A liver storage or in survival on a vitamin A-free diet fed after 6 weeks of age in chicks reared from two to six weeks of age on 750, 1500 or 3000 IU vitamin A per kg of diet at temperatures of 21-25° C versus 28-35° C. However, Kurnick et al. (1964) showed that chicks reared in cool weather had higher hepatic vitamin A levels than those reared on similar diets in hot weather.

Page et al. (1959) subjected twin calves to either high and low ambient temperatures or to solar radiation versus shade. Plasma and liver analyses showed that either high temperatures or solar radiation caused increased expenditures of vitamin A. The calves exposed to solar radiation lost more vitamin A than did the shaded calves, but following administration of vitamin A those exposed to solar radiation retained more. This, of course, is expected since deficient animals always retain more of the missing nutrient than do animals already generously supplied with the nutrient.

McGillivray (1957) of New Zealand found some marked seasonal changes in vitamin A and carotenoid content of the blood plasma and milk fat of cows on pasture. The blood carotene decreased from a high of 1700 μg/100 ml plasma in spring to a low of 800 μg/100 ml in mid-summer. While much of this change probably was due to the changes in carotenoid content of the available forages, some of the reduction may have been due also to the higher environmental temperature, thereby accentuating a possible deficiency state in some areas of poor pasture.

Very little work has been done on the effects of hot climates on vitamin A requirements of humans. The Russian investigators, Smirnova and Golikov (1972), found that serum retinol levels of seamen sailing in the tropical latitutes fell to the "lower limit of normalcy" even though the men were receiving a good diet. As they returned to the temperate zone, retinol levels of plasma returned to normal, with no change in diet. Serum carotenoids did not change with the changes in environmental temperatures.

TEMPERATURE EFFECTS ON VITAMIN C METABOLISM

Man, other primates, guinea pigs, the Indian fruit bat, and the red-vented bulbul lack the enzyme, L-gulonolactone oxidase, and therefore cannot synthesize ascorbic acid. Trout and other fishes also appear to lack a sufficient level of this enzyme to synthesize a sufficient amount of ascorbic acid to meet their metabolic needs for the vitamin.

All other species, including all farm animals and poultry, under normal environmental conditions appear to be able to synthesize sufficient vitamin C for optimal functioning of their metabolic processes.

However, considerable evidence exists which strongly indicates that under certain stresses, particularly under conditions of high environmental temperatures, some mammals and birds may not be able to synthesize sufficient ascorbic acid to replace the severe losses of this vitamin which occur during stress. Under these conditions, supplementation of the diet with ascorbic acid, at levels up to 100-200 ppm, has been shown to produce improvements in growth, egg production and eggshell strength, particularly in broilers and in laying hens.

Vitamin C is found in most tissues of the body, particularly in those organs where intensive biosynthetic processes are taking place (Bourne, 1957). Ascorbic acid is present in the cells forming the bone, teeth and collagen, also in the liver, nerve, white blood cells and the kidney. The adrenal gland, the testes and the ovary contain especially high levels of vitamin C. The particularly high level of ascorbic acid in the adrenals, coupled with the severe drop in adrenal ascorbic acid which occurs during stress, has led to the belief that ascorbic acid must be involved in helping the animal to combat stress (Rumsey, 1969).

Zarrow et al. (1966) and Haltmeyer et al. (1966) found that three minutes after exposure to a high environmental temperature, rats at two days of age showed an increase in plasma corticosterone from 11 to 27 μg/100 ml plasma. A maximum rise occurred with 30 minutes' exposure in two-day-old rats and a 60 minute exposure in 9-day-old rats.

Siddiqui et al. (1972) found that the adrenal depletion in mice was negligible under stress of hypoxia (equivalent to 10,000 ft elevation) when the animals were maintained in a room at 28° C; however, at 32.5° C the same degree of hypoxia produced a pronounced depletion of adrenal ascorbic acid.

ASCORBIC ACID AND EGGSHELL QUALITY

It is general knowledge among poultrymen that high environmental temperatures reduce the thickness of hen's eggs, thus lowering the breaking strength of the eggs. Wilhelm (1940) and Brant et al. (1953) reported studies of the gradual thinning of hen's eggs as summer progresses, and a slow return to improved eggshell strength with onset of autumn and winter.

Thornton and Moreng (1958, 1959) studied the influence of dietary ascorbic acid upon the eggshell quality in hens under conditions of high and moderate environmental temperatures. They reported that supplementation of a practical laying diet with 22 mg of ascorbic acid per kg of diet caused the shell thickness of eggs laid by hens at 29° C to be about equal to that of hens laying eggs in an environmental temperature of 21° C. These reports prompted a flurry of experiments on the possible beneficial effects of ascorbic acid upon eggshell quality under conditions of high environmental temperatures. Several workers (Sullivan and Kingam, 1962; El-Boushy, 1968) obtained confirmatory results indicating a significant improvement in shell quality under conditions of heat stress; however, a number of other workers have been unable to demonstrate any improvement in shell quality with supplemental ascorbic acid under a variety of high-temperature environments (Heywang and Kemmerer, 1955; Hunt and Aitkin, 1962). Perek and Bedrak (1962) showed that cold stress or the stress of debeaking reduced the ascorbic acid levels in the adrenals of laying hens. Perek and Eckstein (1959) found that the adrenal ascorbic acid level of hens was reduced during molting and by administration of ACTH, thereby indicating that the molting of hens is a stress. They did not determine whether or not a dietary supplement of ascorbic acid had any beneficial effect in returning molted hens to maximum egg production. Perek and Kendler (1962) found in hens exposed to Israeli temperatures up to 43.5° C that the high temperatures decreased the plasma ascorbic acid levels in hens and that addition to the diet of 25, 75 or 400 mg ascorbic acid per kg produced a highly significant increase in egg production, but no improvement in eggshell quality. Perek and Snapir (1963) noted a seasonal variation in semen production by several different breeds of cocks, and showed that addition to the diet of 100 mg ascorbic acid per kg produced a highly significant increase in semen production, but no relationship could be shown between this increase and the environmental temperature.

The Colorado group has conducted a number of studies aimed at finding an explanation for the variable effects of ascorbic acid upon eggshell quality.

Ahmad et al. (1967) investigated the possible effects of dietary ascorbic acid upon oxygen consumption and body temperature in several breeds of chickens exposed to elevated environmental temperatures. When Single Comb White Leghorns, New Hampshires and Delawares were first acclimatized at 21° C, then taken through a series of temperature increments (29.4°, 35°, and 43.3° C), egg quality characteristics, body temperatures and oxygen consumption were affected adversely in all three breeds of chickens as the body temperature was increased. The White Leghorns were affected to a lesser extent than were the other two breeds. Dietary ascorbic acid was helpful (though not very significantly so) in maintaining egg quality characteristics, body temperature and metabolic activity.

Thornton and Deeb (1961) found that the levels of ascorbic acid synthesized in the kidneys of laying hens were sufficient for physiological needs at optimum environmental temperatures, but that blood ascorbic acid levels decreased as the environmental temperature was increased from 21° to 31° C. The decrease in blood ascorbic acid levels was postulated to be due in part to exhaustion of endogenous stores (adrenal ascorbic acid, etc.) and also to a reduced synthesis of vitamin C by the kidney.

Lyle and Moreng (1968) found that addition of ascorbic acid to the diet (44 mg/kg) prevented an increase in body temperature as the environmental temperature was increased. Eggshell thickness decreased when the body temperature of the hens rose above normal. This decrease in eggshell strength was prevented only when the ascorbic acid prevented an increase in body temperature.

Freeman (1963) showed that the zone of thermal neutrality in the very young chick is 35° C, but that this moves to lower temperatures as pullets grow older. The upper critical temperature was reported to be 30-33° C at 3 weeks of age and 26-31° C at 4 weeks of age.

The results of Rumsey (1969) and of others strongly indicate that although high environmental temperatures do produce stress in the classical sense, it is not simply the stress which is responsible for the decreased egg production, small egg size and thinning of the eggshells which occur in hens subjected to high environmental temperatures. Some other mechanism, possibly related to thyroid function, is also disturbed by high temperature exposure, and under certain conditions this disturbance is at least partially alleviated by supplementing the diet with ascorbic acid. Wallback and Reineke (1949) showed that injection of rats with thyroxine caused a drop in adrenal ascorbic acid to 6% of normal within four days after the injection, but following this drop the adrenals enlarged and the vitamin C content of the adrenals rose to normal values or above.

Thus, as postulated by Selye (1950), animals have the ability to adapt to stresses. Dietary ascorbic acid may have some value in this adaptation, especially when the stress is one, such as heat stress, which also decreases the biosynthesis of ascorbic acid. Shannon and Brown (1969) found that hens began to adapt to heat stress after three days, and after 12 days showed no further decline in fasting metabolic rate. Possibly, a major effect of dietary ascorbic acid is involved in helping hens to adapt to heat stress. Egg production itself also appears to be a stress.

Jackson and Nalbandov (1969) reported a substance from chicken hypothalamus, similar to vasotocin, which caused depletion of ovarian ascorbic acid. Blanc and Von der Muehll (1967) reported that a flavinoid, trihydroxyethylrutoside, showed a synergistic action in regard to storage of vitamin C in liver and adrenals. The flavinoid appeared to reduce the loss of vitamin C from the adrenals and maintained the weight of guinea pigs during thermal stress.

Tseitina (1965) reported that rats exposed to 35° C over a 30 day period, at 5 hrs per day, showed a severe drop in ascorbic acid of the adrenals and blood serum and an increased excretion of ascorbic acid, and that this excretion was further increased by administration of tea catechines (flavinoids) but at the same time, loss of ascorbic acid from the adrenals was inhibited.

Riker et al. (1967) found that the reduced growth of pigs to six weeks of age at a temperature of 29.4° C as compared to 18.3° C (both at 85% relative humidity) was correlated with a decrease in plasma ascorbic acid at the higher temperature.

Nestor et al. (1972) showed that the addition of 330 mg ascorbic acid per kg of diet prevented a decline in blood ascorbic acid levels which occurred in unsupplemented turkeys during hot weather (highest temperature reached was 30° C), but did not produce a significant improvement in egg production.

VITAMIN C NUTRITION OF HUMANS IN THE TROPICS

Hindson (1970) studied the mean ascorbic acid levels of the white blood cells of European residents in the tropics. For residents that had resided in the tropics less than four months, the average ascorbic acid levels were 29.2 and 32.34 μg per 10^8 white blood cells in males and females, respectively. For those living in the tropics 4-30 months, the average ascorbic acid levels were 22.54 and 29.88 μg per 10^8 white blood cells, respectively. Thus, men showed a highly significant drop while women did not. Although the men produced more sweat, this did not account for the difference since insignificant amounts of ascorbic acid were found in the sweat. Sargent et al. (1944) found negligible amounts of vitamin C or other vitamins in sweat. Hindson postulated that vitamin C is used up during sweating, and that perhaps it is needed for proper functioning of the sweat glands. Shields et al. (1945) reported that men stressed by high environmental temperature and humidity lose much more ascorbic acid than those working in a temperate environment. Similar results were reported by Dluzniewska (1965) and by Dluzniewska et al. (1965).

POSSIBLE EFFECTS OF HEAT STRESS ON VITAMIN D AND E NUTRITION

Conrad (1939) found that high environmental temperatures cause a sharp decrease in blood calcium levels in laying hens. This was confirmed by Warren and Schnepel (1940) and by Mueller (1959), and by Huston and Carmon (1961), who also correlated the decrease in blood calcium at 32.2° C with a decrease in eggshell thickness as compared with eggs from hens housed at a moderate environmental temperature.

In view of the recent evidence reviewed by Scott and Combs (1973) showing that vitamin D_3 must be converted in vivo to 1,25-dihydroxycholecalciferol before it can act in formation of calcium binding proteins, the possibility that heat stress interferes with this conversion is worthy of investigation.

Silber et al. (1969) showed that vitamin E uptake was markedly greater at 37° than at 4° C. The exchange of tocopherol between erythrocytes and plasma was markedly increased at the higher temperature. There was an hourly efflux of 26% of the tocopherol from RBC $\rightarrow$ low density lipoproteins. Vitamin E, like cholesterol, appeared to be a constituent of the erythrocyte membrane which is in dynamic equilibrium with the corresponding plasma compound. High temperatures, therefore, may cause a much more rapid turnover of vitamin E than that which occurs at moderate temperatures. Biswas (1970) found that heat stress stimulates spermatogenesis and causes a marked depletion of lipids in the testes of male toads.

EFFECTS OF HIGH ENVIRONMENTAL TEMPERATURES ON B-VITAMIN REQUIREMENTS

THIAMIN

The best work on the effects of high environmental temperatures on B-vitamin requirements was conducted by Mills et al. in 1947. In critical chick studies with thiamin, pyridoxine, nicotinic acid, folic acid and choline, heat stress appeared to influence only the requirement for thiamin. At 33° C the thiamin requirement was approximately three times that found to be adequate at 21° C. A vitamin B_1 requirement of 3 mg per kg of diet may be difficult to achieve in tropical areas without supplementing the diets of young chicks with thiamin.

A higher requirement for thiamin may be an important contributing factor to the high incidence of beriberi in tropical countries, the main cause of which, of course, stems from the consumption of polished rice and other foods low in thiamin and at times the consumption of raw fish which contains a thiaminase. Itokawa and Cooper (1970) showed that in vitro incubation of fragments of spinal cord, brain and sciatic nerve at high temperatures caused thiamin to be released from the membrane. Possibly high environmental temperatures cause a more rapid turnover of vitamin B_1 in metabolism.

Joslyn and Leichter (1968), studying the effects of traces of SO_2 on thiamin destruction in diets containing commercial casein, found no destruction at either −4° C or 25° C when casein diet was free of SO_2. However, in the presence of residual SO_2, there was much more destruction of vitamin B_1 at 25° C than at −15° C.

RIBOFLAVIN

Seymour et al. (1968) did not find any consistent interactions between dietary riboflavin and temperature in young pigs. Efficiency of feed utilization was improved equally well at both 16° and 32° C when the riboflavin was increased above 3.0 mg per kg of diet. However, Squibb et al. (1954a, b) reported that rats exposed to 34° C for 72 hours lost weight and showed reduced blood levels of riboflavin, vitamin A, vitamin C and alkaline phosphatase. They found no changes in vitamin levels of the blood in 6-week-old cockerels held at 37° C for 7 days.

FOLIC ACID AND VITAMIN B_{12}

Sprue and enterocolitis, endemic in many tropical climates, are often associated with folic acid and vitamin B_{12} deficiency. Baker (1967, 1972) has reviewed the subject and concluded that tropical enteropathy and tropical sprue are human vitamin B_{12} deficiencies. According to Atakhanov (1962, 1963) chronic enterocolitis and similar conditions depress production of gastric mucoprotein intrinsic factor, so that vitamin B_{12} is not absorbed, and B_{12}-deficiency anemia results. In tropical sprue, giving folic acid caused blood vitamin B_{12} values to return to normal, but blood levels of free phenylalanine and tyrosine were returned to normal only by injecting vitamin B_{12}.

Panders and Rupert (1965) showed that chick liver converts more tetrahydrofolic acid (FH_4) to the 5,10 methenyl derivative of FH_4 at 39° than at 37° C. The higher temperature appeared to inhibit conversion of pteroylglutamic acid (PGA) to FH_4. This may be important at temperatures above the critical temperature of the animal. Chanarin and Perry (1970) showed that injected FH_4 was quickly converted to N^5-methyl-FH_4 in four patients with pernicious anemia and one with nutritional aenmia of folic acid deficiency. In contrast, in control subjects the FH_4 was changed to PGA and then to N^5-methyl-FH_4. Ascorbic acid and vitamin B_{12} are essential for normal conversion of PGA to FH_4. Consequently, if high environmental temperatures alter the blood levels of ascorbic acid and vitamin B_{12}, heat stress may influence folic acid metabolism.

Jeejeebhoy et al. (1969) found that sprue patients had low albumin levels in blood serum (1.8% *vs* 4.0% in normals). Since vitamin E and selenium are involved in maintaining plasma albumin levels in chicks, these results may be an indication of a need for these nutrients in addition to folic acid and vitamin B_{12} in people suffering from sprue.

CHOLINE

In 1952, Sellers and You reported that exposure to cold prevented liver fat deposition in choline-deficient rats. Reasoning that higher temperatures might increase choline requirements, Chahl and Kratzing (1966a, b) demonstrated that rats fed a choline-deficient diet did not develop fatty livers at 2° C. However, fatty livers occurred at 21° C when the diet contained 250 mg choline per kg, and no fatty livers occurred at this temperature with a dietary choline level of 500 mg per kg. Fatty livers occurred with 500 mg choline per kg of diet when the environmental temperature was increased to 33° C. In contrast to the findings of Mills et al. (1947), these investigators demonstrated that high environmental temperatures do increase choline requirements of rats when an otherwise good diet is fed. These results appear to explain findings of Radomski and Orme (1967) that at 21° C fatty livers were produced in rats by many hepatotoxic agents, including choline-deficiency, carbon tetrachloride, ethionine, 4-NH_2-pyrazolopyrimidine, orotic acid and etharal, whereas no fatty livers occurred in choline-deficient rats at 4° C and less fatty livers with ethanol at this low temperature. All other hepatotoxic agents produced the same effects at both temperatures.

CONCLUSIONS

The research reviewed in this Section indicates that requirements for several of the vitamins are markedly increased in animals exposed to the high temperatures and humidities typical of tropical countries and also often found in many areas of the temperate zones during summer.

High temperatures, especially temperatures above the upper critical temperatures of animals, affect the metabolism and requirements for vitamins A and C, and for thiamin and choline. Evidence is presented indicating that riboflavin, folic acid and vitamin B_{12} metabolism and requirements also are affected adversely by prolonged exposure of humans and other animals to temperatures at, or somewhat above their critical temperatures.

REFERENCES: see pp. 554-557.

The Redistribution of Cardiac Output in Animals During Heat Stress

by

J. R. S. Hales

INTRODUCTION

Although the primary role of the circulation is to transport metabolic materials to and from the body's cells, an extremely important additional role is the transport of heat from deep to superficial tissues (i.e., from core to shell) for subsequent dissipation in the control of body temperature. Further, the transfer of thermal information from warm or cool tissues to the thermoreceptors may occur via the circulation. In view of the intimate role of the cardiovascular system in body temperature regulation, marked effects of thermal stimuli on cardiovascular function would be expected; surprisingly, cardiovascular physiologists have paid relatively little attention to this. Likewise, thermoregulatory physiologists have largely restricted their studies of blood flow to tissues directly involved in the regulation of heat exchange, despite the fact that the maintenance of cardiac output and blood pressure, and of an adequate blood supply to all tissues is paramount to the well-being of the animal. The concept of a compensatory redistribution of cardiac output during heat stress – from 'non-vital' tissues to tissues important in thermoregulation – (Robinson, 1963; Thauer, 1965) has been based on few and generally non-quantitative estimations of blood flow in very few of the tissues involved. A complete picture of regional blood flow redistribution during heat stress is now emerging (Hales, 1974). The importance of this in relation to the role of the circulation in transport of metabolites, for example, is striking when it is recognised that up to 30% of the cardiac output in sheep may by-pass the capillary bed via arteriovenous anastomoses (AVA's).

The effects of heat on gross cardiovascular function aie dealt with in the following section by G. C. Whittow and therefore, this discussion will be confined to considerations of the regional distribution of blood flow.

2. THERMOREGULATORY INFLUENCES ON THE DISTRIBUTION OF CARDIAC OUTPUT

THERMONEUTRAL ENVIRONMENT: There is a widely varying pattern of changes in metabolic rate of different animal species when exposed to various environmental temperatures. Therefore, the somewhat controversial concept of a zone of thermoneutrality being associated with a range of ambient temperatures within which metabolic rate is minimal, may be better associated with a zone of minimal activity of thermoregulatory mechanisms; this would be the zone of ambient conditions in which body temperature is regulated by alterations in vasomotor function. Thus, even in the zone of thermoneutrality, fluctuating vasodilatation and vasoconstriction in the body shell must involve variations in the cardiac output or in its distribution. As blood vessels of the skin pass from constriction to dilatation, the thermal conductance of body tissues in the dog and ox may show a 20-80% increase (Hammel et al., 1958; Blaxter and Wainman, 1961). The limbs and ears are particularly important in regulating heat loss under such conditions as judged by fluctuations in skin surface temperature (Whittow, 1962), and in sheep, capillary blood flow in the skin of these regions may increase two- to four-fold during vasodilatation (Hales, 1973a; unpubl. observ.). Under such conditions, the proportion of cardiac output received increases from 1% to 2% of the total, and depending upon whether there is cutaneous vasoconstriction or vasodilatation, 0.4-4% of the cardiac output passes through AVA's (Hales, 1973b). Although AVA's are present in many body tissues, those patent under these conditions would most likely be in the extremities or other tissues important in heat exchange such as the tongue and nasal passage (see Hales, 1974, for references). The relatively small and often transient increases in blood flow probably do not involve any decrease in blood flow to other tissues but are probably compensated for by increased venous tone, as the veins constitute the most important blood reservoir immediately available (Alexander, 1963).

MILD HEAT STRESS: Exposure to a moderately hot environment in which deep body temperature either remains unchanged or increases by less than 0.5° C, leads to a marked redistribution of cardiac output. The most comprehensive study of this appears to have been performed in sheep which were exposed to an environment of 40° C dry bulb and 26° C wet bulb until an apparent equilibrium was reached after 2-4 hrs. The results were similar to those illustrated for moderate hyperthermia in Figure 1. Up to 12% of the cardiac output passes through AVA's during heat stress, compared with around 1 or 2% in a thermoneutral environment. An increase in capillary blood flow is seen only in tissues involved in the regulation of body temperature, viz., the skin and nasal region for heat loss, and respiratory muscles to meet the energy requirements of panting. There is no change in cardiac output, and any change in total blood volume or volume of the venous reservoir is unknown. However, it appears that the required increases in blood flow are made possible by decreases to non-respiratory skeletal muscle and abdominal organs (rumen, small intestine, kidneys, spleen). Although blood flow in resting muscle is relatively low (about 6 ml 100 g^{-1} min^{-1}, compared with >400 ml 100 g^{-1} min^{-1} in the kidney for example), approximately 27% of the body weight in sheep (Merino) is muscle, and therefore this must constitute an important blood reservoir. The reduced muscle blood flow in sheep is contrary to the situation in man, in which body heating results in little or no change in muscle blood flow (Edholm et al., 1956). Also, blood flow in the thyroids of sheep decreases and that in the tongue, adrenals, large intestine, heart, brain and spinal cord does not change significantly.

SEVERE HEAT STRESS: The redistribution of cardiac output during heat stress of sufficient severity to evoke marked hyperthermia has been determined in the Merino sheep (Fig. 1; Hales, 1973a) and Greyhound dog (Fig. 2, Hales and Dampney, 1975). Measurements of blood flow were made before exposure to heat, and then at a time corresponding to the peak of rapid shallow panting ("moderate hyperthermia", deep body temperature raised by approximately 1° C), and later during maximal slower deeper panting immediately prior to termination of heat exposure ("severe hyperthermia", deep body temperature raised by approximately 2.4° C). (The reader should refer to Chapter 1, Sect. 5 for a description of these different patterns of panting which occur during severe heat stress). In the sheep, cardiac output remains unchanged and its redistribution with moderate hyperthermia is similar in many respects to that occurring during mild heat stress. However, contrary to the situation in mild heat stress, blood flow is decreased in the large intestine, brain and spinal cord, and increased in the tongue and adrenals. Some changes have also been well established to occur in man. For example, there is a decrease in blood flow to the kidneys (e.g., Smith et al., 1952) and splanchnic areas (e.g., Rowell et al., 1971). With severe hyperthermia in the sheep there is a reversal of some blood flow changes. Thus, blood flow to the extremities which initially shows a 5- to 7-fold increase, ultimately returns to approximately control levels; and blood flow to non-respiratory skeletal muscle ultimately doubles. Also, whereas blood flow to body skin changes insignificantly during mild heat stress and moderate hyperthermia, it ultimately decreases markedly during severe hyperthermia.

Cardiac output in the dog, like that in many other species (e.g., ox and chicken), but in contrast to the sheep, shows a progressive and highly significant increase with progressing hyperthermia (Hales and Dampney, 1975). However, there does not appear to be any important difference in the basic pattern of changes in its distribution induced by severe heat stress (Fig. 2). Qualitatively, the effects of heat stress were generally similar in the dog and sheep, with the exception of blood flow to non-respiratory skeletal muscle. As noted earlier, this decreases with moderate hyperthermia but increases with severe hyperthermia in the sheep, however, in the dog there is a progressive decrease. A very large blood reservoir must be provided by skeletal muscle in the breed of dog studied (racing Greyhound), as it constitutes approximately 50% of the body weight (Hales and Dampney, 1975). Certain quantitative differences between blood flow changes in the two species should be noted. For example, there was a greater fall in blood flow to the brain and spinal cord of the dog than the sheep, and the increase in blood flow to respiratory muscles in the sheep was almost twice that in the dog. Further, the decrease in blood flow to abdominal organs was much less marked in the dog. This latter flow decrease was apparently essential in the sheep to permit the blood flow increase in many tissues and through AVA's, whereas in the dog the necessary flow increases were presumably attained by an increase in cardiac output, in addition to its small redistribution and smaller diversion of blood through AVA's. Finally, there was a much greater increase in blood flow to naso-buccal regions (tongue, nasal mucosa and turbinates) in the dog, an open-mouthed panter, than in the sheep which pants through the nose and only opens the mouth when deep body temperature is raised by about 1° C or more; this was due mainly to differences in tongue blood flow. These interspecies differences in the magnitude of changes in the distribution of cardiac output are even more striking when values for absolute blood flow rate are examined (Hales, 1973a, c; Hales and Dampney, 1975).

In the above studies, the proportion of cardiac output passing through AVA's increased

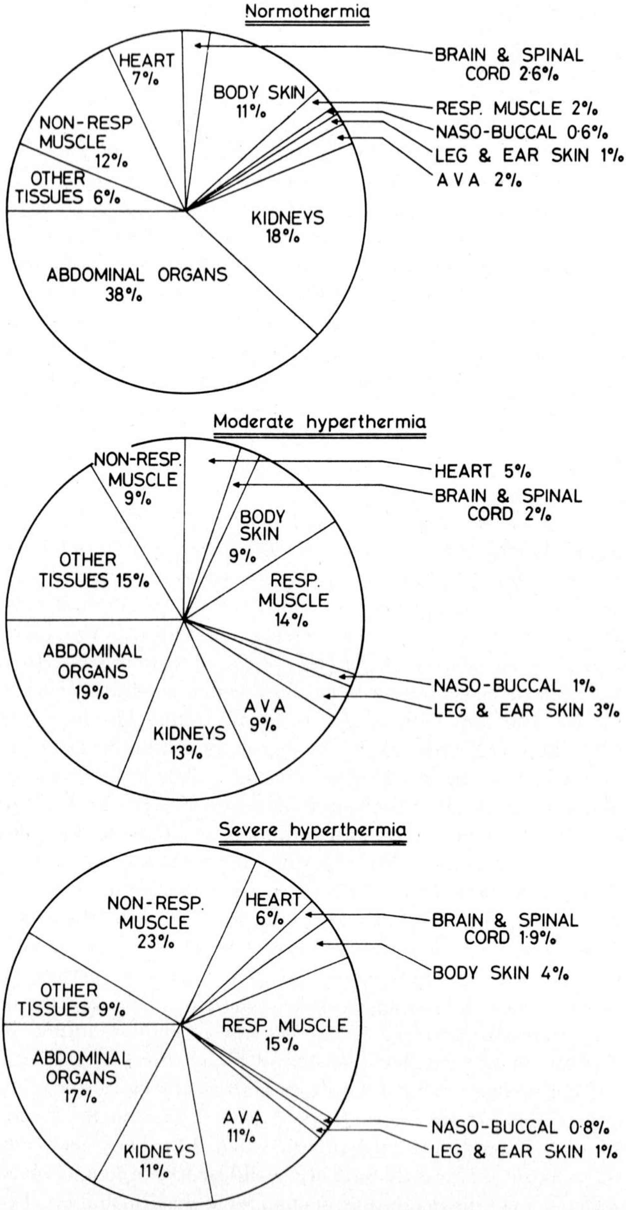

Fig. 1. Redistribution of cardiac output in the conscious Merino sheep during exposure to an environment of 40° C dry bulb and 38° C wet bulb temperature. Normothermia: Resting in thermoneutral environment with normal body temperatures (Tre $\simeq$ 39.5° C). Moderate Hyperthermia: At the peak of rapid shallow panting with a deep body temperature of 40.5° C (ΔTre $\simeq$ 1° C) during heat exposure. Severe Hyperthermia: Near maximal panting with a deep body temperature of approximately 42° C (ΔTre $\simeq$ 2.5° C) during heat exposure. Note that cardiac output, represented by circle area, did not change. Mean results from nine animals. (compiled from data of Hales, 1973a, c, using the radioactive microsphere technique).

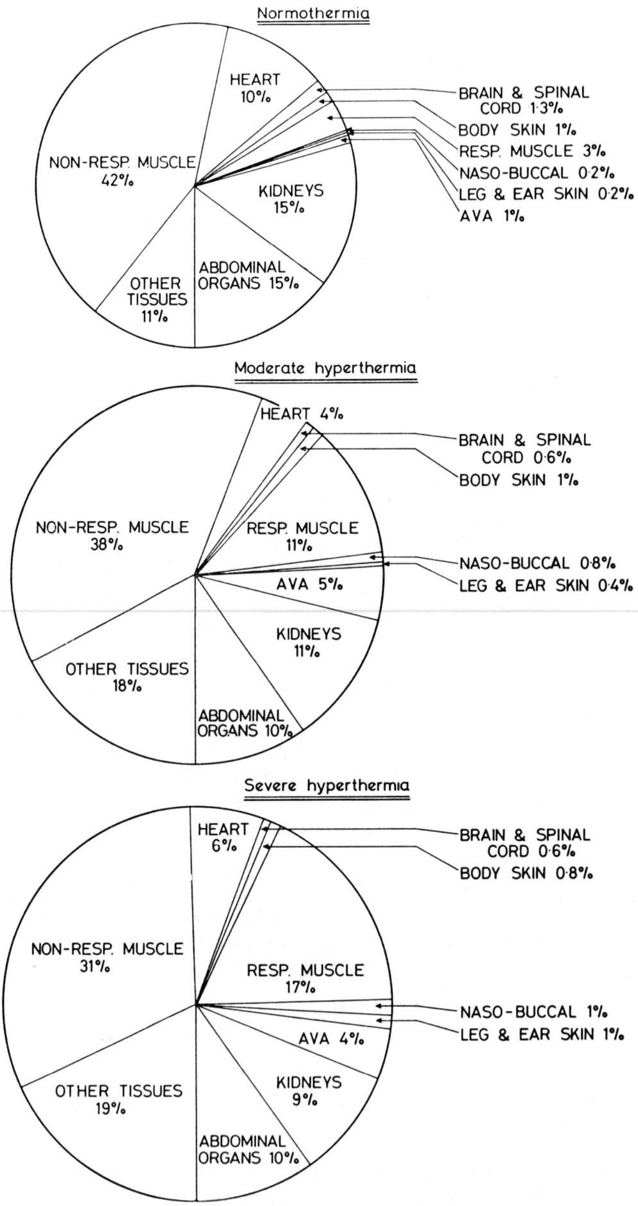

Fig. 2. Redistristribution of cardiac output in the conscious Greyhound dog during exposure to an environment of 40° C dry bulb and 36° C wet bulb temperature. Normothermia: Resting in a thermoneutral environment with normal body temperatures (Tre $\simeq$ 38.7° C). Moderate Hyperthermia: At the peak of rapid shallow panting with a deep body temperature of approximately 40.4° C (ΔTre $\simeq$ 1.7° C) during heat exposure. Severe Hyperthermia: Near maximal panting with a deep body temperature of approximately 42.3° C (ΔTre $\simeq$ 3.6° C) during heat exposure. Note that cardiac output, represented by circle area, had increased by 36% and 50% at moderate and severe hyperthermia, respectively. Mean results from five animals. (Compiled from data of Hales and Dampney, 1975, using the radioactive microsphere technique).

from about 2% in a thermoneutral environment to 10% during severe heat stress in sheep, and respective figures for the dog were 1% and 5%. In individual sheep, up to 30% of the cardiac output has been noted to pass through AVA's >15 μm diameter, and up to 13% through AVA's >50 μm diameter (Hales, 1973b).

INFLUENCES OF ADAPTATION

It is clear that a series of intimately integrated dynamic changes must occur when an animal is confronted with an environmental stress (e.g., altitude or exercise, or thermal), and many physiological systems are involved. Whether the animal continues to thrive depends upon the compatibility of these changes with physiological requirements. The need by different species to cope with different environmental stresses is probably reflected in genetic variations in both structure and function. For example, cardiac output increased in the Greyhound dog but not in the Merino sheep during heat stress; it might therefore be considered that less strain was being imposed on the cardiovascular system of the sheep. The apparent superiority of the Merino over the Greyhound with respect to cardiovascular adjustments to heat may be a result of the evolution of sheep to better cope with challenges of the thermal environment, whereas, the dog, particularly the racing Greyhound, might be expected to exhibit superiority in meeting the requirements of exercise (Hales and Dampney, 1975). The influence of adaptation to continual long-term heat exposure, on regional tissue blood flow, does not appear to have been examined. However, in man, the increase in blood flow to the hand and forearm which is induced by heat becomes less with repeated short-term exposures to heat (Wyndham, 1951) but is higher at a given level of body temperature (Fox et al., 1963). Thus, while the stimulating effect of temperature is enhanced, the demands for raised peripheral blood flow are reduced; presumably this is made possible by the increased sweating rate which occurs with acclimatization. Such a decrease in peripheral blood flow would permit a return toward normal blood flow in the gut and muscle. A restoration of normal blood flow to tissues from which blood had been diverted would also be facilitated by two further factors: first, an increase in blood volume which occurs in cattle (Bianca, 1965) and possibly in man (Robinson, 1963); secondly, a decrease in the volume of peripheral veins as may occur in man sometimes (Whitney, 1954) although not under all conditions (Wood and Bass, 1960). Further, there are indications from histological studies in the sheep (Molyneux, 1965) and pig (Ingram and Weaver, 1969) that increased cutaneous vascularity might be an adaptive mechanism to facilitate heat loss in animals which sweat poorly.

CONTROL OF CARDIAC OUTPUT DISTRIBUTION

Many factors are involved in determining regional tissue blood flow during heat stress. The regulation of blood flow by local factors within a tissue, so that the metabolic requirements of that tissue are met (i.e., autoregulation), is normally considered to be very powerful. This may be the prime factor determining the blood flow changes in some tissues during heat stress, viz., the reduced thyroid blood flow paralleling reduced thyroid activity, the increased adrenal blood flow paralleling increased medullary activity, the increased blood flow to respiratory muscles to meet the energy cost of panting, and the reduced brain and spinal cord blood flow are probably the result of panting-induced hypocapnia (see Chapter 1, Sect. 5).

The most striking changes in blood flow during heat stress seem obviously necessary to facilitate heat loss, and the decreases are necessary to permit those increases. The large arteries, AVA's and venous sections of the vasculature do not exhibit autoregulation but depend almost entirely on extrinsic factors for their control (Burnstock and Prosser, 1960). The specific blood flow responses to thermoregulatory demands are explicable, at least in a qualitative sense, on the basis of a balance between the activity of sympathetic vasoconstrictor and vasodilator nervous activity, autoregulation, and the local effect of temperature on arteries and veins (Thauer, 1965; Shepherd and Webb-Peploe, 1970). Cutaneous blood flow increases as a result of reduced sympathetic vasoconstrictor activity and active vasodilatation and at least under some conditions, the existence of a central nervous system mechanism integrating cardiovascular and

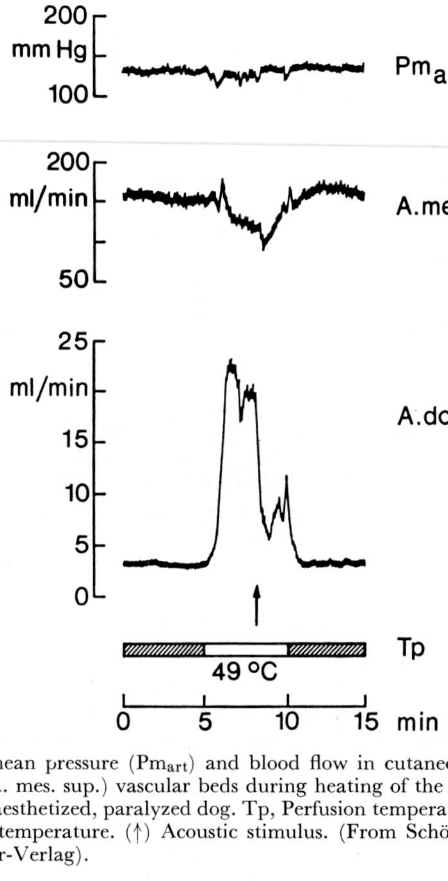

Fig. 3. Arterial mean pressure (Pm_{art}) and blood flow in cutaneous (A. dors. ped.) and intestinal (A. mes. sup.) vascular beds during heating of the anterior hypothalamus of a lightly anaesthetized, paralyzed dog. Tp, Perfusion temperature. Hatched bars: Neutral perfusion temperature. (↑) Acoustic stimulus. (From Schönung et al., 1971, courtesy of Springer-Verlag).

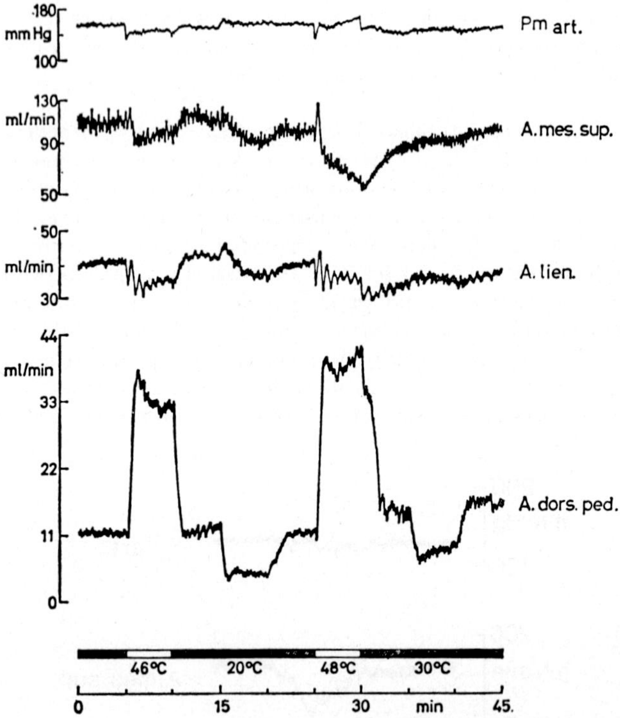

Fig. 4. Arterial mean pressure (Pm_{art}) and blood flow in intestinal (A. mes. sup.) splenic (A. lien.) and cutaneous (A. dors. ped.) vascular beds during thermal stimulation of the spinal cord in an anaesthetized, paralyzed dog. Heating (white bars), cooling (black bars) with perfusion temperatures. (From Kullmann et al., 1970, courtesy of Springer-Verlag).

respiratory activity is implicated. In this regard, Kindermann and Pleschka (1973) have demonstrated an increase in tongue blood flow coinciding with the onset of high frequency phrenic nerve discharges during hyperthermia in anaesthetized, immobilized dogs. Also, Dampney (1974) showed that exercise-induced bursts of panting in the conscious dog were intimately associated with decreases in peripheral vascular resistance; it appears that such changes occur primarily in response to metabolic demands, whereas those occurring during thermally-induced panting are primarily in response to thermoregulatory demands (Dampney and Hales, 1974). Central nervous system drive varies efferent nerve impulses, but alterations in local temperature may modify sensitivity of the blood vessel smooth muscle to incoming impulses (Shepherd and Webb-Peploe, 1968). The interaction between central (thermoreceptor) and peripheral (direct effect and/or thermoreceptor) temperature has been well demonstrated. For example, the magnitude of fall in total peripheral resistance elicited by hypothalamic heating (Whittow, 1968) and the magnitude of rise in ear skin temperature elicited by spinal cord heating (Iriki, 1968) depend upon central and skin temperature. Ingram and Legge (1971) have shown that peripheral blood flow (measured in the pig by plethysmography of the tail) is dependent on hypothalamic, spinal cord and skin temperature.

The influence of deep body thermoreceptor stimulation on the distribution of cardiac output has not been examined, however, reciprocal changes in blood flow to skin and intestines during isolated heating of the hypothalamus (Fig. 3) or spinal cord (Fig. 4) are strongly indicative of a similar pattern to that which occurs during exposure to hot environments. Further, with hypothalamic or spinal cord heating, the activity of sympathetic efferents supplying the skin decreases while that supplying abdominal viscera increases; and it has, therefore, been suggested "that the sympathetic nervous system may perform under certain conditions a sort of "reciprocal innervation" of functionally antagnostic autonomic effector systems" (Walther et al., 1970; Iriki et al., 1971).

CONCLUSIONS

Evidence gained to date on single breeds from two animal species shows that short-term exposure of an animal to heat results in a redistribution of blood flow appropriate to meet the environmental challenge with which the animal is confronted. Thus, there is a redistribution to tissues important in heat exchange (skin and nasal region) and thermoregulatory responses (respiratory muscles), away from what might be considered non-vital tissues for immediate purposes (e.g., non-respiratory skeletal muscle and gut), and with the preservation of flow to vital tissues (heart, brain and spinal cord). During cold stress an appropriate compensatory redistribution of cardiac output also occurs (Alexander et al., 1973). Nevertheless, if an animal is to subsequently perform muscular exercise (such as to search out food or escape from a predator), or even to maintain its well-being or level of productivity, it is clear that rapid adaptation of the cardiovascular system is imperative. For example, it would not be uncommon for livestock to encounter a mild heat stress similar to that which has been noted to result in a halving of blood flow to skeletal muscle in Merino sheep. If this persisted, it is possible that general body growth and meat production would be adversely influenced. It is possible that the higher blood volume in *Bos indicus* than in *Bos taurus* cattle (Bianca, 1965) is a feature of cardiovascular adaptation which allows the former to do so much better in hot climates. There is considerable evidence that development of the cardiovascular system is associated with metabolic demands. For example, there is a higher density of capillaries in muscles of animals of small body size (and high metabolic rate per unit body weight) compared with large animal species (Schmidt-Nielsen and Pennycuik, 1961) and in "fast" compared with "slow" muscles (Hudlická et al., 1973); also the heart of the Greyhound is very large compared with non-sporting dog breeds.

The effects of heat stress on the overall distribution of cardiac output in sheep and dogs were determined using the radioactive microsphere technique. Although quantitative measurements of blood flow were obtained, since microspheres 15 μm in diameter were used, the values represent CAPILLARY or NUTRIENT blood flow. Because of the passage of a large percentage of the cardiac output through AVA's, capillary would often have been less than total blood flow in some tissues. This could be considered a shortcoming of the microsphere technique, however, the observations have demonstrated that temperature may have an extremely potent influence on nutrient blood flow and that a large portion of the cardiac output may be made unavailable for the meeting of metabolic requirements under these conditions. In view of this latter point, the significance of blood flow through AVA's in the regulation of body temperature warrants examination.

Four major lines of research require attention: (1) The effects of heat stress on regional blood flow distribution in species besides the sheep and dog (particularly a non-panting species); (2) The detailed control of this distribution; (3) The consequences of adaptation and interactions with other environmental factors and (4) The significance of blood flow through arteriovenous anastomoses in the regulation of body temperature.

REFERENCES: see pp. 558-559.

Effects of Heat Stress on the Central Circulation in Mammals

by

G. C. Whittow

INTRODUCTION

The cardiovascular system is inevitably involved in the thermoregulatory responses of mammals to heat. An increased heat loss from the skin surface requires an increased delivery of heat to the skin by way of the cutaneous circulation. Active sweat glands also need an increased blood flow to the skin. In those mammals that pant, an increased blood flow to both the respiratory muscles and to the respiratory surfaces from which evaporation occurs is necessary. Changes in heat production may be associated with a diversion of blood flow to or from heat producing tissues. These responses must be effected without seriously disturbing the arterial blood pressure. Consequently, there must be secondary adjustments of cardiac output and peripheral resistance in order to ensure that the arterial blood pressure remains reasonably constant. The emphasis in this section will be on reporting experiments performed on unanesthetized and wherever possible, unrestrained animals. There is abundant evidence that the thermoregulatory responses of anesthetized animals differ significantly from those of conscious animals. For earlier reviews of the basic cardiovascular physiology of domestic animals and their responses to heat, the reader is referred to Whittow (1967, 1968b). The effect of heat on the peripheral circulation is considered elsewhere (Chapt. 1, Sections 4 and 5).

TECHNIQUES

Considerable advances in cardiovascular technology have been made in recent years. It is now possible to measure blood pressure, blood flow, heart rate and cardiac output by telemetry. These techniques have not been used extensively in animals during exposure to heat, but it is likely that they will be used in the future in order to study the responses of animals exercising in the heat and to permit the animals to execute their customary pattern of thermoregulatory behavior. A technique that has proven to be extremely useful in the past for cardiovascular research, is the exteriorizisation of the carotid artery. The application of this procedure to the ox has been discussed by Whittow and Robertshaw (1965).

CARDIAC OUTPUT

A large sustained increase in blood flow to the skin, upper respiratory tract or other region, usually implies an increase in cardiac output. There is good evidence that such an increase occurs in response to heat. For example in the ox, exposure to a hot, dry environment (40° C) resulted in an increased cardiac output which was effected by an increased stroke volume and heart rate (Whittow, 1965). As the deep-body temperature did not increase, it is a reasonable supposition that the increased cardiac output was elicited by warming of the skin. A moderate degree of thermal polypnea was provoked by exposure to dry heat and Whittow speculated that the increased cardiac output was due, in part, to an increased venous return associated with the panting movements. It has subsequently been shown that spontaneous panting in the dog, in the absence of exposure to heat, results in an increase in cardiac output (Jennings et al., 1973).

When the ox became hyperthermic during exposure to heat i.e., when the deep-body temperature increased, a considerable further increase in cardiac output occurred (Whittow, 1965). This was brought about largely by an increased heart rate. At high levels of body temperature, the stroke volume diminished; and it was actually lower at a body temperature of 41.5° C than under control conditions in a cool environment. The reduction in stroke volume might have been related to the increased heart rate and a corresponding diminution in the time available for the ventricles to fill, during diastole. Whittow (1971a) presented evidence to show that the increased heart rate in the hyperthermic ox was probably the direct consequence of the increased temperature of the heart tissue.

However, not all of the effects of hyperthermia can be attributed to the direct effect of an increased temperature of the cardiac pacemaker and other cardiac tissue. Whittow (1968a) was able to produce an increased cardiac output and heart rate in the ox by localized heating of the anterior hypothalamus/pre-optic region of the brain alone. Similar responses have been demonstrated in the dog (Phillips and Jennings, 1973). Both during hyperthermia and during localized heating of the anterior hypothalamus and pre-optic region of the brain, a considerable distortion of the intraventricular and central venous pressure record was observed (Whittow, 1966; 1970). It was believed that the distortion was of respiratory origin, arising from an interaction between the changes in intrathoracic pressure associated with panting and the pressures generated by cardiac contraction. It was postulated that both the filling of the heart and the ejection of blood from it were affected by this phenomenon. Paradoxically, an increase in cardiac output did not occur in the sheep, another panting animal, during exposure to mild heat stress or when the animals became hyperthermic (Hales, 1973).

CARDIAC WORK

The static mechanical work of the heart did not increase significantly during exposure to the ox to a hot dry environment, but it did increase when the animals became hyperthermic (Whittow, 1965).

ARTERIAL BLOOD PRESSURE

The arterial blood pressure of the ox was significantly lower in a hot, dry environment

(40° C) than at 15° C (Whittow, 1965). The diminution in blood pressure, in the face of an increased cardiac output, is consistent with a considerable decrease in total peripheral resistance, which did, in fact, occur (see the next paragraph)). The diminution in arterial blood pressure and total peripheral resistance largely reflect an increased cutaneous blood flow (Chapter 1, Section 11a(1)). When part of the skin surface of the ox was exposed to infra-red irradiation, the change in blood pressure was related to the concurrent change in the blood flow to the skin (Ingram and Whittow, 1963). It is unlikely that the concomitant thermal polypnea contributed to the diminution in blood pressure because in dogs that panted in the absence of exposure to heat, the arterial blood pressure increased (Jennings et al., 1973). When the ox became hyperthermic, in a hot environment, the arterial blood pressure increased (Whittow, 1965). As the total peripheral resistance decreased, the increase in blood pressure was the result of the increased cardiac output (see page 296). The greatly increased respiratory activity could also have contributed to the increased blood pressure (Jennings et al., 1973). An increase in arterial blood pressure during localized heating of the anterior hypothalamus/pre-optic region occurred in the ox, but the magnitude of the effect was variable depending on the environmental temperature and the response of the cutaneous blood flow to heating of the hypothalamus (Ingram and Whittow, 1963; Whittow, 1968a, b). Phillipps and Jennings (1973) reported an increased arterial blood pressure in response to localized heating of the anterior hypothalamus of the dog. In the light of these results, it is likely that the effects of hyperthermia on the arterial blood pressure are mediated in part by an increase in the temperature of the anterior hypothalamus and pre-optic region. Whittow (1970) drew attention to the possible effects of phasic variations in arterial pulse pressure in the hyperthermic animal on the baroreceptive regions of the vascular bed. The arterial blood pressure of the sheep diminished when it became hyperthermic. This presumably was the result of an increased cutaneous blood flow leading to a diminution in total peripheral resistance. The effect of the decrease in total peripheral resistance on arterial blood pressure was not, in the case of the sheep, compensated by an augmented cardiac output (Hales, 1973). The opposing influences on arterial blood pressure during exposure to heat are summarized diagrammatically in Figure 1.

TOTAL PERIPHERAL RESISTANCE

The overall effects of changes in regional blood flow during exposure to heat are represented by changes in the total peripheral resistance. The diminution in total peripheral resistance which occurred when the ox was exposed to a hot, dry environment could not have been elicited by stimulation of warm receptors in the brain or spinal cord because an increase in deep-body temperature did not take place (Whittow, 1965). In view of the results obtained by Jennings et al. (1973), in the dog, the concomitant thermal polypnea also must have contributed to the response. A further decrease in total peripheral resistance occurred in the ox when it became hyperthermic (Whittow, 1965). Part of this decrease may be attributed to an increased temperature of the anterior hypothalamus and pre-optic region because localized heating of this area in the ox and also in the dog, evoked a reduction in peripheral resistance (Whittow, 1968b; Phillips and Jennings, 1973).

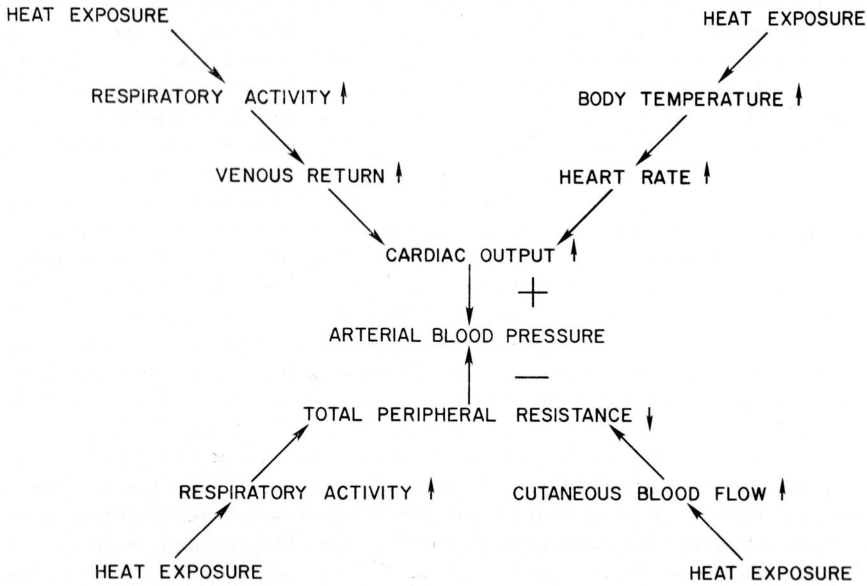

Fig. 1. Opposing influences on arterial blood pressure during exposure to heat. Based on Whittow (1965) and Jennings et al. (1973).

The diminution in total peripheral resistance during hyperthermia was less in the sheep than in the ox (cf. Whittow, 1965; Hales, 1973). It is conceivable that the sheep relies to a large extent on diversion of blood from other tissues, to the skin, rather than on an increased cardiac output. The ox, on the other hand, may provide for an increased cutaneous blood flow (Whittow, 1962, 1971b) by augmenting cardiac output.

BLOOD VOLUME

The blood volume of the ox did not increase during exposure to mild heat or in the hyperthermic animal (Whittow, 1965). The proportion of red cells in the blood consistently increased during hyperthermia; this was related partly to a tendency for the plasma volume to diminish, which in turn was probably related to water loss from the skin, respiratory tract and in the saliva. An increase in the percentage of red blood cells in the blood did not occur when the anterior hypothalamus was heated, in the absence of a general hyperthermia (Whittow, 1968a). Evidence was adduced to show that the increased haematocrit during hyperthermia was mediated by increased blood levels of circulating catecholamines and that an intact innervation of the adrenal glands was a prerequisite for the response (Robertshaw and Whittow, 1966). The absence of increased blood levels of catecholamines during localized heating of the anterior hypothalamus seems to provide a plausible explanation for the failure of the hematocrit to increase (Robertshaw and Whittow, 1966). An increased hemoglobin concentration in the blood of the sheep during hyperthermia was reported by Hales (1973).

PULMONARY CIRCULATION

The pulmonary arterial pressure decreased when the ox was exposed to heat (Whittow 1965). As the cardiac output increased (see p. 295), the pulmonary vascular resistance must have diminished during hyperthermia. The central blood volume increased during exposure to heat and it was postulated that the respiratory changes during thermal polypnea resulted in the transfer of blood from the abdomen into the thorax (Whittow, 1965). Support for this conclusion was adduced by Whittow (1968a) who subjected the ox to localized heating of the anterior hypothalamus and pre-optic region of the brain. This procedure resulted in thermal polypnea and an increased central blood volume. In contrast to the effects of generalized hyperthermia on the circulation, localized heating of the anterior hypothalamus resulted in an increase in pulmonary arterial pressure (Whittow, 1968a). Consequently, the diminution in pulmonary arterial pressure which occurred when the entire animal was exposed to heat must have involved either a direct effect of the increased temperature on the pulmonary vascular bed or the effect of changes in the composition of the blood during generalized hyperthermia. A greater decrease in pulmonary arterial pressure occurred in the sheep during hyperthermia, than in the ox (Hales, 1973). This is perhaps to be expected as the cardiac output did not increase (see page 296). The effects of some of these factors on pulmonary arterial blood pressure are shown in Figure 2.

ROLE OF RESPIRATORY CHANGES IN THE CARDIOVASCULAR RESPONSES TO HEAT

Most of the cardiovascular changes described in the preceding sections occurred against a background of greatly increased respiratory activity associated with panting. Evidence was presented that, in some instances, the cardiovascular changes appeared to be related directly to the variation in respiratory activity. The enhanced operation of the

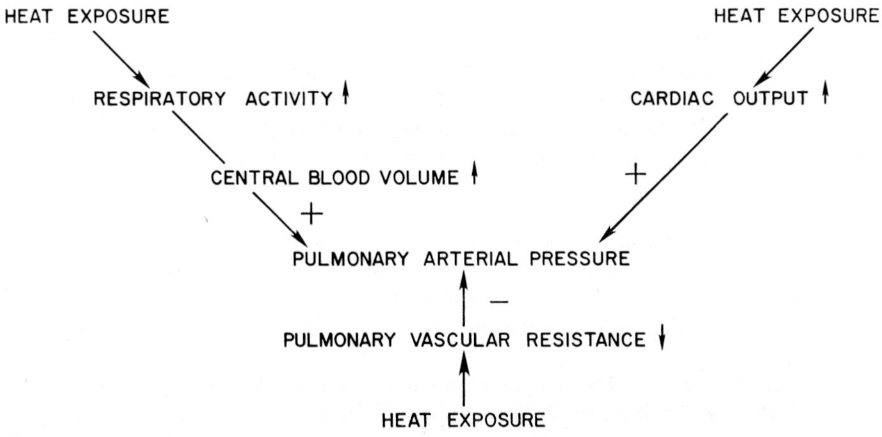

Fig. 2. Factors affecting the pulmonary circulation in an animal exposed to heat. Based on Whittow (1965; 1968a), and Hales (1973).

respiratory pump appeared to increase venous return and lead to an increased stroke volume and central blood volume. In addition, the intrathoracic pressure changes seemed to modulate the arterial, ventricular and venous pulse pressures, probably with important repercussions on the filling and emptying of the heart and on the reflex regulation of arterial blood pressure. Increased respiratory activity also resulted in a diminished total peripheral resistance.

A further consequence of augmented respiratory activity is the alteration in the composition of the blood, particularly with regard to the blood gases. During hyperthermia, the cardinal changes in the blood gases of the ox were a reduction in Pa_{CO_2} and an increase in the pH of arterial blood. The Pa_{O_2} increased during hyperthermia; but the $P\bar{v}_{O_2}$ diminished, indicating a state of tissue hypoxia (Findlay and Whittow, 1966). The diminished $P\bar{v}_{O_2}$ may have had cardiovascular effects by virtue of stimulation of pulmonary chemoreceptors (Duke et al., 1963). Hypocapnia may have contributed to the increase in heart rate but the effects are complex (Jennings and Sparling, 1973). Similarly, a decrease in total peripheral resistance resulted from hypocapnia (Jennings et al., 1973). It is conceivable also that the increased levels of circulating catecholamines in the hyperthermic animal may have contributed to the increased cardiac output, in addition to an effect on the red cell content of the blood.

A strikingly different cardiovascular response to heat, related to the respiratory cycle, is evident in the California Sea Lion. Sea lions, in common with other marine mammals, are natural, spontaneous breath-holders. During the apneic phase of the respiratory cycle, a profound bradycardia occurs. However, if sea lions are made hyperthermic, the apneic bradycardia is less. In other words, the heart rate is higher during apnea. As the heart rate during the inspiratory phase of the respiratory cycle is not altered when the sea lion becomes hyperthermic, the average heart rate of the animals increases in response to heat (Lin et al., 1972; Matsuura and Whittow, 1973).

10. CONCLUSIONS

An increased blood flow to the skin, a diminution in total peripheral resistance and an increased cardiac output characterize the cardiovascular response to heat, at least in panting animals. An increased central blood volume is also a feature of the effects of heat on the circulation, but this may be true only for the panting animal. An increased heart rate is the main determinant of the increased cardiac output. Many of the effects of hyperthermia on the circulation may be produced by localized heating of the anterior hypothalamus and pre-optic region of the brain. There is a great need for information on the effects of heat on the cardiovascular system of an animal that does not pant in order to distinguish between the direct effects of heat and effects secondary to the increased respiratory activity.

ACKNOWLEDGEMENTS: The preparation of this paper was supported in part by a grant (GB-29287X) from the National Science Foundation.

REFERENCES: see pp. 559-560.

Effects of Heat Stress on Cardiovascular Responses in Birds

by

P. C. Harrison

INTRODUCTION

Reports of cardiac and circulatory adjustments by birds in hot environments are contradictory and thus often confusing. This is not surprising, since the different studies seldom involve similar temperature and methodological environments. Hot environments used in studies of cardiovascular responses to heat vary from temperatures that scarcely initiate thermoregulatory responses to those that would render the bird well cooked shortly postmortem. Methodological conditions vary from the subject being strapped and wrapped on a laboratory bench to a free-moving bird telemetering its cardiovascular status. These experiments have yielded important information concerning specific responses of the cardiovascular system to specific hot environments under specific conditions of observation, but interpretation of these data in relation to underlying mechanisms is a perplexing task.

Importance of cardiovascular reflexes for thermoregulatory purposes has been well defined (Thauer, 1965). Cardiovascular adjustments in hot environments aim to redistribute blood flow to facilitate heat loss yet maintain adequate blood supply for local tissue metabolic demands. The ideal adaptive process would be to make these adjustments in the cardiovascular system in a hot environment without adding to the animal's heat load. To accomplish cardiovascular alterations important for thermoregulation of birds in hot environments, during a time of ventilation change due to panting, represents a high degree of integration of peripheral and central cardiovascular control systems. Details of avian cardiac innervation and cardiovascular control systems can be found in Abraham (1969), Jones and Johansen (1972) and Pearson (1972).

VASCULAR RESPONSES TO HOT ENVIRONMENTS

During acute heat exposure and hyperthermia there is a decrease in blood pressure and total peripheral resistance (Whittow et al., 1964). A decrease in peripheral resistance which reflects peripheral vasodilation is the general response observed in most homeotherms and serves to increase surface temperature and thereby increase sensible heat loss and reduce heat input from the environment. Change in surface temperature with changes in environmental temperature on the unfeathered leg area of California quail subjected to hot environments indicates that there are peripheral vascular responses to thermal environments in the uninsulated surface of these birds (Brush, 1965). Considering the relatively small body surface that is comprised by unfeathered parts of most birds, peripheral vascular changes in these parts would have little effect on total blood pressure or peripheral resistance.

Blood pressure and peripheral resistance changes during exposure to hot environments depend upon magnitude of temperature change and length of time of exposure to hot environments. Acute exposure of domestic fowl to hot environments (40 to 45° C) causes a slight increase in blood pressure until maximum heart rates are attained, followed by a fall in diastolic, then systolic, pressure (Frankel et al., 1962; Linsley and Burger, 1964; Weiss et al., 1963). Abrupt increase in temperature from 21° C to 35° C caused a decrease in blood pressure of domestic fowl within the first day and for 10 days following the change (Harrison and Biellier, 1969). Heat-acclimatized birds also showed a decrease in blood pressure at 31° C and a decrease in pulse pressure as a result of lowered systolic pressure (Whittow et al., 1966). These examples clearly demonstrate that the differences in cardiovascular responses to acute heat exposure are dependent upon both degree of thermal change and previous thermal conditioning.

Whittow et al. (1964) measured a decreased blood pressure and calculated a decrease in total peripheral resistance (TPR) 100 min after wrapping hens in a thermal heating pad. This decrease in peripheral resistance is the anticipated response and similar to that reported for mammals (Thauer, 1965). Contrary to data reported for mammals in a hot environment, Vogel and Sturkie (1963) and Sturkie (1967) have reported an increase in peripheral resistance in heat acclimatized chickens. They proposed that the difference in peripheral resistance between heat-acclimated mammals and birds may occur because birds dissipate more heat from the respiratory system than from the surface. However, it would seem logical that blood shunts to heat dissipating areas (if vasodilation occurs), regardless of location, would have the effect of decreasing peripheral resistance. The method of restraint and thermal environment under which the parameters used to calculate TPR [TPR = mean arterial pressure (mm Hg)/ cardiac output (ml/min)] may have an influence on the measurements of the heat-acclimatized animals. The increased TPR observed in heat-acclimatized chickens by Vogel and Sturkie (1963) was observed after the birds had been kept in a 21° C environment for approximately 1 hour. However, Whittow et al. (1966) compared cardiovascular status of heat-acclimatized birds at 22° and 31° C environments and found no significant difference for a 90-min period. Another consideration related to TPR during heat acclimation is the relative changes in the parameters used to calculate resistance. Cardiac output (the denominator) is significantly decreased during heat acclimation, whereas there is a relatively small reduction in blood pressure (the numerator) at the same time (Sturkie, 1967). Therefore, increased TPR may be representative of cardiovascular reflexes separate from peripheral vasodilation.

CARDIAC RESPONSES IN HOT ENVIRONMENTS

Cardiac rate and output of birds in hot environments depends on thermal intensity and treatment of the animal prior to monitoring cardiac status. Hutchinson and Sykes (1953) reported that heart rate was inversely related to the severity of the temperature stress up to environmental temperatures of 37.2° C and rectal temperatures up to

43.3° C. Above 43.3° C rectal temperature, each increase in rectal temperature of about 0.6° C resulted in a 30-beat-per minute increase in heart rate. Wilson (1949) reported that White Leghorn pullets subjected to a change in environmental temperature from 26.6° C to 32.2° C demonstrated a decreased heart rate in conjunction with increased respiratory rate. A reflex bradycardia following exposure of birds to hot environment is in opposition to the response reported for many mammals which show an increase in heart rate as soon as rectal temperature begins to rise. Likewise intense acute exposure of chickens to 45° C air temperature resulted in increases in heart rate that were proportionally related to increased body temperature to a body temperature of about 45° C (Frankel et al., 1962; Linsley and Burger, 1964; Rodbard and Tolpin, 1947). Linsley and Burger (1964) found that heart rate and stroke volume increased for the first half of a 100-min experimental period of exposure to 45° C followed by a decrease in stroke volume for the second half of the experimental period. Whittow et al. (1964) also found that hot environments induced by wrapping birds in a thermal blanket would cause an initial increase in heart rate. These data clearly demonstrate that the intensity of thermal environment greatly affects the cardiac reflexes to hot temperatures.

Domestic fowl acclimated to hot environments (30 to 35° C) tend to exhibit a depressed heart rate (Weiss et al., 1963; Harrison and Bieller, 1969; Sturkie et al., 1970). White Leghorn hens exposed to an increase in temperature from 21° C to 35° C showed a stable heart rate for the first 6 to 8 hrs of exposure which was followed by a decrease in heart rate to a lower stabilized level for 10 days after exposure to the 35° C environment (Harrison and Biellier, 1969). The lowered heart rate of heat-acclimated chickens rapidly returns to normal when removed from hot environments and was not different from cold-acclimated birds when measured at normal room temperatures (22° C) (Whittow et al., 1966). Weiss et al. (1963) found that the depressed heart rate of heat-acclimated chickens reflexly increased in response to additional increases in environmental temperature.

Depression of heart rate as environmental temperature increases has also been demonstrated in quail (*Lophortyx californicus*, *Coturnix coturnix japonica*) and in the morning dove (*Zenaidura macroura*) (Brush, 1965; Chilgren and King, 1973; Harrison et al., 1972; Hudson and Brush, 1964). Hudson and Brush (1964), using 0° C as the constant, calculated heart rate to be, $HR = 218.8(-1.181)^t$ and $HR = 563.6(-1.023)^t$ for the morning dove and the California quail, respectively (t = environmental temperature °C). This relationship of decrease in heart rate with increase in environmental temperature occurred until minimal heart rates were attained around 30 to 35° C for the dove and 35.5° C for the California quail. Chilgren and King (1973) recorded decreased heart rates in Japanese quail subjected to 35 and 40° C environments with relative humidities ranging from 13 to 93% in comparison to heart rates at 25° C and the same range of relative humidities. The depression of heart rate in Japanese quail subjected to abrupt shifts in temperature from 25° C to 36 and 40° C environments occurred within 10 to 15 min following the temperature change (Harrison et al., 1972; Schwalbe, 1973).

The reflex return of heart rates of heat-acclimated birds to normal levels when exposed to thermoneutral room temperatures and increase in heart rate in response to added heat load and hyperthermia indicate that warm environments directly stimulate bradycardia. Adaptive significance of thermal bradycardia as the bird initiates increased physical thermoregulative heat-loss processes is open to speculation. The sensory and neurological integrative mechanisms resulting in heat depression of heart rate is relatively unknown. An increase in vagal tone is the only mechanism that has been related to the depression of heart rate of chickens and quail exposed to hot environments (Sturkie et al., 1970; Harrison and Becker, 1970; Harrison et al., 1972). Similarity between efferent vagal control of diving bradycardia (reviewed by Jones and Johansen, 1972) and thermal bradycardia suggest a common mechanism. However, a common afferent control is difficult to envision.

Thermal bradycardia in hot environments may represent a very beneficial adaptive functional phenomenon in that circulatory adjustments important to heat dissipation can occur without an added heat load. The question arises: Is thermal bradycardia unique to the birds herein reported? Perhaps methodology has masked an important adaptive trait in many homeothermic animals. The occurrence of thermal bradycardia has been reported in cattle and rats (Brody, 1945; Spielman and Lyman, 1971). Morishima and Gale (1972) have observed a decrease in heart rate of baboons following preoptic anterior hypothalamic warming. Spielman and Lyman (1971) have found that more or less restraint than applied in a moderate restraint condition prevented thermal bradycardia from being expressed. In discussion of their results, it is pointed out that thermal bradycardia appears to occur in a number of mammals and birds but that methodological conditions can prevent this phenomena.

SUMMARY AND CONCLUSION

In many respects the cardiovascular responses of birds to hot environments appeared to follow the same general trend as for mammals. Apparent differences in response that are worthy of future research for the elucidation of their adaptive importance and mechanics of occurrence are: (1) Change in total peripheral resistance in birds acutely exposed to hot environments as compared to those chronically exposed to hot environments; (2) vascular reflexes important to shunting blood to heat dissipating areas for thermoregulatory purposes; (3) the mechanisms responsible for thermal bradycardia.

Standard methodological and thermal environments need to be established which will initiate expression of thermoregulatory responses which occur in response to natural heat stress. Once standardized methodology has been established comparative studies of cardiovascular reflexes within and between species of birds and mammals should be initiated in order to elucidate the mechanics and adaptive significances of cardiovascular reflexes to hot environments.

REFERENCES: see pp. 560-561.

Effects of Cold Stress on Cardiovascular Responses in Mammals

by

G. E. Folk, Jr.

The adaptations to meet the challenge of the hostile environment of extreme cold will be considered from the aspect of Comparative Mammalian Physiology. The two main classes of mammalian response to cold and some special difficulties involved in making cardiovascular measurements in the cold will be described.

When a cold snap is experienced by a "summer" mammal, it calls upon certain Darwinian adaptations to counteract the cold exposure. It may vasoconstrict its skin (a physiological mechanism), and it may start to grow a winter coat (an anatomical mechanism). If the cold continues, after about 10 days some changes have taken place such as the replacement of shivering by non-shivering thermogenesis. This change is referred to as ACCLIMATIZATION if the animal is in the free environment, and ACCLIMATION if the animal is in an environmental chamber (Folk, 1974).

The two main classes of response to cold are: (1) hibernation and (2) maintaining the normal body temperature (normothermic). The detailed mechanisms of mammalian hibernation have been reported extensively in recent years; many such references are given in textbooks as well as descriptions of cardiovascular responses of the mammal going into and staying in hibernation (Folk, 1974). For example Lyman (1958) showed that when the internal signal is received by the mammal "calling for hibernation", first the heart rate is reduced, followed by reduction in metabolic rate and body temperature. The core body temperature is gradually reduced to about one degree above the environment of the hibernating burrow. Other examples of cardiovascular responses to hibernation will be cited later as "case histories."

How are measurements such as Lyman's made while the animal is cold exposed? Are there special difficulties in making these recordings? The problem is that restrained mammals are far from normal; this is especially the case when they are exposed to extreme heat or extreme cold. If a rat, rabbit, or cat is exposed to −20° C in a reasonably large cage, they tolerate this easily, even if they are not cold acclimated. However, if they are confined in a tight cylinder of wire, or if the cat is tethered by the hind leg, they will quickly die of hypothermia (restraint-stress). The right enlargement of the cylinder to prevent this restraint-stress has not been determined for each animal. The principle of hypothermic restraint-stress was first described by Barlett (many of his references are given in Magalhaes, 1974). Thus, if an investigator wishes to investigate blood pressure, heart rate, EKG, or vasomotor function of cold-exposed mammals using wires, he must be very ingenious unless he is studying ungulates or horses tied in stalls. The tight cylinder of wire prevents the twisting of the leads from animal to recorders, but is responsible for the death of the animal by hypothermia. Eventually this technical obstacle will be met by studying unrestrained animals by physiological radio-capsules either attached to the body or implanted in the abdominal cavity. Many investigators today are using these devices; an international society of physiologists and engineers using radio-telemetry has been formed by Dr P. Kimmish, but unfortunately their devices often fail if they are not used by the careful fingers of the inventor himself. Because such data are somewhat rare, some examples of cardiovascular recording by telemetry in the cold will be given.

Table 1. Range of heart rate from resting to exercise*

GROUP A (RANGE 1.3–1.8×)	
Hedgehog	1.6× (147–234 b/m)
Snowshoe rabbit	1.3× (137–182 b/m)
Stumptail macaca	1.5× (110–165 b/m)
Wolverine	1.8× (123–218 b/m)
GROUP B (RANGE 2.0–15×)	
Arctic fox	3.5× (64–225 b/m)
Arctic ground squirrel	2.3× (155–348 b/m)
Domestic cat	2.0× (120–241 b/m)
Domestic horse**	13.0× (20–275 b/m)
Lynx, Alaska (Arctic)	2.08× (95–197 b/m)
Polar bear	3.7× (42–156 b/m)
Striped skunk	2.5× (84–209 b/m)
Wolf	2.6× (45–115 b/m)

* For method and original data see Folk and Hedge (1964).
** Hamlin et al., (1970).

First we will consider the circumstances where spontaneous exposure by the animals to cold and to exercise seemed to drive them to high heart rates and in some cases to a maximum rate (Table 1). In most cases the Iowa implantable long-life radio-capsule was used (Folk and Copping, 1972). The only exception is the data from Hamlin (1970) on the domestic horse; in this case cold was not a stimulus for the high rates, nor was a radio-capsule used. The resting heart rates by radio are lower than many handbook values because the animals were always in large enclosures, undisturbed and unaware that heart rates were being recorded. Except for the horse, the animal with the largest change from resting to maximum was the polar bear (3.7×). Probably under these circumstances stroke volume changed very little, while if the animals studied had been lower vertebrates (fish, amphibians, or reptiles) they would have conspicuously increased both stroke volume and heart rate (Johansen, 1965). The convenient breakdown in Table 1 into two groups is artificial and does not seem to have a taxonomic or size basis. In most cases only two or three specimens of each species were studied and so more heart rates on more individuals should be recorded.

The case of recording an instrumented woodchuck (*Marmota monax*) exposed to cold weather will be considered next. This hibernator had first been studied in the laboratory. When she was studied in the field, very different results were obtained; when the temperature of an improvised outdoor den fluctuated with the weather, then the heart beats fluctuated in inverse proportion (Folk et al., 1970). During another season, the animal was given the opportunity to dig her own tunnel (3 m long) and den which was 1.6 m under the soil. Four thermistors were in the wall of her den. Heart rates were recorded every half hour during January and July (this is true of all species discussed subsequently); in this case and the others the nine lowest sleeping heart rates are averaged, each one selected from a different night. For this woodchuck the summer sleeping heart rates of 70 b/m were reduced to 6 b/m during natural hibernation in a natural den (Table 2). The woodchuck was sensitive to vibrations while in outdoor

Table 2. Lowest Daily Heart Rates (± SD) Compared in January and July

Species	Sleeping heart rates (mean) b/m		Place	Conditions	January lowest ambient
	Summer (N = 9)	Winter (N = 9)			
Wolverine					
No. 1	———	85 ± 5	Pt. Barrow	Snow Den	−46° C.
No. 2	———	120 ± 8	Pt. Barrow	Snow Den	−43° C.
No. 3	77 ± 5	94 ± 7	Pt. Barrow	Snow Den	−42° C.
Striped Skunk	83 ± 3	85 ± 6	Iowa	Natural Den (3° to 2° C.)	−10° C.
Arctic Marmot					
No. 1	93 ± 7	7*	Pt. Barrow	Outdoor Dens	−25° C.
No. 2	82 ± 6	8*	Pt. Barrow	Outdoor Dens	−48° C.
No. 3	92 ± 5	7*	Pt. Barrow	Outdoor Dens	−31° C.
Woodchuck	70 ± 9	6*	Iowa	Natural Den (4.2 to 2.6° C.)	−29° C.
Porcupine	85 ± 7	86 ± 11	Pt. Barrow	Outdoor Dens	−21° C.

* Hibernating values.

hibernation. As a special test, on three occasions each a week apart, continuous heart rates were taken in a building 25 m from the burrow, while an observer walked over the underground burrow. The heart rates went up from about 6 b/m to as high as 28 b/m, but within 6 min could return to hibernation level; details are as follows:

Trial
1. (9.30 hr) 6 b/m; (9.45 hr) 15 b/m; (9.58 hr) 7 b/m
2. (13.00 hr) 6 b/m; (13.12 hr) 28 b/m; (13.27 hr) 13 b/m
3. (17.00 hr) 6 b/m; (17.15 hr) 24 b/m; (17.22 hr) 8 b/m

These results from a woodchuck in Iowa were compared with measurements on three Arctic marmots (*M. caligata*). In each case the rodents hibernated in a mound of hay about 2 m across. The den temperature was not measured, but the air temperature outside the hay was recorded (Table 2). These rodents were about the same weight as the woodchuck, their summer rates were higher, but their winter outdoor hibernating values were about the same. Probably these heart rates simulate those which would be obtained on free-living marmots and woodchucks.

These same devices for study of cardiovascular function in the cold have been used on mammals which have been suspected of hibernation, but seldom studied, namely raccoons and skunks. We studied four raccoons in outdoor cages and found that they raised instead of lowered their heart rates in winter; the July values were 97 b/m while January values varied from 99 to 120 b/m (Folk et al., 1968). To ask the same question, a striped skunk was allowed to spend the winter in the burrow dug in the field by the woodchuck (mentioned above). His summer sleeping level (83 b/m) in a relatively warm burrow increased to 85 b/m. Many similar experiments need to be done, especially on badgers which are also suspected of hibernation; the above scanty evidence at least indicates that hibernation in raccoons and skunks is difficult to induce or is not their mechanism to combat cold. It is of interest to compare these results with the largest relative of the skunk, the wolverine. This animal is ever present all winter and has never been associated with hibernation; although the skunk weighs 4 kg and the wolverine about 15 kg, the winter sleeping heart rates of the skunk were similar to those of two of the wolverines (Table 2). In the one case studied both winter and summer, the wolverine increased its winter heart rate markedly, just as had the raccoons and skunk.

Porcupines occur in the arctic and since they are large rodents, may be considered as control animals for the hibernators. One was studied at Point Barrow; he showed no lowering of sleeping heart rate while tolerating long exposure to cold in an outdoor cage.

The foregoing examples of a method of studying cardiovascular responses to cold should be compared with data from the literature; such reports are very scarce except for studies on human subjects. Only the studies of Mühlbock and Nomura (1966) on domestic non-hibernator mice seem to be suitable for comparison here. The mice were acclimated to 25° C. The investigators recorded the response to cold (15° C) of nine strains of mice which were restrained during the experiment; there was usually a decrease in physiological function which varied with the strain as follows (in %):

Heart Rate	36.6 to 77.5
Respiratory Rate	57.5 to 110.0
Blood Pressure	75.2 to 89.5
Body Temperature	59.3 to 80.1

One strain was studied without restraint; it is important to note that in this case the cardiovascular measurements went up as follows:

	Envir. temp.	No. of mice	Measurement
Heart rate (b/m)	15° C	11	746.4
	25° C	16	681.8
Resp. rate (no./m)	15° C	11	217.5
	25° C	16	202.7
Body temperature (°C)	15° C	11	36.47
	25° C	16	36.76

These experiments help to confirm the two concepts emphasized in this chapter: (1) normal cardiovascular responses to cold cannot be measured on restrained animals; and (2) there is a tendency for the resting heart rate of non-hibernator (normothermic) mammals to increase during cold exposure.

Our experiments described in this paper were sponsored by The Arctic Institute of North America with the approval and financial support of the Office of Naval Research under Contract Number N00014-72-A-0375-0002, Subcontract Number ONR-453.

REFERENCES: see p. 561.

Effects of Cold Stress on Cardiovascular Responses in Birds

by

P. C. Harrison

INTRODUCTION

Cardiovascular adjustment represents one of the bird's major mechanisms of defense against heat loss to a cold environment. Peripheral blood-flow adjustment in unfeathered leg and toe areas is a key adaptate favoring survival of aquatic, wading and Arctic birds.

Evidence for cardiovascular adjustment to cold, other than that for peripheral and countercurrent heat-exchange shunts, is scanty. This is understandable since it is difficult to visualize how other cardiovascular responses (with the exception of increased heart rate) could greatly increase heat production or decrease heat loss.

VASCULAR RESPONSES OF BIRDS TO COLD ENVIRONMENTS

Blood pressure follows seasonal trends in chickens. Mean blood pressure is lowest during summer and highest during winter (Weiss et al., 1961). Similarly, in man, blood pressure increases during winter months, reaching its highest level in the spring (Rose, 1961). Increased mean blood pressure is the result of increased systolic and pulse pressures in chickens during winter months (Weiss et al., 1961; Whittow et al., 1966).

Blood-pressure changes in response to acute or abrupt exposure to cold environments depend on the magnitude of decrease in environmental temperature and are closely correlated with decrease in body temperature. Blood pressure of chickens exposed to −1.1° C was not different from birds in a 21° C (thermoneutral) environment (Vogel and Sturkie, 1963). Abrupt decreases in air temperature from 21° C and 35° C to 5° C had little effect on blood pressure (Harrison and Biellier, 1969). In the 35° to 5° C air-temperature shift, blood pressure tended to be lower for the first 12 hrs of exposure to the 5° C temperature. However, after the first 24 hrs at 5° C, blood pressure and body temperature had regained normal levels. A linear relationship between decrease in rectal temperature and decrease in blood pressure has been demonstrated in ice-pack and cold-water-immersion studies (Weiss, 1960; Rodbard and Tolpin, 1947). Blood pressure of chickens decreased linearly with body temperature to a body temperature of around 33° C, below which blood pressure declined precipitously.

In contrast to most cold-acclimatized mammals, cold-acclimatized birds show non-significant changes in total peripheral resistance in cold environments. In mammals, peripheral vasoconstriction may cause a 2- to 3-fold increase in peripheral resistance, even during hypothermia (Thauer, 1965). Chickens exposed to −1.1° C air temperatures had no change in total peripheral resistance when compared to individuals in a 21° C (thermoneutral) environment and it was significantly lower than in individuals at 35° C (Vogel and Sturkie, 1963). However, as mentioned in the previous chapter, an increase in peripheral resistance is observed after acclimatization of birds to hot environments.

Lack of cold-induced change in total peripheral resistance indicates that peripheral vasoconstriction does not occur in cold-stressed birds. Additional information suggesting that cold-induced general peripheral vascular responses are not occurring is that, under feathered body surfaces, skin temperature seldom drops below 30° C even when air temperature is 0° C or lower (Irving and Krog, 1954; Bartholomew and Dawson, 1954; Wilson et al., 1952). Unfeathered areas show drastic decreases in surface temperature in response to decreased environmental temperature, indicating that peripheral vasomotor activity is regionally responding to insulate otherwise uninsulated heat-dissipating surfaces. The thermoregulatory (heat-dissipative and heat-conservative) importance of local vasomotor response in establishing blood-flow shunts in the leg and toe region of many aquatic and wading birds has been recently reviewed by Dawson and Hudson (1970) and Jones and Johansen (1972). Anatomical and physiological evidence of vascular retia which establish countercurrent heat exchange blood flow patterns has been observed in a number of Arctic, aquatic, wading and some terrestrial birds. These vascular retia permit arterial blood cooling as it circulates to the peripheral limb. This maintains a cool extremity and thereby reduces the animal-to-environment thermal gradient. In addition, venous blood is warmed during its return to the animal core. Countercurrent retia are found in numerous animals and serve important functions in heat and salt balance. [See Scholander (1958) for review and diagrams of vascular countercurrent blood flow in animals].

Johansen and Tønnesen (1969) found that as external temperature decreased from 27° to 1.5° C, foot-web temperatures of gulls dropped from 27° to 4° C, while blood-flow rate dropped from 18 to 0 ml/100 g/minute. Similarly, Brush (1965) found a direct linear relationship between ambient temperature and tarsometatarsal and toe temperatures in California quail. Bypasses of leg and foot arteriovenous retia are suggested to aid heat dissipation in birds in positive heat balance and to prevent freeze damage to tissue in cold environments (Steen and Steen, 1965; Johansen and Tønnesen, 1969). Birds in positive-heat-load conditions can dissipate four times more heat to water than to air at the same temperature. Also, leg cooling of birds in positive-heat-load conditions caused an immediate cessation of thermal panting. Evidence for retia bypass and the existence of the hunting reflex [periodic warming of cold-exposed digits caused by local vasodilation and increased blood flow during extended cold exposure (Lewis, 1930)] has been demonstrated in chickens as well as aquatic birds (Grant and Bland, 1931). Blood flow in the foot web of gulls maintained in a 4° C environment for 30 min changed from 1.8 to 0 then 3 ml/100 g/min (Johansen and Tønnesen, 1969). Research of Millard and Johansen, 1972 and Johansen and Millard, 1972 (reviewed by Jones and Johansen, 1972) indicated the blood-flow changes in leg and foot vasculature in response to cold and hot environments are under active adrenergic vasoconstrictor and cholinergic vasodilator nerve tone. Cutting the foot nerves in cold and hot environments resulted in increased and decreased blood flow, respectively. Peripheral vasodilation caused by stimulation of peripheral ends of cut vasodilator foot/nerves was blacked by atropine. Alpha-receptor stimulation with phenylephrine resulted in vasoconstriction, but beta stimulation with isoprenaline had no effect.

CARDIAC RESPONSES TO COLD ENVIRONMENT

Winter-acclimatized birds have increased heart rate, stroke volume, and ventricular work when compared to summer-acclimatized birds (Vogel and Sturkie, 1963; Whittow et al., 1966). An inverse linear relationship between heart rate and environmental temperatures between 30 to 35° C and 0° C has been reported for the morning dove and California quail (Hudson and Brush, 1964) (See previous chapter for equations for the relationship of heart rate to air temperature). Intense body cooling in chickens by cold-water immersion or ice-pack conditions caused transient increases in heart rate, followed by a decrease in heart rate associated with hypothermia (Weiss, 1960; Rodbard and Tolpin, 1947).

Increase in heart rate following abrupt exposure of hot- and thermoneutral-acclimated chickens to 5° C air temperature occurred within the first few minutes in the cooler environment (Harrison and Biellier, 1969). This increase in heart rate was not associated with any detectable change in rectal temperature, thereby indicating peripheral sensory integration with central cardiovascular control systems. The increase in heart rate can be correlated with an increase in circulating catecholamine level in cold-exposed mammals and birds. However, increased heart rate of birds in cold environments appears to be the result of decreased vagal inhibition. Sturkie et al. (1970) found that propranolol blockage of the adrenergic influence on resting heart rate of cold-acclimatized, normothermic-acclimatized and heat-acclimatized chickens caused equal depression of heart rate. However, atropine blockage of cholinergic influence on resting heart rate caused significantly less increase in heart rate in the cold-acclimated birds, thereby indicating decreased vagal cardiac tone. Also in the chicken, pressor effects of epinephrine are the same between body temperatures of 30 and 41° C (Rodbard and Tolpin, 1947). Contrary to this response in cold-exposed birds, increased vagal activity has been found in cold-adapted rats (Leblanc and Cote, 1967).

CONCLUSIONS

Cardiovascular reactions that develop during cold acclimation in birds reflect, but in less magnitude, the adjustments that occur during acute exposure of birds to low temperatures (Fig. 1). Cold-induced hypothermia causes a depression of the levels of cardiovascular function from those that occur in thermoneutral or cold-acclimatized individuals.

Countercurrent or arteriovenous retia blood flow in feet, toe and web areas of birds is an important thermoregulative mechanism for aquatic and long-legged wading birds. The importance (if any) of rete blood flow for thermoregulatory purposes in terrestrial birds that behaviorally insulate the legs with feathered body areas should be investigated.

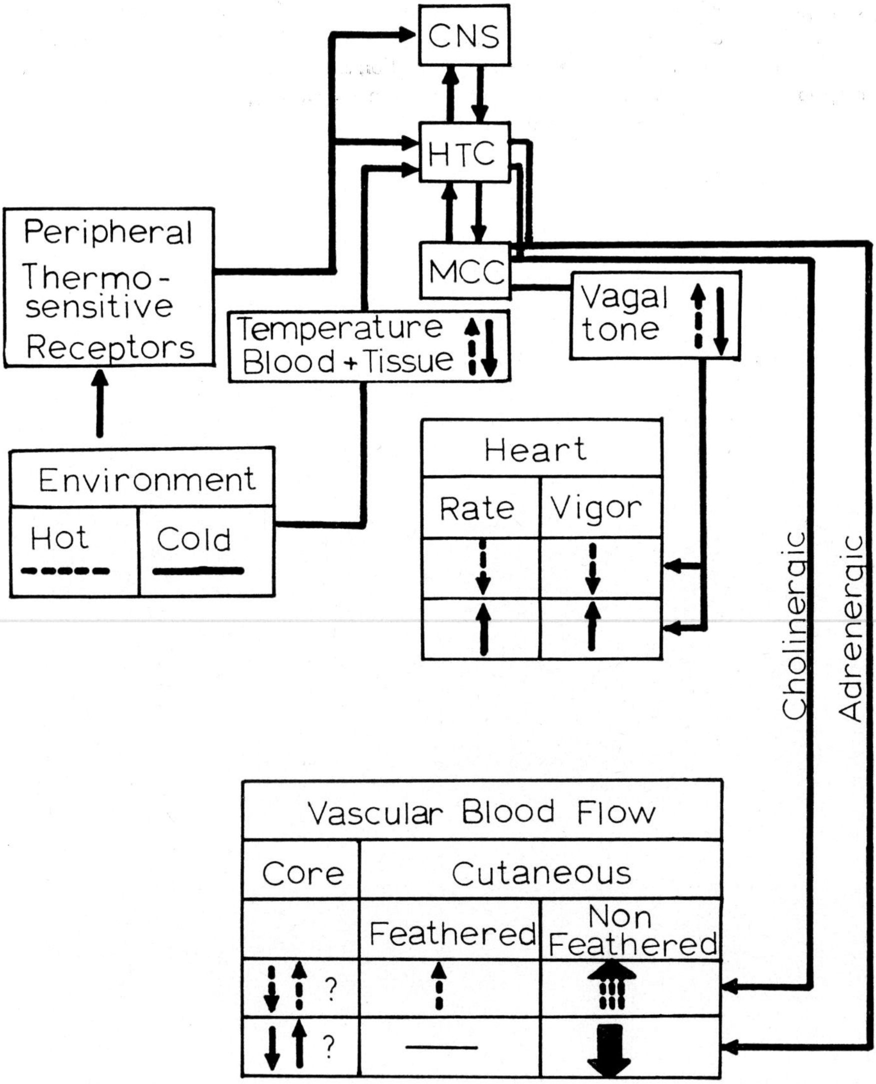

Fig. 1. Cardiovascular adjustments of birds acclimatized or acutely exposed to cold and hot environments that do not induce extreme changes in body temperature. Both cold and hot environments alter surface, blood and tissue temperatures which are centrally sensed at cerebral center (CNS) and hypothalamic thermoregulatory centers (HTC). CNS and HTC efferent response to afferent thermal input acts on medullary cardiovascular centers (MCC), which cause cardiac responses mainly by alteration of vagal tone and vascular reponses via active autonomic motor (cholinergic and adrenergic) control of blood flow by vessel diameter and shunts to countercurrent retia.

Changes in core blood flow and blood volume of birds in both hot and cold environments should be studied. Blood shunts in heat- and cold-acclimatized birds need to be correlated with thermal dissipation and production sites. Blood-shunt data in turn may help explain the greater total peripheral resistance following heat acclimatization as compared to that in cold-acclimatized birds.

REFERENCES: see p. 562.

Effects of Temperature on the Growth of Cattle

by

D. F. Dowling

INTRODUCTION

Growth is not a scientific term with defined and constant meaning, but at no time has it been more relevant to understand its effect on economical food production. At no stage has this protein starved world more needed efficiency in cattle growth. We must breed cattle which make proper use of forage and which can respond to better management situations.

ANIMAL GROWTH

Growth in the biological sense is intimately connected with metabolism and the organisation of living systems in self reproducing units. If the molecule of a chemical compound increases in size, it changes into another chemical compound. An organism, like cattle, for an instance, represents a system consisting of units which may increase in number but remain of the same type within their respective species and so allows for growth as an increase in size of a system which remains specifically identical. Consequently, animal development leads to the establishment of certain shapes or forms. In early development stages, morphogenetic changes take place essentially by segregation and differentiation of tissues. In later stages they are mainly caused by relative growth, that is, different growth rates exhibited by various components of the body. In this sense the problems of organic form is a problem of growth. Thus, animal growth is a complex process which involves not only an increase in SIZE but also changes in form and function of the different parts of the body. What is measured as growth is the outcome of considerable complexity, whether envisioned from the biochemical, physiological, or morphological viewpoint. Furthermore the growing organism undergoes changes in many respects, such as chemical composition, content of water, proteins and other compounds, the ratio between bone, protein and fat and changes in the shape of the body.

The purpose of this Section is to alert all interested in cattle production and research to observe carefully the influence of the environment on the various components in liveweight change.

The increase in body weight tends to follow a characteristic S-shaped curve pattern. However, the course of growth is different in different organisms because the physiological basis of growth is different in each case. Both heredity and environment are important in producing differences among individual animals. Environmental differences include managerial aspects as well as climatic differences (e.g., temperature). For example, growth is clearly dependent upon health and food intake. Hence, the classical sigmoidal growth curve steep at onset of puberty flattening out as maturity comes on (or as fat is laid down) depends on a continuous type growth pattern.

Growth of the organism as a whole e.g., cattle, is estimated by liveweight measurement. But what does liveweight mean, when in beef production the edible portion of the carcass is our goal. Liveweight is related to carcass weight but may not be an accurate measure of beef yield. As carcass weight increases, the dimensions and weights of all component cuts will also increase, but the fat content can vary from 14% in the 364 kg carcass of a 20-22 month old lean beef type e.g., Charolais type, to 44% fat in a 273 kg carcass of the traditional early fattening Bos taurus type at the same age. Liveweight measurement is an inadequate guide to carcass composition and for the variations in the non-carcass factors due to fill of rumen and gastrointestinal tract, blood, viscera, hide, head, hooves, kidney, pelvic, cod and udder fat, diaphragm, tail etc.

Growth, that is, the increase in body mass in time, is not unlimited. As mentioned above, the initial rapid increase gradually slows down until the organism reaches a steady state or is "adult". Changes in weight, once mature, are due mainly to deposition of FAT. Excess fat is wasteful. Recently, Charles (1974) has suggested a simple method of measurement on the live animal to determine the slaughter point. It is important to know when the animal reaches this stage where fat is laid down and muscle growth slows down. From this time on, the high energy cost of depositing fat and the increased maintenance costs of the heavy animal results in low biological efficiency.

MATHEMATICAL PROCEDURES

Hence, a completely accurate picture of growth is difficult to come by where growth is measured as an increase in mass (W) per unit of time.
Mathematical procedure must be based on biological concept to depict the phenomena of change and growth with some simple value which can be simply presented. Mathematical procedures are useful, though limited in describing the relative growth of beef in a hot or cold environm nt.

Certainly, allometric growth coefficient ("b" values) reduces the recorded information to a simple unit. But how may we interprete such figures? Even percentage values, which rely too heavily on accuracy of base calculations, must be avoided in cattle. Simply, it would appear that large muscles grow proportionally faster than small muscles; muscles most closely related to skeleton are smallest and have the lowest growth potential. Complexly, it would seem that assumptions for mathematical models of beef production will often be off beam because the physiological data has not been obtained. For examples of difficulties, Butterfield and Berg (1966) showed that muscle proportions do not change necessarily with age. In fact, muscle weight changes are quite independent of age and are highly total muscle-weight dependent. Naturally, a major function of musculature in cattle is to support total body weight yet muscle weight is a fairly stable 33% of body or live weight in cattle (Callow, 1944).

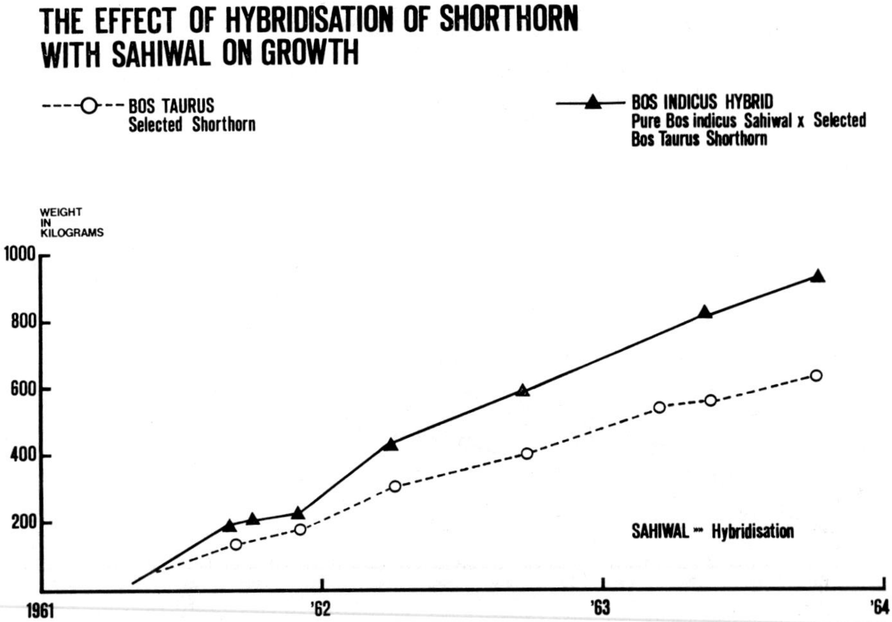

Fig. 1. Shows the typical effect of hybridisation on the growth curve of cattle. The development of the suitable hybrid dam is the first step in efficient beef production. The *Bos indicus* (Sahiwal) × early fattening *Bos taurus* (Shorthorn) is a typical hybrid on the tropical Australian scene.

Mathematical models may seem ideal for cattle, because cattle can be prohibitively expensive to actually use experimentally. Similarly, psychologically contrived negative correlations between maintenance costs and the likely contribution to knowledge from research with cattle are conveniently accepted. Consequently, most research projects shy off "the economic facts of life" the industry must face. Models don't tell you where to go; models are not a substitute for thought at all, if you cannot think about cattle comprehensively. When will we face up to experimenting with cattle based on an understanding of cattle and the influence of environmental factors on the particular genotypes? It is futile to do production experiments unless it is realised how heterogeneous cattle may be and how necessary it is to attack production research problems head-on. Nevertheless, at this stage of our observations, the models which may work is the animal. As ever, the first duty of a geneticist is to make his animals (material) more valuable experimentally. This applies to all researchers on growth in cattle.
That growth and production are presently severely limited due to direct influences of the micro-environment on energy exchange and utilisation has been amply shown (Bianca, 1965; Berg, 1969; Bond, 1974; Dowling, 1974; Hafez and Dyer, 1969; Hahn, 1970; Hammond et al., 1971; Johnson, 1965, 1972; McDowell, 1972; Webster, 1969).

Cattle starting from a single fertilised ovum must grow and be capable of carrying out all life's functions with special adaptations to obtaining and utilising forage, surviving

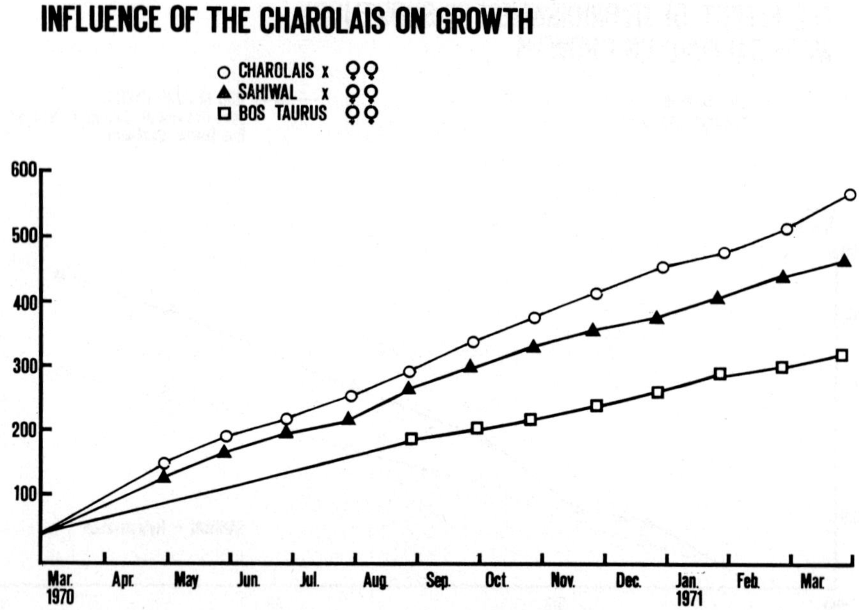

Fig. 2. Liveweight recorded in kilograms. The Figure illustrates the use of the lean beef sire (Charolais) on the selected *Bos indicus* hybrid dams, which makes for ease of calving and livability of calf under hot grazing conditions and results in faster growing, more efficient cattle for the heavier, younger, leaner beef carcass on a high level nutrition.

climatic and environmental stresses and reproducing to increase livability of the strains under conditions to which the cattle are exposed for production. Healthy animals can perform optimally, if genotypically selected, in the tropics (Dowling, 1960, 1974) and the cold (Young, 1974).

SELECTION TO OPTIMISE GROWTH AND PRODUCTIVITY UNDER HOT OR COLD CONDITIONS

Sir John Hammond (Hammond, 1960) saw that the most important animal production technique was A.I., not only for disease and defects control but for genetic improvement. Now that vast genetic resources are available, the improvement i.e., the economic and long range betterment, of existing cattle is practical by this method, if selection on Growth criteria can be affected. Progress will depend on research to evaluate cattle accurately in this regard. National Cattle Recording Schemers carry a heavy responsibility to be meaningful i.e., to continue to propose measurements which are more aligned to optimal efficiency.

The Beef cow is the self generating biological unit in the process. It must be highly efficient in aspects of growth for walking, grazing, breeding, and survival. Further, in our hot environments we are concerned with low-cost beef production under extensive conditions where parasite resistance and hardiness are critical, and fertility, livability of the calf, mothering ability are crucial for success. Our aim is also lower cost production under intensive conditions where cattle gain more rapidly for a larger size at a given compositional age. A better understanding of how cattle grow may enable manipulation of the process so that we can get a heavier carcass, at a younger age, with a desired level of intermuscular and intramuscular fat, but still without excess subcutaneous fat and no waste fat on flank, brisket etc., to comply with high muscle, optimum of fat only and a minimum of bone. In general, any selection for growth must be tested on a herd basis using several "control" sires for comparison to show the differences in adaptability and efficiency. It is time efficiency be taken more seriously e.g., cattle may well be tested for growth for a short period after early weaning or when heavier than 500 kg, because potential for growth may be ascertained early in life and late fattening at an economical market weight.

Thus, the calf must be managed to gain faster, convert feed more efficiently, fatten later, produce a heavier carcass with optimal muscle-weight distribution and minimal fat deposition. Hence it is patently obvious that we must breed and feed animals of each sex and breed type in manner appropriate to their growth characteristics, so as to achieve best combination of carcass weight and carcass composition.

REFERENCES: see pp. 562-563.

Effects of Temperature on the Growth of Pigs

by

J. Steinbach

INTRODUCTION

Although the survival range of adult pigs extends from −30 to +45° C (Porter and Gates, 1969), the zone of thermal indifference is much narrower, 0-15° C (Bianca, 1970). Temperatures above or below this zone elicit in the animal immediate physiological and behavioural responses, which have been reviewed recently in this series (Ingram, 1972) and elsewhere Steinhardt, 1971; Mount, 1968, 1972) and are referred to in other chapters of this volume (Chapt. 1, Sect. 4c, 5d, 8b (3) and 18c (2)).

Extreme ambient conditions, ultimately affecting pig productivity, may be of great economic importance. The following review will discuss the effects of ambient temperatures in the range to be expected under normal management conditions on various traits related to the growth performance and carcass characteristics of pigs.

Since the zone of thermoneutrality of adult pigs covers the lower range of temperatures normally encountered in pig raising, problems of heat may be more critical than those of cold, except for the newborn pig, whose problems have been reviewed recently by Curtis (1970). About one third of the world's pig population is currently found in tropical climates, at temperatures above the zone of thermal indifference. A screening of published results indicates that growth performance is very low indeed, although there can be no doubt that pigs can perform satisfactorily in tropical climates (Van Vaerenbergh, 1960; Oyenuga, 1961; Babatunde and Fetuga, 1974). However, even the best results obtained in the tropics fall about 15% below performance levels which could be expected in a comparable temperate environment. It is the purpose of this review to gather evidence on the effects of climatic changes on the growth performance and carcass characteristics of pigs during the growing and fattening period.

TEMPERATURE EFFECTS ON LIVEWEIGHT GROWTH

The growth rate of animals is interrelated with food intake and the efficiency of food conversion; each of these may be affected by changes in ambient temperature and will in turn effect changes in the other two.

FOOD INTAKE

Growth depends on food intake more than on any other factor. Heitman and Hughes (1949) and Heitman et al. (1958) have shown that food intake decreases with increasing environmental temperatures. Recent studies on younger pigs (Fuller, 1965; Sugahara et al., 1970) and older pigs (Seymour et al., 1964; Hale et al., 1968; Morrison et al., 1969; Morrison and Mount, 1971), confirming this observation, have explored a very wide range of experimental conditions. Yet their results agree well when the daily food intake is expressed in terms of the metabolic body weight in $kg^{0.72}$ (Fuller, 1965) and is plotted against ambient temperatures (t_a). The data of the above-mentioned reports fit the following equation fairly well ($R = 0.783$):

$$\text{Daily food intake (g/kg body weight}^{0.72}) = 145.74 + 1.96t_a - 0.104t_a^2$$

The mean food intake decreases at an accelerated rate rom about 155 g at 10° C to approximately 90 $g/kg^{0.72}$ at 35° C. Similarly, Mangold et al. (1967) found a decrease in food intake of 40 g/°C at temperatures above 23° C in young pigs (34 kg) and above 21° C in older pigs (63 kg). It appears then that even younger pigs exhibit this temperature-dependent decrease in appetite (Sugahara et al., 1964; Fuller, 1965; Mangold et al., 1967).

Food intake may also be influenced by high vapour pressures at high ambient temperatures. Morrison et al. (1969) reported a decrease in appetite due to increasing humidity levels of only 4% at 22° C, as compared to 14% at 33° C.

Food intake may further be affected by the composition of the ration. In particular, a higher protein level may increase food intake (Seymour et al., 1964; Devendra and Parris, 1970). Agarwala and Sundaresan (1956) mention an interaction between season and protein influencing the appetite of pigs. During the summer, pigs preferred the low-protein diet (16% CP), while in winter, intake on the high-protein level was greater.

As in other species, food intake in the pig appears to be regulated in the hypothalamus. Auffray (1969) presented evidence for this by inducing hyperphagia in pigs following bilateral lesions of the ventro-medial hypothalamic nuclei. Thus, ambient or endogenous heat may, in this species as in others, cause the rostral cooling centre of the hypothalamus to stimulate the satiety centre and/or inhibit the appetite centre (Brobeck, 1960).

Endogenous sources of heat are variable. Feeding causes an increase in heat production (Neergaard and Thorbek, 1967; Charlet-Lery, 1970), in heat loss (Bond et al., 1952; Kelly et al., 1963), in rectal temperatures (Ingram and Legge, 1970), and in skin temperatures (Sørensen, 1961); Okebukola and Steinbach (1973) observed a positive correlation between average daily gain and rectal temperature and respiratory frequency as evidence for the statement by Deighton (1929) "that the processes of growth are accompanied by waste of energy as heat."

The chemical composition of the blood may also affect the voluntary intake of food. Marple et al. (1972) reported lower glucose and fatty acid levels at 21° C than at 32° C. Possibly, the higher levels during heat stress are the result of a decrease in nutrient utilization by the body tissues, since Zinn, Weinman and Tumbleson (1968) observed a 40% decrease in the O_2 consumption of heart cell mitochondria in heat-stressed pigs.

GROWTH RATE

Over the past two decades, investigations on the effects of ambient temperatures, pioneered by Heitman and his group from the University of California, have been dominated by agricultural engineers in the U.S.A. In psychrometric chamber studies, Heitman and Hughes (1949) and Heitman et al. (1958) determined the optimum temperatures for growing pigs ranging in weight from 40 to 200 kg, concluding that the ambient temperature conductive to maximum growth rates decreases linearly with increasing body weight, from 24° C at about 40 kg liveweight to 16° C in pigs weighing 160 kg (Morrison et al., 1966). For each °C increase in mean daily temperature which was above the maximum gain temperature, a decrease of 30 g average daily gain was observed (Morrison et al., 1966). These observations have been confirmed by other experiments, carried out under a great variety of conditions, as shown in Table 1.

Table 1. Effect of temperatures above the optimum on the reduction in daily liveweight gain, as calculated from various published sources

Environment*	Temperature range in °C (approx.)	Reduction in ADG (g/°C)	Source
Controlled	20 – 30	30	Heitman et al., 1958
Natural	27 – 30	40	Hazen et al., 1959
Controlled	16 + 32	4	Seymour et al., 1964
Natural	27 + 29	39	Todd and Daniels, 1968
Controlled	22 – 28	28	Morrison et al., 1969
Controlled	28 – 33	41	Morrison et al., 1969
Controlled	23 + 33	21	Sugahara et al., 1970
Natural	24 – 30	24	Nelson et al., 1972

* Controlled – climatic chamber research;
Natural – normal management, natural climatic conditions.

Reductions in growth rate ranging between 20 and 40 g average daily gain/°C temperature increase were reported for studies in both natural and controlled environments, the only exception being the results of Seymour et al. (1964). However, the lower reduction in growth rate per unit temperature increase may be due to the fact that 16° C is well below the maximum gain temperature for pigs of 55 kg body weight. Possibly, the decline in growth rate is accelerating above 28° C, since the study by Morrison et al. (1969) indicates a more rapid decline in the temperature range of 28-33° C than between 22 and 28° C; and Comberg et al. (1971, 1973), who did not observe significantly different growth rates in pigs up to 28° C, found a suppression of daily gains at 30° C. Similar results were obtained in the Irish high-temperature-high-humidity pig houses (Hunter and Jennings, 1961) at least at weights below 75 kg (Tonks and Smith, 1965).

Below the optimum or maximum gain temperatures, results have been more varied. While the graphs of Heitman et al. (1958) and Morrison et al. (1966) indicate a significant rapid decrease in growth rates, other investigators found but little change at temperatures of about 8° C, compared to growth rates at thermoneutrality (Sørensen, 1962; Sugahara et al., 1970; Comberg et al., 1972) or achieved even higher gains around the freezing point than at 16 or 21° C (Seymour et al., 1964; Jensen et al., 1969). In the latter report, only small pigs below 50 kg gained less at an average temperature of −7° C. It appears then, that at temperatures below the optimum, as they are normally found in pig houses throughout the temperate zone, depressions in growth rate are small and negligible, if the food intake is adequate. Fuller and Boyne (1971) have shown that at temperatures below the optimum growth rate, nitrogen retention and carcass fatness are a function of food intake and that this relationship is not affected significantly by falling temperatures.

Food intake and liveweight gain at temperatures below and above optimum within the range of 5-35° C seem to follow a similar pattern, showing little or no sub-optimal temperature effect, but a depression of both appetite and gain above the maximum gain temperature. Indeed, Tonks et al. (1972) reported recently that differences in growth rate due to high ambient temperatures are apparent only when pigs are fed ad libitum and disappear largely under conditions of restricted feeding. A small direct effect of heat stress may, however, persist independent of food intake, since Holmes (1971), when feeding liquid whey at 33 °C, observed a growth reduction of about 12% although food intake decreased by only 6%.

As the dry-bulb temperature increases, physical cooling mechanisms become less effective and evaporative cooling increases from 15% of the total heat loss at 10° C to 90% at 38° C (Bond et al., 1952). This implies that the dissipation of heat becomes increasingly difficult if, at high temperatures, relative humidity rises, which may also affect pig performance. Morrison et al. (1969) have confirmed that, at temperatures above the optimum, an increase in relative humidity from 30 to 80% will cause a further decrease in growth rates, by about 15% at 28° C and by about 30% at 33° C. The decline of 6% recorded at the maximum gain temperature was not significant.

Based on their work on the effects of high ambient temperatures and high relative humidity, Morrison et al. (1968) proposed an index to predict the growth rate of pigs under any set of environmental conditions, the "Gain Reduction Factor", which is the fraction of the gain at given ambient conditions over the gain that can be expected at the maximum gain temperature (22° C for a pig of 70 kg) at low humidity (50%). Nelson et al. (1972) have compared the gain reduction factor of Morrison et al. (1968) with the "Swine Effective Temperature" of Beckett (1964) and the "Wetbulb-Drybulb" Index of Roller and Goldman (1969) under naturally varying conditions. They warned that the gain reduction factor significantly overpredicted the decline in growth rate and feed efficiency; and they favoured the wetbulb-drybulb index, which weighs the wetbulb- and the drybulb-temperature with 25 and 75%, respectively. Since, under natural conditions, relative humidity in the tropics is usually negatively correlated with air temperature, high humidities are not likely to depress growth rates of swine to any great extent.

A few studies have shown an interaction between protein level and ambient conditions.

Jottrand (1956) reported greatly improved gains on a high-protein diet in the equatorial climate of central Africa, and Babatunde et al. (1972) determined the crude protein levels for maximum growth rates as 23 and 20% on a dry matter basis, respectively for weanling and growing pigs reared in a tropical environment. These levels are approximately 15% higher than those recommended for temperate countries. Indeed, Seymour et al. (1964) have been able to show that an increase in protein levels by about 30% improved daily gains but slightly at 7° C (+5%), but to a considerable extent (+20%) under conditions of heat stress (30° C), indicating a higher requirement for protein at higher temperatures.

THE EFFICIENCY OF FOOD CONVERSION

For economic reasons, the efficiency with which the animal organism can convert food into live weight gain is the most important of the three growth parameters affected by heat and cold. This efficiency, the result of food intake and growth rate, can be expected to be highest in the zone of thermoneutrality i.e., between 20 and 25° C in young growing pigs and between 15 and 20° C in finishing pigs; and indeed all published studies on growing pigs confirm a high rate of food conversion for the 15-25° C temperature range (cf list of references). Below 15° C, food conversion efficiency declines on average by 40 g/kg food intake for each °C decrease in temperature (Hazen et al., 1959), ranging from 17 g (Seymour et al., 1964; Jensen et al., 1969) to 70 g (Sørensen, 1962). Above the optimum temperature range, results are more variable: While some investigators (Seymour et al., 1964; Mangold et al., 1967; Riker et al., 1967) did not observe differences in food conversion efficiency between 16 and 29° C, and even up to 33° C in younger (10-35 kg) pigs (Sugahara et al., 1970), others (Todd and Daniels, 1968; Morrison et al., 1969; Holmes, 1971) found a significant deterioration in the efficiency between 28 and 33° C, with an average of about 90 g/kgfood for each °C increase in temperature over the mentioned range. Hazen and Mangold (1960) have suggested that the food efficiency improves with increasing temperatures to the point where rising temperatures reduce the intake of food. The reviewed literature, however, seems to indicate a rapid decline in food intake above 20° C, while the efficiency of food conversion does not appear to be affected significantly below 28° C. The study of Amick and Purcell (1964), correlating data from swine evaluation stations in selected areas of the U.S.A. with climatological records of the U.S. Weather Bureau, gives the temperature range for maximum feed efficiency as between 15 and 20° C and points out an accelerating deterioration below and above this temperature range. In addition, Holmes (1971) presented evidence that the reduction in growth rate and food conversion efficiency at high ambient temperatures may partially be independent of food intake, since the dry matter intake decreased by only 6%, yet growth rate and feed efficiency deteriorated by about 12%.

High ambient humidity depresses the efficiency of food conversion only at very high temperatures (33° C, 80% rh: Morrison et al., 1969).

As changing environmental temperatures affect growth rate and feed efficiency in general, energy and nitrogen retention may be influenced as well. Sørensen (1962), reporting studies in Denmark, indicated that maximum protein deposition coincides with the maximum growth rates between 15 and 23° C and declines at effective temperatures below and above this point. Similarly, Holmes (1973) observed a decrease in energy and nitrogen retention at 34° C as compared to 25° C, although growth rate and food conversion efficiency did not differ significantly. The author attributes these effects of high ambient temperature to an increase in heat production and to urinary nitrogen excretion, respectively. This argument might explain the results of Seymour et al. (1964), who found that, at suboptimal temperatures, an increase of the dietary protein level had little or no effect upon the food conversion rate, but effected a great improvement in food conversion rate in the heat-stressed pig. Charlet-Lery (1970) spoke of a reduction of energy losses following high-protein diets, the saving of energy being proportional to the amount of protein ingested or retained.

SUMMARY

It appears from the discussions outlined in the previous three parts of this Section, that the efficiency of feed conversion deteriorates below 15° C and above 30° C, while the intake of feed, which declines at an accelerated rate above 15° C, shows but little change below 5° C. Between 5 and 25° C, daily growth rates are hardly affected by temperature. Thus, the decreased food intake between 15 and 25° C by about 15% seems to be compensated by decreased maintenance requirements, permitting maximal growth rates. Below the optimum temperature for growth, feed intake cannot be increased sufficiently to compensate increased maintenance requirements. This might explain the decrease in the rate of gain observed in some studies at temperatures below 5° C. Economically then, the temperature range between 20 and 25° C would be optimal, as high growth rates and efficient food conversion rates can be obtained at reduced food intake, but possibly requiring a higher feed quality.

BODY CONFORMATION AND CARCASS QUALITY

Since unfavourable ambient temperatures inhibit the rate of growth and thus produce an animal which is physiologically younger than chronologically, changes in body conformation and carcass quality, particularly in late-maturing organs, may be expected.

BODY CONFORMATION

So far, only one investigation has been carried out under controlled conditions. Weaver and Ingram (1969) reared pigs at 5° C, 20° C and at 35° C at comparable growth rates. Their findings support the causal observation in tropical areas, that animals reared at high ambient temperatures tend to have a relatively larger body surface per unit body weight, are less compact in type than those raised under cool-temperate conditions. Their findings cast doubt on the continued use of body weight for the estimation of body surface (Brody, 1945) under extreme climatic conditions.

CARCASS COMPOSITION

The effects of changing environmental temperatures upon carcass characteristics are apparently much smaller than those related to feeding practice. Several investigators did not observe any real differences in carcass composition over a range of ambient temperatures (Seymour et al., 1964; Jensen et al., 1969; Tonks et al., 1972).

However, restricted feeding may lead to an improved lean-fat ratio at suboptimal temperatures (Comberg et al., 1971) and to an increase in carcass fatness at temperatures above the optimum (Morrow et al., 1957; Zobrisky et al., 1961; Sørensen, 1962), which may reach considerable levels if the temperature-induced reduction in food intake can be contained (Holmes, 1971) or if the feeding level is raised (Holmes, 1973). While at high temperatures (34° C) energy retention was reduced by 15%. the nitrogen retention decreased by 28% (Holmes, 1973).

The reduced protein deposition at temperatures above the optimum has been attributed to the inadequate amino acid intake for maximum protein synthesis while, due to reduced energy requirements, energy is available for increased fat deposition (Sugahara et al., 1970).

It remains to be determined, whether the observed negative relationships between ambient temperature and plasma lipids (Morrow et al., 1963) and plasma cholesterol (Heidenreich et al., 1962) may be the immediate cause for the changes in carcass composition at various temperatures, since Pearson et al. (1966) suggested a positive correlation between growth rate and plasma lipid levels of swine fattened at 27° C.

MEAT QUALITY

In addition to the effects of environmental temperatures on carcass composition – directly or through their effects upon food intake – ambient conditions may affect the meat quality of finishing pigs. These effects, usually referred to as PSE muscle, have been reviewed recently by Judge (1969); and it appears that muscle quality is affected not only by the increased temperatures, but rather by the pig's inability to respond quickly and in a coordinated manner to acute situations of stress, such as fluctuating temperatures. Heat-stress-induced hyperthermia was found to decrease meat quality only in stress-susceptible pigs (Judge et al., 1973). Indeed, Comberg et al. (1971-1973) observed a better waterbinding capacity at 28 and 30° C than at 8° C. These findings indicate that high muscle qualities may well be expected from growing pigs which are adapted to the high ambient temperatures of the tropical climate.

With regard to the fat quality, McGrath et al. (1968) have published evidence that, in addition to the well-known effect of the intake of poly-unsaturated fatty acids on fat consistency, the saturation of backfat in pigs may be influenced by environmental temperatures: pigs reared at 0-5° C exhibited greater unsaturation than those kept at temperatures ranging from 25-30° C.

CONCLUSION

At present, many laboratories throughout the world are engaged in the study of temperature effects upon pig performance; however, relatively few research data have been published. This is particularly true for areas, in which pigs are under a constant heat stress. Many publications from tropical countries do not report climatic data and physiological indicators of heat stress, such as respiratory frequency, when describing the growth performance of their pigs. In view of the interactions between various management procedures and ambient temperatures upon growth rate and feed efficiency e.g., protein level (Seymour et al., 1964), it would be desirable to compile more performance data from pigs born and reared under conditions of high temperatures than hitherto, in order to study the climatic effects in adapted animals under normal management conditions, which may differ considerably from results obtained from unacclimated animals under acute heat stress in psychrometric chambers (Nelson et al., 1972).

REFERENCES: see pp. 563-566.

Effects of Temperature on the Growth of Fish

by

J. R. Brett

INTRODUCTION

The great majority of fishes are poikilothermic*, or cold-blooded, expending no energy on maintaining differential body temperature by specialized metabolic or behavioural means. They are free, however, to use temperature preference within the environmental circumstance available, to select a suitable body temperature. As such, they are *thermal conformers*; every response and every process proceeds within a temperature range dictated by the immediate environment. Consequently growth in temperate climates is strongly seasonal, almost stopping in winter months for some fish (Fig. 1). Yet fish are particularly plentiful in subarctic regions of the ocean, where nutrient-rich water provides abundant food; they may even grow to a fair size while inhabiting ice tunnels under the polar cap e.g., *Trematomus bernacchii* reaches 350 g (Wohlschlag, 1961). Adaptive or compensative mechanisms have evolved to meet the wide range of environmental temperatures. In general, tropical fishes grow fast and mature early; the reverse applies to polar species.

In nature temperature and photoperiod are usually correlated with each other and with seasonal abundance of food, so that the unravelling of how temperature affects

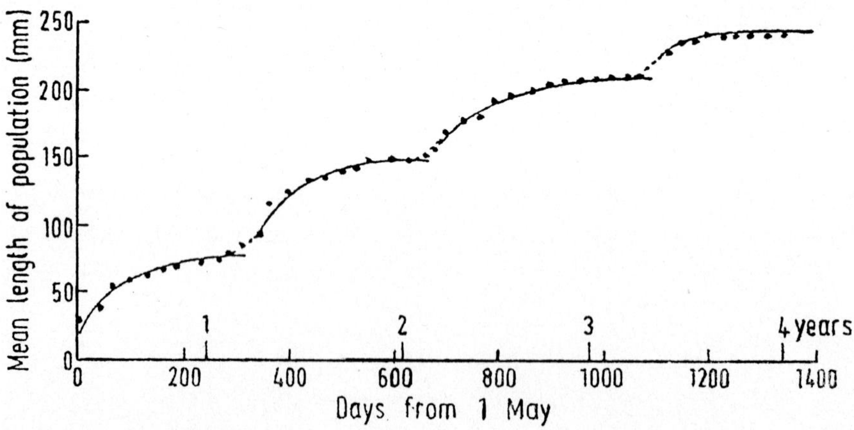

Fig. 1. Seasonal growth curves of stream-dwelling trout (From Lockwood, 1973).

* See also Chapter 1, Section 4e

growth has been sought by controlled experiments. Among the relatively few early studies of this sort are those of Dawes (1930) on plaice, *Pleuronectes platessa*, Hatanaka and Takahashi (1956) on mackeral, *Pneumatophorus japonicus*, Pentelow (1939) and Brown (1946) on brown trout, *Salmo trutta*. Increasing temperature was found to enhance growth up to an optimum, beyond which moderate or rapid diminution of growth occurred; two such optima were reported for brown trout (Table 1). These studies entailed either high feeding levels or the use of maintenance rations. Influenced by an increasing world interest in aquaculture, more extensive studies have commenced

Table 1. Examples of temperatures for optimum growth derived (1) experimentally, (2) observed in nature, and (3) computed from published data. Arranged in order of increasing temperature optima. From Brett (1970) with permission of John Wiley and Sons Ltd.

Species	Optimum (°C)	Salinity (‰)	Size (cm, g)	Comment
Salmo trutta	8	Fresh water	50 g	Lower optimum
Oncorhynchus nerka	10	Fresh water	8 g	3.0% ration*
Pleuronectes platessa	12	Sea water	4.6 mm**	Computed
Pleuronectes platessa	12	Sea water	2.0 cm	Observed
Gadus morhua	13	Sea water	Large	Computed
Pleuronectes platessa	15	Sea water	1.7 cm	Observed
Oncorhynchus nerka	15	Fresh water	25 g	6.0% ration*
Salmo trutta	17	Fresh water	50 g	Higher optimum
Pseudopleuronectes americanus	18	Sea water	10 g	Computed
Cyprinodon macularius	28	15‰	20–30 cm	Experimental
Cyprinodon macularius	30	35‰	20–30 cm	Experimental

* Food ration as percentage of dry body weight.
** Larval stage.

lately involving a wider range of species, rations, diets, and environmental factors (e.g., Andrews and Stickney, 1972; Bardach et al., 1972). Among these a series of experiments performed by the author and associates on sockeye salmon, *Oncorhynchus nerka*, will be used to elaborate the basic principles (cited appropriately).

BASIC GROWTH RELATIONS

The main determinants of fish growth are age, size, ration, and temperature, in the presence of adequate oxygen, and subject to modification by salinity, photoperiod and pH. Age effect has not been separated as yet from its closely correlated size relation. In addition the level of activity, by influencing the metabolic rate, affects the extent to which food-energy is deposited as growth. Spontaneous activity is temperature dependent and can mask temperature effects on growth by its correlated association. Experimentally, this is largely overcome by imposing a fixed, minimum swimming speed on all lots of test fish, using circular or oval tanks in which pressurized waterjets induce similar background velocities thereby damping the fluctuations in spontaneous activity.

Specific growth rate in fish, known as "G" or "K", is determined as the instantaneous increase in relative body weight (or length) and expressed in terms of percent wet weight/day (or dry weight). It is calculated as: $G = \frac{\ln_e w_1 - \ln_e w_0}{t_1 - t_0} \times 100$, where *w* is weight and *t* is time in days (cf Winberg, 1956). It is equivalent to the slope (× 100) of the exponential curve – in any one, short stanza of growth.

A. GROWTH RATE × TEMPERATURE (ONE SIZE, FULL* RATION)

The general shape of the bell curve relating growth rate to acclimation temperature is illustrated in Figure 2. Growth rate of sockeye salmon rises steadily from a depressed level at 1° C to an optimum at 15° C, falling rapidly beyond 20° C to a completely inhibited state at 23.5° C, about 1.5° C below the upper lethal temperature (Brett et al., 1969). For comparison a similar form of curve is depicted for the marine yellowtail (*Seriola quinqueradiata*) extensively cultured in Japan (Harada, 1970). Its optimum temperature occurs 10° C higher than the more northerly distributed salmon. The optimum for both species corresponds roughly to the mean maximum summer temperatures of their normal environments.

B. SIZE EFFECT (FULL RATION)

Growth rate decreases rapidly with increasing size, suppressing the dome of the optimum temperature range progressively, resulting in the relatively flattened configuration illustrated in Figure 3 (Shelbourn et al., 1973). Hatchery stocks of brook trout (*Salvelinus fontinalis*) have been shown to exhibit a similar decrease in growth rate as they grow larger (Cooper, 1961).

C. RATION EFFECT (FOR ONE SIZE)

Growth rate increases almost in direct proportion to the size of the meal, tapering off somewhat at highest rations to provide a maximum growth rate at a maximum ration (Fig. 4). Between the maintenance and maximum rations lies an optimum ration where food conversion efficiency is highest. Because of the depression of food intake at low temperatures and corresponding reduction in growth rate, it is important to note that food conversion efficiency remains high over a fairly wide span of temperatures. This has obvious adaptive value in providing considerable temperature independence within the normal range of temperatures where the species is abundant. Despite wintering under the ice in many northern lakes (with access to the sea) young sockeye salmon retain 75% of their maximum food conversion efficiency at 4° C, the maximum density of water (which underlies the ice).

D. SALINITY × TEMPERATURE INTERACTION

Fishes which migrate from salt to fresh water, like the salmon (anadromous), appear to show enhanced growth at intermediate salinities which are approximately isosmotic with their blood (9-11 ppt). Definitive experiments on the influence of size, stage of development and temperature have yet to be performed on salmon. For the highly saline-tolerant pupfish (*Cyprinodon macularius*) Kinne (1960) demonstrated a counter-clockwise rotation in the growth-temperature bell curve at higher salinities (Fig. 5). Thus, greatest growth occurred at the combination of high temperature (35° C) and high salinity (35 ppt); at low temperatures for this species (15° C) growth was favored by low salinity (fresh water).

In a factorial experiment (temperature × salinity × ration) on the estuarine flatfish, the hogchoker (*Trinectes maculatus*), Peters and Boyd (1972) demonstrated that maximum feeding rate depended primarily on temperature rather than salinity. Growth rate was greatest where intermediate temperature (25° C) and high salinity (30 ppt) combined; the interaction, however, did not follow any simple configuration.

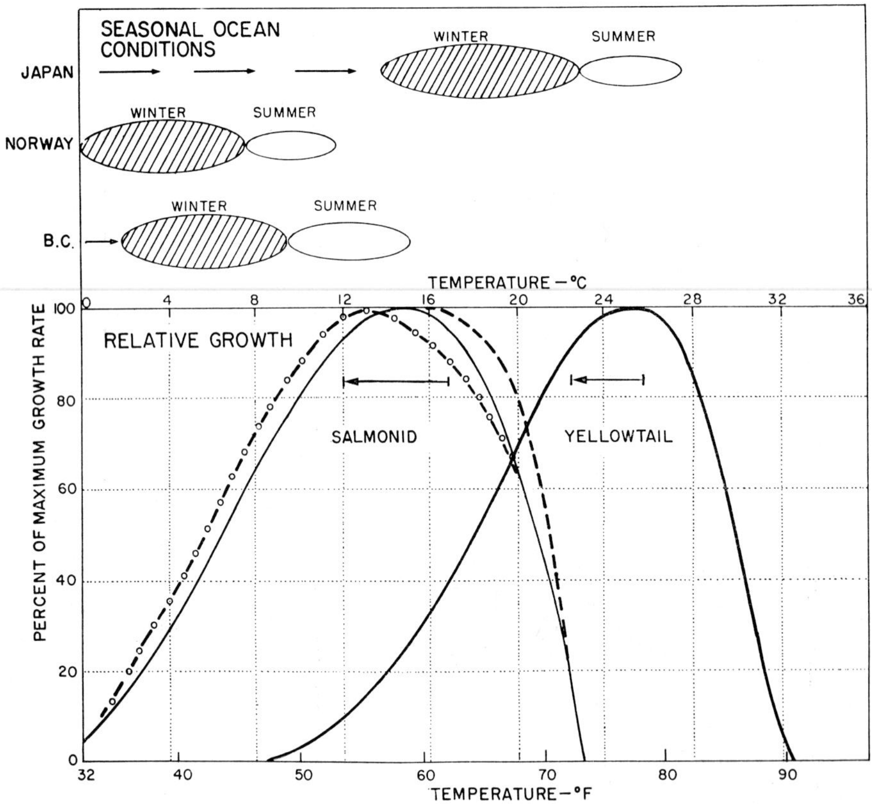

Fig. 2. Relation of temperature to growth rate of yellowtail and a typical salmonid (expressed as a percent of maximum rate for each species). Arrow for yellowtail indicates shift in optimum temperature with increasing size. Arrow for salmonids indicates range for different species – coho salmon and rainbow trout have a slightly higher optimum (curve to the right) than chum salmon and Atlantic salmon (curve to the left). Seasonal ocean temperatures in the natural habitat are shown above. Note the general correspondence of average summer temperatures with optimum temperatures for growth (From Brett, 1974).

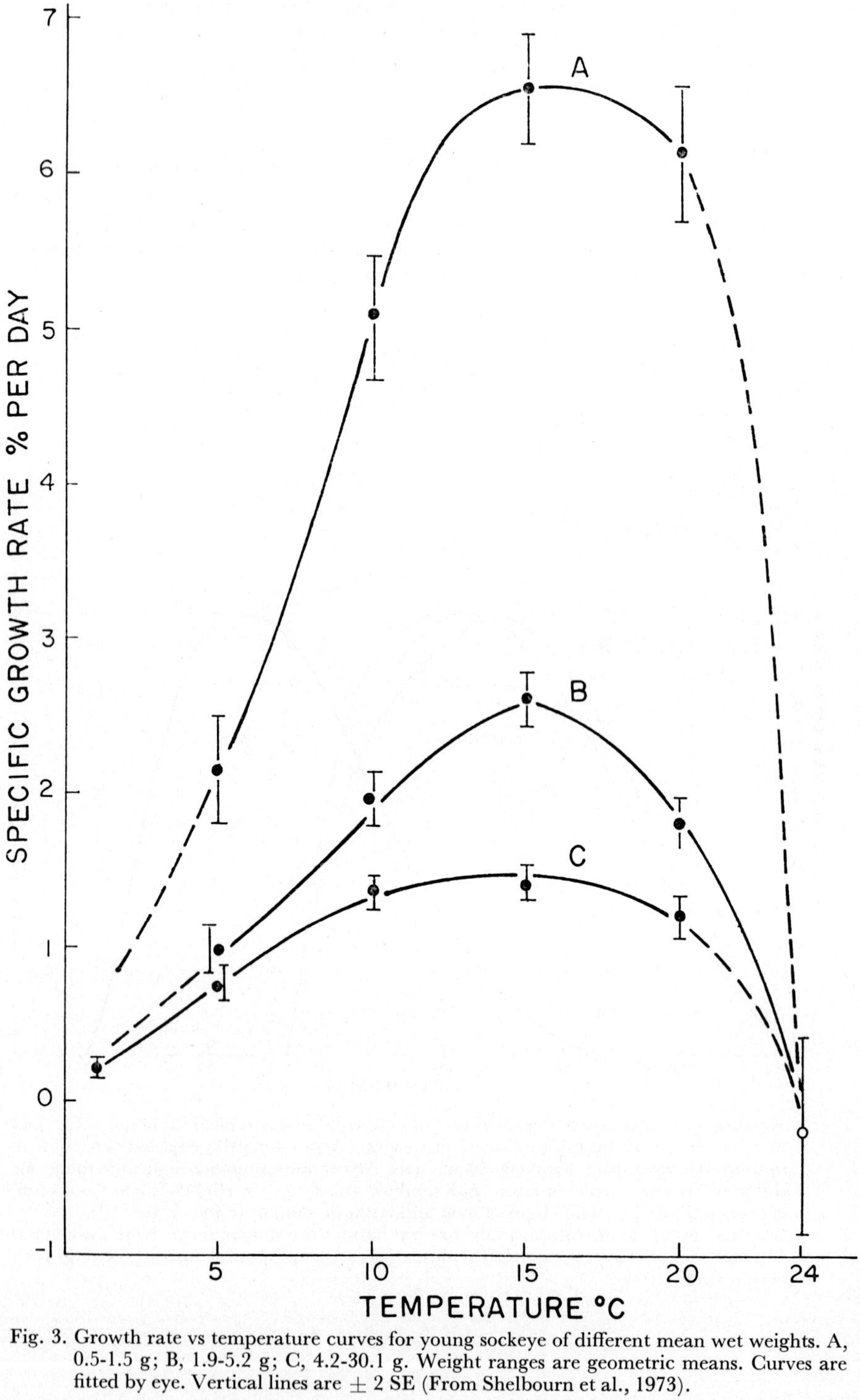

Fig. 3. Growth rate vs temperature curves for young sockeye of different mean wet weights. A, 0.5-1.5 g; B, 1.9-5.2 g; C, 4.2-30.1 g. Weight ranges are geometric means. Curves are fitted by eye. Vertical lines are ± 2 SE (From Shelbourn et al., 1973).

BASIC MECHANISMS

Some of the underlying mechanisms controlling the shape of the temperature-growth curve have been analyzed for sockeye salmon (Brett and Higgs, 1970). Since growth is a metabolic product of a series of processes in which digestion, assimilation, metabolism and excretion all interact to determine the final disposition of body substance, these can be examined in relation to temperature effects on their respective rates. Rate of digestion, maximum meal size (appetite) and gross conversion efficiency (reflecting assimilation and excretion) were studied in relation to the growth rate-temperature response (Fig. 6). For clarity of explanation it is best to consider three temperature ranges: low, intermediate, and high. Unless otherwise stated, the growth relations apply only to daily satiation rations.

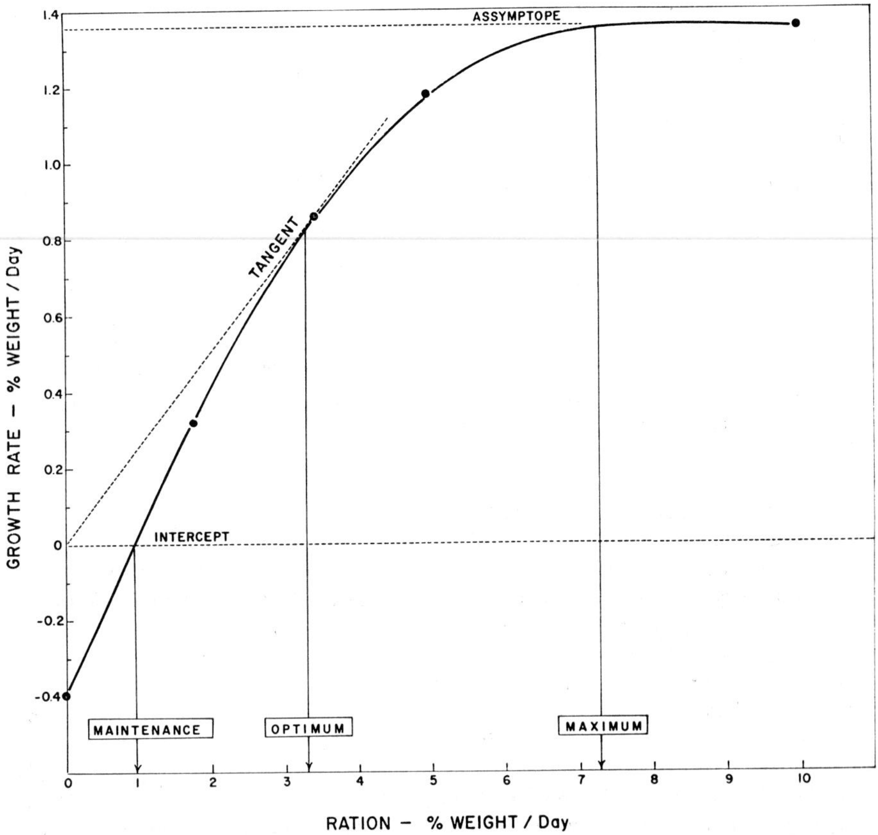

Fig. 4. Geometric derivation of various parameters of growth with accompanying ration. Data for 10°C acclimated sockeye with an average weight of 12 g (From Brett et al., 1969).

* Applied to maximum daily feeding rate, or to ad libitum feeding, resulting in satiation.

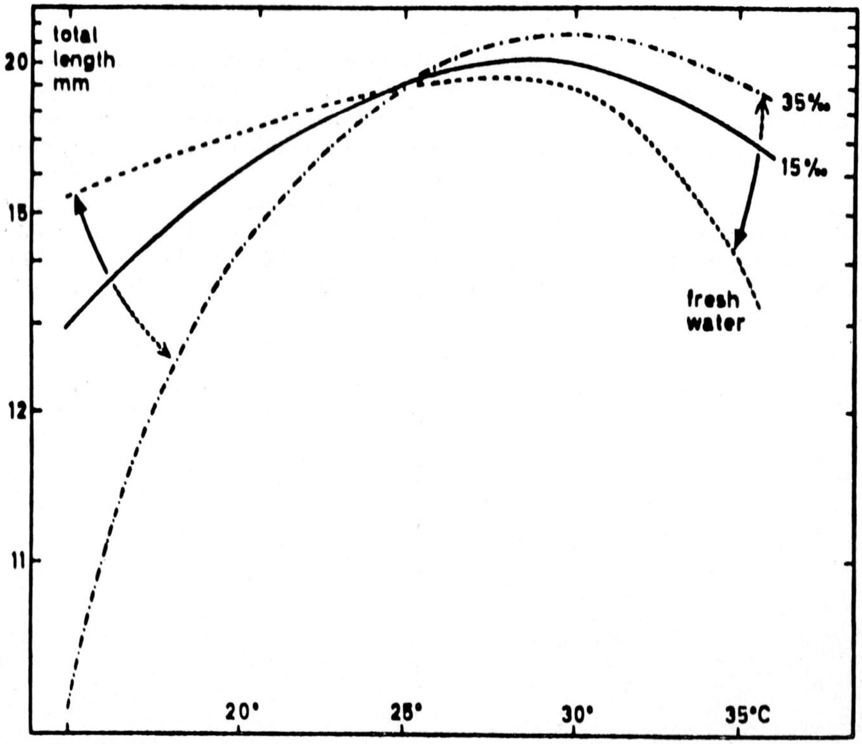

Fig. 5. Diagram of the combined effects of temperature and salinity on the growth process in *Cyprinodon macularius* (After Kinne, 1960).

A. LOW TEMPERATURES: 1 TO 5° C

Rate of digestion is greatly depressed, approaching zero in the region of 0° C. However, digestion shows some compensatory increase; young sockeye will grow at 1° C, consuming 1.5% dry body wt/day. Gross conversion efficiency is only one-half the maximum, which reflects either reduced digestive efficiency or high nitrogenous excretion, or both. Rate of digestion appears to be the chief limiting factor to consumption and hence to growth at low temperatures, with a concomitant reduction in appetite (meal size). Some seasonal increase in food intake may occur, indicating a capacity to take on energy and retain it for gradual digestion.

B. INTERMEDIATE TEMPERATURES: 13 TO 17° C

With rising temperature all systems increase in roughly the same proportion except growth rate, which is relatively more rapid owing to an increased food conversion efficiency coupled with higher intake per meal. At 15° C an optimum is reached for three systems; only rate of digestion continues to rise. Further, at 15° C a significant threshold is reached with daily meals being fully digested within 24 hrs.

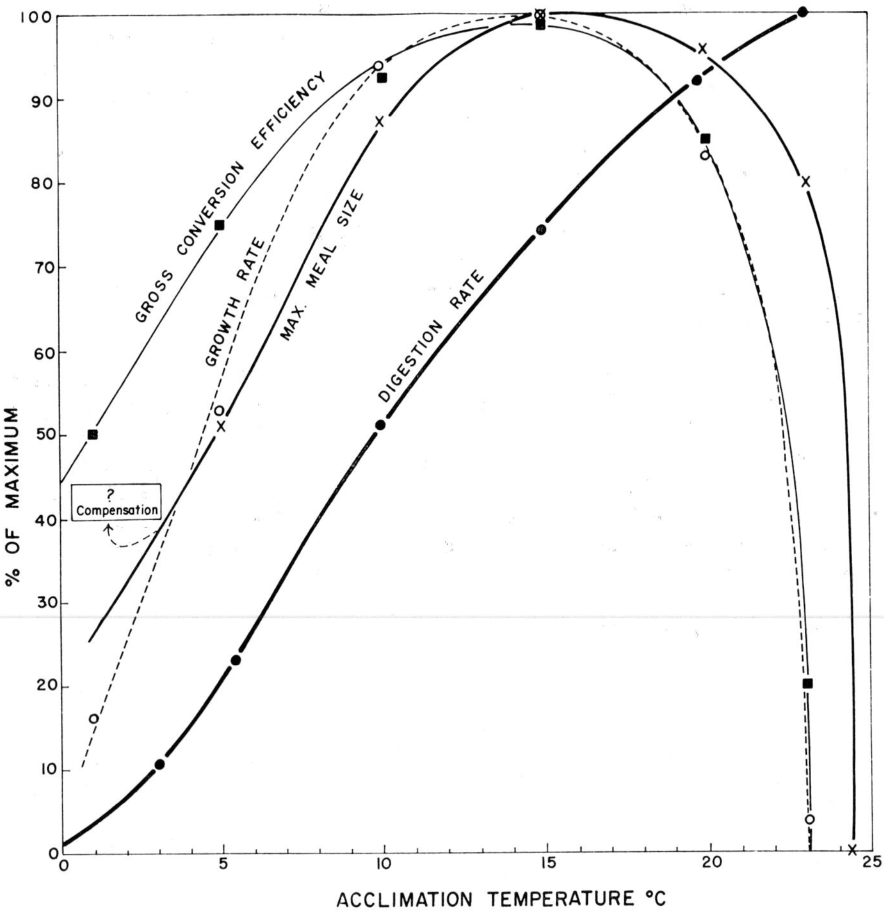

Fig. 6. Relative relation of various growth, food, and digestion parameters to acclimation temperature. Each rate has been expressed as a percentage of the maximum observed (From Brett and Higgs, 1970).

C. HIGH TEMPERATURES: 20 TO 24° C

Rate of digestion continues to increase abetting the ability to deal with multiple daily meals. However, appetite is reduced and conversion efficiency declines precipitously. As a result the sustained capacity to consume and digest is not accompanied by growth, which follows conversion efficiency in a parallel decline. At 23° C high levels of standard metabolism, maintenance ration, and excretion all combine to suppress growth to the point of extinction.

CONCLUSION

As an environmental factor affecting the activities of poikilotherms, temperature was labelled a *controlling* factor by Fry (1947, 1971); it paces the metabolic processes, which provide for growth, at all tolerable levels of the factor. Unlike *limiting* factors which provide thresholds, or *lethal* factors which define fringes of tolerance, temperature as a *controlling* factor governs growth rate in a constantly dependent fashion. This results in minimum, maximum, and optimum relations of temperature. Near or complete cessation in growth occurs at the minimum and maximum temperatures; however, the causes for the cessation are entirely different.

Optimum temperatures for growth are ecologically significant in permitting rapid increase in size during the normal rise and peaking of average seasonal temperatures. The compensatory mechanisms which permit successful growth from tropics to poles are little understood.

It can be anticipated that experimentation will become increasingly active in revealing the complex interaction of temperature × salinity × photoperiod × size × ration, on growth. There will remain the yet more complex question of the effects of fluctuating levels of these factors, based as they are on the fundamental question of metabolic acclimation and its underlying mechanisms. Tracking the nature and rates of excretory products and the compilation of energy budgets as affected by environmental change and quantity of food will undoubtedly assist in understanding the nature of the growth processes in fish.

REFERENCES: see pp. 566-567.

Effects of Temperature on the Growth of the Domestic Chicken

by

F. N. Reece and J. W. Deaton

INTRODUCTION

Although there are different opinions about the exact ancestry of the domestic chicken (Lucas and Stettenheim, 1972), it is generally agreed that the modern chicken arose from the jungle fowl of Southeast Asia. Probably as a result of the many years of domestication (Davenport, 1914, reported that the Malay Fowl has been domesticated for over 3000 yrs), the chicken has become acclimatized to a much broader range of environmental conditions than would be expected in a tropical jungle. Consequently, the chicken has adapted reasonably well to most areas of the Temperate Zones of the world. Man, of course, has assisted in the adaptation process by providing varying degrees of protection from the elements by environmental modification and control.

The thermoneutral range, which is defined by Sturkie (1965) as being the environmental temperature range within which there is little or no change in heat production by the bird, is an indicator of the optimum environmental temperature for the bird. Table 1 gives the thermoneutral range for several species of birds. The range for the

Table 1. Thermoneutral range for some species of birds*

Species		Body weight, g	Thermoneutral range (°C)
Chicken	] 0–1 wks	36	34–36
(Rhode Island]	5 wks	260	32–35
Red, female)]	52 wks	2430	18–24
Pigeon		260	30.2–35.8
Cardinal (winter)		40	18–33
Cardinal (summer)		40	24–33
Abert's Towhee		46.8	25–35
Brown Towhee		43.7	23–33
Yellow Bunting		26.4	25–33
Ortolan Bunting		22	32–38

* Modified from Sturkie, 1965.

young chicken is very narrow. The range for the mature chicken is also rather narrow when it is compared with that for smaller birds that are native to the Temperate Zones.

The thermoneutral ranges shown in Table 1 indicate that the optimum environmental temperature for chickens starts at 35° ± 1° C during the first week and decreases to 33.5 ± 1.5° C by 5 weeks of age. At maturity the optimum temperature is 21° ± 3° C. Barrot and Pringle (1947, 1948, and 1950) measured the temperature required for maximum growth of Rhode Island Red chickens from 1 day to 32 days of age (Table 2) and found that the optimum temperature started at 34.4° C at 1 day and decreased

Table 2. Temperature for maximum growth of Rhode Island Red chickens as determined by Barrot and Pringle (1947, 1948, and 1950), with resulting feed consumption and body-weight gains

Chicken age, days	Temperature for maximum growth, °C	Feed consumption, g/day/chicken	wt gain, g/day/chicken
1–9	34.4, 1st day		
	30.6, 9th day	10	4.3
	(Decrease of 0.475, per day)		
9–18	30.6, 9th day		
	26.7, 18th day	20.5	10
	(Decrease of 0.433, per day)		
18–32	26.7, 18th day		
	18.9, 32nd day	55	20
	(Decrease of 0.557, per day)		

about 0.5° C per day to 18.9° C at 32 days of age. Their work indicated that the optimum temperature for maximum growth for 1-day-old chickens is in the thermoneutral range; but at 5 weeks, the optimum temperature is lower than the thermoneutral range indicated in Table 1.

Because of the rather rapid change in temperature required during the early stages of chicken growth, the domestic growing of chickens, after they hatch, is usually divided into two phases, the BROODING PHASE and the "GROW-OUT" PHASE.

BROODING PHASE OF GROWING CHICKENS

In commercial production, the brooding phase, which includes the chicken age from hatching to 3 or 4 weeks, is usually accomplished by providing a heater that directs heat downward over an area about 2 to 4 m across, where from 500 to 1000 chicks are placed shortly after they hatch. The heater is equipped with controls to permit the temperature to be regulated rather precisely at about 35° C, with provision to reduce the heat according to the following schedule:

Brooding Temperature	
First week	32 to 35° C.
Second week	29 to 32° C.
Third week	26 to 29° C.
Fourth week	23 to 26° C.

The brooding temperatures indicated are the temperatures measured under the heaters about 5 cm from the floor. The brooding area is arranged such that the chickens can move to an area of lower temperature if they so desire. The commercial procedure for brooding chickens essentially duplicates the natural method in which the mother hen provides the correct environmental conditions by hovering over the baby chickens after they hatch.

The heaters used for brooding should produce a relatively large proportion of their heat as infrared energy, because this method is the most practical for directing heat downward to provide heat to the dorsal areas of the baby chickens as the mother hen does. The importance of dorsal heat was demonstrated (Reece and Deaton, 1968) in a study in which a limited quantity of heat was provided from the floor to heat the ventral areas of one group of baby chickens, and the same quantity of heat was provided as infrared energy to heat the dorsal areas of another group of baby chickens. Mortality for the ventrally heated group was 5.5% as compared with 2.5% for the dorsally heated group. The major reason for the higher mortality for the group heated from the floor was that the baby chickens could not locate the warm floor area; thus, they tended to wander to a cold area of the room, where they perished. The dorsally heated group readily located the energy directed from above and were able to stay in a comfortable zone.

The Rhode Island Red chickens used by Barrot and Pringle (1950) weighed approximately 275 g at 4 weeks of age and had a feed conversion of about 2.16 g feed per g weight gain. In a report by Deaton et al. (1973) a commercial stock of meat-type chickens weighed 630 g at 4 weeks of age and had a feed conversion of 1.60 g feed per g weight gain. Theoretically, as shown by Bouchillon et al. (1970), the size of the chicken is an important parameter in the heat exchange between the chicken and its environment. Consequently, one might expect to find that the optimum temperature for the maximum growth for the crosses of strains of chickens with hybrid vigor, which are now in widespread use, might be lower than that found by Barrot and Pringle (1950) for a breed. However, in the absence of supporting data for such a theory, any statement concerning how much lower the optimum temperature might be would be purely speculative.

The work by Barrot and Pringle (1950), which provided the basis for modern brooding temperature recommendations, was designed to determine the optimum temperature regime that should be used during the first 3 to 4 weeks to obtain maximum growth. From a practical standpoint, factors that enter into brooding practices, other than obtaining maximum growth of the chicks during the brooding period, include the amount of energy needed to maintain the desired brooding temperature, the mortality rate of the chickens, their growth rate after the brooding period, and their final feed conversion. For example, in brooding meat-type chickens, Reece and Deaton (1968) used approximately 30% of the energy required to maintain the recommended brooding temperatures. It was found that even though body weights of the chickens were depressed 11% at 3 weeks of age, their mortality rate, body weight at 8 weeks of age, and final feed conversion were not significantly affected by the reduced heat provided during the brooding period.

GROW-OUT PHASE

The grow-out phase of chicken production consists of the period from about 3 or 4 weeks of age (end of brooding phase) to the age at which the chickens have reached the degree of maturity for the intended use. For meat-type chickens the age at slaughter may range from about 6 1/2 weeks for specialty use to about 10 or 12 weeks for general use in areas where a heavy dressed chicken is in demand.

Chickens can be grown over a much wider range of temperatures during the grow-out phase than they can during the brooding phase. However, as indicated in Table 3, the temperature for optimum performance is 18.3° C, which is in the thermoneutral range for mature chickens as indicated in Table 1.

Table 3. Effect of temperature on body weight gain and feed conversion of meat-type chickens from 22 to 56 days of age (Deaton and Reece, 1970)*

Rearing temperature, °C	Body weight at 56 days, g	Feed conversion, 22 to 56 days, g feed/g gain
7.2	1614ab	2.50a
12.8	1602ab	2.39ab
18.3	1641a	2.26bc
23.9	1570b	2.20c

* Different letters in the same column denote significance at the 0.05 level of probability.

As indicated in Table 3, temperatures below the thermoneutral range had no significant effect on growth rate but tended to adversely affect the feed conversion ratio.

In the Temperate Zones of the world the climatic temperatures vary widely, on a seasonal basis, from the optimum temperatures required for chicken growth. For growing chickens commercially, shelter from the elements must be provided by appropriate housing.

During the winter season, poultry house design, insulation, ventilation, stocking rate, and supplementary heat are balanced so that the temperature in the house can be maintained within the range from about 5° to 21° C during the grow-out phase. As indicated in Table 3, as rearing temperatures drop below 18.3° C, feed consumption increases. This increase in feed consumption can be offset by maintaining the house temperature above 18.3° C by improving the housing or heating the house, or a combination of both. The actual temperature to be maintained will be based on an economic analysis that involves housing costs, the cost of fuel to heat the house, and the cost of feed.

During the summer season, poultry reared in the Temperate Zones of the world may be exposed to temperatures ranging from about 15° to 44° C. The diurnal temperature range is quite variable, depending largely on the geographic location. Haurwitz and Austin (1944) indicated that, for areas located at 29° N latitude, the mean annual diurnal temperature range was 5° C for maritime climates and 10.8° C for continental climates. At 9° N latitude, the diurnal temperature range was 5.3° C for maritime climates and 8.3° C for continental climates.

There are two major effects of high temperatures on poultry during summer – depressed growth and decreased survival rate. The excess body heat of homoiotherms in temperatures above the thermoneutral range must be dissipated in two forms – sensible heat, which is lost by conduction, convection, and radiation to the surrounding environment; and latent heat, which is lost through the evaporation of water from the skin surface and the respiratory tract. In the avian species, practically all of the latent heat loss is from the respiratory tract. Therefore, relative humidity of the ambient air has a pronounced effect on poultry growth and survival at high temperatures.

The effect of depressed growth as rearing temperatures are increased above the thermoneutral zone is shown in Table 4 which gives the data for chickens that were grown from 4 to 8 weeks at temperatures ranging from 26.7° C to 35°.

Table 4. Effects of four constant rearing temperatures and two levels of relative humidity on growth of meat-type chickens (Reece and Deaton, 1971)*

Temperature and humidity effects					
Temperature, °C			Relative	8-Week	Feed conversion,
Dry-bulb	Dew-point	Wet-bulb	humidity, percent	body wt, g	g feed/g gain
Experiment 1 – Moderate relative humidity					
35.0	18.3	23.3	38	1344a	2.18a
32.2	19.7	23.3	48	1460b	2.13a
29.4	20.9	23.3	60	1560c	2.10a
26.7	22.1	23.3	76	1615d	2.10a
Experiment 2 – High relative humidity					
35.0	21.1	24.9	44	1230a	2.25a
32.2	22.2	24.9	56	1347b	2.13b
29.4	23.3	24.9	69	1454c	2.08b
26.7	24.3	24.9	87	1458c	2.13b

* Within each experiment, different letters denote significant difference at the 0.05 probability level.
Data are for three trials, each experiment, with equal numbers of males and females.

As the ambient temperature approaches the body temperature of the chicken, relative humidity becomes critical to the survival of chickens. Reece et al. (1972), in working with 8-week-old, meat-type chickens, found that when diurnal temperature reached 40.6° C, the chickens survived at 17% relative humidity and that mortality due to heat prostration became apparent at 33% relative humidity and was severe at 39%. Results of these experiments are summarized in Table 5.

Table 5. Effects of three levels of humidity and 23.9° to 40.6° C. diurnal temperature cycle on heat prostration of 8-week male broilers for three acclimation treatments combined (Reece et al., 1972)*

Weight and No. of dead	Diurnal temperature range		
	23.9°–40.6° C dry-bulb 23.9° C dewpoint	23.9°–40.6° C dry-bulb 21.1° C dewpoint	23.9°–40.6° C dry-bulb 10° C dewpoint
Av. wt, g	1,577a	1,578a	1,566a
Av. dead wt, g	1,704	1,673	–
Av. surv. wt, g	1,505	1,567	1,566
No. dead	43/120	12/120	0/120
Mortality, %	35.8a	10.0b	0c

* Different letters in the same line denote significant difference at the 0.05 probability level.

SUMMARY

The domestic chicken, whose ancestors originated in the jungles of Southeast Asia, is by nature limited to a rather narrow temperature range for optimum growth and feed utilization. This temperature decreases from about 35° C at time of hatch to about 18.3° C at about 3 weeks of age. During the early stages of its growth, the chicken's survival depends on maintaining the brooding temperature near the optimum. After about 3 weeks of age, the chicken can be grown in houses where the temperature may vary from about 5° to 44° C.

At temperatures below 18.3° C, growth rate is not adversely affected; but feed utilization is adversely affected.

At temperatures above 18.3° C, growth rate is depressed and mortality is increased. Relative humidity is an important factor that affects growth rate and mortality at elevated temperatures. As ambient temperature approaches the body temperature of the chicken (about 41° C), relative humidity becomes the critical factor for survival. For 8-week-old male meat-type chickens, mortality due to heat prostration can occur at an environmental temperature of 41° C if the relative humidity is above 30%.

REFERENCES: see p. 567.

Heat Transfer Through Animal Coats

by

K. Cena and J. L. Monteith

INTRODUCTION

This Section is concerned with those physical properties of hair which determine the rate at which heat is lost from the body of an animal to its environment. For many mammals, the thermal insulation provided by the natural coat is a major part of the total insulation between the core and the environment. In electrical terms, the relatively large resistance of the coat ensures that the temperature of the skin surface is decoupled from the air temperature i.e., large changes of air temperature usually induce much smaller changes of skin temperature. The physical processes responsible for heat transfer through the coat are therefore closely related to the extent of the thermoneutral zone within which metabolic rate is minimal and independent of air temperature. Coats also play an important role in protecting the surface of the skin from the heat load imposed by solar radiation.

Heat transfer through coats is usually discussed in terms of conduction but as convection and radiation are also involved, all three processes are considered here. Aspects of vapour transfer through coats are considered in Chapt. 1, Sect. 5a and Chapt. 2, Sect. 2a.

RADIATION

It is important to distinguish between the radiative properties of a coat in bulk and the properties of a single fibre. Radiation which strikes a fibre within the coat of a hairy animal may be absorbed within the fibre, reflected away from the skin, or deflected towards the skin. Cena and Monteith (1975a) showed that the fraction of radiation intercepted by fibres within unit depth of a coat depends on the hair diameter d, on coat depth l and on the length of the stretched hairs l_s. The ratio l/l_s defines the inclination of hairs to the skin surface in straight coats and the degree to which the hairs are crimped in fleecy coats. For a beam of radiation normal to the coat surface, the intercepted fraction p is given by the area of hair projected on a plane parallel to the skin (Fig 1) which is

$$p = nd \tan \arccos (l/l_s)$$

Table 1 shows values of $\bar{p}$ and other coat parameters determined for cleaned samples of hair from several species.

A. LONG-WAVE RADIATION

The behaviour of radiation in a coat depends strongly on its wavelength. For long-wave,

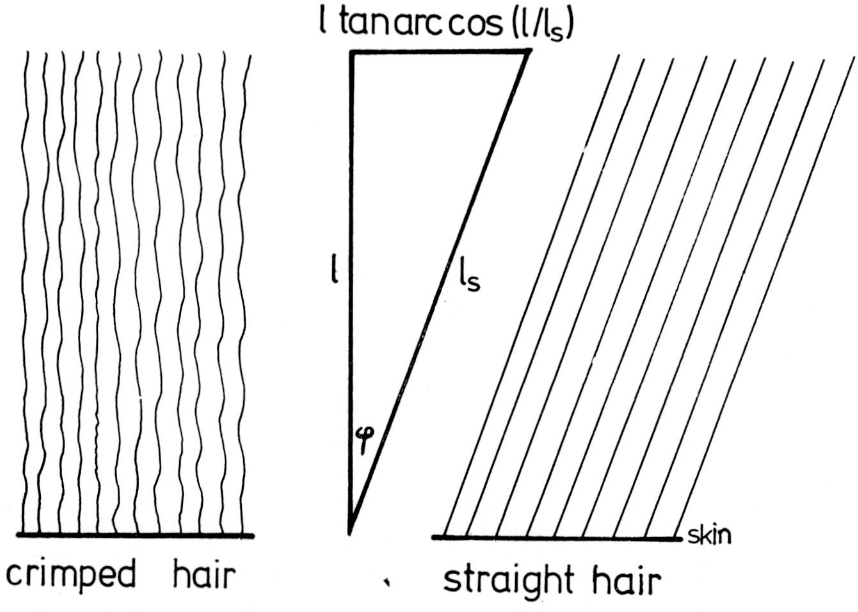

Fig. 1. Simplified geometry for the penetration of radiation in natural coat with crimpled hair and an equivalent coat with stretched hair. l, depth of coat; l_s, length of stretched hairs; tan arc cos (l/l_s) represents the area of hair projected on a plane parallel to the skin for radiation at normal incidence.

thermal radiation in the waveband 3 to 100 μm, hair behaves like a black body absorbing all the radiation that strikes it. It can be shown that within an ISOTHERMAL coat, the fraction of incident long-wave radiation transmitted through a depth x of coat is exp $(= \bar{p}x)$ where $\bar{p} = (\pi/2)$ p is an approximate interception function for diffuse radiation (Cena and Monteith, 1975a). It follows that the depth of coat needed to absorb 95% of incident long-wave radiation ranges from about 1 mm for cattle (p = 18 cm^{-1}) to about 4 mm for sheep (p= 5 cm^{-1}).

In a coat which is not isothermal, long-wave radiation is transmitted by a process of absorption and emission by individual hairs. When the gradient is linear, it is possible to define a radiative conductivity (Cena and Monteith, 1975b) which has the value $16\,\sigma T^3/3p$ where $\sigma = 5.67 \times 10^{-8}$ W m^{-2} K^{-4} is Stefan-Boltzmann constant and T(K) is absolute temperature. At 20°C, the radiative conductivity ranges from 15 mW m^{-1} K^{-1} for fleece to 4 mW m^{-1} K^{-1} for cattle coats. As the thermal conductivity of still air is 25 mW m^{-1} K^{-1} at 20° C, the transfer of heat in coats can be substantially increased by radiative exchange. Some of the values of thermal conductivity quoted in the literature lie in the range 30 to 40 mW m^{-1} K^{-1} and can therefore be accounted for by a combination of radiative transfer and molecular diffusion in still air. Values exceeding 40 mW m^{-1} K^{-1} may be a consequence of free convection, to be discussed later.

Table 1. Typical values for the physical parameters of cleaned coats (From Cena, 1974)

Coat	Number of hairs per unit surface area of skin (n) cm^{-2}	Hair diameter (d) μm	Coat depth (l) cm	Stretched length of hairs (l_s) cm	tan arc cos (l/l_s)	Radiation interception index (p) cm^{-1}	Coat reflection coefficient ($\bar{\varrho}^*$)	Absorption coefficient (α)
Dorset Down sheep	1430	48	8.0	10.0	0.75	5.2	0.79	0.001
Clun Forest sheep	1460	42	5.0	6.0	0.75	4.6	0.60	0.05
Welsh Mountain sheep	850	65	3.5	4.5	0.75	4.1	0.30	0.30
Rabbit	4200	31	2.0	2.5	0.75	9.7	0.81	0.001
Badger	240	71	1.8	4.5	2.30	3.9	0.48	0.11
Cow	1260	44	0.6	2.0	3.20	17.8	0.63	0.04
Goat	110	83	2.3	3.0	0.75	0.7	0.42	0.16
Fox	3600	20	1.4	2.0	1.00	7.2	0.34	0.26
Deer	520	150	1.5	3.7	2.30	17.9	0.69	0.025

In principle, the total flux of radiation EMITTED by a coat is the sum of components from the skin and from individual hairs. In practice, a very large proportion of the flux comes from the hair if the coat is more than a few millimetres deep. If the flux of emitted radiation is Φ, an effective surface temperature T_e can be defined by $\Phi = \sigma T_e^4$. Cena and Monteith (1975b) derived an expression for the value of Φ as a function of skin temperature T_s(K), of the temperature gradient in the coat assumed constant with depth at a value of a (K cm^{-1}), and of the coat depth l. The expression contains the exponential term exp $(-\bar{p}l)$ which has a limiting value of zero for very thick dense coats and an approximate value of $(1 - \bar{p}l + \bar{p}^2l^2/2)$ for thin coats. Approximate values of T_e corresponding to these two cases are

$$\text{thick coats } T_e \simeq T_s - a\,(1 - \bar{p}^{-1})$$
$$\text{thin coats } T_e \simeq T_s - a\bar{p}l^2/2$$

When a directional radiometer is used to measure the effective surface temperature of a coat, the depth of coat contributing to the emitted radiation depends on the angle of view. The scale of this effect was examined by Cena and Clark (1973) who derived a general relation for the interception function p as a function of the angle of view and the arrangement of individual hairs.

B. SHORT-WAVE RADIATION

In the short-wave or solar spectrum from 0.3 to 3 μm, the reflectivity ϱ^* of animal coats in bulk ranges from about 10 to 80%. Cena and Monteith (1975a) derived a complex but explicit expression for the bulk reflection coefficient of a coat as a function of the interception function p, the fractions of radiation absorbed, reflected, and deflected by a fibre, the reflectivity of the skin, and the coat depth l.

A complementary expression for the bulk transmission coefficient τ^* of a coat incorporates the same parameters. Figure 2 shows the relationship between τ^* and coat depth l for representative values of α and p. Assuming a value of $\alpha = 0.001$ for a clean unpigmented fleece, and $\varrho_s = 0.5$, 95% of the incident short-wave radiation would be absorbed in a depth of 10 cm ($\tau^* = 0.05$). For a natural greasy fleece, assuming $\alpha = 0.1$, the corresponding depth is about 2 cm. The latter estimate is consistent with the observation by Macfarlane (1968) that only 2 to 3 cm of fleece are needed to protect the skin of a sheep from solar radiation.

Light coloured coats transmit more solar radiation than dark coats of the same depth, putting a greater radiative load on the skin. The relation between heat load and radiation penetration has been measured experimentally by Hutchinson et al. (1974) and estimated theoretically by Kovarik (1964). It has not yet been possible, however, to test theoretical prediction directly on live animals.

C. THERMOGRAPHY

Measurement of the effective coat temperature, as defined in the last section can be performed with any radiometric instrument and when accurate mapping is not required, hand-held directional radiometers can be used as an alternative to more expensive

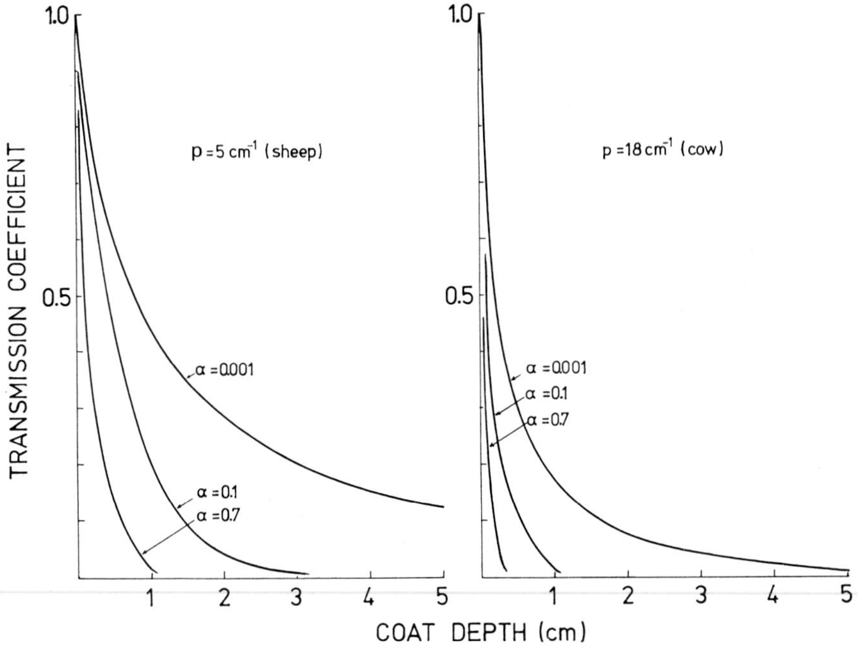

Fig. 2. Relationship between the transmission coefficient τ^* and coat depth l for characteristic values of absorption coefficient α and the radiation interception function p. Values of $p = 5\ cm^{-1}$ and $18\ cm^{-1}$ are appropriate for a sheeps' fleece and cattle coats. The value of $\alpha = 0.001$ represents a very white cleaned fleece and $\alpha = 0.1$ and 0.7 are absorption coefficients for typical natural white and dark coats. Skin reflection coefficient was assumed to be 0.5 for all cases (From Cena, 1974).

thermographic apparatus. Thermography presents the temperature distribution of objects within the field of view of the camera as an image, somewhat like a black-and-white television picture, in which the gradation of grey tone intensity corresponds to the radiative temperatures within a chosen span. Alternatively, an isotherm system may be used for the measurements of temperature and temperature differences with an accuracy of approximately $\pm 1°$ C. This method is particularly suitable for surfaces like animal coats which have such a small heat capacity that measurements with solid probes give false readings. For coated animals, the surface seen by the thermographic camera or a radiometer is some layer within the coat. This radiative surface has the temperature T_e and may be regarded as the site where the various fluxes of energy between the animal and its environment are partitioned. The equilibrium temperature of this surface is determined by the balance between the loss of heat by radiation and convection to the surroundings, the conduction of heat through the coat and the exchange of thermal and short-wave radiation.

During exposure to sunshine, part of the energy intercepted by an animal is re-radiated in the long-wave spectrum and is therefore visible by the thermographic camera. Figure

Fig. 3. Conventional photograph of a cow which appears in Figure 4.

3 shows a conventional photograph of a piebald dairy cow. When the cow was in the shade, thermography showed that its surface temperature was almost uniform. It was then moved into sunshine and temperature differences that developed between the black and white areas are shown by the thermogram (Fig. 4) in which blacker coat areas are light (hot). Within a few minutes of exposure to a weak sunshine of 320 Wm^{-2} irradiance, the maximum difference in radiative surface temperature between irradiated black and white coat areas was 8° C. Cena (1966) showed that skin temperatures beneath black and white areas of cattle coats follow a similar pattern but have a smaller amplitude. It is evident that modification of the energy balance due to variations of colour in the surfaces absorbing short-wave radiation determines subsequent long-wave emission, although the magnitude of the effect will be moderated by prevailing micrometeorological conditions (Clark and Cena, 1972).

CONDUCTION AND CONVECTION

Many workers have reported values for the thermal conductivity of animal coats, determined by mounting samples of coat on a heated plate through which a known flux of heat is passed. The conductivity is defined as the heat flux produced by unit temperature GRADIENT (e.g. $W\ m^{-2}$ per $K\ m^{-1}$ or $W\ m^{-1}\ K^{-1}$), and insulation is the temperature DIFFERENCE needed to produce unit flux (e.g., K per $W\ m^{-2}$ or $K\ m^2\ W^{-1}$).

In a classic paper, Scholander et al. (1950) reported that the coat conductivities of arctic animals ranged from 36 to 94 $mW\ m^{-1}\ K^{-1}$ with an average value of 55 mW

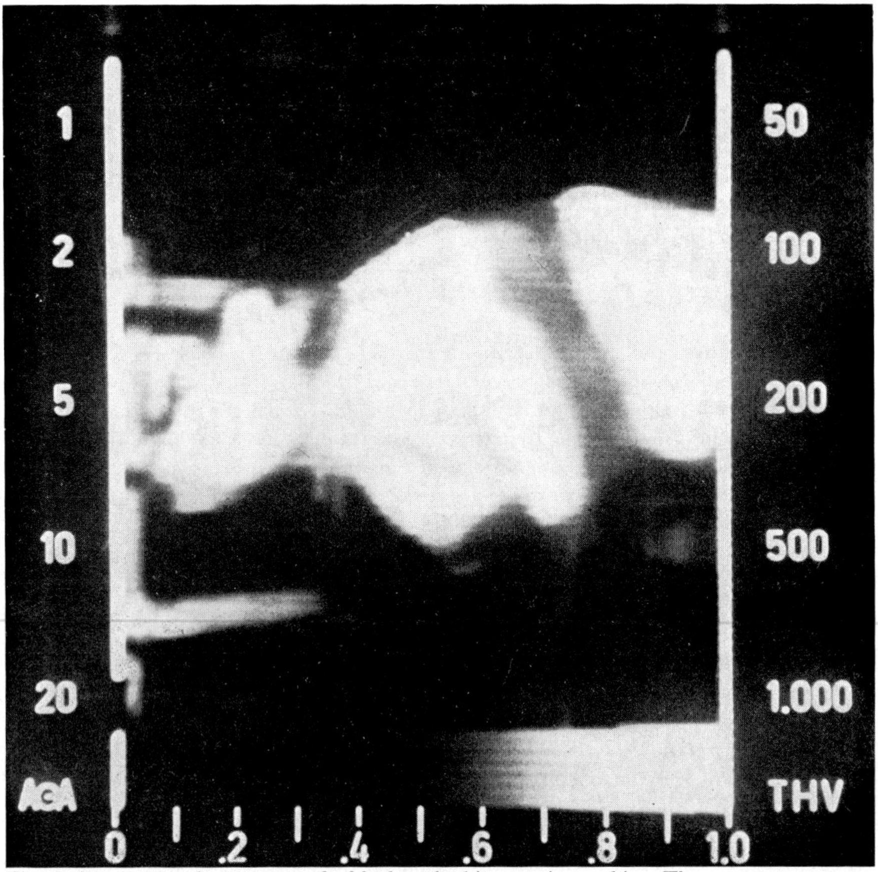

Fig. 4. A grey-tone thermogram of a black and white cow in sunshine. The warmest areas are shown as lightest in the print. The maximum difference between the radiative surface temperature of black and white coat areas is approximately 8°C (From Clark and Cena, 1972).

m^{-1} K^{-1}, more than twice the value for still air. Hammel (1955), Moote (1955) and Lentz and Hart (1960) obtained similar values with a mean of 40 mW m^{-1} K^{-1}. For Merino sheep, Bennett and Hutchinson (1964) reported a range of 37 to 48 mW m^{-1} K^{-1} but for newborn sheep, values between 65 and 107 mW m^{-1} K^{-1} have been reported. Even larger values in the range of 76 to 146 mW m^{-1} K^{-1} were found for cattle (Bennett, 1964; Gonzalez-Jimenez and Blaxter, 1962).

Cena and Monteith (1975b) attached samples of fleece to a heated flat plate mounted in a wind tunnel. They showed that although the thermal conductivity of the fleece was independent of windspeed, it increased with the temperature difference between the skin and the air. Such behaviour is characteristic of a system in which differences of air density associated with temperature differences play an important role in determining air movement and the transfer of heat by free convection.

Free convection can be analysed in terms of non-dimensional groups such as the Nusselt number which is a form of heat transfer coefficient defined by the relation

$$H = Nu\, k\, \Delta T/d$$

where H is the heat loss per unit area from the surface of an object with a characteristic dimension d, ΔT is the difference between mean surface temperature and ambient air temperature and k is the thermal conductivity of still air (see Mitchell, 1974; Monteith, 1973). In free convection, the Nusselt number has the general form

$$Nu = A\, (T_s - T_a)^n$$

where A and n are constants and T_s and T_a are skin and air temperatures respectively. Cena and Monteith found good agreement between values of A and n for a vertical sample of fleece 2 cm deep and for a flat vertical plate. For a 4 cm fleece, they reported $A = 1.66$ and $n = 0.7$ whereas the corresponding figures for a flat plate of the same characteristic dimension are $A = 17.0$ and $n = 0.25$. More measurements with fleece of different depth and different areas are needed to establish values of A and n with general validity. The complexity of the processes involved emphasizes the danger of using values of coat thermal conductivity determined in the laboratory to estimate rates of heat transfer in the coats of live animals with their variable geometry.

Few attempts have been made to quantify the dependence of coat insulation on wind speed. Joyce and Blaxter (1964) exposed sheep to ventilation from fans and showed that the amount of insulation provided by unit depth of coat was (a) independent of coat depth and (b) dependent on the square root of the windspeed. For a fleece 4 cm deep on a sheep standing cross-wind, the insulation in units of K m^2 W^{-1} was $0.48 - 0.11 \sqrt{V}$ where V is the velocity in m/s. Windspeeds between 4 and 5 m/s therefore destroy about half the insulation provided by the fleece in still air. One interpretation of this result is that nearly all the insulation on the windward side of the fleece is removed by air movement whereas insulation on the more sheltered leeward side is little disturbed by ventilation. The angle between the wind direction and the set of the hairs also has an important bearing on insulation as Lentz and Hart (1960) demonstrated for samples of caribou fur, but evidence for other animals is very scarce.

The amount of insulation provided locally by the coat of an animal can be estimated by measuring changes in the temperature of the skin and in the radiative surface temperature of the coat when the external radiative load is changed (Clark et al., 1974; Cena and Clark, 1974). This method has the advantages that the coat is scarcely disturbed during the series of measurements, and that elaborate temperature control of the environment is unnecessary. By combining the method with thermography, it is possible in principle to determine the local heat balance and insulation at a number of representative points over the body of the animal.

SUMMARY OF PROGRESS AND PROSPECTS

This Section has summarised a number of attempts to measure the thermal properties of animal coats in carefully standardised conditions. In some experiments, isolated samples of coat have been exposed in a laboratory or wind tunnel. In others, an animal has been kept in an indoor environment where the air and wall temperatures have been held constant, usually equal. Principles and generalisations emerging from these studies have been reviewed by Tregear (1966) and later references can be found in texts edited by Hardy et al. (1970) and by Monteith and Mount (1974). Graphical methods of estimating heat transfer through animal coats have been developed by Porter and Gates (1969) and by Monteith (1973).

Further progress will depend on a synthesis of physiological evidence about the nature of homeostasis in mammals and physical evidence about the mechanisms responsible for the transfer of heat through the coat (and through the respiratory system) to the environment. There is a wide scope for studies of heat transfer through the coats of animals exposed to natural weather conditions in the field. More information is needed about the complex exchanges of solar and thermal radiation that take place within the coat of an animal in the open and about the effects of wind speed and direction on the insulation of the coat. When this knowledge has been acquired, it will be possible to assess with more confidence and precision the extent to which the metabolic rates and food needs of animals are determined by specific features of the physical environment.

REFERENCES: see pp. 567-568.

Fiber Growth and Insulation Against Heat and Cold in Sheep

by

C. J. Thwaites

INTRODUCTION

In the adult sheep, wool fiber growth exhibits quite a marked seasonal variation, with maximum production occurring in the summer and minimum production in winter. The magnitude of the annual rhythm of wool growth is usually expressed as either the ratio of summer to winter growth rates (useful in describing the amplitude of the curve) or in terms of the difference between maximum and minimum growth rates expressed as a percentage of the mean (useful when partitioning annual production into its seasonal components). This seasonality of wool growth results from changes in both fiber diameter (Stewart et al., 1961) and length (Hart, 1961), and remains in evidence on a constant plane of nutrition (Ferguson et al., 1949). Marked breed differences have been observed in the amplitude of the seasonal rhythm of wool growth, with animals of mutton-type generally showing much greater variation than specialist wool producers such as the Merino (Table 1).

Table 1. The seasonal rhythm of wool growth in ewes on a constant plane of nutrition

Breed and Location	% Rhythm*	Reference
Corriedale (N.Z.)	100	(8)
Cheviot (U.K.)	70–100	(9)
Merino (U.K.)	40–60	(9)
Merino (Aust.)	40	(12)
Corriedale (Aust.)	28	(12)
Merino (Aust.)	17	(4)

* Difference between summer maximum and winter minimum expressed as a percentage of the mean.

Under range conditions a marked seasonality of feed supply (usually maximal in summer and minimal in winter) is normally superimposed upon the basic physiological rhythm of wool growth and exaggerates its amplitude. However, the influence of nutrition on wool growth is sufficiently strong as to enable the basic physiological rhythm of wool growth to be actually reversed in areas where seasonal feed supply patterns are reversed, as occurs in the Mediteranean environment of southern Australia (Stewart et al., 1961). In the case of the ewe, an additional complication arises through the widespread practice of spring lambing. This automatically imposes the stresses of

Information on temperature effects in cattle will be briefly mentioned in Chapter 2, Section 1e.

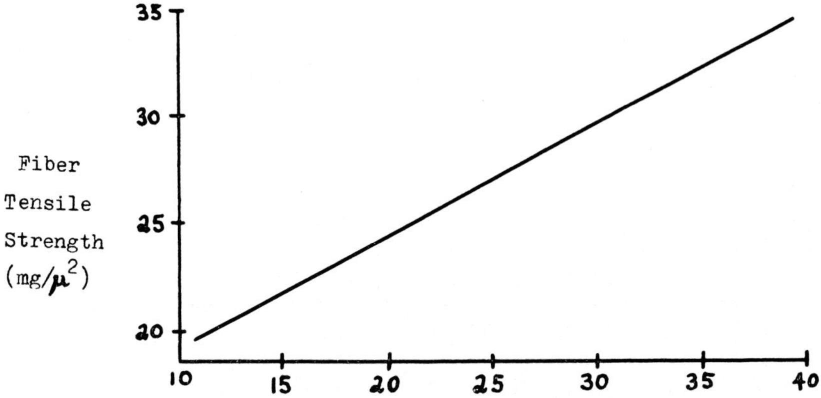

Fig. 1. The relationship between fiber tensile strength and minimum fiber diameter in a range of Merino wools (Hill, 1971).

late pregnancy and early lactation on ewes at a time when seasonal feed supplies and the basic physiological rhythm of wool growth are at their lowest (Story and Ross, 1959). The net result may well be that the diameter and tensile strength of the staple falls to a level which becomes obvious during handling and which interferes with its subsequent spinning characteristics – "tender" wool – or, in extreme cases, which results in an actual "break" in the fleece, part or all of which may then be shed. In Merino fleeces, for example, minimum fiber diameter and fiber tensile strength are significantly related (Fig. 1), and variations in fiber tensile strength between "sound" and "tender" wool (Table 2) thus appear to be almost wholly accounted for by variations in minimum fiber diameter (Hill, 1971).

Seasonal variations in wool growth on a constant plane of nutrition have been linked with changes in both daylength (Coop and Hart, 1953) and temperature (Rudall, 1935). Numerous photoperiod studies have indicated that manipulation of the light environment affects wool growth, but the concensus of opinion is that photoperiodic influences are not able to fully account for the observed seasonal variation (Hutchinson, 1965).

Table 2. The tensile strength of single wool fibers derived from "sound" and "tender" fleeces (After Hill, 1971, Table 2)

Fiber quality	Bradford Quality Number				
	60	64	70	74	80
Tensile strength (Mgm/μ^2)					
– "sound"	25.7	24.9	24.0	23.2	22.5
– "tender"	23.5	22.6	21.8	21.5	20.7
Mean minimum diameter (μ)					
– "sound"	22.0	20.5	18.0	17.0	15.8
– "tender"	17.8	15.9	13.8	13.4	12.2

Reversed photoperiods reverse the seasonal shedding of fibers from the extremities (Rougeot, 1957) and have been reported to both reverse the annual wool growth rhythm on the general body surface (Morris, 1961) and to lead to a bimodal rhythm with peaks in winter as well as summer (Bennett et al., 1962). Hooded sheep exhibit no seasonal rhythm after an initial period of adjustment, but those exposed to continuous light retain their original seasonality (Hart, 1961).

Early attempts to relate environmental temperature and wool growth stemmed either from observations that their seasonal rhythms were approximately in phase, or from the known associations between seasonal temperature patterns and thyroid function on the one hand (Hoersch et al., 1961), and between thyroxine and wool growth on the other (Masqsood, 1955; Hart, 1957). More recently, climate chamber studies have enabled direct evidence to be obtained on the degree to which seasonal temperature cycles are related to those in wool growth (Bennett et al., 1962).

Local effects of skin temperature on wool growth are well known (Rudall, 1935; Ferguson, 1949; Bennett et al., 1962; Doney and Griffiths, 1967; Lyne et al., 1970; Joly and Lyne, 1970), the concensus of opinion being that local skin cooling reduces fiber length growth without greatly affecting fiber diameter, the effect being most likely mediated through alterations in peripheral blood flow and thus in nutrient supply to the wool follicles. Under natural conditions, local skin temperature effects are presumably most relevant to the newly shorn sheep. The fact that apparently quite different fleece types have similar insulative values (Webster and Blaxter, 1966) would suggest that skin temperature is likely to play only a minor role in the regulation of the seasonal wool growth rhythm. However, variations in environmental temperature are paralleled by those of the skin (Bligh, 1959, 1961; Phillips and Raghavan, 1970); and in environments in which wind and rain exacerbate the effects of low winter temperatures, it seems reasonable to expect that the wool production of at least some individuals of some breeds (Blaxter et al., 1966) would be directly affected by low skin temperature and reduced peripheral blood flow.

The depressing effects of the sex steroid oestrogen on wool growth have been known for some time (Slen and Connell, 1958), and it has been suggested that the depressed wool growth rate which characterises the autumn and winter months may be associated, in ewes, with the endogenous oestrogen surges which precede each period of oestrus (Smith, 1973). In the latter work, exogenous oestrogen depressed wool growth by approximately 15%; and wool growth rate during the luteal phase of the oestrous cycle was as much as twice that observed at about the time of oestrus (Fig. 2).

Besides providing man with a valuable textile fiber, the sheep's fleece is of great thermoregulatory significance to the animal itself. A large proportion of the world's sheep graze in hot, dry regions where incident and reflected solar radiation burdens the animal with a thermal load equivalent to several times its endogenous heat production (Priestley, 1957). Animal species have adapted differently to this situation; and thus while *Bos indicus* cattle have developed a short, shiny hair coat which acts as an efficient reflector of radiant energy, sheep have in general developed a thick, insulating barrier – the fleece.

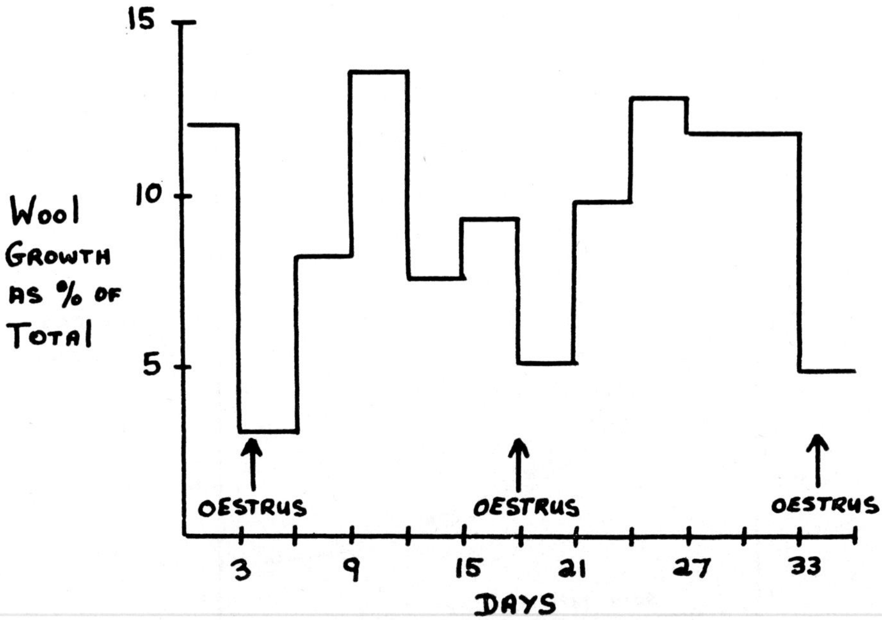

Fig. 2. The wool production of a Border Leicester × Merino ewe during 3-day periods over 2 successive oestrous cycles. (After Smith, 1973).

Standing in the sun on a cloudless day, the fleece-tip temperature of Merino sheep may reach 80-90° C (Macfarlane et al., 1958). Under such conditions a temperature gradient of 40-50° C is established between the skin and the fleece-tip of full-wooled sheep, and the majority of the short-wave solar radiation heat load of approximately 350 kcal/hr is dissipated by the purely physical processes of convection and long-wave re-radiation. The magnitude of the skin – wool tip temperature gradient obviously depends to a large extent on fleece length, and it has been demonstrated experimentally that below 1 cm fleece length heat is conducted very rapidly; while above 4 cm, increasing fleece length affords little additional protection (Fig. 3).

Under severe conditions, the protection afforded by the fleece is probably the single most important factor influencing the sheep's survival; and as such it is a prime determinant of productivity – without the fleece there would be no productivity. However, under less severe circumstances there may be significant physiological and behavioural changes related to fleece length without there being any corresponding effects on productive parameters such as wool growth or liveweight (Parer, 1963), perhaps because the diurnally variable nature of radiant heat stress leads to a shift to predominately night grazing which allows normal feed intake to be maintained.
Besides modifying the sheep's reactions to its environment by the purely physical means described above, the presence or absence of the fleece is also known to have specific

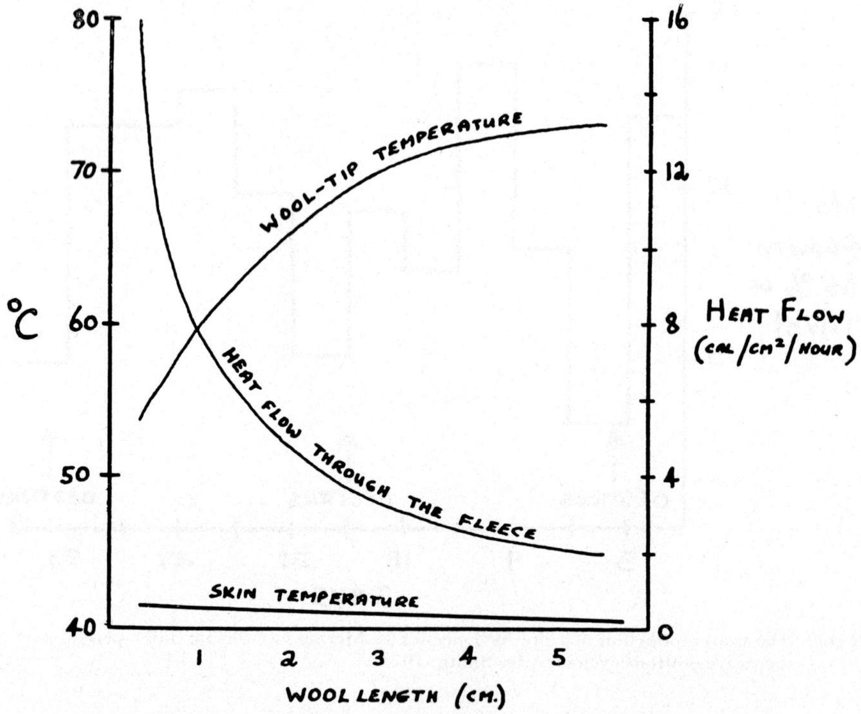

Fig. 3. The relationships between fleece length and skin temperature, fleece-tip temperature, and heat flow through the fleece in Merino sheep exposed to simulated radiant heating at an air temperature of 36°C (After Parer, 1963).

effects with respect to humid atmospheres and the onset of thermal panting. The adsorption of moisture by wool involves an exothermic reaction; and thus the presence of a long fleece is a disadvantage to sheep which are suddenly exposed to hot, wet atmospheres (Klemm, 1962). This reaction is obviously a short-term one, however, and the features of hot wet climates which render them unsuitable for sheep production are related more to impaired efficiency of evaporative cooling and to moisture damage and associated diseases of the skin and fleece.

In the unshorn sheep, physiological responses to severe heat stress precede any change in deep body temperature and are thus apparently triggered by peripheral receptors (Bligh, 1959), which are particularly effective in such areas as the scrotum (Waites, 1961) and mammary glands (Phillips and Raghavan, 1970). Shearing the sheep results in a reduction in skin temperature and respiratory frequency (Bligh, 1961), as well as in a delayed response to thermal stress (Bligh, 1961; Phillips and Raghavan, 1970; Fig. 4).

At the other end of the climatic spectrum, the fleece plays a major role in the survival of sheep under a variety of conditions. The birth coats of lambs vary from the short

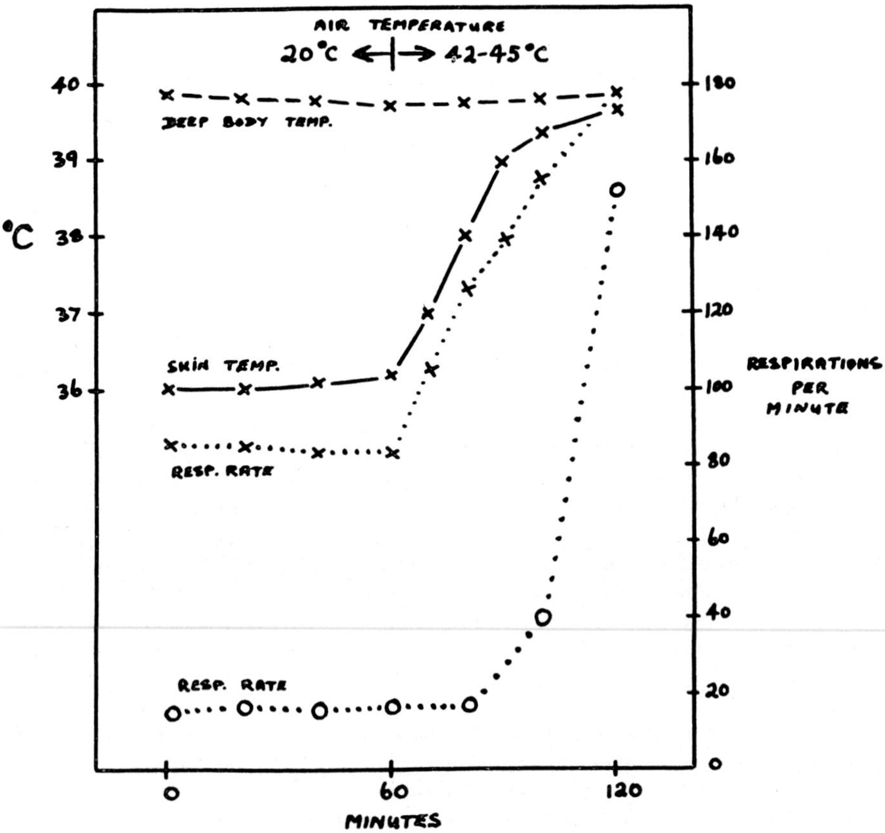

Fig. 4. Respiratory and body temperature responses of shorn (○) and unshorn (×) sheep to acute temperature stress (After Bligh, 1959, 1961; Phillips and Raghavan, 1970).

and fine to the long and hairy, with low rates of heat loss and relatively low thermogenic responses to cold characterising the latter (Alexander, 1961, 1962). Under extremely adverse environmental conditions, involving cold, wind and rain, these thermoregulatory advantages of hairy lambs are presumably of considerable survival significance, although under less stressful circumstances birth coat and survival rate are not significantly related (Mullaney, 1967). Perhaps the most drammatic illustration of the importance of the fleece to the adult sheep is the phenomenon of post-shearing death which has been widely reported, in Australia, to follow the exposure of newly shorn sheep to cold, wet and windy conditions (Hutchinson, 1968). The syndrome is characterised by the sudden (e.g. overnight) death of apparently healthy adult sheep as a result of acute adrenocortical insufficiency brought on by excessive ACTH production (Panaretto and Ferguson, 1969). The adrenals of animals which succumb to cold stress in this way are enlarged and exhibit lipid-laden cortices and in some cases adrenocortical haemorrhages.

REFERENCES: see pp. 568–570.

Effects of Temperature on Lactation of Cattle*

by

H. D. Johnson

INTRODUCTION

Many thousands of years ago in ancient India and Egypt, man discovered a most perfect food in the form of milk secreted from an exocrine gland of domestic cattle. Ancient man soon became aware of environmental influences on the processes of milk secretion when he found it necessary to let the calf nurse the cow for a few minutes before he could extract milk by hand for human consumption. This environmental influence termed psychic-neuro and neuro-endocrine control on the process of lactation is still a most dramatic display of an environmental influence on milk production.

With the development and advances in modern physiology and biometeorology, we are gradually gaining a better understanding of how biometeorological factors such as temperature, humidity, air velocity, radiation and indirect environmental effects influence a mammal's physiological functions including the process of lactation by the mammary gland.

The external and internal environment of a mammal is in a state of continuous change and for survival or for maintenance of an efficient level of productivity in domestic animals, the animal must continually adjust body functions to meet the external environmental changes. These external changes may be cold or hot temperature, high humidity, radiation and wind excesses or deficits, light and other meteorological factors that directly affect thermal balance. These thermal or environmental stimuli incite compensations in energy, hormonal and water regulatory mechanisms which may not be conducive to optimal functioning of the mammary gland and associated reproductive functions. These processes are discussed in other Sections, however, their role in lactation will be cited where appropriate.

The influences of environment, especially climate, on milk production will be summarized and then a brief accounting of physiological mechanisms involved in environment-lactation interactions. The effects of other animals in the surroundings will not be emphasized in this section.

GENERAL BIOMETEOROLOGICAL EFFECTS ON CATTLE

As discussed in other sections, cattle are basically cold tolerant. That is, with adequate

* See also "Progress in Human Biometeorology", Part III, Chapter 10, Section 1a, pp 3-51 (by H. D. Johnson).

protection from extremes of cold, rain, snow and wind and the provision of sufficient energy, dairy cattle will produce milk adequate for commercial operations. However, efficiency of feed conversion is optimal near the thermoneutral zone (Johnson, 1965).

The situation is much different in the tropical and semi-tropical zones of the world, especially in the low-humid tropics. For most of the year in these regions, it is readily apparent that the thermal aspects of the meteorological environment tend to inhibit the optimal functioning of neural and neuro-endocrine systems essential for efficient processes of growth, lactation and reproduction. These factors, combined with other limiting environmental factors or indirect meteorological factors such as lower energy and high fiber feeds and parasites, present a challenge to the development of dairy industries in the tropics. With the increasing efforts of tropical nations and world organizations to develop sound dairy management practices to ameliorate the limiting thermal aspects of the physical environment, control pests and disease and develop forages and concentrates for sufficient energy to maintain good lactation levels, acceptable growth and reproductive efficiency, one may expect the establishment of functional dairy systems in the low tropics in the future.

A. COLD CLIMATES

Early field studies of Davis (1914) reported a lowering of milk yield of cows held in both open and closed barns during cold weather and also observed that cows in the open sheds consumed more roughage.

Popoff (1927) also found that cows produced less milk at a lower temperature, −10 to 4° C (14-39° F), than in a high temperature, 5 to 12° C (41-54° F). In this respect Cobble and Herman (1951) found the fat percentage to increase by 0.2 for each 5.5° C (10° F) decrease over the temperature range of 22 to −3° C (72-27° F).

Kelley and Rupel (1937) observed that changing cows from a warm barn to an open lot (average temperature of −8° C, 18° F) resulted in a daily decrease of 0.38 kg/cow. However, extra feeding minimized this effect. Dice (1935, 1940, 1942) reported similar results on cows and heifers. Witzel and Heizer (1946) in comparing FCM milk of cows kept in cold barns with cows kept in warm barns, showed the cold animals to produce within 2.7% of the warm group.

Sementovskaya and Garkavi (1950) exposed 8 cows to outside winter cold and without supplementary feeding observed a 26% decline in milk yield and a 28.5% increase in butterfat accompanied by a loss in body weight of approximately 0.9 kg/day. All of these reports suggest that cattle may lactate adequately at cold temperatures but with lowered production and greater feed intake.

Extensive studies were initiated on the effects of cold on milk production by Ragsdale et al. (1949). A decline in constant environmental temperature of 10 to −15.5° C (50-4.1° F) resulted in a decline in milk yield and increase in fat percentage – a greater effect being demonstrated on Jersey than Holstein animals. These same experiments also showed the Jerseys to increase feed intake more markedly than the Holsteins at −15.5° C (4.1° F). Cobble and Herman (1951) described the cold effects of increased butterfat percentage, milk total solids, solids non-fat and total nitrogen. McDonald and Bell (1958) similarly observed the milk yield to decline in Holstein cows at −4° C (24.8° F) although the milk yield was not altered presumably due to a free intake of hay feeding program.

B. HOT CLIMATES

Bianca (1965) reviewed much of the hot climate and environmental heat studies on lactation. Recently Thatcher (Personal Comm., Univ. Florida) compared lactating Holstein cattle under Florida field conditions and in air-conditioned barns and observed the air-conditioned animals (21° C, 69.8° F) to produce 9.4% more FCM (4%) fat corrected milk. These losses were similar to the predicted losses by Hahn (1969). These predictions were derived from a comparison of extensive climatic laboratory and air-conditioned barn studies at the University of Missouri.

Mawson and White (1971) presented an interesting and comprehensive study on the effects of the biometeorological environment on lactating cattle in various regions of Australia. Recent data of Maust et al. (1972) compared the response of lactating cows at three stages of lactation and observed that major changes in feed intake and milk yield were seen in the cows in mid-lactation (100-180 days).

Although there are genetic differences in ability to produce milk at high environmental temperature, the level of milk production is a critical factor in assessing the desirability of a specific individual for export to a low tropical environment. A high producer will have a heat production level perhaps 1/3 to 1/2 greater than a lower producer but still must dissipate this heat from the same surface area of the animal (Johnson, 1967).

Most investigations (Johnson et al., 1962; Lee, 1951; Ragsdale et al., 1948-51) showed that heat lowered milk yield and tended to depress butterfat percentage until extremely high temperatures increased the butterfat percentage. Payne (1957) reviewed the effects of climate and particularly tropical climate on milk yield of cattle.

The "comfort zone" for milk production is the optimal zone for maximal milk production, a zone in which the animal is within a normal range of body temperature (Brody, 1948; Brody et al., 1948; Ragsdale et al., 1948). It is not the zone of minimal heat production or thermoneutrality as is often erroneously referred, since the minimal heat production is above the comfort zone (Kibler et al., 1970). The level of wind, radiation, and/or humidity can alter this zone, i.e., at temperatures above the comfort zone increased wind or air velocity will extend this zone, radiation and humidity will lower the critical temperature zone. Furthermore, these zones may vary with acclimation (Johnson et al., 1967) and breed differences (Johnson, 1967).

A comparison of Holstein cattle in the temperate heat of Missouri and subtropical heat of Tabasco, Mexico, when exposed in early August of 1973 to relatively similar "hot days" revealed some interesting but quite different relationships in milk production and rectal temperatures (Fig. 1). The relationship of the change in rectal temperature from early morning (before sunrise) to maximal rectal temperature around 15.00 hr and level of milk production of each animal were studied (DeDios et al., Personal Comm.).

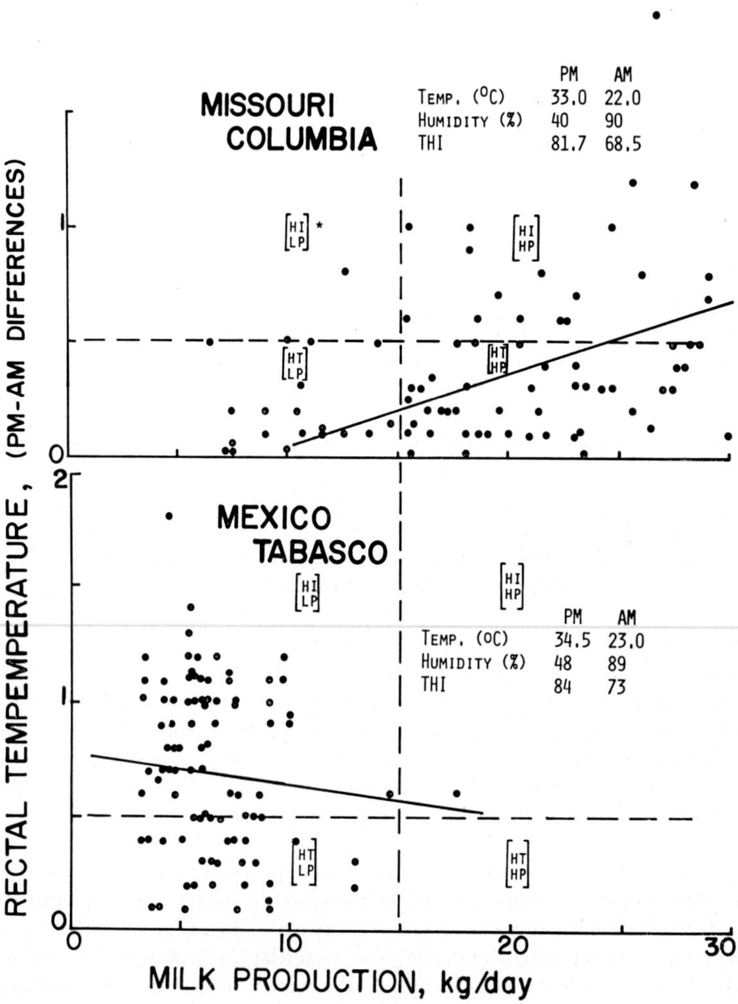

Fig. 1. Milk production, kg/day, as related to AM-PM differences in rectal temperature on animals in Missouri and Mexico. The r values for the Missouri data are r = 0.37 (P < 0.01) and the Mexico data are r = —0.08 (ns).
HI = Heat Intolerant; LP = Low Producer; HT = Heat Tolerant; HP = High Producer.

These data show rather clearly a positive relationship (r = +0.08, P < 0.01) of increase in rectal temperature on a hot day and level of milk production. Thus the combination of the greater endogenous heat load of a higher producer and the external heat load tend to increase the rectal temperature more in higher lactating animals. Of course there are individual differences in rectal temperature rise and level of production which suggest a genetic factor and a production level factor which influence the animal's ability to maintain thermal stability. It is reasonable to assume that 30 kg/day producers whose body temperatures only increase by 0.5° C or less would be most thermally adapted for more adverse low tropical climates. With the enormous capital outlay for exportation of cattle to tropical regions, it seems that an animal with known ability to regulate body temperature would warrant extra premium. Figure 1 generally characterizes the animal represented by each dot into four classes. The heat tolerant, high producer animal has the most potential for success in a low tropical environment. For Holstein cattle in Tabasco, the relationship of change in rectal temperature versus level of production is not significant largely because the production range is too limited in the range of 5-10 kg/day as compared to 10-30 kg range in a temperate zone. That is, the limited energy intake plus the lowered ability to lactate in a hot environment for many months of the year make it difficult to assess these relationships of cattle in a low tropical region. It would be of extreme importance to determine if the relationship of rectal temperature and milk production would become positive with ad libitum availability of high energy feed. This would aid animal biometeorologists and environmental physiologists to better understand the limiting factors for high production in a tropical environment. These relatively straight-forward data reveal several important facts.

1. In temperate zones, higher producing cows have more difficulty maintaining thermal balance.
2. In subtropical zones after prolonged exposure to higher fiber rations, the higher producing cattle are those that appear to have an inherent ability to regulate body temperature.
3. These facts strongly suggest that breeders must give more attention to heat tolerance of heifers destined for export to tropics or sub-tropics at low altitude.
4. One may speculate that progressive breeders in the near future will provide a heat tolerance index as part of the performance or genetic information on a heifer or bull. An approximate estimate of heat regulatory ability need only be a few hours exposure to a high environmental temperature (for example 35 to 38° C (95-100.4° F), 50% relative humidity) for non-lactating dairy heifers.

In summary, it must be re-emphasized that temperate-evolved animals in low tropical zones where seasonal changes are minor undergo long-term acclimitizations which cause a potential high producer to compensate physiologically in such functions as mammary gland development and body size, resulting in production far below their expected or predicted level. De Dios et al. (Personal Comm.) are initiating a project in Tabasco, Mexico, to clarify the thermal and/or energy limitations in purebred Holsteins imported from the United States and Canada. Studies of this nature are long overdue and invaluable and essential to the livestock industry in the tropics and will provide valid, reliable guidelines for rearing and milking of European animals in the low tropics.

METEOROLOGICAL EFFECTS ON MILK PRODUCTION

Meteorological factors influence lactation and related growth and reproductive functions when adverse enough to alter thermal, water, hormonal and energy balances – whether the "productive" functions are altered sufficiently to cause economic losses depends on the duration of time and magnitude of thermal imbalance. Applying this view to the low tropics, one may expect lactation performance below optimal for temperate zones, especially if the European animals are reared under the limitations of thermal imbalance, low energy feedstuffs and possible hormonal compensations.

The lower growth rates, reproductive efficiency and subsequent lower lactation usually reported in the tropics are a result of physiological acclimitization or compensations, especially in feed intake, energy metabolism and hormonal functions (see Chapt. 1, Sect. 12a, 15a and 18c).

The milk production of temperate-evolved Holstein, Brown Swiss, Jersey and the tropically-evolved Brahman cows has been described previously by Johnson (1972). Milk production was plotted versus dry bulb temperature of approximately 50% R.H. and was maximal between 4 and 21° C (40 and 70° F). Milk production generally declined above and below this comfort zone. More specifically at the high environmental temperatures, the yield declined from Holstein animals at approximately 21° C (70° F), Brown Swiss and Jersey at about 24-27° C (75-80° F) and the Brahman at approximately 32° C (90° F). The most pronounced decline in lactation was shown by the Holstein animals.

Johnson (1972) illustrated the effects of various temperature-humidity conditions on animals producing approximately 23, 18, 14 kg/day. Using 18° C as a base condition, the decline in milk production at each THI condition was computed from the following equation: ($THI = 0.55 \times D.B. + 0.2\ D.P. + 17.5$). These values were based on declines in production from base (18° C) values. A switchback procedure was used for the various temperature humidity conditions thus eliminating the persistency effects of lactation. THI values have not been established from conditions below 18° C for cattle. Below 18° C a THI index may be of limited significance; however, an index relating radiation, wind and humidity to temperature would be a most significant contribution of lactating cattle at a cold temperature. The previous application of Wright's (1946) climograph was a valuable contribution in this aspect of shelter evaluation.

An environmental index that quantitates the contributing effects of radiation, humidity, wind and diurnal cycles on lactation is of course more meaningful since dry bulb temperature is always accompanied or qualified by the other variables. Such an environmental index "equivalence index" should be established. It should be re-emphasized that high radiation above the comfort zone would increase the thermal problems for the cow and the available air velocity may be critical. These are also valid concerns below the comfort zone. An index of this nature would relate more directly to the heat balance and comfort of the animal than dry bulb temperature. The application of the temperature-humidity index to milk production was an effort in this direction (Cargill et al., 1962; Johnson et al., 1962).

The relative effects of humidity on milk production and feed consumption are also previously published (Johnson, 1972).

The advantages of air velocity at a high environmental temperature and the relatively

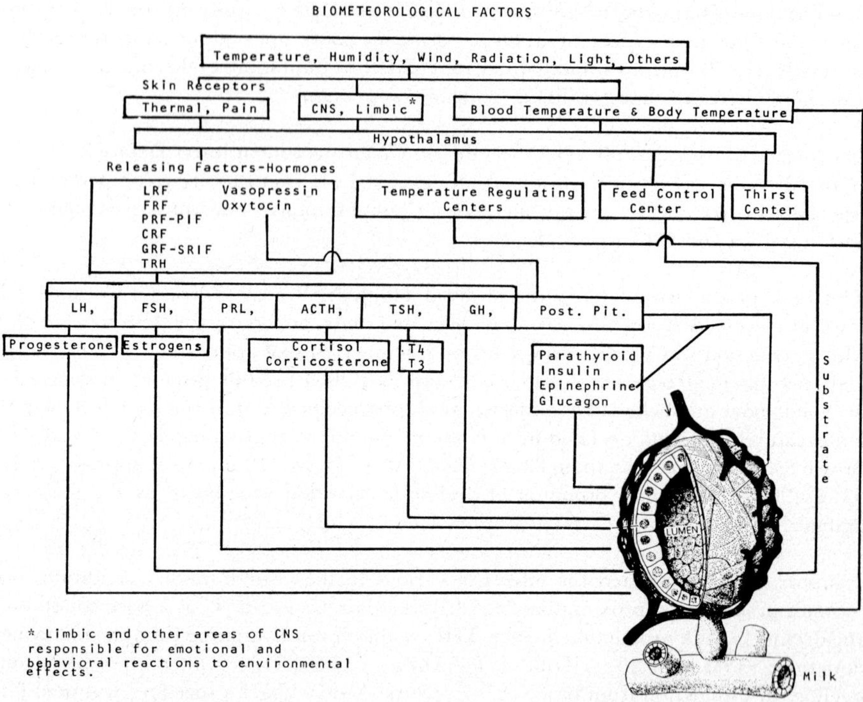

Fig. 2. Meteorological effects on neural, neuro-endocrine and endocrine system and resultant effects on secretion of milk. The scheme is intended to illustrate the principal pathways of meteorological influence on lactation. The specifics of hormonal feed-backs and role of hormonal binding and known compositional changes in milk are important though will not be presented in this brief appraisal of meteorological effects on lactation.

lower thermal neutral or "comfort zone" for Holstein as compared to Brown Swiss or Jersey have been described (Johnson, 1972). A similar conclusion is made from the radiation levels i.e., a higher radiation level limiting production more in the Holstein than other breeds studied (Johnson, 1972).

MODE OF ACTION OF METEOROLOGICAL FACTORS

At the present time, we understand some aspects of the principal regulatory systems of the body which are influenced by meteorological factors and in turn modify the quantity and quality of milk synthesized. These are energy regulatory involving feed intake and subsequent effects on energy available for milk synthesis and temperature regulatory whereby body and surface temperatures alter heat production or heat loss mechanisms and thus modify the other regulatory systems, namely hormonal and water (Fig. 2). It is obvious that these neuro-endocrine systems are intimately interrelated and influenced by meteorological factors. It is this meteorological complex effect on the animals' neural and neuro-endocrine system that requires much intensive investigation to more fully understand the controlling mechanism of milk secretion and the limiting or stimulating role that the meteorological factors may play in these fascinating neuro-hormonal-enzymatic reactions.

The availability of radioimmunoassay techniques to measure the pituitary and target organ hormones and the synthesis of many of the hypothalamic releasing hormones or factors will enable physiologists and biochemists to make rapid advances in the next few years.

To be more specific, we know that biometeorological factors modify the ability of mammals to secrete milk. The pathways (Fig. 2) of this temperature influence are postulated to be direct via (a) cutaneous receptors and the central nervous system* and thus affecting the neuro-endocrine system via the hypothalamus and the pituitary; (b) changes in body temperature and temperature of the blood perfusing the hypothalamic region and (c) temperature changes in the mammary gland, thereby altering the metabolic activity of the mammary tissue.

Indirect effects of temperature or meteorological factors may be regarded to alter the appetite or the nutrient intake of the animals and thus provide less substrate or less available milk precursors for the mammary gland. This scheme points out the major neural and endocrine controls of the mammary gland synthetic activities and suggests the logical sites where biometeorological factors, especially temperature, may affect lactation.

The hypothalamus occupies a key position in the central nervous regulation of processes essential for lactation. Cross (1961) and Meites et al. (1963) have reviewed the influence of the hypothalamus on these functions. Numerous reviews and texts such as those of Meites et al. (1963), Cowie (1961) and Blackwell and Guillemin (1973) document the hormone and neural aspects of these schemes. Figure 2 presents these ideas in a scheme to arrive at a fundamental basis for ascertaining our current state of knowledge of the environmental temperature effects on lactation.

Temperature and especially cold temperature is known to stimulate TSH and ACTH and associated involved functions (Cowie, 1961; Meites et al., 1963). Much less information is available on these two functions at high temperatures. Only by relating the available evidence that thyroid function and lactation are depressed by high temperatures (Johnson and Kibler, 1963) may one begin to speculate on these mechanisms. It was formerly believed by many that ACTH remained elevated at high environmental temperatures but later investigations (Bergman, 1963; Kotby and Johnson, 1967; Christison and Johnson, 1972; Abilay and Johnson, 1973) in the Missouri Laboratory suggest that after a period of heat exposure the glucocortical

* These act by depolarization of unmyelinated nerve fibers with impulses via dorsal root and lateral spino-thalamic tract to limbic and hypothalamic centers. These are controlling areas for thermal regulation, energy intake, hormonal and water regulation. Motor neural and sympathetic stimulation via the ventral roots and sympathetic chain involve the musculature adrenal medulla, etc.

function is lower at the elevated body temperature. Growth hormone (Mitra et al., 1972) and insulin (Kamal et al., 1970) are similarly affected.

Meteorological effects on prolactin and lactation are not documented but Cowie (1961) and Meites et al. (1963) cited evidence that prolactin or lactogenic hormone and ACTH are essential for maintenance of lactation and growth hormone needed to support lactation. The adrenal cortical hormones and insulin are generally regarded as essential for milk secretion but excess hormone is inhibitory. The thyroid gland is not absolutely essential for lactation, but the intensity and duration of lactation are less when this hormone is deficient. (For more information on heat effects on hormone function, see Chapt. 1, Sect. 2a (1)).

Other than data on appetite (Brobeck, 1960; Morgane, 1962) and temperature regulating centers (Hardy, 1961), few facts are available to synthesize specific pathways regarding temperature effects on substrate availability for lactation. To students and workers in environmental physiology, this is a most stimulating challenge.

Cold may have a depressing effect on the mammary gland at three different levels. First, it could result from the fact that there is a greater energy requirement of the whole body because of its increased heat production. Second, there may be a reduced supply of nutrients to the mammary gland by the circulation because in species other than bovines it has been shown that the blood flow of the mammary gland is reduced by lowering its temperature. Third, there may also be a direct effect of cooling the udder on milk synthesis and secretion. Holmes (1971) clipped the hair off one half of the udder of lactating cows and covered the other half. The air temperature was 6 to 10° C (42.8-50° F) and air movement over the clipped udder was artificially raised to 3 to 5 m/s. After 22 hrs of cooling, the milk yield of the exposed half of the udder was lower than that of the covered half; and fat, protein and lactose yields were reduced. Similarly, some of the documented depressive effects of environmental heat on lactation may possibly be a direct action on the mammary gland rather than indirect via the lowered substrate and hormonal deficiencies.

Environmental heat direct effects on the mammary gland are not known. The reduction in feed intake and lowered substrate to the gland could logically be a factor. Forced feeding was of some benefit to lactation (Johnson et al., 1966). Moderate environmental heat may stimulate enzymatic activity, providing circulation is maintained or increased. These academic questions are of importance to fully understand meteorological influences on lactation.

CONCLUSION

Thus, the complex role of the hormonal control, the neural control and integration thereof, the relationship of the responsiveness of the mammary gland cells directly to environmental stimuli and substrate effects on lactation pose many interesting and important challenges and problems which must be resolved before the role of environmental and/or body temperature are understood in the normal functioning of lactation.

REFERENCES: see pp. 570–571.

Breeding Management in Cattle

by

H. A. W. Venter

INTRODUCTION

Environmental conditions under which cattle are produced vary greatly. These environmental conditions influence crop, and animal production. It is however often difficult to separate the direct consequences of climate on animal production from those resulting from the indirect influence of climate on the quality and quantity of food ultimately available for metabolism. Animal productivity depends not only on the flow of nutrients within the environment but also on the adaptability of organisms to the environment and on their interaction (Fig. 1). The animal should be in complete harmony with its total environment to be able to reproduce and produce efficiently.

In an effort to improve cattle reproduction under conditions of high environmental temperatures, producers are often faced with some of the following problems:

a. The formulation of a breeding program that will allow the efficient utilization of the available resources under the existing environment including feed and management;
b. Retarded growth of young breeding stock from the age of weaning up to the age of 2 yrs;
c. The advanced age at first calving of heifers;
d. The occurrence of postpartum anoestrus, especially after the first calf but also in mature cows;
e. Low levels of conception under conditions of both natural and artificial breeding;
f. High mortality around calving time and up to the age of one year;
g. Impaired bull fertility.

From an investigation into the reproductive rates of some world regions the considerable variation according to region and individual operation within the region becomes obvious. This indicates that apart from the direct detrimental effect of high environmental temperatures on reproduction rates, management under the existing environment plays a critical role.

BREEDING MANAGEMENT

Data from several experiments tend to show that in the tropics and sub-tropics, cattle with Zebu breeding, when properly managed, excel and are superior to the exotic breeds in reproduction (Temple, 1972). In all these experiments the herd management was considered to be efficient. The management was not of a completely artificial type

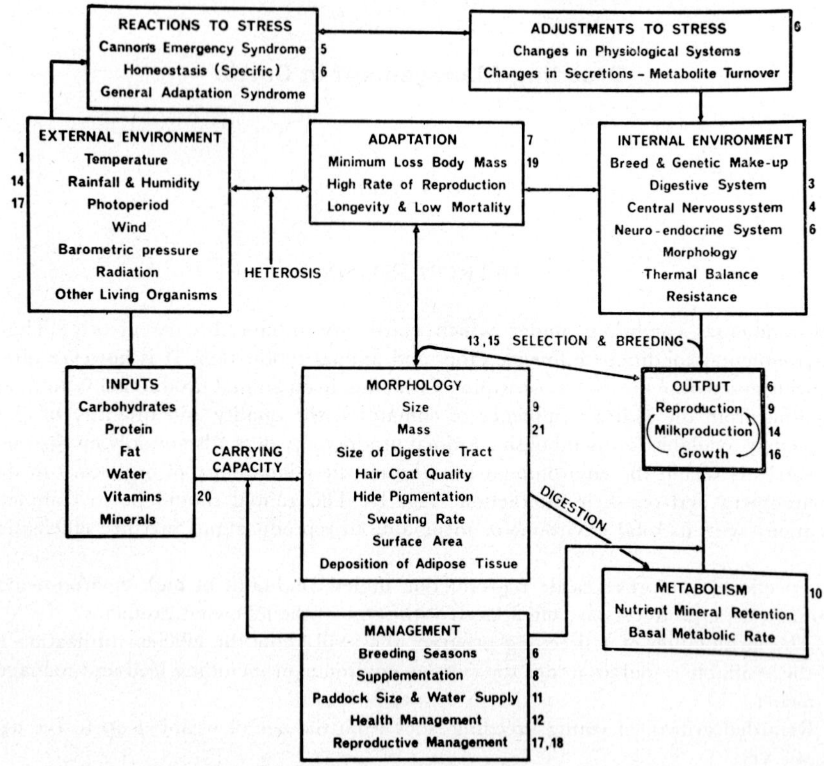

Fig. 1. The interaction between the total environment and the genetic make-up of the animal (The ciphers refer to the ciphers between brackets in the references on pp. 572-573).

which would tend to favour the exotic breeds. Table 1 summarizes the calving percentages and mortality rates on cattle ranches in the Northern Transvaal.

It was further obvious that the maternal influence of Zebu dams is superior to that of the *Bos taurus* dams in the tropics and sub-tropics because the Zebu cow possesses a much greater natural immunity against endemic diseases. Bonsma (1949) maintained that calves suckling Zebu dams probably obtained a greater spectrum of immune bodies through the colostrum of the highly immune cow.

McCarthy and Hamilton (1964) conducted a survey of 28 beef cattle farms in tropical Queensland and compared the performance of British and Zebu cross cattle. From the productivity and management indices used it was found that the only significant difference between the groups was that Zebu cross cattle were marketed at 3 yrs of age compared to 3,9 for British breeds. This, however, is an important aspect of productivity under tropical conditions.

British and European breeds of cattle calve at a younger age than Zebu type cattle

Table 1. Calving percentages and mortality rates on cattle ranches. (Bonsma et al., 1953)

Ranch	Locality	Altitude m	Average Annual temp. °C	Average Annual rainfall mm.	Breed	Number of cows	Calf crop %	Calf Mortality %
A	Long. E 26° 53′ Lat. S 24° 21	853 m	20.1° C	411.5	Herefords	1,477 for 22 yrs	39	18
B	Long. E 28° 21′ Lat. S 24° 56′	1112 m	18.6° C	612.1	Angus	3,143 for 5 yrs	56	23
C	Long. E 29° 45′ Lat. S 20° 45′	1371 m	18.3°	624.8	Sussex	17,400 for 20 yrs	55	14
Mara	Long. E 29° 34′ Lat. S 23° 09′	914 m	19.6° C	419.1	Adaptable tyijes	416 for 6 yrs	83,5	10,6
Zoutpan*	Long. E 28° 51′ Lat. S 25° 24′	1050 m	18, 19	656.23	Africander × Dual purpose	233 for 6 yrs	82,5	6,4

* Calculated by author.

Table 2. The effect of stage of production on Birthrate in Africander and Sussex females (Harwin, 1966).

Type of Cows	Africander		Sussex	
	n	Birthrate %	n	Birthrate %
Non-lactating cows	187	84	129	79
Lactating cows	475	61	571	78
3 Year-old cows	125	51	142	78
Heifers	164	87	175	92
All cows	662	64	700	78

Table 3. Mean body temperatures of Africander and Friesland bulls (Skinner and Louw, 1966)

Breed	At prevailing Atmospheric temp. before exposure °C ISE	During exposure to 40° C °C ± SE		
		After 6 hrs	After 12 hrs	After 14 hrs
Africander	38,00 ± 0,09	38,96 ± 0,11	39,10 ± 0,14	40,05 ± 0,29
Friesland	38,00 ± 0,18	39,66 ± 0,10	40,01 ± 1,37	40,24 ± 1,26

Table 4. Effect of exposure to 40° C for 6 days on spermatozoal characteristics in Africanders and Frieslands (Skinner and Louw, 1966)

Spermatozoal characteristics	Criteria	Breed	
		Africander	Friesland
Mortility index	Mean value before exposure	11,8	12,2
	Mean value after exposure	9,0*	7,3*
	Minimum value after exposure	6,1	3,8
Live sperm % ± SE.	Mean value before exposure	74,8 ± 5,3	71,9 ± 6,2
	Mean value after exposure	59,2 ± 2,3*	46,6 ± 7,2*
	Minimum value after exposure	38,6 ± 5,6	31,0 ± 10,7
Primary abnormalities, % ± S.E.	Mean value before exposure	8,7 ± 2,7	8,8 ± 2,8
	Mean value after exposure	18,6 ± 3,6*	23,5 ± 5,8*
	Maximum value after exposure	29,4 ± 7,0	46,7 ± 13,7
Secondary abnormalities, % ± S.E.	Mean value before exposure	6,4 ± 1,4	8,7 ± 4,3
	Mean value after exposure	19,1 ± 2,8*	16,3 ± 4,2*
	Maximum value after exposure	40,2 ± 4,7	35,5 ± 9,4

* Significantly different from pretreatment mean ($P > 0.01$).

(See Table 6). Tropically adapted cattle are inclined to have a longer lactation anoestrus period (Harwin et al., 1967; Bonsma and Skinner, 1969). Table 2 indicates that the reconception rate of lactating Africander cows was considerably lower than that of the Sussex cows.

High environmental temperatures affects the fertility of bulls of different breeds in a similar way. Skinner and Louw (1966) exposed Friesland and Africander bulls for varying periods to 40° C. The more heat tolerant Africander bulls (Table 3) were less adversely affected than the Friesland bulls (Table 4). A decline in semen quality could be detected as early as a week after treatment commenced, as indicated by decreased sperm motility and an increase in abnormal sperm. It was concluded that short term heat stress may be an important factor under practical conditions. The effects were accentuated three to four weeks after exposure and continued for about six weeks after exposure.

It follows that the selection of livestock breeds, individuals within breeds and breed combinations with the object to optimum reproduction under tropical conditions represents basic management and breeding policy considerations. This aspect has been investigated and reviewed by several authors and recently by McDowell (1972). Table 5 can serve as a broad basis for determining the suitability of types and breeds of cattle for breeding and cross-breeding programs.

GROWTH AND SEXUAL DEVELOPMENT

The early mating of heifers of both dairy and beef breeds is becoming of increasing importance, even under more extensive conditions of production. Friesian heifers fed on high energy rations in Israel had their first estrus at about 177 days of age at an average weight of 225 kg. They required three services per conception and calved down from 13,5 to 17,5 months of age and yielded over 2,700 kg of milk in their first lactation (Amir et al., 1967).

Delayed onset of puberty, particularly in *Bos indicus* types constitutes a major limiting factor in the breeding of yearling heifers. The importance of the level of nutrition has also been demonstrated by several experiments. Table 6 summarises the age and weight at puberty and first conception of different types and breeds of cattle under various environmental conditions.

Maree and Harwin (1971) investigated the sexual development of Bonsmara heifers in the Transvaal Bushveld. Ovarian activity was expressed in terms of a numerical index for each of four treatments for heifers from the age of nine to 17 months (Fig. 2). Creepfeeding had a marked prepubertal residual effect on ovarian activity after the heifers had been weaned. Likewise winter supplementary feeding increased prepubertal ovarian activity in both the creep and the non-creepfed groups. Ovarian activity was reduced in all four groups during August and September. This was attributed to the reduced nutritive value of the grazing, as improved conditions during November and December were reflected in increased ovarian activity. The nutritive effect cannot, however, be completely separated from the effect of increasing temperature and

Table 5. Classification of breed types according to their estimated adaptability to environment and system of production:

Trait	British beef breeds	Dual purpose breeds	Bos Indicus breeds	Specialized beef breeds
Growthrate	Medium	good	Poor	good
Mature weight and size	Early 380 kg Digestive type	Late 580 kg Respiratory type	Medium 420 kg. Respiratory type	Late 500 kg Digestive type except Chianina
Milking ability	Variable	good	Variable	Medium
Fertility Lactation anoestrus	Good	good	medium	medium
Mothering ability	Medium	medium	good	medium
Occurrence of dystocia	Medium Depends on bull used	High	Low	High
Sexual Maturity	Early	Medium	Late	Medium Incidence of dystocia is prohibitive
Muscling	Medium	Good	Poor	Excellent
Grading ability	Good	Medium Depends on Weight at Marketing	Good	Good Depends on weight at Marketing
Feed environment	Semi-intensive	Intensief	Extensive	Intensive
		Adaptability		
Tropics and Subtropics	Poor	Fair in sweetveld areas	Good	Poor
Temperate and cold areas	Good	Good	Medium Increased Maintenance requirements	Medium Depends on soil fertility and quality.
Poor management or extensive management	Poor	Poor	Good	Poor

lengthening photoperiod which have also been shown to increase ovarian activity (Mercier and Salesbury, 1947; Sweetman, 1950; Kiesel and Grein, 1961; Plasse et al., 1969).

In the case of females it has been proved that exposure for 8 hrs daily from 26 weeks

Breeds and Locality (1)	Level of Nutrition: Before weaning	Level of Nutrition: After weaning	No.	Age (days)	Mass (kg)	Remark	Author
Africander	–	0,5 kg A.D.G.*	12	414,7 ± 12,8	245 ± 10,5	First oestrus	Penzhorn and Meintjies (1968)
Hereford × Africander	–	0,6 kg	12	378,7 ± 7,7	259 ± 7,1	First oestrus	
Bonsmara Christiana (2)	Average mass at 7 months	–	35	430	290	First service	Venter (Unpubl.)
Bonsmara Roodeplaat (3)	Creepfed	Supplemented	11	450	346,8	First ovulation	Maree and Harwin (1971)
	non-creep fed	Supplemented	11	450	338,1		
	Creep-fed	Grazing plus lick	11	450	293,6		
	Non-creep	Grazing plus lick	11	450	293,1		
Brahman 50°	Raised	–	3	307	261	First oestrus	Dale et al. (1959)
Brahman 80°		–	3	463	357		
Short-horn 50°	similarly	–	3	303	263	–	
Short-horn 80°	at different temp.	–	3	440	267		
Mara Res. Station (4) Africander	similar conditions		115	1290	–	Age at first calving	Joubert (1959)
Hereford	grazing plus lick		132	1140	–		
Bonsmara			68	1050	–		

* A.D.G. – Average daily gain.

	(1)	(2)	(3)	(4)
Long E.	27°04′	25°10′	28°21′	29°34′
Lat. S.	26°40′	27°55′	25°35′	23°09′
Altitude (m)	1350	1207	1164	914
Average annual rainfall (mm)	615,4	437,5	646,3	419,1
Average annual temp. (°C)	17,04	18,4	18,1	19,6

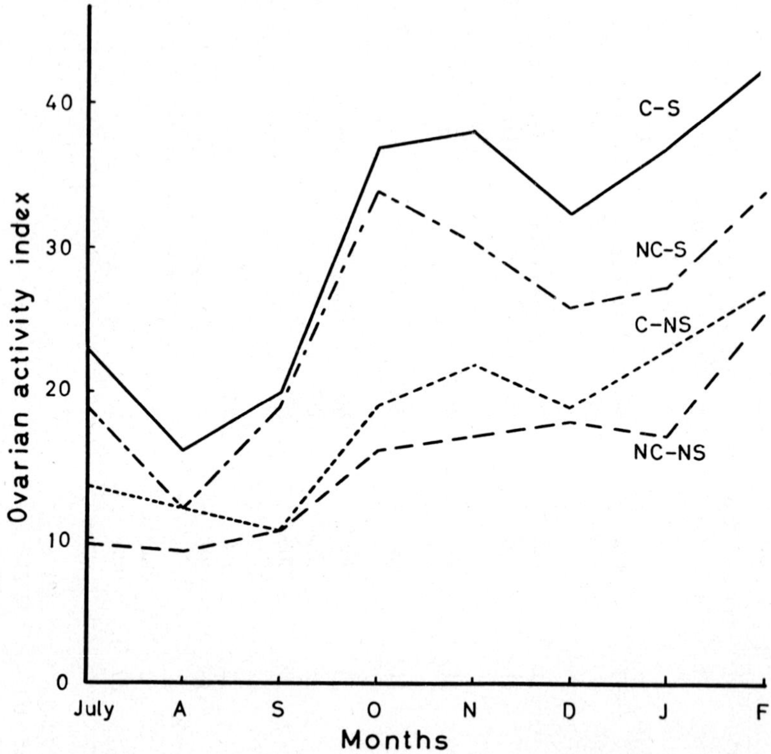

Fig. 2. Relative values of ovarian activity in four groups of heifers in terms of a numerical index value (0-40) C S, Creepfed and supplemented after weaning; N C-S no creepfeeding but supplemented after weaning; C-Ns, creepfed but not supplemented after weaning; NC-NS, no creepfeeding and not supplemented after weaning. (Harwin and Maree, 1971).

of age until puberty at 35° C retarded puberty and lowered semen quality in Jersey bulls (De Alba and Riera, 1966). Libido, however, was unaffected.

REPRODUCTION IN THE COW

The direct effects of high ambient temperatures on reproduction in the cow have been investigated by several researchers (Vincent, 1972; McDowell, 1972; Stott et al., 1972; Stewart, 1973). Depending on the grade of thermal stress, females are usually affected in one or more of the following ways:

1. The length of the estrus cycle is prolonged in both heifers and cows;
2. The duration and intensity of oestrus is decreased;
3. The incidence of clinical anoestrus is increased. The reaction of females depend to some estend on exposure to previous temperature conditions. Beef heifers have been shown to acclimatise to some degree to thermal stress;

4. Increased incidence of uterine infections and delayed post-partum uterine involution;
5. Lowered conception rates;
6. Increased early embryonic mortality;
7. Affected foetal development;
8. The normal hormonal balance of cows are altered. Lactating cows suffer more from reproductive difficulties than non-lactating animals even at equivalent body temperatures, which suggests endocrine involvement.

It follows that the direct effects of high temperature on reproduction are severe and management programs should take cogniance of this and arrange the breeding and management programs accordingly. On investigating the occurrence of estrus in dairy heifers in Louisiana it was estimated that anoestrus would have increased by 20% with twice daily observations only at 06.00 hr and 18.00 hr (Hall et al., 1959). Ameliorization of the environment by artificial means can alleviate the consequence of heat stress. Stott et al. (1972) reported that cows which were cooled under a experimental shade were twice as fertile as a control group in Arizona (Conception – 31 versus 14%).

Delayed onset of estrus and lactation anoestrus are among the most important problems in cattle production in environments with high ambient temperatures. The duration of anoestrus is also influenced by the age of cow, breed, season, date of calving and other environmental factors (Koger et al., 1962; Temple, 1965; Bosman and Harwin, 1969). From studies on postpartum anoestrus, (Jurriaanse et al., 1974; Eden, 1970; Bonsma and Skinner, 1969; Short et al., 1972) it can be concluded that:
1. The suckling effect of the calf is important, as treatments involving weaning at different stages and mastectomy significantly decreased the postpartum interval;
2. The duration of the postpartum anoestrus period and time required for uterine involution were unaffected by the level of milk production;
3. Shorter intervals to first estrus required more services per conception;
4. Breed differences in the duration of postpartum anoestrus intervals are apparent.

Bonsma and Skinner (1969) investigated some factors influencing the distribution of births within calving season and the birth mass of calves. Data on three beef breeds, Africander (926 births), Bonsmara (1,663 births) and Hereford (253 births) on the Mara Research Station (See Table 1 for climatological data), were included.

The calving seasons (Table 7) have been divided into three-weekly periods and the distribution of births within these periods according to breed is illustrated in Figure 3.

Table 7. Division of three-weekly periods in the calving season.

Period No.	Summer	Winter
1	11 December – 31 December	27 May – 16 June
2	1 January – 21 January	17 June – 7 July
3	22 January – 11 February	8 July – 28 July
4	12 February – 4 March	23 July – 18 August
5	5 March – 25 March	

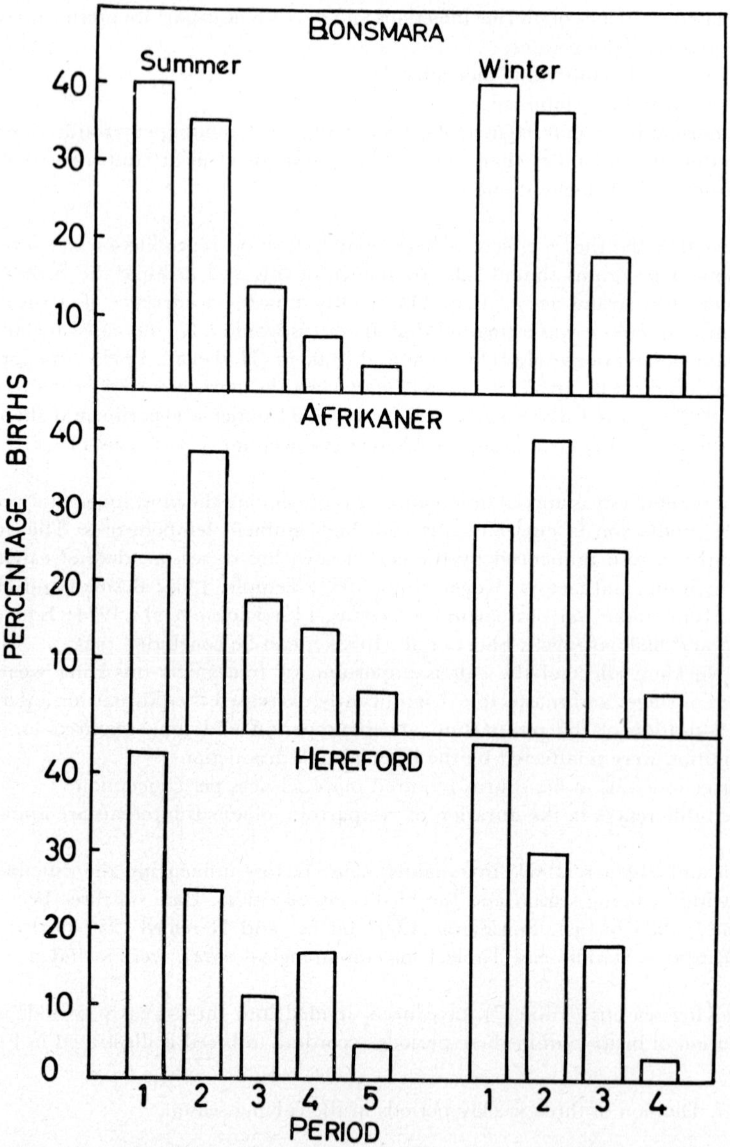

Fig. 3. Distribution of births within the calving seasons. (Bonsma and Skinner, 1969).

Significantly fewer Africander cows calved in the first half of each season than Bonsmaras but there was little difference between Bonsmaras and Herefords. Only 24% of the Africander cows calved in the first 3 weeks of the summer season as against 42 and 44% of the Bonsmaras and Herefords respectively. The overall calving rate of Africander cows was 10% less than that of the other breeds throughout the season. It has been determined that the duration of pregnancy is four to six days longer in the Africander than in the Bonsmara and Hereford, (Bonsma and Joubert, 1959; Skinner and Joubert, 1963) but these differences in the gestation periods were too short to be entirely responsible for the later calving of the Africander cows and a prolonged postpartum anoestrus period is most probably the cause for this phenomenon.

In the same study the birth mass of Africander and Bonsmara bull calves tended to increase the later they are born in the season (Fig. 4). This is in agreement with the findings of Vorster (1954) and Bosman and Harwin (1966) under similar environmental conditions. The birth mass of heifers showed little change in these two breeds. In contrast, the birth mass of Hereford calves tended to decline the later they were born in the season. Early and late summer calves were significantly ($P < 0.01$) heavier than winter calves. The problem of offspring of reduced birth mass has been referred to previously, in British beef breeds in the sub-tropics. (Bonsma, 1949).

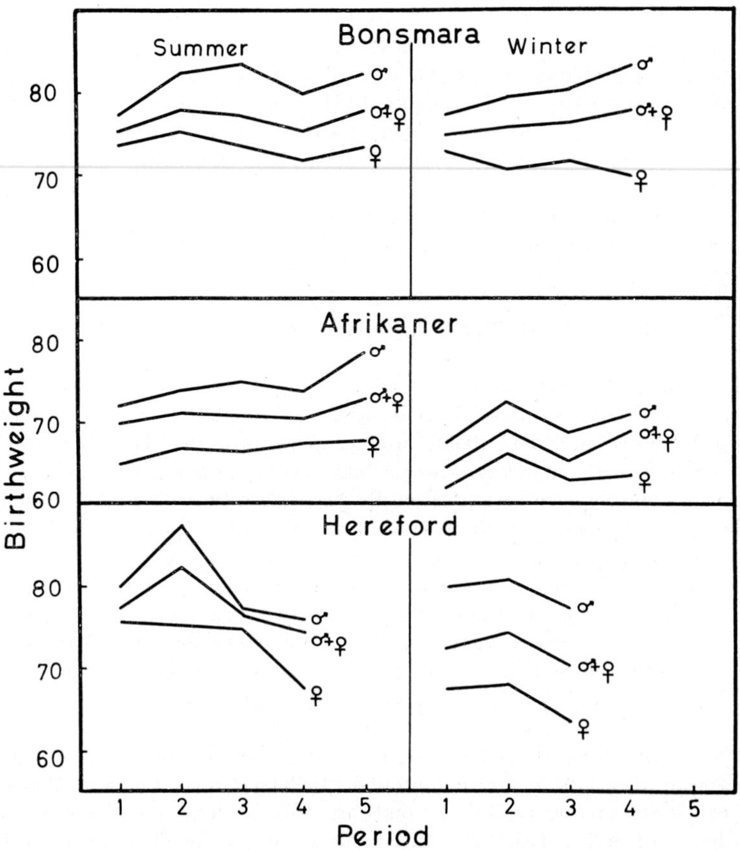

Fig. 4. Variation in birth mass during the calving season. (Bonsma and Skinner, 1969).

Bonsma et al. (1972) reported on a herd of 30 2-yr old Shorthorn heifers imported from the Eastern Cape Province, which is a temperate region where the breed thrives to Messina situated in a sub-tropical region. During their active lifetime at Messina 15 of these 30 cows produced a total of 41 calves. The overall calving percentage was 43,6%. At Messina normal ranching procedure involved the practice of two mating seasons per year, viz. an autumn mating season followed by a winter pregnancy and summer calving, and a spring mating season followed by a summer pregnancy and winter calving. The data on the calving performance of the Shorthorns and their crosses at the two localities are presented in Table 8. In the pregnancies recorded the foetus was carried to term and it would seem that after implantation the foetus was affected by high environmental temperatures. Maximum temperatures were apparently not high enough to cause embryonic loss after implantation had take place.

Under temperate environmental conditions, males are heavier at birth than females. The Messina Shorthorns produced bull calves that were lighter than the heifers at birth. The reason for this is not quite clear but may be due to a higher metabolic rate of the male foetus, which may impose a greater metabolic stress on the cow with a low heat tolerance when she is pregnant during the summer. However, this has never been proved experimently. The foetal adrenal is strongly implicated in terminating gestation (Van Rensburg, 1967; Liggins, 1968, 1969). Gestation length of cows was considerably longer at the Messina research station (12 days) suggesting the possibility of some form of adrenal and thyroid malfunction. The effect of the marked anorexia exhibited by the Messina Shorthorns (Bonsma and Louw, 1966) was not appreciated, however, in sheep it did not appear to be of any importance (Yeates, 1958). According to Joubert (1954) it can vary within a wide range before it affects birth mass of calves.

HIGH TEMPERATURES AND MALE FERTILITY

Very high ambient temperatures 27° C and higher can disrupt spermatogenesis (Bonsma et al., 1940; Skinner and Louw, 1966). Africander bulls were more resistant to a tropical climate and were able to produce more calves than bulls from the British breeds. There is little doubt that lowered male fertility as a result of the influence of climate is responsible for a great deal of the generally poor reproductive performance of cattle in tropical areas. A recent investigation into the fertility problem of Shorthorn bulls in an area where the average annual isotherm exceeds 18° C resulted in the culling of eleven out of twelve Shorthorn bulls, either on the basis of poor semen quality or as a result of clinical abnormalities of the reproductive organs. On the other hand only one out of ten Africander bulls was culled. According to Bonsma (1949) Africander bulls at the Mara Research Station gave an average servicable life of 8 yrs and 10 months compared to 3 yr and 3 months for bulls of the British beef breeds.

Certain tickborne diseases, resulting in an elevated body temperature for some time often adversely affects the fertility of bulls under field conditions. Likewise elephant skin disease can lower calving rates if the bulls are affected. The scrotal skin of affected bulls is left hard and inelastic, and such bulls are unable to maintain a satisfactory testicular abdominal temperature gradient.

Table 8. Comparison of birthmass and gestation lengths of Shorthorn and Shorthorn-Africander calves at Messina Research Station in the subtropics and Dohne Research Station in a temperate region (Bonsma et al., 1972)

Breed	Born in	Sex	No.	Birthweight ± ± S.E. (kg)	Range (kg)	No.	Gestation lenght ± S.E. (days)	Range (days)
Shorthorns at Messina	Winter	Male	9	18,4 ± 2,9	9,5–27,3	6	289,9 ± 1,9	286–199
		Female	6	20,3 ± 10,8	9,5–27,3	2	295,5	281 & 301
	Summer	Male	1	22,7		–	–	–
		Female	4	22,8 ± 2,7	16,4–25,0	–	–	–
Messina								
All Shorthorns			20	20,1 ± 3,2	9,5–27,3	8	290,0 ± 1,9	281–301
Shorthorns at Dohne*	Autumn	Male	23	31,9 ± 0,7				
		Female	22	30,4 ± 0,7				
	Spring	Male	66	30,6 ± 0,5				
		Female	74	29,4 ± 0,4				
All Dohne Shorthorns			185	30,0 ± 0,4	–	121	278,9 ± 1,0	277–286
½ Shorthorn	Winter	Male	1	20,4	20,4	–	–	–
½ Africander		Female	3	28,8 ± 3,1	22,7–31,8	2	297	293 & 301
at	Summer	Male	2	29,5	25,0 & 34,1	–	–	–
Messina		Female	2	27,3	25,0 & 29,5	2	292	283 & 301
¾ Shorthorn:	Summer	Female	2	26,1	25,0 & 27,3	1	288	288
¼ Africander								
Messina			10	27,3 ± 2,6	20,4–34,1	5	293,2 ± 2,5	283–301
All corsses								

* Data provided by E. J. B. Bishop.

Effective performance testing of young beef bulls dictates that bulls should be tested as early as possible on a relatively high energy ration. Although no significant deterious effect of such treatment on the reproductive capacity of young Bonsmara bulls was found, (Venter, 1970) bulls on a high plane of nutrition tended to have lowered concentrations of seminal fructose and citric acid indicating lowered testosterone output. The percentage of abnormal sperm was also consistently higher for the bulls on the high plane of nutrition. Skinner et al. (Unpubl.) noticed that abnormalities increased in the ejaculates of overfed Hereford bulls and were similar to those induced by high temperature. They concluded that the effects of overfeeding could be attributed to fat accumulating in the scrotum which prevented the bulls from maintaining an adequate testicular temperature gradient resulting from insulation of the testis and the veins in the pampimform plescus.

REFERENCES: See pp. 572-573.

Effects of Heat on Reproduction of Sheep

by

C. W. Alliston

INTRODUCTION

The seasonal breeding patterns of mammalian species have been extensively reviewed by Asdell (1946). Among domestic animals, sheep typically exhibit a more pronounced seasonal breeding pattern than most species. Such responses to environmental conditions and seasonal variations of reproductive efficiency no doubt prompted some of the earlier studies concerning effects of the environment upon the reproductive process.

Reviews of the effect of high temperature on reproduction have been compiled (Ulberg, 1958; Dutt, 1960; Moule, 1970) and indicate that temperature-humidity conditions occurring in many geographical areas during the breeding season are of practical concern because of decreased reproductive efficiency. Subsequent reports by a number of independent workers have been directed toward elucidation of physiological mechanisms by which such effects occur. Broadly speaking, the evidence indicates that either the male or female role in the reproductive process may be adversely influenced by hyperthermia and that sensitivity is greatest during the very early stages of gestation. Such influences may occur via (a) "direct effects" of maternal or paternal hyperthermia per se upon the gamete and/or zygote, and (b) "indirect effects" upon the gamete and/or zygote mediated through alterations of maternal or paternal physiology such as the endocrine system (Fig. 1) or a combination of both.

SPECIAL INSTRUMENTATION AND METHODOLOGY

A number of studies has utilized climatic laboratories with varying degrees of sophistication to avoid the vagaries of atmospheric conditions. Design considerations for such facilities have been discussed (Smith et al., 1971) as have bioclimatic factors and their measurement (Kelly and Bond, 1971). Relevant information has also been obtained by in vitro culture techniques alone or in combination with ovum transfer techniques (e.g., Alliston et al., 1965). Such techniques, of course, require careful design, execution and interpretation when extrapolating to naturally occurring circumstances. To facilitate such studies, exogeneous hormones are frequently used for estrous synchronization (e.g., Thwaites, 1969). Such methodology, although frequently necessary, also requires careful application to avoid confounding of results. For example, if gonadotropin levels are altered by high temperature (Miller and Alliston, 1973) but injections of those hormones are utilized for control of ovarian function, the effect would not be detectable.

In general the effects of interest here are associated with hyperthermia of one or more

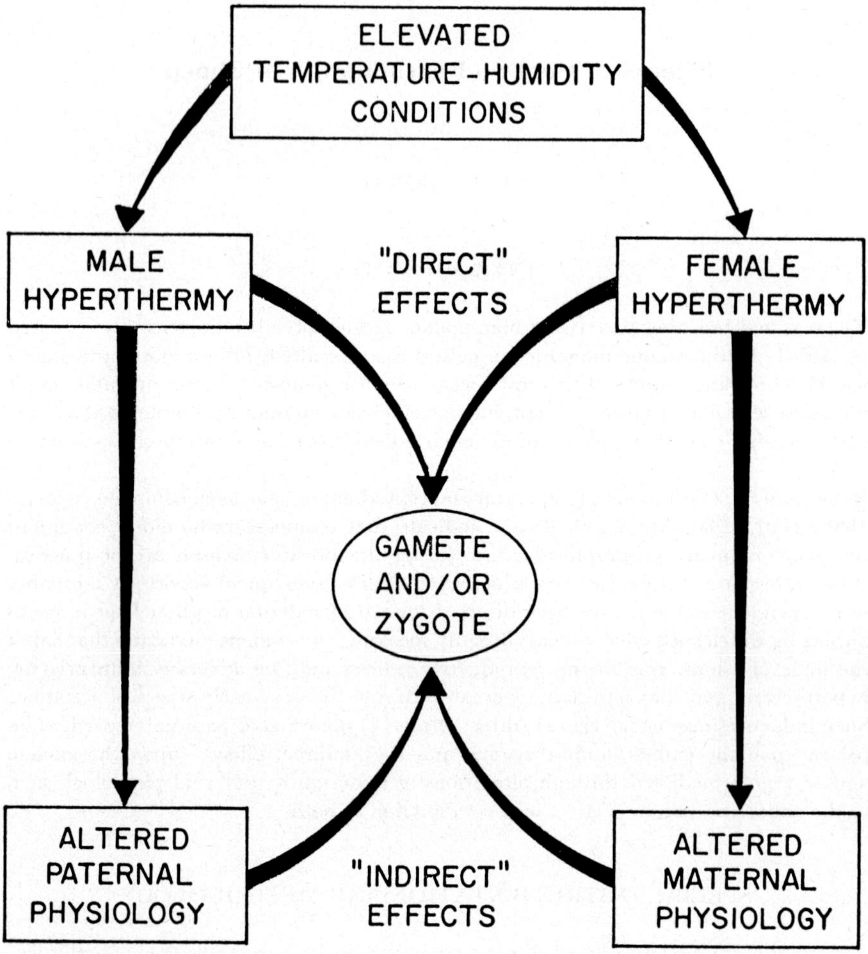

Fig. 1. "Direct" and "indirect" effects of temperature-humidity conditions on the reproductive process.

organs (e.g., testes) or the entire body. Thermoregulatory processes have recently been discussed (Bligh, 1973) and will not be considered here.

INFLUENCE OF TEMPERATURE ON THE MALE

A. SPERMATOGENESIS

Decreased spermatogenesis in the ram during summer months (McKenzie and Berliner, 1937) has also been observed within controlled temperature facilities and studied in detail (Dutt and Hamm, 1957; Moule and Waites, 1963). Other studies have demonstrated that artificial cooling (e.g., Egli et al., 1956) or shearing the ram (Hulet et al., 1956) results in higher quality semen and increased fertilization rates (Dutt and Simpson, 1957; Egli et al., 1956). Temperatures above approximately 26° C appear to reduce semen quality in the ram (Brooks and Ross, 1962). Entwistle (1973) has observed marked variations within groups of rams in hot tropical environment and suggests that the effectiveness of thermoregulatory function by the scrotum may serve as an index for selection.

Evidence from other species indicates that spermatocytes and spermatids are particularly susceptible to high temperature in the guinea pig (Young, 1927) and bull (Casady et al., 1953).

B. EFFECTS VIA FEMALE REPRODUCTIVE SYSTEM

Evidence also indicates that thermal or thermally related conditions encountered within the female reproductive tract adversely influence spermatozoa (Howarth et al., 1965). Rabbit spermatozoa capacitated within normothermic or hyperthermic females were deposited into the oviduct of test females. Fertilization was not impaired, but embryo survival was significantly depressed with semen capacitated within the hyperthermic female. The mechanism for such effects was not established, but it was speculated without experimental evidence that the changes were associated with altered function of nucleic acids. Such evidence appears to be a "direct effect" of elevated temperatures upon the male gamete.

C. ENDOCRINE FUNCTION

Observations supporting an "indirect effect" via altered endocrine function in the ram are less evident. The reproductive consequences of endocrine system responses to high temperature (Collins and Weiner, 1968) have not been elucidated. Whiteman and Brown (1959) suggested that improved performance by "cooled" rams was, in part, because of more matings completed. However, Pepelko and Clegg (1965) observed that the largest number of ejaculations occurred during the fall. The "mounts" per ejaculate were greatest in winter and least during the summer. Steinbach (1973) has discussed the decreased libido in boars and its correlation with effective temperature. However, Wettemann et al. (1973) failed to observe significant decreases of testosterone at temperatures that decreased semen quality of the boar.

INFLUENCE OF TEMPERATURE ON THE FEMALE

A. SENSITIVE STAGES OF REPRODUCTION

Subsequent to the observations that atmospheric conditions depress reproductive performance, considerable effort was directed toward establishing the sensitive stage or stages of reproduction within the female. Yeates (1953) noted adverse effects upon gestation following exposure of ewes to a temperature of 107° F for 7 hr per day beginning the day after mating. The effect was greater when ewes were subjected to the treatment throughout gestation than if only the last two-thirds or one-third of pregnancy. Artificial cooling (65° F) for the first 17 days after mating reduced embryonic mortality within ewes bred during the summer in southern U.S.A. (Alliston et al., 1961).

A series of observations in which ewes were subjected to 32° C at different stages of reproduction has been summarized by Dutt (1960). A significant increase of unfertilized and abnormal eggs occurred when treatment was initiated prior to mating. When exposure to the high temperature was delayed until eight days post-breeding, the embryo loss did not differ significantly from control females. However, in ewes exposed to 32° C immediately after ovulation, a significant increase in embryo mortality was observed. These observations were interpreted to mean the early stages of gestation (i.e., prior to entry into the uterus) were especially sensitive to high ambient temperatures.

Alliston and Ulberg (1961) utilized egg transfer techniques to establish that eggs within ewes at an ambient temperature of 32° C had been adversely affected by three days after mating. When transferred to recipient ewes at 21° C their survival was not enhanced. Woody and Ulberg (1964) did not observe such an effect when one-celled ova were subjected to comparable conditions. Therefore, it appears that in sheep a particularly sensitive interval lies between ovulation and three days after mating.

In vitro culture techniques indicate that in the rabbit, one-cell fertilized eggs are more susceptible to the "direct effect" of maternal hyperthermia than are two-cell eggs. This conclusion was based upon culture at temperatures corresponding to normothermic (38° C) and hyperthermic (40° C) female rabbits at 21 and 32° C ambient temperatures, respectively (Alliston et al., 1965).

Adverse effects of high temperature are also evident during later stages of gestation. Significant depression of birth weight and lamb survival has been reported in a number of observations (Yeates, 1953, 1956, 1958; Shelton and Huston, 1968). Similar effects have been noted in unadapted cattle (Venter et al., 1973). Although such studies define the period of thermal sensitivity rather precisely and provide evidence for the "direct effect", they provide little insight into the physiological mechanisms associated with embryo mortality.

B. PHYSIOLOGICAL MECHANISMS

Studies have been conducted with mice (Elliott, 1968; Bellvé, 1972) and rabbits

Table 1. Incorporation of [^{3}H]uridine by fertilized rabbit eggs during culture in vitro for 4 hr at 39.0 and 40.3° C

Cell stage at recovery	Culture Temperature (°C)	No. of determinations	Uridine incorporation	
			Ct/min/egg	Difference ± S.E.
One-cell	39.0[a]	21	2.88	
	40.3[b]	21	1.68	−1.20 ± 0.26*
Two-cell	39.0[a]	10	3.09	
	40.3[b]	10	3.01	−0.08 ± 0.13

* $P < 0.01$.
[a] Mean rectal temperature of females at 21°C ambient temperature.
[b] Mean rectal temperature of females at 32°C ambient temperature.

(Criblez and Alliston, 1974) to evaluate altered incorporation of ribonucleic acid (RNA) precursors at high temperatures. Decreased uridine uptake has been noted in such studies and the decrease was more pronounced in one-cell rabbit eggs than in two-cell eggs (Table 1) (Criblez and Alliston, 1974). This observation agrees with the greater sensitivity of the one-cell egg to thermal conditions. Metabolic activity (as measured by carbon dioxide production from [^{14}C]pyruvic acid) was not decreased but was greater at the higher temperature as would be expected. Therefore, the observed changes in uptake of precursors are not explained by an overall decrease in metabolic activity. Uridine incorporation subsequent to culture at the higher temperature was also impaired. In mammalian cells approximately 80% of the total amount of RNA formed leaves the nucleus (Shearer and McCarthy, 1967) and has been implicated in the regulation of gene transcription. Impaired cell regulation may therefore be at least a partial explanation for the "direct effect" of maternal hyperthermia. If relatively severe, embryonic mortality as described above may be the result. Perhaps less severe impairment might result in congenital defects in young born subsequent to maternal hyperthermia during gestation as has been observed by Edwards (1969a, 1969b, 1971). Incorporation of [^{3}H]leucine into an acid precipitable material is increased subsequent to heat stressing sperm prior to mating or the embryo during the first cleavage (Sheean et al., 1973).

Many of the early studies were conducted in laboratories utilizing constant temperatures that do not duplicate diurnal temperature patterns. Rich and Alliston (1970b) reported that Columbia and Rambouillet ewes subjected to a diurnal temperature program of 21 to 32° C had significant increases of rectal temperature and respiration rates during the hot portion of the temperature cycle. However, reproductive performance was not significantly decreased. Thwaites (1969) also reported "mitigating effects of a daily respite from high temperatures" in Merino ewes.

C. ENDOCRINE FUNCTION

Alteration of endocrine function has long been hypothesized as at least a contributing factor in hot environments. Evaluation of such endocrine function has not been an objective of most studies related to reproductive performance in hot environments. The relatively recent availability of improved assay techniques has increased interest in such responses. However, the information currently available appears somewhat inconclusive. Dunn et al. (1972) failed to detect significant changes of progesterone levels within ewes in a programmed diurnal cycle of 21 to 32° C. Possible involvement of the adrenal gland is indicated by increased embryonic mortality in ewes injected with hydrocortisone acetate (Howarth and Hawk, 1968). In addition, Tilton et al. (1972) have demonstrated that intact heat stressed ewes had less embryonic death loss than adrenalectomized ewes maintained with daily injections of hydrocortisone. Plasma levels of corticoids were not determined in these studies.

Observations on Angus heifers (Miller and Alliston, 1974a, b, c) did not reveal significant alteration of progesterone or corticoid levels. However, basal and ovulatory levels of luteinizing hormone were significantly depressed. In contrast, Stott and Wiersma (1973) have reported significant decreases of both corticoids and progesterone in dairy cows. Breed differences in response to high temperatures have been reported

(Abilay et al., 1973). Decreased corticoid levels have been reported in thermally stressed Hereford bulls (Rhynes and Ewing, 1973). Level of production and environmental conditions alters plasma thyrotropin (Goret et al., 1973) and glucocorticoids (Vanjonack et al., 1973). Ryle (1961) utilized injections of L-thyroxine to overcome adverse effects of temperature. However, others (Howarth and Hawk, 1968) noted no effect of such therapy. If altered thyroxine level is an adaptive mechanism for high temperatures, therapeutic administration of thyroxine may in fact be undesirable because of a possible elevation of body temperature.

It appears probable that acclimatization is responsible for much of the variation that has been observed in environmental studies related to reproductive or endocrine function. Rich and Alliston (1970a) have reported, for example, some variations of reproduction associated with acclimatization of rabbits.

The complex alterations of endocrine function occurring during exposure to high temperature (Collins and Weiner, 1968) appear to be poorly understood with respect to known and possible interactions with the reproductive process. The status of acclimatization and its probable effects have frequently been disregarded in studies such as those discussed. Such disregard may cause results that appear contradictory whereas one response was chronic and one acute. There appears to be little evidence to adequately access the practical effects of sudden periods of hot weather in contrast to extended periods of similar conditions.

CONCLUSIONS

The available information indicates that the following conclusions can be made with regard to sheep and high temperature:

1. Such conditions adversely influence reproduction in the male and female;
2. Within the reproductive system, maternal or paternal hyperthermia per se exerts a "direct effect" on the gamete and/or zygote;
3. In addition, "indirect effects" of hyperthermia upon endocrine, metabolic or homeostatic mechanisms may alter reproductive processes;
4. Within the female, the initial stages of gestation are particularly sensitive;
5. Additional information is needed to accurately evaluate the reproductive significance of altered endocrine function.

REFERENCES: See pp. 574-575.

Breeding Management of Sheep

by

N. T. M. Yeates

INTRODUCTION

In the previous Section it has already been shown what serious effects high temperature may have on reproduction in both male and female sheep.
In the male, spermatogenesis can be interfered with to varying degrees, according to the severity and duration of heat exposure. Thus very hot weather of several days' duration has long been known to cause complete seminal degeneration for many weeks (Phillips and McKenzie, 1934), while as little as a few hours' exposure to only moderate heat might cause marginal damage – evidenced by slight morphological changes in a proportion of spermatozoa (Rathore, 1970), with perhaps a lowered level of sustained pregnancy following conceptions from such semen (Braden and Mattner, 1970).

As far as the female is concerned, acute exposure to severe heat early in pregnancy is likely to result in embryo resorption (Thwaites, 1971), while chronic exposure to less severe heat over a greater part of pregnancy may lead to the birth of dwarf lambs (Yeates, 1958; Shelton and Huston, 1968).

Fortunately, however, the breeds of sheep which are most susceptible to high temperature are also the ones which, in the course of evolution have become best "protected" against the harmful effect of heat on reproduction. This protection is explained by the animals' now well-known photoperiodic sensitivity (Thwaites, 1965). Thus the higher the latitude of origin of a breed, the more reactive it is to change in daylength; and since decreasing photoperiod initiates the sexual season, the more likely such animals are to conceive and gestate in the autumn and winter (i.e., the cool) months (Hafez, 1952). Under normal circumstances, breeding at a time of year which would be adverse (too hot) to the animals is precluded by their close adherence to a pattern dictated by photoperiod. On the other hand, breeds whose origin is nearer to the equator are less reactive to photoperiod and hence have a more extended breeding season. Although this renders them more liable to conceive and/or gestate in hotter months, the problem is not necessarily too serious because of their relatively good adaptation to heat.

All this means that provided each breed of sheep is maintained in its "natural" geographical environment, and provided breeding is allowed to occur at the season of year physiologically appropriate to that environment, no serious trouble from heat effects in either male or female reproduction should be encountered.

If, on the other hand, man "interferes" with the natural evolutionary pattern in certain ways, hoping thereby to improve production, heat effects may well occur which could offset any apparent benefits of the changes. Since such strategies are all a part of sheep breeding management, we need to examine some of the most likely trouble points; several will now, therefore, be considered.

PROBLEMS OF SHEEP BREEDING MANAGEMENT

A. EXOGENOUS HORMONES FOR OUT-OF-SEASON BREEDING

Techniques are now well enough advanced to allow reasonable efficiency in extra-seasonal lamb production, simply by hormone treatment (Robinson, 1974). This has great appeal in-so-far as it opens the possibility of producing two lamb-crops a year instead of only one, or perhaps supplying lambs in the "off" season when prices are highest. Clearly, however, to the extent that summer pregnancy becomes involved, so does the likelihood of seminal degeneration, embryo resorption and/or foetal dwarfing increase.

B. USE OF LIGHTS FOR OUT-OF-SEASON BREEDING

In earlier years any photoperiodic experiments aimed at modifying the sheep's breeding season involved the expense and trouble of housing the animals (Yeates, 1949). This was mainly because some form of black-out arrangement was used for the believed necessary reduction of daylength during part of the year. Now, however, it appears that lambing frequency may be increased photoperiodically without ever having to use black-out and hence indoor confinement (Ducker and Bowman, 1972). This appears to have enhanced considerably the practical possibilities of photoperiodic manipulation as a technique for intensifying sheep breeding (Fig. 1). However, the same problem as was mentioned for hormone induction of breeding needs to be considered – namely, adverse effects of heat on spermatogenesis or on females caused to gestate during the summer.

C. LATITUDINAL TRANSLOCATION

Two types of problem arise under this heading. One is that of a reasonably heat tolerant and long-breeding-season breed (such as the Merino) being used in a tropical location, where, because of its long breeding season, spring conception is often encouraged and in fact occurs (Entwistle, 1970). Unfortunately, despite the breed's good general heat tolerance, adverse effects of summer reproduction are nevertheless likely owing to the very high temperatures which so often prevail. The other type involves the movement of the much more heat sensitive British breeds of sheep to locations of even moderately warmer climate than that to which the breeds are adapted. There, even though breeding might be confined to the calendar months "normal" for the breed, sperm abnormalities could quite easily be encountered, especially if early-season matings are attempted (Hulet et al., 1956).

AMELIORATIVE MEASURES

Numerous management strategies for combatting the adverse effects of high temperature on reproduction are available. The method or methods of choice will vary according

Fig. 1. Crossbred ewes under out-door night lighting. Advances in knowledge of photoperiodism now make out-of-season breeding a practical possibility; but, to the extent that this involves reproduction during the warmer months of the year, adverse effects of high temperature on male or female fertility become likely.

to particular circumstances, so only some examples will be given here. A decision on which method is "best" in any set of circumstances can only be made against a sound understanding of the physiology involved.

The choice of a heat tolerant breed is important. In this connection, critical studies on relative breed performance in hot climates are lacking. From studies such as those of Lee (1950), Yeates (1956) and Symington (1960), however, as well as from general observation, the following tentative generalisation may be made. The British breeds are least heat tolerant, followed in order among the better-known wool breeds by the Corriedale, Polwarth and Merino; and among the meat breeds by the Dorper, Blackhead Persian and hairy desert breeds (for hot-dry environments), and breeds such as the Barbados (Fig. 2) for tropical maritime or other warm humid environments.

Recognition of the physiologically "correct" time of year to mate particular breeds in particular environments is necessary. If a breed is to be mated for some special reason outside its normal season and/or latitude, then the consequences of any adverse effects on reproduction should be weighed up against the hoped – for advantages – recognising, however, that ameliorative measures of the type mentioned in this section might be helpful in offsetting some of the difficulty.

The provision of shade, either by strategic planting of trees or positioning of artificial

Fig. 2. A Barbados Black-bellied ewe. Of West African origin (Mason, 1969), this breed is being crossed with British breeds in U.S.A. to produce a meat-type sheep of sufficient heat tolerance for unimpaired fertility in even the hottest months of the year.

shelters can be very important, both for the direct relief it affords the animals, and for its cooling effect on the drinking water (Morgan et al., 1972).

Rams should not be driven or even yarded in hot weather at times of the year when overheating from such treatment could be critical to their sperm quality at the mating period. In this regard it should be appreciated that there is generally a lag of about two weeks between overheating and the first appearance of affected sperm in ejaculated semen; also that once sperm damage has resulted, many weeks under heat-free conditions are required for the regenerative processes.

Among the wool breeds, especially if they are to be called upon to reproduce outside their physiologically "normal" period, fleece length (and hence time of year chosen for shearing) can be important. Hot-room trials leave no doubt that the fleece impedes cutaneous evaporation of sweat, and that its removal (shearing) improves heat tolerance (Dutt and Hamm, 1957). In the field, however, close shearing is contra-indicated in hot dry climates where solar radiation is high, especially if shade is either scanty or absent (Macfarlane et al., 1958). The general recommendation, (from the point of view of "safe" reproduction in hot climates), therefore, is to shear rams and ewes prior to mating – removing ALL the wool in hot humid climates (where solar radiation is rarely significant); but ensuring that a depth of at least 4 cm of wool (Parer, 1963) is retained on sheep which will be exposed to strong solar radiation. These arrangements promote maximum evaporation in the humid conditions where heat dissipation by this avenue is essential; while under hot dry ambient conditions the best compromise is reached of facilitating reasonable evaporative opportunity, yet still ensuring protection against skin damage on the one hand and direct radiant heat gain on the other.

The question of closeness of shearing of rams' scrota is an interesting one. A 5-7° C abdomino-testicular temperature gradient is absolutely critical to fertility, and maintenance of that temperature difference under hot conditions is made possible by two processes: evaporative cooling at the scrotal skin surface, and operation of an efficient counter-current heat exchange mechanism in the vasculature of the testis and spermatic cord (Harrison, 1949). Since evaporation should be unimpeded, shearing of woolled scrota would seem to be an obvious management recommendation; and since all but the more pendulous scrota (see below) should be sufficiently shaded by the animals' bodies, close shearing is also probably indicated.

As a longer term policy, selection against rams whose scrota are either too heavily woolled or too pendulous could have merit. On the first of these two points, to avoid any possibility of prejudicing fleece weight, it would need to be confirmed that there is no genetic correlation between scrotal and overall body wool production. Subject to this, and to the character having high enough heritability, selection could be based on a system of scrotal wool scoring such as that proposed by Rathore and George (1969). On the second point (over pendulous scrotum), the character's heritability would require determination before pinning faith in selection against it. In general, however, it can already be said that over-pendulous scrota are disadvantageous for the two reasons that they render the testes more liable to mechanical injury and more exposed to direct solar radiation. Besides this, the condition is probably associated with lack of tone in, or even complete incompetence of the cremaster and dartos muscle retraction systems – possibly resulting from past over-compensation in poorly heat-adapted animals. Culling of such animals is therefore indicated on several grounds, irrespective of any genetic involvement.

Within the Merino breed, selection of plain bodied rather than wrinkle-skinned stock is recommended as one means of lessening susceptibility to seminal degeneration (Fowler and Kennedy, 1968) and increasing lambing percentage (McGuirk, 1969).

FUTURE RESEARCH

Some areas of obvious deficiency in knowledge in the high temperature/reproduction field will now be listed and commented on briefly.

1. The causal mechanism in embryo resorption and foetal dwarfing is poorly understood, and questions such as the following require answers. Are the effects mediated through the maternal system or does the heat act directly on the foetal elements? Is blood supply to the uterus involved? Is some "substance" produced in over- or under-sufficiency – and if so, what is its nature? Are individual differences in animals' susceptibility due merely to corresponding differences in general heat tolerance, or to some other character?

2. The "boundary" between normal and heat affected sperm requires study. Can heat damage occur without ANY morphological change in ANY of the spermatozoa being evident using light microscopy? If so, how can marginally affected semen be identified? What are the marginal changes which permit fertilization, yet are inimical to continuance of pregnancy?
3. The relationship between scrotal character and hot climate ram fertility needs clarification. What is the significance of first, scrotal wool cover and second, the over-pendulous scrotum? What is the definition of "over-pendulousness"? Are both characters heritable – or is the second merely a result or symptom of some physical weakness or physiological deficiency?
4. Comparative breed studies are needed to define the relative suitabilities of the various breeds of sheep for use in regions (or times of the year) of climatic difficulty.

CONCLUSION

Reproductive problems due to heat are unlikely to be encountered if each breed of sheep is managed in accordance with two strict requirements which evolution has set: (a) location in the appropriate climatic milieu, particularly with regard to temperature and light; (b) confining mating to the physiologically normal season of the year.

It is only when man "interferes", either by inappropriate location or by breeding sheep out of season that problems arise. However, since progress in animal husbandry is measured by man's rather than his animals' gains, latitudinal translocation and/or out-of-season breeding are legitimate procedures if they give an overall improvement in production.

To help in achieving such gain in the present context, numerous management strategies are available, which include selecting suitably tolerant stock; treating them gently during the hot hours of the day; and providing conditions (of shade and fleece length, for example) which will assist them to preserve normal body temperature. In the longer term, further help should be available through better understanding of the physiology of heat-induced reproductive wastage, and clearer insight into the inheritance of the factors involved.

REFERENCES: See p. 576.

Effects of Heat on Reproduction of Swine

by

J. Steinbach

INTRODUCTION

One third of the world pig population is found in the tropics, at temperatures constantly above the zone of thermal indifference for this species which, unable to sweat, is highly sensitive to the effects of high ambient temperatures. Nevertheless, there are comparatively few studies about the reproductive performance of heat-stressed pigs. An attempt will be made in this review to present the recent developments in the field of reproductive physiology of the male and female pig under conditions of heat-stress.

HERD FERTILITY

It has been shown (Arganosa et al., 1969; Steinbach, 1973) that high fertility rates, two litters per sow p.a., can be obtained under natural conditions in the tropics. Seasonal effects on farrowing rates, indicating a depressing effect of high summer temperatures, were reported from France (Signoret and Du Mesnil du Buisson, 1968), but were not observed in Italy (Magnani, 1966). In Nigeria, service period and farrowing interval tended to be shorter, when sows farrowed during the dry and hot season, than during the cooler and more humid period of the year, which may indicate that high ambient temperatures depress fertility to a lesser degree in pigs adapted to high ambient conditions (Egbunike and Steinbach, unpubl.), although acutely heat-stressed gilts showed a reduction in pregnancy rate (Teague et al., 1968; Mahoney et al., 1970).

Sow fecundity is but slightly depressed by high ambient temperatures. The total number of piglets born in a litter does not differ significantly between tropical (Gupta et al., 1967; Moulick et al., 1968; Arganosa et al., 1970; Steinbach, 1971) and temperate sow herds (Legault, 1969; Rasbech, 1969), and seasonal changes in the number of piglets born alive have not been observed (Tanhueco and Rigor, 1964; Magnani, 1966; Mathai and Ramachandran, 1972; Steinbach, 1973). Ahlschwede and Robinson (1966) and Steinbach (1968) have shown that only 2-4% of the variability in litter size can be attributed to climatic changes. The number of still-born piglets was found to vary seasonally with the ambient temperature towards the end of pregnancy (Steinbach, 1971). Whatley et al. (1957) were able to reduce the percentage of stillborn piglets at high ambient temperatures considerably by sprinkling the sows with water.

Piglet weights at birth are comparable to those obtainable in temperate countries (Steinbach, 1971), but weaning weights fall far below accepted standards (Abilay and Arganosa, 1962; Arganosa and Penalba, 1971; Steinbach, 1971), possibly due to high temperatures, since the correlation between litter or piglet weight gain from birth to 35 days and ambient temperature was found to be negative (Steinbach, 1971). Whatley et al. (1957) using sprinklers on sows during high ambient temperatures succeeded in reducing rectal temperatures to normal levels and simultaneously improved piglet survival by 4% and litter weight at eight weeks by 39%.

The preweaning mortality of piglets is comparable in tropical and temperate areas (Vandenbranden, 1960; Abilay and Arganosa, 1962; Fraser, 1966a; Rasbech, 1969). It appears then, that only two aspects of the reproductive efficiency of swine herds are affected significantly by high ambient temperatures, the rate of stillborn piglets and the preweaning growth rate.

MALE REPRODUCTION

The reduction in farrowing rate, reported by Thibault et al. (1966) and by Speer (cited by Teague, 1970), may have been due to fertility effects in either the male or the female pig, but subsequent findings of Signoret and Du Mesnil du Buisson (1968), that artificial cooling of the boars will restore fertility to adequate levels, indicate the boar's involvement.

A. SEXUAL DEVELOPMENT

Recently it was shown (Egbunike and Steinbach, 1972) that boars raised in a tropical environment have sigmoid growth curves in testis weight and in several parameters of testis function, which are similar in shape to those in temperate climates, although the growth spurt and sexual maturity were delayed by approximately 2 and 3 months, respectively. Since the rate of liveweight gain (Chapt. 1, Sect. 12b) was severely depressed in this study, the shift in the rate of increase in daily sperm production towards an older age can be explained by an inadequate body development. This contention is supported by the fact that the daily sperm production expressed per unit body weight was even higher in Nigeria than in Germany (Egbunike et al., unpubl.), probably due to the older age at a given body weight of the animals reared in the tropics.

B. SEXUAL ACTIVITY

The libido of boars trained for semen collection has been found to vary with season. Niwa and Mizuho (1954) observed reduced sexual interest of boars during the summer, and Steinbach (1972a) noted a higher incidence of refuses to mount the dummy during the hot season. Ejaculation time was also prolonged under conditions of heat stress both diurnally and seasonally (Steinbach, 1972a). On the other hand, Wettemann et al. (1973) did not note a difference in sexual behaviour between boars kept in climatic chambers at 23 and 32° C for 90 days, but their finding that the plasma testosterone levels were, though not significantly, depressed by 22% at 32° C is of interest in view of the reduced libido discussed above.

C. SPERMATOGENESIS

Few studies have been carried out measuring possible seasonal changes in spermatozoa production in the porcine testis. Niwa and Mizuho (1954) were unable to detect seasonal changes in the spermatogenic activity of the testis, while a study by Egbunike et al. (unpubl.) indicated even a non-significant increase in the daily sperm production/g testicular parenchyma in March and April, the end of the hot season.

Dramatic changes in spermatogenic activity in a naturally varying environment should not be expected in the light of some more recent studies. Mazzarri (1971) reported that acute local heating of the boar's scrotum to 44° C for three hours, thus increasing the intratesticular temperature from 35.5 to 39.5° C, did not affect spermatogenesis, and an ambient temperature of 40° C raised the deep-testis temperature in another study only from 36.4 to 37.0° C (McNitt et al., 1972). Only when the testicular temperature had been raised to more than 40° C by heating the scrotum to 48° C was spermatogenesis impaired (Mazzarri, 1971). Within 24 hrs, the number of spermatocytes in the late pachytene and diplotene phase as well as round spermatids were reduced, leading to a reduction in ejaculated sperm numbers and fertility 2-3 weeks later. The recovery phase did not start until two months after the acute heat treatment.

D. CHARACTERISTICS OF THE EJACULATE

The findings on the effects of acute local heat stress cited above are, however, somewhat in variance with results obtained by collecting semen following artificially or naturally elevated temperatures. Lawrence et al. (1970) and Steinbach (1972b) reported significant seasonal decreases in sperm cell concentrations, and the total number of spermatozoa ejaculated decreased significantly by 4 billion for each °C increase in ambient temperature two months prior to ejaculation (Steinbach, 1972b). These data suggest a time-lag, which has also been observed in studies using acute heat treatment for three days (McNitt et al., 1970; Christenson et al., 1972). In those studies, similar to Mazzarri's experiment (1971), the number of ejaculated spermatozoa was not affected during the first two weeks after the heat treatment, declined thereafter for two weeks, remained low until about eight weeks after the period of heat stress and had completely recovered 2-3 months after the treatment. The reduction in ejaculated sperm ranged between 2.5 (Wettemann et al., 1973) and 3.0 billion spermatozoa (Christenson et al., 1972) for each increase in temperature by one °C.

Motility was generally not affected by heat stress in a naturally varying environment (Wagelie et al., 1959; Lawrence et al., 1970; Steinbach, 1972b), but was considerably suppressed following acute heat treatment at an ambient temperature of about 33° C (McNitt, 1970; Christenson et al., 1972; Wettemann et al., 1973), again reaching lowest levels of about 50 percent between 16 and 45 days post treatment.

Similarly, the percentage of abnormal cells increased rapidly from 4% prior to the heat treatment to 24% 4-6 weeks after the temperature stress in controlled experiments (McNitt et al., 1970; Christenson et al., 1972), but no seasonal variation that could be attributed to high ambient temperatures was found in boars maintained under natural conditions (Lawrence et al., 1970; Olutogun and Steinbach, 1973). Findings by Wettemann et al. (1973) suggest a degree of adaptation, since acrosomal abnormalities were increased only during the period 2-5 weeks after the start of the treatment and returned to a normal level for the remaining experimental period of eight weeks.

Unfortunately, no data on the biochemical composition of ejaculates from heat-stressed boars have been published.
The results compiled above indicate that sperm production rates decrease following prolonged moderate heat stress, even though the intra-testicular temperatures are not likely to reach the critical levels postulated by Mazzarri (1971) under natural conditions (McNitt et al., 1972).

FEMALE REPRODUCTION

As the boar, the sow may be partly responsible for the seasonal changes in farrowing rate observed in France (Thibault et al., 1970) and in the United States (Speer, cited by Teague, 1970). Corteel et al. (1964) have reported a seasonal increase in the frequency of abnormal cycles (up to 45 days) which paralleled the average maximal daily temperature, indicating an increase in embryonic mortality between June and September.

In the sow, reproductive efficiency may be affected through direct or indirect effects of high ambient temperatures on age and weight at puberty, on oestrous cycle characteristics and ovulation and, subsequently, on pregnancy and lactation.

A. SEXUAL DEVELOPMENT

Seasonal changes in the attainment of puberty have been observed in the humid-tropical climate of Southern Nigeria (Steinbach, 1973). Gilts born before the beginning of the cool rainy season gained faster and had their first oestrus at a higher weight (81 kg) and a younger age (224 days) than those born at the onset of the hot and dry season (70 kg and 256 days, respectively), indicating that the gilts which spent their major growth phase at ambient temperatures above 28° C showed retarded body development leading to a delay of sexual maturation. These findings may be attributed to the reduction in food intake and growth rate, as discussed earlier (Chapt. 1, Sect. 12b). One interesting aspect of this investigation was that gilts born during the cool season gained less during the following hot season; but even though their body weight was lower, they attained puberty at 224 days during the following spring. Earlier, Wiggins et al. (1950) observed the lowest proportion of immature gilts at slaughter houses in April (15%) and the highest in October (45%), also pointing to an accelerated maturation at periods of increasing daylight.

There is some evidence of the effect of continuous heat stress retarding directly the expression of oestrus. Jensen et al. (1970) observed oestrous symptoms only in 2 out of 10 heat-stressed gilts of about 10 months of age; however, within two weeks following removal from the heat treatment, the remaining gilts had shown heat and were bred.

B. THE OESTRUS CYCLE AND OVULATION

During the reproductive cycle, high ambient temperatures could disturb behavioural patterns and impair the gametogenic and endocrine functions of the ovary, which may subsequently lead to a change in the intra-uterine environment and have consequences for gamete transport and maintenance.

(1). *Oestrual Behaviour*

The interval between two heat periods is not affected by high ambient temperatures (D'Arce et al., 1970; Steinbach, 1972c), but the duration of oestrus was found to be inversely related to the mean temperature of the environment (Steinbach, 1972c), and the incidence of temporary anoestrus increases in both artificially elevated temperatures above 30° C (Warnick et al., 1965; Teague et al., 1968) and under naturally varying temperatures in the tropics (Steinbach, 1972c).

(2). *Effects on the Ovary and The Genital Tract*

The ovulation rate, as determined by the corpora lutea count, was not significantly affected by heat treatment in most investigations (Warnick et al., 1965; Edwards et al., 1968; Mahoney et al., 1970) although Teague, Roller and Grifo (1968), reporting results on 240 gilts, have observed a significant negative effect of temperatures above 27° C on ovulation rate. There is no indication that naturally varying tropical temperatures (24-30° C) would affect ovulation rate (Steinbach, 1973). Roller and his coworkers (1973) have drawn attention to the fact that ovulation time may be prolonged under heat stress with possible consequences for fertilization rates.

The extensive studies carried out by Roller and his colleagues (1973) did not indicate that a particular period of the cycle preceding ovulation and mating is especially sensitive to heat stress, but the findings of D'Arce show a tendency towards a reduction in ovulation rate with an increased duration of heat stress.

There is, at present, no published evidence that heat stress affects the cyclical changes in the genital tract, such as the endometrial water content (Steinbach, 1972d).

C. PREGNANCY

Three periods during pregnancy have been studied: most extensively the first three weeks after breeding, but also the effects of heat stress during mid- and late pregnancy. Severe heat stress during the first two weeks p.c. causes a reduction in pregnancy rate (Mahoney et al., 1970; Omtvedt et al., 1971) and in the number of viable embryos (Warnick et al., 1965; Tompkins et al., 1967; Mahoney et al., 1970; Omtvedt et al., 1971). Similar results were obtained by other authors (Roller et al., 1967; Edwards et al., 1968), although their results did not differ significantly from the controls. Heat stress after two weeks p.c. did not affect embryo survival (Tompkins et al., 1967; Mahoney et al., 1970). Natural temperatures did not affect the number of live embryos in the tropics (Steinbach, 1973), but Moore et al. (1971) reported a significant reduction in viable embryos between spring and summer in Kentucky.

Heat stress during mid-pregnancy has no effect on reproductive performance (Omtvedt et al., 1971), but heat stress during the final stages of pregnancy causes thermal discomfort, as evidenced by increased rectal temperatures and respiratory rates (Heitman et al., 1951; Steinbach, 1970) and by an increased number of stillborn piglets (Heitman et al., 1951; Omtvedt et al., 1971; Steinbach, 1971) which could be reduced with cooling by sprinkling during the summer (Whatley et al., 1957).

The rapid increase in the susceptibility to heat stress during the terminal weeks of gestation is probably due to the increased metabolic heat production calculated by Mitchell (1962) and observed by Verstegen et al. (1971).

The above discussions indicate that the number of piglets born alive is affected by heat stress during the preimplantation phase and then again during the terminal stages of pregnancy. While it appears obvious that, prior to parturition, the sow cannot cope with the combined effects of exogenous and endogenous heat load, the ultimate causes of early embryonic mortality are not clear.

D. LACTATION

The unsatisfactory weaning weights in tropical regions referred to above and the negative correlation between litter weight gain and ambient temperature (Steinbach, 1971) suggest a direct or indirect effect of high environmental temperatures on the milk yield of sows. Puyaoan and Castillo (1963) and Williams (1971) reported milk yields of 120 and 142 kg respectively, less than 50% of the milk production reported from temperate countries, and Smith (1959) reported higher milk yields in spring than in summer. This reduction in milk yield, as affected by ambient temperatures, may be due to an inadequate development of the mammary glands in the heat-stressed pregnant animal, as suggested by Waites (1968), or due to an insufficient food intake during lactation. Merkel and Hazen (1967) reported increased food consumption and piglet gains in zone-cooled sows during the summer.

CONCLUSION

It has been shown in the preceding sections that high ambient temperatures may affect adversely reproductive behaviour and sexual development in both male and female pigs, that in the boar, gametogenesis may be impaired, while early embryonic mortality, excessive metabolic heat production during the late gestation period and reduced milk yield can reduce fertility in the heat-stressed sow. A considerable portion of the reduced reproductive efficiency observed under conditions of high ambient temperatures may have been mediated indirectly through reduced food intake (Chapt. 1, Sect. 7 and 12b) and an inadequate supply with protein and energy, which under normal temperature conditions play an important role in determining the reproductive rate of pigs (Duncan and Lodge, 1960). This is certainly the case with the effects of heat stress on sexual development and milk yield, but may also influence other aspects of reproduction, such as ovulation rate.

Therefore, investigations into the effects of high ambient temperatures on the various aspects of swine reproduction should be designed in such a way that the food intake and, where possible, the energy and nitrogen balance can be evaluated in the light of the two most important and, in this case, interrelated determinants of fertility, nutrition and ambient temperature.

REFERENCES: See pp. 577-579 .

Effects of Heat on Reproduction of Rat

by

U. A. Sod-Moriah and E. Bedrak

INTRODUCTION

Reproductive processes in the rat are adversely affected by exposure to heat. The effects differ, depending on duration of exposure i.e., acute or chronic. Acute heat exposure which usually lasts for several hours, brings about changes typical of some of the results obtained when applying various stress procedures. The animal reacts according to Selye's general adaptation syndrome (Selye and Heuser, 1955), one of the manifestations being adverse affect on reproduction (Fernandez-Cano, 1958; Chang and Fernandez-Cano, 1959). In the female rat it may temporarily block ovulation, cause the death of blastocysts, or cause embryonic malformations or death in utero, depending on timing, degree and duration of heat exposure (Shah, 1956; Chang and Fernandez Cano, 1959; Edwards, 1968). In the male rat it may induce testicular degeneration and infertility, the extent depending upon degree and duration of heat exposure (Elfving, 1950).

This discussion deals with the effects of chronic heat exposure on the reproductive performance of heat-acclimatized animals. The heat-acclimatized animals (CH) were kept in a hot room at 35° C and relative humidity of 25-40% for at least 30 days prior to initiation of the experiments. Control animals (C) are "acclimatized" to 22° C and relative humidity of 35-50%. An animal is considered heat-acclimatized when its metabolic rate (Gelineo, 1964) is consistently about 70% of that of controls, and body weight (Brobeck, 1960) some 60-70% of that of controls, while the post-acclimatization growth rate of both groups remains unchanged (Bedrak et al., 1971b; Sod-Moriah, 1971b). The CH animals consume 30% less food than controls, although when normalized according to body weight, the difference is not significant. CH rats consume significantly larger quantities of water (Sod-Moriah and Pollack, 1970). The male is apparently more susceptible than the female to high ambient temperature since males resume their normal growth rate only after 9 weeks while females do so after 5 weeks. Male mortality rate is significantly higher.

PHYSIOLOGY OF REPRODUCTION

A. DEEP BODY AND INTRASCROTAL TEMPERATURE

Rats exposed to high ambient temperature become hyperthermic. After acclimatization the hyperthermia disappears (Gelineo, 1964). In contrast, Hamilton (1963), Sod-Moriah and Yagil (1973) and Sod-Moriah et al. (1974) have shown increased deep body temperature in both male and female rats soon after the animals are placed in

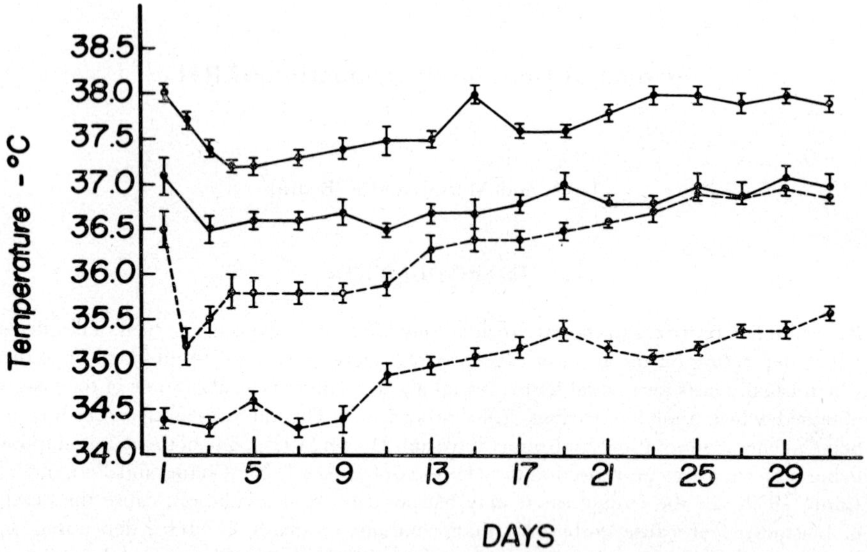

Fig. 1. Deep body (———) and intrascrotal (------) temperatures of control (curves A) and heat-acclimatized (curves B) animals. Means and standard errors.

the hot room. The animals remain hyperthermic as long as they are kept in the hot room. Body temperature must be measured in the hot room, since it quickly reverts to normal when returned to control temperature (Hamilton, 1963; Sod-Moriah and Yagil, 1973). Hence CH rats, despite their reduced metabolic rate, are incapable of "acclimatizing" as far as their body temperature is concerned (Sod-Moriah and Yagil, 1973).

Figure 1 demonstrates the differences in body temperature of male CH and C rats. CH intrascrotal temperature is higher than C, and is probably similar to that of cryptorchid testis. However, body-to-scrotum temperature gradient is maintained, being 1.4° C and 1.8° C in C and CH animals, respectively.

B. TESTICULAR FUNCTION

In vitro incorporation rate of labelled lysine into testicular proteins, at temperature of incubation of 32° C, is higher in CH than in C animals (Fig. 2), similar to that reported by Davis et al. (1964; 1965) for cryptorchids shown to be associated with increased gonadotropins secretion rate (Amatayakul et al., 1971). However, uptake is maximal at 36° C (Sod-Moriah and Bedrak, 1974) as opposed to 32° C in cryptorchids. Despite some similarities in temperature, gonadal size and lysine incorporation rate into proteins, there is an essential difference between CH and cryptorchid testes. Histological examination reveals that in CH testes necrobiotic tubules are scattered among normal seminiferous tubules with active spermatogenesis. These necrobiotic tubules resemble those of cryptorchid testis cross-sections. As opposed to cryptorchid males, the CH rats are capable of mating and reproducing although the reproductive perfor-

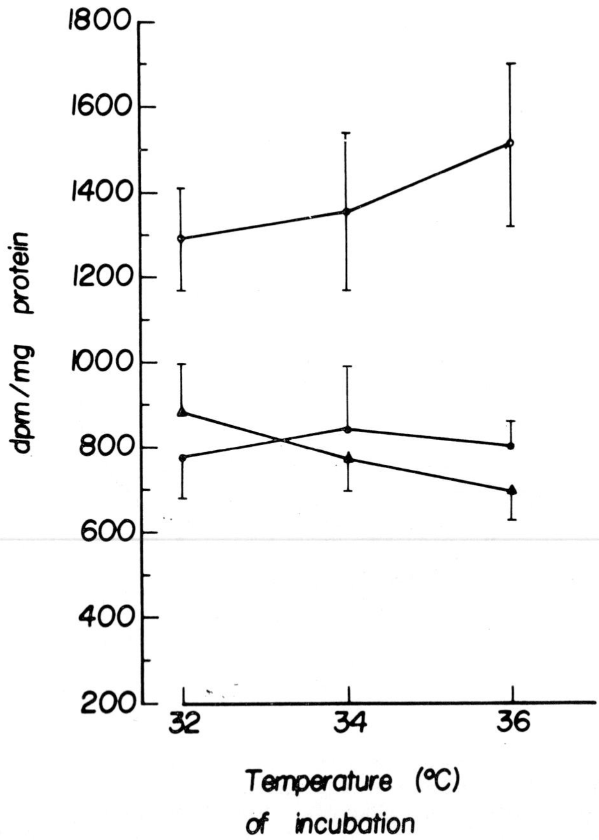

Fig. 2. Incorporation rate of labelled lysine in incubation experiments applying different temperatures. Means and standard errors. (C—●; CH—○; 4HE—▲).

mance is lower than that of control. This is largely due to the lower mating and conception rates of females mated with CH males. After conception pregnancy proceeds normally (Sod-Moriah et al., 1974).

C. FEMALE REPRODUCTION

Estrous cycle length in CH differs from that of C whether assessed in terms of mean cycle length or number of cycles occurring during a given period of time. Our results resemble those of Sundstroem (1930) and Macfarlane et al. (1959). Figure 3 demonstrates that the number of days spent at the metestrus and diestrus stages of the cycle is markedly greater ($P < 0.001$) in CH animals with significantly fewer days for possible impregnation. Thus the effect of heat would tend to reduce the size of the population.

The mating rate is estimated by the proportion of proestrus females caged overnight

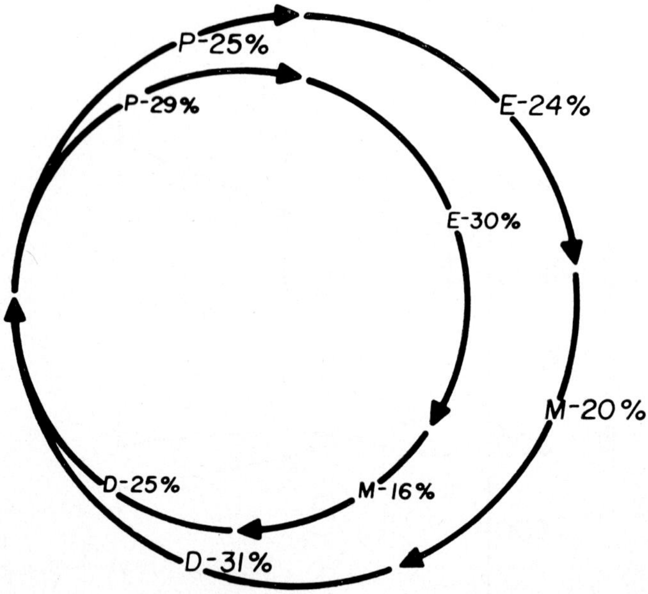

Fig. 3. Days of proestrus (P), estrus (E), metestrus (M), and diestrus (D), as percentage of total number of days, in control (inner circle, mean cycle length of 5.0 days) and heat-acclimatized rats (outer circle, mean cycle length of 6.2 days) (From Sod-Moriah and Pollack 1970).

with fertile males which show spermatozoa in the vaginal smear on the following morning. It indicates a 53% reduction in CH animals as compared with controls (Sod-Moriah and Pollack, 1970; and unpubl. data). The validity of this finding is uncertain.

Previous reports (Macfarlane et al., 1957) indicate a reduced ovulation rate as estimated by the number of corpora lutea (CL) on the sixteenth day of gestation. The validity of this criterion is discussed by Macfarlane et al. (1957). Our results are at variance and there is no difference between the groups in the number of CL counted either on day 4 of pregnancy or on parturition day (Sod-Moriah and Pollack, 1970; Sod-Moriah, 1971). The different results could be due to differences in strain (Hafez, 1968) or diet (Gupta and Lacy, 1967; Short et al., 1968).

The number of ova of day 1 of pregnancy that could be flushed from the oviducts in the CH animals is much smaller than the number of CL (Table 1) suggesting entrapped ova in luteinizing follicles (for discussion and other explanatory options see Sod-Moriah, 1971). Table 1 also demonstrates that the fertilizing rate is the same for both groups. All day 1 ova present are fertilized.

The preimplantation period demonstrates (Table 1) some degeneration of fertilized ova in CH animals up to day 4 of gestation. However, all ova appearing to be normal

Table 1. Losses (mean/rat) of various parameters in control and heat-acclimatized rats expressed also as a percentage of the number of corpora lutea

	Control		Heat-acclimatized		%* difference	Level of significance
	Means	S.E.	Means	S.E.		
No. of CL	12.3 (35)	0.312	11.9 (28)	0.242	− 3.3	NS
No. of Day-1 ova recovered	11.1 (16)	0.273	9.3 (10)	0.471	−26.2	0.01
as % of CL	90.2 (16)	5.427	78.2 (10)	4.920	−13.4	0.01
No. of fertilized Day-1 ova	10.9 (16)	0.310	8.8 (10)	0.468	−19.3	0.001
as % of CL	88.6 (16)	4.250	74.8 (10)	5.392	−15.6	0.01
No. of Day-4 ova recovered	10.9 (15)	0.342	9.0 (11)	0.381	−17.4	0.001
as % of CL	90.2 (15)	5.427	75.6 (11)	4.853	−16.2	0.01
No. of developing Day-4 ova	10.9 (15)	0.284	8.1 (11)	0.639	−25.7	0.001
as % of CL	88.6 (15)	4.341	68.1 (11)	5.886	−23.2	0.001
No. of metrial glands	11.0 (20)	0.319	8.2 (14)	0.883	−25.5	0.01
as % of CL	89.4 (20)	4.421	68.9 (14)	5.859	−22.9	0.001
No. of young born, total	8.0 (20)	0.896	5.2 (14)	1.115	−35.3	0.01
as % of CL	65.0 (20)	3.146	43.7 (14)	5.624	−32.8	0.001
No. of young born alive	6.9 (20)	1.064	3.1 (14)	1.040	−55.1	0.05
as % of CL	56.1 (20)	4.225	26.1 (14)	5.720	−53.6	0.001

* Control values = 100%.
( ) Number of animals.
(From Sod-Moriah, 1971a).

Table 2. Proportionate* preimplantation, postimplantation and perinatal losses in control and heat-acclimatized rats

	Control		Heat-acclimatized		%† difference	Level of significance
	Means	S.E.	Means	S.E.		
Preimplantation losses (%)	10.6 (20)	0.891	31.1 (14)	1.208	+194.1	0.001
Postimplantation losses (%)	27.3 (20)	1.845	36.6 (14)	2.120	+ 34.1	0.001
Perinatal losses	10.0 (20)	3.100	25.6 (14)	3.270	+156.0	0.001

* Pre-implantation losses = the difference between the number of CL and metrial glands expressed as percentage of CL. Postimplantation losses = number of metrial glands less total number of young born as percentage of metrial glands. Perinatal losses = number of young born dead as percentage of metrial glands.

† Control values = 100%.

( ) Number of animals.

(From Sod-Moriah, 1971 a)

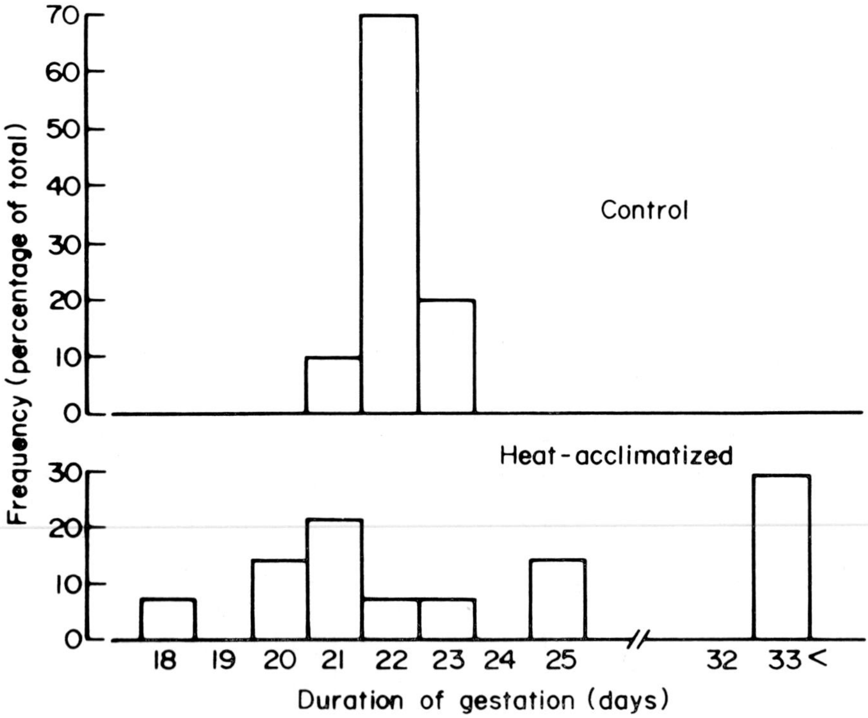

Fig. 4. Frequency distribution of pregnancies of various durations (From Sod-Moriah, 1971a).

on day 4 implant. In the postimplantation period there is increased intrauterine death and resorption of fetuses in the CH animals (Tables 1 and 2).

Macfarlane et al. (1959) and Pennycuik (1964) noted prolonged pregnancies in the heat. With each day of prolongation the number of dead fetuses increased, so that by day 25 of gestation all fetuses were born dead. Results obtained in our laboratory differ somewhat (Fig. 4). Some of the CH animals (70%) delivered spontaneously. However, the process of parturition is extremely irregular in timing and duration, possibly contributing to the markedly increased proportion of stillbirths (Tables 1, 2). In the remaining 30% of CH animals parturition fails completely (Sod-Moriah, 1971). The different results may be due to strain differences of rats used or to diet offered. For further discussion, see Sod-Moriah (1971).

ENDOCRINOLOGY

Figure 5 demonstrates that the testosterone concentration in plasma of CH male rats is lower both than that of C animals and that of rats acclimatized to a hot environment by daily four-hour exposure to 35° C (4HE). This process of heat acclimatization was

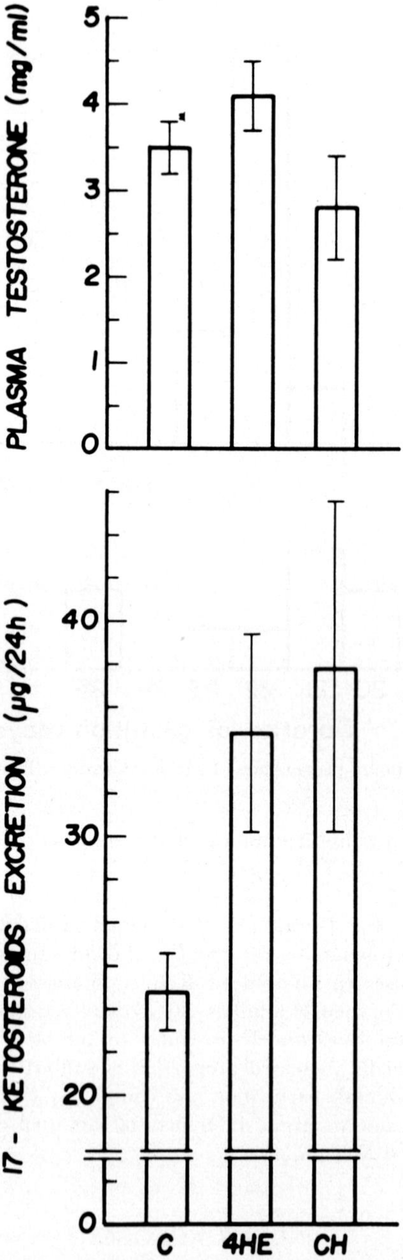

Fig. 5. Peripheral blood plasma testosterone concentration and urinary excretion of 17-ketosteroids in control (C) and in rats which were heat-acclimatized either by daily 4 hr exposure to 35°C (4HE) or by continuous exposure to 35°C (CH).

adopted in order to simulate natural conditions (Bedrak et al., 1971b). The lower testosterone concentration could be partially explained by (a) a significantly smaller gonadal mass in CH rats as compared with 4HE and C ($P < 0.01$) i.e. 0.54%, 0.82% and 0.84% of body weight, respectively and (b) the presence of degenerating tubules in CH testes accompanied by slight hyperplasia of the interstitium. The significantly higher excretion rate of urinary 17-ketosteroids of CH animals compared with that of C ($P < 0.01$) suggests an increased ACTH secretion rate and possibly steroid-inactivation processes in CH and 4HE groups.

Another phenomenon characterizing the CH rat is its changed pattern of testicular capacity of in vitro synthesis of testosterone. With some exceptions, there is increased activity of the enzymes metabolizing progesterone to testosterone via the 4-ene pathway (Bedrak et al., 1971b) and those metabolizing pregnenolone to testosterone via the 5-ene route (Bedrak, Samoiloff and Sod-Moriah, 1973). Moreover, the activity of the converting enzymes, 3β-hydroxysteroid dehydrogenase and 5-ene-hydroxysteroid isomerase, is increased. Consequently a higher proportion of testosterone is metabolized via the 4-ene pathway (Fig. 6). Since the activity of the converting enzymes is stimulated by HCG (Shikita and Hall, 1967) and that of 20α-hydroxysteroide dehydrogenase is enhanced by FSH (Kariem and Samuels, 1974), it is suggested that gonadotropin secretion increases in the heat-acclimatized rats. The increased activity of 20α-hydroxysteroid dehydrogenase (Bedrak et al., 1971b) may be regarded as the rate-limiting step of 17, 20 lyase operting in the 5-ene pathway (Slaunwhite and Burgett, 1965; Bedrak et al., 1973), and the lower activity of 17β-hydroxysteroid dehydrogenase as a regulatory mechanism for testosterone production in the heat-acclimatized rat.

The pattern of these enzymic reactions brought about by heat acclimatization is similar to those observed in mice whose natural habitat is the desert (Bedrak et al., 1971a; Bedrak, 1972).

The extended proportionate time occupied by the metestrus and diestrus stages in the estrus cycle, the prolonged and/or irregular timing of the onset of labour, and the disturbed process of parturition, suggest that the high ambient temperature may adversely affect the processes involved in luteolysis thus leading to an inadequate, high level of progesterone. This hypothesis is supported by the fact that the duration of pseudopregnancy in CH females is significantly longer than in C animals. In addition, while 3 CL will maintain normal pregnancy in CH animals, 4 CL are required in C females (Sod-Moriah and Bedrak, 1973). Preliminary plasma progesterone level determinations in cycling rats indicate slightly increased hormonal levels in CH animals (unpubl. data).

HYPOTHESIS OF MODE OF ACTION

The basic concept underlying this discussion involves the participation of the hypothalamo-pituitary-gonadal axis. The changes observed in the reproductive performance of the animal must be a result of changes occurring in the controlling areas in the central nervous system in response to exposure to high ambient temperature. Consequently,

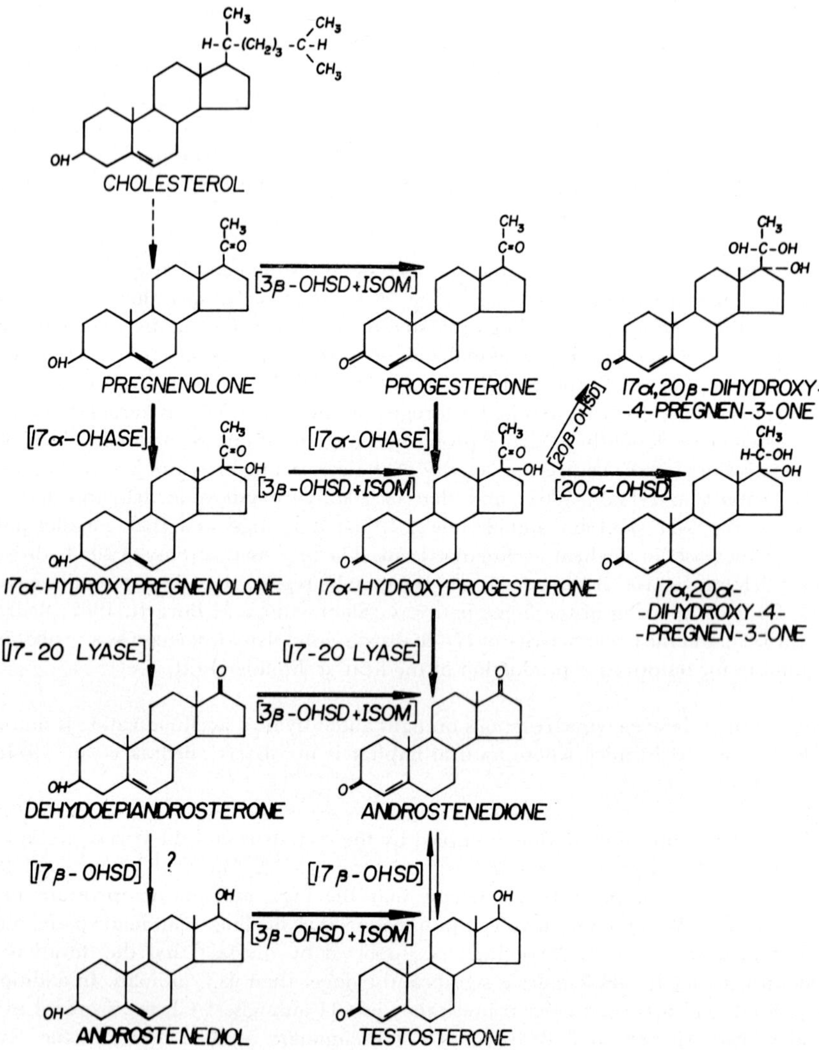

Fig. 6. Proposed relative activities of enzyme systems involved in androgen metabolism in testicular tissue of heat-acclimatized rats. Arrow thickness is proportionate to activity. Dashed arrow indicates that more than one step in involved. Query (?) indicates that the rate of the reaction has not been determined in heat-acclimatized animals. Trivial names of steroids are used. In parentheses are the enzyme systems: 3β-OHSD, 3β-hydroxysteroid dehydrogenase; 17α-OHASE, 17α-hydroxylase; 17β-OHSD, 17β-hydroxysteroid dehydrogenase; 17-20 lyase, C_{17}—C_{20} lyase; 20α-OHSD, 20α-hydroxysteroid dehydrogenase; 20β-OHSD, 20β-hydroxysteroid dehydrogenase; Isom., 5-ene-hydroxysteroid isomerase.

a new endocrine balance is reached: the animal is said to be acclimatized (see review by Collins and Weiner, 1968). It is capable of reproducing although its performance is lower than before acclimatization. While some information is available on the involvement of the hypothalamus in several physiological processes such as hunger or body temperature regulation, there are no experimental data on the effect of high ambient temperature on the hypothalamo-pituitary interactions with regard to reproduction.

The information available, although limited, indicates that the first responses to heat exposure include an immediate increase of ACTH secretion with concomitantly decreased gonadotropin output, thus reducing the secretion rate of gonadal hormones. In the male this reduces testosterone production (Rhynes and Ewing, 1973) and in the female ovulation is inhibited and estrus cycle prolonged (Chang and Fernandez-Cano, 1959).

Acclimatization premises a new endocrine equilibrium and the gonadotropin level reverts to normal. Gonadotropin secretion rate increases and possibly even exceeds normal levels. This is supported by the fact that the activities of some enzymes known to be stimulated by gonadotropins are enhanced in heat-acclimatized animals. In consequence, plasma testosterone levels in such males return to normal, and in animals which are acclimatized by simulation of nature (4HE) and which do not show testicular damage, testosterone levels are higher than normal. This suggests that the chain of events controlling the secretion of gonadotropins has been altered. The commonly accepted negative feedback mechanism does not function, thus allowing an increased level of gonadotropins to exist simultaneously with an increased level of gonadal hormones.

In the female, the estrous cycle is re-established, thus indicating the presence of adequate levels of ovarian hormones. The observed deviations from normal in the reproductive function could be attributed to excessive progesterone levels. The same enzyme systems involved in testosterone biosynthesis in the male and which are stimulated by gonadotrcpins, are also key enzymes in progesterone biosynthesis in the ovary.

To simplify the working hypothesis presented here, the mechanism suggested excludes other hormonal interactions that may play an important role in heat-acclimatization and steroidogenesis. Some of those not mentioned are, for example, thyroxine and hydrocortisone which are known to be permissive hormones in the biosynthesis of estrogens and participate in the process of acclimatization.

CONCLUSION

Rats are capable of acclimatizing to a continuous high ambient temperature. This acclimatization involves a reduction in body mass, and lowered metabolic rate. Food intake per unit body weight is unchanged and water consumption is increased. Such an animal manages to survive and reproduce. However in comparison with other mammals for example, man, the rat is incapable of maintaining normal body temperature and remains hyperthermic while exposed to high ambient temperature.

Despite increased intrascrotal temperature and some testicular damage, the level of hormones involved in reproduction, albeit slightly lower than the controls, enables spermatogenesis to continue and reproduction to take place. In the female, acclimatization includes some hormonal changes. These, together with increased body temperature, apparently cause some adverse effects at various stages of the reproductive cycle, resulting in a smaller number of viable offspring (Table 2).

To elucidate the consequences of heat exposure and acclimatization, the following points need investigation:

1. Hypothalamic response, and production and secretion of gonadotropin-releasing facotrs.
2. Interrelationship between the CRF and gonadotropin-RF.
3. The secretory control mechanism of gonadotropins including hypothalamic receptors and gonadal hormones.
4. Meteorological effects on the properties of seminal and uterine secretions.
5. Biochemistry, physiology and histology of the uterus and oviduct.
6. Fetal parameters e.g. endocrinology and biochemistry, involved in maintenance ot pregnancy and parturition.

In the discussion presented here it should be borne in mind that the extreme experimental conditions described are not encountered in nature. Hence the findings on the simulated desert group (4HE) indicate that the effects of high environmental temperature on reproduction should be investigated under natural desert conditions, or laboratory conditions simulating hot desert areas.

As existing areas of the world become overpeopled and overworked, man will have to consider introducing his lifestyle into that remaining one-third which is still uninhabitable. This is the essential reason for investigating the effects of heat on mammalian physiology, and here specifically reproductive physiology, and so to promote the development of those vast desert areas.

REFERENCES: See pp. 579-580

Effects of Temperature on Oviposition and Egg Formation in Poultry

by

W. O. Wilson

INTRODUCTION

The processes of egg formation including ovulation, oviposition as well as egg production has been reviewed by Biellier and Ostmann (1960) and by Gilbert and Wood-gush (1971). Due to the lack of knowledge of this topic it would appear that the ambient temperature effects the processes only indirectly. In both heat and cold, feed intake becomes the major factor relating to egg production. In turn, feed intake is governed by air temperature and according to newer concepts put forth by Lepkovsky (1973), fat stores. The hypothalamus appears to use many mechanisms in regulating the size of the fat depots in turn, the adipose tissue play a role in the regulation of food intake and the hypothalamus acts centrally in regulating food intake. Further, it appears that adipose tissue is involved in the regulation of food intake. Mechanisms appear to influence the rate of absorption of nutrients, especially of calories from the intestine. The set-point concept applies only to chickens that are fed ad libitum. If force fed, a level of fat in the adipose tissue mass is reached, beyond which, little or no fat can be deposited in the depots. Concomitantly, the rate of absorption of food from the small intestine falls and little food leaves the crop. Consequently, little food can be forced into the crop because it is full most of the time.

The depressing effect of high temperatures on egg production has been investigated by a number of workers e.g., Wilson (1949) and Smith (1971). The effect of high temperature on feed intake is curvilinear and affect heavy birds adversely to a greater extent than Leghorns. Smith concluded from his data that the reduction in food intake at high temperatures could be the cause of lowered egg production. That is, there simply was not enough energy consumed for sustained high production of large sized eggs.

The amount of feed consumed by layers may be predicted from the mean body weight, changes in body weight and egg mass. Byerly's (1941) original prediction equation has been modified by Combs (1968) to use a correction factor for high temperatures. From the results obtained in calorimeters, Longhouse et al. (1960) has shown that the Byerly prediction equation overestimates feed consumption at the high temperature above 24° C and underestimates it at the low temperature below 5° C.

The starvation of chickens and Japanese quail laying hens results in immediate cessation of egg production (Morris and Nalbandov, 1969; Wilson et al., 1965). However, by the use of exogenous gonadotropins and dried avian pituitary powders, it was possible to extend the time before egg production stopped. However, the egg size and yolk size decreased.

The discussion will follow in the following order: (a) Effect of temperature on time of oviposition; (b) length of egg formation time as a result of temperature; (c) egg components affected by temperature and the relation of ambient temperature to rate of egg production and feed efficiency.

EFFECT OF TEMPERATURE ON TIME OF OVIPOSITION

The methods of studying effect of temperature on the egg formation processes need some refinement. At the present time oil-soluble dyes are given to the birds at daily intervals. The eggs are boiled and sliced. The number of rings of dyed yolk is used as an indication of the time spent in its formation.

There is no suitable technique for recording ovulation time available at the present time. Most of the information available on egg formation comes from slaughter trials where the bird is killed at different times of the day and the position of the ova in the oviduct are observed and the time calculated back from the last oviposition. Such procedures leave much to be desired. The time of egg laying or ovipositing is usually done with birds that lay on sloping wire floors and the egg hits against a switch and the event is recorded electrically.

The parts of the ovary and oviduct are represented in Figure 1. About 20 to 30 min after an ovum is released from the ovary it is picked up by the infundibulum. It takes about 5 hrs to pass down the oviduct where the albumen and membranes are added. Normally the stay in the uterus or shell forming gland is 19.5 hrs.

In considering the time of oviposition, it must be recognized that ovulation time determines oviposition time and not vice versa. The evidence for this is found in the review by Biellier and Ostmann (1960). Hypothermia (a reduction of about 4° C) did not hinder ovulation. But acceleration in the time of oviposition can result from low ambient temperature (hypothermia) according to Sturkie (1946). When the hen has a hard shelled egg in the uterus and is exposed to hyperthermic conditions, the eggs are usually aborted. Other conditions that result in the acceleration of time of oviposition to be expected have been discussed by Van Middelkoop (1972). A normal egg laying results from an ovulation occurring at least 21 hrs before or after another ovulation and the laying of normal eggs is associated with a short ovulation interval. Soft-shelled eggs appear to be laid 6 ± hrs prematurely and membranous eggs approximately 13 ± hrs premature.

On the other hand, high ambient temperature 32° C (compared with 21° C) results in a slowing down of passage of eggs through the oviduct with an additional 2 hrs being spent in the uterus according to Nordstrom (1973). Other factors known to delay time of oviposition have been reported by Grau and Kamei (1949), due to feeding of purified diets and by fright of coturnix by Opel (1966).

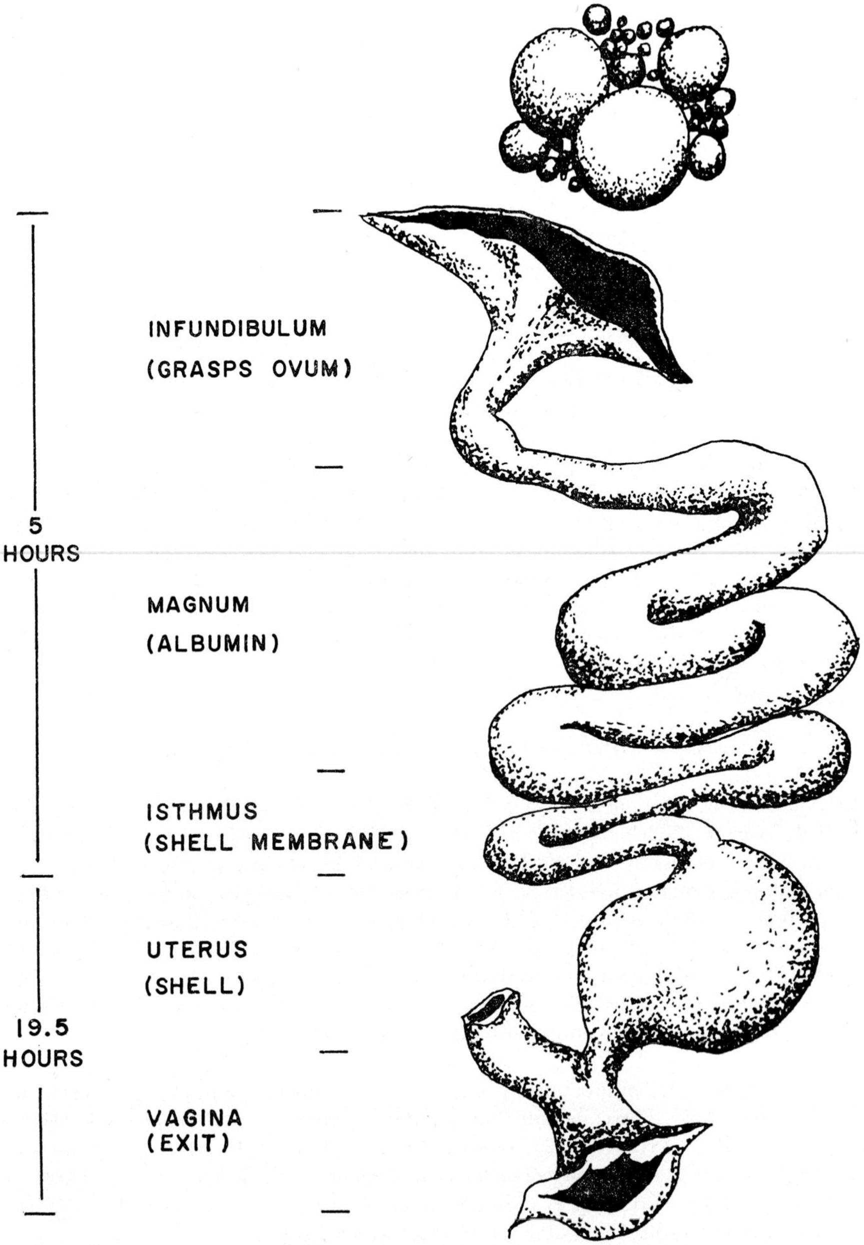

Fig. 1. Ovary and oviduct of a hen. Segments and their function as indicated by center horizontal lines. On the left is given the approximate duration of time spent in the oviduct. Ambient temperature modifies the time spent in the uterus. The ovary sheds an ovum $\frac{1}{4}$ to $\frac{1}{2}$ hr after an egg is laid in coupled sequences.

When continuous light is provided, high ambient temperatures may act as a Zeitgeber or synchronizing agent which determines the time of egg laying. These results have been reported by Payne et al. (1965). Cain (1971) confirmed these results using cyclic temperature regimens of 31 and 10° C for 10 hrs each with a 2 hr change over period between extremes.

The activity associated with oviposition usually results in a rise in body temperature according to Winget et al. (1965) in chickens and Woodard et al. (1970) in coturnix. By using telemetric techniques and computers to reduce the noise of the system, Winget et al. (1965) found that the time of ovulation also resulted in a slight rise in body temperature.

LENGTH OF EGG FORMATION TIME AS A RESULT OF TEMPERATURE

The length of egg formation time is affected by temperatures mentioned previously. Nordstrom (1973) found that additional time, amounting to 2 hrs, was spent in the uterus of the chicken before laying. This was a surprising discovery because many had suggested that thin shells resulting from high temperatures was due to premature expulsion of eggs. The timing of eggs is much more responsive to the photoperiod as has been discussed by Morris (1973). He was able to show that by using ahemeral daylengths, of 27 hrs, the egg that was retained in the oviduct for a longer period of time weighed more than when the same photoperiod was a 24 hrs day.

EGG COMPONENTS AFFECTED BY TEMPERATURE

The various egg components as a result of the high ambient temperature have been studied. Warren (1933) reported that the egg size was decreased when the ambient temperature was in excess of 21° C. Later, Warren and Schnepel (1940) showed that high ambient temperatures (>32° C) depressed shell thickness of the egg. Conrad (1939) reported that the blood calcium of chickens was decreased 20 to 30% as a result of increasing ambient temperature from 21 to 32° C and this possibly could be the cause of thin-shelled eggs which were observed during the summer time. Smith (1969) confirmed earlier work by Warren that the yolk weight decreased as a result of high ambient temperature. He compared temperature of 21° C and 32° C.

The depression of egg production by high ambient temperature has been reported by many investigators. These include Payne (1967), Romijn and Vreugdenhil (1969), Sukki et al. (1971), Sykes (1972), Wilson (1949), Wilson et al. (1957) and Wilson et al. (1971). From a regression equation it was reported by Francis et al. (1972) at temperatures of 21° C the energy requirement for a gram of egg was 0.4 kcal/g whereas, at a temperature of 35° C only 0.31 kcal/g egg was required.

Current concepts of the role of the environment on egg formation and laying are illustrated in Figure 2. Space limitations prevent a thorough discussion in this section.

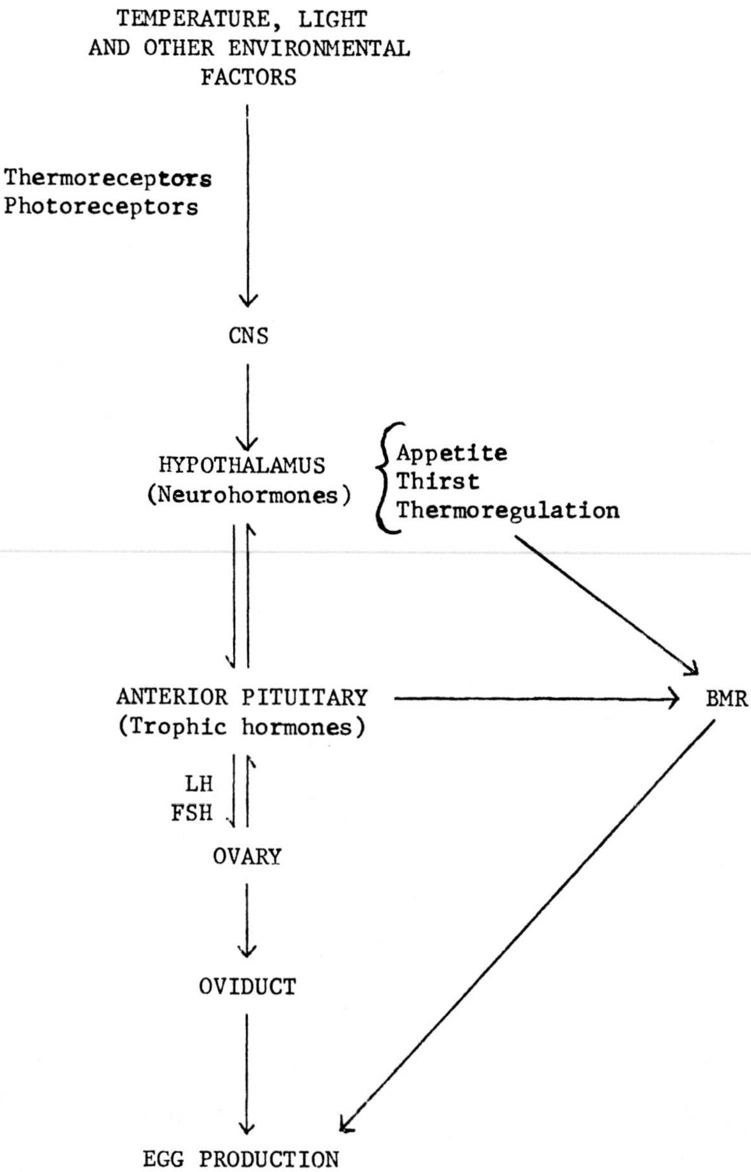

Fig. 2. Steps in the regulation of egg production by the environment. Not all feedback and interrelations are shown in this simplified diagram.

From the above review, it is apparent that egg laying, measured by the time of oviposition, may be affected by ambient temperature. It is hastened by cold temperatures and retarded by high ambient temperatures, that is, eggs are retained in the oviduct longer under conditions of high ambient temperature. Ambient temperature may act as a synchronizer of time of lay under conditions of continuous light. The effect of high temperatures are evident in reduction in egg size, egg shell thickness, yolk weight, as well as decrease in egg production. The optimum mean temperature for egg production appears to be in the range of 10 to 16° C. Hens benefit from cyclic temperature in contrast with continuous high temperature. However, the most efficient production is obtained when the ambient temperature is of the order of 20 to 30° C.

Further investigations as needed to determine with greater accuracy, the time of ovulation. Studies of egg yolk and egg formation should be investigated further. A study of multiple ovulations, yolk reabsorption should also be investigated. The location and timing of the egg yolk as it transverses the oviduct without having to kill a bird is a technique that needs to be developed. It would be very desirable to have information on the effect of other environmental effects that are related to heat loss. For example, relative humidity, air movement and radiation. It is apparent that the environment affects the neural and hormonal control of the processes of ovulation and egg formation. Therefore, further information is needed on the endocrine control in order that we may relate the physiology of egg formation to the biometeorological effects.

REFERENCES: See pp. 541-582

Chapter 1, Section 17

Effects of Temperature as a Function of Age in Mammals

by

W. Bianca

See Progress in Human Biometeorology, Part IA, Chapt. 6, Sect. 6a, pp. 273-284.

Morphological and Anatomical Adaptations

by

W. G. Bradley and M. K. Yousef

INTRODUCTION

The Glossary Committee of the International Union of Physiological Sciences has defined adaptation as "a change which reduces the physiological strain produced by a stressful component of the Total Environment. This change may occur within the lifetime of an organism (phenotypic) or be the result of genetic selection in a species or subspecies (genotypic)" (Bligh and Johnson, 1973). The many relationships whereby form and function interrelate to produce patterns of physiological adaptation are described and discussed in various Sections of this book. Indeed, basic morphological patterns or body plans which favor survival in varied ecological habitats and niches are the features which characterize higher taxonomic groups. Morphological types relating to phylogeny are well illustrated in texts of comparative anatomy and vertebrate biology. Basic body plans are ultimately refined at the specific and population level.

This Section is restricted to a consideration of morphological adaptations of mammals at lower taxonomic levels, specifically morphological and anatomical features that relate to thermoregulation. This subject has received a great deal of attention, both historical and recent, and is an area of interest to our laboratory. Further evaluation is needed due to the numerous misunderstandings and controversy involving ecological and physiological interpretations of morphological and anatomical features.

After publication of Darwin's "Origin of Species", naturalists were preoccupied with the development of rules which explained the geographic and ecologic distribution of plants and animals on an evolutionary basis. Numerous rules relating body size, shape, proportions of appendages and internal organs, and body coloration were proposed by early naturalists and ecologists (for review see Rensch, 1960). Ecological and physiological literature emphasize some of these rules, generally called ecogeographic rules, which are based on the pioneer work of Gloger (1833), Bergmann (1847), Allen (1877), and Semper (1881). The major ecogeographic rules are summarized here as applied to endothermic animals and refer to changes in closely related species, or as interpreted by Mayr (1963) only subspecies and/or localized populations. These changes occur as morphological north-south clines in relation to latitude, or as related to elevational zonation.

A. BERGMANN'S RULE: In general terms this rule states that body size is greater in colder climates than in warmer.
B. ALLEN'S RULE: As a generalization, changes in surface areas due to increase or

decrease in appendage size are correlated with geographic distribution. The length of appendages increases in warm climates.

C. WILSON'S RULE: Morphological changes in characteristics of the animal's coat such as hair length, density, and shape are correlated with climatic conditions i.e., dense and longer hair in colder regions.

D. GLOGER'S RULE: Melanin production resulting in darker coloration is greater in warm and humid environments.

The above rules and certain related phenomena are evaluated and discussed throughout this Section as are various functional aspects of adaptation on an anatomical level.

MORPHOLOGICAL ADAPTATIONS

A. BERGMANN'S RULE: BODY SIZE

According to Bergmann's Rule, body size in mammals is believed to increase in populations which inhabit higher latitudes as well as higher elevations. The functional basis for this rule generally has been related to changes in ratio of body surface to volume, an important consideration in the heat exchange between the organism and its environment. In other words, a large size animal has a reduced surface/volume ratio which aids in conservation of heat in a cold environment. Conversely, reduced body size and increased surface area facilitates heat loss, thus preventing hyperthermia in a hot environment.

Two major problems exist: (1) to what extent does body size of mammalian populations follow the basic rule and (2) do other environmental factors such as availability of food, relation of body size to feeding niche, and predator-prey relationships modify body size?

The validity of Bergmann's rule is quite controversial and has been debated in the literature (Mayr, 1963; Scholander, 1955; McNab, 1971). A question fundamental to the debate is to which taxonomic level does the rule apply. Mayr (1963) emphasized that while there are good examples of related species which change in body size in accordance with the rule, there are also numerous examples where it does not. He concluded that the rule should apply only to subspecies, races, or localized populations of the same species. Furthermore, in many instances, size changes are statistically significant only at the population level. Rensch (1936) analyzed various groups and found that 81% of North American mammals and 60% of Western European mammals corresponded to Bergmann's rule. However, McNab (1971) has criticized Rensch's conclusions as he found that only 32% of the North American species showed a statistically positive correlation of head-body length with latitude. Also the rule was considered to be valid for certain mammals in relation to elevational zonation (Rümmler, 1938).

Undoubtedly even with restriction of the rule to subspecific levels there are numerous exceptions. Some of these exceptions are related to highly specific habitat requirements whereby certain species do not greatly interact with prevailing climate during periods of heat or cold stress. For example, avoidance of thermal stress and in the case of small mammals the selection of favorable microenvironments are of particular importance (Bradley and Yousef, 1972). Availability and quality of food play an important role in modification of body size. Niche segregation and its relationship to species diversity have been an area of extensive research in recent years (MacArthur, 1965; Ricklefs, 1973). A common phenomenon related to niche segregation of sympatric species is that they usually differ in size of body and trophic structures. Probably such differences are related both to microhabitat and food nich selection, which in turn would reduce competition for shelter and food resources (Rosenzweig, 1968; Schoener, 1969; Bradley and Mauer, 1973). In addition, McNab (1971) has shown that size differences of predators are correlated with size of prey and this is more important as a factor modifying size than competition for food as was suggested by Rosenzweig. The available evidence indicates that body size is modified by a wide array of biotic and abiotic factors which differ widely, according to the basic structure of any particular biotic community.

The interpretation of Bergmann's rule as related to heat exchange was criticized by Scholander (1955) who concluded that there is no physiological evidence to support phylogenetic pathways of heat conserving adaptation. He pointed out that insulation is of more importance in conservation of heat than is body size. In our opinion, it is not only insulation or body size independently that govern heat conserving adaptation but both synergistically facilitate either heat loss or conservation more efficiently.

B. APPENDAGES AND ACCESSORY STRUCTURES

Allen's Rule represents the major statement concerning appendage shape and size in relation to thermal environment. As generally stated this rule indicates that extremities such as limbs, tails and ears are longer in warm climates than in cold. It has been suggested by numerous investigators that these structures function as heat radiating organs.

The classical example pertaining to ears is for rabbits (hares) (genus *Lepus*) in which four species representing environments ranging from Arctic tundra to Sonoran Desert show a general increase in ear length and size from northern to southern latitudes (Nelson, 1918). Similarly Saharan Desert populations of *Lepus capenis* are distinguished by larger ears and longer limbs than populations from cooler climates (Petter, 1952). Longer and/or larger ears are common in numerous southern or desert mammals when compared to similar forms from more temperate or arctic regions (Allee and Schmidt, 1951). Specific examples are desert foxes, several African antelopes and the African elephant.

Several examples of changes in tail length in relation to climate are discussed in the literature. One example is tail length in pocket gophers (*Thomomys* and *Geomys*) which varies with latitudinal and elevational zonation (Merriam, 1895; Bradley et al., 1974) and is of particular interest due to the closed burrow system which provides a stable, hot humid environment.

The role of extremities as heat radiators does not depend on size alone as Allen's rule implies but on vascularization of these areas. Specifically, blood flow and the density of the arterio-venous anastomoses are the primary factors related to heat exchange. Based upon observations of peripheral circulation in heat stress experiments, it has been suggested that the feet and tail of fossorial rodents are especially effective in heat loss (McNab, 1966; Bradley et al., 1974; Bradley and Yousef, in press). Another example is the observed peripheral circulation of the bats wing, accompanied by fanning, under conditions of thermal stress (Reeder and Cowles, 1951; Bradley and O'Farrell, 1969). Similarly, it has been suggested that fanning of the large ears of the African elephant represents an important avenue of heat loss under hot conditions (Folk, 1974). In addition to vascularization of the ear of the jack rabbit, behavioral traits such as shading contribute to their effectiveness as heat exchangers (Schmidt-Nielson, 1964). In general however, the heat exchange function assigned to Allen's rule is largely inoperative as many mammals do not display the degree of vascularization or associated behavior which allows extremities to function efficiently in heat dissipation.

Accessory structures which may play a role in heat loss but not commonly considered as part of Allen's rule include: the dewlap of Indian cattle, the horns of Bovidae and the antlers of Cervidae. Although these structures increase the surface area, their role in heat loss is questioned. Removal of the dewlap of Indian cattle during heat stress showed no effect on their ability to thermoregulate effectively (McDowell et al., 1953). Similarly, Krog and Wika (1974) have reported that heat loss from reindeer antlers appears to be a function of the large local blood flow necessary for the rapid growth and not due to a special thermoregulatory adjustment as has previously been suggested. However, Taylor (1966) has suggested that the horns of goats may play a role in thermoregulation, but he found no correlation between horn size and habitat.

C. WILSON'S RULE: INSULATION OF ANIMAL COAT

The high insulative qualities of hair and feathers are well known and considered a necessary adaptation for development of endothermy. The classical study of the insulative characteristics of the mammal pelt has clearly shown that pelage represents a major morphological adaptation which varies greatly depending on body size and climatic conditions (Scholander et al., 1950). Morphological insulation i.e., physical insulation includes the various properties of the pelt such as hair shape, length, density and color. According to Wilson's rule insulative quality of the pelt is related to climatic conditions i.e., the colder the environment the thicker the pelt. Scholander and his coworkers (1950) found that small arctic mammals (weasels, lemmings) have much less insulation than large ones and overlap many of the tropical forms. The conclusions reached by Scholander et al. (1950) on small mammals contradict Wilson's rule. However, since few species were used in that study further examination of hair characteristics of many species, including large ones, is warranted. Evidence supporting Wilson's rule is available for *Thomomys* in which pelage weight and length of hair was related to elevational zonation i.e., desert and mountain environments (Bradley et al., 1974). Also laboratory studies have shown that hair weight and density are greater in rats and mice acclimated to cold when compared to controls (Heroux, 1961).

D. GLOGER'S RULE: MELANIN PRODUCTION

Concerning pelage color, Gloger's Rule states that darker forms are characteristic of warm humid climates. Whereas melanin pigmentation is reduced in dry warm and cool areas, it has been postulated that the lighter pelage coloration found in many desert mammals is of adaptive significance for heat exchange. This generalization is particularly meaningful in a consideration of energy exchange (heat reflection) of large mammals exposed to considerable solar radiation in semiarid or desert regions (Finch, 1972).

Medium sized mammals, such as North American carnivores, are chiefly crepuscular or nocturnal and the benefits of light coloration as related to thermoregulation is minimal. Most small mammals, such as rodents, are nocturnal and semifossorial. Although in general their coloration is light in desert regions, there are numerous examples of melanistic forms which match their surroundings (Sumner, 1925). The major advantage of coat color to crepuscular or nocturnal forms is the matching of their substrate. This camouflage reduces detection which is of advantage to both predator and prey. Experimental evidence for this concept was the study of owl predation on rodents in relation to pelage and soil color (Dice, 1947).

In the arctic dark and white coloration are found in both large and small mammals. Several mammalian species exhibit seasonal dichromatism (Hock, 1965). White coloration in the arctic appears to be of little significance in heat exchange but, as in the desert, it does provide protective coloration.

In conclusion, not only are there numerous exceptions to Gloger's rule; but when operative it is of adaptive significance as protective coloration and/or unknown ecologic factors. With the exception of some large desert mammals, there is no evidence that coloration plays a significant role in heat exchange adaptation.

While hair is characteristic of mammals there are several groups which are naked skinned or largely devoid of hair. These naked mammals fall into three ecologic groups: (1) cetaceans, earless seals and walrus which are aquatic where lack of hair facilitates movement through water, (2) large mammals such as elephant, rhinoceros and some swine which inhabit warm environments and (3) the naked mole rat (*Heterocephalus glaber*) which is completely fossorial. In the aquatic group, insulation and streamlining of the body is provided by a thick layer of subcutaneous fat or blubber which makes up between 25 to 45% of the total body weight of cetaceans (Kanwisher and Sundnes, 1966). It appears that the naked skin of rhinoceros and elephant functions primarily in heat loss. In the naked mole rat heat exchange at high, humid temperatures is greater than for any other mammal (McNab, 1966).

The topographic variability of insulation that can be attributed to hair as a physical insulator was demonstrated by Morrison (1966) on the guanaco. Differences in distribution of hair on body and appendages demonstrate the flexibility in the employment of hair in regulating the emission of heat from the animal. This flexibility allows the animal to stand with bare surfaces exposed to varying degrees depending upon the nature of energy exchange with the environment. This provides an effective regulatory mechanism of heat flow to and from the animal. Such a mechanism is operative in large arctic ungulates, for example in reindeer, which are effectively insulated during activity in the winter or moderate temperatures during the arctic summer. This flexibility in hair distribution should be further investigated in desert ungulates.

ANATOMICAL ADAPTATIONS

Morphological adaptations are complemented by functional anatomical features. Recent studies on anatomical adaptations are directed to a better understanding of interrelationships between macroscopic and microscopic structures and the environment of the organism. These studies have established a broader base for understanding the function of morphological features.

Available anatomical data on adaptation are incomplete and hence generalizations or rules relating these adaptations to lower taxonomic levels have not been attempted as they have been for morphological adaptation. Anatomical models appear to be used in various unrelated phylogenetic groups and these may warrant considerations as general rules of anatomical adaptations. Available anatomical data relating to thermoregulatory functions are restricted to specific vascular arrangements in different organs and some unique anatomical features of the kidneys and lungs. Vascular adaptation is either related to: (a) formation of a rete system, (b) close proximity of veins and arteries in an organ; or (c) selective shunting of blood-borne body heat through a cutaneous vascular plexus. Regardless of the type of anatomical vascularization, this arrangement results in formation of a countercurrent heat exchange model where heat is exchanged between arteries and veins. This concept is utilized efficiently by mammals in heat loss, heat conservation or keeping a specific organ cool regardless of body temperature. Some examples illustrating this principle as related to thermoregulation are summarized below.

A. THERMAL GRADIENT OF THE NASAL PASSAGEWAY

Jackson and Schmidt-Nielsen (1964) have described a model of heat exchange in the nasal passageways of Merriam's kangaroo rat, *Dipodomys merriami*. A longitudinal gradient of temperature inside the nasal passageway allows the animal to reduce its respiratory evaporation. As exhaled air leaves the lung, it is saturated and its temperature is equal to that of the body. On its passage to the outside, exhaled air undergoes substantial cooling and water vapor is recondensed on the walls of the respiratory tract. In addition, recent studies on histological peculiarities of the pulmonary tract of some desert rodents show the presence of some anatomical features which reduce pulmonary water loss. For example, Babero et al. (1973) found that mucous secreting cells combined with reduction of respiratory surface of the lung are anatomical features present in some desert rodents. They suggest these features are related to water economy.

B. THERMAL GRADIENT OF APPENDAGES

The unique vascularization in the extremities of some mammals constituted the principle of the countercurrent heat exchange model proposed by Scholander and Krog (1957). Whales swimming in water near freezing point are well insulated with blubber, except in the region of the flukes and flippers which are highly vascularized and uninsulated. To prevent these extremities from freezing and to conserve heat when arterial blood flows to these structures, it exchanges heat with the returning cold venous blood which is reheated before it enters the body. These arteries and veins occur in close proximity, in fact, the veins completely surround the artery (Schmidt-Nielsen, 1970). Similarly, in the limbs of large terrestrial arctic mammals, arteries and veins branch into a large number of parallel, intermingled vessels forming a vascular rete which allows recovery of heat from arterial blood and reheating of venous blood.

C. A COOL BRAIN

Hyperthermia is used as a physiological strategy by many large desert mammals. Recently, Taylor (1972) reported that African gazelles may reach body temperatures of 46° C without apparent ill effects for as long as 6 hrs. It is inconceivable that heat sensitive organs i.e., brain, can tolerate such high temperature. This paradox was resolved by selective cooling of the brain through countercurrent heat exchange as demonstrated in sheep, gazelles and dogs (Hayward, 1968; Taylor, 1972; Baker and Hayward, 1968). Blood going to the brain is cooled by the unique vascularization of the brain. The external carotid artery passes through the cavernous sinus which is filled with venous blood returning from the nasal passages where it was cooled by evaporation; thus arterial blood to the brain exchanges its heat and is cooled on its way to the brain.

D. COOL TESTES

In some ungulates it is necessary to keep the testes cooler than the body temperature for formation of fertile sperm. This is accomplished by a countercurrent heat exchange in the vessels supplying the testes. Evaporation from the scrotal surface cools the testes while the vascular heat exchanger prevents heat gain from the body.

E. HYPERTONIC URINE

The ability of mammals to conserve water by formation of highly concentrated urine is well known as a mechanism for survival. Production of hypertonic urine is associated with the unique anatomical arrangement of the kidneys. The significance of these peculiar structures received little attention until 1942 when Kuhn and Ryffel suggested that the kidney operates as a countercurrent multiplier system. Their suggestion has been confirmed experimentally and is well presented in general physiology textbooks. The maximum concentration of urine depends primarily on three factors (Schmidt-Nielsen, 1961): (a) the relative length of the multiplier system, (b) the concentration difference between the two limbs of loop of Henle and (c) the rate of flow through the system. Maximum concentration is directly related to the first two factors and inversely related to the third one. Although Schmidt-Nielsen (1964) has correlated the length of renal papilla, numbers and length of the loop of Henle to habitat and diet in a few species of mammals, comparative studies involving subspecies from different habitats are needed to develop a general model relating kidney features to their concentrating ability.

SUMMARY AND CONCLUSIONS

Evaluation of the major ecogeographic rules as applied to mammals has been presented. In many instances, Bergmann's and Allen's rules which relate to body size and length of extremity apply to related species or subspecies. Exceptions can be found to either of these rules. Their value in studies of geographic and ecologic variation remain but their functional interpretation as physiological adaptation is limited and still subject to debate.

Validity of Wilson's is questioned by Scholander and his co-workers for small mammals. Some evidence, however, for small mammals conforms to the rule. Gloger's rule appears to be valid for many mammals although there are numerous exceptions to the rule.

Exceptions to these rules are not surprising and are to be expected since body size, form, pelage thickness and coloration are all modified by selective forces emanating from the total environment and not just in response to prevailing climate.

Ecogeographical rules will undoubtedly be retained in ecological literature, however, the numerous exceptions should be noted and the selective forces must be identified and studied. Although these rules have historical significance for the student of thermoregulation, it should be appreciated that in most instances they have little thermoregulatory significance. Examples which illustrate a thermoregulatory function of the rules usually deal with mammals inhabiting specialized environments such as fossorial, aquatic or large ungulates. Also there appear to be associated anatomical features such as countercurrent mechanisms which augment the morphological features as they relate to thermoregulation. In conclusion, survival of mammals, particularly those inhabiting extreme environments, depends on interaction of varying behavioral traits.

REFERENCES: See pp. 582-583

Adaptation to Climatic Zones and the Ecophysiology of Ruminants

by

W. V. Macfarlane

Ruminants originated from several stocks which evolved in different climatic zones. Since it probably takes 3 to 5 million years for a generic level of structure to be stabilized by natural selection (Simpson, 1953) the present ruminants of the world are largely late Pleiocene or Pleistocene products. Part of their recent distribution is due to human intervention but the earlier wanderings were induced by other pressures (Miller, 1969). Physiological ecotypes have clearly dispersed widely but basic patterns of function suited to original habitats can still be discerned (Macfarlane et al., 1971, 1972).

From earth satellites it is now easy to see the main climatic zones of the world – jungle, monsoonal savanna, hot desert, temperate winter rain and cold desert. The belt of tropical clouds yields high rainfall which washes salts from the earth (Fig. 1). In this region there is no shortage of water or heat; but metals, sodium and iodide are leached from soils (Howard et al., 1962). Between latitudes 5° and 15° there is a monsoonal sequence of hot wet and hot dry seasons. Rain falls in the summer so that water for evaporative cooling is readily available, but food and water are scarce in the dry period. The great deserts of the world lie in the 15° and 30° belt between the wet tropical and wet temperate air masses. In these hot dry zones there may be accumulations of salt in soil, water and also in halophytic plants. Animals in the desert, with intermittent food and water, need reserves of energy and water and they can cope with high salt intake. Between latitudes 30° and 45° there is a predominantly winter rainfall with hot dry summers. Salt is not a problem, but supplies of water and of energy-yielding foodstuffs may be low in summer, so that fat storage has survival value. In the colder zones there is usually shortage of food in winter, but water is freely available through the year. In cold temperate latitudes with a short growing season, fat storage is essential, together with insulation against the winter cold. Beyond latitude 60° there is a cold desert, where neither food nor water is readily available.

Most ruminant groups were well defined in the early Pleiocene. Extant genera are probably 5 to 10 million years old but the palaeontology is patchy. During the Pleistocene there was considerable migration between Africa, Asia and the Americas in both directions. But the functions which came to be associated with the different environments can be surmised from present forms and their zones of origin.

CATTLE like guar, banteng and buffalo are still tropical animals, heavily dependent upon water, which moves rapidly through them (Macfarlane and Howard, 1970). *Bos indicus* breeds will cross with the banteng and they are differentiated into wet land breeds like the Hariana, or into more arid environmental types such as the Sahiwal or the Boran (Table 1). European offshoots, through the Aurochs, maintain a high rate

Fig. 1. Climatic zones of the earth seen from 180,000 km distant. The equatorial cloud and rain belt allowed the evolution of high water and energy consuming types of mammal – cattle, eland, pig.
In the semiarid zones sheep, goats, horses, and wildebeest developed. Hot deserts gave rise to camel, oryx and rodents or marsupials like Notomys Dipodomys or Dasyuroides.
In the wet temperate zones deer evolved with high water demands.
The dry arctic gave rise to desert forms like the musk ox with a camel-like water turnover.
Probably Pleistocene migrations caused many genera to move well beyond their eco-physiological limits.

of water turnover in cooler country, while breeds such as the Jersey have coat characteristics and genetic red cell muco-protein characters which relate them closely to more tropical *Bos indicus* breeds. Throughout these bovids there is not only high water turnover but also a high rate of energy turnover; and metabolism runs at about twice the standard rate-for-size on the Kleiber (1961) metabolic $kg^{0.75}$ (Macfarlane and Howard, 1972). Forms like *Bos taurus* use energy at the highest rate while the more desert types of *Bos indicus* have lower metabolic rates. Amongst all these animals the kidneys are not designed for high levels of urinary concentration, so that water flows through readily. Concentrating power does not rise above 2.5 osmol/l. The kidney is

Table 1. Ecophysiological hierarchy of water turnover among grazing cattle, antelope, goats and sheep

Breed species or genus		Body weight kg	Body solids % wt	Water turnover/24 hr		Environment
				ml/kg	$ml/kg^{0.82}$ ± ± SD	
Hereford cattle	(5)	616	39	104	328 ± 51	Kenya
Zebu cattle	(4)	262	28	105	284 ± 40	highlands
Donkey	(2)	160	22	126	314	lush *Cynodon*
Masai sheep	(8)	29	23	106	193 ± 35	grass
Somali goat	(7)	22	23	79	136 ± 23	lat. 1°S
Boran cattle	(6)	417	28	76	224 ± 14	Dry
Eland	(5)	247	20	78	213 ± 43	grasslands
Wildebeest	(1)	175	27	53	137	Kenya
Kongoni	(2)	88	15	52	116	lat. 1°S
Oryx	(1)	136	30	29	70	
Buffalo	(3)	354	21	212	535 ± 33	Wet
Shorthorn	(4)	322	26	168	461 ± 48	tropics
Santa Gertrudis	(2)	523	37	141	373 ± 60	Arnhem Land
Banteng	(4)	372	23	132	348 ± 22	Summer
Zebu	(3)	532	39	121	350 ± 36	lat. 12°S
Shorthorn cattle	(2)	447	36	149	415	
Leicester sheep	(2)	62	39	166	362	Spring
Donkey	(1)	375	38	75	217	pasture
Goats	(2)	36	40	89	170	Temperate
Camel	(1)	716	46	39	126	lat. 35°S
Reindeer	(6)	100	23	128	293 ± 63	Late winter
Moose	(1)	186	23	111	284	taiga Alaska
Goats	(3)	70	33	52	112 ± 10	lat. 65°N
Musk oxen	(2)	324	34	35	99	T° −40 to −5° C

Among ungulates there are hierarchies of water turnover rates. The tropically evolved types have high rates of water use, while desert adapted forms have low rates of turnover. In different environments the rank order is maintained, so tropical cattle in hot desert do not change from their high demand for water.

relatively insensitive to vasopressin, about 50 times less than the sensitivity for water reabsorption achieved by camels (Macfarlane et al., 1964, 1967). In a similar category is the tolerance to salt (NaCl), which is poor in all cattle so that 1% of salt in drinking water is near the limit of physiological tolerance (Macfarlane, 1971).

This series of physiological rate constants (water, energy, salt handling) is characteristic of these species and makes them basically suitable for grazing in wet areas or jungle where water is not limiting and where the food supply is consistent through the year, with high water content (Table 2). The muzzle of cattle is covered with hairless

Table 2. Adaptive functions among desert, intermediate and jungle types of ruminant

	CATTLE	SHEEP	CAMEL
Energy turnover	+++	++	+
Water turnover	+++	++	+
Renal concentration	+	++	++
Vasopressin sensitivity	+	++	+++
Salt tolerance	+	++	++++

squamous epithelium as a further adaptation to wet grazing. Through polygonal segments of epithelium, ducts lead in to acinous glands of salivary type. These glands secrete not only mucus, but also a high-potassium, low-sodium fluid which is ejected by myoepithelial cells. Acetylcholine from parasympathetic nerves induces secretion and ejection from the glands (Brewer and Macfarlane, 1967). The ducts are glandular, also, and reabsorb sodium while secreting potassium into the fluid, before it finally reaches the nose. It appears that this type of muzzle is a protection against water on marshy vegetation. Long-term exposure to water usually leads to excoriation and breakdown of the epidermis, but with mucus protection cattle graze freely in wet environments.

SHEEP and GOATS appear to have originated in semi-arid and probably upland areas of Eurasia such as those in Kazakhskoya (Zeuner, 1953; Ucko and Dimbleby, 1969). Sheep and goats can interbreed, and their skeletons are virtually indistinguishable. They have lower requirements for water, better renal concentrating power and lower metabolic rates than cattle. These characteristics suit them for more arid environments (Macfarlane and Howard, 1972). Sheep have been most successful in the middle latitudes between 25° and 55° from the equator. They adapt fairly well to aridity because of lower water turnover and correspondingly reduced metabolic rates than cattle. Kidneys can concentrate to 3 osmol/l and in some breeds up to 3.8 osmol/l (Macfarlane, 1964). Salt tolerance is not high but it is above that of cattle. Sheep and goats survive with 1.3 to 1.5% sodium chloride in drinking water (Pierce, 1968). The renal sensitivity to vasopressin is greater than that of cattle by a factor of 25. Sheep differentiated into high mountain breeds (such as *Ovus dalli* which migrated to America), wet temperate breeds such as Leicester and Romney; into forms suited for warm arid climates such as the Merino, Awassi, or Bikaneri, and into tropical breeds such as the Ogaden, Meidob or Dorper which resist heat and aridity. Each type has characteristic functions which suit it to the environment. Leicesters use more water and energy than

Merinos and they are rather low in renal concentrating ability. When exposed to high concentrations of salt in the diet the Leicester uses more water per metabolic kg to wash through the excess electrolyte, than Merinos grazing with them (Macfarlane et al., 1967). The intermittency of food and water supplies in the environments where sheep appear to have evolved has led to several adaptations. In areas with winter rainfall and summer drought there is a close link between water and food intake. When there is succulent food, up to 3 or 4 times more dry matter can be ingested than during the dry summer periods. The fat which is deposited during periods of plenty may be used in dry summers when feed is scarce (Macfarlane and Howard, 1970). At the same time both food and water turnover are reduced when the food supply is low (Fig. 2). The amount of fat laid down by different breeds is not the same even though they graze the same pasture. Dorset × Merino or Leicester × Merino cross sheep, for instance, will put down 2 parts of fat to 1 of other tissue during times of adequate feed, while Merinos deposit 2 of tissue to 1 of fat.

Among goats there is a similar range of ecophysiological adaptation. Goats are less water dependent than sheep, and wider ranging in the vegetation they graze or browse. They penetrate to and survive in deserts better than most sheep. The Sinai goat has a 30% lower metabolic rate and a lower water turnover than the usual middle-eastern goat of Palestine and Egypt (Shkolnik et al., 1972). These physiological differences would be equivalent to those between Leicester and Merino sheep, but in a lower register.

CAMELIDS evolved more than 5 million years ago under arid conditions in Nebraska-Dakota. The genus *Camelus* walked to Asia and then to Africa 3 million years ago. *Lama* and its relatives, the guanaco and vicuna, moved south to the hills, sometimes in the jungle and sometimes to high wastelands, about 3 million years ago. The camel is characterised by a low rate of water turnover, a low metabolic rate and high tolerance to salt (it can live while drinking 5% sodium chloride, Maloiy, 1972). The oval red cells of camels have very high resistance to osmotic rupture. In addition they carry the highest level of Na K ATP-ase activity found amongst ruminants. This is probably associated with ability to pump electrolytes. The precisely known duration of separation of the genera *Camelus* and *Lama* could be expected to have lead to some changes of rates of functioning in very different environments. It was interesting, therefore, to be able to compare camels, llamas and guanacos together on the same feed in the same environment. There was near identity of water turnover rates in the genera, with the llama having a slightly lower rate of turnover per metabolic kg than the camel. The red blood cell of the llama was also resistant to haemolysis and had an even higher level of Na K ATP-ase activity than that of the camel. In these respects no physiological drift had occurred in the course of 3 million years. Yet the desert adaptation of the camel is of a high order, presumably implanted and sustained for something over 3 million years. Similarly the llama seems to have maintained its initial adaptations in non-desert environments.

Amongst these ruminants there is a hierarchy of linked functions (Macfarlane et al., 1971). The jungle syndrome of high energy and water turnover, with low salt tolerance and poor kidney concentration, persists in animals which subsequently may be depastured in arid areas (Table 2). Cattle on the Saharan fringe or in the central Australian desert maintain high water demands like the banteng or buffalo. Cattle die after 3 to 5 days in summer on the desert if water is not available. After such a period of dehydration, they rehydrate rapidly. *Bos taurus* cattle replace 75% of water lost within 15 min. They are more avid for water than similarly dehydrated camels who replace only 60% of loss in the initial drink (Siebert and Macfarlane, 1969). The rate of dehydration of the different ruminants during summer in dry conditions is a mirror of the rates of water turnover. With daily air temperatures of 35° to 40° (under solar radiation, adult *Bos taurus* will lose 20% of body weight in 4 days, Merino sheep in 6 to 8 days and camels in 12 to 15 days). During dehydration, plasma and extracellular volume is reduced more in cattle and sheep than in camels. Sheep can sustain a loss of 45% of extracellular fluid without dying (Macfarlane et al., 1961). When camels are dehydrated the rate of loss of plasma volume is slow so that they maintain good circulation, partly apparently because they synthesise more albumin to increase the oncotic pressure of the plasma, partly because their capillaries are not so permeable to albumin as in other species (Macfarlane, 1964; Siebert and Macfarlane, 1969).

During water lack in hot environments sheep and cattle continue to excrete sodium (as the extracellular fluid is diminished) so that the ratio of sodium to potassium in the urine remains high. The sodium in these animals rarely rises above 175 meq/l during dehydration. The camel, however, does not excrete all the sodium that would be lost from extracellular fluid as it dries out, so that sodium is retained and may rise to over 200 meq/l in the plasma without serious consequences. This is associated, also, with retention of urea which may reach more than 500 mg/dl (Macfarlane, 1964).

On rehydration cattle take in a large volume of water rapidly and this moves quickly across the rumen as well as the intestinal walls, to plasma. The rate of passage (as in sheep) may sometimes be so fast that haemolysis occurs and red urine is produced. Even if that is not so there is an initial period of movement of some salt and urea from the plasma to the intestine, but within 2 hr water dilutes the electrolytes of the blood and a diuresis occurs. In the camel, although the proportion of water lost is not fully replaced in the first drink, the fluid taken is slowly absorbed and does not equilibrate with plasma solutes for more than 24 hr. During this time urine outflow of sodium, potassium and water is halved and at the same time plasma urea and sodium move into the fresh water in the gut (Macfarlane, 1968; Blair-West et al., 1970). Blair-West et al. (1972) have analysed this electrolyte behaviour further, and found that during dehydration sheep do not secrete more aldosterone although plasma renin rose. But when the urinary water and sodium output falls after drinking, there is a rise of both aldosterone and renin, which would account for the sodium reabsorption. This secretion of aldosterone presumably follows a fall of plasma sodium as electrolyte moves to the gut to raise the osmolality of the newly ingested water and the stimulus to renin release is presumed to be the sodium concentration at the macula densa near glomeruli. Slowly this isoosmotic fluid is absorbed. After passing the intestinal wall it appears to move into the cellular space, and is not quickly excreted, in contrast with cattle. The passage of urea from the blood through the salivary glands to the rumen then allows new synthesis of protein by the bacteria of the rumen. This can be looked on as a mechanism by which urea is stored in the extracellular space during dehydration, and brought into use again when salivation begins after hydration. In a similar way, to expand plasma volume there is sodium waiting in the extracellular space to bring water towards osmotic balance with cell fluids. This same process occurs to some extent in sheep but it is less well developed in cattle than in any of the other genera. The extraordinarily high tolerance of camel cells for sodium is a useful adaptation to quick restoration of plasma volume when water becomes available after a long period of deprivation (Siebert and Macfarlane, 1969). Similarly the slow turnover rate for water determines the rate of water loss and thus the time in the summer desert that a camel can travel without water (15 days or more), while high turnover rate animals like cattle have a much more limited range.

In parallel with water, the metabolic rates determine survival without food. Although when water is not available there is depression of metabolic rate, each unit mass of an ox is using about twice as much energy as a unit of camel so that any fat reserves will last about twice as long in the camel as in cattle. A camel with 150 kg of fat in its hump could last half a year without eating.

Rate functions therefore determine the generic levels of survival in extreme conditions as well as the rates of recovery from deprivation of water.

Other hierarchies. Amongst the ANTELOPE there is a range of water and energy turnover rates equal to that found amongst the bovids (Table 1). The largest antelope, the eland, has water turnover and metabolic rates very similar to those of *Bos indicus* in the same environment (Macfarlane and Howard, 1972). The ELAND is found near the equator and it is possible that with its high metabolism and fast rate of water usage, it originated in more wet or jungle habitats. Its survival potential without water is not known. The KONGONI and the WILDEBEEST measured at the same time and place as the eland and with access to the same food, have water turnover rates about half those of the eland. The metabolic rates were also lower (Rogerson, 1968). The ORYX and the ADDAX both live in hot desert areas beyond zones where camels normally would persist. Measurements on these are few but estimates on one oryx studied in parallel with *Bos indicus* (Boran) cattle indicated that these rate functions were lower than in the camel (Macfarlane et al., 1971). It appears, therefore, that modern antelope probably had ancestral forms evolved in different environments, so that the eland would be like cattle; kongoni and wildebeest like sheep or goats, and the oryx like the camel in overall functions.

A similar hierarchy was examined in Alaska. REINDEER use energy at an even higher rate than cattle and have an equally high rate of water turnover (McEwan, 1970; Macfarlane and Howard, 1972). MOOSE are not far behind; but the MUSKOX uses water at a rate similar to that of the camel. DEER appear to have evolved in temperate climates and the reindeer presumably has been translated from a cool wet zone of evolution to the arctic desert where it survives by considerable food intake and by making use of melted snow for water. It is the physiological analogue of cattle displaced to arid lands. Muskoxen, on the other hand, are arctic in evolution and their low turnover rates for water presumably mirror the desert conditions of their evolution.

THYROXINE AND METABOLIC RATE. There is a general impression that thyroxine is related to metabolic rate and that the plasma level of thyroxine or tri-iodothyronine is proportional to metabolic rate. This derives mainly from studies on man and rat.

Amongst ruminants in the field, however, it is clear that the plasma level of thyroxine is not specifically related to metabolic rate. Similarly, plasma free thyroxine is no clear indicator of metabolic rate since camels have as much free thyroxine as cattle and yet camels have much lower rates of metabolism than cattle.

Table 3. Plasma bound iodine and thyroxine secretion rates of ruminants in the arid tropics

	PBI	Plasma	Thyroxine secretion rate	
	μgI/dl	thyro-xine μg/dl	μg/kg/24 hr	μg/kg$^{0.75}$/24 hr
Goat (1)	4.7	7.2	8.3	19.9
Merino sheep (3)	3.9	6.1	6.9	18.5
Buffalo (2)	4.6	7.0	3.3	14.4
Banteng (1)	3.5	5.4	2.3	9.7
Cattle (4)	4.4	6.7	1.6	7.5
Camel (2)	7.8	11.9	1.9	7.7

All animals were together and fed lucerne hay in the open, at Alice Springs lat. 24°S, during May (autumn, dry but not hot). Camels and goats have the lowest metabolic rates and water turnover of these genera, but these bear no relation to circulating thyroxine concentration, nor to the secretion rate of thyroxine. There is, however, an inverse relationship between body weight (size) and the thyroxine secretion rate, so large animals (camel, cattle) have lower rates than small ruminants (sheep, goats).

The rise and fall of thyroxine in the plasma is not necessarily related to the rate of use of thyroxine. In sheep there is often a high level of plasma thyroxine when this hormone is being turned over at the highest rate. Similarly with low secretion rates for thyroxine there may be high plasma levels. When ^{125}I thyroxine is injected and used to estimate turnover rates (Table 3) several findings emerge:

1. The secretion rate of thyroxine is raised when the food supply and food intake increase. In sheep the basal utilization rate is about 1.5 μg/kg/24 hr, and this rises by 0.1 μg/kg for each additional 100 g of dry matter eaten daily.
2. Thyroxine secretion rate is proportional to body mass and not to metabolic rate. Thus a camel and a cow of the same mass have the same thyroxine secretion rate on a given pasture, but a cow may have a metabolic rate over twice that of the camel. Sheep and goats have the highest ruminant secretion rates, more than twice those of the camel.
3. Although there is some seasonal fluctuation in rate of thyroxine secretion the rise during the cold season is associated with increased food intake rather than and in addition to any direct effect of cold on ruminant thyroid activity.

In each generic group there is a characteristic relation between plasma thyroxine set-point and the turnover rate for the hormone. Among the larger mammals, thyroxine secretion rate lies between 7 and 10 μg/kg$^{0.75}$/24 hr, but the camelids have high PBI levels of 7 to 9 μg/dl, the cattle (zebu, banteng, buffalo, *B. taurus* breeds) between 4 and 7 μg/dl, and donkeys 1.5-2 μg/dl. Sheep secrete at 10 to 15 μg/kg$^{0.75}$/24 hr, goats at 12-20 μg, while their PBI ranges are 5-8 μg/dl. These quantities show some relation to size, but little to metabolic rate. The ranges were determined over a variety of pastures and seasons.

Thyroxine may be regarded as a hormone which is mobilized to help metabolise increments of food intake, but it is not the primary determinant of metabolic rate (Macfarlane et al., 1974).

SUMMARY. The rates at which energy and water are used by different genera of ruminants vary by a factor of three, and the rates for water and energy are closely linked. Other functions such as renal concentrating power, sensitivity to vasopressin and tolerance of salt are linked with water turnover rate. Desert animals use water and energy sparingly, concentrate urine and tolerate salt, while jungle-type animals respond in the opposite fashion.

It seems likely that evolution of functions for a desert or wet habitat took place millions of years ago, and the functions do not change rapidly. Cattle (with a wet tropical origin) can survive in desert areas but because of their high metabolic rates and rapid water turnover they die in a few days when deprived of water. They have been moved to arid grasslands but they do not change their rate functions to those of the camel, which evolved in dry lands over 3 million years ago. The persistence of this rate function is illustrated by the retention by llama and guanaco of water turnover rates as low as those of the camel, even though they parted 3 million years ago and *Lama* has lived in wet, cold and diverse habitats over that period.

The relative metabolic rates of ruminants are not fixed by the secretion rate of thyroxine. Genera of the same size, but differing by a factor of at least 2 in metabolic rate (cattle and camel) have the same secretion rate when in the same environment. Thyroxine is secreted proportionally to body mass rather than metabolic rate. The amount of food eaten is the chief determinant of the rate of using thyroxine. And the plasma thyroxine concentration is no indicator of the metabolic rate.

REFERENCES: See p. 584.

Adaptability of Different Breeds of Cattle to Heat

by

J. E. Vercoe

INTRODUCTION

In tropical environments, productivity of cattle can be increased by deriving adapted genotypes. Tropical environments are characterized by high temperatures and often high humidity, parasitic and other diseases, and by large seasonal variations in food quality and quantity. Cattle must be able to cope with these factors in order to survive, to be fertile, and to produce fast growing calves. Some breeds can do this better than others and by cross-breeding and selection, high producing adapted cattle can be obtained.

In this Section, the reactions to heat of contrasting breeds of cattle (Brahman and Africander representing tropically adapted breeds and Hereford and Shorthorn representing temperate breeds) will be considered. Information will be drawn mainly from research work being done at Rockhampton, Queensland, Australia, the aim of which is to define the make-up of an animal that will produce beef most efficiently under the climatic and other environmental conditions that prevail. Attributes of the Brahman and Africander breeds, and the Hereford and the Shorthorn breeds, are being combined in the following cross-breeds: Brahman × Shorthorn × Hereford (BX), Africander × Shorthorn × Hereford (AX), and Shorthorn × Hereford (SH). The contribution of "blood" from the zebu-types (Brahman and Africander) to the respective crossbreeds is approximately 50%. The effects of heat stress on growth, feed utilization, and on other aspects of metabolism and physiological function are being studied under field conditions and in intensive climate-room experiments. Work has recently commenced to find out how high temperatures and heat stress influence the relative fertility of the three breed-types which are markedly different to each other (Seebeck, 1973). However, this aspect is covered elsewhere in this book.

FIELD OBSERVATIONS ON THE EFFECT OF HEAT

The average weights at various stages of growth of the three breed-types of cattle under investigation are shown in Table 1. Details of the conditions and herd management procedures under which these data were collected have been published elsewhere (Kennedy and Chirchir, 1971).

The importance of heat tolerance in determining the growth of young cattle has been demonstrated (Turner, 1962). The heat stress experienced under grazing conditions during the summer months in northern Australia significantly depresses the growth of

Table 1. Liveweights (kg) of breeds of cattle grazed together at various stages of growth (From Kennedy and Chirchir, 1971)

Breed[a]	Birth	Weaning (8-9 months)	Weaning + 4 months[b]	Weaning + 9 months[b]
BX	29.7	199.2	211.7	294.8
AX	31.0	185.5	204.2	282.8
SH	31.8	170.3	181.0	244.2

[a] See text for symbols.
[b] Weights for females only

susceptible animals. When the heat stress of Hereford yearlings was partly relieved during the summer months by clipping, so that the rectal temperature was lowered by 0.7° C, gains in weight increased from 90 to 102 kg. From the relationships between rectal temperatures and gains, Turner (1962) has estimated that differences in susceptibility to heat accounted for approximately half the difference in growth rate between locally adapted Hereford cattle and a strain of the same breed introduced from a more temperate region. In addition, the difference in rectal temperature could account for 20 to 25% of differences between animals of the locally adapted British breeds, 25 to 30% of the difference between SH and AX and for about 15% of the difference between SH and BX.

The AX and BX crossbreeds maintain lower rectal and skin temperatures under hot conditions in the field or in yards. The sweating rates of BX are lower than SH under mild conditions but become higher as the ambient temperature increases. In Shorthorn and Hereford steers, it has been observed that animals with high sweating rates have lower body temperatures and that differences between animals are consistent under stressful conditions (Schleger and Turner, 1965). Related to these observations are measurements of water turnover in grazing animals. Africander and AX steers have consistently lower turnovers than SH, but the water turnover in the zebu-breeds increases relatively more than that of the SH during the summer months (Springell, 1968).

It is evident that the way in which various breed-types react to hot conditions influences their growth rates. The mechanisms by which this occurs are being studied under experimental conditions in which the ambient temperature and humidity can be controlled.

EFFECTS OF HEAT MEASURED UNDER CONTROLLED CONDITIONS

In many of the published papers relating to air temperature and loss of production, the contribution of the specific effect of heat in the overall effect has not been clearly defined. Cattle have been exposed to a variety of environmental conditions which have often resulted in depression of food intake and variable effects on the rectal temperatures of the breed-types under investigation. The experiments at Rockhampton have, as a rule, titrated each breed-type to the same elevated rectal temperature (*c.* 1.3° C above normal) keeping food intake constant at a restricted level. By this method the specific effect of heat (i.e., the effect independent of changes in food intake) has been assessed on some aspects of food utilization and metabolism. The advantages of this experimental procedure can be seen in Tables 2-4; air temperature is a variable which reflects the relative heat tolerance of the breed-types and the effect of the same increase in rectal temperature on aspects of metabolism in each breed can be directly assessed.

1. EFFECTS OF INCREASED TEMPERATURE ON FOOD INTAKE

In equilibrium conditions, heat produced equals the heat lost by evaporative and other means, and as air temperatures increase, thermal equilibrium can be maintained by reducing food intake and consequently heat production.

In the three breed-types of cattle under study at Rockhampton, increase in air temperature results in a decline in food intake. However, the air temperature at which food intake starts to decline varies for the three breeds and is also thought to depend on the quality of the feed offered. In general terms, the intake of SH starts to decline at a lower air temperature than either the BX or the AX. A hypothetical model is represented diagrammatically in Figure 1. The model indicates that the rate of decline in food intake with increasing air temperature may be faster for SH than for the zebu-crossbreed cattle, which is consistent with evidence that BX steers have a lower fasting heat production per unit weight and a lower heat increment of feeding (Vercoe, 1970; Vercoe and Frisch, 1974). The correctness of this model is being studied.
The effect of high air temperatures on food intake is not only significant but can be an important criterion in the overall assessment of an animal's heat tolerance. There are considerable differences in the reaction among animals within a breed to an increase in air temperature. It has been observed that animals which suffer the smallest percentage reduction in ad libitum intake when exposed to high air temperatures are those which have small increases in rectal temperature, and, more significantly, high weight

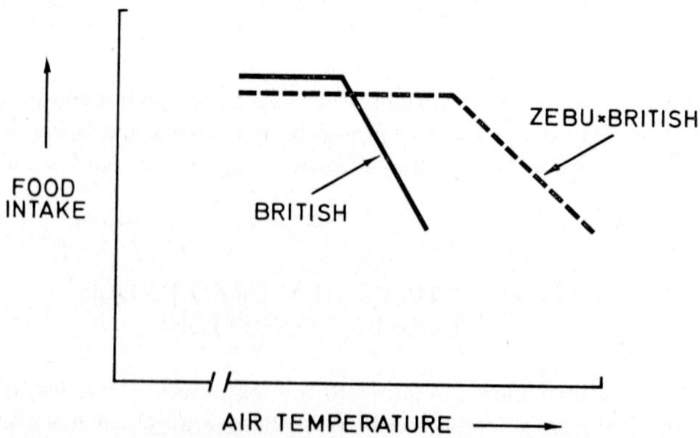

Fig. 1. Hypothetical model representing the effect of continuous exposure to a range of air temperatures on food intake in British and zebu × British cattle.

gains under grazing conditions. In some animals, the decline in food intake that occurs as air and rectal temperature increase is very marked. Usually these animals show obvious signs of distress, such as high respiration rate, excess salivation, high rectal temperature, but sometimes an animal stops eating before the onset of these extreme symptoms.

2. THE EFFECT OF INCREASED RECTAL TEMPERATURE ON FOOD DIGESTIBILITY

The effect of raising the rectal temperature of mature cattle by approximately 1.3° C has little effect on the dry matter digestibility of high or low nitrogen roughage diets. Any increase is of the order of one percentage unit or less (Vercoe, 1969; Vercoe and Frisch, 1970; Vercoe, Frisch and Moran, 1972). However, in young calves fed a concentrate calfmeal and a small amount of lucerne hay, dry matter digestibility is significantly depressed by approximately 4 percentage units (O'Kelly, 1973a). This is illustrated in Table 2.

Table 2. Apparent dry matter and nitrogen digestibilities, nitrogen balance in unstressed and heat-stressed cattle

Breed	Treatment (Rectal Temperature in Parenthesis)	Apparent Digestibility Dry Matter %	Apparent Digestibility Nitrogen %	Nitrogen Balance g/day	Room Temperature °C
	Mature Steers (From Vercoe et al., 1972)				
BX	Stressed (40.0)	62.8	71.3	−1.8	39.4
	Normal (38.2)	60.8	71.1	9.0	25.8
AX	Stressed (39.6)	61.2	70.5	−2.7	39.5
	Normal (38.3)	60.7	70.2	11.3	27.5
SH	Stressed (39.8)	59.9	69.1	0.4	31.7
	Normal (38.7)	59.4	68.7	6.8	26.8
	Calves (3-4 months) (From O'Kelly, 1973)				
AX	Stressed (39.5)	71.5	72.0	−4.2	38
	Normal (38.1)	74.0	72.5	6.4	23
SH	Stressed (39.5)	70.9	70.2	0.5	33
	Normal (38.4)	75.8	74.4	9.8	23

In mature steers, the air temperature required to increase the rectal temperatures of the BX, AX and SH cattle by 1.3° C (with vapour pressure remaining similar) is approximately 40.5° C, 40° C and 32° C respectively, but in younger calves (3 to 4 months old) the corresponding values are 38° C for AX and 32° C for SH.

3. THE EFFECT OF INCREASED RECTAL TEMPERATURE ON APPARENT NITROGEN DIGESTIBILITY AND METABOLISM

Raising the rectal temperature of either mature cattle or calves has little effect on apparent nitrogen digestibility. However, total urinary nitrogen is markedly increased;

and the increased amount of nitrogen excreted is similar for AX, BX and SH cattle when the rectal temperature increase is similar. The increase in total urinary nitrogen is largely due to both an increase in urea nitrogen and creatinine excretion (Vercoe, 1969; Vercoe and Frisch, 1970; O'Kelly, 1973a).

On a high nitrogen roughage diet (lucerne hay), urinary urea increases within 24 hrs of the rectal temperature increase, but on a low nitrogen roughage diet (tropical-grass hay), the increase in urea does not occur until 4 to 5 days after the increase in rectal temperature. The increase in urinary urea is independent of changes in plasma urea which is unaffected by the rectal temperature increase. However, on the low nitrogen diet, the increase may be partly associated with an increase in urine volume since water loading is known to increase urinary urea and lower plasma urea concentration (Vercoe, 1971).

The nitrogen content of the diet has no effect on the time taken for creatinine excretion to increase following a rise in rectal temperature; the increase is immediate. Also, the amount of the increase is directly related to the increase in rectal temperature and is of the order of 2 g/day per °C for steers weighing 250 kg. Diurnal variation in creatinine excretion parallels diurnal variation in rectal temperature.

It has also been observed that plasma α-amino-nitrogen concentration is increased by an increase in rectal temperature but plasma total protein concentration remains unaltered (Table 3).

The overall effect of an increased rectal temperature is to reduce nitrogen balance. In fact cattle, which in an unstressed state may be in a positive balance of 10 g/day, may have a negative balance of 2 g/day when their rectal temperatures are increased by approximately 1.3° C (Table 2).

These effects of increased rectal temperature on nitrogen metabolism are consistent with the view that an increase in muscle catabolism occurs when rectal temperature is raised. Part of this increase in catabolism may be contributed by the van't Hoff effect. A major influence may be due to hormonal disturbances, in particular an increase in the secretion of adrenal corticoids. Thyroid activity is depressed in cattle at high ambient temperatures, but the effect of such a deficiency on protein metabolism is not clear.

4. EFFECTS OF INCREASED RECTAL TEMPERATURE ON LIPID METABOLISM

Continuous exposure of the three breed-types under consideration to environmental temperatures which cause an increase in rectal temperature of 1.3° C, produces marked changes in plasma lipid constituents (O'Kelly, 1973a, b). An increase in rectal temperature in mature steers lowers the concentration of total cholesterol, phospholipid, and alkaline-phosphatase activity, and increases the ratio of free to total cholesterol. Non-esterified fatty acids and glucose levels are unaffected. Similar changes have been observed when the animals are subjected to diurnal variations in ambient temperature. In three-month old calves, hyperthermia is associated with reduced plasma concentrations of cholesterol and phospholipid and increases in the ratio of free to total cholesterol

Table 3. Plasma lipid and protein constituents in unstressed and heat-stressed cattle (From O'Kelly, 1973)

Breed	Treatment (Rectal Temperature in Parenthesis)	Total Cholesterol mg/l	Free Cholesterol mg/l	Phospholipid mg/l	Non-esterified Fatty Acid μ Equiv./l	Total Protein g/l	α-Amino Nitrogen mg/l
AX	Stressed (39.5)	682	164	1,221	317	65.3	40.0
	Normal (38.1)	1,021	217	1,482	222	65.8	38.2
SH	Stressed (39.5)	437	108	880	179	68.7	40.0
	Normal (38.4)	743	160	1,218	185	68.7	37.6

(Table 3). No effect has been observed on the concentration of circulating non-esterified fatty acids, but approximately twice the amount of fat is excreted in the faeces by hyperthermic calves (O'Kelly, 1973a). The direct significance of these findings to the reduced production associated with high temperatures has yet to be fully explored. However, the fact that the plasma supplies the bulk of the metabolically active cholesterol of the adrenal gland may be relevant to the changes in activity of the adrenal gland associated with climatic stress. In this connection, it is worth noting that the increase in circulating lymphocytes and eosinophils associated with hyperthermia is opposite to those changes normally associated with the intravenous administration of ACTH.

5. THE EFFECT OF INCREASED RECTAL TEMPERATURE ON WATER METABOLISM

The total evaporative water losses in the three breed-types have been estimated from measurements of water intake and faecal and urinary water losses. Water balance, equal to water intake less faecal and urinary water, measured over a ten-day period is closely related to evaporative water loss. Liveweight is relatively constant during this time, and the change in bodywater content is therefore likely to be small.

At the same food intake and with rectal temperatures elevated by a similar amount, mature BX and AX steers have a higher water balance than SH steers (Table 4). Not only is there this marked difference between breeds in apparent evaporative losses

Table 4. Water intake, urine volume, water balance[a] and respiration rate of unstressed and heat-stressed cattle (From Vercoe et al., 1972)

Breed	Treatment (Rectal temperature in parenthesis)	Water Intake l/day	Urine Volume l/day	Water Balance l/day	Respiration Rate per min.
BX	Stressed (40.0)	39.4	16.2	18.6	104
	Normal (38.2)	34.4	24.5	4.0	25
AX	Stressed (39.6)	33.1	10.7	17.0	99
	Normal (38.3)	21.2	11.2	4.2	25
SH	Stressed (39.8)	28.8	12.9	10.3	95
	Normal (38.7)	23.1	10.9	5.3	47

[a] Water balance = Water intake – (faecal water + urine volume).

but it is also probable that the breeds differ in how they derive the water for evaporative cooling. The BX sometimes show only a slight increase in water intake but a reduced urine volume when rectal temperature is raised, whereas the AX increase intake and the urine volume remains similar (Vercoe, Frisch and Moran, 1972). The SH apparently do not possess the capacity to utilize as much water for evaporative cooling purposes as the zebu-cross breeds presumably because of reduced sweat gland activity or function (Schleger and Turner, 1965; Johnson, 1970). In younger animals, the increase in water balance in SH and AX steers is similar for similar increases in rectal temperature in contrast to the more mature animals. It is encouraging, from a genetic view point, that the marked differences between animals in water turnover is highly repeatable.

In non-stressful conditions, the differences between breeds are small in relation to the variation within breeds. When SH bulls are exposed to high air temperatures, the measured water balances are correlated negatively with both the increase in rectal temperature and the percentage reduction in ad libitum food intake associated with the rise in rectal temperature.

CONCLUSIONS

The studies on the breed-types of cattle show that as far as the major aspects of food utilization are concerned, there are no differences between breeds in the reaction to increased rectal temperature. The difference between the breed-types lies in the vastly different air temperatures that are required to produce a similar rise in rectal temperature. This difference, as well as the variations among rectal temperatures within a breed-type at a stressful air temperature, is related to different capacities for evaporative water loss.

No mention has been made in this brief outline of some of the effects of heat stress on the possible interaction that exists with the susceptibility to parasitic and non-parasitic diseases. It is possible that heat susceptible animals are also those which are susceptible to diseases. Immunological responses may be impaired either by direct effects of heat on the immune system or by the increased turnover of body proteins with which the immunoglobulin fractions are associated. Studies on the effect of heat on body protein metabolism, with particular emphasis on the various immunoglobulin fractions and lymphocyte activity which are associated with ecto- and endo-parasitic infections, should be a high priority in any programme aimed at producing efficient beef cattle for tropical and sub-tropical environments.

It is apparent that any attempt to identify individuals which tolerate high rectal temperatures is likely to be futile. Individuals that maintain a normal rectal temperature without reducing their food intake are those which are desirable to identify. The simultaneous determination of rectal temperature and water turnover under grazing conditions may provide such an identification; the measurement of rectal temperature alone is insufficient since this may merely reflect a reduced food intake.

REFERENCES: See p. 585.

Adaptability and Thermoregulatory Behaviour of Pigs

by

D. L. Ingram

INTRODUCTION

Behavior plays a large part in the regulation of body temperature in many species (Baldwin and Ingram, 1968a), and it is particularly significant for the pig. In the heat pigs have a limited capacity for evaporative heat loss (see Chapt. 1, Sect. 5d) and must therefore wallow in mud in order to keep cool. In the cold the absence of substantial external thermal insulation makes the young pig dependent on huddling or the use of nest material to keep warm (Mount, 1968).

Studies of behavior have tended to fall into two classes: (1) observation of natural behavior and (2) the use of operant conditioning. Observation of natural behavior can often be made simply by looking through the window of a calorimeter or thermocline, or photographing the animal's posture (Mount, 1968). When the animals are allowed to move over larger areas, or into places which cannot be approached without disturbing them, closed circuit television has proved to be of use (Ingram and Legge, 1970). The technique of operant conditioning involves the use of animals which have been trained to perform a response such as pushing against a switch in order to obtain a short period of heating (Baldwin and Ingram, 1968a). Variations of this basic method have also been used in which the animal breaks a light beam by putting its snout into an opening. The advantage of operant techniques is that, although they do not involve natural thermoregulatory behavior such as huddling, they do allow quantitative measurements to be made under closely controlled conditions.

NATURAL BEHAVIOUR

A. NATURAL BEHAVIOUR OF PIGS UNDER HOT CONDITIONS

Under conditions of high ambient temperature, pigs, particularly large ones, are liable to become hyperthermic and may die. Wild pigs in warm climates shelter from the heat during the day and become active at night, while the domestic pig is active during the day. This change to a nocturnal pattern of activity is also seen in other species on exposure to heat but if the temperatures are extreme, some sort of mud or water wallow is essential for the pig (Heitman et al., 1962). In the absence of water the pig has been observed to wallow in its own urine and faeces under hot conditions (Heitman and Hughes, 1949; Mount, 1968). The behavioral wetting of the skin with either mud or water can be very efficient as a means of cooling the body and gives the animal a similar capacity for evaporative heat loss as that of a sweating man (Ingram, 1965).

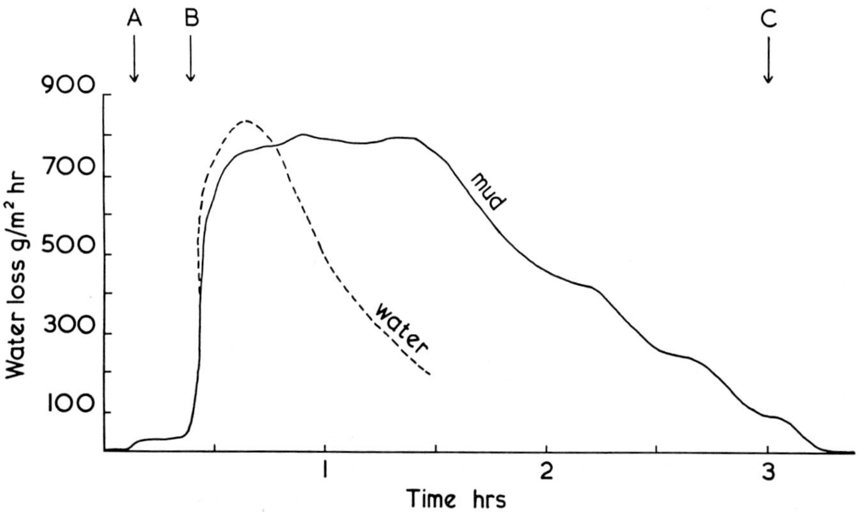

Fig. 1. Water loss from the trunk of a pig at 35°C DB. At A the sweat cup was placed on the skin. At B the skin was smeared with either mud or water. At C the capsule was removed from the skin (From Ingram, 1965).

Moreover, mud which sticks to the body and bristle holds enough water to enable the animal to make only infrequent visits to the wallow (Fig. 1).

A measure of the effect of providing a wallow is supplied by the effects on food intake and growth. Under hot conditions food intake declines particularly in adult pigs (Heitman and Hughes, 1949; Morrison and Mount, 1971) but the provision of a wallow or even a drinking bowl from which water can be splashed, allows sufficient heat loss for food intake to increase again to the control level (Bray and Singletary, 1948).

The extent to which high temperatures have an adverse effect thus depend to a considerable degree on whether the conditions allow the pig to use behavioral means of thermoregulation.

B. NATURAL BEHAVIOUR OF PIGS UNDER THERMONEUTRAL AND COLD CONDITIONS

The term "cold" is used in thermal physiology to indicate ambient temperatures below the thermally neutral zone. For this reason a given environment may be cold to one animal and warm to another. This is true of pigs where the sow and new-born litter share the same environment, since the critical temperature of a new-born pig is about 34° C (Mount, 1968) and that for a lactating sow probably near to 5° C. Under these conditions the litter members huddle together and effectively form a "group animal" with a much lower critical temperature. Mount (1960) measured the oxygen consumption of pigs in the post natal period and estimated the saving in metabolism represented by this huddling behavior (see Chapt. 1, Sect. 8b (3)). He found that the group of animals responded as a single animal of an equivalent weight (Fig. 2). Growing pigs

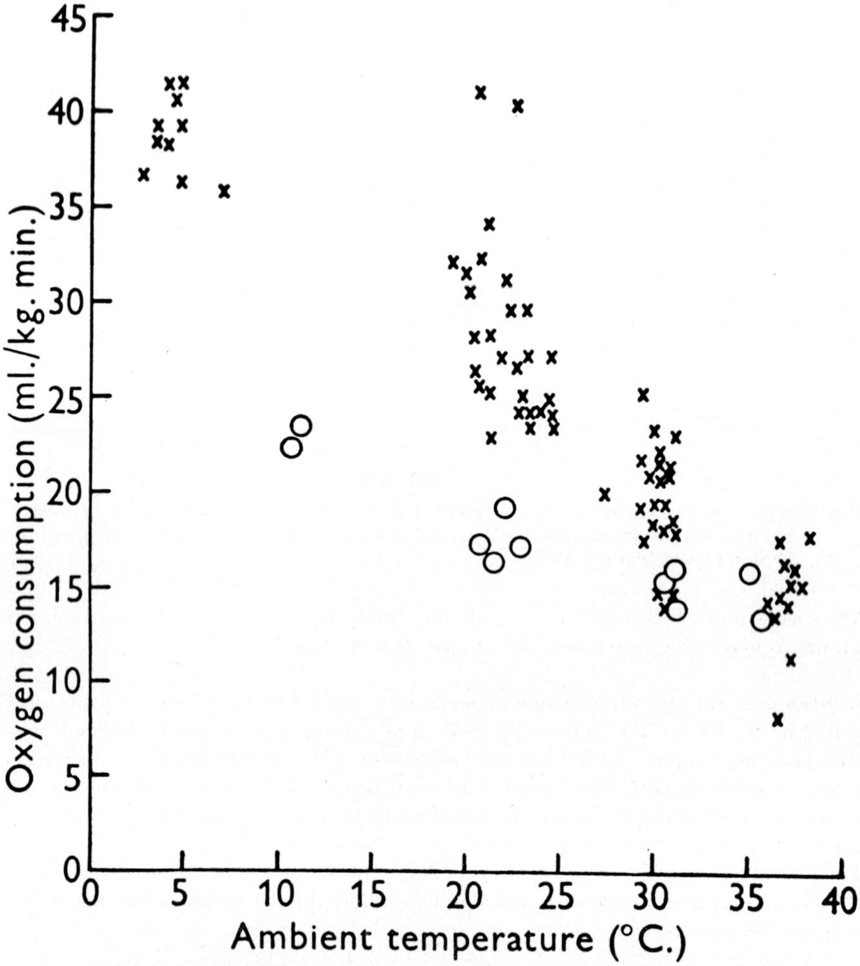

Fig. 2. Oxygen consumption of single new-born pigs × and groups of pigs ○ at different ambient temperatures (From Mount, 1960).

were investigated by Holmes and Mount (1967) who found that even when pigs were 60 kg in weight there was still a saving in metabolic rate if the animals huddled at an environmental temperature of 10° C and similar findings have been reported by Verstegen (1971). The presence of bedding material also influences the effect of a given cold environment and Stephens (1971) has obtained measurements of the reduction in oxygen consumption when straw is placed in the calorimeter during exposure of new-born pigs to a cold air temperature.

In the absence of other pigs, a single animal can still modify the effects of the environmental temperature by alterations in its posture. In thermally neutral or warm environments pigs lie on their side and stretch the legs out sideways; but under cold conditions they adopt a cold defensive position with the limbs held close to the body, and the minimum of contact is made between the body and the floor. These effects of changes in posture on heat loss have been investigated by Mount (1964, 1968). He exposed single pigs in a calorimeter in which the temperature of the walls and the air temperature could be varied independently while oxygen consumption was measured. From these studies it was estimated that the effective radiating surface area of the pig's body decreased from 76% to 67% of the total surface area as the air temperature changed from 30° C to 20° C and the animal altered its posture. The effective convective area decreased as the pig took up a cold defensive position from 86% to 76% of the total surface area over the same range of temperature. At the same time as the posture of the animal changes there are also modifications in peripheral blood flow which influence the rate of heat loss. Using comparisons with physical models, however, Mount (1968) calculated that when the ambient temperature fell from above to below thermal neutrality a third of the change in heat loss could be attributed to alterations in posture.

Because of the capacity of the pig to modify the effects of its environment by behavioural patterns the performance of pigs under husbandry conditions is much less affected over a temperature range of 8° to 20° C than might have been expected (Verstegen, Close, Start and Mount, 1973).

C. CHOICE OF ENVIRONMENTAL CONDITIONS

Two methods of studying the pig's preference for different environments have been used. In one (Mount, 1963) a thermocline consisting of a series of interconnected chambers which were controlled at different air temperatures was used. New-born pigs were introduced either singly or in groups into the thermocline and a note was made of the temperature of the compartment in which they chose to stay. As can be seen (Table 1) the choice of temperature was similar for all the pigs except single new-born animals, but the spread of temperature choices over the range investigated was wide. The inference is that the ambient temperature does not necessarily need to be strictly controlled for a pig to be thermally comfortable. Probably, within certain limits, any

Table 1. Ambient temperatures preferred by young pigs, both singly and in groups of five; numbers choosing different temperature (From Mount, 1963)

Age of pigs days	Ambient temperature °C				Mean ± S.E. temperature °C
	23–28	29–31	32–34	35–37	
Single pigs					
<1	1	2	5	2	32.3 ± 0.94
1–7	8	16	4	0	29.3 ± 0.44
8–41	16	33	11	6	29.8 ± 0.64
Groups					
2–7	2	3	3	0	30.1 ± 1.03
8.41	3	9	4	1	30.4 ± 0.66

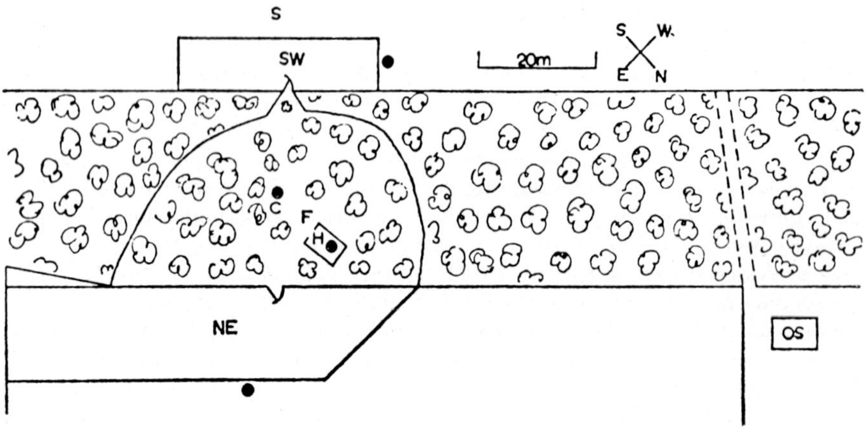

Fig. 3. Area of ground in which pigs were observed. H = hut, F = food, C = television camera, ● = points at which ambient conditions were measured, OS = observation hut, NE and SW = north east and south west paddocks. (From Ingram and Legge, 1970).

departure of ambient temperature from the optimal can be compensated by changes in posture.

A second method of studying the preference of pigs for certain environmental conditions has been used by Ingram and Legge (1970). In their study groups of pigs were kept outdoors in an enclosure which consisted of two grass paddocks on either side of a belt of trees (Fig. 3). A hut containing bedding was placed in the wood and a closed circuit television camera was arranged so that the pigs could be observed among the trees. The grass paddocks were visible from the observation hut, or by watching through the belt of trees, so that the pigs could always be seen without being disturbed by the presence of the observer. Climatic conditions (radiant temperature, air movement and air temperature) were recorded at various points in the enclosure and the animals received a normal farm ration of food once a day. Groups of pigs just after weaning were watched during all seasons of the year and it was found that on dry days the pigs displayed no tendency to shelter in the hut until the air temperature was below 5° C. On days when it was raining there was an increased tendency to shelter, but on many occasions the pigs would be seen out in light rain particularly during a spell of prolonged wet weather. From the experiments using operant conditioning, which will be discussed later, it is known that the animals outside were below the preferred temperature for most of the time. In spite of this the pigs chose to shelter only when the air temperature was well below the critical temperature. In wet weather the additional loss of heat by vaporization of water would have been expected to present a considerable degree of discomfort. The most likely explanation for these results is simply that in an environment which was rich in stimuli, thermal comfort was given a low priority (just as it is by man engaged in some sports).

One outstanding feature of the pig's behavior seen in this study was the consistent way in which they avoided the most windy part of the enclosure. The choice of position depended almost entirely on the direction of the wind, the pigs spending most of the time in the paddock on the least windy side of the wood and when the direction of the wind was parallel with the belt of trees the pigs stayed in the wood. These effects remained even after the differences in the sizes of the areas had been taken into account, and were true in both winter and summer. It is possible that avoidance of the wind was associated with thermal discomfort since the pig is particularly susceptible to heat loss in a wind (Mount and Ingram, 1965); but there may also be other components of discomfort for the pig in a windy environment.

The pigs in Ingram and Legge's (1970) study did not appear to select the areas with the highest radiant temperatures, and except for a few occasions in spring did not lie down in the sun.

OPERANT CONDITIONING

A. OPERANT CONDITIONING TECHNIQUES

The use of operant conditioning techniques in the study of thermoregulatory behavior was first used on rats by Carlton and Marks (1958) and Weiss and Laties (1961). Later it was used to study the thermal preferences in the pig by Baldwin and Ingram (1966, 1967a, 1967b). The pigs were placed in a cage with a switch panel which could be operated by the pig pushing its snout against it. This action turned on infra-red heaters above the pig for a few seconds after which the pig had to make a further response. Once the pigs had learned this response animals were exposed to a range of ambient temperatures and the frequency with which the heaters were turned on was recorded. At temperatures above the critical temperature some responses were made when the pig was first placed in the cage, but usually the pig then went to sleep. Below the critical temperature the pigs obtained more and more heat as the temperature became colder (Fig. 4). In very cold environments of 0° and −5° C, however, the relationship between temperature and the number of times the heaters were turned on was sometimes equivocal because the animals spent time trying to get out of the cage, or they adopted a cold defensive posture with the limbs held close to the body and were then unable to operate the switch. Measurements of oxygen consumption made at the same time as the pig was obtaining heat suggested that the pig adjusted the effective temperature to correspond to a thermally neutral environment (Baldwin and Ingram, 1966). Later studies using more refined measurements of oxygen consumption, however, revealed that the adjustment of the effective temperature by the pig is not perfect, and oxygen consumption still increases slightly as the air temperature is decreased (Ingram, unpubl.). If the time for which the heaters remained on was 6 sec in some experiments and 12 in others then the pig adjusted the number of responses it made and so it obtained roughly the same amount of heating (Baldwin and Ingram, 1967a). Pigs which had been fed on a low plane of nutrition obtained more heat than controls on a high plane when exposed to the same ambient temperature (Baldwin and Ingram, 1967b) as would be expected on the basis of nutritional studies which have shown the critical temperature to be lowered on a high plane of nutrition (Close et al., 1971).

Similar experiments have also been made on groups of pigs housed outside in a hut

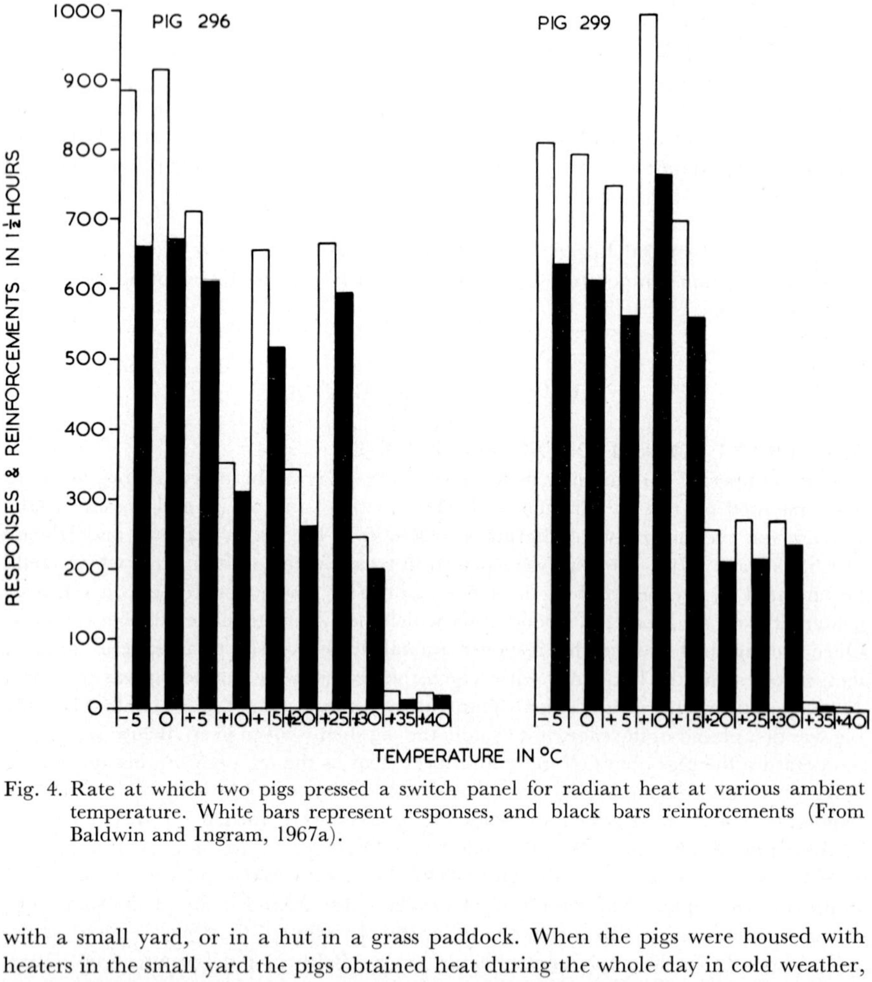

Fig. 4. Rate at which two pigs pressed a switch panel for radiant heat at various ambient temperature. White bars represent responses, and black bars reinforcements (From Baldwin and Ingram, 1967a).

with a small yard, or in a hut in a grass paddock. When the pigs were housed with heaters in the small yard the pigs obtained heat during the whole day in cold weather, but in spring heat was no longer obtained in the middle of the day when air temperature was high. At night the pigs always returned to the hut. When the pigs were housed in the grass paddock very few reinforcements of heat were obtained even in cold weather (Baldwin and Ingram, 1967a). Just as in the studies involving the pig's choice of a wood or a field, less attention was paid to thermoregulatory behavior when the animals were kept in an environment which was comparatively rich in stimuli.

Pigs have also been trained to perform the response of pushing a switch panel in order to turn off a draught of air (Baldwin and Ingram, 1967a), but the relation between the magnitude of the reinforcement and the number of responses made was less clear-cut than with infra-red heat. The possibility is that a draught of air is aversive to the pig for some other reason, as well as that connected with leat loss, and that this complicates the pattern of behavior.

B. MECHANISMS CONTROLLING BEHAVIOUR

The technique of operant conditioning has also been used in the investigation of the mechanism by which thermoregulatory behavior is controlled. Baldwin and Ingram (1968) implanted thermodes into the preoptic region of the hypothalamus of pigs and demonstrated that cooling this region while the pig was in a thermally neutral environment led to an immediate demand for infra-red heating which was sustained for as long as the hypothalamus was cooled. If the hypothalamus was cooled while the pig was in a cool room and already making responses for heat reinforcement, then the rate at which the pig responded increased. Cooling the hypothalamus in a hot room affected behavior in only half the instances.

Warming the hypothalamus while the pig was in a cold environment tended to decrease the rate at which the pigs demanded heat, but the effect was much less pronounced than the effect of cooling. These studies by themselves suggest that deep body temperature alone may control thermoregulatory behavior, but it is known that exposure to a change in ambient temperature can result in a shift of body temperature in the opposite direction to the ambient temperature (Ingram and Legge, 1970b). The effects of skin temperature in modifying the effects changing the temperature of the hypothalamus have been investigated by Baldwin and Lipton (1973). In some experiments heaters could be turned on for as long as a beam of light was interrupted by the pig placing its head in a certain position, and in other experiments the heaters remained on unless the light beam was interrupted. In both situations, "heat reinforcement" and "heat escape", the amount of heat which the pig contrived to receive at a given ambient temperature was the same. During some short periods the pigs switched the heaters on in such a way that the skin temperature on the back appeared to be controlled between 40 and 45° C. Such periods were, however, punctuated by intervals when the pigs ceased to respond and skin temperature fell, to below 30° C, before another period followed in which heat was again obtained and skin temperature controlled at about 43° C. Cooling the hypothalamus during these experiments resulted in the pig maintaining its skin temperature at a higher value of between 45-47° C. It would thus appear that the skin and hypothalamic temperatures interact in the control of behavior, but that signals from a cold skin may sometimes be ignored just as they are when pigs are kept outside.

Another region of the nervous system which is sensitive to changes in temperature is the spinal cord. The role of this region in behavior was investigated by Carlisle and Ingram (1973) using an operant conditioning technique in which the pig could operate the heaters while standing up or lying down. They found that cooling the cervical region of the cord influenced the demand for heat just as did cooling the hypothalamus, but there were important differences. When the hypothalamus was cooled the increased demand for heat was sustained for as long as the cooling continued; but when the cord was cooled there was an initial burst of activity which subsequently declined (Fig. 5). A second difference was that cooling the hypothalamus had no obvious effect on the pig's posture, but cooling the cord led to the pig taking up a cold defensive position with the limbs held close to the body and this position tended to be maintained for as long as the cooling continued. It would therefore seem likely that different aspects of behavior in relation to temperature depend on different mechanisms.

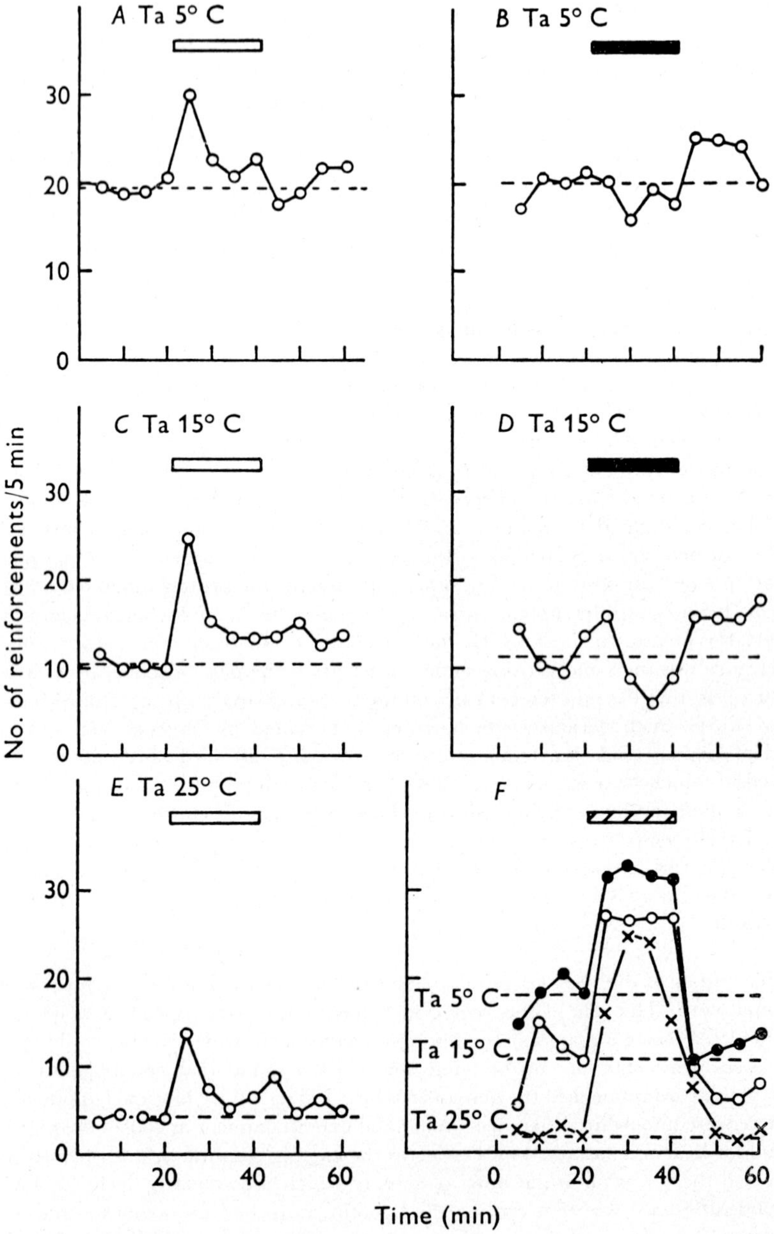

Fig. 5. Mean number of reinforcements per 5 min period before, during and after cooling of the spinal thermode to 20°C (open bar) at three ambient temperatures (A, C, E), heating of the spinal thermode to 43°C (black bar) at two ambient temperatures (B, D) and cooling of the hypothalamic thermode to 20°C (hatched bar) at three ambient temperatures (F). The dashed lines indicate the base line rate of reinforcement (From Carlisle and Ingram, 1973).

CONCLUSIONS

Behavior plays an important part in thermoregulation of pigs particularly at both extremes of environmental conditions. Under less severe conditions, however, the extent to which the pig uses its behavior to overcome the effects of the climate seem to depend on the competition from other stimuli. In a restricted environment in a cage, or a pig pen, the animal may direct its behavior almost entirely in relation to the ambient conditions; but in an open field thermoregulatory behavior occupies relatively little time. The control of thermoregulatory behavior seems to depend on similar but not exactly the same factors as those involved in autonomic thermoregulation.

REFERENCES: See pp. 585-586

Genetic Differences in the Adaptability of Rodents to Temperature Extremes

by

Pamela R. Pennycuik

INTRODUCTION

Races of wild rodents differ in their abilities to survive and breed at extreme temperatures (Brown, 1968). Their methods of adapting to these environments vary: spiny mice, *Acomys russatus*, adjust their level of heat production (Chaim and Borut, 1973), wood rats, *Neotoma cinerea* and *N. albigula*, adjust heat loss by varying surface mass ratio, hair cover and the efficiency of vascular heat exchange (Brown, 1968) and *Peromyscus* spp adjust the size of their nests (King et al., 1964). While these differences must be genetically determined, little is known about the amount of variability in these characters within a population, about the mode of interaction between genes determining adaptive characters and environmental temperature, or about the number of generations necessary for the evolution of measurable differences between strains. Experiments on laboratory rodents provide some answers to these questions. These experiments are of two types (1) those involving unrelated stocks and lines selected for growth rate and tail length at 21° C, and (2) those in which lines have been selected at temperatures extremes. In almost all instances mice have been used, but one report on rats is also available.

UNRELATED LABORATORY STOCKS AND THEIR CROSSES

There is good evidence that at least some of the established laboratory lines differ in the ability to survive and reproduce at extreme temperatures (Ogilvie, 1970; Barnett, 1965). They therefore provide useful material for investigating the characters associated with adaptation to these conditions, namely, heat production, heat loss and behaviour.

1. HEAT PRODUCTION

The simplest way of assessing heat production is to measure O_2 consumption. At low temperatures, if body temperature is to remain constant, consumption must be high; at thermal neutrality heat production for essential functions must balances heat loss to the environment and consumption is at a minimum; at high temperatures (especially short exposures) where the animal is panting excessively to cool itself, consumption is again high.

Murine strains differ significantly in their consumptions both at thermal neutrality (30-32° C) (Schlesinger and Mordkoff, 1963; Pennycuik, 1967) and at temperature extremes (Barnett, 1965; Gorecki and Krzanowska, 1970; Pennycuik, 1967, 1972). Differences at extreme temperatures could be due to differences in body size, for small animals lose heat more readily than large ones (see below). However A2G mice, which weighed as much as C57BL mice at −3° C (Barnett and Scott, 1963) had higher consumption at −3° C than C57BL animals (Barnett, 1965) and Wild mice, which were lighter than R70 mice, had higher O_2 consumptions at 34 and 36° C than R70 animals (Pennycuik, 1967). Differences could also be due to differences in the mass of heat producing organs (e.g., liver) to differences in the respiratory chain of the liver (cf. Cassuto, 1968; Cassuto, Chayoth and Rabi, 1970) or to behavioural differences (cf. Clarke, 1971), but strain differences in these characters still remain to be demonstrated.

The relationship between the consumptions of F_1 hybrids of strains with different consumptions, and the consumptions of the parental strains themselves is very variable. Only in one instance were consumptions intermediate between those of the parental strains at all temperatures investigated: Gorecki and Krzanowska (1970) found that the consumptions of F_1 hybrids of a cross between the inbred strains KP and KE were intermediate between the parental strains at 10, 20 and 30° C. Barnett (1965) who measured consumptions at −3° C found that the F_1 offspring of an A2G × C57 cross had consumptions lower than those of either of the parental strains, possibly because the hybrids grew better at −3° C than either of the parental strains (Barnett and Scott, 1963). Pennycuik (1972) measured the consumptions of mice from two random bred strains and their F_1 offspring at temperatures between 26 and 34° C. The R70 animals had consumptions lower than the Wild mice at all temperatures, the F_1 mice had consumptions similar to those of the R70 mice at 26, 28 and 30° C but, at temperatures above thermal neutrality, consumptions of the F_1 mice increased rapidly in a manner similar to that of their Wild parents. Apparently the factors determining consumption below and above thermal neutrality were different, and the F_1 mice exhibited the characteristics of one parent below 30° C and those of the other parent above 30° C. In addition to differences in consumption, the Wild and R70 mice differed in their response to acclimatization to high temperature; following 2 weeks at 33° C, the R70 animals had significantly lower consumptions at 32 and 34° C than they did when acclimatized to 21° C, the Wild mice and the F_1 offspring of the R70 X Wild cross, failed to show this response to acclimatization to 33° C. F_2 animals were also examined in this experiment; these were found to have consumptions very similar to F_1 animals and variances in consumption were very similar in both filial generations.

Control of O_2 consumption must be polygenic, a fact reflected in the similarity of the variances in the consumption of the F_1 and F_2 mice in the experiment described above. The results of the R70 X Wild cross suggest that some of the genes involved in determining consumption below thermal neutrality are probably without effect above 30° C, and that some determining consumption above 30° C may be inactive below 30° C. Some of the genes involved in determining consumption below 30° C may also be involved with determining body size, for Barnett found that the heavy hybrid offspring of the A2G × C57 cross had lower consumptions than those of either parent (Barnett, 1965). The genes involved in determining whether an animal will respond to acclimatization to high temperature with a fall in O_2 consumption may be related to those determining consumption above 30° C, for response to acclimatization and rate of consumption at temperatures above thermal neutrality appeared to be inherited together in the offspring of the R70 × Wild cross reported above.

2. HEAT LOSS

Characters which affect heat loss at extreme temperatures include, body surface: body weight ratios, insulation, and the efficiency of vascular heat exchange mechanisms.

(a) *BODY SURFACE: BODY WEIGHT RATIOS*

These are important because heat production is proportional to body mass (linear dimension) while heat loss is proportional to surface area (linear dimension). The ratio between surface and weight is therefore affected by (1) the weight of the animal and (2) the length of its appendages.

(1) *Body weight:* At normal animal house temperatures (usually 21° C) weight at a given age depends upon body weight at birth, rate of growth during the rapid growth phase and the duration of the rapid growth phase (Crary and Sawin, 1960; Laird and Howard, 1967). Growth during the first 4 or 5 weeks is due largely to protein and water deposition. After this period, fat contributes more and more to weight increase, particularly in some of the strains selected for high body weight (Roberts, 1965). Strain differences in growth rate reflect differences in the amount of food consumed and differences in the efficiency with which protein and fat are absorbed and stored (Fowler, 1962).

Animals exposed to low temperatures are usually lighter at a given age than those reared at 21° C (Barnett and Scott, 1963; Bigham and Cockrem, 1969). In "new stock" A2G mice exposed to −3° C this depression in body weight at −3° C was due to a slowing of the growth rate during the rapid growth phase (Barnett and Scott, 1963). This resulted from a reduction in both protein and fat deposition, the decline in body fat content being particularly noticable in females (Barnett and Widdowson, 1965). This decline in fat storage suggests that the lowered growth rates of these animals may have been due to inability to process sufficient food to keep pace with increased energy expenditure at −3° C. The results of Bigham and Cockrem (1969) were similar to those of Barnett and co-workers: these authors exposed mice from two lines selected for high weight gain at 21° C and mice from two lines selected for low weight gain at 21° C, to 7° C. In all cases, gain between 3 and 6 weeks of age was depressed by exposure to this low temperature. This was due to depressed protein deposition although depressed fat deposition also played a part.

Not all strains show a decline in growth rate at low temperatures: C57BL mice maintained at −3° C were as heavy at 16 weeks of age as those maintained at 21° C (Barnett and Scott, 1963) and their fat stores were similar at the two temperatures (Barnett, 1965) i.e., their food intake and absorption kept pace with energy expenditure at −3° C. When C57BL mice were crossed with A2G mice (which grew slowly at −3° C) the weight of the F_1 offspring were as great at −3° C as at 21° C (Barnett and Scott, 1963). Apparently these animals inherited the characterisation of their C57BL parents, which enabled them to perform as well at −3° C as at 21° C.

Animals exposed to high environmental temperatures are also usually lighter than those reared at 21° C (Harrison, 1963; Bigham and Cockrem, 1969). Harrison (1963) exposed C57 and R111 mice to 21 and 30° C. In both lines the animals at 30° C grew better than those at 21° C between 4 and 5 weeks of age, but the duration of the rapid growth phase was shortened and weights at 12 weeks of age were below those of controls kept at 21° C. Exposure to 30° C depressed the growth rates of F_1 offspring of the C57 × R111 cross slightly, but these animals were less severely affected by high temperatures than animals of either of the parental strains. Bigham and Cockrem (1969) investigated the effects of exposure to 30° C on growth rate by exposing mice from 2 lines selected for high body weight and 2 selected for low body weight to this temperature when the animals were between 3 and 6 weeks of age. Weight gains were lower at 30° C than at 21° C in 3 of the 4 lines investigated. This reduction in body size was due to depressed protein deposition; fat stores were either unchanged or increased. The fourth line, one of the high body weight lines, grew as well at 30 as at 21° C, but carcase composition was changed: fat deposition increased significantly, and fat free carcase weight declined. The presence of good fat stores in the mice from these selection lines exposed to 30° C, suggests that weight depression at high temperatures is not due to an imbalance between intake and energy expenditure. (They also suggest that exposure to high temperature has less effect on appetite than on energy expenditure). Harrison's results (1963) suggest that small body size at high temperatures is due rather to a shortening of the duration of the rapid growth phase.

Body weight of mice is determined by a large number of loci of approximately equal effect (Falconer, 1960). There is some evidence that genes favouring large body size confer advantages at low environmental temperatures, for mice from lines selected for large body size at 21° C survived better at 7° C than lines selected for small body size (Bigham and Cockrem, 1969). There is also evidence that some strains are more sensitive than others to the depressing effects of low temperature on growth (Barnett and Scott, 1963). These observations suggest that adaptation to low temperature favours the accumulation of genes associated with rapid growth, a long period of rapid growth, and a low sensitivity to the depressing effects of low temperature on growth.

Mice from lines selected for large body size survive as well at 30° C as animals selected for small body size (Bigham and Cockrem, 1969). This might not have been the case if animals from these lines had been tested at temperatures above thermal neutrality. However, such tests still remain to be carried out. If small size should prove to be an advantage at high environmental temperature one might expect strains adapted to this temperature to accumulate genes associated with slow growth, and a short growth phase at all temperatures and a high sensitivity to the effects of high environmental temperature on growth.

(2) *Appendage length:* Although differences in the sizes of the limbs and ears could also affect heat exchange, the tail is the only appendage which has been examined in any detail in the mouse. There is some evidence that the size of this structure is of importance at high environmental temperatures. (Harrison, 1958; Pennycuik, 1967; Bigham and Cockrem, 1969) but at low temperatures, tail length appears to be of little importance to well being (Barnett, 1965; Bigham and Cockrem, 1969).

Mouse strains show considerable variation in the lengths of their tails (Cockrem, 1959; Barnett, 1965) and exposure to extreme temperatures causes further variations; cold causes a shortening of the tail, while heat causes it to lengthen (Barnett, 1965; Bigham and Cockrem, 1969).

Control of tail length appears to be polygenic. Some of the genes involved are the same as (or linked to) those affecting body weight (Falconer, 1954), others are unconnected with general body growth (Cockrem, 1959; Baker and Cockrem, 1970). In some mouse strains the developmental system determining tail length is more sensitive to temperature change than it is in others: Bigham and Cockrem (1969) found that lines selected for long tails showed a greater increase in tail length when exposed to high temperature than those selected for short tails.

(b) *INSULATION*

Hair is of little insulative value in small animals (Barnett, 1959, 1965). However, even in mice the amount of hair/unit surface area has been found to be inversely related to environmental temperature (Barnett, 1959; Pennycuik, 1969b) and these variations in hair quantity have been found to have a small but significant effect on the insulative value of the skin (Barnett, 1959).

In inbred strains housed at 21° C, Barnett (1959) observed small differences in the amount of hair/unit surface area. Moreover two of the three strains investigated (A2G and A) showed a much larger increase in hair weight on exposure to −3° C than the third (C57BL). Whether or not these differences in hair cover confer a selective advantage at temperature extremes, still remains to be investigated: Pennycuik (1971), who exposed bare mutants (N, hr) and their normal sibs to 33° C, found that the mutant animals had no advantages over the normal mice at this temperature; however these results could reflect qualities of these mutant mice other than their lack of hair cover.

(c) *VASCULAR HEAT EXCHANGE*

Differences in the response of the vascular system to exposure to extreme temperatures also affect the amount of heat lost to the environment. Brown (1968) found that wood rats (*Neotoma albigula*) from different habitats, differed in their vascular responses at 0 and 10° C, but no reports have been made of similar differences in laboratory rodents. Pennycuik (1969b) was unable to find any differences in the morphology of the sub-epidermal capillary plexus of lines selected at 21 and 34° C.

3. BEHAVIOUR

Behavioural characters which could affect heat exchange include nest building, burrow construction, sizes of groups sharing a nesting area and postural responses to temperature change. Of these, the only character known to show genetic variability is nest building (Lee, 1970; Lynch and Hegman, 1972). Unfortunately, no information is available about strain differences in nest building behaviour at extreme temperatures, although it is well known that environmental temperature has a profound effect on the size of the nest built (Kinder, 1927).

LINES SELECTED AT EXTREME ENVIRONMENTAL TEMPERATURES

In addition to closed colonies of mice and lines selected at 21° C, lines have been selected for reproductive productivity and for growth rate at temperature extremes. Such selection gives some measure of the speed with which changes can be achieved in these environments.

1. SELECTION FOR REPRODUCTIVE PRODUCTIVITY

The most extensive experiments at low temperatures are those of Barnett (see review, 1965) who maintained both a genetically mixed stock and two inbred lines, A2G and C57BL, at −3° C for a number of generations. In all lines, reproductive productivity at −3° C was considerably lower than that at 21° C. This was due to combinations of the following factors: increase in the number of barren pairs, decline in the number of litters per pair, reduced litter size and increased mortality between birth and weaning. The genetically mixed stock was selected for productivity by using only the offspring of pairs which had reared a litter of 5 or more young within 5 weeks of pairing. After 12 generations, this stock showed significant improvement in litter size and significant decline in nesting mortality. The inbred strains maintained at −3° C were not subjected to deliberate selection; the size of their litters showed no change with increasing generation number, but unexpectedly, nesting mortality declined in a fashion similar to that of the genetically mixed stock. Barnett (1961, 1965) presents evidence that this was not due to genetic changes but rather to cumulative maternal changes, possibly in the uterus, and/or in the nest provided for the new-born pups. These results emphasise the need for caution in interpreting the changes observed in new environments.

In addition to changes in reproductive productivity, the genetically mixed stock selected at −3° C showed an improvement in growth rate, adding support to the suggestion that increased size is an advantage in cold environments (Barnett, 1961). Tail length increased in the 18 generations at −3° C, suggesting that the size of this structure is of little importance in the cold (Barnett, 1965).

A genetically mixed stock has also been selected for reproductive productivity at 34° C. The productivity of mice at this temperature was only one quarter of that at 21° C. This was due to an increase in the number of barren pairs, a decline in the number of litters born over a 12 week mating period, reduced litter size and increased pup mortality. For the first 12 generations only a weak selection pressure was applied to the line at 34° C (all fertile pairs contributed at least one offspring to the next generation) and no improvement was observed in the number of young reared by each pair (Pennycuik, 1969a). At generation 12, the "hot" line was split: some animals were used to maintain the line, (these were selected on the same basis as those in earlier generations), others were crossed to animals from a Wild stock and the offspring of the cross were used to establish a new line. These were subjected to a stronger selection pressure than the old line (progeny from the best 10, out of 20, matings were used as parents for the next generation). The reproductive productivity of the old line improved significantly over the next 12 generations of selection but the number of young reared to 6 weeks was still significantly below that of the control line at generation 24. The new line, on the other hand, showed a rapid improvement in productivity and by generation 12, the animals in the incubator were rearing as many 6 weeks old pups as those at 21° C (Pennycuik, unpubl. data). Because both 34° C lines were genetically mixed at the start of the experiment it seems most likely that improvement in productivity was due to selection of favourable genotypes. However, cumulative maternal changes, similar to those observed by Barnett, may also have contributed to the improvement in productivity.

Although selection for productivity for 12 generations at 34° C had no detectable effects on the number of pups reared to 6 weeks of age, it did affect growth rate: body weight at 6 weeks showed a gradual decline between generations 4 and 12 (Pennycuik, 1969a). This decline in growth was not apparent when mice from the same stock were reared at 21° C (Pennycuik, 1969b). Hair cover and the sub-epidermal capillary plexus were unaffected by selection at 34° C (Pennycuik, 1969b).

Both the experiments of Barnett and Pennycuik show that a significant improvement in productivity in an extreme environment can be achieved in 12 to 24 generations of selection. This improvement probably involves the accumulation of genes favouring productivity directly, but it also involves the accumulation of genes favouring large size in the "cold" stocks and small size at 34° C (but not at 21° C) in the "hot" stocks. Clearly, these stocks should now be examined in other characters involved in adapting animals to temperature extremes.

2. SELECTION FOR GROWTH RATE

In addition to lines selected for productivity at extreme temperatures, two experiments have been carried out in which animals were selected for growth rate following exposure to high and low temperatures for limited periods during growth. Both these experiments were designed to investigate the effects of selection in one environment on performance in another, with the objective of choosing the environment which would give the greatest progress for the character in question.

Baker and Cockrem (1970) established 3 lines of mice which they maintained at 7, 21 or 31° C for the first 6 weeks of life. Selection for high 6 week weight was practiced over 4 generations in all lines. In all environments body weight increased but realized heritability was greatest at 31° C. When samples of mice were switched between environments at 4 weeks of age, the animals from the "hot" line were the ones which grew best between 4 and 7 weeks of age at 7 and 21° C, i.e. maintenance at 30° C was more effective in exposing genetic variability in growth rate and more effective in promoting successful selection for this character than maintenance at 21° C or 7° C. Selection for high weight gain had no effect on the reproductive productivity of the 3 selection lines over the 4 generations of selection.

Roubicek and Ray (1971) established four lines of rats which they selected at 21 and 35° C (i.e., at a temperature 5° C higher than that used by Baker and Cockrem). Two of these were selected for high weight gain between 3 and 13 weeks of age and two were selected for low weight gain over the same period. In the high weight gain lines, only those housed at 21° C showed any advance at the selection temperature during the 10 generations of selection, but when animals were switched between environments the animals from the "hot" line grew as well at 21° C as those selected at 21° C. That is, some genetic advance was made at 35° C in spite of the absence of a phenotypic response, but advance was not as great as it was in Baker and Cockrem's experiment where selection for increased weight gain at 30° C had a measurable effect on weight gain at that temperature. Both of Roubicek and Ray's high weight gain lines showed a marked decline in fertility after 5 generations of selection i.e., one generation after the termination of Baker and Cockrem's experiment. It would be interesting to know if the lines selected at 7, 21 or 30° C by the latter authors showed a similar decline in fertility with continued selection for improved growth rate. On theoretical grounds one might expect the line maintained at 30° C, where small size is an advantage, to show a decline in fertility before either of the lines maintained at lower temperatures.

Selection for low weight gain was equally effective at 21 and at 35° C (Roubicek and Ray, 1971) when animals were switched between environments growth rate at the new temperature was indistinguishable from that of the animals selected at that temperature i.e., there was no evidence that maintenance at 21° C uncovered genetic variability in growth rate more successfully than maintenance at 35° C. Neither of these selection lines showed a decline in reproductive productivity over the 10 generations of selection.

CONCLUSIONS

Table 1 summarizes the characters thought to be connected with adaptation to temperature extremes and the species which have been found to show significant strain differences in these characters when representatives of these strains were exposed to the same extreme environment. It demonstrates that, all the physiological and morphological characters investigated so far, do show significant variability in one species or another. It also demonstrates the paucity of our knowledge about strain differences in behavioural characters, a serious omission in a group of such small body size, where behaviour is probably of major importance in survival at temperature extremes.

During pre-natal development the young mammal is cushioned against temperature extremes by the homeostatic mechanism of the mother. After birth, when it is small but fully formed, it is exposed directly to the environment and its subsequent success will depend upon the interaction between its genetic complement and the environment itself. In some instances this interaction will produce changes in the characters which adapt the animal to its environment, for example, the reduced body size, increased tail length and possibly the reduced insulation observed in animals exposed to high temperatures, or the high O_2 consumptions, and possibly the increased insulation observed in animals exposed to low temperatures. In other cases interaction may produce changes which are not adaptive, for example, high O_2 consumption at high temperatures or small body size at low temperatures. Evolution of races adapted to extreme environments will depend upon the accumulation of genes promoting favourable interactions and genes promoting cushioning of interactions which are unfavourable. The lines selected for productivity at −3° C and at 30° C showed signs of accumulating genes of this kind during the first 12 generations of selection: the random breeding

Table 1. Characters, connected with adaptation to temperature extremes, which show significant differences between strains when representatives of these are exposed to low or to high temperatures

Character	Low Temperature		High Temperature	
	Species	Reference	Species	Reference
HEAT PRODUCTION	*Acomys russatus*	Chaim and Borut (1973)	*Mus musculus*	Pennycuik (1972)
O_2 consumption	*Mus musculus*	Barnett (1965)		
		Gorecki and Krzanowska (1970)		
HEAT LOSS				
Body weight	*Mus musculus*	Barnett (1965)	*Mus musculus*	Bigham and Cockrem (1969)
				Pennycuik (1969a)
Appendage size	*Mus musculus*	Barnett (1965)	*Mus musculus*	Baker and Cockrem (1970)
		Baker and Cockrem (1970)		
Insulation	*Mus musculus*	Barnett (1959)	–	
Vascular heat exchange	*Neotoma albigula*	Brown (1968)	–	
Behaviour	–		–	

strain selected at −3° C increased in body weight at −3 and at 21° C (Barnett and Scott, 1963) and the strain selected at 34° C decreased in weight at 34° C, but not at 21° C (Pennycuik, 1969a, 1969b).

Unfortunately, very little is known about the site of action of the genes producing changes in the characters associated with temperature adaptation. They could be in the hypothalamus, which integrates information about the temperature of the environment, for example, genes at this site could be responsible for regulation of food intake and therefore O_2 consumption, fat storage and growth. Or they could be in the endocrine glands, for example, genes at this site could determine the amount of growth hormone produced or the age at which growth hormone levels declined. Or they could be in the tissues where they could determine sensitivity to hormonal levels or sensitivity to temperature changes in the tissues themselves, for example in the peripheral tissues and organs.

In spite of these wide gaps in our knowledge about genetic control of temperature adaptation, the information available to date suggests that rodents display considerable variation in many of the characters associated with temperature adaptation and that it is possible to produce measurable changes in some of these characters after relatively few generations of selection.

REFERENCES: See pp. 586-587.

Chapter 1, Section 18d (1)

Adaptation of Mammals to Thermal Stresses

by

W. V. Macfarlane

See Progress in Human Biometeorology, Part IB, Chapt. 6, Sect. 19a, pp. 468–473

Seasonal Adaptations of Birds to Cold

by

G. C. West

INTRODUCTION

Most areas of the earth inhabited by birds exhibit seasonal changes in ambient temperatures. In high latitude and high altitude regions, these changes may be profound while in tropical regions, they may be negligible. Birds that reside outside of the equatorial quasi constant temperature regions have either evolved behavioral mechanisms to avoid the seasonal changes in temperature in one region by migrating to another region or have evolved a combination of morphological, physiological, and behavioral adaptations to maintain homeothermy in a single region of residence (West and Norton, 1974).

The problem of maintenance of homeothermy in birds has been discussed in Chapt. 1, Sect. 4f and 5e and in several recent reviews (Calder, 1974; Dawson and Hudson, 1970; King and Farner, 1961; West, 1962). Except for members of a few orders, birds maintain a relatively constant body temperature which fluctuates on a diurnal basis. Absolute body temperatures are related to size and phylogenetic group averaging somewhat lower in the larger non-passerines than in the smaller passerines (Dawson and Hudson, 1970; McNab, 1966; Rodbard, 1950; Wetmore, 1921). Maintenance of homeothermy requires a constant expenditure of energy at any ambient temperature below body temperature (T_b). The energetic cost of maintenance of homeothermy increases as the ambient temperature decreases below the critical temperature (T_c). It is remarkable that most small wild birds from temperate, subarctic, and arctic regions live at ambient temperatures (T_a) below their T_c for most of the year. Therefore the problem to be solved by these species is how to maintain a constant T_b when T_a is far below T_b for long periods. Nocturnal hypothermia in adult northern birds noted by Steen (1958) and Budd (1972) could serve as a means of reducing energy expenditure during winter when days are short. The natural circadian rhythm of body temperature, lowering from one to several degrees at night in all birds studied results in a small but significant energy saving at night (Aschoff and Pohl, 1970b). Sandpiper (*Calidris*) chicks undergo hypothermia, but feed and are active on the tundra in summer where air temperatures are near 0° to +10° C, but apparently this is an exceptional situation (Norton, 1973). Further investigations on hypothermia are needed; some are in progress (Grossman, 1974).

Some migrants may encounter equivalent or colder temperatures on their summer breeding grounds than they faced in winter e.g., *Calidris* spp. sandpipers migrating from South America to the high arctic of Canada or Alaska (Norton, 1973), the Pacific Brant (*Branta bernicla*) which leaves the Pacific coast of California for the Kuskokwim delta in Alaska (Morehouse, 1974). Thus the seasonal alteration of residency between northern and southern regions may in some cases result in an equalization of the environmental temperature to which a population is exposed throughout the year (West, 1960). However, the majority of species populations which either migrate from high latitude breeding grounds to temperate or south temperate regions or remain resident in winter on their breeding grounds encounter a marked seasonal difference in thermal environment.

This Section discusses the mechanisms that birds have evolved which have adapted them to cold. Adaptation to cold involves combinations of mechanisms to increase heat production and to reduce heat loss.

MORPHOLOGICAL ADAPTATIONS TO COLD

Morphological adaptations involve primarily seasonal changes in feather insulation and to a lesser extent, restriction of evaporative heat loss in the cold and alteration in peripheral blood flow. Birds are protected by an insulating shell of feathers which are arranged over the skin surface in most species in tracts (pterylae) leaving intervening areas bare of feathers (apteria). By regulation of feather position, portions of the apteria can be exposed thus facilitating heat loss. Feathers are perhaps the most efficient naturally occurring insulation material and by positioning of feathers uniformly over all skin surfaces, heat loss from the body can be greatly reduced. Regions such as the eyes, bill, and feet of some species are not protected by feathers and are avenues for heat loss in the cold (Veghte and Herreid, 1965).

Down feathers and the basal portions of contour feathers trap air between their barbules thus reducing convective heat loss while the distal portions of the contour feathers together form a smooth aerodynamic surface which acts as a windproof coat. This outer layer forms an ideal shape for movement through air in most species and through water in diving birds. In aquatic species, the waterproof shell serves to retain air in the down thus maintaining an efficient insulation in cold water.

The core temperature of birds is only a degree or two above that under or at the skin surface which implies that the feather layer acts as the major insulator (Dawson and Hudson, 1970; Dawson and Tordoff, 1959; Irving and Krog, 1955; Pohl and West, 1973; West, 1962; West and Hart, 1966). Changes in peripheral blood flow to the feathered portion of the skin surface probably exert only minor influence in overall thermal conductance. However, changes in blood flow to exposed appendages such as the bill or legs may result in alteration of body temperatures (Dawson and Hudson, 1970; Kahl, 1963; Steen and Steen, 1965). Seasonal changes in either the ability or propensity to alter peripheral blood flow have not been demonstrated and the anatomical and physiological mechanisms for such alteration have been adequately described for only a few species (Hyrtl, 1863; Irving and Krog, 1955; Steen and Steen, 1965).

It is reasonable to expect that birds living in cold climates or those adapting to seasonal cold would have greater plumage insulation than those from warmer regions or seasons.

Total plumage insulation however is very difficult to measure since positioning of the plumage by the bird itself is important in achieving maximum heat retention. Attempts to measure heat flow through removed patches of skin and feathers have therefore not yielded reliable data (Irving, 1972; Scholander et al., 1950). Similarly cooling curves determined on dead birds have not always given conductance values comparable to those determined through body temperature and metabolic rate measurements on living birds (Herreid and Kessel, 1967; West, 1972a; 1972b). Some investigators have either counted feathers or weighed portions of the plumage in order to estimate changes in insulation, but unless these measurements are correlated with direct determination of thermal conductance or metabolism on live birds, differences in numbers or weight of feathers need not necessarily reflect changes in insulation. For example, West and Pohl (1975) showed that the plumage (exclusive of flight feathers) of redpolls is heavier in winter than in summer indicating increased insulation. However, there was no corresponding reduction in winter oxygen consumption at comparable temperatures (West, 1972b; Pohl and West, 1973). In contrast, the Willow Ptarmigan (*Lagopus lagopus*) increases its plumage density in winter as inferred from changes in the slope of the metabolic rate versus T_a regression line (West, 1972a). The small increase in plumage insulation has a large effect on the ability of this bird to exist in the cold (see beyond). Direct estimation of plumage insulation for different body parts, for a variety of species in varying body positions and configurations and in different levels of activity remains an area open for investigation.

Terrestrial birds apparently do not utilize body fat as an insulator even though winter weight and fat levels are generally, but not always, higher than in summer (Kendeigh, 1974; West, 1960; West and Meng, 1968). Fat is not only distributed under the skin surface but is also in the folds of the intestines, around the heart, and in other locations not suggesting its effectiveness as an insulator. Aquatic species, however, such as ducks, geese, alcids, and penguins do have significant layers of subcutaneous fat. As in seals, this layer of "blubber" may participate in the insulative gradient (Irving, 1972).

Although not unique to cold weather enduring species, many arctic permanent residents have crops (gallinaceous species) or esophageal diverticulae (redpolls) in which food can be gathered and stored during daylight and subsequently digested during darkness.

PHYSIOLOGICAL ADAPTATIONS TO COLD

Physiological adaptations to cold include the ability to increase the rate and/or duration of sustained heat production and the ability to increase maximal heat production. The seasonal changes in rates of heat production are manifest both in the zone of thermal neutrality and below T_c. Many investigators have shown that the basal metabolic rate [resting oxygen consumption of a fasted bird measured in the dark during its normal rest time (ϱ) at thermal neutrality (Aschoff and Pohl, 1970)] increases in winter over summer values (Goldfinch, *Carduelis carduelis*, and Canary, *Serinus canaria*: Gelineo, 1969; Redpoll, *Acanthis* spp.: West, 1972b; Northwest crow, *Corvus caurinus*: Irving et al., 1955; Brambling, *Fringilla montifringilla*: Pohl, 1971). This result is unexpected as it appears not to be adaptive. The increased resting rate at thermal neutrality in winter birds may be explained by the fact that birds in winter are rarely exposed to the high T_a that define the thermal neutral zone. Therefore, in the short duration of tests at these unseasonably high T_a, the birds' heat production system does not return to its minimal level during the course of the test (West, 1972b). Other authors have not seen the increase in winter basal metabolic rates over those from summer (Black-headed gull, *Larus ridibundus*: Davydov, 1971; Cardinal *Cardinalis cardinalis*: Dawson, 1958; Pigeon, *Columba domestica*: Dontcheff and Kayser, 1934; Riddle et al., 1934; Brambling: Rautenberg, 1957; Yellow Bunting, *Emberiza citronella*: Wallgren, 1954).

Seasonal changes in the slope of the regression line describing the function of metabolic rate vs ambient temperature and the lower T_c have been demonstrated. In small birds (under 100 g), winter rates at cold T_a may be either higher or the same as those from summer, with changes in T_c resulting from the increased resting metabolic rate at thermal neutrality (Fig. 1A; Redpolls: West, 1972b; Pohl and West, 1973; Bramblings: Pohl, 1971; Starling, *Sturnus vulgaris* and Evening Grosbeak, *Hesperiphona vespertina*: Hart, 1962; House Sparrow, *Passer domesticus*: Kendeigh, 1949; Davis, 1955; Goldfinch and Canary: Gelineo, 1969). One exception is the limited data on Yellow Buntings cited by Wallgren (1954). These 25-30 g birds showed a significantly lower CO_2 production at $-11°$ C in winter over summer acclimatized birds.

Larger birds (over 250 g) such as the Willow Ptarmigan demonstrate substantially decreased winter rates at low T_a (West, 1972a). The single Pacific Brant studied by Irving et al. (1955) showed little difference between summer and winter metabolic rates and had a high T_c despite its thick plumage. A larger sample of brant investigated by Morehouse (1974) however indicate that the metabolic rate is higher in winter than in summer. This implies increased thermal conductance in winter birds. As mentioned above, this may be a result of the fact that the population studied is subjected to cool summers in comparison to relatively warm winters and reverse acclimatization has occurred (Morehouse, 1974). Fasted pigeons had lower rates at cold temperatures in winter, indicating increased insulation (Riddle et al., 1934). Decreases in T_c in the larger birds are due to increased feather insulation and a concomitant decreased thermal conductance in winter (West, 1972b). Assuming that the Newtonian model (Kleiber, 1961) applies, the slope describing metabolic rate vs T_a must extrapolate to T_b at zero metabolism. When T_c is lowered, the slope is also decreased. Without altering the absolute level of heat production, large birds such as ptarmigan can exist at significantly lower T_a in winter than in summer (Fig. 1B) (or the reverse situation for brant).

Small birds on the other hand are apparently not able to carry a significantly greater feather insulation in winter. Their small size and requirement for flight imposes a limit on the density of their plumage. Therefore T_c is not lowered as a function of insulation and the metabolic rate vs T_a slope is little changed. This results in little or no change in ability to exist at lower T_a in winter over summer with the same metabolic output (West, 1960; Fig. 1A).

During molt, energy requirements are increased, primarily due to the added cost of feather production, but partly due to the reduction in insulation when old feathers are

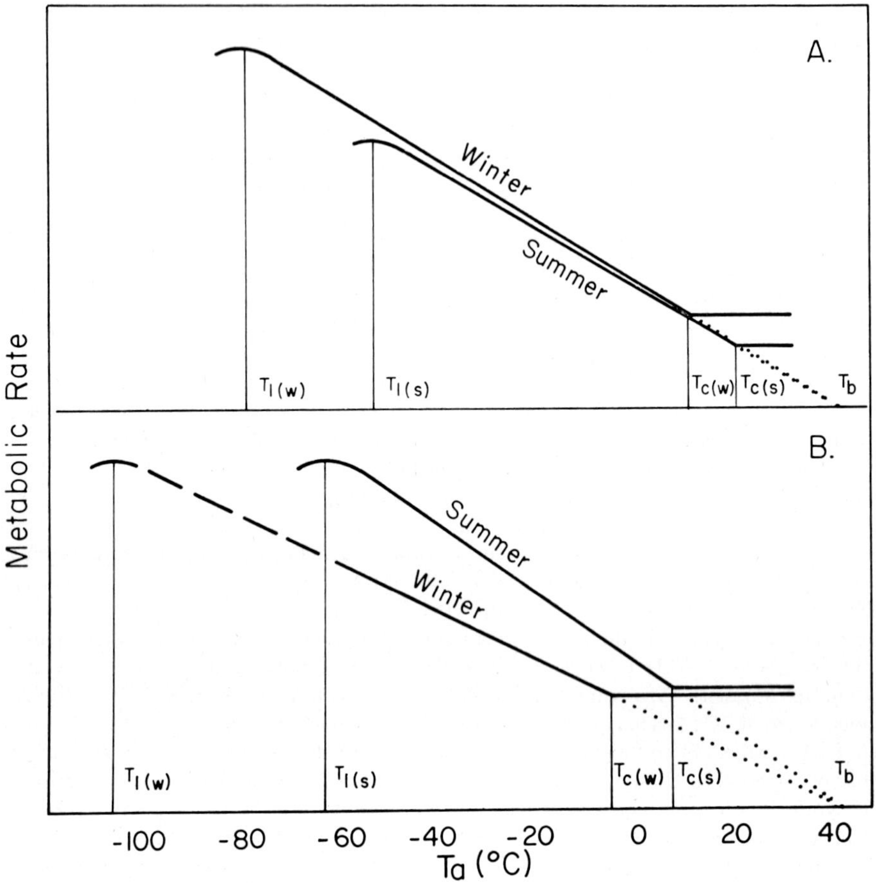

Fig. 1. Metabolic response of birds of two size classes to ambient temperature (T_a). Dotted lines extrapolate metabolic rates to zero at body temperature T_b. Lower lethal temperatures (T_l) are indicated for summer (s) and winter (w) as are lower critical temperatures (T_c).
A. small birds, generally under 100 g.
B. large birds generally over 250 g. Dashed line projects metabolic rate for birds acclimatized to winter (see text).

lost (Blackmore, 1969). However, on a long term basis, it has been suggested that the added insulation provided by the fresh plumage decreases heat loss and saves energy to about the same extent as the energetic cost of molt (Blackmore, 1969; Davis, 1955; Kendeigh, 1974). Molt is a seasonal process that occurs in late summer and fall in most species prior to the onset of winter cold weather and thus may be considered in part as a seasonal adaptation to cold. Willow Ptarmigan however molt continuously all summer changing plumage coloration, except when females are laying eggs (West, 1968) and thus do not have a complete insulative coat in summer.

Some species can exist at lower temperatures by increasing their maximum metabolic rate in winter over summer (Fig. 1A). This mechanism is not clearly understood. House sparrows increase their maximum sustained daily metabolic rate along the same slope so that they can exist at −35° C in winter but only near 0° in summer (Barnett, 1970; Kendeigh, 1974; Davis, 1955). Redpolls also increase their maximum metabolic rate in winter. At −50° C, winter redpolls required a 17% increase in metabolic output to maintain a 1.4° C higher T_b in winter over summer (Pohl and West, 1973). Evening grosbeaks, pigeons and starlings tested by Hart (1962) all showed increased cold resistance in winter as measured by mortality rates. This demonstrates an increase in metabolic capacity by winter birds which enables them to survive at lower temperatures in winter. Some migrant species do not extend their lower limits of temperature tolerance from summer to winter presumably because they are exposed to equivalent ambient temperature ranges both in summer and winter regions (Tree Sparrow, *Spizella arborea*: West, 1960; Dickcissel, *Spiza americana*: Zimmerman, 1965). Larger species such as the Snowy Owl (*Nyctea scandiaca*) and Willow Ptarmigan which increase their insulation significantly in winter do not require additional heat production ability in winter over summer because their summer metabolic level coupled with the increased insulation enable them to exist at the lowest temperatures encountered (Gessaman, 1970; West, 1972a). Tests have not been made to see if an increased metabolic capability in winter indeed exists. Redpolls acclimatized to summer, when tested in a He-O_2 atmosphere, were able to maintain homeothermy for several minutes at an equivalent temperature of −75° C (Rosenmann and Morrison, 1974). Tests on winter acclimatized redpolls have not been made but according to our observations on their metabolic ability during exercise (Pohl and West, 1973), they probably could extend their limit to lower T_a.

Thermoregulatory heat production in adult birds appears to be limited to muscular activity through gross exercise or shivering below T_c (Hart, 1962; Steen and Enger, 1957; West, 1965). There is no evidence in the limited studies available that birds significantly alter their shivering activity seasonally at any given temperature (Hart, 1962; West, 1965). In fact winter shivering rates are higher than summer rates. This indicates that in contrast to many mammalian species, adult birds do not have a non-shivering thermogenic ability which is augmented with cold acclimatization (West, 1965). Therefore, the increase in metabolic heat producing capability of winter birds must be a result of an increased ability to derive heat from more vigorous shivering. Non-shivering thermogenesis has been documented in chicks of some precocial species (Freeman, 1970; Wekstein and Zolman, 1968; Murrish and Guard, 1973) but tests with various blocking agents and injections with catecholamines and other hormones have not yielded conclusive evidence regarding the mechanisms of heat production (Freeman, 1971). Chickens show increased catecholamine levels in winter (Lin and Sturkie, 1972) but the increase does not appear to be directly related to stimulation of heat production mechanisms.

Increases in specific dynamic action (SDA) (King and Farner, 1961) or exercise (West and Hart, 1966) probably do not substitute for thermoregulatory heat production except at extremely low temperatures (Pohl and West, 1973). Problems of the utilization of the heat produced by means other than the resting BMR and shivering to maintain homeothermy need detailed study. This includes SDA, normal exercise in feeding activity and flight (Berger and Hart, 1972).

One area little studied in birds is that of seasonal changes in appetite drive regulating feed intake and digestive efficiency or metabolized energy coefficient (MEC) (Kendeigh, 1974). It is well known that MEC varies with diet, usually within the range of 45 to over 90% (Kendeigh, 1974; West, 1968; Moss, 1973). MEC is also influenced by ambient temperature often being higher at higher T_A (West, 1960; Seibert, 1949; Kendeigh, 1949) but not always (Brooks, 1968; West and Hart, 1966; Willson and Harmeson, 1973). Since most species change their diet seasonally, seasonal changes in MEC could reflect direct influence of the diet or could result from an innate change in digestive or assimilation ability. In caged birds held under naturally varying outdoor conditions on a single diet throughout the year, the MEC does vary seasonally (Owen, 1970; West, 1968). This suggests that enzymatic or microbial changes, gut morphology, or other parameters affecting digestion and assimilation in birds are controlled by innate circannual rhythms as has been shown for other seasonal functions (Gwinner, 1971), and which in turn are synchronized by temperature and/or photoperiodic Zeitgebers rather than by the direct effect of diet. It is also doubtful if temperature or light is the direct causative agent since the observed seasonal changes in MEC have been in opposite directions in different species. How winter cold temperatures influences food intake through appetite drive at a time when energy requirements except for thermoregulation are low, have not been investigated.

BEHAVIORAL ADAPTATION TO COLD

The most dramatic behavioral response to cold is avoidance through migration to warmer regions. In arctic regions, most species take this course of action and depart in fall thus escaping the extreme short days, long nights, and cold temperatures of arctic and subarctic winter (Irving, 1972; West and Norton, 1974). This escape phenomenon is not unique to high latitudes. Because of a multitude of evolutionary factors, there appears to be a continuum of species tolerance to environmental conditions from arctic to subtemperate latitudes so that many species from northern regions migrate and winter in areas vacated by other migrants which move further south and thus are exposed to cold temperate winters which the summer residents of that region could not tolerate.

Short distance migrations to avoid cold or to gain other advantage also occur. Examples include altitudinal movements of breeding birds down mountain slopes to warmer valleys in winter e.g., chickadees, *Parus* spp. (Tanner, 1952), or to higher slopes where there is special food supply (Blue Grouse, *Dendragapus obscurus*: Mussehl, 1960) or where, due to thermal inversion, the air is warmer at higher elevations (chickadees: pers. observ.); local movements in cities to take advantage of shelter and warmth provided by buildings (e.g., Starlings, Pigeons, House Sparrows); latitudinal shifts in portions of the population during winter (Willow Ptarmigan: Irving et al., 1967; Rock Ptarmigan, *Lagopus mutus*: Weeden, 1964).

Several northern species are nomadic, apparently moving in response to food supply rather than to temperature; crossbills, *Loxia* spp. and waxwings, *Bombycilla* are examples. Winter residents can escape cold air temperatures at night and prevent radiation to the cold night sky thus conserving energy, by roosting in shelters such as man-made nest boxes (Kendeigh, 1961) hollows in trees, under snow laden spruce branches, in caves, rock crevices, and in snow tunnels (Irving, 1972; Sulkava, 1969; Braun and Rogers, 1971). By roosting in cavities, birds create a warmer microclimate which can result in considerable energetic savings (Kendeigh, 1961). Convective heat loss is significantly increased by increasing wind velocity (Gessaman, 1972) and can be avoided by obtaining shelter in cavities or by resting on the lee side of vegetation breaks, land or snow irregularities, etc.

A few species have been observed to huddle which reduces surface area available for heat loss (Dawson and Hudson, 1970) but this is not a common mechanism among northern residents. In contrast, some species may bask in the sun which may provide additional warmth that substitutes for the energy requirement of temperature regulation (Norton, 1966; Lustick, 1969).

Individuals reduce the opportunity for heat loss through non-insulated body parts by postural adjustments such as pulling the legs and feet into the plumage and tucking the heat into the axilla.

In order to remain operable, the legs and feet must be maintained at temperatures above 0 when air temperatures reach −40 to −50° C. Yet they cannot be heated much above 0 since this would not only result in excessive heat loss but would also cause the snow and ice substrate to melt. The water would then refreeze around the foot and render it dysfunctional. The peripheral blood flow mechanisms are partly described to account for this ability, but more work, some of which is in progress, is needed. Also despite the observations of Irving and Krog (1955) and others (Baudinette et al., 1974; Chatfield et al., 1953), the biochemical and physiological mechanism of the operation of heterothermous tissues in birds, such as those in the legs, have not received adequate attention.

Seasonal adaptation to cold includes adaptation to the short day lengths of winter. Although little studied, there is evidence that indicates winter acclimatized arctic species have a lower light intensity threshold in winter than in summer which permits them to be active in the predawn and evening dim twilight of high latitude winters (Pohl, 1972). Extension of feeding activity into dusk decreases reliance on stored food in crops or caches, a decided advantage for arctic winter residents.

REFERENCES: See pp. 588-590.

Indices for Heat Adaptability of Domestic Animals*

by

T. H. Kamal

INTRODUCTION

Environmental heat is a systemic stressor that evokes in the homeotherms a series of a generalized intensified or depressed biological changes. Most domestic animals, a nonprofusely sweating species of homeotherms, are greatly affected in this respect, with resultant an impairment in their productive and reproductive performances, as reviewed by Macfarlane (1963), Kamal (1965) and Johnson (1965). Therefore the native cattle evolved in the tropics have become inferior to those evolved in temperate zone. Moreover, when exotic European-evolved cattle of similar milk production are transferred to tropical and subtropical areas their milk production is adversely affected but with great individual variation, reaching in some cases one-half of the average yield of such breeds in temperate countries (Abdel-Ghani and Hathout, 1966).

Animal breeders tackle this problem by selecting from the high productive cattle those which maintain their high milk production or high body weight gain under hot climate relatively unimpaired. However, the production level is a rather time-consuming index to use in the tropics for heat adaptability as it takes 2 yrs to know the mature body weight and 4 yrs to know the milk yield of a cow. Besides, it is not economical to cull many of these expensive animals after they are imported in the developing tropical countries because they fail to hold their high level of production under hot climate.

Nevertheless the use of heat-induced changes in economic traits as indices of heat tolerance, especially for dairy cattle is questionable. There are cows that maintain their high production at the expense of high body temperature and of tissue destruction. Eventually, such cows develop a lowered resistance to disease and other stressful agents. Moreover, body weight gain in summer (Cartwright, 1955) is a misleading index of heat adaptability because of the interaction that exists under heat stress between the varied increase in body water and varied decrease in body protein and fat (Kamal and Seif, 1969; Kamal and Johnson, 1971). Two animals might have a similar body weight gain in summer, but one increased its body water at the expense of body tissue loss while the other actually increased its body solids. It would be erroneous to consider the first animal is as adapted to heat as the second animal, though both had similar apparent body weight gain in summer.

Therefore, it is quite obvious that a quick and accurate physiological, physical or biochemical index for heat adaptability related to the productive performance is

---

* See also Progress in Human Biometeorology, Part IA, Chapt. 4, Sect. 2, pp 138-145.

urgently needed to help predicting the ability of young animals to express their high productive potential rather unimparied after they are transferred to and raised under tropical and semitropical climatic conditions.

PHYSIOLOGICAL INDICES

Several attempts have been made to develop physiological indices for heat adaptability of domestic animals. The physiological indices so far developed are based on rectal temperature, respiration rate and moisture vaporization rate. Unfortunately these indices, although it helps identifying the animals that are capable of maintaining the thermal balance fairly well or those which are more efficient than others in evaporative cooling, are not useful in predicting the productive performance of the animals under hot climate, which is the character that domestic animals are bred for.

The reason for the difficulty of developing an index for heat adaptability related to productive performances is the fact that heat adaptability is a complex character that depends on the integrity of various systems such as the respiratory, circulatory, excretory, nervous, endocrine, and enzymatic systems. The coordination of all these systems under thermal stress is different, not only between species, but also between breeds and even between individuals within breed. Some animals may combat heat most efficiently through depressing their energy metabolism to a greater extent than other animals. Others may attain the same goal by an efficient evaporative or non-evaporative cooling system. Other animals may have an enzymatic system that is capable of reacting normally at a relatively high body temperature. Heat adaptability may also be achieved in animals by low sensitivity of their warm receptors to heat. Such variation among animals in the methods used to combat heat stress makes unsatisfactory the attempts to attribute heat adaptability in domestic animals to one particular mechanism.

1. PHYSIOLOGICAL INDICES BASED ON RECTAL TEMPERATURE AND RESPIRATION RATE RESPONSES:

Rhoad (1944) devised the Iberia Heat Tolerance Test which has been used most extensively to compare different breeds and individuals within breeds in heat adaptability. The test is carried out by keeping the animal in a cattle chute exposed to direct sunlight on a bright calm summer day for three consecutive days with ambient temperature in shade between 29° and 35° C and the daily rectal temperatures and respiration rate taken at 10.00 hr and 15.00 hr of the three days are averaged. The Iberia heat tolerance coefficient = 100 – (10 (average rectal temperature – 101), where 101 is considered the average normal rectal temperature of cattle. If two groups show the same coefficient, the one which has lower respiration rate is considered higher in heat tolerance.

The use of one average rectal temperature in the formula (101° F) results in obtaining unrealistic coefficient values because rectal temperature varies due to factors other than ambient temperature such as age, lactation, breed, etc. Besides, the relationships between the cow's Iberia heat tolerance coefficient and each of her birth, 6-month and 5-yr body weights and the corresponding weights of her offspring were found too low for use in predicting performance at the Iberia Livestock Experiment Station (Vernon et al., 1959).

Lee and Philips (1948) made some improvement in the Iberia test by using a heat chamber in their test (R- value heat tolerance test) to obtain a standard series of climatic conditions of ambient temperature and humidities for 6 hrs, to which the animals are exposed. Benezra (1954) added the respiration rate response to the rectal temperature response in his test (Benzra's index of adaptability). The formula he used, body temperature, °C/38.3 + respirations per minute/23 = 2, where values above and below 2 represent lower and higher heat adaptability than normal. The results obtained by this test are somewhat different from those obtained by rectal temperature alone (Bianca, 1963).

McDowell et al. (1955) also used the rectal temperature and respiration rate responses (6-hr hot room test) in estimating the relative responses of cattle. The trapzoidal mean rectal temperature or respiration rates over a 6-hr period (t_m) = 0.5 $t_0 + t_1 + t_2 + t_3 + t_4 + t_5 +$ 0.5 t_6/6, where t_1, t_2, etc. are temperatures observed after 1, 2, etc. hrs of exposure. Unfortunately, this test similarly to the Iberia test, showed no relationship to the production traits. McDowell (1966) reported that the within heifer-breed-season-station correlation of trapzoidal mean rectal temperature and growth rate was non-significant. He also stated that the relationship of the average response to all tests as a heifer versus first lactation milk yield was low for rectal temperature and for respiration rate. Independent of level of production, the within animal correlation of response as a dry cow versus lactating was non-significant, for rectal temperature and for respiration rate.

Final rectal temperature was proposed by Bianca (1963) as a promising indicator of heat tolerance related to body weight gain in hot climate. He subjected four Ayrshire calves to ten different hot environments of 5 hrs each. The lower the final rectal temperature was, the higher gain in body the animals had by the end of the 39-days period. The question arised as to whether the same would hold true when comparing animals of different age groups, different breeds, or different climatic conditions. While the average final rectal temperature of the two Santa Gertrudis groups reared at 10° C and 26.7° C were 40° C and 39.8° C, respectively at 43° C (Kibler, 1957), the body weight gain during the same period as computed from the data presented by Kibler and Yeck (1959) were 26 kg and 20 kg, respectively.

Brown et al. (1969) suggested that the rate of rectal-temperature rise is definitely related to heat sensitivity and could be used as a heat sensitivity test provided that multiple tests are required to determine the degree of heat sensitivity of individual animals. This indicates that variations from test to test are considerably large. Furthermore it seems that the differences between animals in the rate of rectal-temperature rise have to be great enough to be detected. When five cows in this study were selected in 1967 to be at nearly the same age, stage of lactation, and level of production and were tested 14 times each no significant differences in the rate of rectal temperature rise were detected between these cows. No consistent differences in the rate of rise of rectal temperature were also found between the three Agrshire calves studied by Bianca (1963).

2. PHYSIOLOGICAL INDICES BASED ON WATER LOSS

Water loss is carried out through skin and respiratory vaporization, urine, feces, milk, salivation, tears and nasal tract secretions. In hot climate water loss plays a major role in heat dissipation through evaporative cooling and thermal conduction between consumed water, body core and excreted water. But vaporization is considered, especially at high ambient temperature, the sole mean of dissipating the body heat. Implying this phenomenon Yeck and Kibler (1958) computed the ratios of total evaporative rates of 36 female calves of 3 beef and 3 dairy breeds at high ambient temperature (26.7° C) to those at 10° C and related these ratios to heat tolerance. They showed that Brahman calves which are considered the most heat tolerant had the highest ratio (2.75) followed by Santa Gertrudis, Brown Siwss, Jersey, Holstein, and Shorthorn, in a descending order. Evidently, such order of heat tolerance does not seem to follow these breeds order in production.

Since water turnover rate is closely related to water loss, and it is more accurate in determination than total vaporization it has been used for comparing species and breeds in heat adaptability. Kamal et al. (1972) showed that the percentage increase in water turnover rate, using tritiated water, was highest in Friesians reaching 96%, followed by goats 90%, sheep 77% and buffaloes 72%, when ambient temperature of the climatic chamber was increased from 18° C to 32° C. This indicates that the native fat-tailed sheep and water Buffaloes were less affected by heat and 12 hrs of water deprivation than Friesians and goats.

PHYSICAL INDICES

Coat and skin characteristics have recived more attention in connection with heat adaptability since Bonsma (1949) first demonstrated such a relationship. The Hair coat score proposed by Turner and Schleger (1960), seems promising for predicting the ability of cattle to adapt to hot climate. These workers showed that coat scores were correlated with body temperature and respiration rate in Shorthorns, Herfords and their crosses with Zebu, but genetic correlation between coat score and growth rate was found only in british breeds and not in Zebu crosses. They reported that a sleek coat is important in favoring heat dissipation and woolly coat favoring cold climate. It is apparent, however that sleek coat is also associated with some qualities which might be advantageous in a cool climate.

Percent capillary layer thickness of the total skin thickness in cattle was reported to have marked inverse relationship with heat tolerance, but the results are contradictory as reviewed by Kamal (1965).

The mechanism underlying the relationship between heat adaptability and coat or skin characteristics is not clear. Research along this point and further investigation to develop a numerical expression for coat characteristics related to productive performance in hot climate is worth undertaking.

BIOCHEMICAL INDICES

Biochemical approaches to the problem of devicing an index for heat adaptability related to productive performance was made by Kamal et al. (1962), Kamal and Johnson (1970) and Kamal and Johnson (1971). The premise on which these authors based their proposed indices is that the most productive and heat tolerant animal should show the least depression in the amount of energy it retains (net energy) during heat exposure. This because a heat tolerant animal that maintains its high production level under hot climate is expected to be energetically efficient. It should have the ability of trapping most of the energy produced in the cells from the consumed feed substrates and convert it into useful products rather than heat that augments the heat load on the animal under hot climate. Determination of the depression in net energy at a specific high temperature would consequently provide a quantitative estimation of heat adaptability in domestic animals. This was done in a simpler way by determining the changes in nitrogen retention (Kamal et al., 1962), in lean mass (Kamal and Johnson, 1970) or body solids (Kamal and Johnson, 1971), which are related to the change in energy retention.

The nitrogen retention index for heat adaptability proposed by Kamal et al. (1962) was carried out by determining nitrogen intake at a comfort (10° C) and elevated (35° C) temperatures, and multiplying it by the digestion coefficient of the ration, which does not change markedly with change in temperature. The urinary nitrogen at both temperatures were subtracted from the corresponding values of digestible nitrogen consumption to obtain nitrogen retention at both temperatures. The percentage decline in nitrogen retention at high temperature was then computed as the index proposed. These workers showed that the heat-induced decrease in nitrogen retention reached a negative nitrogen balance indicating a cellular breakdown in all the three dairy breeds studied especially those heifers reared at low temperature. The percentage decline in nitrogen retention ranged from 49 to 254 percent. The occurrence of such a wide individual variation in the heat-induced decrease of nitrogen retention was demonstrated by Kamal et al. (1970) in Holstein breed, where the nitrogen balance was positive in calves and negative in cows. Significant negative nitrogen balance was also observed in heat-stressed sheep in our laboratory (El-Fouly, 1974). This author, noted that the urea entry rate (urea pool size × ^{14}C-urea exponential disappearance rate in blood), increased variably in heat-stressed rams indicating furthermore that protein catabolism takes place in heat stress. Since the entry rate is quicker and more accurate to determine than nitrogen retention it was suggested to use as an index for heat adaptability. Evidently for practical application one would have to use the non-radioactive ^{15}N-urea instead of the hazardous ^{14}C-urea.

Kamal and Johnson (1970) realizing the difficulty encountered in determining nitrogen retention, developed a simpler numerical index for heat adaptability (^{40}K-predicted heat tolerance index). In this method, the amount of radioactivity of the naturally occurring ^{40}K in the body is counted for few minutes in the whole body counter before and after 3 days of heat exposure in the climatic chambers at 32° C. From the ^{40}K loss the amount of lean mass loss can be known. The percentage decrease in body ^{40}K is subtracted from 100 to obtain a coefficient for heat adaptability. The authors showed that among the 11 cows used the coefficient varied from 75 to 102 and it was possible to arrange the animals into 6 classes the lowest class ranged from 70 to 75 and the highest ranged from 96 to 100. The test was repeated on 3 cows at different time and they came out in the same classes they first fitted into.

The animal, which destroys its own lean mass at high ambient temperature is considered a heat intolerant animal because such an animal will eventually become incapable of tolerating other stresses – such as those of disease, pregnancy, and lactation – whether or not it maintains its milk production under hot climate temporarily.

The last method measures only the change in lean mass, but not in adipose tissue, which may precede the protein loss or in growing animals may be the only tissue to be catabolized. Therefore, Kamal and Johnson (1971) proposed another index which determines both lean mass and body fat losses, with the use of the tritiated water. In this test total body water is determined before and 3 days after heat exposure (32° C). Each value is then subtracted from the corresponding live body weight to obtain the total body solids at comfort (10° C) and hot (32° C) climates. The percentage decrease in total body solids may be subtracted from 100 to obtain a coefficient for heat adaptability related to productive performance i.e., body gain or loss of fat and protein. The 12 Holstein calves showed wide variation in the percentage loss in total body solids, ranging from 0.2% to 24.5%.

REFERENCES: See pp. 590-591

Effects of Heat and Cold on Cattle Disease*

by

R. W. Dougherty

INTRODUCTION

In a recent National Academy of Sciences publication, "A Guide to Environmental Research on Animals", Dougherty and Olsen gave a brief review of the effects of environmental changes on disease resistance and susceptibility in farm animals.

In this publication, they listed the losses due to infectious and noninfectious livestock diseases from 1951-1960 as being approximately 1.4 billion dollars (USDA 1965). Of the 78 most important infectious diseases of domestic livestock, at least 53 diseases are affected by environmental (seasonal) factors.

There are relatively few installations throughout the world equipped for the study of the effect of the environment on the physiology, metabolism, or productivity of domestic animals. There are practically no facilities with the capabilities of objectively studying the effects of environmental changes on disease resistance or susceptibility. This puts the onus of our meager information largely on field observations and heresay. Some of the design requirements for such a facility will be described later.

EFFECTS OF COLD STRESS

Milk yields are adversely affected by low ambient temperatures only when relative humidity reaches 90-100%. Butterfat percentage also varies inversely with humidity at low temperatures.

European breeds of cattle feel comfortable and are most productive within a temperature range of 0-15° C. Below this range, they are adversely affected physiologically. There appears to be no definite experimental proof that these adverse physiological effects make them more susceptible to disease.

Olsen (1969), in a limited number of experiments, has shown that the cardiovascular system of cattle undergoes pronounced changes in subzero (F) temperatures. In most of his experimental cattle, there were periodic elevations in systolic aortic blood pressure every 5 to 7 min. These elevations were as much as 100 mm/Hg above normal and

* See also Medical Biometeorology, 1963, pp. 681-703 and Progress in Human Biometeorology, Part III (August 1972), Chapter 10, Section 6 (paper by W. Ferguson and D. Branagan, pp. 93-108) and Section 7 (paper by W. O. Haufe, pp. 109-117).

lasted 5-6 min. In these experiments, no attempt was made to associate these cardiovascular changes with susceptibility or resistance to disease. It is not unreasonable that these changes are of potential survival value to the animals.

Olsen and Trenkel (1969) had 2 groups of cattle before and during 3 days of cold exposure. The first group of 4 cows showed an increase in growth hormone during the exposure. Plasma glucose and free fatty acids were also significantly increased in both groups. The second group of 3 cows showed a more significant increase of growth hormone after cold exposure of −19 to −26° C.

McClurkin exposed 6 Holstein calves to zero (F°) temperatures for 1 month to determine the effects of cold exposure on calf scours and pulmonary infections common to calves reared in the winter months in Wisconsin. The calves remained perfectly healthy until he exposed them to contaminated aerosols. Cold exposure did not augment the aerosol effects.

Ewbank (1968) using 2 six-year-old cows in the fifth and sixth month of lactation cooled both the forequarters of each animal with thin-walled plastic packs, filled with 10-11° C water. One-third of the surface area of each quarter was exposed for 7-12 hrs. The procedure was repeated 3-6 times. The average heat loss was 11 times greater than that of the uncooled parts of the mammary glands. Cooling did not influence milk yield or bacterial colony counts. There was an increase in body cell counts after 5-6 periods of cooling.

Brown moved 2 cows from the mastitis study herd at the National Animal Disease Center into the environmental facility. The purpose of the experiment was to determine if udder contact with a cold ambient temperature or with a cold floor increased the animal's susceptibility to acute mastitis. The first of the 2 cows had rather serious acute mastitis before any chamber temperature changes were effected. Prior to this time, this cow had been in good health and had no recent history of mastitis. Apparently the change from her accustomed place in the herd to an isolated room in the environmental facility was sufficient to influence the occurrence of acute mastitis.

Mikajlovic et al. (1972) using 32 calves of nearly equal weight and age and of native stock were exposed to microclimates with high concentrations of ammonia (80-150 ppm). The CO_2 concentration and humidity were also above the tolerance level. After 5 days in these microclimates, some of the animals were inoculated with 3 different respiratory virus strains. Clinical and postmortem changes were more distinctive and severe in the exposed animals than in the control animals that were inoculated but not exposed. The microclimatic conditions alone did not produce detectable changes in the calves.

In other experiments, Puhae et al. (1972) studied the effects of high concentrations of ammonia (104-116 ppm) and high humidity on the susceptibility of steers to artificial inoculations with IBR virus. High ammonia concentrations within the immediate environment produced slightly greater respiratory symptoms in the inoculated steers than in the control group which was inoculated with the same virus.

Shaw and Nichols (1964) measured the circulating (plasma) 17 hydroxycorticosteroids in calves shipped from Montana to Wisconsin. Two-thirds of the animals had values lower or higher than the 2-8 μg range observed in indigenous, undisturbed calves. Both the low and high 17-OHCS values were indicative of adrenal response. These values did not return to the normal level until 3 weeks after their arrival in Wisconsin. New developments in methodology indicate that their 17-OHCS levels were about 10-fold too high; however, the relative levels might be significant.

Witzel and Collier tested the role of chill stress in the induction of cattle respiratory disease. Calves weighing 300-500 pounds each were confined in a wind tunnel. Each of 6 pairs of beef calves was subjected to chill stress for 10 days with a 20 mph wind. Cold stress alone induced a slight leukocytosis, a leukopenia, a neutrophilia, and a 3-5% drop in the packed cell volume. Some of the calves were treated with extracellular bacterial cell substance and viral agents. Of the control calves 16% died of stress alone and 50% died of induced respiratory disease. Probably, the main point in these studies was the wind chill factor, differing from the quiet air at zero degrees exposure of McClurkin's calves previously described.

In New Zealand when cold rains occur in the high pasture lands, severe losses of sheep occur in flocks that have been recently shorn. This is due entirely to the chilling effects of the cold rains on the exposed skin of the sheep after shearing (Dougherty).

EFFECTS OF HEAT STRESS AND ULTRAVIOLET

A rather interesting study was that of Hughes et al. (1965) on the additive effects of ultraviolet radiation on the keratoconjunctivitis produced by infecting the eyes of cattle with *Moraella bovis* organisms. This may help explain the high incidence of keratoconjunctivitis during the hot months of the year or during the peak months of ultraviolet radiation.

Temperature changes can indirectly affect many other disease of cattle. Parasite loads and infectivity are influenced by temperature and humidity. Warmer and more humid climates have pronounced effects on the hatchability and viability of parasite eggs. Conversely extremely cold climates have deleterious effects on the viability and hatchability of parasite eggs.

Plant growth can be adversely affected by dry hot weather so that grazing range animals may be forced to eat poisonous plants such as *Veratrum californicum*, which in certain areas and at certain times contain teratogenic compounds and if ingested at a certain time in the gestation period may cause fetal malformations.

In Nairobi, Shifrine heat stressed a group of 5 cattle in a 40° C room for 2 weeks. Another group of 5 were not heat stressed. At the end of 2 weeks both groups were subjected to an aerosol of virulent Gladysdale contagious bovine pleuropneumonia (CBPP). The stressed group was returned to the hot room for an additional 23 days. Ninety days later both groups were killed. There were no serological differences in 2 groups; however, one of the heat stressed animals had a slight rise in the agglutination titers 30 days after the aerosol. Necropsy findings were negative.

THE NEED FOR EXPERIMENTAL FACILITIES TO STUDY ENVIRONMENTAL EFFECTS ON DISEASE SUSCEPTIBILITY AND RESISTANCE

In a complete search of available literature, it would be possible to find more experimental work on this particular phase of disease study; however, there is a striking weakness in much of the work. This weakness is caused by the inadequate facilities used in the experiments and not by the inabilities of the research workers.

Facilities for the study of the effects of the environment on disease susceptibility and resistance have requirements that differ drastically from some of the very good environmental facilities that are used to study effects on production and metabolism. These latter facilities can be set at certain temperatures and humidity levels and left in narrow ranges of variability for prolonged periods of time. Examples of these types of facilities can be found at the University of Missouri; ARS Laboratories, Beltsville, Maryland; Rowett Research Institute, Aberdeeshire, Scotland; Hannah Research Institute, Ayrshire, Scotland; and the Ian Clunies Ross Animal Research Laboratory, Prospect, Australia, and others. These facilities are not concerned with the rapid changes in temperature, humidity, and wind velocity that occur under natural conditions, especially in many parts of the world's temperate belts.

No evidence can be found of pulmonary or other diseases in cattle following severe temperature changes. This does not mean that they do not occur.

Besides the capabilities of making rapid changes in temperature, humidity, and wind velocity in the environmental facilities, it should be possible to contaminate the rooms with bacterial or viral containing aerosols. This would require special air filtration systems containing either high efficiency or ultra high efficiency (high efficiency particle arrestants, HEPA) filter media. For further suggestions the reader is referred to "A Guide to Environmental Research on Animals," National Academy of Sciences, Washington, DC., 1971.

REFERENCES: See pp. 591-592

Effects of Heat and Cold on Diseases in Sheep

by

M. E. Hugh-Jones

INTRODUCTION

While this Section will discuss the effects of extremes of temperature on the health of sheep, in nature these extremes rarely occur on their own. Wind and rain will modify the effects of temperature and usually synergistically and not in a cumulative manner. These climatic effects will be referred to when they are relevant. The term "ill-health" has been purposefully used because the classical and traditional definitions of "disease" become less and less applicable when one considers healthy but dead animals and how they reached that state. On the one hand there are the sick animals with various collections of clinical signs and symptoms and pathological lesions which are clearly recognisably suffering from a specific "disease" and on the other hand there are the normal "healthy" animals which through mismanagement have low production rates and non-economic returns. In between there are those animals in which the components of "disease" and mismanagement are inextricably entwined in differing amounts and complexity. Animals in this middle area are of interest both to the veterinarian and the animal production expert. It is certainly easier and possibly more accurate to describe the animals as suffering from "ill-health" than being diseased.

This Section is divided into three parts: (1) conditions where temperature has a direct effect, (2) conditions produced by simple management changes and temperature and (3) multi factoral conditions following indirectly upon temperature changes.

DIRECT EFFECTS OF TEMPERATURE ON SHEEP DISEASES

There are a limited number of sheep diseases in which temperature plays a significant part. The major conditions associated with cold weather appear to be pregnancy toxaemia, hypomagnesaemia and swayback. Pasteurellosis can occur in epizootic proportions if sheep are driven in cold weather but the mechanism is obscure.

Pregnancy toxaemia mainly occurs among ewes bearing twin-lambs in cool temperate climates, where flocks are at pasture throughout the year and with limited grazing during the winter (pregnancy) months. Severe losses can be produced by two sets of circumstances: (1) when the ewes have suffered a prolonged and progressive under-nourishment during pregnancy, (2) when well-nourished ewes face a sudden and almost complete lack of food usually as a result of a harsh environmental stress. In addition twin-bearing ewes have a mild voluntary anorexia in late pregnancy, possibly due to the compression of the rumen by the enlarging uterus and possibly as a result of the depressing effects of the increasing amounts of oestrogens being produced at this time. Essentially, the disease is caused by a shortage of readily available dietary energy at a time when demands are high. The amount of serum free fatty acids is an index of the rate of mobilisation of body fat and indicates the extent to which the energy demand exceeds that available from the current nutrient intake; these demands can be severe in very cold weather and grazing is absent or covered by snow (Slee, 1971). However acclimatized sheep are able to survive while mobilising less free fatty acids than non-acclimatised sheep (Halliday et al., 1969). This can explain the observation that hill-sheep with little or no supplementary feeding rarely suffer from this condition (Watt, 1965). Smith (1970) has produced a semi-empirical formula to forecast the incidence of disease in North Wales

$$V = 13.5\,xy - 0.11\,v + 45$$

where V is the number of cases per 20,000 ewes, x the November to March rainfall minus 16 inches, y the November to March accumulated temperature deficit below 43° F minus ten degree-months, and v the number of cases the previous year; the latter parameter allows for husbandry changes resulting from last year's experience.

Swayback in newborn lambs is associated with a progressive copper deficiency in the ewe. In the United Kingdom there is a wide geographical distribution of this disease, which is associated with a variety of soils, some of which have higher than average molybdenum content. The variation in numbers of cases from year to year shows that the disease is more prevalent after a mild winter than after a winter with many days of snow cover. On this basis a qualitative forecast (high, above average, average, below average, and low) is possible using the November-February snow cover, corrected for the available hay quality (Smith, 1970). The explanation is not completely elucidated but in severe winters with a subsequent low prevalence the pregnant flock will receive extra supplementary feed which may contain more utilisable copper than is present in the available winter herbage. But it now seems possible that the severe winters also limit the involuntary ingestion of soil as well as the consumption of herbage molybdenum (Alloway, 1973).

Hypomagnesaemia has been admirably reviewed by Voisin (1963). Ewes with twin lambs are particularly liable to this disease and it is the older ewes that are most affected; old ewes have significantly less serum magnesium (1.62 mg/100 ml) than young ewes (2.26 mg/100 ml). The period of greatest risk is from shortly before lambing to six weeks afterwards, with the highest incidence of disease just after lambing when the ewe's lactation is at its maximum. There are two forms of the disease – Spring grass tetany, and Autumn and Winter tetany; the former is caused mainly by the imbalance in the herbage composition and the latter form from under-feeding. Spring grass tetany occurs when, following a cold spell of weather, there is a sudden rise in temperature triggering a growth of young grass. This grass can have 15% or less dry matter, which itself is a component of under-feeding and will accentuate any other hypomagnesaemic effects. It can also cause excessive amounts of rumenal acid to be produced which will reduce the amount of magnesium that will be absorbed. The temperature rise is associated with an increase of the potassium content of the grazing. This increase of the magnesium antagonist will upset the K: Mg ratio and set off a clinical attack of disease. While there is no forecasting facility for hypomagnesaemia as yet, it has been noted that the lower the absolute temperature, the more marked is the effect of the temperature rise. A delay of some five warm days has been noted between the temperature rise and the outbreak of tetanic cases.

In Autumn and Winter tetany the seasonal diet of poor quality and quantity with little or no magnesium slowly robs the animal of its few mobilisable reserves of magnesium; this is occasionally accompanied by hypocalcaemia. Any digestive upset with anorexia will exacerbate the seasonal magnesium loss. The cold temperature will increase the thyroid activity and the magnesium requirements and consumption; thyroxine itself has an antagonistic effect on magnesium. Sheep experimentally exposed to an 8° C environment had a 12% reduction in serum magnesium while those acutely exposed to a −20° C temperature had a reduction of both magnesium and calcium (Sykes et al., 1969). Inglis et al. (1959) noted that the serum magnesium of cold underfed sheep fell during five days from 2.1 mg to 1.3 mg/100 ml. As soon as the diet was supplemented by forage the serum magnesium level returned to almost normal in three days. The major factors in this form of hypomagnesaemia are acute and chronic lack of feed and sudden and prolonged cold. No forecasting facility exists for this type of hypomagnesaemia either but Voisin commented on the tetanogenic effects of sudden temperature drops to 0° C, showers and cold wind, and snowfalls preventing grazing.

Although certain diseases are claimed to be associated with hot weather or seasons there is as yet scant evidence to support them. Until the multitudinous problems of poor husbandry and management, heavy parasitism and inadequate nutrition, with extremes of humidity, approach solution in the tropics, it might be wiser to delay any discussion as to the importance of high temperatures in frank diseases of sheep to another day. That the Australian workers in Queensland have not identified any specific disease strongly associated with high temperatures would indicate their infrequency. However the current development of sheep feed lots with thousands of sheep in the Middle East and North Africa may result in the identification of specific disease conditions in the future.

SHEEP DISEASES DUE TO CHANGES IN MANAGEMENT AND TEMPERATURE

IATROGENIC AND GEORGOGENIC CONDITIONS: Certain diseases follow upon the activities of the veterinarian and the farmer. Although not frequently reported, significant excess deaths have been associated with drenching sheep with carbon tetrachloride four days after colder than average weather (Setchell, 1962). The mechanism for this is obscure. The effect might be due either to the carbon tetrachloride or to bringing together and handling sheep suffering from a temperature associated condition, such as pneumonia, irrespective of or exacerbated by the drenching.

Housing sheep to minimise the effects of winter weather and to conserve pastures can result in severe problems until the management can resolve them. Thus "cold" can indirectly result in housed sheep suffering pneumonia (due to faulty ventilation), acidosis (following overeating), rumenitis (low roughage intake), pregnancy toxaemia, copper poisoning (improper diets), periorbital eczema, urolithiasis, pot-bellied lambs (irregular feeding), parasitism (helminthiases, coccidiosis and lice), increased numbers of prolapsed vaginas (breeding young stock and too little exercise) and nutritional muscular dystrophy. Because of the increased stocking density and shared air space, the following infectious diseases can become realities if care is not taken, contagious pustular dermatitis, abortion (enzootic, vibrionic, salmonellosis and toxoplasmosis), colibacillosis, navel infections, contagious ophthalmia, mastitis, Johne's disease, scrapie and pulmonary adenomatosis (Watt, 1966; Sharman, 1973; Cooper, 1966). The Moredun Institute at Edinburgh suffered a small epizootic of jaagsiekte after housing some experimental sheep (MacKay et al., 1971). The sheep were in cold stress experiments and were housed together for 9-10 months. Not only were 11/62 affected but the sheep were all less than 20 months old. The normal attack rate at the Institute was 14/1,200 for similarly aged sheep and 41/1,200 for older sheep. No disease had been recorded among the 5,000 sheep on the farm where the sheep originated. However, many of the disease problems of housing sheep in winter are more potential than real and with good management will not occur.

Although deaths after shearing are usually low, an average of 0.7% has been quoted by Hutchinson and McCrae (1969) and Hutchinson (1968), losses over 2% are associated with wind and rain on the newly shorn sheep. Geyterbeck (1963) reported an 8% mortality on 17 different properties after a single severe storm. The situation is made worse if the sheep are either underweight or if there has been a general loss of condition in the months before shearing (Hutchinson, 1966). Half the deaths occur during the week immediately following shearing. While the immediate effect of shearing on some sheep is to decrease their feed intake, which will affect the animal's summit metabolism capability, rapid acclimatisation takes place enabling the animal to increasingly maintain its body temperature and regain its normal feed intake during the following weeks (Bennett, 1972; Wheeler et al., 1963; Slee, 1971).

The form of death can be presented as "pneumonia" and later cases will have all the lesions typical of pneumonia. The actual initial cause of death might be either hypothermia and/or an endocrine crisis precipitated by cold exposure (Panaretto and Vickery, 1970). While one considers post-shearing chilling to be a problem limited to cooler latitudes, it must be remembered that in dry hot climates there are extreme day-night temperature fluctuations and post-shearing deaths are not unknown. The fleece insulates and resists heat dissipation by convection, conduction and radiation. It also reduces the inward transmission of solar radiation absorbed on the fleece's outer surface. The unshorn sheep is more heat tolerant than the shorn sheep (MacFarlane et al., 1958). The sheep's critical temperature is affected by the fleece length, nutritional status and body size (Blaxter, 1962). Armstrong and his co-workers (1959) found that close clipped sheep on a high level of nutrition had a critical temperature 24° C while those on a sub-maintenance diet had a critical temperature of 38° C. On a maintenance diet and with a fleece length of 0.1 cm the critical temperature was 32° C, 2.5 cm and 13° C, and 12 cm and 0° C. Below the critical temperature the lighter fleeced sheep's body temperature also falls more quickly into a normally irreversible hypothermia. Losses can be prevented either by leaving about half an inch of fleece on the sheep (though a "snow-comb" is not suitable for fine woolled merinos), or by providing the sheared sheep with plastic coats (only economically worthwhile if the expected losses are over 3-6%) or by regularly patrolling the flock and placing any "downer" sheep, head and all, overnight in an empty feed-sack to recover her body temperature (Hutchinson et al., 1960; Hutchinson and Bennett, 1969, 1970).

INDIRECT EFFECTS OF TEMPERATURE AND OTHER FACTORS ON "ILL-HEALTH" AND PRODUCTIVITY OF SHEEP

While excess heat and cold will play their separate parts in affecting growth and disease susceptibility, there are certainly many more important effects resulting from nutrition and management. Therefore it would be foolish to claim that temperature is the main determinant of lowered productivity, but there are a number of indirect effects which can result in lamb deaths or poor fertility.

Sheep appear to be more susceptible to excess heat than cold during spermatogenesis, ovulation and conception. The effects on the ram's physiology has been described by Watts and Ortavant (1968) and Hales and Hutchinson (1971), and reviewed by Moule (1970) and Waites (1970), while the effect on his behaviour and libido has been reviewed by Entwhistle (1973). The effect of heat on the ram is held to be more severe than on the ewe (Braden and Mattner, 1970). Various acclimatization experiments have isolated the various effects on the ewe (Ryle, 1961; Dutt, 1963; Thwaites, 1967, 1969, 1971 and 1972) and the conceptus (Moule, 1954; Yeates, 1958; Shelton, 1964; Alexander and Williams, 1966; Shelton and Houston, 1968; Edey, 1969; Alexander and Williams, 1971). It is a moot point whether sheep ever meet a natural climate similar to the constant stress characteristic of some chamber experiments. In the field a significant drop in conception rates has been noted following severe drops in air temperature (27° C to 16° C) followed by rain (Davis, 1973). But as with excess heat the effect is the result of both lowered libido and fertility (Webster, 1939).

Both heat and cold will affect the nutrition of the pregnant ewe; the former restricts her to shade and the latter increases her energy demands and occasionally forces her to seek shelter instead of food. During both times of the year the grazing is at its least nutritious. While the energy requirements of the pregnant ewe have been reported (Russel et al., 1967a) the workers commented that such levels were seldom available "on the hill". Thus with "normal" undernourishment during the last six weeks of pregnancy, single lambs can be 10% underweight and twins 25% under their proper total birthweight (Russel et al., 1967b). The same effect can also follow upon prolonged underfeeding, though the full effect is mitigated by the lamb foetus growing at the expense of the ewe (Robinson, 1973). If the ewe is properly fed after lambing there will be no real effect on her lactation but small lambs seem unable to fully exploit her potential (Peart, 1967). However, both cold and heat will affect the lactation and development of the normal udder (Johnson, 1965). A malnourished foetus, if it survives, will be born with a small body-weight and therefore a relatively large surface area, a reduced birth coat density with low thermal insulation and poor vigour.

Alexander (1972a, b) has listed climatic stress as only one of the many causes of neonatal lamb losses, pre and post natal infections, congenital abnormalities, undernutrition, qualitative nutritional disorders, birth accidents, deficient milk supply, ewe misbehaviour, predators, multiple births and extreme birth weights. While one must not overemphasise climate during the lambing season it can kill a significant number of lambs. Lambs are born wet and therefore the body temperature will immediately fall after birth until the necessary respiratory and circulatory adjustments for an independant existence are complete. These usually occur within a few hours of birth. The newborn lamb has limited energy resources, which may have already been depleted by disease in utero, such as by "Border disease" or congenital hypomyelinogenesis. The ability to maintain the body temperature or summit metabolism decreases progressively as the cold exposure continues. The maintenance of summit metabolism depends on the milk intake and is proportional to the body weight; thus the small neonate is especially liable to chilling (Alexander, 1970; Gemell et al., 1972; Alexander 1972). Under mild conditions the neonate's energy may last for three days but with severe cold conditions it will be expended within 24 hrs. Hypothermia reduces the lamb's teat seeking activity which is stopped completely below 37° C; the ewe loses interest if the lamb is cold and inert. Without milk the lamb's energy reserves, and especially the rapid available brown fat, soon fail the lamb dies by hypothermia and starvation (Alexander and Williams, 1966; Alexander et al., 1959; Alexander and Peterson, 1961). The situation facing a premature lamb is worse even if it survives birth and establishes spontaneous respiration as the premature lamb is reluctant to suck and has a thermogenic response 60% of the full term lamb. At the same time its mother has a retarded onset of what will be a reduced lactation. She will be slow to rise and make the udder available to the lamb. It is not surprising that less than 10% of premature lambs survive (Alexander et al., 1972).

Lambs can be chilled by the heat lost in wet windy weather. Mild wind speeds and rainfalls can have marked synergistic effects upon the numbers of lambs surviving until "marking" (Egan, 1972; Obst and Day, 1968). The type of birthcoat is important. Lambs with long hairy coats have twice the insulation of short curly coats; even wet hairy coats are advantageous (Alexander, 1962; Obst and Evans, 1970). The total number of secondary follicles in the skin at birth is markedly affected by poor nutrition during late pregnancy and by high ambient temperatures (Cartwright, 1972). The finer the birthcoat the higher is the subsequent neonatal mortality in wet windy weather. However, the disadvantages of fine wool coats in early life must be balanced by the better later value of the fleece and the better dressing-out percentages at slaughter (Purser and Karam, 1967).

Winds can chill older animals by opening the fleece and destroying the insulating microclimate as well as by driving water into the fleece. There are breed differences due to fleece type in ability to withstand climatic stress (Blaxter et al., 1966). The flock responding to wind by actively seeking shelter has been described by Munro (1962) and all sheep seek shelter if the temperature falls below freezing. Some protection is provided by geomorphic shelter near the ground (Nichols, 1960) as well as by permiable windbreaks. But strong winds will drive sheep before them to seek impermiable windbreaks which should be sited on the leeside of paddocks. The particularly stressful times of excessive winds are the same as for cold; in the adult, late pregnancy, parturition, lactation and shearing, and for the young stock, birth, the neonatal period and weaning.

Lambs are as susceptible to excess heat as to cold. Moule (1954), Smith (1961) and Morgan et al. (1972) have reported their observations. Lambs are most susceptible to hyperthermia at birth and tolerance improves with age. With time there is a progressive drop in rectal temperatures. Merino lambs in shaded pens with an air temperature of 32° C can have rectal temperatures of 40.6°. Further rises in temperature follow short periods of exercise or exposure to direct sunlight. With air temperature over 37° C, half or more of the newborn lambs in unshaded pens will die, while none or a few will die in nearby shaded pens. In Northern Australia, where there is frequently little natural shade or water, ewes may walk long distances to find grazing and water. Lambs can collapse with heat prostration after following their mothers for some 200 yards in the sun. The provision of shade and watering points can markedly reduce lamb mortality.

REFERENCES: See pp. 592-594

Effects of Heat and Cold on Diseases in Birds

by

E. Simensen and L. D. Olson

INTRODUCTION

High and low environmental temperatures appear to influence the pathogenesis of disease in birds by acting as stress factors which apparently affect the preventive mechanisms for disease. Environmental temperature may also influence other climatic conditions such as humidity, content of gases, dust particles, etc., in the environment.

High and low temperatures also affect disease by influencing the survival time of pathogenic microorganisms outside the bird. For example, Newcastle disease virus can survive on chick down for 3 months at 37° C as compared to 6 months or longer at room temperature. The virus survived in chicken feces for 25 days at 37° C as compared to more than 1 1/2 years at 3-6° C (Olesiuk, 1951).

In poultry husbandry today, birds are often confined where the climate is modified by housing, ventilation, etc. Under such conditions, the birds are seldom exposed to temperature extremes, and the effects of temperature on disease is not readily recognized. Therefore, reports on the relationship of disease to environmental temperature are few.

VIRAL DISEASES

One of the first reported viral diseases of birds which appeared to be related to weather, was INFECTIOUS LARYNGOTRACHEITIS (ILT). Hudson (1931) found that mortality in chickens inoculated with ILT and maintained at room temperature was higher than when infected birds were confined at 37° to 39° C. Sinovic (1970) observed a higher mortality in growing chickens exposed to ILT at 5° C than those exposed at 20° or 35° C. The mortality from ILT also was higher in chickens housed at 35° C as compared to birds maintained at 20° C. Jordan (1966) was unable to establish a relationship between weather or seasonal factors and the severity of ILT.

NEWCASTLE DISEASE virus (NDV) has been reported to occur principally during the fall and winter where it is endemic, and to disappear during the higher temperatures of the summer months (Brandly et al., 1946). Hanson (1972) has suggested that environmental temperature and air pollution associated with poor ventilation in the houses during the winter can decrease the host response of the bird to NDV.

It has been observed, with NDV in chickens at different environmental temperatures, that a higher mortality occurs at the higher temperatures. After infecting 6 to 7-week-old chickens, acclimatized to different environmental temperatures with NDV aerosols, the mortality was 100% at 29.5-32° C, 90% at 21-23.5° C, 75% at 10-13° C, and 55% at 0° C. The incubation period was shorter and nervous signs were more marked at the higher temperatures, whereas, respiratory signs were more pronounced at the lower temperatures (Sinha, 1957).

High environmental temperature has been considered as a possible contributing factor to the cause of AVIAN MONOCYTOSIS (BLUE COMB DISEASE). Van Nes (1953) observed that the disease occurred most frequently 3 to 4 days after the environmental temperature exceeded 29.5° C. Cold weather appeared to be insignificant in the development of EPORNITICS of avian monocytosis. Maas and Voute (1961) related epornitics of avian monocytosis to sudden changes in temperature.

BACTERIAL DISEASES

Fowl cholera (*Pasteurella multocida* infections) in turkeys has been reported to be related to environmental temperature. Albers and Graham (1948) observed a relationship between changes in weather from warm and dry to cold and wet and epornitics of fowl cholera. An increase in the number and severity of the epornitics during a period of high environmental temperature in the summer months has been observed. This relationship was most evident after a decrease in the temperature when the incidence decreased markedly (Olson, et al., 1968).

Under experimental conditions, heat stress increased the mortality among chickens inoculated intramuscularly with a virulent strain of *P. multocida* (Juszkiewic, 1966).

In an attempt to further determine the influence environmental temperature has on the pathogenesis of fowl cholera in turkeys, experiments were conducted at the University of Missouri where turkeys were maintained in two temperature controlled chambers at temperatures of 35° and 20° C and 5° and 20° C. Turkeys were exposed to *P. multocida* either by injecting approximately 1000 organisms (isolate 8964) into the palatine sinuses or by administering 3×10^{10} organisms (isolate 9481) in 4 l of drinking water. It appeared from the preliminary data that high temperatures (35° C) increased the resistance of turkeys, thus resulting in a decrease in mortality as compared to birds maintained at 20° C. This increase in resistance was observed both when the turkeys were simultaneously infected and exposed to high temperature, and when the turkeys were exposed to heat for several days prior to infection. In contrast, a higher mortality was observed when turkeys were exposed to low temperatures (5° C) for several days prior to infection.

With all turkeys exposed to *P. multocida*, except the turkeys maintained at 5° C, there was a rise in the cloacal temperature for the first 24 hrs after exposure. This rise appeared to be higher in the turkeys maintained at 35° C than those at 20° C.

There was a higher incidence of TORTICOLLIS in turkeys maintained at 35° C than in those kept at 20° C. Joint infections were more prevalent at 5° C. High or low temperatures did not significantly affect the phagocytic activity of the reticuloendothelial system as measured by the rate of clearance of colloidal carbon injected into the blood. Yet to be determined, are the levels of corticosteroids in the plasma of infected-noninfected birds at the different temperatures.

CONCLUSION

From the disagreement between the experimental data and the field observations in the investigations cited, it appears that temperature stress produces responses in birds that are complex and difficult to reproduce in simple experimental models. The mechanisms involved in determining the effects high and low environmental temperatures have on the pathogenesis of disease still are not clearly understood.

REFERENCES: See pp. 594-595.

Effects of Heat and Cold on Diseases of Arthropods*

by

W. O. Haufe

INTRODUCTION

The potential for biological control of arthropods has been widely recognized in agriculture and forestry for many years. It was largely responsible for the rapid development of insect pathology as a scientific discipline during the fifth and sixth decades of this century. Advances, however, in the last decade have not kept pace with the recent emphasis on biological, cultural and integrated approaches to pest control in response to the restricted use of chemical methods (Haufe, in press).

The role that physical factors play in the environment of pathogens varies widely with the type of organism and its particular characteristics. Temperature by itself appears to have a twofold effect on entomogenous microorganisms. One effect is on survival of the pathogen outside the host. This is important since microorganisms are known to vary widely in their inherent ability to withstand unfavourable temperatures. For example, viruses that form resistant polyhedra or capsules are able to survive extreme temperatures that are lethal to viruses lacking resistive stages. Usually species with these or similar survival characteristics that withstand adverse temperatures are considered to be more suitable as biological control agents. The second effect is on the activity of the microorganism within the body of the host. The actions of some pathogens such as crystalliferous sporeforming bacteria do not appear to be influenced by variations in temperature. Growth of some microorganisms such as insect viruses, however, are generally affected within reasonable limits so that incubation periods are inversely related to temperature.

WEATHER AND CLIMATE

General recognition of the importance of weather and climate in biological control of arthropods has been evident in the literature for many years. Very meagre studies, however, have been reported to date to investigate precise functions of weather and climate in inhibiting or promoting the pathogenicity of microorganisms in insect pests. This lack of knowledge is noteworthy in view of the importance of biological control agents as practical alternatives to chemical pesticides. It is not surprising, therefore, that references to effects of heat and cold on insect pathogens have been limited to

* See also Progress in Human Biometeorology, Part III (Aug. 1972), Chapt. 10, Sect. 7, pp. 109-117 by W. O. Haufe.

the requirements of laboratory protocols in maintaining cultures of experimental strains of some well studied species.

A few references are available on the importance of weather in epizootics of protozoan diseases, particularly in relation to nosema disease in the honey bee. Weiser (1956) has concluded that weather generally has no effect on microsporidian infections. There are exceptions such as those reported by Kramer (1959) in which the European Corn Borer (*Pyrausta nubilalis*) infected with microsporidian organisms has lower survival rates than uninfected populations in years with low winter and high summer temperatures. Goetze and Zeutzchel (1959) found that the susceptibility of bee colonies to epizootics of *Nosema apis* Zander is increased by unfavourable environmental conditions including past and present poor weather, unsuitable locations for hives and defective housing.

No direct relation has been observed so far between protozoan infections such as *Nosema apis* in honey bees and climatic factors in a region. Incidence of infection, however, is definitely related to season of the year and Burnside and Revell (1948) have correlated rise and fall of the disease with seasonal changes in weather. Average temperature of broodless bees in winter is well below the most favourable temperature (31.0° C) for development of an infection. Infections occur and spread most rapidly in spring when temperature is favourable. According to Burnside and Revell, the upper temperature threshold for disease is 37.0° C and the lower one is between 10.6 and 14.0° C. Diseased bees recover from infections if they are held at 37.2 for 14 days.

GENERAL ETIOLOGY AND EPIZOOTOLOGY OF ARTHROPOD PATHOGENS

The influence of environmental factors such as heat and cold on diseases is expected to vary with the properties and characteristics of primary conditions of infection. Latent as well as active infections are affected by environmental factors. Numerous authors have noted the importance of environmental factors in epizootics of arthropods, but quantitative data on specific factors such as heat and cold and related physiological mechanisms are very meager. Undoubtedly this is due to the complexity of the epizootic relationship of microorganisms to their hosts in nature and to the need for studies on physical factors in controlled environments. Some observations indicate that temperature conditions within the normal range of growth in insect hosts have only a limited effect on epizootics but they are of greater importance in association with other factors such as humidity in the etiology of resultant diseases. Generally, high temperatures accelerate the progress of disease. There are a few exceptions with viruses and protozoa in which insects become resistant rather than susceptible when reared under high temperature. Certain insects are more susceptible to fungi, bacteria and viruses when reared at low temperatures (Franz, 1961). The impression from a number of reports is that humidity may be more important than temperature as a single physical factor in epizootics among the fungi, nematodes and viruses.

Coliform bacteria cause widespread dysentry and septicemia in grasshoppers and locusts. One causative agent *Aerobacter aerogenes* var. *acridiorum* (d'Her.) has been credited with the termination of locust outbreaks in Mexico and has been extensively studied and applied in various regions of the world as a biological control agent. Various authors have emphasized the importance of weather conditions such as moderately warm temperatures and high humidity in influencing epizootics of this disease. Less than satisfactory success of the pathogen as a control agent so far may be indicative of an obvious neglect of studies to determine the role of temperature and other physical factors in the etiology of the disease.

RESISTANCE TO DISEASE

Resistance to pathogenic infection in arthropods, as in other animals, is considered to depend on the activity of lysins. Lysins function as antibodies and act on foreign cells to cause their dissolution or lysis. Bacteriolysins, which kill and dissolve bacteria, have been the most easily demonstrated antibodies found in the blood of insects. Although data are still meager, bacteriolysins respond differently to temperature in insects than they do in vertebrates. Unlike the bacteriolysins of vertebrates, those of insects cannot be separated into two fractions by heating. Furthermore, the bacteriolysins of insects are not completely inactivated by temperatures as high as 75° C. This has been demonstrated with the blood of the cutworm *Euxoa segetum* Schiff., which does not lose its bacteriolytic power against *Bacterium melolonthae nonliquefaciens* Pail. when heated at 72° C, although it may decrease slightly up to 75° C after which it is destroyed (Paillot, 1933). Zernoff (1939) found that immune haemolymph of larvae of the wax moth and the European Corn Borer diluted in physiological saline could be heated to 115° C for 20 min without destruction of the bacteriolysins contributing to immunization of the insects to bacteria.

Phagocytosis has received little attention in relation to temperature and other extrinsic conditions for arthropods. Marked variations in temperature cause corresponding variations in the speed with which phagocytosis takes place. The beginning of phagocytosis is usually retarded at temperatures as high as 15 to 18° C. At 10 to 12° C it is delayed by one-two hours. Practically no phagocity activity is considered to occur below 10° C.

INDUCTION OF INFECTIONS

VIRUSES

Occurrence of virus diseases in certain species of insect has been closely related to either abnormal temperature or a change in temperature. Experiments on the role of high and low temperatures in laboratory infections, especially in the silkworm *Bombyx mori*, indicate that temperature influences induction of the disease. Both negative and positive results have been obtained by controlling temperature to induce virus infections in experiments with several other kinds of insects.

Only a few experiments have indicated that heat stress increases incidence of death from virus diseases. Rearing larvae at a high temperature and moderate relative humidity did not increase the incidence of either polyhedrosis or granulosis in the imported cabbage worm, *Pieris rapae* (Linnaeus) (Tanada, 1954). An increase in infection was observed in caterpillars raised at high temperature combined with high humidity (Vago, 1951). It has been recognized by Japanese researchers that, in the case of *B. mori*, conditions of hatching and incubating eggs play a part in later occurrence of some diseases. The occurrence of cytoplasmic polyhedrosis in *B. mori* can be controlled to some extent by the temperature of incubation of eggs with higher incidence occurring at 30° C than at 25° C (Yokokawa and Yamaguchi, 1960). There are numerous strains of *B. mori* differing in their susceptibility to high or low temperatures, especially among uni-, bi-, tetra and polyvoltine races. Generally the tetra- and polyvoltine are more resistant to abnormal environmental conditions including temperature than are the uni- and bivoltine races.

RICKETTSIAE

In contrast to viruses, rickettsial pathogens in arthropods appear to have thermal limits that are related to the temperature ranges of their hosts. Growth of *Rickettsiella* spp. pathogenic for coleopterous larvae has been found to be minimal between 14 and 20° C, optimal between 25 and 28° C, and maximal at 32° C, the average upper limit of temperature ranges for insects. By comparison, rickettsial pathogens in man have approximately the same minimum temperatures (16 to 20° C) but their optima are between 32 and 35° C and their maxima are near 37° C.

BACILLI AND AGENTS OF MILKY DISEASES

Milky diseases have been studied extensively as control agents of the Japanese beetle, *Popillia japonica* Newman. The large-scale use of the type A strain in pest control has been one of the more impressive successes of biological control with an insect pathogen. All of the milky-disease organisms studied have a narrow range of temperature permitting development. All have about the same lower limit near 15° C. The upper limit for *Bacillus popilliae* and its related strains is about 36° C. Both the time required for the onset of gross symptoms i.e., assumption of a milky appearance in the larvae, and the longevity of the diseased larvae after symptoms first appear, increase under reduced temperature. There is an indication at the upper extreme of temperature, and possibly also at the lower, that sporulation is reduced but there is no change in numbers of spores throughout the middle of the range in temperature.

The precise relationship between the time required for onset of symptoms and temperature has not been fully established although studies have been reported by several workers (Dutky, 1963). Data in many cases unfortunately are not comparable because of differences in dosage, host species and environmental factors. There is some contention whether the relationship between temperature and measurable periods of infection such as onset of symptoms, sporulation and time of death is hyperbolic (Beard, 1945), or exponential (Dutky, 1959). Plots of Beard's data (Dutky, 1963) show that straight-line relationships between temperature and the reciprocal of time have a common intercept for various phases of infection which demonstrates a valid temperature threshold at 15.8° C. Tests measuring the number of doublings in the infection rate equivalent to a ten-fold increase in the dosage indicate a doubling time of about six hours at 30° C for *B. popilliae*. At the same level of inoculation the median time of infection is 10.2 days at 30° C and about 21 days at 22° C.

Development of milky diseases in arthropods appears to be more than a simple relationship between the development of the pathogen and temperature. A recent paper (Julian and Hall, 1968) suggests that, contrary to expectations, the mass production of *B. popilliae* seems to be slow in progress. The repeatedly demonstrated increase in activity and germination in heated spore preparations in the laboratory still begs the question whether such heat activation is a phenomenon likely to occur in the field. The host's response to temperature and its nutritional state at time of inoculation are also involved. Interrelation of temperature with these factors appears to have been neglected in studies of infection.

Certain strains of the crystalliferous bacteria such as *Bacillus thuringiensis* var. *thuringiensis* and *B. entomocidus* var. *entomocidus* have been the subject of intensive study for their pesticidal properties attributed to their production of an apparently common "thermostable toxin active per os" (Heimpel, 1967). Organisms in this group are pathogenic to lepidopterous insects because they produce a toxic, proteinaceous, parasporal body known as the crystal when incubated on acceptable nutrients between 28 and 35° C in the presence of a liberal supply of oxygen. However, the role of temperature and the concept for the activity of the thermostable toxin in strains pathogenic for insects is still not clear.

NEOPLASMS

Studies in comparative pathology have produced an extensively descriptive literature on insect neoplasms and tumors, mostly referring to *Drosophila* spp. (Harshbarger and Taylor, 1968). The early literature on lymph gland and blood cell lesions describes the influence of the environment on inherited melanotic tumors. The effect of temperature, for example, on tumor incidence varies with the tumor genes present; it is quite striking in some strains whereas changes in temperature produce little or no effect in others (Burdette, 1959). It has been inferred that temperature enhances or inhibits tumor pigmentation and therefore counting black spots may belie the actual number of tumors present (Kroman, 1957). In any event, the role of temperature as well as that of other factors is not understood since differentiation between true neoplasia and various inflammatory reactions so far has had to be based on insufficient information on insect response to specific causative factors (Harshbarger and Taylor, 1968).

CONCLUSIONS

Steinhaus (1949) has pointed out that without any question the success or failure of man's efforts to use disease organisms to control insects is related to climate and weather. Nevertheless, only occasional or scant experimental attention has been given to basic factors such as temperature even in experiments in practical application of insect pathology in biological control. References to climatic influence on pathogenic organisms in insect disease cite temperature and humidity as principal but relatively unstudied factors in the etiology and epidemiology of infections. Moreover, both factors are considered to operate interdependently both on the disease organism and on the defensive mechanisms of the host.

In most cases, insect diseases develop and spread most rapidly within ranges of temperature and humidity that provide for the growth of the pathogenic organism within the host during the infective stage. As a general rule the probability of disease outbreaks in insects depends on a high density of susceptible individuals in the population, average daily temperatures falling in the range for infective incubation, high relative humidity (90% or more) and adequate concentration of the infective agent. Descriptions of various entomogenous diseases in nature indicate without doubt that weather conditions, primarily those related to temperature and moisture, are of primary importance during the incubation periods of pathogens among protozoa, bacteria, fungi, viruses and nematodes.

It is significant perhaps that in reviewing the current status of pathophysiology for insects, Martignoni (1964) deplored the neglect of quantitation in studies of insect disease in contrast to insect toxicology and to the procedures normally applied in human medicine. Whether for the lack of reports in the particular field or for the failure of studies to quantitate temperature relationships according to his own definition, he made no reference to temperature in his assessment of pathophysiology for insects. Weiser (1970) also expressed his concern for the general lack of attention to the "classical" insect pathology. It can only be concluded that with recent emphasis in biochemistry and other experimental areas of insect pathology, including pathophysiology, there is a serious neglect of temperature relationships and of the important environmental associations that are essential for a sound etiology in the knowledge of arthropod diseases.

REFERENCES: See: pp. 595-596

Shelter Engineering for Cattle and Other Domestic Animals

by

G. L. Hahn

INTRODUCTION

Economic constraints and lack of technology have in the past largely limited shelter engineering to a "protective" role of keeping animals alive during adverse conditions, perhaps with a small incidental benefit to production. Development of the "productive" function – which can benefit not only production, but reproduction, health and comfort – involves consideration of environmental modification practices. To evaluate these modification practices, we must be able to assess the penalties for departing from maximum production environments in view of potential economic returns and energy availability.

This Section discusses available means for environmental modification, acceptable environments, production responses, management factors, and current applications of livestock shelters. A section on future needs includes areas for further research that will permit reasonable appraisal of livestock shelter needs. Cost and efficiency of labor, mechanization of feeding and milking, and waste handling are not considered.

ENERGY TRANSFER

Energy transfer between animals and their surroundings has recently been reviewed and summarized by several authors. The "Guide to Environmental Research on Animals" (National Research Council, 1971) discusses the physical laws that govern the four modes of heat transfer: radiation, convection, evaporation and conduction. Gates (1962) describes the many aspects that must be considered in heat transfer between an animal and its surroundings. Table 1 summarizes the factors that influence such heat exchanges. To effectively alter the microclimate of an animal through shelter or environmental modification, we must alter one or more of these factors: temperature of the surroundings (e.g., by providing shades or other thermal radiation shields); air temperature (e.g., by utilizing animal body heat or by providing auxiliary heating or cooling); air velocity (e.g., by augmenting natural airflow by fans or by reducing airflow by windbreaks); air vapor pressure (e.g., by evaporating water); radiation shape factors; and conductivity of surfaces that an animal might contact (e.g., by providing bedding materials).

OPTIMUM ENVIRONMENTS

An optimum environment for maximum animal productivity and/or efficiency is not

Table 1. Factors that influence heat transfer from the surface of an animal

Factor	Mode of Heat Transfer			
	Radiation	Convection	Conduction	Evaporation
Surface area of animal	1)	×	2)	3)
Temperature of animal surface	×	×	×	4)
Temperature of surroundings	×		5)	
Temperature of air		×		×
Velocity of air		×		×
Vapor pressure of air				×
Shape factor of radiation source or sink	×			
Emissivity of animal surface	×			
Conductivity of surroundings			5)	

1) Area of animal directly exposed to the radiation source or sink.
2) For standing animals, conduction heat transfer is negligible; for animals lying down, the area of animal surface in contact with the supporting structure becomes a factor.
3) The wetted area of the animal surface, including the respiratory passages.
4) The temperature of the animal surface is an indirect factor, since vapor pressure is a function of temperature.
5) Only that portion of the surroundings actually in contact with the animal.

Table 2. Average daily temperatures for nominal losses in production and efficiency of livestock (based primarily on ASAE D249.2, 1974; and Sainsbury, 1967)

Animal	Acceptable[1]) Temperature Range
Dairy cattle,	
Lactating or within 2 weeks of breeding	4–24° C
Calves	10–26° C
Beef cattle	4–26° C
Sheep	4–24° C
Hogs	
Weaning to market weight	10–24° C[2])
Farrowing sows	10–16° C
Poultry	
Growing (over 10 days old)	13–27° C
Laying hens	7–21° C

1) Acceptable average daily temperatures for long-term exposure (concurrent relative humidity less than 75%). The range should be shifted downward at least 3°C for high radiant heat loads (greater than 1 cal/cm^2-min). Nutrition, management, housing and other factors can also alter the acceptable range.
2) Optimum temperature shifts downward within this range as weights increase.

necessarily the optimum in terms of money and energy inputs. Other factors such as low genetic potential for production, low nutritional levels, or inadequate management may restrict production so much that the direct effects of environmental factors (temperature, solar radiation, insects, etc.) are minimal. However, protective shelters are sometimes necessary to permit the existence of livestock production enterprises in adverse climates. If livestock are exposed for long periods to climatic conditions outside the acceptable ranges listed in Table 2, protective shelters (such as solar radiation shades or winter windbreaks) should be considered regardless of animal production level.

In regions with adverse climates, as production potential and nutrition levels increase, environmental factors become limiting to livestock production. Figure 1 illustrates the

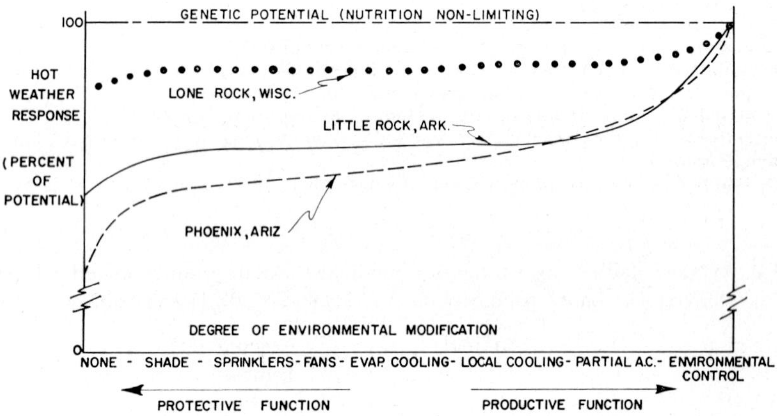

Fig. 1. Production response of livestock to environmental modification practices varies with location, genetic potential for production of the animal, individual differences in sensitivity to heat or cold, energy levels of feed, etc. This schematic illustrates, for hot weather, the functions of various practices together with the generalized benefits to production at three locations.

protective versus the productive function for hot-weather livestock shelters. If the protective function predominates, inexpensive shelters are sufficient; for a predominantly productive function, more costly shelters and environments can provide greater profits over their useful life (Stuber, 1969).

Shelters require properly designed ventilation systems to minimize the undesirable effects of heat, moisture and wastes as by-products of animal production. Presently, the effects on livestock of chronic exposure to low gas concentrations in confinement livestock housing is largely unknown. However, animals have died during mixing operations preparatory to removal of stored wastes (Baxter, 1969; Lillie, 1970).

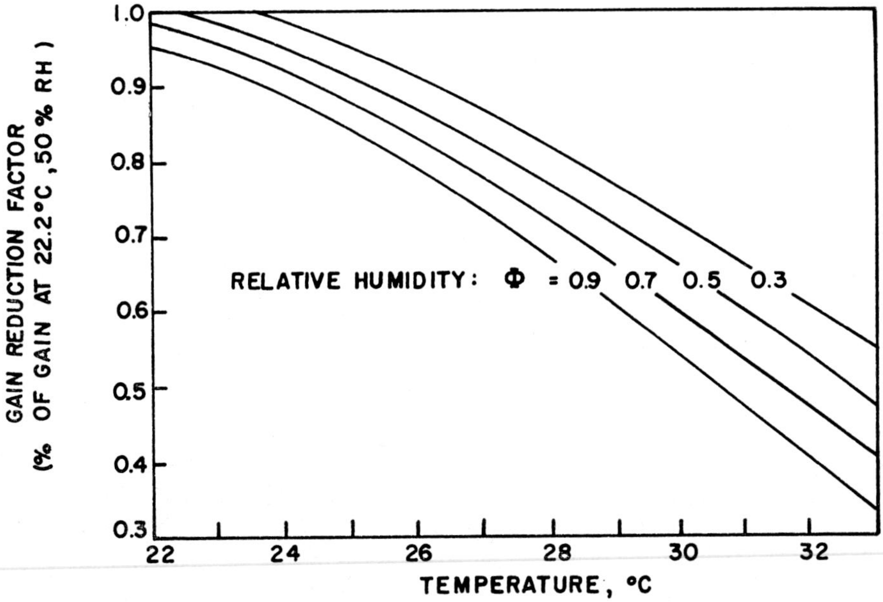

Fig. 1. Ratio of weight gain at conditions above optimum to weight gain at 22.2° C and 50% relative humidity ($\Phi = 0.5$) for 68 kg pigs. (Morrison, Bond and Heitman, 1968). The ratio has been designated as the Gain Reduction Factor (GRF).

RATIONAL SELECTION OF SHELTER NEEDS

Rational decision-making requires a basis for selection among alternatives. Rational decisions by livestock managers concerning the type, design and operation of livestock housing are dependent on several interacting factors. Duckham (1967) includes: (a)

Table 3. Currently available functional relationships between weather events and livestock production responses

Dairy Cows (adapted from Berry et al., 1964):
$M\ Dec = -1.075 - 1.736\ NL + 0.02474\ (NL)\ (THI)$
where M Dec = absolute decline in milk production (kg/day-cow)
NL = normal level of production (kg/day)
THI = daily mean Temperature-Humidity Index value[1])

Laying Hens (adapted from Ota, Garver and Ashby, 1953):
$E = +0.556 + 0.0389\ t_{db} - 0.00179\ t_{db}^2 + 0.0000186\ t_{db}^3$
where E = number of eggs/day-hen (based on 2.25 kg Rhode Island Red Hens)
t_{db} = ambient dry-bulb temperature (°C)
[1] THI is a derived statistic computed from the relation:

[1]) THI is a derived statistic computed from the relation: $THI = t_{db} + 0.36\ t_{dp} + 41.2$.
where t_{db} = dry-bulb temperature (°C).
t_{dp} = dew-point temperature (°C).

livestock response to one or more weather factors; (b) the probability of occurrence of critical weather events affecting response and (c) the ability of the manager to decide among all the alternatives and act upon those decisions for gain (or to avoid loss).

Functional relationships between production (or related animal responses) and weather factors are extremely limited. Hahn and McQuigg (1970) have summarized the currently available empirically-developed livestock production response functions (Table 3 and Fig. 2). Livestock managers can benefit greatly from such information.

Simulation models of animal systems, based on information concerning animal physiology, are being developed at various research stations. For example, in Nebraska, Teter et al. (1973) simulated operational characteristics of animals and in Oklahoma, Paine et al., (1972) developed a simulation model of energy metabolism in beef animals. These models provide a better understanding of why the animal responds as it does to a given environment.

Production response functions for dairy cows (Table 3) and fattening pigs (Fig. 2) have been combined with the probability of occurrence of the related weather events (developed from climatological records) to predict production losses* for June through September (Hahn and Osburn, 1969; Morrison et al., 1970). Figure 3 illustrates

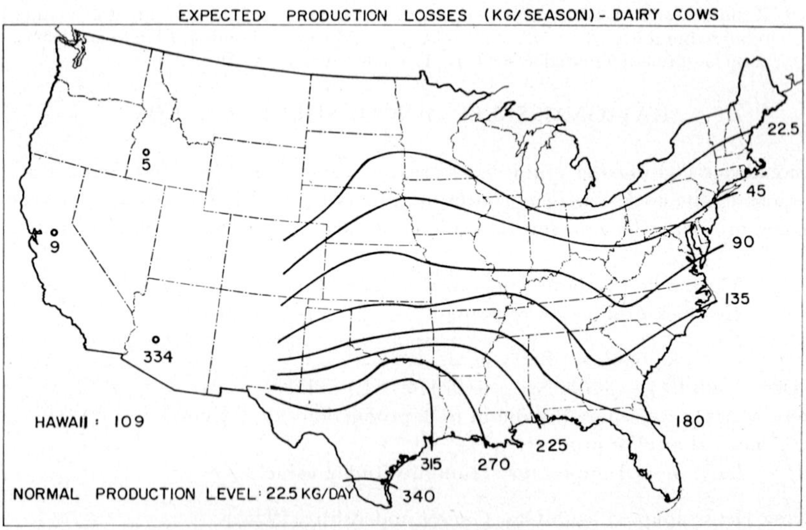

Fig. 3. Expected summer-season (June 1 – September 30) milk production losses (kg/cow-season) for cows with a normal production level of 22.5 kg/day (Hahn and Osborn, 1969). (Solar radiation shades, hot-weather rations and adequate water are assumed available; if not provided, the production losses would be larger.)

* "Losses" in the sense that the animal is not able to produce according to its genetic potential that would occur under optimum conditions, even though it may produce as much as possible under existing environmental constraints.

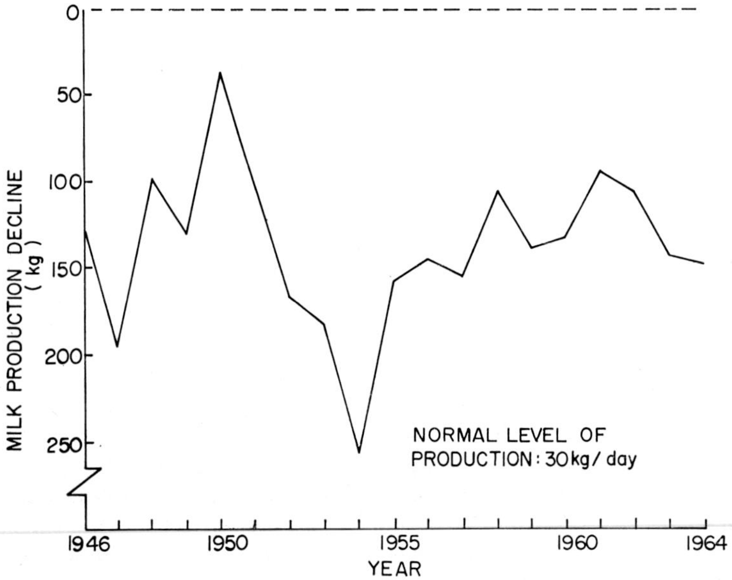

Fig. 4. Year-to-year variability in predicted summertime production decline for cows with a normal production level of 30 kg/day (based on the functional relationships in Table 3) when exposed to the naturally-varying environment at Columbia, Missouri (Hahn, 1974). The values indicate some decline in each of the 19 years used as the climatological data base; the coolest summer (1950) shows an expected decline of about 40 kg/cow and the hottest (1954) an expected decline of over 250 kg/cow.

expected production losses for dairy cows (normal production level, 22.5 kg/day) exposed to the naturally-varying environment within the United States. When predicted and measured losses were compared in Ohio, Missouri, Louisiana and Florida, they agreed within 5 to 17%, which indicates that a reasonable prediction model has been obtained (Hahn, 1969; Thatcher et al., 1974). Estimates of year-to-year variability of milk production due to the direct influence of climate can also be developed for a specific location (Fig. 4).

The intensity of the farming operation also affects the impact of weather and hence the shelter requirements for livestock. Intensive livestock production requires more capital investment, a higher degree of management skill and greater availability of energy resources. However, Taylor (1967) has stated that "farmers operating more intensively near the extreme limit [of weather risks affecting production] maximize resources and profits in the greater number of seasons." Taylor also discusses the exceptionally conservative farmer whose lack of ambition, management skills or capital results in a less intensive livestock operation. Although he obtains lower profits, these profits tend to remain relatively stable from year to year partially because the operation is well within the limits of weather factors.

Even if all livestock managers were provided with precise information for rational nor a uniform level of management. Each perceives the research results demonstrating improved technology on the basis of his own experience. His willingness to experiment with potentially beneficial practices is based on the technique's requirements, the changes in his practices necessary to realize its potential, the hardship that failure might cause, etc. As Wharton (1968) has stated, "... each farmer must live with his individual risk."

CURRENT APPLICATIONS OF LIVESTOCK SHELTERS

McDowell (1974) has pointed out that application of environmental modification practices for improving livestock production is quite limited. In some cases, where factors other than environment are limiting to production, this limited application may reflect the livestock managers' realistic conclusion that such practices are unnecessary. In many cases, however, it indicates his lack of strategic facts for rationally selecting shelters and environmental modification practices.

The economic and technical feasibility of modifying livestock environment is largely unproven, as discussed by Hahn (1974). For example, using mechanical refrigeration for maintaining dairy cow production during the summer in temperate climates has indicated little or no economic feasibility, although it is technically feasible. Conversely, tests with dairy cows breathing air cooled by mechanical refrigeration have given strong indications of economic feasibility, while technical feasibility for commercial applications is still unproven.

When adequate facts are available for rational decision making, adoption of innovations has occurred relatively quickly. For example, evaporative cooling for dairy cows in Arizona in hot summer weather has expanded rapidly. This practice has been shown to be economically feasible through favorable benefit/cost ratios and technically feasible through adequate research and improved design criteria for the coolers (Stott et al., 1972).

FUTURE NEEDS

Increased consideration of shelters and environmental modification practices for livestock production requires improved knowledge of the risk of application. This knowledge can aid the decision making of farmers in their allocation of resources.

Many studies in the past – both laboratory and field – have been directed toward a single factor. Likewise, many past studies have compared only two factors (such as shade vs no shade, or growth at one condition compared with another). Such studies do not provide the rate of change – a fact of strategic value. No general application of such results exists, even though they may indicate adverse effects on the animals. Future research will undoubtedly refine existing biological response relationships and simulation models, but even relatively crude models permit reasonable assessment of the economic feasibility of shelters and environmental modification systems.

Laboratory research under controlled environments is needed to further refine existing hot-weather production response functions for dairy cows, laying hens and fattening hogs to incorporate other climatic factors such as air velocity and thermal radiation effects. Response functions for reproduction and health are also essential for valid evaluations of environmental effects. Biological response models are also needed for cold weather effects, particularly utilization of feed energy. The interacting effects of other environmental factors, such as insects, disease, etc., with climate must be determined. Trade-offs may exist between altering the physical environment of an animal and other factors (e.g., hormone levels in the blood stream). Physiological simulation models and production response models for beef cattle, broilers and sheep need to be developed, especially for aspects of the physiologic response which may alter or invalidate observed short-term reactions (e.g., the compensatory ability of growing beef animals as reported by Hahn et al., 1974). Field research is needed to determine technical feasibility of practices developed from laboratory studies or tests that involve limited numbers of animals. Laboratory and field experiments must be coordinated to validate laboratory results and to avoid restricted geographic applicability of field results. Multidisciplinary research teams are essential in developing comprehensive results to satisfy these needs. Animals of high genetic potential are needed for all studies to realistically appraise the potential for applications of shelter and environmental modification. Especially important is the advocation of Weiss (1945): "The primary aim of research must not just be more and more facts, but more facts of strategic value. The implication, of course, is the need for deeper insight that may lead to the linking of previously unrelated facts and ideas".

REFERENCES: See pp. 596-597

Management, Including Breeding and Nutrition of Cattle

by

W. F. Y. Mawson

INTRODUCTION

Management may be defined as the manipulation and/or modification of factors to achieve a desired result. While in this context it may be assumed that the "desired result" is more efficient animal production the type of production may vary considerable from place to place. For example, in some areas, the desired result may be a dairy cow capable of high milk production in a modified environment while in another it may be a cow capable of a lower level of milk production without a great deal of environmental modification.

This Section is related to cattle management in the out-of-doors or grazing situation and it is necessary to outline the factors which it is feasible to modify, recognising that with modern technology the determinant of feasibility may be decided more by the economist than by the physical scientist.

Although modifications such as fertilizer application, terracing and cloud-seeding may be employed in specific situations it is practical to regard soil type, landform and climate as relatively fixed in location and as such provide the framework of the environment.

Substantial modifications are technically feasible in terms of changing the vegetation and hence the forage supplies for animals and in the selection and/or development of animals adapted to the particular environment. Arising from studies of the characteristics of the forage supply and of the nutrition and habits of the grazing animal is developed a set of management principles for both plants and animals to attain the objectives within the practical limitations of the environmental rigidities.

Climatic factors such as temperature and humidity have a direct effect on cattle while indirect effects are manifest by the limitations imposed on plant growth through lack of moisture, excess moisture or extremes of temperature. Both direct and indirect factors may operate together, for example in modifying the grazing behaviour of animals in a drought situation in mid summer. Climate also affects the distribution of important animal parasites, such as the tsetse fly (*Glossima* spp.) and cattle tick (*Boophilus* and *Rhicpicephalus* spp.) which are responsible for the spread of blood parasites. The incidence of diseases such as ocular carcinoma also appears to be strongly correlated with climatic factors.

Recent reviews with bibliographies are by Mawson and White (1971) concerning certain aspects of dairy cattle, Turner (1972) on differences in animal adaptability and Brown and Hutchinson (1973) for the pastoral situation. A comprehensive approach to animal production in warm climates, with a diagrammatic presentation of the influence of direct and indirect climatic factors was previously given by McDowell (1972, Chapt. 2).

In terms of scientific research the study of the effect of climate on the grazing animal is complex. Brown and Hutchinson (1973) in discussing the difficulties of comparing natural climates to conditions in climate rooms point out that in the latter the air is still, there is no sunshine and humidity may be very high. The distribution of the thermal load around the animal is different in outdoor conditions being modified by such factors as wind speed, wetting by rain, topography, presence of vegetation and solar radiation. For example Murray (1966) has produced evidence that under temperate field conditions cutaneous evaporation rates from the same animals were almost twice those obtained during climate-laboratory exposures. The suggestion is that sweat glands are stimulated to a greater extent by direct solar radiation than by artificial heat.

In addition, the unrestained animal out-of-doors can stand, lie, stretch and make various movements to alleviate stress. Brown and Hutchinson (1973: p. 357) conclude that "studies in climate rooms are difficult to relate to natural conditions but are employed because of difficulties in making measurements on unrestricted animals in the field".

Because of the biological complexities involved progress in the quantification and understanding of the inter-relationships of factors influencing the performance of the grazing animal is not likely to be rapid. However, using the temperature-humidity index (T.H.I.) described by Johnson (1965) Mawson and White (1971) demonstrated that a (T.H.I.) of 75 corresponded to the level at which a dairy production decline could be expected in sub-tropical and tropical North Eastern Australia. This coincides with experience in other centres and its use could be extended to classify areas on a climatic basis. Advances are being made in the use of radio telemetry to obtain physiological data from stock in the grazing situation, scales with automatic recording devices for weighing animals without their grazing pattern being disrupted and identification discs to facilitate behaviour studies at long distances or during night observations. Infra red ocular equipment is also in use. Oesphageal and rumenal fistulas are being employed in nutritional and utilisation studies with grazing animals.

Grazing behaviour studies, particularly where animals are under continual scrutiny for a continuous period are most useful but are necessarily confined to fairly intensive units within a relatively small area.

Because of the difficulties imposed by the interaction and duration of climatic factors and problems of correlating performance with single factors (McIntyre, 1967; Hancock and Payne, 1955) quantification has been in terms of measuring the magnitude of animal response to the total situation in terms of litres of milk or kilograms of mass increase.

In explanation of terms, among types of Indian cattle (*Bos indicus*) the terms "Zebu" and "Brahman" tend to be used interchangeably although the Brahman was developed in U.S.A. from Zebu stock there is some chance of a small fraction of Bos taurus breeding in the Brahman (Payne, 1970: p. 179). British or European stock are termed *Bos taurus.* The Criollo on Creole stock of the Central American region are derived from *Bos taurus* animals but a substantial degree of regional adaptability has been achieved. In dairy breeds the terms Holstein and Friesian are used by different writers to describe what may be regarded as the same breed.

GRAZING BEHAVIOUR

In a review, Hancock (1953) considered that temperatures on the range in the temperate zone has little effect on the time cattle spent grazing. Sheppard et al. (1957) also quoted evidence that grazing time in summer in temperate climates varied considerably but showed no correlation with temperatures. Hancock (1953) states that cattle prefer to graze in daylight but probably his main conclusion is "that cattle, given time, will change their grazing habits to a considerable extent in order to meet changes in their environment". This is borne out in studies in tropical environments by Lampkin and Quartermain (1962), Larkin (1954) and Smith (1959) who report that animals will graze for considerable periods at night under tropical conditions. There are conflicting reports concerning the relationship between time spent grazing and quality and quantity of available fodder.

PARASITE INFESTATION

Control of parasites is an essential feature of management in commercial cattle production and the types of parasites of economic importance vary according to the environment. For example, in portions of sub-tropical and virtually all of tropical Australia the cattle tick (*Boophilus microplus*) is by far the most important because of the direct effect of infestation and the transmission of blood parasites (*Babesia* spp.). A recent enquiry by the Cattle Rick Control Commission, (Anon., 1973) places the estimated annual cost of the cattle tick in Australia at close to $A 42 million. Acaracides, with or without pasture spelling and inoculation have been used as control measures, with the use of acaracide playing a key role. In a recent review and forecasting paper Wharton (1972) states that "the need to apply acaricides at frequent intervals to present *B. microplus* from increasing to damaging levels on herds of European (*Bos taurus*) cattle has led to crises, particularly in Argentina, Brazil and Australia". Resistance has developed to each successive acaricide and some are no longer used because of unacceptable residues in food for human consumption.

Wharton (1972) states that resistance occurs in all breeds of cattle but is associated primarily with the *Bos indicus* types. Although aware of the opposition to the use of *Bos indicus* animals by some producers he considers that there is, amongst scientists working in this field, "general acceptance that they (resistant cattle) provide a logical long-term solution to the cattle tick problem".

ANIMAL ADAPTATION

1. DAIRY CATTLE

Climatic stresses, particularly of lowland tropical areas, have resulted in attempts to develop a tropical dairy breed which would combine the tropical adaptability of the *Bos indicus* with the productive capacity of the *Bos taurus*.

Brief reference is made here to two examples of tropical dairy breeds. One highly productive animal is the Jamaica Hope, developed in Jamaica from Sahiwal and Jersey stock. Lecky (1972) indicates how the breed traces back to a Zebu cow "Norbrook" purchased in 1919.

Wellington (1966) records production trials from 15 years of records (1950-1964) of the government farm at Bodles and seven years of records from six private herds. Selected data are tabulated.

Table 1. Production characteristics of Jamaica Hope cattle

Characteristic	Bodles Herd		Private Herds	
	No. of Records	Mean	No. of Records	Mean
Age at first calving (months)	903	34.2	368	35.9
Length of lactation (days)	2153	332	1634	324
Length of calving interval (days)	1751	439	1394	405
305 day milk yield (litres)	2153	3221	1665	2757
Butterfat percentage	879	4.88	–	–
Gestation period (days)	2471	281	–	–
Birth mass of calves (kg)	1882	24.5	–	–

Compared with temperate regions age at first calving is high and could be reduced by a faster growth rate in heifers and high conception rates at first mating. However, short lactations were not a serious problem. At Bodles 93% of all first lactations exceeded 200 days. The longer calving interval at Bodles compared to private farms is influenced by the fact that the Bodles herd was not exposed to rebreeding until 90 days after calving.

On the sub-tropical coastal region of Australia a new breed, the Australian Milking Zebu (A.M.Z.) has been developed. As with the Jamaica Hope, this breed is based on Sahiwal (25%) and Jersey (75%) and bulls were progeny testing in commercial herds by means of artificial insemination. In breed development, and in contrast to line breeding, crossbred animals were inter-mated and subsequent selection made on the basis of production, adaptability to a tropical climate and tick resistance (Anon, 1965). Semen from A.M.Z. bulls is now available from the Queensland Department of Primary Industries.

From results of testing in Jamaica, Australia and Kenya (Payne, 1970) the Sahiwal has been the outstanding *Bos indicus* breed for dairy purposes. Work with Red Sindhi crosses in U.S.A. and Australia has given disappointing results. The Jersey has also emerged as a suitable *Bos taurus* animal for crossing. Currently, at the Ayr Field Station, located in the tropical coastal region of Queensland, an attempt is being made to develop a tropical dairy breed based on Sahiwal and Friesian.

2. BEEF CATTLE

Early attempts to develop animals for tropical conditions were influenced by the need to have stock capable of surviving the harsh seasonal conditions of the dry tropics (Kelley, 1943), and a great deal of success has been achieved in this direction. However, the widespread distribution of emerging breeds with *Bos indicus* and *Bos taurus* parentage reflects also the desire to exploit heterosis in beef cattle production. North America has seen the development of the Santa Gertrudis, Beefmaster (selected purely on economic traits), Brangus, Braford, Charbray and others; Jamaica has the Jamaica Black and Jamaica Red; South Africa the Drakensberger and the Bonsmara and Australia the Droughtmaster and Belmont Red. Generally, animals of *Bos indicus* inheritance exhibit a greater capacity to survive and produce in a hot harsh environment while some, at least, are capable of high production in environments conducive to high individual performance.

NUTRITION

Emphasis is given to the nutritional aspects of pasture production, recognizing that cropping and grain production may also be practized in suitable agronomic situations. Various meals (peanut, cocoanut, soybean, linseed, cottonseed), by-products (brewer's grains, citrus and pineapple waste, molasses) and crop residues (sugar cane tops, crop stubbles) may also be used in addition to transported concentrates but these are not necessarily specific to the climatic situation.

In the wet tropical situation, where rainfall for plant growth is available for most of the year, certain species of grasses and legumes have been identified and developed into very productive species, frequently away from their native habitat. Among the grasses, species of *Panicum*, *Digitaria*, *Setaria*, *Pennisetum* and *Brachiaria* are prominent while *Desmodium*, *Centrosema*, *Calopogonium*, *Glycine*, *Pueraria* and *Stylosanthes* are legume species which contain highly productive cultivars (Cassidy, 1967; Teitzel, 1969). Edgeley and Tucker (1973), in a tropical elevated area report an average annual herd production of over 4200 l from a herd of 80 Friesian cows grazing tropical grass – legume pasture with a supplement of molasses of 1.4 kg per cow daily. Quinlan et al. (1973) using irrigation to maintain continual growing conditions state that similar pastures will support over seven cows per hectare using a grazing cycle of 3-4 weeks and a stocking density of 100 cows per hectare. In a sub-tropical environment Holder (1967) quotes production of up to 7400 l per hectare from heavily fertilised *Pennisetum* pasture.

Under coastal conditions, long period mass increase in steers grazing pastures of this type is commonly 0.7 to 0.8 kg per day at a stocking rate approaching two animals per hectare.

Under irrigation on the tropical coast Allen and Cowdry (1961) summarise seven years' data in recording beef cattle mass increase of 440-500 kg/ha over a fattening period of 300 days.

A high level of nutrition, moisture and management is necessary to achieve high productivity. Tropical pastures have some deficiencies in comparison with the best temperate species. Holder (1967) states that digestibility of organic matter is some 10 units lower while Minson (1971) points to a depression of digestibility with increasing maturity and the role of nitrogen in influencing the rate of digestion and hence forage intake. Minson and McLeod (1970) conclude that differences in digestibility between temperate and tropical species are closely associated with climate.

In the tropical dryland areas, outstanding introduced grass species are *Cenchrus, Chloris* and to some extent cultivars of *Panicum* with *Stylosanthes, Dolichos* as tropical legumes and *Medicago sativa* as essentially a temperate species which is adaptable to some sub-tropical and tropical situations. Data on productivity appear in articles by Young et al. (1959) who indicate a threefold increase in animal productivity compared with native pastures. Robinson and Winks (1970) indicate that productivity of native pastures can be more than doubled by the introduction of a suitable legume.

Over wide areas of the dry tropics, however, native pastures will continue to be the major source of nutrition for pastoral animal production. These pastures are characterised by rapidly growing, quick maturing plants which have a high nutritive value for four weeks or so then a rapid decline in digestibility as fibre content increases and then a long period as a rather unpalatable, low quality roughage. The growth phase is initiated by rain in late spring or early summer and can be extended through the summer months if sufficient grazing pressure can be maintained – a procedure which is difficult under extensive range conditions. Droughts of varying lengths are common in this environment and stock losses can be heavy in the absence of managerial action.

The performance of cattle grazing such pastures closely parallels the quality status of the plants and is documented by Sutherland (1959), Alexander and Chester (1956), Norman (1965), Davidson (1967) and Alexander and Carraill (1973). Performance is characterised by rapid mass increase (up to 1.5 kg/day) as pastures commence to grow in spring/summer slowing down to a stationary position by the end of autumn, followed by a decline in late winter compounding into spring until grass-producing rains occur. As male castrate animals increase in age the period of mass decrease tends to both shorten and be less pronounced. Lactating cows and cows advanced in pregnancy, together with weaners are the most vulnerable groups in the dry season conditions (Alexander, 1968).

CONCLUSIONS

1. DAIRY CATTLE

The major factor is to improve the quality and availability of forage on a continual basis (Mawson, 1959). Irrigation is an obvious means of achieving this where it is feasible. Norman (1971) in a tropical coastal location quotes a herd average of 6060 l for Friesian cows grazing irrigated pastures plus a small supplement of molasses at milking. This was improved by 450 l when the herd was confined and forage taken to it and later when a cooled barn was used for housing.

Vohnout and Bateman (1972) indicated the need for increasing the calorific density of feed and McDowell (1972: p. 110) points to lower feed efficiency under high temperatures. Where supplementary fodder is supplied to high producing lactating animals on pasture the need is for a high energy, low fibre material, given that nitrogen is adequate.

In breeding, two issues dominate – firstly the degree of specialization required in the industry and secondly the efficiency of acaricides to protect the current high producing animals from parasites. *Bos indicus* strains have been used successfully to a relatively small extent and it is clear that a tropical type dairy breed can be developed. In many areas close to the equator it would seem that the dairy cow may need to be a multi-purpose animal producing in a sub-optimal environment.

Hancock and Payne (1955) in a comparative study involving the use of identical twins concluded that there was evidence of a genotype – climate interaction indicating that selection within existing cattle could be effective.

As indicated from grazing behaviour studies the provision of adequate supplies of drinking water for stock, availability of ample supplies of forage for night grazing and provision of natural shade (McDaniel and Roark, 1956) are useful ameliorants under hot conditions. Where forage supply is seasonal, some manipulation of parturition times of cows makes best use of the available forage (Pegg, 1955).

2. BEEF CATTLE

The nutrition of grazing animals may be improved by:

a. *Use of sown pastures and forage crops* to extend quality and availability of native pastures. (Alexander and Chester, 1956; Ebersohn, 1970; Shaw and 't Mannetje, 1970).

b. *Use of supplements to grazing animals* for maintenance purposes, particularly nitrogen-rich materials to offset low N content of mature grass. A recent review by Alexander et al. (1970) records that the use of a urea-molasses mixture available on a restricted basis resulted in a sustained advantage to yearling steers in a drought year and in fewer mortalities and better performance by cows and calves in drought. In some cases the benefit from supplementation of steers was not sustained subsequently. *Medicago sativa* hay proved superior to urea-molasses as an N supplement in two experiments.

c. *Use of mineral supplements*, of which phosphorus is probably of most significance. Hart and Michell (1965) report increase in mass and improvement in fertility in lactating cows. In many reports response from mineral supplementation alone has been disappointing until combined with a nitrogen supplement and management practices such as controlled mating and weaning.

Turner in reporting comparative performance of Africaner cross, Brahman cross and Hereford × Shorthorn cross under the same conditions in a tropical near-coastal location produced evidence that when exposed to moderate tick infestations there was negligible effect on Brahman cross, a reduction of 20 kg in mass of Africaner cross and 40-50 kg/year in Hereford × Shorthorn cross. A similar order of ranking applied when the groups of animals were exposed to gastro-intestinal helminths.

Susceptibility to ocular carcinoma (a problem in animals lacking pigmentation in the eye region) was almost entirely limited to the pure *Bos taurus* animals where the incidence of contagious ophthalmia was six times greater than in the *Bos indicus* crosses. The comparison also supported the data of Francis and Little (1964) and Daly and Hall (1955) that *Bos indicus* type cattle are relatively resistant to the blood parasite *Babesia argentina*. In feeding studies Turner concluded that the heat tolerant animals show better utilization under hot conditions and Brahman cows have a lower requirement for maintenance. From these studies it was deduced that the Brahman was heat tolerant, tick resistant, efficient at feed utilization and had a general immunity to pathogens. Under the conditions of the observations the crosses showed a reduced reproductive efficiency and some undesirable temperament. Africaners were both highly fertile and docile. It was calculated that the mass increase of the *Bos taurus* crosses was some 14% lower than the *Bos indicus* crosses under free grazing conditions. Gee et al. (1971) in a comparative study under extensive grazing conditions in the dry tropics reported a superior performance by *Bos indicus* crosses, as have Mawson (1956) and Arbuckle (1958). In a property survey of the Queensland subtropical and tropical coast and hinterland McCarthy and Hamilton (1964) found that managers of *Bos indicus* type stock turned off slaughter animals at 3.0 yrs compared with 3.9 yrs of age for *Bos taurus* animals.

Management of an extensive beef cattle herd centres around improving growth and reproductive performance and, in drought, minimising losses. In the dry tropics Davidson (1967) recommends that heifers be withheld from mating until at least two years of age and that calving times be controlled to avoid late autumn and winter calves. Burns (1964) recommends early weaning (say, mid to late autumn for spring and summer calves) to improve the body condition of cows. Sparke and Lamond (1968) stress the influence of adequate bodyweight in cows as a means of improving fertility of range animals. Stating that "nutritin is a key factor in poor reproductive performance" Alexander (1965) advocates early weaning of calves coupled with pregnancy diagnosis as a means of identifying animals with poor breeding performance. Alexander et al. (1964) found that milk production of the cow exerts a major influence on suckling gain and there is evidence (Arbuckle, 1959) that little milk is produced by the dams after autumn, even with a late spring/early summer calving.

Management integrates the many meteorological, biological, physical and economic factors which comprise animal production. For research to be management-oriented consideration of management objectives is required. While research teams are necessarily confined to a particular facet their conclusions will be more meaningful if they are aware of the relationship of their research to that of other teams and to the overall objective, which may have both short and long term characteristics.

REFERENCES: See pp. 597-599

REFERENCES

CHAPTER 1

Effects of Temperature on Animals*

Chapter 1, Section 1a: Effects of Heat on Cellular Metabolism: Enzymes and Other Cellular Changes (R. R. J. CHAFFEE)

ALLEN, J. R. and CHAFFEE, R. R. J. (1964): The alteration of lactic dehydrogenase isoenzyme patterns of brown fat in the golden hamster by cold exposure. Physiologist, 7: 80.

ARINE, R. M., PLATNER, W. S., ROBERTS, J. C. and CHAFFEE, R. R. J. (1973): Evidence of kidney chemical thermosuppression in temperature acclimated hamsters. Comp. Biochem. Physiol., 45: 149-155.

BEYER, R. E., NOBLE, W. M. and HIRSCHFELD, T. J. (1962): Coenzyme Q (ubiquinone) levels of tissues of rats during acclimation to cold. Canad. J. Biochem. Physiol., 40: 511-518.

BEYER, R. E. (1963): Regulation of energy metabolism during acclimation of laboratory rats to a cold environment. Fed. Proc., 22: 874-880.

BEYER, R. E. (1971): Biochemical aspects of acclimation to a cold environment. In: Biochemical Responses to Environmental Stress. I. A. Bernstein (ed.): 94-118.

BLATT, W. F., WALKER, J. and MEGER, M. (1965): Tissue lactic dehydrogenase isozymes: variations in rats during prolonged cold exposure. Amer. J. Physiol., 209: 785-89.

BUTCHER, T. and KLINGENBERG, M. (1958): Wege des Wasserstoffs in der lebendigen Organisation. Angew. Chem., 70: 552-570.

CASSUTO, Y. and CHAFFEE, R. R. J. (1963): The thermogenic role of the liver in the heat-acclimated hamster (*Mesocricetus auratus*). Canad. J. Biochem., 41: 1840-1842.

CASSUTO, Y. and CHAFFEE, R. R. J. (1966): Effects of prolonged heat exposure on cellular metabolism of hamsters. Amer. J. Physiol., 210: 423-426.

CASSUTO, Y. (1968): Metabolic adaptations in chronic heat exposure in the golden hamster. Amer. J. Physiol., 214: 1147-1151.

CASSUTO, Y. and AMIT, Y. (1968): Thyroxine and norepinephrine effects on the metabolic rates of heat-acclimated hamsters. Endocrinology, 82: 17-20.

CASSUTO, Y., CHAYOTH, R. and RABI, T. (1970): Thyroid hormone in heat acclimated hamsters. Amer. J. Physiol., 218: 1287-1290.

CHAFFEE, R. R. J. (1957): Mitochondrial oxidative and phosphorylative enzyme systems in mammals during cold exposure hibernation. Ph. D. Thesis, Cambridge, Harvard University.

CHAFFEE, R. R. J. (1968): A comparison of environmentally induced physiological and biochemical changes in generalized and advanced primates. U.S.A.F. techn. Rep. ARL-TR-68-13. 6571st Aeromed. Res. Lab., Aerospace med. Div., Air Force Systems Command, Holloman A. F. Base, New Mexico.

CHAFFEE, R. R. J. and ALLEN, J. R. (1973): Effects of ambient temperatures on the resting metabolic rate of cold- and heat-acclimated *Macaca mulatta*. Comp. Biochem. Physiol., 44: 1215-1225.

CHAFFEE, R. R. J., ALLEN, J. R., ARINE, R. M., FINEG, J., ROCHELLE, R. H. and ROSANDER, J. (1974): Studies on thermoregulation in brown adipose tissue in temperature acclimated *Macaca mulatta*. In press.

CHAFFEE, R. R. J., ALLEN, J. R., CASSUTO, Y. and SMITH, R. E. (1964): Biochemistry of brown fat and liver of cold-acclimated hamsters. Amer. J. Physiol., 207: 1211-1214.

CHAFFEE, R. R. J., ALLEN, J. R., KELLEY, C., LOVE, J., McCLINTOCK, J. T. and ROCHELLE, R. H. (1969a): Studies on the effects of temperature acclimation on the whole-body protein turnover rate in *Macaca mulatta*. Physiologist, 12: 195.

CHAFFEE, R. R. J., CLARK, R. T., REYNAFARJE, B., CUNNINGHAM, M. D. and BARTLETT, W. L. (1963): Some effects of cold-acclimation on the biochemistry and histology of the hamster kidney. Proc. Soc. exp. Med., 113: 115-121.

* P.H.B. = Progress in Human Biometeorology (Ed. S. W. Tromp).

CHAFFEE, R. R. J., HOCH, F. L. and LYMAN, C. P. (1961): Mitochondrial oxidative enzymes and phosphorylations in cold exposure and hibernation. Amer. J. Physiol. 201: 29-32.

CHAFFEE, R. R. J., KAUFMAN, W. C., KRATOCHVIL, C. H., SORENSON, M. W., CONAWAY, C. H. and MIDDLETON, C. C. (1969b): Comparative chemical thermoregulation in cold- and heat-acclimated rodents, insectivores, protoprimates, and primates. 28: 1029-1034.

CHAFFEE, R. R. J., PLATNER, W. S., PATTON, J. and JENNY, C. (1968b): Fatty acids of RBC ghosts, liver mitochondria and microsomes of cold-acclimated hamsters. Proc. Soc. exp. Biol., 127: 102-106.

CHAFFEE, R. R. J. and ROBERTS, J. C. (1971): Temperature acclimation in birds and mammals. Ann. Rev. Physiol., 33: 155-202.

CHAFFEE, R. R. J. and SMITH, R. E. (1963): Some effects of prolonged cold and heat exposure on the brown fat of the hamster (*Mesocricetus auratus*). Amer. Zool., 3: 274.

CHAFFEE, R. R. J., YOUSEF, M. K. and JOHNSON, H. D. (1968a): Possible role of protein recycling in chemical thermogenesis. Fed. Proc., 27: 633.

COLLINS, K. J. and WEINER, J. S. (1968): Endocrinological aspects of exposure to high environmental temperatures. Physiol. Rev., 48: 785-839.

DEPOCAS, F. (1966): Concentration and turnover of cytochrome c in skeletal muscles of warm- and cold-acclimated rats. Canad. J. Physiol. Pharmacol., 44: 875-880.

EDELMAN, I. S. and ISMAIL-BEIGI, F. (1972): Role of ion transport in thyroid calorigenesis. In: Bioenergetics. Smith, Hannon, Shields and Horwitz (eds). Proc. int. Symp. environm. Physiol.: 67-70, FASEB, Washington.

ESTABROOK, R. W. and SACKTOR, B. (1958): α glycerophosphate oxidase of flight muscle mitochondria. J. biol. Chem., 233: 1014-1019.

FREGLY, M. J. (1968): Metabolic responses to heat. Proc. 3rd Midwest Conf. Thyroid, 1967. Univ. Missouri Columbia: 1-20.

GALE, C. C. (1973): Neuroendocrine aspects of thermoregulation. Ann. Rev. Physiol., 35: 391-430.

GIRARDIER, L., SEYDOUX, J. and CLAUSEN, T. (1968): Membrane potential of brown adipose tissue. A suggested mechanism for the regulation of thermogenesis. J. gen. Physiol., 52: 925-940.

HANNON, J. P. (1963a): Current status of carbohydrate metabolism in the cold-acclimatized mammal. Fed. Proc., 22: 856-861.

HANNON, J. P. (1963b): Cellular mechanisms in the metabolic acclimatization to cold. In: Temperature, its Measurement and Control in Science and Industry. Proc. Amer. Inst. Physic Symp., 3. Rheinhold Publ. Co. New York: 469-484.

HIMMS-HAGEN, J. (1969): The role of brown adipose tissue in the calorigenic effect of adrenaline and noradrenaline in cold-acclimated rats. J. Physiol., 205: 393-404.

HIMMS-HAGEN, J. (1972): Lipid metabolism during cold exposure and during cold acclimation. Lipids, 7: 310-323.

HIMMS-HAGEN, J., BUKOWIECKI, L., BEHRENS, W. and BONIN, M. (1972): Mechanisms of non-shivering thermogenesis in rats. In: Bioenergetics, Smith, Hannon, Shields and Horwitz (eds). Proc. int. Symp. environm. Physiol.: 127-133.

HLUSZKO, M. T., PORTET, R. and CHEVILLARD, L. (1969): Bilan énergétique chez les rats adaptés à 5°C ou 30°C et soumis à des régimes contenant 5% ou 30% de lipides. C. R. Soc. Biol., 163: 344-348.

HOCH, F. L. (1971): Energy Transformation in Mammals. W. B. Saunders Co., Pa.: 45-48.

HORWITZ, B. A. (1974): Physiological and biochemical characteristics of adrenergic receptors and pathways in brown adipocytes. Proc. 2nd Symp. Pharmacol. Thermoregul., Paris: 147-154.

HORWITZ, B. A., HOROWITZ, J. M. Jr. and SMITH, R. E. (1969): Norepinephrine-induced depolarization of brown fat cells. Proc. nat. Acad. Sci. (Wash.), 64: 113-120.

HORWITZ, B. A. and SMITH, R. E. (1972): Function and control of brown fat thermogenesis during cold exposure. In: Bioenergics. Smith, Hannon, Shields and Horwitz (eds). Proc. int. Symp environm. Physiol., FASEB, Washington: 134-140.

HSIEH, A. C. and CARLSON, L. D. (1957): Role of adrenaline and noradrenaline in chemical regulation of heat production. Amer. J. Physiol., 190: 243-246.

ISMAIL-BEIGI, F. and EDELMAN, I. S. (1971): The mechanism of the calorigenic action of

thyroid hormone. Stimulation of Na^+ and K^+ activated adenosinetriphosphatase activity. J. gen. Physiol., 57: 710-722.
JANSKY, L. (1966): Body organ thermogenesis of the rat during exposure to cold and at maximal metabolic rate. Fed. Proc., 25: 1297-1302.
JANSKY, L. (1973): Non-shivering Thermogenesis and its thermoregulatory significance. Biol. Rev., 48: 85-132.
KLAIN, G. J. (1963): Alterations of rat-tissue cytochrome c levels by a chronic cold exposure. Biochim. biophys. Acta (Amst.), 74: 778-780.
LARDY, H. A. and WELLMAN, H. (1952): Oxidative phosphorylations: role of inorganic phosphate and acceptor systems in control of metabolic rates. J. biol. Chem., 195: 215-224.
LEDUC, J. and RIVEST, P. (1969): Effets de l'ablation de la graisse brune interscapulaire sur l'acclimatation au froid chez le rat. Rev. Canad. Biol., 28: 249-266.
LEE, Y-P. and LARDY, H. A. (1965): Influence of thyroid hormones on L α- glycerophosphate dehydrogenase and other dehydrogenases in various organs of the rat. J. biol. Chem., 240: 1427-1436.
LIANIDES, S. P. and BEYERS, R. E. (1960): Oxidative phosphorylation in liver mitochondria from cold exposed rats. Amer. J. Physiol., 199: 836-840.
LUSENA, C. V. and DEPOCAS, F. (1967): Fragility and turnover of liver mitochondria of cold-acclimated rats. Canad. J. Physiol. Pharmacol., 45: 683-687.
MASORO, E. J. (1973): Factors influencing intermediary metabolism. In: Physiology and Biophysics, 20th Ed. T. C. Ruch and H. D. Patton (eds). W. B. Saunders Co. Pa.: 136-147.
MORRISON, G. R., BROCK, F. E., SORBAL, D. T. and SHANK, R. E. (1966): Cold-acclimation and intermediary metabolism of carbohydrates. Arch. Biochem., 114, 494-501.
PANGOS, S., BEYER, R. E. and MASORO, E. J. (1958): Oxidative phosphorylation in liver mitochondria prepared from cold-exposed rats. Biochim. biophys. Acta (Amst.), 29: 204-205.
PATTON, J. F. and PLATNER, W. S. (1970): Cold acclimation and thyroxine effects on liver and liver mitochondrial fatty acids. Amer. J. Physiol., 218: 1417-1422.
PENNER, P. E. and HIMMS-HAGEN, J. (1968): Gluconeogenesis in rats during cold acclimation. Canad. J. Biochem., 46: 1205-1213.
REYNAFARJE, B. and CHAFFEE, R. R. J. (1960): Pyridine nucleotide-cytochrome C reductases in acclimatization to cold. Proc. Soc. esp. Biol., 103: 225-226.
RUEGAMER, W. R., WESTERFELD, W. W. and RICHERT, D. A. (1964): α glycerophosphate dehydrogenase response to thyroxine in thyroidectomized, thiouracil-fed and temperature adapted rats. Endocrinology, 75: 908-916.
RUBNER, M. (1902): Die Gesetze des Energieverbrauchs bei der Ernährung. Franz Deuticke, Leipzig.
SMITH, R. E. (1960): Comparative effects of thyroxin and cold acclimation on metabolic activity of cell fractions from rat liver. Fed. Proc. 19: 64-70.
SMITH, R. E. and FAIRHURST, A. S. (1958): A mechanism of cellular thermogenesis in cold-adaptation. Proc. nat. Acad. Sci. (Wash.), 44: 705-711.
SMITH, R. E. and HOIJER, J. J. (1962): Metabolism and cellular function in cold acclimation. Physiol. Rev., 42: 60-142.
SMITH, R. E. and HORWITZ, B. A. (1969): Brown fat and thermogenesis. Physiol. Rev., 49: 331-425.
SOKOLOFF, L., FRANCIS, C. M. and CAMPBELL, P. L. (1964): Thyroxine stimulation of amino acid incorporation into protein independent of any action on messenger RNA synthesis. Proc. nat. Acad. Sci. (Wash.), 52: 728-736.
SUTHERLAND, E. W. (1972): Studies on the mechanism of hormone action. Science, 177: 401-408.
TESKEY, N., HORWITZ, B. A. and HOROWITZ, J. (1974): Personal communication.
THERRIAULT, D. G., MORNINGSTAR, Jr. J. F. and WINTERS, V. G. (1969): Adenylcyclase activity in isolated fat cells from cold exposed rats. Life Sci., 8: 1353-1358.
YOUSEF, M. K. and CHAFFEE, R. R. J. (1970): Studies on protein-turnover rates in cold-acclimated rats. Proc. Soc. exp. Biol., 133: 801-803.
YOUSEF, M. K. and JOHNSON, H. D. (1968): Effects of heat and feed restriction during growth on thyroxine secretion rate of male rats. Endocrinology, 82: 353-358.

Chapter 1, Section 1b: Effects of Cold on Cellular Metabolism: Enzymes and Other Cellular Changes (E. J. MASORO)

BRENGELMANN, G. (1973): Temperature regulation. In: Physiology and Biophysics, 20th Ed. T. C. Ruch and H. D. Patton (eds). W. B. Saunders Co., Pa.: 105-135.
CARLSON, L. D. (1966): The role of catecholamines in cold adaptation. Pharmacol. Rev., 18: 291-301.
CHAFFEE, R. R. J. and ROBERTS, J. C. (1971): Temperature acclimation in birds and mammals. Ann. Rev. Physiol., 33: 155-202.
HANNON, J. P. and LARSON, A. M. (1962): Fatty acid metabolism during norepinephrine-induced thermogenesis in the cold-acclimatized rat. Amer. J. Physiol., 203: 1055-1061.
HANNON, J. P. (1963): Current status of carbohydrate metabolism in the cold-acclimatized mammal. Fed. Proc., 22: 856-861.
HIMMS-HAGEN, J. (1972): Lipid metabolism during cold-exposure and during cold acclimation. Lipids, 7: 310-323.
HOCH, F. L. (1971): Energy Transformations in Mammals: Regulatory Mechanisms. W. B. Saunders Co., Pa.: 45-53.
JANSKY, L. and HART, J. S. (1963): Participation of skeletal muscle and kidney during non-shivering thermogenesis in cold-acclimated rats. Canad. J. Biochem., 41: 953-964.
MASORO, E. J. (1966): Effect of cold on metabolic use of lipids. Physiol. Rev., 46: 67-101.
MASORO, E. J. (1973): Factors influencing intermediary metabolism. In: Physiology and Biophysics, 20th Ed. T. C. Ruch and H. D. Patton (eds). W. B. Saunders Co., Pa.: 136-147.
SMITH, R. E. and HORWITZ, B. A. (1969): Brown fat and thermogenesis. Physiol. Rev., 49: 330-425.
THERRIAULT, D. G., MORNINGSTAR, J. F. and WINTERS, S. U. G. (1969): Adenyl cyclase activity in isolated fat cells from cold exposed rats. Life Sci., 8, II: 1353-1358.

Chapter 1, Section 2a(1): Effects of Heat on Endocrine Responses of Domestic and Laboratory Animals (H. D. JOHNSON)

ALVAREZ, M. B. and JOHNSON, H. D. (1973): Environmental heat exposure on cattle plasma catecholamine and glucocorticoids. J. Dairy Sci., 56: 189-194.
BLACKWELL, R. E. and GUILLEMIN, R. (1973): Hypothalamic control of adenohypophysical secretions. Ann. Rev. Physiol., 35: 357-390.
CHRISTISON, G. I., MITRA, R. and JOHNSON, H. D. (1970): Glucocorticoids in acutely heat-stressed steers. J. Animal Sci., 31: 219.
CHRISTISON, G. I. and JOHNSON, H. D. (1972): Cortisol turnover in heat-stressed cows. J. Animal Sci., 35: 1005-1010.
CONVEY, E. M. (1973): Neuroendocrine relationships in farm animals. J. Animal Sci., 37: 745.
GALE, C. C. (1973): Neuroendocrine aspects of thermoregulation. Ann. Rev. Physiol., 35: 391-430.
A GUIDE TO ENVIRONMENTAL RESEARCH ON ANIMALS (1971): Nat. Acad. Sci. (Wash.).
HAYWARD, J. N. and BAKER, M. A. (1968): Diuretic and thermoregulatory response to preoptic cooling in the monkey. Amer. J. Physiol., 214: 843-850.
JOHNSON, H. D., RAGSDALE, A. C., BERRY, I. L. and SHANKLIN, M. D. (1963): Temperature, humidity effects including influence of acclimation in feed and water consumption of Holstein cattle. Univ. Miss. agric. exp. Stat. Res. Bull., 846.
JOHNSON, H. D. (1974): Tracer studies in environmental adaptation. In: Tracer Techniques in Tropical Animal Production. Int. Atom. Ener. Ag., Vienna.
KAMAL, T. H., IBRAHIM, I. I., SEIF, S. M. and JOHNSON, H. D. (1970): Plasma insulin and free amino acids changes with heat exposure in bovine. J. Dairy Sci, 53: 651.
KÖPPEN, W. and GEIGER, R. (1936): Handbuch der Klimatologie. Vol. 1, Part C, Borntraeger, Berlin.
LEE, D. H. K. (1953): Physiological objectives in hot weather housing. U.S. Governm. Print. Off., Washington.
MADAN, M. L. and JOHNSON, H. D. (1973): Environmental heat effects on bovine luteinizing hormone. J. Dairy Sci., 56: 1420-1423.

McDOWELL, R. E. (1972): In: Improvement of Livestock Production in Warm Climates. G. W. Salisbury and E. W. Crampton (eds). W. H. Freeman Com., San Francisco, California.

MITRA, R. and JOHNSON, H. D. (1972): Growth hormone response to acute thermal exposure in cattle. Proc. Soc. exp. Biol., 139: 1086-1089.

MITRA, R., CHRISTISON, G. I. and JOHNSON, H. D. (1972): Effect of prolonged thermal exposure on growth hormone (GH) secretion in cattle. J. Animal Sci., 34: 776-779.

RAND, H. R., KIDDY, C. A. and ODELL, W. D. (1971): The effect of stress upon the determination of serum prolactin by radioimmunoassay. Proc. Soc. exp. Biol., 136: 689-693.

RHYNES, W. E. and EWING, L. L. (1973): Plasma corticoids in Hereford bulls exposed to high ambient temperature. J. Animal Sc., 36: 369-373.

SEIF, S. M. (1973): Water kinetics, heat adaptability and ADH of bovine. Ph.D. Thesis, Univ. Missouri.

STOTT, G. H., THOMAS, M. R. and GLENN, L. W. (1967): Blood progesterone in thermally stressed bovine. J. Dairy Sci., 50: 966.

THOMPSON, G. E. (1973): Review of the progress of dairy science climatic physiology of cattle. J. Dairy Res. 40: 441-473.

THORNTHWAITE, C. E. (1948): An approach toward a rational classification of climate. Geogr. Rev., 38: 55-94.

TREWARTHA, G. T. (1954): An Introduction to Climate. 3rd Ed. McGraw-Hill, New York.

ULBERG, L. C. (1967): Effects of macro- and microenvironment on the biology of mammalian reproduction. In: Ground level climatology. R. H. Shaw (ed). Amer. A. A. Sci. Publ. 86 (Wash.).

WEBSTER, C. C. and WILSON, P. N. (1966): Agriculture in the Tropics. Longmans, Green & Co., London.

WILLIAMSON, G. and PAYNE, W. J. A. (1965): An Introduction to Animal Husbandry. 2nd Ed. Longmans, Green & Co., London.

YOUSEF, M. K. and JOHNSON, H. D. (1965): Time course of thyroxine I^{131} disappearance rates in cattle during exposure to hot and cold environments. Life Science, 4: 1531.

Chapter 1, Section 2b(1): Effects of Cold on Endocrine Responses of Ruminants and Other Domestic Animals (A. J. F. WEBSTER)

ANDERSSON, B., EKMAN, L., HÖKFELT, B., JOBIN, M., OLSSON, K. and ROBERTSHAW, D. (1967): Studies of the importance of the thyroid and the sympathetic system in the defence to cold of the goat. Acta. physiol. scand., 69: 111-118.

ASHMORE, J. and MORGAN, DOLORES (1967): Metabolic effects of the adrenal glucocorticoid hormones. In: The Adrenal Cortex. A. B. Eisenstein (ed), Churchill Ltd., London: 249-268.

CARLSON, L. D. (1966): The role of catecholamines in cold adaptation. Pharmacol. Rev., 18: 291-301.

FALCONER, I. R. (1963): The exteriorisation of the thyroid gland and measurement of its function. J. Endocrinol., 26: 241-247.

HALES, J. R. S. (1973): Radioactive microsphere measurement of cardiac output and regional tissue blood flow in the sheep. Pflügers. Arch. ges. Physiol., in press.

HEROUX, O. (1969): Catecholamines, corticosteroids and thyroid hormones in nonshivering thermogenesis under different environmental conditions. In: Physiology and Pathology of Adaptation Mechanisms. E. Bajusz (ed.). Pergamon Press, Oxford.

HIMMS-HAGEN, Jean (1967): Sympathetic regulation of metabolism. Pharmacol. Rev., 19: 367-461.

JANSKY, L. (1970). Non-Shivering Thermogenesis. Swets and Zeitlinger Publ. Co., Amsterdam.

KOKKA, N., GARCIA, J. F., GEORGE, R. and ELLIOTT, H. W. (1972): Growth hormone and ACTH secretion: evidence for an inverse relationship in rats. Endocrinology, 90: 735-743.

LANDON, J. (1968): The basis of radioimmunoassay with particular reference to adrenocorticotrophic hormone. In: Recent Advances in Endocrinology. V. H. T. James (ed). Churchill, London: 240-270.

McDONALD, I. R., GOLDING, J. R. and WRIGHT, R. D. (1958): Transplantation of the adrenal gland of the sheep to provide access to its blood supply. Austr. J. exp. Biol., 36: 83.

PANARETTO, B. A. and VICKERY, M. R. (1970): The rates of plasma cortisol entry and clearance in sheep before and during their exposure to a cold, wet environment. J. Endocrinol., 47: 273-285.

RYDER, M. L. and STEPHENSON, S. K. (1968): Wool Growth. Academic Press, London: 617-625.

SEJRSEN, P. (1969): Blood flow in cutaneous tissue in man studied by washout of radioactive Xenon. Circulation Res., 25: 215-229.

SHUM, A., JOHNSON, G. E. and FLATTERY, K. V. (1969): Influence of ambient temperature on excretion of catecholamines and metabolites. Amer. J. Physiol., 216: 1164-1169.

TRENKLE, A. (1971): Postprandial changes in insulin secretion rates in sheep. J. Nutr., 101: 1099-1104.

TRENKLE, A. (1972): Growth hormone metabolism in cattle and sheep. J. Anim. Sci., 35: 256.

WEBSTER, A. J. F., HEITMAN, J. H., HAYS, F. L. and OLYNYK, G. P. (1969): Catecholamines and cold thermogenesis in the sheep. Canad. J. Physiol. Pharmacol. 47, 719-724.

WEBSTER, A. J. F. (1973): Physiological effects of cold exposure. M.T.P. Int. Rev. Sci. In: Physiology, 7. D. Robertshaw (ed). Butterworths, London, Chapt. 2.

YOUSEF, M. K. and JOHNSON, H. D. (1967): A rapid method for estimation of thyroxine secretion rate in cattle. J. Anim. Sci., 26: 1108-1112.

YOUSEF, M. K., CAMERON, R. D. and LUICK, J. R. (1971): Seasonal changes in hydrocortisone secretion rate of reindeer, *Rangifer tarandus*. Comp. Biochem. Physiol., 40: 495-501.

Chapter 1, Section 2b(2): Effects of Acute and Chronic Exposure to Cold on the Endocrine System of Laboratory Animals (V. M. PETROVIC)

BOOKER, W. M. (1960): Relation of ascorbic acid to adrenocortical function during cold exposure. Fed. Proc., 19: 94-96.

BOULOUARD, R. (1966): Adrenocortical activity during adaptation to cold in the rat: role of Porter-Silber chromogens. Fed. Proc., 25: 1195-1199.

BROWN-GRANT, VON EULER, C., HARRIS, G. W. and REICHLIN, S. (1954): The movement and experimental modification of thyroid activity in the rabbit. J. Physiol. (Lond.), 126: 1-28.

CADOT, M., JULIEN, M. F. and CHEVILLARD, L. (1969): Estimation of thyroid function in rats exposed or adapted to environments at 5° or 30°C. Fed. Proc., 28: 1228-1233.

CARLSON, L. D. (1960): Non-shivering thermogenesis and its endocrine control. Fed. Proc., 19: 25-30.

CHAFFEE, R. R. and ROBERTS, J. C. (1971): Temperature acclimation in birds and mammals. Ann. Rev. Physiol., 33: 155-203.

CHOWERS, J., CONFORTI, N. and FELDMAN, S. (1970): Effect of changing levels of glucocorticosteroids on body temperature on exposure to cold. Amer. J. Physiol., 218: 1563-1567.

CHOWERS, J., CONFORTI, N. and FELDMAN, S. (1972): Body temperature and adrenal function in cold-exposed hypothalamic-disconnected rats. Amer. J. Physiol., 223: 341-345.

COTTLE, W. H. and CARLSON, L. D. (1956a): Turnover of thyroid hormone in cold-exposed rats determined by radio-active iodine studies. Endocrinology, 59: 323-330.

COTTLE, W. H. and CARLSON, L. D. (1956b): Regulation of heat production in cold adapted rats. Proc. Soc. exp. Biol. (N.Y.), 92: 845-849.

COTTLE, W. H. (1960a): Role of thyroid secretion in cold acclimation. Fed. Proc., 19: 59-63.

COTTLE, W. H. (1960b): Urinary noradrenaline in cold acclimated rats. Proc. Canad. Fed. Biol. Soc., 3: 20-25.

D'ANGELO, S. A. (1960a): Hypothalamus and endocrine function in persistent estrus rats at low environmental temperature. Amer. J. Physiol., 199: 701-706.

D'ANGELO, S. A. (1960b): Adrenohypophysal function in the guinea pig at low environmental temperature. Fed. Proc., 19: 51-56.
DEPOCAS, F. (1960): Calorigenesis from various organ systems in the whole animal. Fed. Proc., 19: 19-24.
DESMARAIS, A. (1957): L'acide ascorbique dans l'acclimation au froid. Rev. canad. Biol., 16: 189-195.
DUPONT, A., BASTARACHE, E., ENDRÖCZI, E. and FORTIER, C. (1972): Effect of hippocampal stimulation on plasma thyrotropin (TSH) and corticosterone responses to acute cold exposure in the rat. Canad. J. Physiol. Pharmacol., 50: 364-367.
ENDRÖCZI, E. and LISSAK, K. (1962): Interrelations between palaeo-cortical activity and pituitary-adrenocortical function. Acta Physiol. Acad. Sci. hung., 21: 257-263.
FONTAIN, M. and LACHIVER, F. (1955): Thyroïd et thermorégulation chimique. Arch. Sci. physiol., 9: C63-70.
FORTIER, C. (1966): The Pituitary Gland. Butterworth, London: 195- 234.
HARRIS, G. W. (1955): Neural Control of the Pituitary Gland. Arnold, London.
HARRIS, G. W. and DONOVAN, B. T. (1966): The Pituitary Gland, Vol. 2 Anterior Pituitary. Butterworths, London.
HART, J. S. (1963): Physiological responses to cold in nonhibernating homeotherms. In: Temperature, its Measurement and Control. Reinhold Publ. Co., New York: 373-406.
HÉROUX, O. and HART, J. S. (1954a): Cold acclimation and adrenal cortical activity as measured by eosinophil levels. Amer. J. Physiol., 178: 453-456.
HÉROUX, O. and HART, J. S. (1954b): Adrenal cortical hormone requirement of warm and cold acclimated rats after adrenalectomy. Amer. J. Physiol., 178: 449-452.
HÉROUX, O. (1955): Acclimation of adrenalectomized rats to low environmental temperature. Amer. J. Physiol., 181: 75-78.
HÉROUX, O. (1960): Adjustments of the adrenal cortex and thyroid during cold acclimation. Fed. Proc., 19: 82-85.
HÉROUX, O. and BRAUER, R. (1965): Critical studies on determination of thyroid secretion rate in cold-adapted animals. J. appl. Physiol., 20, 597-606.
HÉROUX, O. and PETROVIĆ, V. (1969): Effect of high and low-bulk diets on the thyroxine turnover rate in rats with acute and chronic exposure to different temperatures. Canad. J. Physiol. Pharmacol., 47: 963-968.
HOIJER, D. J. (1957): Metabolic function of ascorbic acid in acclimation to cold. Fed. Proc., 19: 90-95.
HSIEH, A. C. L. and CARLSON, L. D. (1957): Role of the thyroid in metabolic response to low temperature. Amer. J. Physiol., 188: 40-44.
ITOH, S., HIROSHIGE, T., KOSEKI, T. and NAKOTSUGAWA, T. (1969): Release of thyrotropin in relation to cold exposure. Fed. Proc., 25: 1187-1192.
JANSKY, L. (1973): Non-shivering thermogenesis and its thermoregulatory significance. Biol. Rev., 48: 85-132.
KAWAKAMI, M., SETO, K., TERASAWA, E., YOSHIDA, K., MIYAMOTO, T., and HATTATORI, Y. (1968): Influence of electrical stimulation and lesion in limbic structure upon biosynthesis of adreno-corticoid in the rabbit. Neuroendocrinology, 3: 337-348.
KNIGGE, K. M. and BIERMAN, S. M. (1958): Evidence of central nervous system influence upon cold-conditioned acclimation of thyroidal I^{131} release. Amer. J. Physiol., 192: 625-630.
KNIGGE, K. M. (1960): Neuroendocrine mechanisms influencing ACTH and TSH secretion and their role in acclimation. Fed. Proc., 19: 45-51.
KVETNANSKY, R., GEWIRTZ, P. G., WEISE, V. K. and KOPIN, I. J. (1971): Catecholamine-synthesizing enzymes in the rat adrenal gland during exposure to cold. Amer. J. Physiol., 220: 928-931.
LACHIVER, F. and PETROVIĆ, V. M. (1960): Activité thyroïdienne comparée de deux hibernants: le lérot (*Eliomys quercinus*) et le spermophile (*Citellus citellus*) et d'un homéotherme: le rat au cours de l'exposition au froid pendant l'été. C. R. Acad. Sci. (Paris), 250: 3883-3885.
LE BLANC, J. A. and NADEAU, G. (1961): Urinary excretion of adrenaline and noradrenaline in normal and cold adapted animal. Canad. J. Biochem., 39: 215-217.
LE BLANC, J. and VILLEMAIRE, A. (1970): Thyroxine and noradrenaline on noradrenaline sensitivity, cold resistence, and brown fat. Amer. J. Physiol., 218: 1742-1745.

LEDUC, J. (1961a): Excretion of catecholamines in rats exposed to cold. Acta physiol. scand., 51: 94-95.
LEDUC, J. (1961b): Catecholamine production and release in exposure and acclimation to cold. Acta physiol. scand., 53, suppl. 183: 1-101.
LEVIN, L. (1945): The effects of several varieties of stress on the cholesterol content of the adrenal glands and of the serum of rats. Endocrinology, 37: 34-39.
LUTHERER, L. O., FREGLY, M. J. and ANTON, A. H. (1969): An interrelationship between theophylline and catecholamines in the hypothyroid rat acutely exposed to cold Fed. Proc., 28: 1238-1245.
NICHOLS, D. and ROSSITER, R. J. (1955): The role of the pituitary and thyroid glands in the phosphorus metabolism of the adrenal gland during cold stress. Endocrinology, 56: 547-559.
NICHOLS, D. and ROSSITER, R. J. (1956): Phosphorus metabolism of the adrenal gland of the rat. Amer. J. Physiol., 187: 11-14.
PETROVIĆ, V. M. and ANDJUS, R. K. (1960): Métabolisme de base et adaptation thermique du rat; role de la thyroïde et des surrénales. J. Physiol. (Paris), 52: 191-192.
PETROVIĆ, V. M., ANDJUS, R. K. and DUKIĆ, S. (1961): Role de la thyroïde et des surrénales dans l'adaptation thermique du rat. Arch. Biol. Sci., (Beograd), 13: 7-20.
PETROVIĆ, V. M. and JANIĆ, V. (1964): Comparative study of the effect of cold on 17, 21-hydroxi-20-ketosteroids and adrenal ascorbic acid in rats and ground squirrels. Arch. Biol. Sci. (Beograd), 16: 145-153.
PETROVIĆ, V. M., JANIĆ, V. and DUKIĆ, S. (1964): L'action du froid sur l'activité corticosurrénalienne chez le spermophile. J. Physiol. (Paris), 56: 631-632.
PETROVIĆ, V. M., RAJČIĆ, O. and HUDNIK-PLEVNIK, T. (1966): Étude comparée de l'action du froid et de l'ACTH, sur le taux des acides nucléiques et des protéines dans la surrénale et dans le foie chez le rat et le spermophile. J. Physiol. (Paris), 58: 590-591.
PETROVIĆ, V. M. and RAJČIĆ, O. (1969): Effect of prolonged cold exposure and ACTH administration on RNA, DNA and protein content in adrenals and liver of rat and ground squirrel. Fed. Proc., 28: 1247-1250.
PETROVIĆ, V. M., RAJČIĆ, O., VRBAŠKI, M. and ŠEVALJEVIĆ, Lj. (1973a): Incorporation of labelled amino acids into the liver proteins of rats exposed or adapted to cold. Experientia (Basel), in press.
PETROVIĆ, V. M., RAJČIĆ, O. and VRBAŠKI, M (1973b): Incorporation of labelled amino acids into the adrenals proteins of rats exposed to cold. Arch. Biol. Sci., in press.
PETROVIĆ, V. M., GRIPOIS, D., JANIĆ, V. and ROFFI, J. (1973c): Phenylethanolamine-N-methyl transferase activity in the ground squirrel (*Citelluc citellus*). Experientia (Basel), 29: 876-877.
PETROVIĆ, V. M. and MARKOVIĆ-GIAJA, L. (1973): A comparative study of the calorigenic action of noradrenaline in rat and ground squirrel adapted to different temperatures. Experientia (Basel), 29, in press.
SCHOENBAUM, E., CASSELMAN, W. G. B. and LARGE, R. E. (1959): Studies on the time course of the response of the adrenal cortex to histamine and cold. Canad. J. Biochem., 37: 399-404.
SCHOENBAUM, E. (1960): Adrenocortical function in rats exposed to low environmental temperatures. Fed. Proc., 19: 85-88.
SELLERS, E. A., FLATTERY, R. V., SHUN, A., and JOHNSON, G. E. (1970): Thyroid status in relation to catecholamines in cold and warm environment. Canad. J. Physiol. Pharmacol., 49: 268-275.
SMITH, R. E. (1960): Comparative effects of thyroxin in vivo and cold acclimation on metabolic activity of cell fractions from rat liver. Fed. Proc., 19: 64-70.
STEVENS, C. E., D'ANGELO, A., PASCHKIS, K. E., CANTAROW, A. and SUNDERMAN, F. W. (1955): The response of the pituitary-thyroid system of the guinea pig to low environmental temperature. Endocrinology 56: 143-156.
SWANSON, H. (1957): The effect of temperature on the potentiation of adrenaline by thyroxine in the albino rat. Endocrinology, 60: 205-213.
THIBAULT, O. (1949): Hormonal control of chemical thermoregulation. Rev. Canad. Biol., 8: 3-131.
WOODS, R. and CARLSON, L. D. (1956): Thyroxine secretion in rats exposed to cold. Endocrinology, 59: 323-330.

YOUSEF, M. K. and JOHNSON, H. D. (1968): Effects of heat and feed restriction during growth on thyroxine secretion rate of male rats. Endocrinology, 82: 353-358.

Chapter 1, Section 3a: Water Metabolism in Sheep in Relation to Heat and Cold (W. V. MACFARLANE)

ALEXANDER, G. (1970): Thermogenesis in young lambs. In: Physiology of Digestion and Metabolism in the Ruminant. A. T. Phillipson (ed). Oriel Press, Newcastle on Tyne: 199-210.

BLAIR-WEST, J. R., BOTT, E., BOYD, G. W., COGHLAN, J. P., DENTON, D. A., GODING, J. R., WELLER, S., WINTOUR, M. and WRIGHT, R. D. (1965): General biological aspects of salivary secretion in ruminants. In: Physiology of Digestion in the Ruminant. R. W. Dougherty (ed.). Butterworths, Washington, D.C.: 198-220.

BLAXTER, K. L. (1962): The energy metabolism of ruminants. Hutchinson, London.

BROWN, G. D. and HUTCHINSON, J. C. D. (1973): Climate and animal production. In: The Pastoral Industries of Australia. G. Alexander and O. B. Williams (eds). Sydney Univ. Press: 336-370.

MACFARLANE, W. V., MORRIS, R. J. H. and HOWARD, B. (1956): Water economy of tropical Merino sheep. Nature (Lond.), 178: 304-305.

MACFARLANE, W. V., MORRIS, R. J. H., HOWARD, B., McDONALD, J. and BUDTZ-OLSEN, O. E. (1961): Water and electrolyte changes in tropical Merino sheep exposed to dehydration during summer. Aust. J. agric. Res., 12: 889-912.

MACFARLANE, W. V. (1964): Terrestrial animals in dry heat: ungulates. In: Handbook of Physiology. Environment. D. B. Dill, E. F. Adolph and C. G. Wilber (eds). Amer. physiol. Soc. (Wash.), 4: 509-530.

MACFARLANE, W. V., HOWARD, B. and MORRIS, R. J. H. (1966): Water metabolism of Merino sheep shorn during summer. Aust. J. agric. Res., 17: 219-225.

MACFARLANE, W. V., HOWARD, B. and SIEBERT, B. D. (1967): Water metabolism of Merino and Border Leicester sheep grazing saltbrush. Aust. J. agric. Res., 18: 947-958.

MACFARLANE, W. V. and HOWARD, B. (1970): The ecophysiology of water. In: The Physiology of Digestion and Nutrition in the Ruminant. pp. 362-374. A. T. Phillipson (ed.). Oriel Press, Newcastle: 362-374.

MACFARLANE, W. V. (1971): Salinity and the whole animal. In: Salinity and Water Use. T. Talsma and J. R. Philip (eds). Macmillan & Co., London: 161-178.

MACFARLANE, W. V. and HOWARD, Beth (1972): Comparative water and energy economy of wild and domestic mammals. Symp. zool. Soc. (Lond.), 31: 261-296.

MORRIS, R. J. H., HOWARD, B. and MACFARLANE, W. V. (1962): Interaction of nutrition and air temperature with water metabolism of Merino wethers shorn in winter. Aust. J. agric. Res., 13: 320-334.

MOULE, G. R. (1945): Salt poisoning of sheep following evaporation of saline waters. Aust. vet. J., 15: 37-42.

PORTER, W. P. and GATES, D. M. (1969): Thermodynamic equilibria of animals with environment. Ecol. Monogr., 39: 227-244.

PEIRCE, A. W. (1968): Salt tolerance studies of sheep. VII. The tolerance of ewes and their lambs in pens for drinking waters of types obtained from underground sources in Australia. Aust. J. agric. Res., 19: 577-587.

POTTER, B. J. (1963): The effect of saline water on kidney tubular function and electrolyte excretion in sheep. Aust. J. agric. Res., 14: 518-528.

POTTER, B. J. (1968): The influence of previous salt ingestion on the renal functions of sheep subjected to intravenous hypertonic saline. J. Physiol. (Lond.), 194: 435-455.

PRIESTLEY, C. H. B. (1957): The heat balance of sheep standing in the sun. Aust. J. agric. Res., 8, 271-280.

WEBSTER, A. J. F. and PARK, C. (1967): The effect of jute coats on the heat losses of two breeds of sheep exposed to different environments. Anim. Prod., 9: 483-490.

WILSON, A. D. (1966): The tolerance of sheep to sodium chloride in food or drinking water. Aust. J. agric. Res., 17: 503-514.

WILSON, A. D. and HINDLEY, N. L. (1968): Comparison of voluntary water intakes of some Australian sheep. CSIRO Field Stat. Rec., 7: 25-36.

Chapter 1, Section 3b: Water and Salt Metabolism in the Camel (G. M. O. MALOIY)

MACFARLANE, W. V., MORRIS, R. J. H. and HOWARD, B. (1962): Water metabolism of Merino sheep and camels. Aust. J. Sci., 25: 112.

MACFARLANE, W. V., MORRIS, R. J. H. and HOWARD, B. (1963): Turnover and distribution of water in desert camels, sheep, cattle and kangaroos. Nature (Lond.), 197: 270-271.

MACFARLANE, W. V. (1964): Terrestrial animals in dry heat: Ungulates. In: Handbook of physiology, 4. Amer. physiol. Soc. (Wash.): 509-540.

MACFARLANE, W. V., KINNE, R., WALMSLEY, C. M., SIEBERT, B. D. and PETER, D. (1967): Vasopressins and the increase of water and electrolyte excretion by sheep, cattle and camels. Nature (Lond.), 214: 979-981.

MACFARLANE, W. V. and SIEBERT, B. D. (1967): Hydration and dehydration of desert camels. Aust. J. exp. Biol. med. Sci., 45: 29.

MACFARLANE, W. V. and HOWARD, B. (1972): Comparative water and energy economy of wild and domestic mammals. Symp. zool. Soc. (Lond.), 31: 261-294.

MALOIY, G. M. O. and TAYLOR, C. R. (1971): The water requirements of African goats and haired-sheep. J. agric. Res. Sci. (Camb.), 77: 203-208.

MALOIY, G. M. O. (1972): Renal salt and water excretion in the camel (*Camelus dromedarius*). Symp. zool. Soc. (Lond.), 31: 243-259.

SCHMIDT-NIELSEN, B., SCHMIDT-NIELSEN, K., HOUPT, T. R. and JARNUM, S. A. (1956): Water balance of the camel. Amer. J. Physiol., 185: 185-194.

SCHMIDT-NIELSEN, B., SCHMIDT-NIELSEN, K., HOUPT, T. R. and JARNUM, S. A. (1957a): Urea excretion in the camel. Amer. J. Physiol., 188: 477-484.

SCHMIDT-NIELSEN, K. (1964). Desert Animals. Oxford Univ. Press.

SCHMIDT-NIELSEN, K., CRAWFORD, E. C., NEWSOME, A. E., RAWSON, K. S., and HAMMEL, H. T. (1967a): Metabolic rate of camels: effect of body temperature and dehydration. Amer. J. Physiol., 212: 341-346.

SCHMIDT-NIELSEN, K., SCHMIDT-NIELSEN, B., JARNUM, S. A. and HOUPT, T. R. (1957b): Body temperature of the camel and its relation to water economy. Amer. J. Physiol., 188: 103-112.

SIEBERT, B. D. and MACFARLANE, W. V. (1971): Water turnover and renal function of dromedaries in the desert. Physiol. Zool., 44: 225-240.

TAYLOR, C. R. (1968): The minimum water requirements of some East African bovids. Symp. zool. Soc. (Lond.), 21: 195-206.

Chapter 1, Section 3c: Water Metabolism in Rodents (R. E. MACMILLEN and D. E. GRUBBS)

BAUDINETTE, R. V. (1972): The physiological ecology of a burrowing rodent, *Spermophilus beecheyi*. Unpubl. Ph.D. Thesis, Univ. Calif., Irvine, 99 pp.

CARPENTER, R. E. (1966): A comparison of thermoregulation and water metabolism in the kangaroo rats *Dipodomys agilis* and *Dipodomys merriami*. Univ. Calif. Publ. Zool., 78: 1-36.

CHURCH, R. L. (1966): Water exchanges of the California vole, *Microtus californicus*. Physiol. Zool., 39: 326-340.

CHURCH, R. L. (1969): Evaporative water loss and gross effects of water privation in the kangaroo rat, *Dipodomys venustus*. J. Mammal., 50: 514-523.

COLLINS, B. G. (1973a): Physiological responses to temperature stress by an Australian murid, *Rattus lutreolus*. J. Mammal., 54: 356-368.

COLLINS, B. G. (1973b): Studies on the metabolism, thermoregulation, and evaporative water losses of two species of Australian rats, *Rattus villosissimus* and *Rattus rattus*. Physiol. Zool., 46: 1-21.

COULOMBE, H. N. (1970): The role of succulent halophytes in the water balance of salt marsh rodents. Oecologia, 4: 223-247.

HUDSON, J. W. (1962): The role of water in the biology of the antelope ground squirrel *Citellus leucurus*. Univ. Calif. Publ. Zool., 64: 1-56.

HUDSON, J. W. (1964): Temperature regulation in the round-tailed ground squirrel, *Citellus tereticaudus*. Ann. Acad. Sci. fenn. A4, 71: 219-233.
HUDSON, J. W. and RUMMEL, J. A. (1966): Water metabolism and temperature regulation of the primitive heteromyids, *Liomys salvani* and *Liomys irroratus*. Ecology, 47: 345-354.
JACKSON, D. C. and SCHMIDT-NIELSEN, K. (1964): Countercurrent heat exchange in the respiratory passages. Proc. nat. Acad. Sci. (Wash.), 51: 1192-1197.
KENAGY, G. J. (1973): Daily and seasonal patterns of activity and energetics in a heteromyid rodent community. Ecology, 54: 1201-1219.
MACFARLANE, W. V., HOWARD, B., HAINES, H., KENNEDY, P. J. and SHARPE, C. M. (1971): Hierarchy of water and energy turnover of desert mammals. Nature (Lond.), 234: 483-484.
MACMILLEN, R. E. (1965): Aestivation in the cactus mouse, *Peromyscus eremicus*. Comp. Biochem. Physiol., 16: 227-248.
MACMILLEN, R. E. and LEE, A. K. (1970): Energy metabolism and pulmocutaneous water loss of Australian hopping mice. Comp. Biochem. Physiol., 35: 355-369.
MACMILLEN, R. E. (1972): Water economy of nocturnal desert rodents. In: Comparative physiology of desert animals. G. M. O. Maloiy (ed). Symp. zool. Soc. (Lond.), 31: 147-174.
MACMILLEN, R. E., BAUDINETTE, R. V. and LEE, A. K. (1972): Water economy and energy metabolism of the sandy inland mouse, *Leggadina hermannsburgensis*. J. Mammal., 53: 529-539.
RAAB, J. L. and SCHMIDT-NIELSEN, K. (1972): Effect of running on water balance of the kangaroo rat. Amer. J. Physiol., 222: 1230-1235.
SCHMIDT-NIELSEN, K. (1964): Desert Animals: *Physiological Problems of Heat and Water*. Oxford Univ. Press, London, 277 pp.
SCHMIDT-NIELSEN, K., HAINSWORTH, F. R. and MURRISH, D. E. (1970): Countercurrent heat exchange in the respiratory passages: effect on water and heat belance. Resp. Physiol., 9: 263-276.
SCHMIDT-NIELSEN, K. (1972): How Animals Work. Cambridge Univ. Press, London, 114 pp.
SHKOLNIK, A. and BORUT, A. (1969): Temperature and water relations in two species of spiny mice (*Acomys*). J. Mammal., 50: 245-255.
WUNDER, B. A. (1970): Temperature regulation and the effects of water restriction on Merriam's chipmunk, *Eutamias merriami*. Comp. Biochem. Physiol., 33: 385-403.

Chapter 1, Section 4a: Temperature Regulation in Cattle (J. BLIGH)

BERMAN, A. (1971): Thermoregulation in intensively lactating cows in near-natural conditions. J. Physiol. (Lond.), 215: 477-489.
BIANCA, W. (1965): Cattle in a hot environment. J. dairy Sci., 32, 291-345.
BIANCA, W. (1971): Environmental physiology of domesticated animals in ecological perspective. Proc. 25th int. Congr. physiol. Sci., 8: 157-158.
BLIGH, J. and HARTHOORN, A. M. (1965): Continuous radiotelemetric records of the deep body temperature of some unrestrained African mammals under near-natural conditions. J. Physiol. (Lond.), 176: 136-144.
BLIGH, J. and LAMPKIN, G. H. (1965): A comparison of the deep-body temperature of Hereford and Zebu cows recorded continuously by radiotelemetry under similar field conditions. J. agric. Sci., 64: 221-227.
BLIGH, J. (1973): Temperature Regulation in Mammals and Other Vertebrates. North Holland, Publ. Co., Amsterdam.
BLIGH, J. and HENSEL, H. (1974): Modern theories on the location and function of the thermoregulatory centres in mammals including man. In: Progress in Human Biometeorology, Div. A, Vol. 1, Part I B. S. W. Tromp and J. J. Bouma (eds), Swets & Zeitlinger Publ. Co., Amsterdam.
BLIGH, J., BAUMANN, I., SUMAR, J. and POCCO, F. (1974): Studies of body temperature patterns in South American Camelidae. Comp. Biochem. Physiol., 50A, in press.
CHQUILOFF, M. A. G. (1964): Estudo comparativo da tolerância de novilhas das raças Gir, Schwyz, Jersey, Guernsey e Holandesa P.B., às condições climáticas de Pedro Leopoldo, Minas Gerais. Arquivos Esc. Vet. (Minas Gerais), 16: 19-95.

HALES, J. R. S. (1974): Physiological responses to heat. In: MTP Intl. Review of Science, Physiology Series 1, Vol. 7, D. Robertshaw (ed.). Butterworths, London: 107-162.
HENSEL, H. (1973): Neuronal processes in thermoregulation. Physiol. Rev., 58: 948-1017.
MORAN, J. B. (1973): Heat tolerance of Braham cross, buffalo, Banteng and Shorthorn steers during exposure to sun and as a result of exercise. Aust. J. agric. Res., 24: 775-782.
RHOAD, A. O. (1944): The Iberia heat tolerance test for cattle. Trop. agric. Trin., 21: 162-164.
RICHARDS, S. A. (1973): Temperature Regulation. Wykeham Publ., Lond.
SCHMIDT-NIELSEN, K., SCHMIDT-NIELSEN, B., JARNUM, S. A. and HOUPT, T. B. (1957): Body temperature of the camel and its relation to water economy. Amer. J. Physiol., 188: 103-112.
SCHMIDT-NIELSEN, K. (1964): Desert Animals: Physiological Problems of Heat and Water. Oxford Univ. Press, Oxford.
THOMPSON, G. E. (1973): Climatic physiology of cattle. J. dairy Sci. 40: 441-473.
VERNON, E. H., DAMON, R. A., HARVEY, W. R., WARWICK, E. J. and KINCAID, C. M. (1959): Relation of heat tolerance determinations to productivity in beef cattle. J. anim. Sci., 18: 91-94.

Chapter 1, Section 4b: Temperature Regulation in Sheep (R. O. RAWSON)

BAKER, M. A. and HAYWARD, J. N. (1968a): Intracranial heat exchange and regulation of brain temperature in sheep. Life Sci., 7: 349-354.
BAKER, M. A. and HAYWARD, J. N. (1968b): The influence of the nasal mucosa and the carotid rete upon hypothalamic temperature in sheep. J. Physiol. (Lond.), 198: 561-579.
BAKER, M. A. (1972): Influence of the carotid rete on brain temperature in cats. J. Physiol. (Lond.), 220: 711-728.
BLIGH, J. (1961): Possible temperature sensitive elements in or near the vena cava of sheep. J. Physiol. (Lond.), 159: 85-86P.
BLIGH, J. (1963): The receptors concerned in the respiratory response to humidity in sheep at high ambient temperature. J. Physiol (Lond.), 168: 747-763.
DANIEL, P. M., DAWES, J. D. K. and PRITCHARD, M. L. (1953): Studies of the carotid rete and its associated arteries. Phil. Trans. roy. Soc. B, 237: 173-208.
JACOBSON, F. H. and SQUIRES, R. D. (1970): Thermoregulatory responses of the cat to pre-optic and environmental temperatures. Amer. J. Physiol., 218: 1575-1582.
RAWSON, R. O. and QUICK, K. P. (1970): Evidence of deep-body thermoreceptor response to intra-abdominal heating of the ewe. J. appl. Physiol., 28: 813-819.
RAWSON, R. O. and QUICK, K. P. (1971): Unilateral splanchnotomy: Its effect on the response to intra-abdominal heating in the ewe. Pflügers. Arch. ges. Physiol., 330: 362-365.
RAWSON, R. O. and QUICK, K. P. (1972): Localization of intra-abdominal thermoreceptors in the ewe. J. Physiol. (Lond.), 222: 665-677.
RIEDEL, W., SIAPLAURAS, G. and SIMON, E. (1973): Intra-abdominal thermosensitivity in the rabbit as compared with spinal thermosensitivity. Pflüg. Arch. ges. Physiol., 340: 59-70.
WOOD, J. D. (1970): Electrical activity from single neurones in Auerbach's plexus. Amer. J. Physiol., 219: 159-169.

Chapter 1, Section 4c: Temperature Regulation in Pigs (D. L. INGRAM)
See references in this volume, chapter 1, Section 5d and 18c(2)

Chapter 1, Section 4d: Temperature Regulation in Marine Mammals (G. C. WHITTOW)

ELSNER, R. (1969): Cardiovascular adjustments to diving. In: The Biology of Marine Mammals. H. T. Andersen (ed.). Academic Press, New York.
FAY, F. H. and RAY, C. (1968): Influence of climate on the distribution of Walruses, *Odobenus rosmarus* L. 1. Evidence from thermoregulatory behavior. Zoologica, 53: 1-14.

GENTRY, R. L. (1973): Thermoregulatory behavior of eared seals. Behavior, in press.
HAMPTON, I. F. G., WHITTOW, G. C., SZEKERCZES, J. and RUTHERFORD, S. (1971): Heat transfer and body temperature in the Atlantic Bottlenose Dolphin, *Tursiops truncatus*. Int. J. Biometeor., 15: 247-253.
HARRISON, R. J. and KOOYMAN, G. L. (1968): General physiology of the pinnipedia. In: The Behavior and Physiology of Pinnipeds. R. J. Harrison, R. C. Hubbard, R. S. Peterson, C. E. Rice and R. J. Schlusterman (eds). Appleton - Century - Crofts, New York.
HART, J. S. and IRVING, L. (1959): The energetics of Harbor Seals in air and in water with special consideration of seasonal changes. Canad. J. Zool., 37: 447-457.
IRVING, L. (1969): Temperature regulation in marine mammals. In: The Biology of Marine Mammals. H. T. Andersen (ed.). Academic Press, New York.
IRVING, L. (1973): Aquatic mammals. In: Comparative Physiology of Thermoregulation, Vol. III, G. C. Whittow (ed.). Academic Press, New York.
KANWISHER, J. and SUNDNES, G. (1966): Thermal regulation in cetaceans. In Whales, Dolphins and Porpoises. K. S. Norris (ed.). Univ. Calif. Press, Berkeley.
MATSUURA, D. T. (1972): Temperature regulation in the California Sea Lion (*Zalophus californianus*): Responses to heat. Ph.D. Thesis. Univ. Hawaii, Honolulu.
MATSUURA, D. T. and WHITTOW, G. C. (1973a): Oxygen uptake of the California Sea Lion and Harbor Seal during exposure to heat. Amer. J. Physiol., 225: 711–715.
MATSUURA, D. T. and WHITTOW, G. C. (1973b): Evaporative heat loss in the California Sea Lion and Harbor Seal. Comp. Biochem. Physiol., in press.
McGINNIS, S. M., WHITTOW, G. C., OHATA, C. A. and HUBER, H. (1972): Body heat dissipation and conservation in two species of dolphins. Comp. Biochem. Physiol., 43A: 417-423.
MILLER, K., ROSENMANN, M., IRVING, L. and MORRISON, P. (1973): Metabolism and temperature regulation in newborn and developing Harbor Seals. Fed. Proc., 32: 391.
OHATA, C. A. and WHITTOW, G. C. (1973): Conductive heat loss to sand in California Sea Lions and a Harbor Seal. Comp. Biochem. Physiol., in the press.
PETERSON, R. S. and BARTHOLOMEW, G. A. (1967): The Natural History and Behavior of the California Sea Lion. Amer., Soc. Mammal., Spec. Publ., 1.
PIERCE, R. W. (1970): Design and operation of a metabolic chamber for marine mammals. Ph.D. Thesis, Univ. Calif., Berkeley.
RIDGWAY, S. H. (1972): Mammals of the sea. C. C. Thomas, Springfield, Ill.
WEBB, P. (1970): Body heat loss in undersea gaseous environments. Aerospace Med., 41: 1282-1288.
WHITE, F. N. and ODELL, D. (1971): Thermoregulatory behavior of the Northern Elephant Seal, *Mirounga augustirostris*. J. Mammal., 52: 756-774.
WHITTOW, G. C., MATSUURA, D. T. and LIN, Y. C. (1972): Temperature regulation in the California Sea Lion (*Zalophus californianus*). Physiol. Zool., 45: 68-77.
WHITTOW, G. C., OHATA, C. A. and MATSUURA, D. T. (1971): Behavioral control of body temperature in the unrestrained California Sea Lion. Comm. Behav. Biol., 6: 87-91.
WHITTOW, G. C., HAMPTON, I. F. G., MATSUURA, D. T., OHATA, C. A., SMITH, R. M. and ALLEN, J. F. (1974): Body temperature of three species of whales. Marine Behav. Physiol., in press.

Chapter 1, Section 4e: Temperature Regulation in Fish (F. E. J. FRY)

BARANS, C. A. and TUBB, R. A. (1973): Temperatures selected seasonally by four fishes from western Lake Erie. J. Fish. Res. Bd. Can., 30: 1697-1703.
BENNETT, D. H. (1971): Preliminary examination of body temperatures of largemouth bass (*Micropterus salmoides*) from an artificially heated reservoir. Arch. Hydrobiol., 68: 376-381.
BONE, Q. (1966): On the function of the two types of myotomal muscle fibre in elasmobranch fish. J. Marine Biol. Assn. U.K., 46: 321-349.
BRETT, J. R. (1952): Temperature tolerance in young Pacific salmon, genus Oncorhynchus. J. Fish. Res. Bd. Can., 9: 265-323.
BRETT, J. R. (1971): Energetic responses of salmon to temperature. A study of some thermal relations in the physiology and freshwater ecology of sockeye salmon (*Oncorhynchus nerka*). Amer. Zool., 11: 99-113.

CAREY, F. G. and TEAL, J. M. (1966): Heat conservation in tuna fish muscle. Proc. nat. Acad. Sci. (Wash.), 56: 1464-1469.

CAREY, F. G. and TEAL, J. M. (1969): Regulation of body temperature by the bluefin tuna. Comp. Biochem. Physiol., 28: 205-213.

CAREY, F. G. and LAWSON, K. D. (1973): Temperature regulation in free-swimming bluefin tuna. Comp. Biochem. Physiol., 44A: 375-392.

CRAWSHAW, L. I. and HAMMEL, H. T. (1973): The regulation of internal body temperature in the brown bullhead *Ameiurus nebulosus*. In: The Pharmacology of Thermoregulation. E. Schönbaum and P. Lomax. (eds). Proc. Symp., San Francisco, 1972. Karger Verlag, Basel: 142-145.

DENDY, J. S. (1945): Fish distribution, Norris Reservoir, Tennessee, 1943. II. Depth distribution of fish in relation to environmental factors, Norris Reservoir. Rep. Reelfoot Lake Biol. Stat., Vol. 9: 114-135.

DENDY, J. S. (1948): Predicting depth distribution of fish in three TVA storage-type reservoirs. Trans. Amer. Fish. Soc., (1945) 75: 65-71.

FERGUSON, R. G. (1958): The preferred temperature of fish and their midsummer distribution in temperate lakes and streams. J. Fish .Res. Bd. Can., 15: 607-624.

FRAENKEL, G. S. and GUNN, D. L. (1961): The orientation of animals. 2nd Ed. Dover, New York.

FRY, F. E. J. (1964): Animals in aquatic environments: fishes. In: Handbook of Physiology, 4: Adaptation to the environment. D. B. Dill, E. F. Adolph and C. G. Wilber (eds). Amer. Physiol. Soc. (Wash.): 715-728.

FRY, F. E. J. and HOCHACHKA, P. W. (1970): Fish. In: Comparative Physiology of Thermoregulation, 1: Invertebrates and nonmammalian vertebrates. G. C. Whittow (ed.), Academic Press, New York, 79-134.

FRY, F. E. J. (1971): The effect of environmental factors on the physiology of fish. In: Fish Physiology, 6. W. S. Hoar and D. J. Randall (eds). Academic Press, New York: 1-98.

IVLEV, V. S. (1960): Analiz mekhanizma raspredelniia ryb v usloviiakh temperaturnovo gradienta. Zool. Zh. 39: 494-499. Transl in: Fish. Res. Board Can., Transl. Ser., 364, 1961.

LEIM, A. H. and SCOTT, W. B. (1966): Fishes of the Atlantic Coast of Canada. Fisheries Res. Bd. Can. Bull., No. 155, 485 pp.

MORROW, J. E. and MAURO, A. (1950): Body temperature of some marine fishes. Copeia, 108-116.

NEILL, W. H., MAGNUSON, J. J. and CHIPMAN, G. G. (1972): Behavioral thermoregulation by fishes: A new experimental approach. Science, 176: 1443-1445.

OLLA, B. L. and STUDHOLME, A. L. (1971): The effect of temperature on the activity of bluefish, *Pomatomus saltatrix* L. Biol. Bull., 141: 337-349.

PETERSON, R. H. and ANDERSON, J. M. (1969): Influence of temperature change on spontaneous locomotor activity and oxygen consumption of Atlantic salmon, *Salmo salar*, acclimated to two temperatures. J. Fish. Res. Bd. Can., 26: 93-109.

PITT, T. K., GARSIDE, E. T. and HEPBURN, R. L. (1956): Temperature selection of the carp (*Cyprinus carpio* L.). Canad. J. Zool., 34: 555-557.

ROZIN, P. N. and MAYER, J. (1961): Thermal reinforcement and thermoregulatory behaviour in the goldfish, *Carassius auratus*. Science, 134: 942-943.

SCHMEIN-ENGBERDING, F. (1953): The temperature preference of some bony fishes and its physiological importance. Z. Fischerei, 2: 125-155.

SPIGARELLI, S. A. and THOMMES, M. M. (1973): Sport fishing survey at the Point Beach Nuclear Power Plant. Argonne Nat'l. Lab., Radiological and Environmental Res. Div. Ann. Rep. Pt. III, Ecology Jan.-Dec. 1972. Rep. No. ANL-7960: 102-110.

SULLIVAN, C. M. (1954): Temperature reception and responses in fish. J. Fish. Res. Bd. Can., 11: 153-170.

STEVENS, E. D. and FRY, F. E. J. (1971): Brain and muscle temperatures in ocean caught and captive skipjack tuna. Comp. Biochem. Physiol., 38A: 203-211.

ZAHN, M. (1963): Jahreszeitliche Veränderungen der Vorzugstemperaturen von Scholle (*Pleuronectes platessa* L.) und Bitterling (*Rhodeus sericeus Pallas*). Verh. dtsch. zool. Ges. München: 562-580.

Chapter 1, Section 4f: Temperature Regulation in Birds (Effects of Cold on Metabolism and Endocrinal Reactions of Birds) (J. D. RISING)

ASCHOFF, J. and POHL, H. (1970): Der Ruheumsatz von Vögeln als Funktion der Tageszeit und der Körpergröße. J. Orn., 111: 38-47.
BROOKS, W. S. (1968): Comparative adaptations of the Alaskan redpolls to the arctic environment. Wils. Bull., 80: 253-280.
CHAFFEE, R. R. J., MAYHEW, W. W., DREBIN, M. and CASSUTO, Y. (1963): Studies on thermogenesis in cold-acclimated birds. Canad. J. Biochem., 41: 2215-2220.
CHAFFEE, R. R. J. and ROBERTS, J. C. (1971): Temperature Acclimation in birds and mammals. Ann. Rev. Physiol., 33: 155-202.
CHAPLIN, Susan B.: Daily Energetics of the Black-capped Chickadee, *Parus atricapillus*, in winter. J. comp. Physiol., 89: 321-330.
EL HALAWANI, M. E., WILSON, W. O. and BURGER, R. E. (1970): Cold-acclimation and the role of catecholamines in body temperature regulation in male Leghorns. Poultry Sci., 49: 621-632.
FREEMAN, B. M. (1966): The effects of cold, noradrenaline and adrenaline upon the oxygen consumption and carbohydrate metabolism of the young fowl (*Gallus domesticus*). Comp. Biochem. Physiol., 18: 369-382.
FREEMAN, B. M. (1970): Thermoregulatory mechanisms of the neonate fowl. Comp. Biochem. Physiol., 33: 219-230.
HART, J. S. (1962): Seasonal acclimatization in four species of small wild birds. Physiol. Zool., 13: 224-236.
HISSA, R. and PALOKANGAS, R. (1970): Thermoregulation in the titmouse (*Parus major* L.). Comp. Biochem. Physiol., 33: 941-953.
IRVING, L. (1960): Birds of Anaktuvuk Pass, Kobuk, and Old Crow. Bull. 217, U.S. nat. Mus. (Wash.).
IRVING, L. (1964): Terrestrial animals in cold: birds and mammals. In: Handbook of Physiology, IV. Adaptation to the Environment. Amer. physiol. Soc., (Wash.): 361-377.
JANSKY, L. (1973): Non-shivering thermogenesis and its thermoregulatory significance. Biol. Rev., 48: 85-132.
KENDEIGH, S. C. (1973): Monthly variations in the energy budget of the House Sparrow throughout the year. In: Productivity, population dynamics and systematics of granivorous birds. Pol. sci. Publ. (Warszawa): 17-44.
LASIEWSKI, R. C. (1964): Body temperatures, heart and breathing rate, and evaporative water loss in hummingbirds. Physiol. Zool., 37: 212-223.
LASIEWSKI, R. C. and DAWSON, W. R. (1967): A re-examination of the relation between standard metabolic rate and body weight in birds. Condor, 69: 13-23.
LIN, Y.-C. and STURKIE, P. D. (1968): Effect of environmental temperatures on the catecholamines of chickens. Amer. J. Physiol., 214: 237-240.
MILLER, Dorothea S. (1939): A study of the physiology of the sparrow thyroid. J. exp. Zool., 80: 259-286.
McNAB, B. K. (1966): An analysis of the body temperatures of birds. Condor, 68: 47-55.
RISING, J. D. and HUDSON, J. W. (1974): Seasonal variation in the metabolism and thyroid activity of the black-capped chickadee (*Parus atricapillus*). Condor, 76, in press.
STEEN, J. and ENGER, P. S. (1957): Muscular heat production in pigeons during exposure to cold. Amer. J. Physiol., 191: 157-158.
STEEN, J. (1958): Climatic adaptation in some small northern birds. Ecology, 39: 625-629.
VEGHTE, J. H. and HERREID, C. F. (1965): Radiometric determination of feather insulation and metabolism of arctic birds. Physiol. Zool., 38: 267-275.
WEKSTEIN, D. R. and ZOLMAN, J. F. (1968): Sympathetic control of homeothermy in the young chick. Amer. J. Physiol., 214: 908-912.
WEST, G. C. (1965): Shivering and heat production in wild birds. Physiol. Zool., 38: 111-120

Chapter 1, Section 4g: Thermoregulation in Wild Ungulates (M. K. YOUSEF)

ALLEN, T. E. and BLIGH, J. (1961): A comparative study of the temporal patterns of cutaneous water vapor loss from some domesticated mammals with epitrichial sweat glands. Comp. Biochem. Physiol., 31: 347-363.

BAKER, M. A. and HAYWARD, J. N. (1968): The influence of the nasal mucosa and the carotid rete upon hypothalamic temperature in sheep. J. Physiol. (Lond.). 198: 561-579.

BIANCA, W. (1961): Heat tolerance in cattle – its concept, measurement and dependence on modifying factors. Int. J. Biometeor., 5: 5-30.

BIANCA, W. (1963): Rectal temperature and respiratory rate as indicators of heat tolerance in cattle. J. agric. Sci., 60: 113-120.

BIANCA, W. (1968): Thermoregulation. In: Adaptation of Domestic Animals. E. S. E. Hafez (ed). Lea & Febiger, Philadelphia, 97.

BLIGH, J. (1967): A thesis concerning the process of secretion and discharge of sweat. Environm. Res., 1: 28-45.

BLIGH, J. and ROBINSON, S. G.: Radiotelemetry in a veterinary research project. Med. biol. Ill., 15: 94-99.

BULLARD, R. W., DILL, D. B. and YOUSEF, M. K. (1970): Responses of the burro to desert heat stress. J. appl. Physiol., 29: 159-167.

BUXTON, P. A. (1924): Heat, moisture and animal life in deserts. Proc. roy. Soc. B, 673; 96: 123-131.

CAMERON, R. D. and JUICK, J. R. (1972): Seasonal changes in total body water, extracellular fluid, and blood volume in grazing reindeer. Canad. J. Zool., 50: 107-116.

DANIEL, P. M., DAWES, J. D. K. and PRITCHARD, M. M. L. (1953). Studies of the carotid rete and its associated arteries. Phil. Trans. roy. Soc. B, 237: 173-215.

DILL, D. B. (1938). Life, Heat and Altitude. Harvard Univ. Press, Cambridge, Mass.

HAFEZ, E. S. E. (1968): Adaptation of Domestic Animals. Lea & Febiger, Philadelphia.

HAMMEL, H. T., HOUPT, T. R., ANDERSON, K. L. and SKJENME, S. (1962): Thermal and metabolic measurements on a reindeer at rest and in exercise. Techn. Rep. Arctic aeromed. Lab., Fairbanks, Alaska: 61-64.

HART, J. S., HEROUX, O., COTTLE, W. H. and MILLS, C. A. (1961): The influence of climate on metabolic and thermal responses of infant caribou. Canad. J. Zool., 39: 845-856.

IRVING, L. and KROG, J. (1954): Body temperatures of arctic and subartic birds and mammals. J. appl. Physiol., 6: 667-680.

IRVING, L., SCHMIDT-NIELSEN, K. and ABRAHAMSEN, N. (1956): On the melting points of animal fat in cold climates. Physiol. Zool., 30: 93-105.

IRVING, L. (1964): Terrestrial animals in cold: birds and mammals. In: Handbook of Physiology, 4. Adaptation to the Environment. D. B. Dill et al. (eds). Amer. Physiol. Soc. (Wash.): 361.

JOHNSON, H. D. (1965): Responses of animals to heat. Meteor. Monogr., 6: 109-122.

JOHNSON, K. G., MALOIY, G. M. O. and BLIGH, J. (1972): Sweat gland function in the red deer (*Cervus elaphus*). Amer. J. Physiol., 223: 604-607.

JENKINSON, D. M. (1972): Evaporative temperature regulation in domestic animals. In: Comparative Physiology of Desert Animals. G. M. O. Maloiy, (ed.), Academic Press, New York: 345.

KLEIBER, M. (1961): The Fire of Life. Wiley, New York.

KROG, J. O. and WIKA, M. (1974): The Circulation in the growing reindeer antlers. In: Proc. 1st International Reindeer/Caribou Symposium. Inst. arct. Biol., Univ. Alaska, Fairbanks, in press.

LEE, D. H. K. (1972): Large mammals in the desert. In: Physiological Adaptations: Desert and Mountain. M. K. Yousef et al. (eds). Academic Press, New York: 109.

LUICK, J. R., PERSON, S. J., CAMERON, R. D. and WHITE, R. G. (1973): Seasonal variations in glucose metabolism of reindeer (*Rangifer tarandus* L.) estimated with [U-^{14}C] glucose and /3-^{3}H/ glucose. Brit. J. Nutr., 29: 245-259.

MALOIY, G. M. O. (1970): Water economy of the Somali donkey. Amer. J. Physiol., 219: 1522-1527.

MALOIY, G. M. O. (1971): Temperature regulation in the Somali donkey (*Equus asinus*). Comp. Biochem. Physiol. 39A: 403-412.

MALOIY, G. M. O., and HOPCRAFT, D. (1971): Thermoregulation and water relations of two East African Antelopes: the hartebeest and impala. Comp. Biochem. Physiol., 38A: 525-534.

MALOIY, G. M. O. (1972): Renal salt and water excretion in the camel (*Camelus dromodarius*).

In: Comparative Physiology of Desert Animals. G. M. O. Maloiy (ed.). Academic Press, New York: 243.
MACFARLANE, W. V. (1964): Terrestrial animals in dry heat: ungulates. In: Handbook of Physiology. Adaptation to the Environment. D. B. Dill et al. (eds). Amer. physiol. Soc., (Wash.): 509.
ROBERTSHAW, D. and TAYLOR, C. R. (1969): A comparison of sweat gland activity in eight species of East African Bovids. J. Physiol. (Lond.), 203: 135-153.
ROSENMANN, M. and MORRISON, P. (1963): The physiological response to heat and dehydration in the guanaco. Physiol. Zool., 36: 45-51.
ROSENMANN, M. and MORRISON, P. (1967): Some effects of water deprivation in reindeer. Physiol. Zool., 40: 134-142.
SCHMIDT-NIELSEN, K. (1964): Desert Animals. Oxford Univ. Press, New York.
SCHOLANDER, P. F., WALTERS, V., HOCK, R. and IRVING, L. (1950): Body insulation of some arctic and tropical mammals and birds. Biol. Bull., 99: 225-236.
SCHOLANDER, P. F. and KROG, J. (1957): Countercurrent heat exchange and vascular bundles in sloths. J. appl. Physiol., 10: 405-411.
TAYLOR, C. R. (1968): Hydroscopic food: a source of water for desert antelope. Nature (Lond.), 219: 181-182.
TAYLOR, C. R. (1969): The leand and the oryx. Sci. Amer., 220: 88-95.
TAYLOR, C. R., ROBERTSHAW, D. and HOFMANN, R. (1969): Thermal panting: a comparison of wildebeest and zebu cattle. Amer. J. Physiol., 217: 907-910.
TAYLOR, C. R., SPINAGE, C. A. and PLYMAN, C. (1969): Water relations of the water buck, an East African antelope. Amer. J. Physiol., 217: 630-634.
TAYLOR, C. R. (1970): Dehydration and heat: effects on temperature regulation of East African ungulates. Amer. J. Physiol., 219: 1136-1139.
TAYLOR, C. R. (1972): The desert gazelle: A paradox resolved. In: Comparative Physiology of Desert Animals. G. M. O. Maloiy (ed.). Academic Press, New York, p. 215.
TAYLOR, C. R. and LYMAN, C. P. (1972): Heat storage in running antelopes: independence of brain and body temperatures. Amer. J. Physiol., 222: 114-117.
YOUSEF, M. K., CAMERON, R. D. and JUICK, J. R. (1971): Seasonal changes in hydrocortisone secretion rate of reindeer, *Rangifer tarandus*. Comp. Biochem. Physiol., 40A: 495-501.
YOUSEF, M. K. and DILL, D. B. (1969a): Resting energy metabolism and cardiorespiratory activity in the burro *Equus asinus*. J. appl. Physiol., 27: 229-232.
YOUSEF, M. K. and DILL, D. B. (1969b): Energy expenditure in desert walks; man and burro *Equus asinus*. J. appl. Physiol., 27: 681-683.
YOUSEF, M. K., DILL, D. B. and MAYES, M. G. (1970): Shifts in body fluids during dehydration in the burro, *Equus asinus*. J. appl. Physiol., 29: 345-349.
YOUSEF, M. K., DILL, D. B. and FREELAND, D. V. (1971): Energetic cost of grade walking in man and burro, *Equus asinus*: desert and mountain. J. appl. Physiol., 33: 337-340.
YOUSEF, M. K. and LUICK, J. R. (1971): Estimation of thyroxine secretion rate in reindeer, *Rangifer tarandus*: effects of sex, age and season. Comp. Biochem. Physiol., 40A: 789-795.
YOUSEF, M. K. and LUICK, J. R. (1975): Responses of reindeer, *Rangifer tarandus*, to heat stress. In: First International Reindeer/Caribou Symposium. Inst. arct. Biol., Univ. Alaska, Fairbanks, in press.
ZERVANOS, S. M. and HADLEY N. F. (1973): Adaptational biology and energy relationships of the Collared peccary (*Tayassu tajacu*). Ecology, 54: 759-774.

Chapter 1, Section 5a: Interaction Between Respiratory rnd Thermoregulatory Systems of Domestic Animals in Hot Enrironments (J. R. S. HALES)

ALBERS, C. (1961): Der Mechanismus des Wärmehechelns beim Hund. I. Die Ventilation und die arteriellen Blutgase während des Wärmehechelns. Pflügers Arch. ges. Physiol., 274: 125-147.
BIANCA, W. and FINDLAY, J. D. (1962): The effect of thermally-induced hyperpnea on the acid-base status of the blood of calves. Res. vet. Sci., 3: 38-49.
CHAI, C. Y. and LIN, M. T. (1972): Effects of heating and cooling the spinal cord and medulla oblongata on thermoregulation in monkeys. J. Physiol. (Lond.), 225: 297-308.

CHAPOT, G. (1967): Action de la température et de F_{ACO_2} sur les décharges du nerf phrénique. Pflügers Arch. ges. Physiol., 296: 196-201.
CLARK, F. J. and VON EULER, C. (1972): On the regulation of depth and rate of breathing. J. Physiol. (Lond.), 222: 267-295.
COHEN, M. I. (1968): Discharge patterns of brain stem respiratory neurons in relation to carbon dioxide tension. J. Neurophysiol., 31: 142-165.
EYZAGUIRRE, C. and LEWIN, J. (1961): Effect of different oxygen tensions on the carotid body in vitro. J. Physiol. (Lond.), 159: 238-250.
FINDLAY, J. D. and INGRAM, D. L. (1961): Brain temperature as a factor in the control of thermal polypnoea in the ox (*Bos taurus*). J. Physiol. (Lond.), 155: 72-85.
FINDLAY, J. D. and WHITTOW, G. C. (1966): The role of arterial oxygen tension in the respiratory response to localized heating of the hypothalamus and to hyperthermia. J. Physiol. (Lond.), 186: 333-346.
FINDLAY, J. D. and HALES, J. R. S. (1969): Hypothalamic temperature and the regulation of respiration of the ox exposed to severe heat. J. Physiol. (Lond.), 203: 651-653.
FRANKEL, H., HOLLANDS, K. G. and WEISS, H. S. (1962): Respiratory and circulatory responses of hyperthermic chickens. Arch. int. Physiol., 127: 997-999.
FRANKEL, H. M., ELLIS, J. P. and CAIN, S. M. (1963): Development of tissue hypoxia during progressive hyperthermia in dogs. Amer. J. Physiol., 205: 733-737.
GRUNSTEIN, M. M., FISK, W. M., LEITER, L. A. and MILIC-EMILI, J. (1973): Effect of body temperature on respiratory frequency in anesthetized cats. J. appl. Physiol., 34: 154-159.
HALES, J. R. S. (1967): The partition of respiratory ventilation of the panting ox. J. Physiol. (Lond.), 188: 45-46P.
HALES, J. R. S. and WEBSTER, M. E. D. (1967): Respiratory function during thermal tachypnoea in sheep. J. Physiol. (Lond.), 190: 241-260.
HALES, J. R. S., FINDLAY, J. D. and MABON, R. M. (1967): Tissue hypoxia in oxen exposed to severe heat. Resp. Physiol., 3: 43-46.
HALES, J. R. S. and FINDLAY, J. D. (1968a): Respiration in the ox: Normal values and the effects of exposure to hot environments. Resp. Physiol., 4: 333-352.
HALES, J. R. S. and FINDLAY, J. D. (1968b): The oxygen cost of thermally-induced and CO_2-induced hyperventilation in the ox. Resp. Physiol., 4: 353-362.
HALES, J. R. S. (1969): Changes in respiratory activity and body temperature of the severely heat-stressed ox and sheep. Comp. Biochem. Physiol., 31: 975-985.
HALES, J. R. S. and BLIGH, J. (1969): Respiratory responses of the conscious dog to severe heat stress. Experientia (Basel), 25: 818-819.
HALES, J. R. S., BLIGH, J. and MASKREY, M. (1970): Cerebrospinal fluid acid-base balance during respiratory alkalosis in the panting animal. Amer. J. Physiol., 219: 469-473.
HALES, J. R. S. (1973): Effects of exposure to hot environments on the regional distribution of blood flow and on cardiorespiratory function in sheep. Pflügers Arch. ges. Physiol., 344: 133-148.
HALES, J. R. S. (1974): Physiological responses to heat. In: MTP International Review of Science, Physiology Series 1, vol. 7. D. Robertshaw (ed.), Butterworths & Co. Ltd, London: 107-162.
HEISEY, S. R., ADAMS, T., HOFMAN, W. and RIEGLE, G. (1971): Thermally induced respiratory responses of the unanaesthetized goat. Resp. Physiol., 11: 145-151.
HELLON, R. F. (1971): Central thermoreceptors and thermoregulation. In: Handbook of Sensory Physiology, 3. 1; Enteroceptors. E. Neil (ed.). Springer-Verlag, Berlin: 161-186.
HOLMES, R. L., NEWMAN, P. P. and WOLSTENCROFT, J. H. (1960): A heat sensitive region in the medulla. J. Physiol. (Lond.), 152: 93-98.
INGRAM, D. L. and WHITTOW, G. C. (1962): The effect of heating the hypothalamus on respiration in the ox (*Bos taurus*). J. Physiol. (Lond.), 163: 200-210.
INGRAM, D. L. and LEGGE, K. F. (1969): The effect of environmental temperature on respiratory ventilation in the pig. Resp. Physiol., 8: 1-12.
JOELS, N. and WHITE, H. (1968): The contribution of the arterial chemoreceptors to the stimulation of respiration by adrenaline and noradrenaline in the cat. J. Physiol. (Lond.), 197: 1-23.
KAPPEY, F., ALBERS, C. and SCHMIDT, R. (1962): Die ventilatorische CO_2- Reaktion des Hundes während der Wärmetachypnoe. Pflügers Arch. ges. Physiol., 275: 312-326.

KOSAKA, M., SIMON, E., THAUER, R. and WALTHER, O.-E. (1969): Effect of thermal stimulation of the spinal cord on respiratory and cortical activity. Amer. J. Physiol., 217: 858-863.
LIPTON, J. M. (1973): Thermosensitivity of the medulla oblongata in control of body temperature. Amer. J. Physiol., 224: 840-897.
OTIS, A. B. and GUYATT, A. R. (1968): The maximal frequency of breathing of man at various tidal volumes. Resp. Physiol., 5: 118-129.
PLESCHKA, K., ALBERS, C. and HEERD, E. (1965): Der Einfluss der Temperatur auf die CO_2-Schwelle des Atemsystems. Pflügers Arch. ges. Physiol., 286: 142-158.
PLESCHKA, K. (1969): Der Einfluss der Temperatur auf die elektrische Aktivität des Nervus phrenicus. Pflügers Arch. ges. Physiol., 308: 357-381.
PURVES, M. J. (1970a): The effect of hypoxia, hypercapnia, and hypotension upon carotid body blood flow and oxygen consumption in the cat. J. Physiol. (Lond.), 209: 395-416.
PURVES, M. J. (1970b): The role of the cervical sympathetic nerve in the regulation of oxygen consumption of the carotid body of the cat. J. Physiol. (Lond.), 209: 417-431.
RICHARDS, S. A. (1970a): The biology and comparative physiology of thermal panting. Biol. Rev., 45: 223-264.
RICHARDS, S. A. (1970b): Physiology of thermal panting in birds. Ann. Biol. anim., 10: 151-168.
ROBERTSHAW, D. and WHITTOW, G. C. (1966): The effect of hyperthermia and localized heating of the anterior hypothalamus on the sympathoadrenal system of the ox (*Bos taurus*). J. Physiol. (Lond.), 187: 351-360.
SCHLÄFKE, M. and LOESCHCKE, H. H. (1967): Lokalisation eines an der Regulation von Atmung und Kreislauf beteiligten Gebietes an der ventralen Oberfläche der Medulla oblongata durch Kälteblockade. Pflügers Arch. ges. Physiol., 297: 201-220.
SCHMIDT-NIELSEN, K., KANWISHER, J., LASIEWSKI, R. C., COHN, J. E. and BRETZ, W. L. (1969): Temperature regulation and respiration in the ostrich. Condor, 71: 341-352.
SCHOENER, E. P. and FRANKEL, H. M. (1972): Effect of hyperthermia and P_{ACO_2} on the slowly adapting pulmonary stretch receptor. Amer. J. Physiol., 222: 68-72.
SIEMON, G., PLESCHKA, K. and ALBERS, C. (1966): Die alveolar-arterielle Sauerstoffdifferenz ($AaDO_2$) und die alveolare Ventilation beim wachen Hund in Abhängigkeit von der Umgebungstemperatur. Pflügers Arch. ges. Physiol., 289: 255-266.

Chapter 1, Section 5b: Evaporative Temperature Regulation in Ruminants (D. ROBERTSHAW)

AHLQUIST, R. P. (1948): A study of adrenotropic receptors. Amer. J. Physiol., 153: 586-600.
ALLEN, T. E. and BLIGH, J. (1969): A comparative study of the temporal patterns of cutaneous water vapour loss from some domesticated mammals with epitrichial sweat glands. Comp. Biochem. Physiol., 31: 347-363.
BLIGH, J. (1957): A comparison of the temperature of the blood in the pulmonary artery and in the bicarotid trunk of the calf during thermal polypnoea. J. Physiol. (Lond.), 136: 404-412.
BLIGH, J. (1961): The synchronous discharge of apocrine sweat glands of the Welsh Mountain sheep. Nature (Lond.), 189: 582-583.
BLIGH, J. (1967): A thesis concerning the processes of secretion and discharge of sweat. Environm. Res., 1: 28-45.
BIANCA, W. and HALES, J. R. S. (1970): Sweating, panting and body temperature of newborn and one-year-old calves at high environmental temperatures. Brit. vet. J., 126: 45-53.
BLATT, C. M., TAYLOR, C. R. and HABAL, M. B. (1972): Thermal panting in dogs: the lateral nasal gland, a source of water for evaporative cooling. Science, 177: 804-805.
BLAXTER, K. L. and WAINMAN, F. W. (1961): Environmental temperature and the energy metabolism and heat emission of steers. J. agric. Sci., 56: 81-90.
CRAWFORD, E. C. (1962): Mechanical aspects of panting in dogs. J. appl. Physiol., 17: 249-251.
FINDLAY, J. D. and ROBERTSHAW, D. (1965): The role of the sympathoadrenal system in the control of sweating in the ox (*Bos taurus*). J. Physiol. (Lond.), 179: 285-297.

GONZALEZ, R. R., KLUGER, M. J. and HARDY, J. D. (1971): Partitional calorimetry of the New Zealand white rabbit at temperatures 5-35°C. J. appl. Physiol. 31: 728-734.
HALES, J. R. S. and FINDLAY, J. D. (1968): The oxygen cost of thermally-induced and CO_2-induced hyperventilation in the ox. Resp. Physiol., 4: 353-362.
HALES, J. R. S. (1973): Effects of heat stress on blood flow in respiratory and non-respiratory muscles in the sheep. Pflügers Arch. ges. Physiol., 345: 123-130.
HALES, J. R. S. (1974): Physiological responses to heat. In: MTP International Review of Science, Physiology Series 1, Vol. 7, Ed. D. Robertshaw (ed.). Butterworths & Co. Ltd., London.
HALES, J. R. S. and BROWN, G. D. (1974): Net energetic and thermoregulatory efficiency during panting in the sheep. Comp. Biochem. Physiol, in press.
INGRAM, D. L. and WHITTOW, G. C. (1962): The effects of variations in respiratory activity and in the skin temperatures of the ears on the temperature of the blood in the external jugular vein of the ox (*Bos taurus*). J. Physiol. (Lond.), 163: 211-221.
JENKINSON, D. McE. and ROBERTSHAW, D. (1971): Studies on the nature of sweat gland "fatigue" in the goat. J. Physiol. (Lond.), 212: 455-465.
JOHNSON, K. G., MALOIY, G. M. O. and BLIGH, J. (1972): Sweat gland function in the red deer (*Cervus elaphus*). Amer. J. Physiol., 223: 604-607.
JOYCE, J. P. and BLAXTER, K. L. (1964): Respiration in sheep in cold environments. Res. vet. Sci. 5: 506-516.
McLEAN, J. A. (1963a): The partition of insensible losses of body weight and heat from cattle under various climatic conditions. J. Physiol. (Lond.), 167: 427-447.
McLEAN, J. A. (1963b): The regional distribution of cutaneous moisture vaporization in the Ayrshire calf. J. agric. Sci., 61: 275-280.
McLEAN, J. A. and CALVERT, D. T. (1972): Influence of air humidity on the partition of heat exchanges of cattle. J. agric. Sci., 78: 303-307.
ROBERTSHAW, D. and WHITTOW, G. C. (1966): The effect of hyperthermia and localized heating of the anterior hypothalamus on the sympathoadrenal system of the ox (*Bos taurus*). J. Physiol. (Lond.), 187: 351-360.
ROBERTSHAW, D. (1968): The pattern and control of sweating in the sheep and the goat. J. Physiol. (Lond.), 198: 531-539.
ROBERTSHAW, D. and TAYLOR, C. R. (1969a): A comparison of sweat gland activity in eight species of East African bovids. J. Physiol. (Lond.), 203: 135-143.
ROBERTSHAW, D. and TAYLOR, C. R. (1969b): Sweat gland function of the donkey (*Equus asinus*). J. Physiol. (Lond.), 205: 79-89.
ROBERTSHAW, D. (1974): Neural and humoral control of apocrine glands. J. invest. Dermatol., 63: 160-176.
SCHMIDT-NIELSEN, K. (1972): How animals work. Cambridge Univ. Press, London: 15.
TAYLOR, C. R. (1969): Metabolism, respiratory changes and water balance of an antelope, the eland. Amer. J. Physiol., 217: 317-320.
TAYLOR, C. R. (1974): Exercise and thermoregulation. In MTP International Review of Science. Physiology Series 1, Vol. 7, D. Robertshaw (ed.), Butterworths & Co. Ltd., London.
VENDSALU, A. (1960): Studies on adrenaline and noradrenaline in human plasma. Acta physiol. scand., 49, Suppl. 173.
WHITTOW, G. C. and FINDLAY, J. D. (1968): Oxygen cost of thermal panting. Amer. J. Physiol., 214: 94-99.

Chapter 1, Section 5c: Evaporative Temperature Regulation in Sheep (K. G. JOHNSON)

ALEXANDER, G. and WILLIAMS, D. (1962): Temperature regulation in the new-born lanb VI. Heat exchange in lambs in a hot environment. Aust. J. agric. Res., 13: 122-143.
ALLEN, T. E. and BLIGH, J. (1969): A comparative study of the temporal patterns of cutaneous water vapour loss from some domesticated mammals with epitrichial sweat glands. Comp. Biochem. Physiol., 31: 347-363.
AMES, D. R., NELLOR, J. E. and ADAMS, T. (1971): Energy balance during heat stress in sheep. J. anim. Sci. 32: 784-788.

BAKER, M. A. and HAYWARD, J. N. (1968): The influence of the nasal mucosa and the carotid rete upon hypothalamic temperature in sheep. J. Physiol. (Lond.), 198: 561-579.
BIANCA, W. (1962): Tolerance to severe heat and behaviour of respiratory minute volume in cattle. Nature (Lond.), 195: 1208-1209.
BLATT, C. M., TAYLOR, C. R. and HABAL, M. B. (1972): Thermal panting in dogs: the lateral nasal gland, a source of water for evaporative cooling. Science, 177: 804-805.
BLAXTER, K. L., GRAHAM, N. McC. and WAINMAN, F. W. (1959): Environmental temperature, energy metabolism and heat regulation in sheep III. The metabolism and thermal exchanges of sheep with fleeces. J. agric. Sci., 52: 41-49.
BLIGH, J. (1959): The receptors concerned in the thermal stimulus to panting in sheep. J. Physiol. (Lond.), 146: 142-151.
BLIGH, J. (1961a): The synchronous discharge of apocrine sweat glands of the Welsh Mountain sheep. Nature (Lond.), 189: 582-583.
BLIGH, J. (1961b): Possible temperature-sensitive elements in or near the vena cava of sheep. J. Physiol. (Lond.), 159: 85-86P.
BLIGH, J. (1963a): The receptors concerned in the respiratory response to humidity in sheep at high ambient temperature. J. Physiol. (Lond.), 168: 747-763.
BLIGH, J. (1963b): Inhibition of thermal polypnoea in the closely shorn sheep. J. Physiol. (Lond.), 168: 764-781.
BROCKWAY, J. M., McDONALD, J. D. and PULLAR, J. D. (1965): Evaporative heat-loss mechanisms in sheep. J. Physiol. (Lond.), 179: 554-568.
BROOK, A. H. and SHORT, B. F. (1960): Sweating in sheep. Aust. J. agric. Res., 11: 557-569.
BROWN, W. H. and SHANKLIN, M. D. (1970): Respiratory fraction of total insensible heat loss from shorn and unshorn sheep. Trans. Amer. Soc. agric. Eng. 13: 505-507.
CENA, K. and MONTEITH, J. L. (1975): Transfer processes in animal coats III. Water vapour diffusion. Proc. roy. Soc. B., in press.
HALES, J. R. S. and WEBSTER, M. E. D. (1967): Respiratory function during thermal tachypnoea in sheep. J. Physiol. (Lond.), 190: 241-260.
HOFMAN, W. F., RIEGLE, G. D. and DOW, P. (1971): Respiratory evaporative cooling during heat-induced hyperventilation in sheep. Fed. Proc., 30: 319 (Abstract).
HOFMEYR, H. S., GUIDRY, A. J. and WALTZ, F. A. (1969): Effects of temperature and wool length on surface and respiratory evaporative losses of sheep. J. appl. Physiol., 26: 517-523.
JENKINSON, D. McE. and BLACKBURN, P. S. (1967): The distribution of nerves, monoamine oxidase and cholinesterase in the skin of the sheep and goat. J. Anat., 101: 333-341.
JENKINSON, D. McE. and ROBERTSHAW, D. (1971): Studies on the nature of sweat gland "fatigue" in the goat. J. Physiol. (Lond.), 212: 455-465.
JOHNSON, K. G. (1971a): The discharge of sweat in Welsh Mountain sheep. J. Physiol. (Lond.), 215: 743-753.
JOHNSON, K. G. (1971b): Sweating and panting in Welsh Mountain sheep. Int. J. Biometeor., 15: 281-285.
JOHNSON, K. G. (1973). Sweat storage as a factor influencing sweat discharge in sheep. J. Physiol. (Lond.), in press.
JOYCE, J. P. and BLAXTER, K. L. (1964): Respiration in sheep in cold environments. Res. vet. Sci., 5: 506-516.
KERSLAKE, D. McK. (1972): The Stress of Hot Environments. Cambridge Univ. Press, London.
KNAPP, B. J. and ROBINSON, K. W. (1954): The role of water for heat dissipation by a Jersey cow and a Corriedale ewe. Aust. J. agric. Res., 5: 568-577.
LEE, D. H. K. (1950): Studies of heat regulation in the sheep, with special reference to the Merino. Aust. J. agric. Res., 1: 200-216.
LINZELL, J. L. and BLIGH, J. (1961): Polypnoea evoked by heating the udder of the goat. Nature (Lond.), 190: 173.
MASKREY, M. M. and BLIGH, J. (1971): Interactions between the thermoregulatory responses to injections into a lateral cerebral ventricle of the Welsh Mountain sheep of putative neurotransmitter substances, and of local changes in anterior hypothalamic temperature. Int. J. Biometeor., 15: 129-133.
MUTO, K. (1916): Über die Wirkung des Adrenalins auf die Schweissekretion; zugleich ein

Beitrag zur Kenntnis der doppelsinnigen Innervation der Schweissdrüse. Mitteilungen med. Fak. Tokyo, 15: 365-386.
RAWSON, R. O. and QUICK, K. P. (1971): Thermoregulatory responses to temperature signals from the abdominal viscera of sheep. J. Physiol. (Paris), 63: 399-402.
ROBERTSHAW, D. (1968): The pattern and control of sweating in the sheep and the goat. J. Physiol. (Lond.), 198: 531-539.
SCHMIDT-NIELSEN, K., BRETZ, W. L. and TAYLOR, C. R. (1970): Panting in dogs: unidirectional air flow over evaporative surfaces. Science, 169: 1102-1104.
SCHMIDT-NIELSEN, K., HAINSWORTH, F. R. and MURRISH, D. E. (1970): Counter-current heat exchange in the respiratory passages: effect on water and heat balance. Resp. Physiol., 9: 263-276.
WAITES, G. M. H. (1962): The effect of heating the scrotum of the ram on respiration and body temperature. Quart. J. exp. Physiol., 47: 314-323.
WAITES, G. M. H. and VOGLMAYR, J. K. (1963): The functional activity and control of the apocrine sweat glands of the scrotum of the ram. Aust. J. agric. Res., 14: 839-851.
WEBSTER, M. E. D. and JOHNSON, K. G. (1968): Some aspects of body temperature regulation in sheep. J. agric. Sci., 71: 61-66.
WOOL SCIENCE REVIEW (1956): Physical properties of wool fabrics Wool-water relations, No. 16.
WOOL SCIENCE REVIEW (1963): The physiological properties of wool clothing, No. 22, Part 2.

Chapter 1, Section 5d: Evaporative Temperature Regulation in Pigs (D. L. INGRAM)

ALLEN, T. E. and BLIGH, J. (1969): A comparative study of the temperal patterns of cutaneous water vapour loss from some domesticated mammals with epitrichial sweat glands. Comp. Biochem. Physiol. 31, 347-363.
BIANCA, W. (1962): Relative importance of dry and wet bulb temperatures in causing heat stress in cattle. Nature (Lond.), 195: 251-252.
BOND, T. E., KELLY, C. F. and HEITMAN, Jr. H. (1952): Heat and moisture loss from swine. Agric. Eng. St. Joseph, Mich., 33: 148-152.
BRODY, S. (1948): Environmental Physiology. I Physiological Backgrounds. Res. Bull. Mo. agric. exp. Stat., 423: 3-43.
BUTCHBAKER, A. F. and SHANKLIN, M. D. (1964): Partitional heat losses of new-born pigs as affected by air temperature, absolute humidity, age and body weight. Trans. Amer. Soc. agric. Eng., 7: 380-383.
FINDLAY, J. D. (1957): The respiratory activity of calves subjected to thermal stress. J. Physiol. (Lond.), 136: 300-309.
HALES, J. R. S. and WEBSTER, M. E. D. (1967): Respiratory function during thermal tachypnoea in sheep. J. Physiol. (Lond.), 190: 241-260.
HEITMAN, Jr. H. and HUGHES, E. H. (1949): The effects of air temperature and relative humidity on the physiological well being of swine. J. anim. Sci., 8: 171-181.
HEMINGWAY, A. (1938): The panting response of normal anaesthetized dogs to measured dosages of diathermy heat. Amer. J. Physiol., 121: 747-754.
HOLMES, C. W. and MOUNT, L. E. (1967): Heat loss from groups of growing pigs under various conditions of environmental temperature and air movement. Anim. Prod., 9: 435-452.
HOLMES, C. W. (1968): Heat losses from young pigs at three environmental temperatures, measured in a direct calorimeter. Anim. Prod., 10: 135-147.
INGRAM, D. L. (1964): The effect of environmental temperature on heat loss and thermal insulation in the young pig. Res. vet. Sci., 5: 357-364.
INGRAM, D. L. (1965a): The effect of humidity on temperature regulation and cutaneous water loss in the young pig. Res. vet. Sci., 6: 9-17.
INGRAM, D. L. (1965b): Evaporative cooling in the pig. Nature (Lond.), 207: 415-416.
INGRAM, D. L. (1967): Stimulation of cutaneous glands in the pig. J. comp. Path., 77: 93-98.
INGRAM, D. L. and LEGGE, K. F. (1970): The effect of environmental temperature on respiratory ventilation in the pig. Resp. Physiol., 8: 1-12.

INGRAM, D. L. and LEGGE, K. F. (1972a): The influence of deep body temperature and skin temperatures on respiratory frequency in the pig. J. Physiol. (Lond.) 220: 283-296.
INGRAM, D. L. and LEGGE, K. F. (1972b): The influence of deep body and skin temperature on thermoregulatory responses to heating of the scrotum in pigs. J. Physiol. (Lond.), 224: 477-487.
JOYCE, J. P. and BLAXTER, K. L. (1964): Respiration in sheep in cold environments. Res. vet. Sci., 5: 506-516.
McLEAN, J. A. (1963a): Measurement of cutaneous moisture vaporization from cattle by ventilated capsules. J. Physiol. (Lond.), 167: 417-426.
McLEAN, J. A. (1963b): The partition of insensible losses of body weight and heat from cattle under various climatic conditions. J. Physiol. (Lond.), 167: 427-447.
MORRISON, S. R., BOND, T. E. and HEITMAN, H. (1967): Skin and lung moisture loss from swine. Trans. Amer. Soc. agric. Eng., 10: 691-692.
MOUNT, L. E. (1962): Evaporative heat loss in the new-born pig. J. Physiol. (Lond.), 164: 274-281.
MOUNT, L. E. (1968): The Climatic Physiology of the Pig. Edward Arnold, London.
PROVINS, K. A., HELLON, R. F., BELL, C. R. and HIRONS, R. W. (1962): Tolerance to heat of subjects engaged in sedentary work. Ergonomics, 5: 93-97.
RANDALL, W. C. and HIESTAND, W. A. (1939): Panting and temperature regulation in the chicken. Amer. J. Physiol., 127: 761-767.
ROBINSON, K. and LEE, D. H. K. (1941): Reactions of the pig to hot atmospheres. Proc. roy. Soc. Queensland, 53: 145-158.
SCHMIDT-NIELSEN, K., BRETZ, W. L. and TAYLOR, C. R. (1970): Panting in dogs unidirectional flow over evaporative surfaces. Science, 169: 1102-1104.
SCHMIDT-NIELSEN, K., HAINSWORTH, F. R. and MURRISH, D. E. (1970): Countercurrent heat exchange in the respiratory passage and effect on water and heat balance. Resp. Physiol., 9: 263-276.
WAITES, G. M. H. (1961): Polypnea evoked by heating the scrotum of the ram. Nature (Lond.), 190: 172-173.

Chapter 1, Section 5e: Evaporative Temperature Regulation in Birds (M. VAN KAMPEN)

CADE, T. J. (1964): Thirst. M. J. Wayner (ed.). Pergamon Press, Oxford: 237-256.
CRAWFORD, E. C. AND SCHMIDT-NIELSEN, K. (1967): Temperature regulation and evaporative cooling in the ostrich. Amer. J. Physiol., 212: 347-353.
LASIEWSKI, R. C., ACOSTA, A. L. and BERNSTEIN, M. H. (1966): Evaporative water loss in birds. I. Characteristics of the open flow method of determination, and their relation to estimates of thermoregulatory ability. Comp. Biochem. Physiol., 19: 445-457.
LASIEWSKI, R. C., BERNSTEIN, M. H. and OHMART, R. D. (1971): Cutaneous water loss in the roadrunner and poor-will. Condor, 73, 4: 470-472.
MARDER, J. (1973): Body temperature regulation in the brown-necked raven (Corvus corax ruficollis). I. Metabolic rate, evaporative water loss and body temperature of the raven exposed to heat stress. Comp. Biochem. Physiol., 45A: 412-430.
SCHMIDT-NIELSEN, K. (1972): How Animals Work. Cambridge Univ. Press.
SCHMIDT-NIELSEN, K., KANWISHER, J., LASIEWSKI, R. C., COHN, J. E. and BRETZ, W. L. (1969): Temperature regulation and respiration in the ostrich. Condor, 71: 341-352.
SCHMIDT-NIELSEN, K., HAINSWORTH, F. R. and MURRISH, D. E. (1970): Countercurrent heat exchange in the respiratory passages: effect on water and heat balance. Resp. Physiol., 9: 263-276.
VAN KAMPEN, M. (1971): Some aspects of thermoregulation in the White Leghorn Fowl. Int. J. Biometeor., 15, 2-4: 244-246.
VAN KAMPEN, M. (1973): Energy metabolism and heat regulation in the White Leghorn hen. Thesis, Univ. Utrecht.
VAN KAMPEN, M. (1974): Physical factors affecting energy expenditure. Brit. poult. Sci., in press.

Chapter 1, Section 6: Possible Mechanisms Involved in Death in Laboratory Animals due to Thermal Stresses (Heat and Cold) (G. L. WRIGHT)

ASAHINA, E. (1966): Freezing and frost resistance in insects. In: Cryobiology, H. T. Meryman (ed.). Academic Press, New York: 451-486.

BATTISTA, A. F. (1957): Effects of cold on evoked cortical potentials in the rat. Amer. J. Physiol., 191: 209-212.

BELEHRÁDEK, J. (1931): Le mécanisme physicochronique de l'adaptation thermique. Protoplasma, 12: 406-434.

BOGERT, C. M. (1959): How reptiles regulate their body temperature. Sci. Amer. 200: 105-120.

BRATTSTROM, B. H. and LAWERENCE, P. (1962): The rate of thermal acclimation in anuran amphibians. Physiol. Zool., 35: 148-156.

BUCHSBAUM, R., PECHERSKY, J. and McKIBBEN, D. (1970): Effects of heat stress on cellular structure and function. AMRL-TR-70-1 Aerospace Med. Res. Lab. Wright-Patterson AFB, Ohio: 1-15.

BURGER, F. J. and FUHRMAN, F. A. (1964): Evidence of injury to tissues after hyperthermia. Amer. J. Physiol., 206: 1062-1964.

CHAN, M., HIMES, R. H. and AKAGI, J. M. (1971): Fatty acid composition of thermophilic, mesophilic and psychrophilic clostridia. J. Bact., 106: 876-881.

CHERRY, L. M. (1959): Fat metabolism and temperature acclimatization in the fly. *Phormia terraenovae R.-D.* Ent. exp. appl., 2: 68-76.

COWLES, R. B. and BOGERT, C. M. (1944): A preliminary study of the thermal requirements of desert reptiles. Bull. Amer. Mus. nat. Hist., 83: 261-296.

DAVISON, T. F. and BOWLER, K. (1971): Changes in the functional efficiency of flight muscle sarcosomes during heat death of adult *Calliphora erythrocephala*. J. cell. comp. Physiol., 78: 37-48.

DAVISON, T. and BOWLER, K. (1971): Ultrastructural changes in sarcosomes after exposure of adult *Calliphora erythrocephala* to lethal and sub-lethal high temperatures. J. cell. comp. Physiol., 78: 49-58.

DUNLAP, D. G. (1968): Critical thermal maximum as a function of temperature of acclimation in two species of Hylid frogs. Physiol. Zool., 41: 432-439.

FAIRFIELD, J. (1948): Effects of cold on infant rats: Body temperature, oxygen consumption, electrocardiograms. Amer. J. Physiol., 155: 355-365.

FRAENKEL, G. and HOPF, H. S. (1940): The physiological action of abnormally high temperatures on poikilothermic animals I. Temperature adaptation and the degree of saturation of phosphatides. Biochem. J., 34: 1085-1091.

FRANKEL, H. M. (1959): Effects of restraint on rats exposed to high temperatures. J. appl. Physiol., 14: 997-999.

FRANKEL, H. M., ELLIS, J. P. Jr. and CAIN, S. M. (1963): Development of tissue hypoxia during progressive hyperthermia in dogs. Amer. J. Physiol., 205: 733-737.

GRANGER, J. N. R. (1969): Heat death in *Arianta arbustorum*. Comp. Biochem. Physiol., 29: 665-670.

HEATWOLE, H., MERCADO, N. and ORTIZ, E. (1965): Comparison of CTM of Two species of Puerto Rican frogs of the genus *Eleuthrodactylus*. Physiol. Zool., 38: 1-8.

HOAR, W. S. (1955): Seasonal variation in the resistance of goldfish to temperature. Trans. roy. Soc. Can. 49: 25-34.

HOUSE, H. L., RIORDAN, D. R. and BARLOW, J. S. (1958): Effects of thermal conditioning and of degree of saturation of dietary lipids on resistance of an insect to a high temperature. Canad. J. Zool., 36: 629-632.

HUTCHISON, V. H. (1961): Critical thermal maxima in salamanders. Physiol. Zool., 34: 92-125.

KOSH, R. J. and HUTCHISON, V. H. (1968): Daily rhythmicity of temperature tolerance in the eastern painted turtle, *Chrysemys picta picta*. Copeia, 2: 244-246.

KOUR, E. L. and HUTCHISON, V. H. (1970): Critical thermal tolerances and heating and cooling rates of lizards from diverse habitats. Copeia, 2: 219-229.

LEVITT, J. (1967): Molecular mechanisms of temperature adaptations. Horm-Shafer C. Baltimore: 41-51.

LOVELOCK, J. E. (1953): The haemolysis of human red cells by freezing and thawing. Biochim. biophys. Acta (Amst.), 10: 414-426.
LOWE, C. H. and VANCE, V. J. (1955): Acclimation of critical thermal maximum of the reptile, *Urosauris ornatus*. Science, 122: 73-74.
LOWE, C. H., LARDNER, P. J. and HALPERN, E. A. (1971): Supercooling in reptiles and other vertebrates. Comp. Biochem. Physiol., 39: 125-135.
LUYET, B. J. (1949): Preservation of the formed elements and of the proteins of the blood. Amer. nat. Red Cross.
MAHONEY, J. J. and HUTCHISON, V. H. (1969): Photoperiod acclimation and 24-hour variations in the critical thermal maxima of a tropical and a temperate frog. Oecologia, 2: 143-161.
MAYER, A. G. (1914): The effects of temperature on tropical marine birds. Pap. Tortugas Lab., 6: 1-24.
MERYMAN, H. T. (1968): A modified model for the mechanism of freezing injury in erythrocytes. Nature (Lond.), 218: 333-336.
ROSENBERG, D., KEMERY, G., SWITZER, R. C. and HAMILTON, T. C. (1971): Quantitative evidence for protein denaturation as the cause of thermal death. Nature (Lond.), 232: 471-473.
SALT, R. W. (1961): Resistance of poikilotherm animals to cold. Brit. med. Bull., 17: 5-8.
SCHERAGA, H. A. (1963): The Proteins. Academic Press, New York: 477-594.
SHANES, A. M. (1954): Effect of temperature on potassium liberation during nerve activity. Amer. J. Physiol., 177: 377-382.
SMITH, A. U. (1961): Biological effects of freezing and supercooling. Williams and Wilkins, Baltimore.
STEARN, A. E. (1949): Kinetics of biological reactions with special reference to enzymic processes. Advanc. Enzymol., 9: 25-74.
SULLIVAN, B. J. and LeBLANC, M. F. (1957): Effect of inositol and rapid rewarming on extent of tissue damage due to cold injury. Amer. J. Physiol., 189: 501-503.
UMMINGER, B. L. (1971): Chemical studies of cold death in the Gulf killifish, *Fundulus grandis*. Comp. Biochem. Physiol., 39: 625-632.
USHAKOV, B. P. (1964): Thermostability of cells and proteins of poikilotherms and its significance in speciation. Physiol. Rev., 44: 518-560.
WILLIAMS, R. J. (1970): Freezing tolerance in *Mytilus edulis*. Comp. Biochem. Physiol., 35: 145-161.
WOODHEAD, P. M. J. (1964): The death of North Sea fish during the winter of 1962-63, particularly with reference to the sole. *Solea vulgaris*, Helgoland. Wiss. Meeresuntersuch., 10: 283-300.
WRIGHT, G. L., LINDSEY, W. D., JOHNSON, H. L. and KRZYWICKI, H. J. (1972): The effect of high temperatures on tissue lactate, pyrurate and venous blood properties in the rat. Int. J. Biometeor., 16: 71-77.
ZWEIFEL, R. G. (1957): Studies on the critical thermal maxima of salamanders. Ecology, 38: 64-69.

Chapter 1, Section 7: Environmental Temperature and Feeding Behaviour (C. L. HAMILTON)

ANDIK, I., DONHOFFER, S., FARKAS, M. and SCHMIDT, P. (1963): Ambient temperature and survival on a protein-deficient diet. Brit. J. Nutr., 17: 257-261.
APPLEMAN, R. and DELAUCHE, J. C. (1958): Behavioral, physiological and biochemical responses of goats to temperature. J. Anim. Sci., 17: 326-335.
BERNARD, C. (1949): An Introduction to the Study of Experimental medicine. Henry Schuman Co., New York (Transl. from French).
BROBECK, J. R. (1945): Effects of variations in activity, food intake and environmental temperature on weight gain in albino rat. Amer. J. Physiol., 143: 1-5.
BRODY, S. (1964): Bioenergetics and Growth. Hafner Publ. Co., New York (Reprint edit.).
CANNON, W. B. (1939): The Wisdom of the Body. W. W. Norton Publ. Co., New York.
DONHOFFER, S. and VONOTZKY, J. (1947): The effect of environmental temperature on food selection. Amer. J. Physiol., 150: 329-333.

DUGAL, L. P., LEBLOND, C. P. and THERIEN, M. (1945): Resistance to extreme temperatures in connection with different diets. Ann. J. Res., 23: 244-258.

DURRER, J. L. and HANNON, J. P. (1962): Seasonal variations in caloric intake of dogs living in an arctic environment. Amer. J. Physiol., 202: 375-378.

FUSCO, M., HARDY, S. D. and HAMMEL, H. T. (1961): Interaction of central and peripheral factors in physiological temperature regulation. Amer. J. Physiol., 200: 572-580.

GASNIER, A. and MAYER, A. (1939): Recherches sur la régulation de la nutrition. Part III: Mécanismes régulateurs de la nutrition chez le lapin domestique. Ann. Physiol. Physicochim. biol., 15: 186-194.

HALE, E. B. and SCHIEN, M. W. (1962): The behavior of turkeys. In: The behavior of Domestic Animals. E. S. E. Hafez (ed.). Williams and Wilkins, Baltimore.

HAMILTON, C. L. and BROBECK, J. R. (1962): Temperature response of tube fed rats. Amer. J. Physiol., 203: 383-384.

HAMILTON, C. L. (1963): Interactions of food intake and temperature regulation in the rat. J. comp. physiol. Psychol., 56: 476-488.

HAMILTON, C. L. and BROBECK, J. R. (1964): Food intake and temperature regulation in rats with rostral hypothalamic lesions. Amer. J. Physiol., 207: 291-297.

HARRIS, N. R. (1959): Factors Influencing the Food Intake of a Poikilothermic Vertebrate, the Toad, *Bufo wookhausei fowleri*. Ph.D. Thesis, Univ. Pennsylvania Pa.

HEITMAN, H., Jr. and HUGHES, E. H. (1949): The effects of air temperature and relative humidity on the physiological well-being of swine. J. Anim. Sci., 8: 171-181.

KLAIN, G. J. and WINDERS, R. L. (1964): Metabolic studies of an amino acid imbalance in cold-exposed rats. J. Nutr., 82: 333-337.

KLEIBER, M. (1961): The Fire of Life. Wiley and Sons, New York.

MAGOUN, H. W., HARRISON, F., BROBECK, J. R. and RANSON, S. W. (1938): Activation of heat loss mechanisms by local heating of the brain. J. Neurophysiol., 1: 101-114.

McMAHON, J. W. (1965): Some physical factors influencing the feeding behavior of *Daphnia magna straus*. Canad. J. Zool., 43: 603-611.

MEYER, J. H. and HARGUS, W. A. (1959): Factors influencing food intake in rats fed low-protein rations. Amer. J. Physiol., 197: 1350-1352.

NAKAYAMA, T., HAMMEL, H. T., HARDY, J. D. and EISENMAN, J. S. (1963): Thermal stimulation of electrical activity of single units of the pre-optic region. Amer. J. Physiol., 204: 1122-1126.

SPECTOR, N. H., BROBECK, J. R. and HAMILTON, C. L. (1968): Feeding and core temperature in albino rats: Changes induced by preoptic heating and cooling. Science, 161: 286-287.

SPECTOR, N. H., PERVALL, J. H. F. and GROVE, F. T. (1972): Effects of low ambient temperature and shaving upon food-intake in rats eating two hours. Pahlavi med. J. (Iran), 3: 569-571.

Chapter 1, Section 8a (1): Effects of Heat and Exercise on Energy Metabolism of Cattle (J. A. McLEAN)

BIANCA, W. (1962): Relative importance of dry- and wet-bulb temperatures in causing heat stress in cattle. Nature (Lond.), 195, 251-252.

BIANCA, W. (1965): Reviews of the Progress of Dairy Science – Cattle in a hot environment. J. Dairy Res., 32, 291-345.

BLAXTER, K. L. (1962): The Energy Metabolism of Ruminants. Hutchinson & Co. Ltd., London.

BLINCOE, C. and BRODY, S. (1955): The influence of ambient temperature, air velocity, radiation intensity and starvation on thyroid activity and iodide metabolism in cattle. Univ. Mo. agric. exp. Stat. Res. Bull., 576.

BRODY, S. (1945): Bioenergetics and Growth. Reinhold Publ. Co., New York.

BROUWER, E. (1965): Report on sub-committee on constants and factors. In: Energy Metabolism K. L. Blaxter (ed.), Europ. Ass. Anim. Prod. Publ. No. II. Academic Press, London: 441-443.

CHRISTISON, G. I. and JOHNSON, H. D. (1972): Cortisol turnover in heat-stressed cows. J. Anim. Sci., 35, 1005-1010.

E.A.A.P. (1958): Symposium on Energy Metabolism. Europ. Ass. Anim. Prod., Pub. No. 8, Rome.
HALES, J. R. S. and FINDLAY, J. D. (1968): The oxygen cost of thermally induced and CO_2-induced hyperventilation in the ox. Resp. Physiol., 4, 353-362.
KIBLER, H. H. (1960): Oxygen consumption in cattle in relation to rate of increase in environmental temperature. Nature (Lond.), 186; 972-973.
KIBLER, H. H., JOHNSON, H. D., HAHN, L. and SHANKLIN, M. D. (1970): Thermal regulation in cattle at 2° to 35° as influenced by controlled feeding, ad libitum feeding and fasting. Univ. Mo. agric. exp. Stat. Bull., 975.
MACFARLANE, W. V. (1964): Terrestial animals in dry heat: ungulates. In: Handbook of Physiology, Sect. 4. Adaptation to the Environment. W. O. Fenn and H. Rahn (eds). Amer. physiol. Soc. (Wash.), 509-539.
McLEAN, J. A. (1963): The partition of insensible losses of body weight and heat from cattle under various climatic conditions. J. Physiol. (Lond.), 167, 427-447.
McLEAN, J. A. (1971): A gradient layer calorimeter for large animals. J. Inst. Heat. Vent. Engrs., 39: 1-8.
McLEAN, J. A. (1972): On the calculation of heat production from open-circuit calorimetric measurements. Brit. J. Nutr., 27: 597-600.
McLEAN, J. A. and CALVERT, D. T. (1972): Influence of air humidity on the partition of heat exchanges of cattle. J. agric. Sci. (Camb.), 78: 303-307.
THOMPSON, G. E. (1973): Review of the Progress of Dairy Science – Climatic Physiology of cattle. J. Dairy Res., 40: 441-473.
VERCOE, J. E. (1973): The energy cost of standing and lying in adult cattle. Brit. J. Nutr., 30: 207-210.
WHITTOW, G. C. and FINDLAY, J. D. (1968): Oxygen cost of thermal panting. Amer. J. Physiol., 214: 94-99.
YOUSEF, M. K. and JOHNSON, H. D. (1966): Blood thyroxine degradation rate of cattle as influenced by temperature and feed intake. Life Sciences, 5: 1349-1363.
YOUSEF, M. K. and JOHNSON, H. D. (1967): Calorigenesis of cattle as influenced by hydrocortisone and environmental temperature. J. Anim. Sci., 26: 1087-1093.
YOUSEF, M. K., KIBLER, H. H. and JOHNSON, H. D. (1967): Thyroid activity and heat production in cattle following sudden ambient temperature changes. J. Anim. Sci., 26: 142-148.

Chapter 1, Section 8a (2): Effects of Heat on Energy Metabolism of Sheep (K. L. KNOX)

BHATTACHARYA, A. N. and HUSSAIN, F. (1974): Intake and utilization of nutrients in sheep fed different levels of roughage under heat stress. J. Anim. Sci. 38: 877.
BLAXTER, K. L. (1962): The Energy Metabolism of Ruminants. C. C. Thomas Publ., Springfield: 329.
BLAXTER, K. L., GRAHAM, N. McC. and ARMSTRONG, D. G. (1959): Environmental temperature, energy metabolism and heat regulation in sheep. III. The metabolism and thermal exchanges of sheep with fleeces. J. agric. Sci., 52: 41.
BLAXTER, K. L., GRAHAM, N. McC., WAINMAN, F. W. and ARMSTRONG, D. G. (1959): Environmental temperature, energy metabolism and heat regulation in sheep. II. The partition of heat losses in closely clipped sheep. J. agric. Sci., 52: 25.
BLIGH, J. (1959): The receptors concerned in the thermal stimulus to panting in sheep. J. Physiol. (Lond.), 146: 142.
BRAY, G. A. (1974): Endocrine factors in the control of food intake. Fed. Proc., 33: 1140.
BROCKWAY, J. M. (1965): Posturally associated changes in the heat content of sheep. In: Energy Metabolism. Academic Press, New York: 395.
COOK, C. W. (1970): Energy budget of the range and range livestock. Colo. St. Univ. expt. Stat. Bull. TB109: 28.
GRAHAM, N. McC., WAINMAN, F. W., BLAXTER, K. L. and ARMSTRONG, D. C. (1959): Environmental temperature, energy metabolism and heat regulation in sheep. I. Energy metabolism in closely clipped sheep. J. agric. Sci., 52: 13.
HOFMEYR, H. S., GUIDRY, A. J. and WALTZ, F. A. (1969): Effects of temperature and

wool length on surface and respiratory evaporative losses of sheep. J. appl. Physiol., 26: 517.
KLEIBER, M. (1961): The Fire of Life. John Wiley & Sons, Inc., New York: 454.
KNOX, K. L. and SODERQUIST, H. G. (1969): Effect of environmental temperature on energy utilization in lambs. J. Anim. Sci., 29: 115.
KNOX, K. L. and HANDLEY, T. M. (1973): The California Net Energy System: theory and application. J. Anim. Sci., 37: 190.
KOTB, A. R. and PFANDER, W. H. (1964): Metabolism of sheep under cold and hot conditions. J. Anim. Sci., 23: 1226.
SODERQUIST, H. G. (1967): Effects of environmental temperature on energy retention in fattening lambs. M. S. Thesis, Colo. State Univ.
WALLACE, J. D. (1969): Nutritive value of forage selected by cattle on sandhill range. Ph.D. Thesis, Colo. State Univ.
WALLACE, J. D. and DENHAM, A. H. (1970): Digestion of range forage by Sheep collected esophageal fistulated cattle. J. Anim. Sci., 30: 605.

Chapter 1, Section 8a (3): Effects of Heat on Energy Metabolism and Related Endocrine Functions in Rodents (M. K. YOUSEF)

AMES, R. G. and VAN DYKE, H. B. (1950): Antidiuretic hormone in the urine and pituitary of the kangaroo rat. Proc. Soc. exp. Biol. (N.Y.), 75: 417-420.
BARTHOLOMEW, G. A. and DAWSON, W. R. (1968). Temperature regulation in desert mammals. In: Desert Biology. 1. G. W. Brown, Jr. (ed.). Academic Press, New York: 396.
BLIGH, J. and JOHNSON, K. G. (1973). Glossary of terms for thermal physiology. J. appl. Physiol., 35: 941-961.
BRADLEY, W. G. and YOUSEF, M. K. (1972): Small mammals in the desert. In: Physiological Adaptations: Desert and Mountain. M. K. Yousef et al. (eds). Academic Press, New York: 127.
BRODY, S. (1945): Bioenergetics and Growth. Reinhold Pub. Co., New York.
CASSUTO, Y. and AMIT, A. (1968): Thyroxine and norepinephrine effects on the metabolic rates of heat acclimated hamsters. Endocrinology, 82: 17-20.
CHAFFEE, R. R. J. and ROBERTS, J. C. (1971): Temperature acclimation in birds and mammals. Ann. Rev. Physiol., 33: 155-202.
COLLINS, K. J. and WEINER, J. S. (1968): Endocrinological aspects of exposure to high environmental temperatures. Physiol. Rev., 48: 785-839.
EL-HUSSEINI, M. and HAGGAG, G. (1974): Antidiuretic hormone and water conservation in desert rodents. Comp. Biochem. Physiol., 47A: 347-350.
EVANS, E. S., CONTOPOULOS, A. N. and SIMPSON, M. E. (1957). Hormonal factors influencing calorigenesis. Endocrinology, 60: 403-419.
HAGGAG, G. and EL-HUSSEINI, M. (1974): The adrenal cortex and water conservation in desert rodents. Comp. Biochem. Physiol., 47A: 351-359.
HAINSWORTH, F. R. and STRICKER, E. M. (1970): Salivary control by rats in the heat. Physiological and Behavioral Temperature Regulation. In: J. O. Hardy et al. (eds). C. C. Thomas, Springfield, Ill.: 611.
HART, J. S. (1971): Rodents. In: Comparative Physiology of Thermoregulation. G. C. Whittow (ed.). Academic Press, New York: 1.
HEROUX, O. (1960): Adjustments of the adrenal cortex and thyroid during cold acclimation. Fed. Proc., 19: 82-85.
HUDSON, J. W. and BARTHOLOMEW, G. A. (1964): Terrestrial animals in dry heat: estivators. In: Adaptation to the Environment. Handbook of Physiology, Sect. 4. D. B. Dill et al. (eds). Amer. physiol. Soc. (Wash.): 493.
HUDSON, J. W. and DEAVERS, D. R. (1973): Metabolism, pulmocutaneous water loss and respiration of eight species of ground squirrels from different environments. Comp. Biochem. Physiol., 45A: 69-100.
JOHNSON, H. D., WARD, M. W. and KIBLER, H. H. (1966): Heat and aging effects on thyroid function of male rats. J. appl. Physiol., 21: 689-694.
KIBLER, H. H. and JOHNSON, H. D. (1967): Growth and metabolism in rats as affected by dietary restriction, environmental temperature and endocrine therapy. Growth, 31: 205-216.

KOTBY, S. and JOHNSON, H. D. (1967): Rat adrenal cortical activity during exposure to a high (34C) ambient temperature. Life Sci., 6: 1121-1132.
PARKHIE, M. R. and JOHNSON, H. D. (1973): Hypothalamic growth hormone releasing factor: Release and synthesis after exposure to a high ambient temperature. Proc. Soc. exp. Biol. (N.Y.), 142: 311-315.
ROBINSON, K. W. and MACFARLANE, W. V. (1956): The influence of environmental temperature on the level of plasma antidiuretic substances in the rat. Aust. J. biol. Sci., 9: 130-138.
SCHMIDT-NIELSEN, K. (1964): Terrestrial animals in dry heat: desert rodents. In: Adaptation to the Environment. Handbook of Physiology, Sect. 4. D. B. Dill et al. (eds). Amer. physiol. Soc., (Wash.): 493.
SCHMIDT-NIELSEN, K. (1972): How Animals Work. Cambridge Univ. Press., London.
SHKOLNIK, A. and BROUT, A. (1974): Physiological adaptations to the desert environment. In: Environmental Physiology. D. Robertshaw (ed.). Butterworths Publ. Co., London: 185.
THORINGTON, R. W. Jr. (1966): The biology of rodenta tails: A study of form and function. AAL-TR-65-8.
YOUSEF, M. K. and JOHNSON, H. D. (1967): Time courses of oxygen consumption in rats during sudden exposure to high environmental temperature. Life Sci., 6: 1221-1228.
YOUSEF, M. K. and JOHNSON, H. D. (1968a): Effects of heat and feed restriction during growth on thyroxine secretion rate of male rats. Endocrinology, 82: 353-358.
YOUSEF, M. K. and JOHNSON, H. D. (1968b): Iodine compounds in plasma of rats: effect of exposure to high environmental temperature. Nature (Lond.), 217: 182-183.
YOUSEF, M. K., ROBERTSON, W. D., DILL, D. B. and JOHNSON, H. D. (1970): Energy expenditure of running kangaroo rats, *Dipodomys merriami*. Comp. Biochem. Physiol., 36: 387-393.
YOUSEF, M. K. and DILL, D. B. (1971a): Daily cycles of hibernation in the kangaroo rats, *Dipodomys merriami*. Cryobiology, 3: 122-129.
YOUSEF, M. K. and DILL, D. B. (1971b): Responses of Merriam's kangaroo rat to heat. Physiol. Zool., 44: 33-39.
YOUSEF, M. K., ROBERTSON, W. D., DILL, D. B. and JOHNSON, H. D. (1973): Energetic cost of running in the antelope ground squirrel, *Ammospermophilus leucurus*. Physiol. Zool., 46: 139-147.
YOUSEF, M. K., JOHNSON, H. D., BRADLEY, W. G. and SEIF, S. M. (1974): Tritiated water turnover rate in rodents: desert and mountain. Physiol. Zool., in press.

Chapter 1, Section 8a (4): Effects of Heat on Energy Metabolism in Birds (H. S. SIEGEL)

ALLEN, D. J. and MARLEY, E. (1967): The effect of sympathomimetic and allied amines on temperature and oxygen consumption in chickens. Brit. J. Pharmacol., 31: 290-312.
AMIN, A. H., CRAWFORD, T. B. B. and GADDUM, J. H. (1954): The distribution of substance P and 5-hydroxytryptamine in the central nervous system of the dog. J. Physiol. (Lond.), 126: 598-618.
AULIE, A. (1971): Body temperatures in pigeons and budgerigars during sustained flight. Comp. Biochem. Physiol., 39A: 173-176.
BALDWIN, S. P. and KENDIEGH, S. C. (1932): Physiology of the temperature of birds. Sci. Publ. Cleveland Mus. nat. Hist., 3: 1-196.
BARTHOLOMEW, G. A. and DAWSON, W. R. (1954): Body temperature and water requirements in the mourning dove, *Zenaidura macroura margiuella*. Ecology, 35: 181-187.
BARTHOLOMEW, G. A. and CADE, T. J. (1957): The body temperature of the American kestrel, *Falco sperverius*. Wilson Bull., 69: 149-154.
BARTHOLOMEW, G. A. and DAWSON, W. R. (1958): Body temperatures in California and Gambel's quail. Auk., 75: 150-156.
BENEDICT, F. G. and LEE, R. C. (1937): Lipogenesis in the animal body with special reference to the physiology of the goose. Carnegie Inst. (Wash.), Publ. No., 489: 1-232.
BERMAN, A. and SNAPIR, N. (1965): The relation of fasting and resting metabolic rates to heat tolerance in the domestic fowl. Brit. Poultry Sci., 6: 207-216.

BRODY, S. and PROCTOR, R. C. (1932): Growth and development with special reference to domestic animals. XXIII. Relation between basal metabolism and mature body weight in different species of mammals and birds. Mo. agric. expt. Stat. Res. Bull., 166: 89-101.
CALDER, W. A. (1964): Gaseous metabolism and water relations of the Zebra Finch, *Taeniopygin castanotis*. Physiol. Zool., 37: 400-411.
CALDER, W. A. and SCHMIDT-NIELSEN, K. (1967): Temperature regulation and evaporation in the pigeon and the roadrunner. Amer. J. Physiol., 213: 883-889.
CRAWFORD, E. C., Jr., and SCHMIDT-NIELSEN, K. (1967): Temperature regulation and evaporative cooling in the ostrich. Amer. J. Physiol., 212: 347-353.
DAWSON, W. R. (1954): Temperature regulation and water requirements of the brown and Albert Towhees, *Piplio fuscus* and *Piplio aberti*. Univ. Calif. Berkeley Publ. Zool., 59:81-124.
DAWSON, W. R. (1958): Relation of oxygen consumption and evaporative water loss to temperature in the cardinal. Physiol. Zool., 26: 37-48.
DAWSON, W. R. and TORDOFF, H. B. (1959). Relation of oxygen consumption to temperature in the Evening Grosbeak. Condor, 61: 388-396.
DAWSON, W. R. and HUDSON, J. W. (1970): Birds. In: Comparative Physiology of Thermoregulation I. G. C. Whittow (ed.). Academic Press, New York.
DRUMMOND, G. I. (1967): Muscle metabolism. Forsch. Zool., 18: 359-429.
EGGE, A. S. and CHAISSON, R. B. (1963): Endocrine effects of diencephalic lesions in the White Leghorn hen. Gen. comp. Endocr., 3: 346-361.
ELIASSEN, E. (1963): Preliminary results from new methods of investigating the physiology of birds during flight. Ibis., 105: 234-237.
ENGER, P. S. (1957): Heat regulation and metabolism in some tropical mammals and birds. Acta physiol. scand., 40: 161-166.
EULER, C. VON (1961): Physiology and pharmacology of temperature regulation. Pharmac. Rev., 13: 361-398.
FARNER, D. S. (1956): Body temperature of the Fairy Prion *Pachyptila turtur* in flight and at rest. J. appl. Physiol., 8: 546-548.
FELDBERG, W. and MYERS, R. D. (1963): A new concept of temperature regulation by amines in the hypothalamus. Nature (Lond.), 200: 1325.
FELDMAN, S. E., LARSSON, S., DIMICK, M. K. and LEPKOVSKY, S. (1957): Aphagia in chickens. Amer. J. Physiol., 191: 259-261.
GELINEO, S. (1964): Organ systems in adaptation: The temperature regulation system. In: Handbook of Physiology. Sect. 4 Adaptation to the Environment. D. B. Dill, E. F. Adolph, and C. G. Wilbur (eds). Amer. physiol. Soc. (Wash.).
GEORGE, J. C. and BERGER, A. J. (1966): Avian Mytology. Academic Press, New York.
HART, J. S. and ROY, O. Z. (1966): Respiratory and cardiac responses to flight in pigeons. Physiol. Zool., 39: 291-306.
HART, J. S. and ROY, O. Z. (1967): Temperature regulation during flight in pigeons. Amer. J. Physiol., 213: 1311-1316.
HIESTAND, W. A. and RANDALL, W. C. (1942): Influence of proprioceptive vagal efferents on panting movements in mammals and birds. Amer. J. Physiol., 138: 12-15.
HOFFMAN, E. and SHAFFNER, C. S. (1950): Thyroid weight and function as influenced by environmental temperature. Poultry Sci., 29: 365-376.
HUSTON, T. M. and CARMON, J. L. (1962): The influence of high environmental temperature on thyroid size of domestic fowl. Poultry Sci., 41: 175-179.
HUSTON, T. M., COTTON, T. E. and CARMON, J. L. (1962): The influence of high environmental temperature on the oxygen consumption of the mature domestic fowl. Poultry Sci., 41: 179-183.
HUTCHINSON, J. C. D. and SYKES, A. H. (1953): Physiological acclizatimation of fowls to a hot humid environment. J. agric. Sci., 43: 294-322.
HUTCHINSON, J. C. D. (1954): Evaporative cooling in fowls. J. agric. Sci., 45: 48-59.
JOHNSTON, D. W. and McFARLANE, R. W. (1967): Migration and bioenergetics of flight in the Pacific Golden Plover. Condor, 69: 156-168.
JUORIO, A. V. and VOGT, A. (1967). Monoamines and their metabolites in avian brain. J. Physiol. (Lond.), 189: 489-518.
KAYSER, C. (1930): Contribution à l'étude de la régulation thermique. L'émission d'eau et le rapport $H_2O:O_2$ chez quelques espèces homéothermes adultes et en cours de croissance. Ann. Physiol. Physicochim. Biol., 6: 721-744.

KING, J. R. and FARNER, D. S. (1961): Energy metabolism, thermoregulation, and body temperature. In: Biology and Comparative Physiology of Birds. II. A. J. Marshall (ed.). Academic Press, New York.
KING, J. S. and FARNER, D. S. (1964): Terrestrial animals in humid heat: Birds. In: Adaptation to the Environment. Handbook of Physiology. Sect. 4. D. B. Dill, E. F. Adolph and C. G. Wilbur (eds). Amer. physiol. Soc. (Wash.).
KING, J. S. and FARNER, D. S. (1965): Studies of fat deposition in migratory birds. Ann. N.Y. Acad. Sci., 131: 422-440.
LASIEWSKI, R. C. (1962): The energetics of migrating hummingbirds. Condor, 64: 324.
LASIEWSKI, R. C. (1963). Oxygen consumption of torpid, resting, active, and flying hummingbirds. Physiol. Zool., 36: 122-140.
LASIEWSKI, R. C. and DAWSON, L. R. (1964): Physiological response to temperature in the common nighthawk. Condor, 66: 477-490.
LASIEWSKI, R. C., ACOSTA, A. L. and BERNSTEIN, M. H. (1966): Evaporative water loss in birds. I. Characteristics of the open flow method of determination and their relations to estimates of thermoregulatory ability. Comp. Biochem. Physiol., 19: 445-457.
LASIEWSKI, R. C. and BARTHOLOMEW, G. A. (1966): Evaporative cooling in the Poor-will and the Tawny Frogmouth. Condor, 68: 253-262.
LASIEWSKI, R. C. and DAWSON, L. R. (1967): A re-examination of the relation between standard metabolic rate and body weight in birds. Condor, 69: 13-23.
LEFEBVRE, E. A. (1964): The use of D_2O^{18} for measuring energy metabolism in *Columbia livia* at rest and in flight. Auk, 81: 403-416.
LEPKOVSKY, S., SNAPIR, N. and FURUTA, F. (1968): Temperature regulation and appetitive behavior in chickens with hypothalamic lesions. Physiol. Behav., 3: 911-915.
MARLEY, E. and STEPHENSON, J. D. (1969): Effects of some catecholamines infused into the hypothalamus of young chickens. Brit. J. Pharmacol. Proc., 36: 194P-195P.
MILLS, S. H. and HEATH, J. E. (1970): Thermoresponsiveness of the preoptic region in the brain in house sparrows. Science, 168: 1008-1009.
NISBET, I. C. T., DRURY, W. H. Jr., and BAIRD, J. (1963): Weight loss during migration. Part I: Deposition and consumption of fat by the Blackpoll Warbler, *Dendroica striata*. Bird-Banding, 34: 107-138.
PEARSON, O. P. (1964): Metabolism and heat loss during flight in pigeons. Condor, 66: 182-185.
RANDALL, W. C. (1943a): Factors influencing the temperature regulation of birds. Amer. J. Physiol., 139: 46-63.
RANDALL, W. C. (1943b): Alterations in response to changing body temperature following artificial fever and chilling. Proc. Soc. exp. biol. (N.Y.), 52: 240-242.
RAUTENBERG, W. (1969): Untersuchungen zur Temperaturregulation wärme-und kälteakklimatisierter Tauben. Z. vergl. Physiol., 62: 221-234.
REINEKE, E. P. and TURNER, C. W. (1945): Seasonal rhythms in the thyroid secretion of the chick. Poultry Sci., 24: 299-504.
ROGERS, F. T. (1923): The relations between lesions of the brain stem, water elimination and body temperature. Amer. J. Physiol., 66: 284-287.
ROGERS, F. T. (1928): Studies on the brain stem. XI. The effects of artificial stimulation and traumatization of the avian thalamus. Amer. J. Physiol., 86: 639-650.
SCHOLANDER, P. F., HOCK, R., WALTERS, V., JOHNSON, F. and IRVING, L. (1950): Heat regulation in some arctic and tropical mammals and birds. Biol. Bull., 99: 237-258.
TUCKER, V. A. (1968): Respiratory exchange and evaporative water loss in the flying budgerigar. J. exp. Biol., 48: 67-87.
TUCKER, V. A. (1970): Energetic cost of locomotion in animals. Comp. Biochem. Physiol., 34: 841-846.
TUCKER, V. A. (1971): Flight energetics in birds. Amer. Zool., 11: 115-124.
VOGT, M. (1954): The concentration of sympathin in different parts of the central nervous system under normal conditions and after administration of drugs. J. Physiol., 123: 451-481.
VON SAALFELD, E. (1936): Untersuchungen über das Hecheln bei Tauben. Z. vergl. Physiol., 23: 727-743.
WALLGREN, H. (1954): Energy metabolism of two species of the genus *Emberiza* as correlated with distribution and migration. Acta zool. fenn., 84: 1-110.

WASHBURN, K. W., SIEGEL, P. B., FREUND, R. J. and GROSS, W. B. (1962): Effect of thiouracil on the body temperature of White Rock females. Poultry Sci., 41: 1354-1356.
WEIS-FOGH, T. (1952): Weight economy of flying insects. Trans. 9th int. Congr. Entomol., 1: 341-347.
WHITTOW, G. C., STURKIE, P. D. and STEIN, G., Jr. (1964): Cardiovascular changes associated with thermal polypnea in the chicken. Amer. J. Physiol., 207: 1349-1353.
WHITTOW, G. C. (1965): Regulation of body temperature. In: Avian Physiology. 2nd Ed. P. D. Sturkie (ed.). Comstock Publ., Ithaca, N.Y.
WILSON, W. O. (1948): Some effects of increasing environmental temperatures on pullets. Poultry Sci., 27: 813-817.
YEATES, N. T. M., LEE, D. H. K. and HINES, H. J. G. (1941): Reaction of domestic fowl to hot atmospheres. Proc. roy. Soc. Queensl., 53: 105-128.
ZEUTHEN, E. (1942): The ventilation of the respiratory tract in birds. Kgl. Danske Videnskab. Selskab. Biol. Medd., 17: 1-51.

Chapter 1, Section 8b (1): Effects of Cold on Energy Metabolism of Cattle (G. E. THOMPSON)

ALEXANDER, G., BENNETT, J. W. and GEMMEL, R. T. (1975): Brown adipose tissue in the new-born calf. J. Physiol., 244: 223-234.
ALVAREZ, M. B. and JOHNSON, H. D. (1970): Urinary excretion of adrenaline and noradrenaline in cattle during heat and cold exposure. J. Dairy Sci., 53: 928-930.
BELL, A. W., THOMPSON, G. E. and FINDLAY, J. D. (1974): The contribution of the shivering hind leg to the metabolic response to cold of the young ox (*Bos taurus*). Pflügers Arch. ges. Physiol., 346: 341-350.
BHATTACHARYA, A. N. and WARNER, R. G. (1968): Influence of varying rumen temperature on central cooling or warming and on regulation of voluntary feed intake in dairy cattle. J. Dairy Sci., 51: 1481-1489.
BLAXTER, K. L. and WAINMAN, F. W. (1961): Environmental temperature and the energy metabolism and heat emission of steers. J. agric. Sci., Camb. 56: 81-90.
BLAXTER, K. L. and WAINMAN, F. W. (1964): The effect of increased air movement on the heat production and emission of steers. J. agric. Sci. (Camb.), 62: 207-214.
BLAXTER, K. L. (1967): The Energy Metabolism of Ruminants. 2nd Ed. Hutchinson & Co. Ltd., London, 140 pp.
BOND, T. E., GARRETT, W. N., GIVENS, R. L. and MORRISON, S. R. (1970): Comparative effects of mud, wind and rain on beef cattle performance. Presented at the Annual Meeting Amer. Soc. agric. Eng., Minneapolis, U.S.A.
BROUWER, E. (1965): In: Energy Metabolism. K. L. Blaxter (ed.). 3rd Symp., Troon, 1964. Academic Press, London, 441 pp.
CALVERT, D. T., CLOUGH, D. P., FINDLAY, J. D. and THOMPSON, G. E. (1972): Hypothalamic cooling heat production and plasma lipids in the ox. Life Sci., II: 223-229.
CLOUGH, D. P., KENNY, J. D. R., SCOBIE, A. and THOMPSON, G. E. (1971): Dopamine and mast cells in the adipose tissue of the ox and the effect of noradrenaline and dopamine infusions on the plasma unesterified fatty acids. Res. vet. Sci., 12: 228-233.
COBBLE, J. W. and HERMAN, H. A. (1951): The influence of environmental temperatures on the composition of milk of the dairy cow. Res. Bull. Mo. agric. exp. Stat., No. 485.
DIETZ, W. and YOUNG, B. A. (1972): Plasma metabolites in beef cows over winter. Proc. W. Sect. Amer. Soc. Anim. Sci., 23: 265-269.
FINDLAY, J. D. and THOMPSON, G. E. (1968): The effect of intraventricular injections of noradrenaline, 5-hydroxytryptamine, acetylcholine and tranylcypromine on the ox (*Bos taurus*) at different environmental temperatures. J. Physiol. (Lond.), 194: 809-816.
FISHER, L. J., LISTER, E. E., JORDAN, W. A., BROSSARD, G. A., WAUTHY, J. M., COMEAU, J. E. and PROULX, J. (1972): Effects of plane of nutrition, confinement system and forage preservation on supplemental mineral intake and content of minerals in the blood of pregnant cows. Canad. J. Anim. Sci., 52: 693-702.
GONZALEZ-JIMENEZ, E. and BLAXTER, K. L. (1962): The metabolism and thermal regulation of calves in the first month of life. Brit. J. Nutr., 16: 199-212.
HAMADA, T. (1971): Estimation of lower critical temperatures for dry and lactating dairy

cows. J. Dairy Sci., 54: 1704-1705.
HIDIROGLOU, M., JENKINS, K. J., LESSARD, J. R. and CARSON, R. B. (1970): Copper and molybdenum status of growing beef cattle in a selenium deficient area of northern Ontario. Canad. J. Anim. Sci., 50: 279-284.
HIDIROGLOU, M. and LESSARD, J. R. (1971): Some effects of fluctuating low ambient temperatures on beef cattle. Canad. J. Anim. Sci., 51: 111-120.
HIRONAKA, R. and PETERS, H. F. (1969). Energy requirements for wintering mature pregnant beef cows. Canad. J. Anim. Sci., 49: 323-330.
HOLMES, C. W. (1971a): The effect of milk given at various temperatures on the oxygen consumption of young calves. Anim. Prod., 13: 619-625.
HOLMES, C. W. (1971b): Local cooling of the mammary gland and milk production in the cow. J. Dairy Res., 38: 3-7.
JESSEN, C., McLEAN, J. A., CALVERT, D. T. and FINDLAY, J. D. (1972): Balanced and unbalanced signals generated in spinal cord of the ox. Amer. J. Physiol., 222: 1343-1347.
JOHNSON, H. D. and YOUSEF, M. K. (1966): Effect of short-term fasting on thyroid activity of cattle at various environmental temperatures. J. Anim. Sci., 25: 1069-1075.
JORDAN, W. A., LISTER, E. E. and ROWLANDS, G. J. (1968a): Effect of planes of nutrition on wintering pregnant beef cows. Canad. J. Anim. Sci., 48: 145-154.
JORDAN, W. A., LISTER, E. E. and ROWLANDS, G. J. (1968b): Effect of varying planes of winter nutrition of beef cows on calf performance to weaning. Canad. J. Anim. Sci., 48: 155-161.
KEENER, H. A., BECHDEL, S. I., GUERRANT, N. B. and THORP, W. T. S. (1942): Carotene in calf nutrition. J. Dairy Sci., 25: 571-583.
KELLEY, R. O., MARTZ, F. A. and JOHNSON, H. D. (1967): Effect of environmental temperature on ruminal volatile fatty acid levels with controlled feed intake. J. Dairy Sci., 50: 531-533.
KHAN, M. A., DICKSON, W. M. and MEYERS, K. M. (1970): The effect of low environmental temperature on plasma corticosteroid and glucose concentrations in the newborn calf. J. Endocr., 48: 355-363.
KLEIBER, M. (1965): In: Energy Metabolism. Proceedings 3rd Symp., Troon, 1964 (K. L. Blaxter ed.). Academic Press, London, 427 pp.
LAWSON, J. E. (1967): Subsequent performance of bull calves wintered on two planes of nutrition and two feeding methods. Canad. J. Anim. Sci., 47: 181-186.
LISTER, E. E., JORDAN, W. A., WAUTHY, J. M., COMEAU, J. E. and PIGDEN, W. J. (1972): Effects of housing and type of forage on the response of pregnant beef cows to dietary energy intake in winter. Canad. J. Anim. Sci., 52: 671-679.
LISTER, E. E., FISHER, L. J., JORDAN, W. A., WAUTHY, J. M., COMEAU, J. E. and PROULX, J. (1973): Influence of shelter, level of feeding and method of forage conservation on packed cell volume and plasma metabolite levels in pregnant beef cows. Canad. J. Anim. Sci., 53: 81-88.
LUNGU, A., MITRACHE, L. and TEODORU, V. (1972): Serum cholesterol variations in bovines in relation to environmental temperature. Rev. roum. Endocr., 9: 249-252.
MacDONALD, M. A. and BELL, J. M. (1958): Influence of ambient air temperature on feed intake of lactating Holstein-Friesian cows. Canad. J. Anim. Sci., 38: 148-159.
McLEAN, J. A., HALES, J. R. S., JESSEN, C. and CALVERT, D. T. (1970): Influences of spinal cord temperature on heat exchange of the ox. Proc. Aust. Physiol. Pharmacol. Soc., I: 32-33.
McLEAN, J. A. (1972): On the calculation of heat production from open-circuit calorimetric measurements. Brit. J. Nutr., 27: 597-600.
MARKOVIČ, P. (1963): Thermoregulation in young cattle. Věd. Pr. výzk. Úst. živoč Výroby Uhřiněvsi, 6: 181-215.
MILLIGAN, J. D. and CHRISTISON, G. I. (1973): The influence of winter on the productivity of feedlot steers. Univ. Sask. 18th Ann. Stockman's Day Rep., Dept. Anim. Sci., Univ. Sask., Saskatoon, Canada.
NATIONAL RESEARCH COUNCIL (1963): Nat. Acad. Sci. Nat. Res. Counc. Publ., No. 1137, Rev. Ed. (Wash.).
OLSEN, J. D. and TRENKLE, A. (1973): Exposure of cattle to controlled sub zero temperature: growth hormone glucose and free fatty acid concentration in plasma. Amer. J. vet. Res., 34: 747-751.

PETERS, H. F. and SLEN, S. B. (1966): Range calf production of cattle and bison Cattalo and Hereford cows. Canad. J. Anim. Sci., 46: 157-164.
RAGSDALE, A. C., WORSTELL, D. M., THOMPSON, H. J. and BRODY, S. (1949): Influence of temperature, 50° to 0°F and 50° to 95°F on milk production, feed and water consumption and body weight in Jersey and Holstein cows. Res. Bull. Mo. agric. exp. Stat., No. 449.
ROGAL, I. G. (1955): Cold method of rearing young farm animals and its physiological premises. Zh. obshch. Biol., 16: 275-284.
SMOLIAK, S. and PETERS, H. F. (1955): Climatic effects on foraging performance of beef cows on winter range. Canad. J. Anim. Sci., 35: 213-216.
SZEKELY, H. (1967): Body temperature changes during the first seven days of life in calves delivered by extraction or by means of Caesarian section. Wien. tierärztl. Mschr., 54: 333-335.
THOMPSON, G. E. and CLOUGH, D. P. (1970): Temperature regulation in the new-born ox. Biol. Neonate, 15: 19-25.
THOMPSON, G. E. and CLOUGH, D. P. (1971): The effect of intraventricular noradrenaline on plasma unesterified fatty acid concentration in the ox. Life Sci., 10: 415-420.
THOMPSON, G. E. and CLOUGH, D. P. (1972): The effect of cold exposure on plasma lipids of the new-born and adult ox. Quart. J. exp. Physiol., 57: 192-198.
WEBSTER, A. J. F., CHLUMECKY, J. and YOUNG, B. A. (1970): Effects of cold environments on the energy exchanges of young beef cattle. Canad. J. Anim. Sci., 50: 89-100.
WEBSTER, A. J. F. (1971): Prediction of heat losses from cattle exposed to cold outdoor environments. J. appl. Physiol., 30: 684-690.
YOUNG, B. A. and BERG, R. T. (1970): Energy requirements for maintenance of pregnant beef cows. Univ. Alb. Feeders Day June 15th. Dept. Anim. Sci., Univ. Alb., Edmonton, Canada.
YOUNG, B. A. (1972): Influence of adaptation to cold on resting heat production of beef cows. Proc. W. Sect. Amer. Soc. Anim. Sci., 23: 270-274.
YOUSEF, M. K. and JOHNSON, H. D. (1965a): Some blood constituents of dairy cattle: influence of thyroxine and high environmental temperature. J. Dairy Sci., 48: 1074-1078.
YOUSEF, M. K. and JOHNSON, H. D. (1965b): Time course of thyroxine I^{131} disappearance rates in cattle during exposure to hot and cold environments. Life Sci., 4: 1531-1543.
YOUSEF, M. K. and JOHNSON, H. D. (1966a): Calorigenesis of dairy cattle as influenced by thyroxine and environmental temperature. J. Anim. Sci., 25: 150-156.
YOUSEF, M. K. and JOHNSON, H. D. (1966b): Blood thyroxine degradation rate of cattle as influenced by temperature and feed intake. Life Sci., 5: 1349-1363.
YOUSEF, M. K. and JOHNSON, H. D. (1966c): Calorigenesis of cattle as influenced by growth hormone and environmental temperature. J. Anim. Sci., 25: 1076-1082.
YOUSEF, M. K., KIBLER, H. H. and JOHNSON, H. D. (1967): Thyroid activity and heat production in cattle following sudden ambient temperature changes. J. Anim. Sci., 26: 142-148.

Chapter 1, Section 8b (2): Effects of Cold on Energy Metabolism of Sheep (A. J. F. WEBSTER)

ALEXANDER, G. (1961): Temperature regulation in the new born lamb. Aust. J. agric. Res., 12: 1139-1151.
ALEXANDER, G. and WILLIAMS, D. (1968): Shivering and non-shivering thermogenesis during summit metabolism in young lambs. J. Physiol. (Lond.), 198: 251-276.
ALEXANDER, G. (1974): Heat losses from sheep. In: Heat Loss from Animals and Man, Assessment and Control. Proc. 20th Easter School, Univ. Nottingham. Butterworths, London, in press.
BENNETT, J. W. (1972): The maximum metabolic response of sheep to cold. Effects of rectal temperature, shearing, feed consumption, body posture and body weight. Aust. J. agric. Res., 23: 1045-58.
BLAXTER, K. L. (1967): The Energy Metabolism of Ruminants. Hutchinson Publ. Co., London.

BROCKWAY, J. M., BOYNE, A. W. and GORDON, J. G. (1971): Simultaneous calibration of gas analysers and meters. J. appl. Physiol., 31: 296-297.
BURTON, A. C. and EDHOLM, O. G. (1955): Man in a Cold Environment. Arnold Ltd., London.
CORBETT, J. L., LENG, R. A. and YOUNG, B. A. (1969): Measurement of energy expenditure by grazing sheep. In: Energy Metabolism of Farm Animals. K. L. Blaxter, J. Kielanowski and G. Thorbek (eds). Oriel Press, Newcatle: 117.
CORBETT, J. L., FARRELL, D. J., LENG, R. A., McCLYMONT, G. L. and YOUNG, B. A. (1971): Determination of the energy expenditure of penned and grazing sheep from estimates of carbon dioxide entry rate. Brit. J. Nutr., 26: 277-291.
JOYCE, J. P. and BLAXTER, K. L. (1964): The effect of air movement, air temperature and infra red radiation on the energy requirements of sheep. Brit. J. Nutr., 18: 5-27.
JOYCE, J. P., BLAXTER, K. L. and PARK, C. (1966): The effect of natural outdoor environments on the energy requirements of sheep. Res. vet. Sci., 7: 342-359.
McLEAN, J. A. (1972): On the calculation of heat production from open-circuit calorimetric measurements. Brit. J. Nutr., 27: 597-600.
MOUNT, L. E. (1968): The Climatic Physiology of the Pig. Arnold Ltd., London.
PULLAR, J. D. (1969): Methods of calorimetry (A) Direct. In: Nutrition of Animals of Agricultural Importance, 1. Pergamon Press, Oxford: 471.
ROYAL SOCIETY (1972): Report of the Royal Society's British National Committee for Nutritional Sciences. Proc. Nutr. Soc., 31: 239-247.
SYKES, A. R. and SLEE, J. (1968): Acclimatization of Scottish Blackface sheep to cold. Anim. Prod., 10: 17-35.
WEBSTER, A. J. F. and BLAXTER, K. L. (1966): The thermal regulation of 2 breeds of sheep exposed to air temperatures below freezing point. Res. vet. Sci., 7: 466-479.
WEBSTER, A. J. F. (1967): Continuous measurement of heart rate as an indicator of the energy expenditure of sheep. Brit. J. Nutr., 21: 769-785.
WEBSTER, A. J. F. and PARK, C. (1967): The effect of jute coats on the heat losses of two breeds of sheep exposed to different environments. Anim. Prod., 9: 483-490.
WEBSTER, A. J. F., HEITMAN, J. H., HAYS, F. L. and OLYNYK, G. P. (1969): Catecholamines and cold thermogenesis in the sheep. Canad. J. Physiol. Pharmacol., 47: 719-724.
WEBSTER, A. J. F. (1971): Prediction of heat losses from cattle exposed to cold outdoor environments. J. appl. Physiol., 30: 684-690.
WEBSTER, A. J. F. (1974): Physiological effects of cold exposure. In: Physiology, 7. M.T.P. Int. Rev. Sci. D. Robertshaw (ed.).

Chapter 1, Section 8b (3): Effects of Heat and Cold on Energy Metabolism of the Pig (L. E. MOUNT)

BARNETT, S. A. and MOUNT, L. E. (1967): Resistance to cold in mammals. In: Thermobiology. A. H. Rose (ed.). Academic Press, London/New York: 411-477.
BRODY, S. (1945): Bioenergetics and growth. Reinhold Publ. Co., New York.
CAIRNIE, A. B. and PULLAR, J. D. (1959): An investigation into the efficient use of time in the calorimetric measurement of heat output. Brit. J. Nutr. 13: 431-439.
CAPSTICK, J. W. (1921): A calorimeter for use with large animals. J. agric. Sci. (Camb.), 11: 408-431.
CURTIS, S. E. (1970): Environmental-thermoregulatory interactions and neonatal piglet survival. J. Anim. Sci., 31: 576-587.
FULLER, M. F. (1965): The effect of environmental temperature on the nitrogen metabolism and growth of the young pig. Brit. J. Nutr., 19: 531-546.
GIAJA, J. (1938): L'homéothermie. Actualités sci. industr., 576. IX, 1.
INGRAM, D. L. and WEAVER, M. E. (1969): A quantitative study of the blood vessels of the pig's skin and the influence of environmental temperature. Anat. Rec., 163: 517-524.
IRVING, L. (1964): Terrestrial animals in cold: birds and mammals. In: Adaptation to Environment. D. B. Dill (ed.). Amer. physiol. Soc.: 361-377.
KLEIBER, M. (1961): The fire of life. Wiley and Sons, New York.
MOUNT, L. E. (1968): The climatic physiology of the pig. Edward Arnold, London.

MOUNT, L. E. and INGRAM, D. L. (1971): The pig as a laboratory animal. Academic Press, London.
MOUNT, L. E. (1972): Environmental physiology in relation to pig production. In: Pig Production. D. J. A. Cole (ed.). Butterworths, London: 71-89.
STEPHENS, D. B. (1971): The metabolic rates of newborn pigs in relation to floor insulation and ambient temperature. Anim. Prod., 13: 303-313.
VERSTEGEN, M. W. A., CLOSE, W. H., START, I. B. and MOUNT, L. E. (1973): The effects of environmental temperature and plane of nutrition on heat loss, energy retention and deposition of protein and fat in groups of growing pigs. Brit. J. Nutr., 30: 21-35.
VERSTEGEN, M. W. A. and VAN DER HEL, W. (1974): The effects of temperature and type of floor on metabolic rate and effective critical temperature in groups of growing pigs. Anim. Prod., in press.

Chapter 1, Section 8b (4): Effects of Cold and Exercise on Energy Metabolism of Small Mammals (L. JANSKÝ)

BARTŮŇKOVÁ, R. and JANSKÝ, L. (1970): Cardiac output and organ blood flow in cold-acclimated rats exposed to cold or work. Physiol. bohemoslov., 19: 314.
BRÜCK, K. and WÜNNENBERG, W. (1967): Eine kälteadaptative Modifikation: Senkung der Schwellentemperaturen für Kältezittern. Pflügers Arch. ges. Physiol., 293: 226-235.
BRÜCK, K. and WÜNNENBERG, W. (1970): "Meshed" control of two effector systems: Nonshivering and shivering thermogenesis. In: Physiological and Behavioral Temperature Regulation. J. D. Hardy, A. P. Gagge and J. A. J. Stolwijk (eds). C. C. Thomas, Springfield, Ill.: 562-580.
COTTLE, W. H. and CARLSON, L. D. (1956): Regulation of heat production in cold-adapted rats. Proc. Soc. exp. Biol. (N.Y.), 92: 845-849.
GIAJA, J. (1938): Homéothermie et Thermorégulation. II. La Thermorégulation. Hermann et Cie, Paris.
GOLENHOFEN, K. (1965): Das Reaktionmuster der menschlichen Muskulatur im Rahmen der Thermoregulation. Pflügers Arch. ges. Physiol., 285: 124-146.
HART, J. S. (1950): Interrelations of daily metabolic cycle, activity and environmental temperature of mice. Canad. J. Res., 28: 293-307.
HART, J. S. (1953): The relation between thermal history and cold resistance in certain species of rodents. Canad. J. Zool., 31: 80.
HART, J. S. and HEROUX, O. (1955): Exercise and temperature regulation in lemmings and rabbits. Canad. J. Biochem., 33: 428-435.
HART, J. S. (1956): Seasonal changes in insulation of the fur. Canad. J. Zool., 34: 53-57.
HART, J. S., HEROUX, O. and DEPOCAS, F. (1956): Cold acclimation and the electromyogram of unanesthetized rats. J. appl. Physiol., 9: 404-408.
HART, J. S. (1957): Climatic and temperature induced changes in the energetics of homeotherms. Rev. Canad. Biol., 16: 133-174.
HART, J. S. (1963a): Surface cooling versus metabolic response to cold. Fed. Proc., 22: 940-943.
HART, J. S. (1963b): Physiological responses to cold in non-shivering homeotherms. In: Temperature – Its Measurements and Control in Science and Industry, 3. J. D. Hardy (ed.). Reinhold Publ. Co., New York: 373-406.
HART, J. S. and JANSKÝ, L. (1963): Thermogenesis due to exercise and cold in warm- and cold-acclimated rats. Canad. J. Biochem., 41: 629-634.
HART, J. S. (1971): Rodents. In: Comparative Physiology of Thermoregulation. G. C. Whittow (ed.). Academic Press, New York and London: 1-149.
HAYWARD, J. S. and LYMAN, C. P. (1967): Nonshivering heat production during arousal from hibernation and evidence for the contribution of brown fat. Proc. III int. Symp. natur. mamm. Hibern. Oliver and Boyd Publ. Co., Edinburgh: 346-355.
HAYWARD, J. S. (1968): The magnitude of noradrenaline-induced thermogenesis in the bat (*Myotis lucifugus*) and its relation to arousal from hibernation. Canad. J. Physiol., 46: 713-718.
HEIM, T. and HULL, D. (1966): The blood flow and oxygen consumption of brown adipose tissue in the new-born rabbit. J. Physiol. (Lond.), 186: 42-55.

HELDMAIER, G. (1971): Relationship between nonshivering thermogenesis and body size. In: Nonshivering Thermogenesis. Proc. Symp. Nonshivering Thermogenesis. L. Janský (ed.). Swets & Zeitlinger, Amsterdam, 73-81.
HEMINGWAY, A. and STUART, D. (1963): Shivering in man and animals. In: Temperature – Its Measurements and Control in Science and Industry. J. D. Hardy (ed.). Reinhold Publ. Co., New York, Vol. 3: 407-427.
HIMMS-HAGEN, J. (1969): The role of brown adipose tissue in the calorigenic effect of adrenaline and noradrenaline in cold-acclimated rats. J. Physiol. (Lond.), 205: 393-403.
HIMMS-HAGEN, J. (1972): Effects of Catecholamines on Metabolism. In: Catecholamines. H. Blaschko and E. Muscholl (eds). Springer-Verlag Berlin/Heidelberg/New York: 363-462.
HORVATH, S. M., SPURR, G. B., HUTT, B. K. and HAMILTON, L. H. (1956): Metabolic cost of shivering. J. appl. Physiol., 8: 595-602.
HSIEH, A. C. L. and CARLSON, L. D. (1957a): Role of the thyroid in metabolic response to low temperature. Amer. J. Physiol., 188: 40-44.
HSIEH, A. C. L. and CARLSON, L. D. (1957b): Role of adrenaline and noradrenaline in chemical regulation of heat production. Amer. J. Physiol., 190: 243-246.
HSIEH, A. C. L., CARLSON, L. D. and GRAY, G. (1957): Role of the sympathetic nervous system in the control of chemical regulation of heat production. Amer. J. Physiol., 190: 247-251.
JANSKÝ, L. (1959a): Einfluss der Arbeit und der niedrigen Temperaturen auf den Sauerstoffverbrauch des Goldhamsters. Acta Soc. Zool. bohemoslov., 23: 266-274.
JANSKÝ, L. (1959b): Oxygen consumption in white mice during physical exercise. Physiol. bohemoslov., 8: 464-471.
JANSKÝ, L. (1959c): Working oxygen consumption in two species of wild rodents (*Microtus arvalis, Clethrionomys glareolus*). Physiol. bohemoslov., 8: 472-478.
JANSKÝ, L. (1963): Body organ cytochrome oxidase activity in cold- and warm-acclimated rats. Canad. J. Biochem., 41: 1847-1854.
JANSKÝ, L. and HART, J. S. (1963): Participation of skeletal muscle and kidney during nonshivering thermogenesis in cold-acclimated rats. Canad. J. Biochem., 41: 953-964.
JANSKÝ, L. (1965): Adaptability of heat production mechanisms in homeotherms. Acta Univ. Carol. Biol.: 1-91.
JANSKÝ, L. (1966): Body organ thermogenesis of the rat during exposure to cold and at maximal metabolic rate. Fed. Proc., 25: 1297-1302.
JANSKÝ, L., BARTŮŇKOVÁ, R. and ZEISBERGER, E. (1967): Acclimation of the white rat to cold: noradrenaline thermogenesis. Physiol. bohemoslov., 16: 366-372.
JANSKÝ, L. and HART, J. S. (1968): Cardiac output and organ blood flow in warm- and cold-acclimated rats exposed to cold. Canad. J. Physiol., 46: 653-659.
JANSKÝ, L., VOTÁPKOVÁ, Z. and FEIGLOVÁ, E. (1969): Total cytochrome-oxidase activity of the brown fat and its thermogenetic significance. Physiol. bohemoslov., 18: 443-451.
JANSKÝ, L., FUNKE, E. and HART, J. S. (1970): Analysis of electromyograms in shivering rabbits. Physiol. bohemoslov., 19: 397-402.
JANSKÝ, L. (1971): Participation of body organs during nonshivering heat production. In: Nonshivering Thermogenesis. Proc. Symp. Nonshivering Thermogenesis. L. Janský (ed.). Swets & Zeitlinger, Amsterdam, 159-170.
JANSKÝ, L. (1973): Non-shivering thermogenesis and its thermoregulatory significance. Biol. Rev., 48: 85-132.
JANSKÝ, L., LUKÁŠOVÁ, G., VYBÍRAL, S. and BAŽENOV, J. (1973): Thermoregulatory response of the rabbit to local peripheral cooling. Physiol. bohemoslov., 22: 125-136.
KAWAMURA, Y. (1961): Neuromuscular organization of shivering. In: Neural aspects of Temperature Regulation. Symp. Arct. Biol. Med. I. J. P. HANNON and E. Vilreck (eds): 193-224.
KOČKOVÁ, J. and JANSKÝ, L. (1968): Cold acclimation in the rabbit. Physiol. bohemoslov., 17: 309-316.
LARDY, H. A. and WELLMAN, H. (1952): Oxidative phosphorylations: role of inorganic phosphate, and acceptor systems in control of metabolic rates. J. biol. Chem., 195: 215.
LEDUC, J. (1961): Excretion of catecholamines in rats exposed to cold. Acta physiol. scand., 51: 94-95.

LINDBERG, O., DE PIERRE, J., RYLANDER, E. and AFZELIUS, B. A. (1967): Studies of the mitochondrial energy-transfer system of brown adipose tissue. J. Cell Biol., 34: 293-310.

MEJSNAR, J. and JANSKÝ, L. (1970): Shivering and nonshivering thermogenesis in the bat (*Myotis maotis* Borkh.) during arousal from hibernation. Canad. J. Physiol., 48: 102-106.

MEJSNAR, J. and JANSKÝ, L. (1971a): Methods for estimating of nonshivering thermogenesis. In: Nonshivering Thermogenesis. Proc. Symp. Nonshivering Thermogenesis. L. Janský (ed.). Swets & Zeitlinger, Amsterdam, 27-36.

MEJSNAR, J. and JANSKÝ, L. (1971b): Nonshivering thermogenesis and calorigenic action of catecholamines in white mouse. Physiol. bohemoslov., 20: 157-162.

SCHOLANDER, P. F., HOCK, R., WALTERS, V., JOHNSON, F. and IRVING, L. (1950): Heat regulation in some arctic and tropical mammals and birds. Biol. Bull., 99: 237-271.

SWANSON, H. E. (1956): Interrelations between thyroxin and adrenalin in the regulation of oxygen consumption in the albino rat. Endocrinology, 59: 217-225.

THAUER, R. V. (1939): Der Mechanismus der Wärmeregulation. Ergebn. Physiol., 41: 607.

TKACHENKO, E. Y. and IVANOV, K. P. (1971): On the physiological mechanisms of the chemical thermoregulation after the cold-adaptation. Sechenov physiol. J. U.S.S.R., 57: 111-115.

TWENTE, J., TWENTE, J. W. and JANSKÝ, L. (1970): The effects of environmental temperature upon the physiological "preparation" for hibernation. In: Adaptation to Arid Zone Conditions: Hibernation. Novosibirsk: 188-192.

VYBÍRAL, S. and JANSKÝ, L. (1974): Non-shivering thermogenesis in the golden hamster. Physiol. bohemoslov., in press.

ZEISBERGER, E., BRÜCK, K., WÜNNENBERG, W. and WIETASCH, C. (1967): Das Ausmass der zitterfreien Thermogenese des Meerschweinchens in Abhängigkeit vom Lebensalter. Pflügers Arch. ges. Physiol., 296: 276-288.

Chapter 1, Section 8b (5): Effects of Cold on Energy Metabolism in Birds (H. S. SIEGEL)

ALLEN, D. J., GARG, K. N. and MARLEY, E. (1969): Hypothermia due to α-methylnoradrenaline in young chickens. Brit. J. Pharmacol., 36: 195P-197P.

AULIE, A. (1971): Body temperatures in pigeons and budgerigars during sustained flight. Comp. Biochem. Physiol., 39A: 173-176.

BALL, E. G. and JUNGAS, R. L. (1961): On the action of hormones which accelerate the rate of oxygen consumption and fatty acid release in rat adipose tissue *in vitro*. Proc. nat. Acad. Sci. (Wash.), 47: 932-941.

BARTHOLOMEW, G. A., HOWELL, T. R. and CADE, T. J. (1957): Torpidity in the white-throated swift, Anna hummingbird, and the poorwill. Condor, 59: 145-155.

BARTHOLOMEW, G. A., HUDSON, J. A. and HOWELL, T. R. (1962): Body temperature, oxygen consumption, evaporative water loss, and heart rate in the poor-will. Condor, 64: 117-125.

BARTHOLOMEW, G. A. and TROST, C. H. (1970): Temperature regulation in the speckled mousebird, *Colius striatus*. Condor, 72: 141-146.

BAROTT, H., FRITZ, J. C., PRINGLE, E. M. and TITUS, H. W. (1938): Heat production and gaseous metabolism of young male chickens. Condor, 15: 145-167.

BAROTT, H. G. and PRINGLE, E. M. (1947): Effect of environment on growth and feed and water consumption of chickens. I. The effect of temperature of environment during the first nine days after hatch. J. Nutr., 34: 53-68.

BAROTT, H. G. and PRINGLE, E. M. (1949): Effect of environment on growth and feed and water consumption of chickens. II. The effect of temperature of environment during the period from 18 to 32 days of age. J. Nutr., 37: 153-161.

BERGER, M., HART, J. S. and ROY, O. Z. (1970): Respiration, oxygen consumption, and heart rate in some birds during rest and flight. Z. vergl. Physiol., 66: 201-214.

CALDER, W. A. and BOOSER, J. (1973): Hypothermia of broad-tailed hummingbirds during incubation in nature with ecological correlations. Science, 180: 751-753.

CARPENTER, F. L. (1974): Torpor in an Andian hummingbird: Its ecological significance. Science, 183: 545-547.

COWLES, R. B. and BOGERT, C. M. (1944): A preliminary study of the thermal requirements of desert reptiles. Bull. Amer. Mus. nat. Hist., 83: 261-296.
DAVIS, E. A., Jr. (1955): Seasonal changes in the energy balance of the English sparrow. Auk, 72: 385-411.
DAWKINS, M. J. R. and HULL, D. (1964): Brown adipose tissue and the response of new-born rabbits to cold. J. Physiol., 172: 216-238.
DAWSON, W. R. and HUDSON, J. W. (1970): Birds. In: Comparative Physiology of Thermoregulation, I. G. C. Whittow (ed.). Academic Press, New York.
DUKES, H. H. (1955): The Physiology of Domestic Animals. 7th Ed. Comstock Publ. Ass., Ithaca, N.Y.
EISENTRAUT, M. (1961): Heat regulation in primitive mammals and in tropical species. Bull. Mus. comp. Zool., Harvard Coll., 124: 31-43.
EL-HALAWANI, M. E., WILSON, W. O. and BURGER, R. E. (1970): Cold-acclimation and the role of catecholamines in body temperature regulation in male Leghorns. Poultry Sci., 49: 621-632.
FREEMAN, B. M. (1967): Some effects of cold on the metabolism of the fowl during the prenatal period. Comp. Biochem. Physiol., 20: 179-193.
FREEMAN, B. M. (1969): Effect of noradrenaline on the plasma free fatty acid and glucose levels of *Gallus domesticus*. Comp. Biochem. Physiol., 30: 993-996.
FREEMAN, B. M. (1970): Thermoregulatory mechanisms of the neonate fowl. Comp. Biochem. Physiol., 33: 219-230.
FREEMAN, B. M. (1971): Impaired thermoregulation in the thiouraciltreated neonate fowl. Comp. Biochem. Physiol., 40A: 553-555.
GELINEO, S. (1955): Température d'adaptation et production de chaleur chez les oiseaux de petite taille. Arch. Sci. physiol., 9: 225-243.
GELINEO, S. (1964): Organ systems in adaptation: The temperature regulation system. In: Adaptation to the Environment Handbook of Physiology, Sect. 4. D. B. Dill, E. F. Adolph, and C. G. Wilbur (eds). American physiological Soc. (Wash.).
GROEBBALS, F. (1928): Fortgesetzte Untersuchungen über den Stoffwechsel der Vögel und das Stoffwechselproblem im allgemeinen. Pflüger's Arch. ges. Physiol., 218: 99-114.
HART, J. S. (1957): Climatic and temperature induced changes in the energetics of homeotherms. Rev. Canad. Biol., 16: 133-174.
HART, J. S. (1962): Seasonal acclimation in four species of small wild birds. Physiol. Zool., 35: 224-236.
HART, J. S. and ROY, O. Z. (1967): Temperature regulation during flight in pigeons. Amer. J. Physiol., 213: 1311-1316.
HEATH, J. E. (1962): Temperature fluctuation in the turkey vulture. Condor, 64: 234-235.
HEIM, T. and HULL, D. (1966): The effect of propranolol on the calorigenic response in brown adipose tissue of new-born rabbits to catecholamines, glucagon, corticotrophin, and cold exposure. J. Physiol. (Lond.), 187: 271-283.
HOWELL, T. R. and BARTHOLOMEW, G. A. (1959): Further experiments on torpidity in the poorwill. Condor, 61: 180-185.
HULL, D. and SEGALL, M. M. (1965): The contribution of brown adipose tissue to heat production in the new-born rabbit. J. Physiol. (Lond.), 181: 449-457.
HUTCHINSON, J. C. D. (1954): Heat regulation in birds. In: Progress in the Physiology of Farm Animals, I. J. Hammond (ed.). Butterworths Publ. Co., London.
IRVING, L. (1960): Birds of Anakutuvuk Pass, Kabuk and Old Crow. U.S. nat. Mus. Bull., 217.
IRVING, L. (1964): Terrestrial animals in cold: birds and mammals. In: Adaptation to the Environment. Handbook of Physiology, Sect. 4. D. B. Dill, E. F. Adolph, aud C. G. Wilber (eds.). Amer. physiol. Soc. (Wash.).
ITOH, S., KUROSHIMA, A., DOI, K., MORIYA, D., SHIRATO, H. and YASHIMURA, K. (1969): Lipid metabolism in relation to cold adaptability in man. Fed. Proc., 28: 960-964.
JOEL, C. D., NEAVES, W. B. and RABB, J. M. (1968): Mitochondria of brown fat: oxidative phosphorylation sensitive to 2,4-dinitrophenol. Biochem. biophys. Res. Commun., 29: 490-495.
KAYSER, C. (1929): Régulation thermique après section médullaire dorsale chez le pigeon. C. R. Soc. Biol. (Paris), 100: 286-288.

KENDIEGH, S. C. (1949): Effect of temperature and season on the energy resources of the English sparrow. Auk, 66: 113-127.
KING, J. R. and FARNER, D. S. (1961): Energy metabolism, thermoregulation and body temperature. In: Biology and Comparative Physiology of Birds, II. A. J. Marshall (ed.). Academic Press, New York.
LANGSLOW, D. R. and HALES, C. N. (1969): Lipolysis in chicken adipose tissue, in vitro. J. Endocr., 43: 285-294.
LASIEWSKI, R. C. (1963): Oxygen consumption of torpid, resting, active, and flying hummingbirds. Physiol. Zool., 36: 122-140.
LASIEWSKI, R. C. (1964): Body temperatures, heart, and breathing rate, and evaporative heat loss in hummingbirds. Physiol. Zool., 37: 212-223.
LASIEWSKI, R. C. and DAWSON, L. R. (1964): Physiological response to temperature in the common nighthawk. Condor, 66: 477-490.
LASIEWSKI, R. C. and LASIEWSKI, R. J. (1967): Physiological responses of the blue-throated and Rivoli's hummingbirds. Auk, 84: 34-48.
LASIEWSKI, R. C., WEATHERS, W. W. and BERNSTEIN, M. H. (1967): Physiological responses of the giant hummingbird, *Patagona gigas*. Comp. Biochem. Physiol., 23: 797-813.
LEFEBVRE, E. A. (1964): The use of D_2O^{18} for measuring energy metabolism in *Columbia livia* at rest and in flight. Auk, 81: 403-416.
MacMILLAN, R. E. and TROST, C. H. (1967): Thermoregulation and water loss in the Inca dove. Comp. Biochem. Physiol., 29: 263-273.
MARLEY, E. and STEPHENSON, J. D. (1969): Effects of some catecholamines infused into the hypothalamus of young chickens. Brit. J. Pharmacol. Proc. 36: 194P-195P.
MARSHALL, J. T. (1955): Hibernation in captive goatsuckers. Condor, 57: 129-134.
MILLER, D. S. (1939): A study of the physiology of the sparrow thyroid. J. exp. Zool. 80: 259-281.
ODUM, E. P. (1941): Variations in the heart rate of birds: a study in physiological ecology. Ecol. Monogr., 11: 299-326.
ODUM, E. P. (1942): Muscle tremors and the development of temperature regulation in birds. Amer. J. Physiol., 136: 618-622.
PEARSON, O. P. (1950): The metabolism of hummingbirds. Condor, 58: 222-228.
POHL, H. (1969): Some factors influencing the metabolic response to cold in birds. Fed. Proc. 28: 1059-1064.
RANDALL, W. C. (1943): Factors influencing the temperature regulation of birds. Amer. J. Physiol., 139: 46-63.
SCHOLANDER, P. F., WALTERS, V., HOCK, R. and IRVING, L. (1950): Body insulation of some arctic and tropical mammals and birds. Biol. Bull., 99: 225-236.
STEEN, J. and ENGER, P. S. (1957): Muscular heat production in pigeons during exposure to cold. Amer. J. Physiol., 191: 157-158.
STEEN, J. (1958): Climatic adaptations in some small northern birds. Ecology, 39: 625-629.
THAXTON, P., WYATT, R. O. and HAMILTON, P. B. (1974): The effect of environmental temperature on parathyroid infection in the neonatal chick. Poultry Sci. 53: 88-94.
TUCKER, V. A. (1966): Oxygen consumption of a flying bird. Science, 154: 150-151.
TUCKER, V. A. (1968): Respiratory exchange and evaporative heat loss in the flying budgerigar. J. exp. Biol., 48: 67-87.
TUCKER, V. A. (1970): Energetic cost of locomotion in animals. Comp. Biochem. Physiol., 34: 841-846.
TUCKER, V. A. (1971): Flight energetics in birds. Amer. Zoologist 11: 115-124.
WEKSTEIN, D. R. and ZOLMAN, J. F. (1969): Ontogeny of heat production in chicks. Fed. Proc. 28: 1023-1028.
WEST, G. C. (1960): Seasonal variation in the energy balance of the tree sparrow in relation to migration. Auk, 77: 306-329.
WEST, G. C. (1965): Shivering and heat production in wild birds. Physiol. Zool., 38: 111-120.
WEST, G. C. and HART, J. S. (1966): Metabolic responses of evening grosbeaks to constant and fluctuating temperatures. Physiol. Zool. 39: 171-184.

Chapter 1, Section 9: Effects of Cold on Electrolyte Metabolism in Rodents and Laboratory Animals (W. S. PLATNER)

BADER, R. A., ELIOT, J. W. and BASS, D. E. (1952): Hormonal and renal mechanisms of cold diuresis. J. appl. Physiol., 4: 649-658.
BAKER, D. G. and SELLERS, E. A. (1957): Electrolyte metabolism in the rat exposed to a low environmental temperature. Canad. J. Biochem., 35: 631.
BASS, D. E. and HENSCHEL, A. (1956): Responses of body fluid compartments to heat and cold. Physiol. Rev., 36: 128-144.
BAUMAN, T. R. and TURNER, C. W. (1967): The effect of varying temperatures on thyroid activity and the survival of rats exposed to cold and treated with L-thyroxine or corticosterone. J. Endocr., 37: 355-359.
BRUCE, D. S. and WIEBERS, J. E. (1970): Calcium and phosphate levels in bats (*Myotis lucifungus*) as function of season and activity. Experientia (Basel), 26: 625-627.
CADOT, M., JULIEN, M. F. and CHEVILLARD, L. (1969): Estimation of thyroid function in rats exposed or adapted to environments at 5° or 30°C. Fed. Proc., 28: 228-233.
FENN, W. O. (1940): The role of potassium in physiological processes. Physiol. Rev., 20: 377-415.
FERRAND, R. L. (1959): Cold acclimatization in the golden hamster. Univ. Iowa Stud. nat. Hist., 20: 3-29.
FISHER, K. C. and MANERY, J. F. (1967): Water and electrolyte metabolism in heterotherms. In: Mammalian Hibernation, III. K. C. Fisher, A. R. Dowe, C. P. Lyman, E. Schönbaum and F. E. South (eds). Oliver and Boyd, Edinburgh: 235-279.
FISHER, K. C. and MROSOVSKY, N. (1970): Effectiveness of KCl and NaCl injections in arousing 13-lined ground squirrels from hibernation. Canad. J. Zool., 48: 595-596.
GALTON, V. A. and NISULA, B. C. (1969): Thyroxine metabolism and thyroid function in the cold-adapted rat. Endocrinology, 85: 79-86.
HALL, V. E. and WHALEN, W. J. (1951): Mechanism of magnesium poly prea in the urethanized cat. Fed. Proc., 10: 59.
HANEGAN, J. L. and WILLIAMS, B. A. (1973): Brain calcium: Role in temperature regulation. Science, 181: 663-664.
HANNON, J. P., LARSON, A. M., and YOUNG, D. W. (1958): Effect of cold acclimatization on plasma electrolyte levels. J. appl. Physiol., 13: 239.
HEAGY, F. C. and BURTON, A. C. (1948): Effect of intravenous injection of magnesium chloride on the body temperature of the unanaesthetized dog with some observations on magnesium levels and body temperature in man. Amer. J. Physiol., 152: 407.
HEGSTED, D. M., VITALE, J. J. and McGRATH, H. (1956): The effect of low temperature and dietary calcium upon magnesium requirement. J. Nutr., 58: 175.
HEMINGWAY, A., PRICE, W. M. and STUART, D. (1964): The calorigenic action of catecholamines in warm acclimated and cold acclimated non-shivering cats. Int. J. Neuropharmacol., 3: 495-503.
HEROUX, O and BRAUER, R. (1965): Critical studies on determination of thyroid secretion rate in cold-adapted animals. J. appl. Physiol., 20: 597-606.
HEROUX, O. and PETROVIC, V. M. (1969): Effect of high- and low-bulk diets on the thyroxine turnover rate in rats with acute and chronic exposure to different temperatures. Canad. J. Physiol., 47: 963-968.
HSIEH, A. C. L. (1966): Thyroid hormone requirement in rats exposed to cold. Gunma Symp. Endocri., 3: 239-248.
JANSKY, L., BARTUNKOVA, R. and ZEISBERGER, E. (1967): Acclimation of the white rat to cold: noradrenaline thermogenesis. Physiol. bohemoslov. 16: 366-372.
KNIGGE, K. M. (1963): Thyroid function and plasma binding during cold exposure of the hamster. Fed. Proc., 22: 755-760.
KOCKOVA, J. and JANSKY, L. (1968): Cold acclimation in the rabbit. Physiol. bohemoslov., 17: 309-316.
LACHIVER, F. and PETROVIC, V. (1963): Statement made in the discussion of paper presented by K. M. Knigge. Fed. Proc. int. Sym. Temp. Reg., 22: 760.
LE BLANC, J. and CÔTÉ, J. (1967): Increased vagal activity in cold-adapted animals. Canad. J. Physiol., 45: 745-48.
LECOQ, R., CHANCHARD, P. and MAZOUE, H. (1960): Protective and corrective effect

of Ca and Mg on nervous disequilibrium from thermoregulation. Ann. pharm. franc., 18-101-111.

LOCKER, A. and WEISH, P. (1966): Quantitative aspects of cold-adaptation and its thyroxine model in cold and warm-blooded animals. Helgoländer wiss. Meeresunters., 14: 503-513.

LUSTIG, B. T., ERNST, T. and REUSS, E. (1937): Die Zusammensetzung des Blutes von *Helix pomatia* bei Sommer- und Wintertieren. Biochem. Z., 290: 95-98.

MASORO, E. J. (1966): Effect of cold on the metabolic use of lipids. Physiol. Rev., 46: 67-101.

MEYER, D. K., WESTFALL, B. A. and PLATNER, W. S. (1956): Water and electrolyte balance of goldfish under conditions of anoxia, cold and inanition. Amer. J. Physiol., 184: 553-556.

NAGASAKA, T. and CARLSON, L. D. (1965): Responses of cold and warm adapted dogs to infused norepinephrine and acute body cooling. Amer. J. Physiol., 209: 227-230.

NEUBEISER, R. E., PLATNER, W. S. and SHIELDS, J. L. (1961): Magnesium in blood and tissues during cold acclimation. J. appl. Physiol., 16: 247-249.

NUNGESSER, W. S. (1955): Acute cold and renal function in dogs. Fed. Proc., 14: 108.

PATTON, J. F. and PLATNER, W. S. (1970): Cold acclimation and thyroxine effects on liver and liver mitochondrial fatty acirs. Amer. J. Physiol., 218: 1417-1422.

PATTON, J. F. and PLATNER, W. S. (1971): Effects of thyroxine and cold-acclimation on hepatic fatty acid metabolism in the hamster. Proc. Soc. exp. Biol., 137: 196-201.

PLATNER, W. S. (1952): Rate of change in concentration of serum magnesium upon cooling rewarming. Fed. Proc., 11: 124.

PLATNER, W. S. and HOSKO, M. F. (1953): Mobility of serum magnesium in hypothermia. Amer. J. Physiol., 174: 273-276.

PLATNER, W. S. (1960): Electrolytes in acclimation to cold. Fed. Proc., 19 :130-131.

PLATNER, W. S. (1970): Effects of low temperature on magnesium content of blood, body, fluids and tissues of goldfish and turtle. Amer. J. Physiol., 161: 399-405.

POHL, H. and HART, J. S. (1965): Thermoregulation and cold acclimation in a hibernator *Citellus tridecemlineatus*. J. appl. Physiol., 20: 398-404.

PROSSER, C. L. (1967): Molecular mechanisms of temperature adaptation. Publ. 84 Amer. Ass. Advanc. Sci. (Wash.).

RAO, K. P. (1967): Biochemical correlates of temperature acclimation. In: Molecular Mechanisms of Temperature Adaptation. C. Ladd Prosser (ed.). Proc. Symp. A.A.A.S. Berkeley. California. 1965. Publ. 84 Amer. Ass. Advanc. Sci. (Wash.).

SOTHER, B. T. and ROGERS, T. A. (1969): Studies on rat water and electrolyte balances during acute fasting and cold exposure; including estimations of changes inextracellular fluids. Bull. New Jersey Acad. Sci. Sym. Influence of Cold on Metabolic Regulation, March 1969.

SUOMALAINEN, P. N. (1938): Magnesium and calcium content of hedgehog serum during hibernation. Nature (Lond.) 141: 471.

UMMINGER, B. L. and GIST, D. (1972): Effects of thermal acclimation on physiological responses to handling stress, cortisol and aldosterone injections in the goldfish *Carassius auratus*.

WILLIAMS, D. D. (1962): Magnesium, calcium and water in peripheral tissues of the cold exposed rat. M.S. Thesis Univ. Missouri, Columbia.

WILLIS, J. S. (1972): The possible roles of cellular K for survival of cells at low temperature. Cryobiology, 9: 351-366.

WOODS, R. and CARLSON, L. D. (1956): Thyroxine secretion in rats exposed to cold. Endocrinology, 59: 323-330.

Chapter 1, Section 10: Effects of Heat on Vitamin Metabolism (M. L. SCOTT)

AHMAD, M. M., MORENG, R. E. and MULLER, H. D. (1967): Breed responses in body temperature to elevated environmental temperature and ascorbic acid. Poultry Sci., 46: 6-15.

ANDERSON, T. A., HUBBERT, F., Jr. and ROUBICEK, C. B. (1964): Effect of thyroxine, thiouracil and ambient temperature on the utilization of vitamin A by vitamin A-deficient rats. J. Nutr., 82: 457-462.

ASCARELLI, I. and BARTOV, I. (1963): Vitamin A requirement of chicks at moderately elevated temperature. Poultry Sci., 42: 232-235.
ATAKHANOV, E. I. (1962): The effect of chronic intestinal disease on the blood system. Proc. 8th int. Congr. Hematol., Tokyo, 2: 1145-1150.
ATAKHANOV, E. I. (1963): Enterocolitis in hot climate and associated changes in some metabolic indices. Vestn. Akad. med. Nauk, 18: 42-48.
BAKER, S. J. (1967): Human vitamin B_{12} deficiency. Wld Rev. Nutr. Diet., 8: 62-126.
BAKER, S. J. and MATHAN, V. I. (1972): Tropical enteropathy and tropical sprue. Amer. J. clin. Nutr., 25: 1047-1055.
BENNION, N. L. and WARREN, D. C. (1933): Temperature and its effect on egg size in the domestic fowl. Poultry Sci., 12: 69-82.
BISWAS, N. M. (1970): Histochemical changes in tests of the heat-exposed toad. Acta histochem. (Jena), 37: 363-368.
BLANC, B. and VON DER MUEHLL, M. (1967): Interaction of AP factor (flavonoid) and of vitamin C, its effect on guinea pig weight and the vitamin C content of its organs-effect of thermal variations. Int. Z. Vitamin-forsch., 37: 145-169.
BRAEKKAN, O. R., INGEBRIGSTEN, O. and MYKLESTAD, H. (1969): Uptake and storage of vitamin A in rainbow trout. Int. J. Vit. Res., 39: 121-130.
BRANT, A. W., OTTE, A. W. and CHIN, C. (1953): A survey of egg quality at two egg-laying tests. Techn. Bull. No. 1066, USDA, Beltsville.
CHAHL, J. C. and KRATZING, C. C. (1966): Environmental temperature and choline requirement in rats. I. Choline deficiency in rats at various temperatures. J. Lipid Res., 7: 17-21.
CHAHL, J. S. and KRATZING, C. C. (1966): Environmental temperature and choline requirements in rats. II Choline and methionine requirements for lipotropic activity. J. Lipid Res., 7: 22-26.
CHANARIN, I. and PERRY, J. (1970): Exchange between parenteral and tissue folate in man. Brit. J. Haemat., 18: 691-698.
CONRAD, R. M. (1939): The effect of high temperature on the blood calcium of the laying hen. Poultry Sci., 18: 327-329.
DLUZNIEWSKA, K. (1965): Studies on vitamin C metabolism in workers employed in hot industrial departments. I. Folia Med. Cracov., 7: 199-207.
DLUZNIEWSKA, K., OBTULOWICZ, A. and SIENKO, S. (1965): Studies on vitamin C metabolism in workers employed in hot industrial departments. II. Folia med. (Cracov.), 7: 209-219.
EL-BOUSHY, A. R., SIMONS, P. C. M. and WIERTZ, G. (1968): Structure and ultrastructure of the hen's egg shell as influenced by environmental temperature, humidity and vitamin C additions. Poultry Sci., 47: 456-467.
FREEMAN, B. M. (1963): Gaseous metabolism of the domestic chicken. IV. The effect of temperature on the resting metabolism of the fowl during the first month of life. Brit. Poultry Sci., 4: 275-278.
HALTMEYER, G. C., DENENBERG, V. H., THATCHER, J. and ZARROW, M. X. (1966): Response of the adrenal cortex of the neonatal rat after subjection to stress. Nature (Lond.), 212: 1371-1373.
HEITMAN, H. and HUGHES, E. H. (1949): The effects of air temperature and relative humidity on the physiological well-being of swine. J. Anim. Sci., 8: 171.
HEYWANG, B. W. (1952): The level of vitamin A in the diet of laying and breeding chickens during hot weather. Poultry Sci., 31: 294-301.
HEYWANG, B. W. and KEMMERER, A. R. (1955): The effect of procaine penicillin and ascorbic acid on egg weight and shell thickness during hot weather. Poultry Sci., 34: 1032-1036.
HINDSON, T. C. (1970): Ascorbic acid status of Europeans resident in the tropics. Brit. J. Nutr., 24: 801-802.
HUNT, J. R. and AITKIN, J. R. (1962): Studies on the influence of ascorbic acid on shell quality. Poultry Sci., 41: 219-226.
HUSTON, T. M. and CARMON, J. L. (1961): The influence of high environmental temperature on specific gravity and albumen quality of hen eggs. Poultry Sci., 40: 1060-1062.
ITOKAWA, Y. and COOPER, J. R. (1970): Ion movements and thiamine. II. The release of the vitamin from membrane fragments. Biochim. biophys. Acta (Amst.), 196: 274-284.

JACKSON, G. L. and NALBANDOV, A. V. (1969): Ovarian ascorbic acid depleting factors in the chicken hypothalamus. Endocrinology, 85: 113-120.

JEEJEEBHOY, K. N., SAMUEL, A. M., SINGH, B., NADKARNI, G. D., DESAI, H. G., BORKAR, A. V. and MANI, L. S. (1969): Metabolism of albumin and fibrinogen in patients with tropical sprue. Gastroenterology, 56: 252-267.

JOSLYN, M. A. and LEICHTER, J. (1968): Thiamine instability in experimental wet diets containing commercial casein with sulfur dioxide. J. Nutr., 96: 89-93.

KURNICK, A. A., HEYWANG, B. W., HULETT, B. J., VAVICH, M. G. and REID, B. L. (1964): The effect of dietary vitamin A, ambient temperature and rearing location on growth, feed conversion and vitamin liver storage of White Leghorn pullets. Poultry Sci., 43: 1582-1586.

KUROKAWA, G. (1941): Relation between the high temperature environment and vitamin A. I. Influence of high temperature environment upon the appearance of vitamin A avitaminosis. Kokumin. Esei, 18: 123-134; 135-155.

LYLE, G. R. and MORENG, R. E. (1968): Elevated environmental temperature and duration of post exposure ascorbic acid administration. Poultry Sci., 47: 410-417.

MANGOLD, D. W., HAZEN, T. E. and HAYS, V. W. (1967): Effect of air temperature on performance of growing-finishing swine. Trans. ASAE, 10: 370.

McGILLIVRAY, W. A. (1957): Factors influencing the vitamin A content of milk fat. III. Seasonal changes in the vitamin A and carotenoid content of the blood plasma and milk fat of cows on pasture. J. Dairy Res., 24: 95-101.

McGILLIVRAY, W. A. (1959): Factors influencing the vitamin A content of milk fat. N. Z. J. agric. Res., 2: 694-701.

MILLS, C. A., COTTINGHAM, E. and TAYLOR, E. (1947): Influence of environmental temperature on dietary requirements for thiamine, pyridoxine, nicotinic acid, folic acid and choline in chicks. Amer. J. Physiol., 149: 376-382.

MUELLER, W. J. (1959): The effect of environmental temperature and humidity on the calcium balance and serum calcium of laying pullets. Poultry Sci., 38: 1296-1301.

NESTOR, K. E., TOUCHBURN, S. P. and TREIBER, M. (1972): The influence of dietary ascorbic acid on blood ascorbic acid level and egg production in turkeys. Poultry Sci., 51: 1676-1680.

NIR, I. and ASCARELLI, I. (1966): Effect of dietary protein and level of thyroxine on vitamin A depletion from liver of chicks. Brit. J. Nutr., 20: 41-53.

NIR, I. and ASCARELLI, I. (1966): Seasonal influence on vitamin A alcohol concentration in chick liver. Brit. Poultry Sci., 7: 297-300.

NOGUCHI, T., CANTOR, A. H. and SCOTT, M. L. (1973): Mode of action of selenium and vitamin E in prevention of exudative diathesis in chicks. J. Nutr., 103: 1502-1511.

PAGE, H. M., ERWIN, E. S. and NELMS, G. E. (1959): Effect of heat and solar radiation on vitamin A utilization by the bovine animal. Amer. J. Physiol., 196: 917-918.

PANDERS, J. T. and RUPERT, M. S. E. (1965): The effect of temperature on folic acid metabolism. Brit. J. Haematol., 11: 518-524.

PEREK, M. and ECKSTEIN, B. (1959): The adrenal ascorbic acid content of molting hens and the effect of ACTH on the adrenal ascorbic acid content of laying hens. Poultry Sci., 38: 996-999.

PEREK, M. and BEDRAK, E. (1962): The effect of cold and debeaking upon the adrenal ascorbic acid concentration of chickens fed aureomycin supplement. Poultry Sci., 41: 1149-1156.

PEREK, M. and KENDLER, J. (1962): Vitamin C supplementation to hen's diets in a hot climate. Poultry Sci., 41: 677-678.

PEREK, M. and KENDLER, J. (1963): Ascorbic acid as a dietary supplement for White Leghorn hens under conditions of climatic stress. Brit. Poultry Sci., 4: 19-26.

RADOMSKI, M. W. and ORME, T. (1967): Effect of environmental temperature on fatty livers produced by various hepatotoxic agents in rats. J. Nutr., 92: 19-22.

RIKER, J. T., III, PERRY, T. W., PICKETT, R. A. and HEIDENREICH, C. J. (1967): Influence of controlled temperatures on growth rate and plasma ascorbic acid values in swine. J. Nutr., 92: 99-103.

RUMSEY, G. L. (1969): Studies on the effects of simulated stress and ascorbic acid upon avian adreno-cortical function and eggshell formation. Ph.D. Thesis, Cornell Univ.

SARGENT, F., ROBINSON, P. and JOHNSON, R. E. (1944): Water-soluble vitamins in sweat. J. biol. Chem., 153: 285.

SCOTT, M. L. and COMBS, Jr., G. F. (1973): The effect of vitamin D on eggshell quality. In: Proc. Maryland Nutr. Conf.: 44-48.
SELLERS, E. A. and YOU, S. S. (1952): Effects of cold environment on deposition of fat in the liver in choline deficiency. Biochem. J., 51: 573.
SELYE, H. (1950): The Physiology and Pathology of Exposure to Stress. Acts, Inc., Montreal, 1025 pp.
SEYMOUR, E. W., SPEER, V. C. and HAYS, V. W. (1968): Effect of environmental temperature on the riboflavin requirement of young pigs. J. Anim. Sci., 27: 389-393.
SHANNON, D. W. F. and BROWN, W. O. (1969): The period of adaptation of the fasting metabolic rate of the common fowl to an increase in environmental temperature from 22°C to 28°C. Brit. Poultry Sci., 10: 13-18.
SHIELDS, J. B., JOHNSON, B. C., HAMILTON, T. S. and MITCHELL, H. H. (1945): The excretion of ascorbic acid and dehydroascorbic acid in sweat and urine under different environmental conditions. J. biol. Chem., 161: 351.
SIDDIQUI, H. H., ARORA, R. B. and MALHOTRA, N. K. (1972): Effect of hypoxia on blood glucose, liver glycogen and adrenal ascorbic acid in mice at varying environmental temperature. Indian J. med. Res., 60: 153-158.
SILBER, R., WINTER, R. and KAYDEN, H. J. (1969): Tocopherol transport in the rat erythrocyte. J. clin. Invest., 48: 2089-2095.
SMIRNOVA, A. N. and GOLIKOV, K. K. (1972): Retinol requirements of sailors in hot climates. Rus. Vop Pitan., 31(1): 55-57.
SMITH, J. E. and BORCHERS, R. (1972): Environmental temperature and the utilization of β-carotene by the rat. J. Nutr., 102: 1017-1024.
SQUIBB, R. L., GUZMAN, M. A. and SCRIMSHAW, N. S. (1954a): Effect of high environmental temperatures. I. Growth and blood constituents of rats exposed to 94 F for 72 hours. J. Nutr., 52: 155-163.
SQUIBB, R. L., GUZMAN, M. A. and SCRIMSHAW, N. S. (1954b): Relation of high environmental temperature, starvation and coryza to several blood constituent levels of New Hampshire chickens. Fed. Proc., 13: 478.
SQUIBB, R. L., BRAHAM, J. E. and ARROYAVE, G. (1958): Rate of vitamin A absorption in New Hampshire laying hens subjected to 20° and 36°C temperature. Poultry Sci., 37: 932-934.
SULLIVAN, T. W. and KINGAN, J. R. (1962): Effect of dietary calcium level, calcium lactate and ascorbic acid on the egg formation of Single Comb White Leghorn hens. Poultry Sci., 41: 1596-1602.
THORNTON, P. A. and MORENG, R. E. (1958): The effect of ascorbic acid on egg quality factors. Poultry Sci., 37: 691-698.
THORNTON, P. A. and MORENG, R. E. (1959): Further evidence on the value of ascorbic acid for maintenance of shell quality in warm environmental temperature. Poultry Sci., 38: 594-599.
THORNTON, P. A. and DEEB, S. S. (1961), The influence of thyroid regulators on blood ascorbic acid levels in the chicken. Poultry Sci., 40: 1063-1067.
TSEITINA, A. Ya. (1965): Effects of vitamin P on the ascorbic acid metabolism in rats during long-term exposure to high temperatures. Vop. Pitan., 24: 35-40.
UMANSKY, S. Sl. (1972): Seasonal variations in the acidity of gastric juice, uropepsin and of figures reflecting partial deficiency of A, C and B-complex vitamins in healthy subjects. Vop. Pitan., 31(1): 35-38.
WALLBACK, D. P. and REINEKE, E. P. (1949): The effect of varying levels of thyroidal stimulation on the ascorbic acid content of the adrenal cortex. Endocrinology 45: 75-81.
WARREN, D. C. and SCHNEPEL, R. L. (1940): The effect of air temperatures on egg shell thickness in the fowl. Poultry Sci., 19: 67-72.
WASSERMAN, R. H. and TAYLOR, A. N. (1969): Some aspects of the intestinal absorption of calcium, with special reference to vitamin D. In: Mineral Metabolism, 3. C. L. Comar and F. Bronner (eds). Academic Press, New York: 321-403.
WILHELM, L. A. (1940): Some factors affecting variations in egg shell quality. Poultry Sci., 19: 246-253.
ZARROW, M. X., HALTMEYER, G. C., DENENBERG, V. H. and THATCHER, J. (1966): Endocrinology, 79: 631-634.

Chapter 1, Section 11a (1): The Redistribution of Cardiac Output in Animals During Heat Stress. (J. R. S. HALES)

ALEXANDER, R. S. (1963): The peripheral venous system. In HANDBOOK OF PHYSIOLOGY, Sect. 2, CIRCULATION, vol. II. W. F. Hamilton and P. Dow (eds). Amer. physiol. Soc., (Wash.): 1075-1098.

ALEXANDER, G., BELL, A. W. and HALES, J. R. S. (1973): Effect of cold exposure on tissue blood flow in the new-born lamb. J. Physiol. (Lond.), 234: 65-77.

BIANCA, W. (1965): Reviews of the progress of dairy science. Section A. Physiology. Cattle in a hot environment. J. Dairy Res., 32: 291-345.

BLAXTER, K. L. and WAINMAN, F. W. (1961): Environmental temperature and the energy metabolism and heat emission of steers. J. agric. Sci. (Camb.), 56: 81-90.

BURNSTOCK, G. and PROSSER, C. L. (1960): Responses of smooth muscles to quick stretch; relation of stretch to conduction. Amer. J. Physiol., 198: 921-925.

DAMPNEY, R. A. L. (1974): Cardiovascular alterations associated with bursts of panting in the exercising dog. J. Physiol. (Lond.), in press.

EDHOLM, O. G., FOX, R. H. and MacPHERSON, R. K. (1956): The effect of body heating on the circulation in skin and muscle. J. Physiol. (Lond.), 134: 612-619.

FOX, R. H., GOLDSMITH, R., KIDD, D. J. and LEWIS, H. E. (1963): Blood flow and other thermoregulatory changes with acclimatization to heat. J. Physiol. (Lond.), 166: 548-562.

HALES, J. R. S. (1973a): Effects of exposure to hot environments on the regional distribution of blood flow and on cardiorespiratory function in sheep. Pflügers Arch. ges. Physiol., 344: 133-148.

HALES, J. R. S. (1973b): Radioactive microsphere evaluation of thermal influences on blood flow through arteriovenous anastomoses. Proc. Aust. physiol. pharmacol. Soc., 4, 1: 55.

HALES, J. R. S. (1973c): Effects of heat stress on blood flow in respiratory and non-respiratory muscles in the sheep. Pflügers Arch. ges. Physiol. 345: 123-130.

HALES, J. R. S. (1974): Physiological responses to heat. In: MTP International Review of Science, Physiology. Ser. 1, Vol. 7. D. Robertshaw (ed.). Butterworths Publ. Co., London: 107-162.

HALES, J. R. S. and DAMPNEY, R. A. L. (1975): The redistribution of cardiac output in the dog during heat stress. J. of therm. Biol. 1, in preparation.

HAMMEL, H. T., WYNDHAM, C. H. and HARDY, J. D. (1968): Heat production and heat loss in the dog at 8-36°C environmental temperature. Amer. J. Physiol., 194: 99-108.

HUDLICKÁ, O., PETTE, D. and STAUDTE, H. (1973): The relation between blood flow and enzymatic activities in slow and fast muscles during development. Pflügers Arch. ges. Physiol., 343: 341-356.

INGRAM, D. L. and WEAVER, M. E. (1969): Quantitative study of the blood vessels of the pig's skin and the influence of environmental temperature. Anat. Rec., 163: 517-524.

INGRAM, D. L. and LEGGE, K. F. (1971): The influence of deep body temperatures and skin temperatures on peripheral blood flow in the pig. J. Physiol. (Lond.), 215: 693-707.

IRIKI, M. (1968): Änderung der Hautdurchblutung bei unnarkotisierten Kaninchen durch isolierte Wärmung des Rückenmarkes. Pflügers Arch. ges. Physiol., 299: 295-310.

IRIKI, M., RIEDEL, W. and SIMON, E. (1971): Regional differentiation of sympathetic activity during hypothalamic heating and cooling in anesthetized rabbits. Pflügers Arch. ges. Physiol., 328: 320-331.

KINDERMANN, W. and PLESCHKA, K. (1973): Local blood flow and metabolism of the tongue before and during panting in the dog. Pflügers Arch. ges. Physiol., 340: 251-262.

KULLMANN, R., SCHÖNUNG, W. and SIMON, E. (1970): Antagonistic changes of blood flow and sympathetic activity in different vascular beds following central thermal stimulation I. Blood flow in skin, muscle and intestine during spinal cord heating and cooling in anaesthetized dogs. Pflügers Arch. ges. Physiol., 319: 146-161.

MOLYNEUX, G. D. (1965): Observations on the structure, distribution and significance of arterio-venous anastomoses in sheep skin. In: Biology of the Skin and Hair Growth. A. G. Lyne and B. F. Short (eds). Angus and Robertson, Sydney: 591-602.

ROBINSON, S. (1963): Circulatory adjustments of men in hot environments. In: Temperature, Its Measurement and Control in Science and Industry, 3, 3. J. D. Hardy (ed.). Reinhold Publ. Co., New York/London: 287-297.

ROWELL, L. B., DETRY, J. M., PROFANT, G. E. and WYSS, C. (1971): Splanchnic vasoconstriction in hyperthermic man-role of falling blood pressure. J. appl. Physiol., 31: 864-869.
SCHMIDT-NIELSEN, K. and PENNYCUIK, P. (1961): Capillary density in mammals in relation to body size and oxygen consumption. Amer. J. Physiol., 200: 746-750.
SCHÖNUNG, W., WAGNER, H., JESSEN, C. and SIMON, E. (1971): Differentiation of cutaneous and intestinal blood flow during hypothalamic heating and cooling in anaesthetized dogs. Pflügers Arch. ges. Physiol., 328: 145-154.
SHEPHERD, J. T. and WEBB-PEPLOE, M. M. (1970): Cardiac output and blood flow distribution during work in heat. In: Physiological and Behavioral Temperature Regulation. J. D. Hardy, A. P. Gagge and J. A. J. Stolwijk (eds). C. C. Thomas, Springfield, Ill.: 237-253.
SMITH, J. H., ROBINSON, S. and PEARCHY, M. (1952): Renal responses to exercise heat and dehydration. J. appl. Physiol., 4: 659-665.
THAUER, R. (1965): Circulatory adjustments to climatic requirements. In: Handbook of Physiology, Sect. 2, Circulation, III. W. F. Hamilton and P. Dow (ed.). Amer. physiol. Soc., (Wash.): 1921-1966.
WALTHER, O.-E., IRIKI, M. and SIMON, E. (1970): Antagonistic changes of blood flow and sympathetic activity in different vascular beds following central thermal stimulation 11. Cutaneous and visceral sympathetic activity during spinal cord heating and cooling in anaesthetized rabbits and cats. Pflügers Arch. ges. Physiol., 319: 162-184.
WHITNEY, R. J. (1954): Circulatory changes in the forearm and hand of man with repeated exposure to heat. J. Physiol. (Lond.), 125: 1-24.
WHITTOW, G. C. (1962): The significance of the extremities of the ox (*Bos taurus*) in thermoregulation. J. agric. Sci. (Camb.), 58: 109-120.
WHITTOW, G. C. (1965): The effect of hyperthermia on the systemic and pulmonary circulation of the ox (*Bos taurus*). Quart. J. exp. Physiol., 50: 300-311.
WHITTOW, G. C. (1968): Cardiovascular response to localised heating of the anterior hypothalamis, J. Physiol. (Lond.), 198: 541-549.
WOOD, J. E. and BASS, D. E. (1960): Responses of the veins and arterioles of the forearm to walking during acclimatization to heat in man. J. clin. Invest., 39: 825-833.
WYNDHAM, C. H. (1951): Effect of acclimatization on circulatory responses to high environmental temperatures. J. appl. Physiol., 4: 383-395.

Chapter 1, Section 11a (2): Effects of Heat Stress on the Central Circulation in Mammals (G. C. WHITTOW)

DUKE, H. N., GREEN, J. H., HEFFRON, P. F. and STUBBENS, V. W. J. (1963): Pulmonary chemoreceptors. Quart. J. exp. Physiol. 48: 164-175.
FINDLAY, J. D. and WHITTOW, G. C. (1966): The role of arterial oxygen tension in the respiratory response to localized heating of the hypothalamus and to hyperthermia. J. Physiol. (Lond.), 186: 333-346.
HALES, J. R. S. (1973): Effects of exposure to hot environments on the regional distribution of blood flow and on cardiorespiratory function in sheep. Pflügers Arch. ges. Physiol., in press.
HALES, J. R. S. (1974): Regional tissue blood flow. Chapter this volume.
INGRAM, D. L. and WHITTOW, G. C. (1962): The effects of variations in respiratory activity and in the skin temperatures of the ears on the temperature of the blood in the external jugular vein of the ox (*Bos taurus*). J. Physiol. (London.), 163: 211-221.
INGRAM, D. L. and WHITTOW G. C. (1963): Changes of arterial pressure and heart rate in the ox (*Bos taurus*) with changes of body temperature. J. Physiol. (Lond.), 168: 736-746.
JENNINGS, D. B. and SPARLING, J. (1973): Effects of low O_2 and high CO_2 on cardiorespiratory function in conscious resting dogs. Amer. J. Physiol., in press.
JENNINGS, D. B., PHILLIPS, H. H., CHEN, C. C. and SPARLING, J. (1973): Cardiovascular function in panting and nonpanting resting conscious dogs. Amer. J. Physiol., 224: 1059-1065.
LIN, Y. C., MATSUURA, D. T. and WHITTOW, G. C. (1972): Respiratory variation of heart rate in the California sea lion. Amer. J. Physiol., 222: 260-264.

MATSUURA, D. T. and WHITTOW, G. C. (1973): Oxygen uptake of the California Sea Lion and Harbor Seal during exposure to heat. Amer. J. Physiol., 225: 711-715.
PHILLIPS, H. H. and JENNINGS, D. B. (1973): Cardio-respiratory effects of hypothalamic heating in conscious dogs. Amer. J. Physiol., in press.
ROBERTSHAW, D. and WHITTOW, G. C. (1966): The effect of hyperthermia and localized heating of the anterior hypothalamus on the sympatho-adrenal system of the ox (*Bos taurus*). J. Physiol. (Lond.), 187: 351-360.
WHITTOW, G. C. (1962): The significance of the extremities of the ox (*Bos taurus*) in thermoregulation. J. agric. Sci. (Camb.), 58: 109-120.
WHITTOW, G. C. (1965): The effect of hyperthermia on the systemic and pulmonary circulation of the ox (*Boc taurus*). Quart. J. exp. Physiol., 50: 300-311.
WHITTOW, G. C. and ROBERTSHAW, D. (1965): The use of an exteriorized carotid artery for cardiovascular studies in the ox (*Bos taurus*). Amer. J. vet. Res., 26: 183-186.
WHITTOW, G. C. (1966): Cardiovascular responses of the ox (*Bos taurus*) to hyperthermia. Biometeorology, 2, 1. S. W. Tromp and W. H. Weihe (eds). Pergamon Press, Oxford: 460-463.
WHITTOW, G. C. (1967): Climatic effects on physiological functions. In: Ground Level Climatology. R. H. Shaw (ed.). Amer. Ass. Advanc. Sci.
WHITTOW, G. C. (1968a): Cardiovascular response to localized heating of the anterior hypothalamus. J. Physiol. (Lond.), 198: 451-548.
WHITTOW, G. C. (1968b): Cardiovascular regulation. In: Adaptation of Domestic Animals. E. S. E. Hafez (ed.), Lea and Febiger, Pa.
WHITTOW, G. C. and FINDLAY, J. D. (1968): Oxygen cost of thermal panting. Amer. J. Physiol., 214: 95-99.
WHITTOW, G. C. (1970): Respiratory variations of blood pressure in the ox (*Bos taurus*) during panting. Brit. vet. J., 126: 652-655.
WHITTOW, G. C. (1971a): Cardioacceleration in the ox (*Bos taurus*) during hyperthermia. Res. vet. Sci., 12: 495-496.
WHITTOW, G. C. (1971b): Ungulates. In: Comparative Physiology of Thermoregulation, 2. G. C. Whittow (ed). Academic Press, New York.

Chapter 1, Section 11a (3): Effects of Heat Stress on Cardiovascular Responses in Birds (P. C. HARRISON)

ABRAHAM, A. (1969): Microscopic Innervation of the Heart and Blood Vessels in Vertebrates Including Man. Pergamon Press, Oxford.
BRODY, S. (1945): Bioenergetics and Growth. Reinhold Publ. Co., New York, Chapt. 11: 281.
BRUSH, A. H. (1965): Energetics, temperature regulation and circulation in resting, active and defeathered California quail, *Lophortyx californicus*. Comp. Biochem. Physiol., 15: 399-421.
CHILGREN, J. D. and KING, J. R. (1973): Thermoregulatory responses of Japanese quail to various air temperatures and humidities. Int. J. Biometeor., 17: 29-40.
FRANKEL, H., HOLLAND, K. and WEISS, H. S. (1962): Respiratory and circulatory responses of hypthermic chickens. Arch. int. Physiol., 70: 555-563.
HARRISON, P. C. and BIELLIER, H. V. (1969): Physiological response of domestic fowl to abrupt changes of ambient air temperature. Poultry Sci., 48: 1034-1045.
HARRISON, P. C. and BECKER, W. C. (1970): Vagal influence on heart rate of Coturnix quail in a hot environment. Poultry Sci., 49: 1393.
HARRISON, P. C., BECKER, W. C. and BECKER, W. A. (1972): Vagal influence on respiratory and cardiac responses of Coturnix quail to hot environments. Int. J. Biometeor., 16: 45-52.
HUDSON, J. W. and BRUSH, A. R. (1964): A comparative study of the cardiac and metabolic performance of the dove, *Zenaidura macroura* and the quail *Lophortyx californicus*. Comp. Biochem. Physiol., 12: 157-170.
HUTCHINSON, J. C. D. and SYKES, A. H. (1953): Physiological acclimatization of fowls to a hot humid environment. J. agric. Sci. (Camb.), 43: 294-321.
JONES, D. R. and JOHANSEN, K. (1972): Avian Biology, 2. D. S. Farner and J. R. Kind (eds). Academic Press, New York/London, Chapt. 4: 157.

LINSLEY, J. G. and BRUGER, R. E. (1964): Respiratory and cardiovascular responses in the hypthermic domestic cock. Poultry Sci., 43: 291-305.
MORISHIMA, M. S. and GALE, C. C. (1972): Relationship of blood pressure and heart rate to body temperature in baboons. Amer. J. Physiol., 223: 387-395.
PEARSON, R. G. (1972): The Avian Brain. Academic Press, New York/London.
RODBARD, S. and TOLPIN, M. (1947): A relationship between body temperature and blood pressure in the chicken. Amer. J. Physiol., 151: 509-515.
SCHWALBE, C. R. (1973): Physiological responses and cost of thermal panting of Japanese quail to a hot environment. M.S. Thesis, Wash. State Univ.
SPIELMAN, R. S. and LYMAN, C. P. (1971): Thermal bradycardia in the mildly stressed rat. Amer. J. Physiol., 221: 948-951.
STURKIE, P. D. (1967): Cardiovascular effects of acclimatization to heat and cold in chickens. J. appl. Physiol., 22, 1: 13-15.
STURKIE, P. D., LIN, Y. and OSSORIO, N. (1970): Effects of acclimatization to heat and cold on heart rate in chickens. Amer. J. Physiol., 219: 34-36.
THAUER, R. (1965): In: Handbook of Physiology. Sect. 2. Circulation, III. W. F. Hamilton and P. Dow (eds). Amer. physiol. Soc. (Wash.), Chapt. 55: 1921.
VOGEL, J. A. and STURKIE, P. D. (1963): Cardiovascular responses of the chicken to seasonal and induced temperature changes. Science, 140: 1404-1406.
WEISS, H. S., FRANKEL, H. and HOLLANDS, K. G. (1963): The effect of extended exposure to a hot environment on response of the chicken to hyperthermia. Canad. J. Biochem., 41: 805-815.
WHITTOW, G. C., STURKIE, P. D. and STEIN, Jr., G. (1964): Cardiovascular changes associated with thermal polypnea in the chicken. Amer. J. Physiol., 207: 1349-1353.
WHITTOW, G. C., STURKIE, P. D. and STEIN, Jr, G. (1966): Cardiovascular differences between cold-acclimatized and heat-acclimatized chickens. Res. vet. Med., 7: 296-301.
WILSON, W. O. (1949): High environmental temperature as affecting the reaction of laying hens to iodized casein. Poultry Sci., 28: 581-592.

Chapter 1, Section 11b (1): Effects of Cold Stress on Cardiovascular Responses in Mammals (G. E. FOLK, Jr.)

FOLK, G. E., Jr. and HEDGE, R. S. (1964): Comparative physiology of heart rate of unrestrained mammals. Amer. Zool., 4: 43.
FOLK, G. E., Jr., COADY, K. B. and FOLK, M. A. (1968): Physiological observations on raccoons in winter. Iowa Acad. Sci., 75: 301-305.
FOLK, G. E., Jr., FOLK, M. A. and KREUZER, F. (1970): Initial stages of hibernation: Is cold acclimatization necessary. Acta Theriol., 15: 373-380.
FOLK, G. E., Jr. and COPPING, J. R. (1972): Telemetry in animal Biometeorology. In: Biometeorology, 5, 2. S. W. Tromp, W. H. Weihe and J. J. Bouma (eds). Proc. 6th int. Congr. Biometeor.: 153-170.
FOLK, G. E., Jr. (1974): Textbook of Environmental Physiology. Lea & Febiger, Philadelphia, 465 pp.
HAMLIN, R. L. et al. (1970): Atrial activation paths and P waves in horses. Amer. J. Physiol., 219: 306-309.
JOHANSEN, K. (1965): Cardiovascular dynamics in fish, reptiles, and amphibians. Ann. N.Y. Acad. Sci., 127: 414-442.
LYMAN, C. P. (1958): Oxygen consumption, body temperature, and heart rate of woodchucks entering hibernation. Amer. J. Physiol., 194: 83-91.
MAGALHAES, H. (ed.) (1974): Environmental Variables in Animal Experimentation. Bucknell Univ. Press, Lewisberg, 147 pp.
MÜHLBOCK, O. and NOMURA, T. (eds) (1966): Experimental Animals in Cancer Research. Maruzen Co. Ltd., Tokyo.

Chapter 1, Section 11b (2): Effects of Cold Stress on Cardiovascular Responses in Birds (P. C. HARRISON)

BARTHOLOMEW, G. A. and DAWSON, W. R. (1954): Body temperature in the morning dove *Zenaidura macroura marginella*. Ecology, 35: 2: 181-187.

BRUSH, A. H. (1965): Energetics, temperature regulation and circulation in resting, active and defeathered California quail, *Lophortyx californicus*. Comp. Biochem. Physiol., 15: 399-421.

DAWSON, W. R. and HUDSON, J. W. (1970): Comparative Physiology of Thermoregulation. Vol. I. Invertebrates and Nonmammalian Vertebrates (G. C. Whittow, ed.). Academic Press, New York/London, Chapt. 6: 240-242.

GRANT, R. T. and BLAND, E. F. (1931): Observations on arteriovenous anaestomosis in human skin and in the bird's foot with special reference to the reaction to cold. Heart, 15: 385-411.

HARRISON, P. C. and BIELLIER, H. V. (1969): Physiological response of domestic fowl to abrupt change in ambient air temperature. Poultry Sci., 48: 1034-1045.

HUDSON, J. W. and BRUSH, A. R. (1964): A comparative study of the cardiac and metabolic performance of the dove, *Zenaidura macroura* and the quail, *Lophortyx californicus*. Comp. Biochem. Physiol., 12: 157-170.

IRVING, L. and KROG, J. (1954): Temperature of skin in the arctic as a regulator of heat. J. appl. Physiol., 7: 355-364.

JOHANSEN, K. and TØNNESEN, K. H. (1969): Blood flow in the interdigital web of sea gull at low temperatures. Acta physiol. scand., 76: 21-22 (Abstract).

JONES, D. R. and JOHANSEN, K. (1972): Avian Biology, Vol. II (D. S. Farner and J. R. King, eds.). Academic Press, New York/London, Chapt. 4: 264-270.

LEBLANC, J. and COTE, J. (1967): Increased vagal activity in cold-adapted animals. Canad. J. Physiol., 45: 745-748.

LEWIS, T. (1930): Observations on the reactions of the vessels of the human skin to cold. Heart, 15: 177-208.

RODBARD, S. and TOLPIN, M. (1947): A relationship between body temperature and blood pressure in the chicken. Amer. J. Physiol., 151: 509-515.

ROSE, G. (1961): Seasonal variation in blood pressure in man. Nature (Lond.), 189: 235.

SCHOLANDER, P. F. (1958): Counter current exchange. A principle in biology. Hvalrad. Skr. Nar. Vindenaki.-Akad. Oslo, 44: 1-24.

STEEN, I. and STEEN, J. B. (1965): The importance of the leg in the thermoregulation of birds. Acta physiol. scand., 63: 285-291.

STURKIE, P. D., LIN, Y. and OSSORIO, N. (1970): Effect of acclimatization to heat and cold on heart rate in chickens. Amer. J. Physiol., 219: 34-36.

THAUER, R. (1965): In: Handbook of Physiology. Sect. 2, Circulation. III. W. F. Hamilton and P. Dow (eds). Amer. physiol. Soc. (Wash.), Chapt. 54: 1909-1910.

VOGEL, J. A. and STURKIE, P. D. (1963): Cardiovascular responses of the chicken to seasonal and induced temperature changes. Science, 140: 1404-1406.

WEISS, H. S. (1960): Cardiovascular reactions to immersion hypothermia with a note on prefeeding reserpine. Poultry Sci. 39: 1304 (Abstract).

WEISS, H. S., FISHER, H. and GIMINGER, P. (1961). Seasonal changes in avian blood pressure related to age, sex, diet, confinement and breed. Amer. J. Physiol., 201: 655-659.

WHITTOW, G. C., STURKIE, P. D. and STEIN, G. Jr., (1966): Cardiovascular differences between cold-acclimatized and heat-acclimatized chickens. Res. vet. Med., 7: 296-301.

WILSON, W. O., HILLERMAN, J. P. and EDWARDS, W. H. (1952): The relationship of high environmental temperature to feather and skin temperature of laying pullets. Poultry Sci., 31: 843-846.

Chapter 1, Section 12a: Effects of Temperature on the Growth of Cattle (D. F. DOWLING)

BERG, R. T. (1969): Cattle Breeding Plan. Feeders Day Rep., Univ. Alberta, 45: 23-28.

BIANCA, W. (1965): Cattle in a hot environment. J. Dairy Res. 32: 291-345.

BOND, T. W. (1974): The Influence of Environmental Factors on Non-Ruminant Production. In: Proc. 1st Int. Livestock Environment Symposium, Lincoln, Ne., in press.

BUTTERFIELD, R. M. and BERG, ROY (1966): A nutritional effect on relative growth of muscles. In: Proc. Aust. Soc. Anim. Prod., 6: 298.
CALLOW, E. H. (1944): The food value of beef from steers and heifers, and its relation to dressing-out percentage. J. agric. Sci. (Camb.), 34: 177-189.
CHARLES, D. D. (1974): A Method of Estimating Carcase Components in Cattle. Res. vet. Sci., 16: 89-94.
DOWLING, D. F. (1960): Comparative Growth Rates of Different Strains of Cattle in a Hot Tropical Environment. In: Proc. Aust. Soc. Anim. Prod., 3: 184-191.
DOWLING, D. F. (1974): Biometeorology: as it relates to Beef Production in Tropical Australia. Brit. vet. J., in press.
HAFEZ, E. S. E. and DYER, I. A. (1969): Animal Growth and Nutrition. Lea and Febiger. Philadelphia.
HAHN, G. L. (1974): The Influence of Environmental Factors on Ruminant Production. In: Proc. 1st int. Livestock Environm. Symp., Lincoln, Ne., in press.
HAMMOND, J. (1960): Farm Animals. Edward Arnold Ltd, London.
HAMMOND, J. Jnr., MASON, I. L., ROBINSON, T. J. (1971): Hammond's Farm Animals. Edward Arnold Ltd., London.
JOHNSON, H. D. (1967): Climatic Stress Responses and Indices in Domestic Animals. Biometeorology, 2. 3rd int. biometeor. Congr. S. W. Tromp and W. H. Weihe (eds). Pergamon Press, Oxford.
JOHNSON, H. D. (1972): Meteorological Effects on Milk Production. In: Progress in Biometeorology, Swets & Zeitlinger Publ. Co., Amsterdam: 43-51.
McDOWELL, R. E. (1972): Improvement of Livestock Production in Warm Climates. W. H. Freeman & Co., S.F.
YOUNG, B. A. and CHRISTOPHERSON, R. J. (1974): The effect of prolonged cold exposure on digestion and metabolism in ruminants. In: Proc. 1st int. Livestock Environm. Symp., Lincoln, Ne., in press.

Chapter 1, Section 12b: Effects of Temperature on the Growth of Pigs (J. STEINBACH)

AGARWALA, O. P. and SUNDARESAN, D. (1956): Cheaper high-protein versus expensive low-protein rations for growth and fattening of pigs in India. Indian J. vet. Sci., 26: 185-192.
AMICK, R. J. and PURCELL, J. C. (1964): Influence of temperature on feed conversion in swine. J. Farm Econ., 46: 1227-1231.
AUFFRAY, P. (1969): Effets des lésions des noyaux ventromédians hypothalamiques sur la prise d'aliment chez le porc. Ann. Biol. anim. Bioch. Biophys. 9: 513-526.
BABATUNDE, G. M., OLOMU, M. J. and OYENUGA, V. A. (1972): Determination of the optimum crude protein requirement of pigs in a tropical environment. Anim. Prod., 14: 57-67.
BABATUNDE, G. M. and FETUGA, B. L. (1974): Pig production in Nigeria: possibilities and problems. Proc. Agric. Soc. Nigeria. Nig. agric. J., 10, in press.
BECKETT, F. E. (1964): Effective temperature as a means of evaluating or designing hog environments. ASAE, Paper, 64-911.
BIANCA, W. (1970): Animal response to meteorological stress as a function of age. Biometeorology 4, 1. S. W. Tromp and W. H. Weihe (eds.). Proc. 5th int. biometeor. Congr. Swets & Zeitlinger Publ. Co., Amsterdam: 119-131.
BOND, T. E., KELLY, C. F. and HEITMAN, H. (1952): Heat and moisture loss from swine. Agric. Eng., 33: 148-152, 154.
BOND, T. E., KELLY, C. F. and HEITMAN, H. (1967): Physiological response of swine to cycling environmental conditions. Anim. Prod., 9: 453-462.
BROBECK, J. R. (1960): Food and temperature. Rec. Progr. Hormone Res., 16: 439-466.
BRODY, S. (1945): Bioenergetics and Growth. Reinhold Publ. Corp., New York.
CHARLET-LERY, G. (1970): Variations d'extra-chaleur induites par l'ingestion de repas très riches en protéines chez le porc en croissance. Ann. Biol. anim. Bioch. Bipohys., 10: 123-141.
COMBERG, G., SPÄTH, H. und STEPHAN, E. (1971): Zur Problematik der Klimaempfindlichkeit von Haustieren. IV. Die Reaktion von Ebernachkommengruppen der Deutschen

Landrasse auf verschiedene stallklimatische Bedingungen (2. Mitteilung). Züchtungskunde 43: 253-267.
COMBERG, G., PLISCHKE, R., WEGNER, W. und FEDER, H. (1972): Die Auswirkungen suboptimaler Temperaturen (+8 °C) auf die Leistungen von Piétrains, Belgische Landrasse und deren Kreuzungen. 1. Mitteilung: Mastleistungseigenschaften und Schlachtkörperbewertung. Züchtungskunde, 44: 91-97.
COMBERG, G., WEGNER, W., STEPHAN, E., PLISCHKE, R., FEDER, H. und REETZ, I. (1973): Die Auswirkungen von Temperaturen oberhalb des Optimums (+30°C) auf Piétrains, Belgische Landrasse und deren Kreuzungen. Züchtungskunde, 45: 366-374.
CURTIS, S. E. (1970): Environmental-thermoregulatory interactions and neonatal piglet survival. J. Anim. Sci., 31: 576-587.
DEIGHTON, T. (1929): A study of the metabolism of two breeds of pig (with some remarks on a third). J. agric. Sci., 19: 140-184.
DEVENDRA, C. and PARRIS, E. C. C. (1970): Optimum crude protein levels for growing finishing pigs in a tropical environment. Trop. Anim. Hlth. Prod., 2: 162-165.
FULLER, M. F. (1965): The effect of environmental temperature on the nitrogen metabolism and growth of the young pig. Brit. J. Nutr., 19: 531-546.
FULLER, M. F. and BOYNE, A. W. (1971): The effects of environmental temperature on the growth and metabolism of pigs given different amounts of food. 1. Nitrogen metabolism, growth and body composition. Brit. J. Nutr., 25: 259-272.
HALE, O. M., JOHNSON, J. C. and WARREN, E. P. (1968): Influence of season, sex and dietary energy concentration on performance and carcass characteristics of swine. J. Anim. Sci., 27: 1577-1582.
HAZEN, T. E., CURRY, H. N. and GIESE, H. (1959): Swine growth and efficiency in a naturally varying environment. Agric. Eng., 40: 210-213.
HAZEN, T. E. and MANGOLD, D. W. (1960): Functional and basic requirements of swine housing: Production approaching conditions of controlled confinement. Agric. Eng., 41: 585-590.
HEIDENREICH, C. J., GARWOOD, V. A. and CARTER, M. W. (1962): Seasonal influence on serum cholesterol levels in swine. Amer. J. vet. Res., 23: 457-459.
HEITMAN, H. and HUGHES, E. H. (1949): The effects of air temperature and relative humidity on the physiological well-being of swine. J. Anim. Sci., 8: 171-181.
HEITMAN, H., KELLY, C. F. and BOND, T. E. (1958): The ambient temperature and weight gain in swine. J. Anim. Sci., 17: 62-67.
HOLMES, C. W. (1971): Growth and backfat depth of pigs kept at a high ambient temperature. Anim. Prod., 13: 521-527.
HOLMES, C. W. (1973): The energy and protein metabolism of pigs growing at a high ambient temperature. Anim. Prod., 16: 117-133.
HUNTER, A. A. and JENNINGS, J. W. (1961): The effects of high-temperature, high-humidity environmental conditions on the performance of bacon pigs. Res. Exp. Rec. Ministr. Agric. N. Irel., 10: 121-137.
INGRAM, D. L. and LEGGE, K. F. (1970): Variations in deep-body temperature in the young unrestrained pig over the 24-hour period. J. Physiol. (Lond.), 210: 989-998.
INGRAM, D. L. (1972): Meteorological effects on pigs. In: Progress in Biometeorology. S. W. Tromp (ed.). Swets & Zeitlinger B.V., Amsterdam, Div. A, Vol. 1, III: 71-79.
JENSEN, A. H., KUHLMAN, D. E., BECKER, D. E. and HARMON, B. G. (1969): Response of growing-finishing swine to different housing environments during winter seasons. J. Anim. Sci., 29: 451-456.
JOTTRAND, M. (1956): L'engraissement du porc dans la région d'Elisabethville. Bull. Inf. INEAC 5: 351-376.
JUDGE, M. D. (1969): Environmental stress and meat quality. J. Anim. Sci., 28: 755-760.
JUDGE, M. D., EIKELENBOOM, G., ZUIDAM, L. and SYBESMA, W. (1973): Blood acid-base status and oxygen binding in stress-induced hyperthermia in pigs. J. Anim. Sci., 37: 776-784.
KELLY, C. F., BOND, T. E. and HEITMAN, H. (1963): Direct "air" calorimetry for livestock. Trans. ASAE, 6: 126-128.
MANGOLD, D. W., HAZEN, T. E. and HAYS, V. W. (1967): Effect of air temperature on performance of growing-finishing swine. Trans. ASAE, 10: 370-375, 377.
MARPLE, D. N., ABERLE, E. D., FOREST, J. C., BLAKE, W. H. and JUDGE, M. D.

(1972): Effects of humidity and temperature on procine plasma adrenal corticoids, ACTH and growth hormone levels. J. Anim. Sci., 34: 809-812.
McGRATH, W. S., VAN DER NOOT, G. W., GILBREATH, R. L. and FISHER, H. (1968): Influence of environmental temperature and dietary fat on back fat composition of swine. J. Nutr., 96: 461-466.
MORRISON, S. R., HEITMAN, H., BOND, T. E. and FINN-KELCEY, P. (1966): The influence of humidity on growth rate and feed utilization of swine. Int. J. Biometeorol., 10: 163-168.
MORRISON, S. R., BOND, T. E. and HEITMAN, H. (1968): Effect of humidity on swine at high temperature. Trans. ASAE, 11: 526-528.
MORRISON, S. R., HEITMAN, H. and BOND, T. E. (1969): Effect of humidity on swine at temperatures above optimum. Int. J. Biometeor., 13: 135-139.
MORRISON, S. R. and MOUNT, L. E. (1971): Adaptation of growing pigs to changes in environmental temperatures. Anim. Prod., 13: 51-57.
MORROW, R. E., REINEKE, E. P., PEARSON, A. M. and HOEFER, J. A. (1957): The effect of two environmental temperatures and fasting on blood fat levels of swine. J. Anim. Sci., 16: 1047 (Abstracts).
MORROW, R. E., PEARSON, A. M., REINEKE, E. P. and HOEFER, J. A. (1963): Effects of feeding, fasting and environmental temperature upon blood plasma lipid levels of swine. J. Anim. Sci., 22: 466-470.
MOUNT, L. E. (1968): The Climatic Physiology of the Pig. Edward Arnold (Publ.) Ltd., London.
MOUNT, L. E. (1972): Environmental physiology in relation to pig production. In: Pig Production. D. J. A. Cole (ed.) Proc. 18th Easter Sch. agric. Sci. Butterworths Publ. Co., London.
NEERGAARD, L. and THORBEK, G. (1967): Variation in heat production in growing pigs. 1. Some observations on the relationship between feed intake and heat production in pigs fed barley and skimmilk. Europ. Ass. Anim. Prod., 4th Symp. Energy Metabolism.
NELSON, D. P., READ, C. H., BARFIELD, B. J., WALKER, J. N., HAYES, V. and CROMWELL, G. (1972): The performance of swine under warm environments. Trans. ASAE, 15: 133-136.
OKEBUKOLA, O. E. and STEINBACH, J. (1973): Effect of weight gain on respiratory rate and rectal temperature of growing pigs in a tropical environment. Unpubl. Rep., Univ. Ibadan.
OYENUGA, V. A. (1961): Nutritive value of cereal and cassava diets for growing and fattening pigs in Nigeria. Brit. J. Nutr., 15: 327-338.
PEARSON, A. M., REINEKE, E. P., HOEFER, J. A. and MORROW, R. E. (1966): Effect of environmental temperature and thiouracil feeding upon growing-fattening pigs. J. Anim. Sci., 25: 994-999.
PORTER, P. and GATES, D. M. (1969): Thermodynamic equilibria of animals with environment. Ecol. Monogr., 39: 227-244.
RIKER, J. T., PERRY, T. W., PICKETT, R. A. and HEIDENREICH, C. J. (1967): Influence of controlled temperatures on growth rate and plasma ascorbic acid values in swine. J. Nutr., 92: 99-103.
ROLLER, W. L. and GOLDMAN, R. F. (1969): Response of swine to acute heat exposure. Trans. ASAE, 12: 166-169.
SEYMOUR, E. W., SPEER, V. C., HAYS, V. W., MANGOLD, D. W. and HAZEN, T. E. (1964): Effects of dietary protein level and environmental temperature on performance and carcass quality of growing-finishing swine. J. Anim. Sci., 23: 375-379.
SØRENSEN, P. H. (1961): Staldklimaets inflydelse på vaekst, foderudnyttelse og slagtekvalitet hos svin. Årsberetn. Inst. Sterilitetforsk., 1961: 185-201.
SØRENSEN, P. H. (1962): Influence of climatic environment on pig performance. In: Nutrition of Pigs and Poultry. J. T. Morgan and D. Lewis (eds). Butterworths, Publ. Co., London.
STEINHARDT, M. (1971): Der Wärmehaushalt beim Schwein. In: Der Wärmehaushalt landwirtschaftlicher Nutztiere. L. Lyhs (ed.). VEB Gustav Fischer Verlag, Jena: 105-185.
SUGAHARA, M., BAKER, D. H., HARMON, B. G. and JENSEN, A. H. (1970): Effect of ambient temperature on performance and carcass development in young swine. J. Anim. Sci., 31: 59-62.

TODD, A. C. E. and DANIELS, L. J. (1968): The influence of housing on the performance of growing pigs in a subtropical environment. Proc. Aust. Soc. Anim. Prod., 7: 285-288.
TONKS, H. M. and SMITH, W. C. (1965): The effect of a hightemperature, high-humidity indoor environment on the performance of bacon pigs. Anim. Prod. 7: 280 (Abstract).
TONKS, H. M., SMITH, W. C. and BRUCE, J. (1972): The influence of a high-temperature, high-humidity indoor-environment under farm conditions. Vet. Rec., 90: 531-537.
VAN VAERENBERGH, R. (1960): Note préliminaire sur l'économie de l'engraissement des porcs de la race "Large White" nés et élevés dans la cuvette congolaise. Bull. Inf. INEAC, 9: 7-19.
WEAVER, M. E. and INGRAM, D. L. (1969): Morphological changes in swine associated with environmental temperature. Ecology, 50: 710-713.
ZINN, G. M., WEINMAN, D. E. and TUMBLESON, M. E. (1968): Hyperthermic stress effects in swine. J. Anim. Sci., 27: 1782 (Abstract).
ZOBRISKY, S. E., MOODY, W. G., TRIBBLE, L. F. and NAUMANN, H. D. (1961): Meatiness of swine as influenced by breed, season and sex. J. Anim. Sci. 20: 923 (Abstract).

Chapter 1, Section 12c: Effects of Temperature on the Growth of Fish (J. R. BRETT)

ANDREWS, J. W. and STICKNEY, R. R. (1972): Interactions of feeding rates and environmental temperature on growth, food conversion, and body composition of channel catfish. Trans. Amer. Fish Soc., 101: 94-99.
BARDACH, J. E., RYTHER, J. H. and McLARNEY, W. O. (1972): Aquaculture. The farming and husbandry of freshwater and marine organisms. Wiley-Interscience, John Wiley and Sons Ltd., London/New York, 868 pp.
BRETT, J. R., SHELBOURN, J. E. and SHOOP, C. T. (1969): Growth rate and body composition of fingerling sockeye salmon, *Oncorhynchus nerka*, in relation to temperature and ration size. J. Fish. Res. Bd. Can., 26: 2363-2394.
BRETT, J. R. (1970): 3. Temperature. 3.3. Animals. 3.32. Fishes. In: Marine Ecology, 1. Environmental Factors, Part 1. O. Kinne, Wiley Interscience, John Wiley and Sons Ltd., London/New York: 515-560.
BRETT, J. R. and HIGGS, D. A. (1970): Effect of temperature on the rate of gastric digestion in fingerling sockeye salmon, *Oncorhynchus nerka*. J. Fish. Res. Bd. Can., 27: 1767-1779.
BRETT, J. R. (1974): Marine fish aquaculture in Canada. Bull. Fish. Res. Bd. Can., 188: 53-84.
BROWN, M. E. (1946): The growth of brown trout (*Salmo trutta* Linn.) III. The effect of temperature on the growth of two-year-old trout. J. exp. Biol., 22: 145-155.
COOPER, E. L. (1961): Growth of wild and hatchery strains of brook trout. Trans. Amer. Fish. Soc., 90: 424-438.
DAWES, B. (1930): Growth and maintenance in the plaice (*Pleuronectes platessa* L.). Part I. J. Marine Biol. Ass., 17: 103-174.
FRY, F. E. J. (1947): Effects of the environment on animal activity. Univ. Toronto Stud. Biol., Ser. No. 55, Publ. Ontario Fish. Res. Lab. No. 68: 1-62.
FRY, F. E. J. (1971): The effect of environmental factors on the physiology of fish. In: Fish Physiology. W. S. Hoar and D. J. Randall (eds). Academic Press, New York/London: 1-98.
HARADA, T. (1970): The present status of marine fish cultivation research in Japan. Helgoländer Wiss. Meeresunters, 20: 594-601.
HATANAKA, M. and TAKAHASHI, M. (1956): Utilization of food by mackerel *Pneumatophorus japonicus*. Tohoku J. agric. Res., 7: 51-57.
KINNE, O. (1960): Growth, food intake, and food conversion in a euryplastic fish exposed to different temperatures and salinities. Physiol. Zool., 33: 288-317.
LOCKWOOD, S. J. (1973): The use of the Von Bertalanffy growth equation to describe the seasonal growth of fish. J. Cons. int. Expl. Mer., 35, in press.
PENTELOW, F. T. K. (1939): The relation between growth and food consumption in the brown trout (*Salmo trutta*). J. exp. Biol., 16: 446-473.
PETERS, D. S. and BOYD, M. T. (1972): The effect of temperature, salinity, and availability of food on the feeding and growth of the hogchoker, *Trinectes maculatus* (Bloch and Schneider). J. exp. Marine Biol. Ecol., 7: 201-207.

SHELBOURN, J. E., BRETT, J. R. and SHIRAHATA, S. (1973): Effect of temperature and feeding regime on the specific growth rate of sockeye salmon fry (*Oncorhynchus nerka*), with a consideration of size effect. J. Fish. Res. Bd. Can., 30: 1191-1194.
WINBERG, G. G. (1956): Rate of metabolism and food requirements of fishes. Belorussian State Univ., Minsk, 251 p. (Fish. Res. Bd. Can. Transl. Ser. 194).
WOHLSCHLAG, D. E. (1961): Growth of an Antarctic fish at freezing temperatures. Copeia, 11-17.

Chapter 1, Section 12d: Effects of Temperature on the Growth of Chickens (F. N. REECE and J. W. DEATON)

BARROT, H. G. and PRINGLE, EMMA M. (1947): Effect of environment on growth and feed and water consumption of chickens. I. The effect of temperature of environment during the first nine days after hatch. J. Nutr., 34: 53-58.
BARROT, H. G. and PRINGLE, EMMA M. (1949): Effect of environment on growth and feed and water consumption of chickens II. The effect of temperature and humidity of environment during the first eighteen days after hatch. J. Nutr., 37: 153-161.
BARROT, H. G. and PRINGLE, EMMA M. (1950): Effect of environment on growth and feed and water consumption of chickens III. The effect of temperature of environment during the period from 18 to 32 days of age. J. Nutr., 41: 25-30.
BOUCHILLON, C. W., REECE, F. N. and DEATON, J. W. (1970): Mathematical modeling of thermal homeostasis in a chicken. Trans. Amer. Soc. agri. Eng., 13, 5: 648-652.
DAVENPORT, C. B. (1914): The origin of domestic fowl: J. Hered. 5: 313-315.
DEATON, J. W. and REECE, F. N. (1970): Implications of environmental control for broilers. Feedstuffs, 42, 30: 34.
DEATON, J. W., REECE, F. N., KUBENA, L. F., LOTT, B. D. and MAY, J. D. (1973): The ability of the broiler chicken to compensate for early growth depresssion. Poultry Sci., 52, 1: 262-265.
HAURWITZ, BERNHARD and AUSTIN, J. M. (1944): Climatology. McGraw-Hill Book Co., New York: 35.
LUCAS, A. M. and STETTENHEIM, P. R. (1972): Avian Anatomy Integument, Part I. U.S. Govt. Print. Off.: 2-3.
REECE, F. N. and DEATON, J. W. (1968): Implications of reduced heat during brooding of broilers. Proceedings, Association of Southern Agricultural Workers: 218-219.
REECE, F. N. and DEATON, J. W. (1971): Use of evaporative cooling for broiler chicken production in areas of high relative humidity. Poultry Sci., 50, 1: 100-104.
REECE, F. N., DEATON, J. W. and KUBENA, L. F. (1972): Effects of high temperature and humidity on heat prostration of broiler chickens. Poultry Sci., 51, 6: 2021-2025.
STURKIE, PAUL D. (1965): Avian Physiology, 2nd Ed. Cornell University Press: 200-201.

Chapter 1, Section 13a: Heat Transfer Through Animal Coats (K. CENA and J. L. MONTEITH)

BENNETT, J. W. (1964): Thermal insulation of cattle coats. Proc. Aust. Soc. Anim. Prod., 5: 160-166.
BENNETT, J. W. and HUTCHINSON, J. C. D. (1964): Thermal insulation of short lengths of Merino fleece. Aust. J. agric. Res., 15: 427-445.
CENA, K. (1966): Absorption of solar radiation by cattle and horses with various coat colours. Acta agric. silv., 6: 93-138.
CENA, K. and CLARK, J. A. (1973): Thermal radiation from animal coats: coat structure and measurements of radiative temperature. Phys. in Med. Biol., 18: 432-443.
CENA, K. (1974): Radiative loss from animals and man. In: Heat Loss From Animals and Man. J. L. Monteith and L. E. Mount (eds). Proc. 20th. Easter Sch. agric. Sci. Butterworths Publ. Co., London.
CENA, K. and CLARK, J. A. (1974): Heat balance and thermal resistances of sheep's fleece. Phys. in. Med. Biol., 19: 51-65.
CENA, K. and MONTEITH, J. L. (1974a): Transfer processes in animal coats. I. Radiative transfer. Proc. roy. Soc. B.

CENA, K. and MONTEITH, J. L. (1974b): Transfer processes in animal coats. II. Conduction and convection. Proc. roy. Soc. B.
CLARK, J. A. and CENA, K. (1972): The application of thermal imaging techniques to animals. Dtsch. tierärztl. Wschr., 79: 292-296.
CLARK, J. A., CENA, K. and MONTEITH, J. L. (1973): Measurements of the local heat balance of animal coats and human clothing. J. appl. Physiol., 35: 751-754.
GONZALEZ-JIMENEZ, E. and BLAXTER, K. L. (1962): The metabolism and thermal regulation of calves in the first month of life. Brit. J. Nutr., 16: 199-212.
HAMMEL, H. T. (1955): Thermal properties of fur. Ann. J. Physiol., 182: 369-376.
HARDY, J. D., GAGGE, A. P. and STOLWIJK, J. A. J. (1970): Physiological and Behavioural Temperature Regulation. C. C. Thomas, Springfield.
HUTCHINSON, J. C. D., BROWN, G. D. and ALLEN, T. E. (1973): Effect of the coat on the heat dissipation of cattle in a radiant environment. In: Proc. 1st Aust. Conf. Heat and Mass Transfer, 1973.
JOYCE, J. P. and BLAXTER, K. L. (1964): The effect of air movement, air temperature and infrared radiation on the energy requirements of sheep. Brit. J. Nutr., 18: 5-27.
KOVARIK, M. (1964): Flow of heat in an irradiated protective cover. Nature (Lond.), 201: 1085-1087.
LENTZ, C. P. and HART, J. S. (1960): The effect of wind and moisture on heat loss through the fur of newborn caribou. Canad. J. Zool., 33: 679-688.
MACFARLANE, W. V. (1968): Weather, climate and domestic animals. In: Proc. W.M.O. Seminar, Melbourne, 1: 119-161.
MITCHELL, D. (1974): Convective heat transfer from man and other animals. In: Heat Loss from Animals and Man. J. L. Monteith and L. E. Mount (eds). Butterworths Publ. Co., London.
MONTEITH, J. L. (1973): Principles of Environmental Physics. Edward Arnold, London.
MONTEITH, J. L. and MOUNT, L. E. (Eds) (1974): Heat Loss from Animals and Man. Proc. 20th Easter Sch. agric. Sci. Butterworths Publ. Co., London.
MOOTE, I. (1955): The thermal insulation of Caribou pelts. Text. Res. J., 25: 832-837.
PORTER, W. P. and GATES, D. M. (1969): Thermodynamic equilibria of animals with environment. Ecol. Monogr., 39: 245.
SCHOLANDER, P. F., WALTERS, V., HOCK, R. and IRVING, L. (1950): Body insulation of some Arctic and tropical mammals and birds. Biol. Bull., 99: 225-236.
TREGEAR, R. T. (1966): Physical functions of skin. Academic Press, London.

Chapter 1, Section 13b: Fiber Growth and Insulation Against Heat and Cold in Sheep (C. J. THWAITES)

ALEXANDER, G. (1961): Temperature regulation in the newborn lamb. III. Effect of environmental temperature on metabolic rate, body temperatures and respiratory quotient. Aust. J. agric. Res., 12: 1152-1174.
ALEXANDER, G. (1962): Temperature regulation in the newborn lamb. IV. The effect of wind and evaporation of water from the coat on metabolic rate and body temperature. Aust. J. agric. Res., 13: 82-99.
BENNETT, J. W., HUTCHINSON, J. C. D. and WODZICKA-TOMASZEWSKA, M. (1962): Climate and wool growth. Proc. Aust. Soc. Anim. Prod., 4: 32-33.
BENNETT, J. W., HUTCHINSON, J. C. D. and WODZICKA-TOMASZEWSKA, M. (1962): Annual rhythm of wool growth. Nature (Lond.), 194: 651-652.
BLAXTER, K. L., CLAPPTERTON, J. L. and WAINMAN, F. W. (1966): The extent of differences between six British breeds of sheep in their metabolism, feed intake and utilization, and resistance to climatic stress. Brit. J. Nutr., 20: 283-294.
BLIGH, J. (1959): The receptors concerned in the thermal stimulus to panting in sheep. J. Physiol. (Lond.), 146: 142-151.
BLIGH, J. (1961): The delayed polypnoeic response to heat in the shorn sheep. J. Physiol. (Lond.), 159: 84-85P.
COOP, I. E. and HART, D. S. (1953): Environmental factors affecting wool growth. Proc. N.Z. Soc. Anim. Prodc., 13: 113-119.

DONEY, J. M. (1966): Breed differences in response of wool growth to annual nutritional and climatic cycles. J. agric. Sci. (Camb.), 67: 25-30.
DONEY, J. M. and GRIFFITHS, J. G. (1967): Wool growth regulation by local skin cooling. Anim. Prod., 9: 393-397.
FERGUSON, K. A. (1949): The effect of sympathectomy on wool growth. Aust. J. sci. Res., B2: 438-443.
FERGUSON, K. A., CARTER, H. B. and HARDY, M. H. (1949): Studies of comparative fleece growth in sheep. 1. The quantitative nature of inherent differences in woolgrowth rate. Aust. J. sci. Res., B2: 42-81.
HART, D. S. (1957): Stimulation of wool growth by thyroxine implantation. N.Z.J. sci. Techn. 38: 871-877.
HART, D. S. (1961): The effect of light-dark sequences on wool growth. J. agric. Sci. (Camb.), 56: 235-242.
HILL, M. K. (1971): Studies on the growth and properties of Merino wool with particular reference to wool tenderness. Ph.D. Thesis, Univ. New England, Australia.
HOERSCH, T. M., HENDERSON, H. E., REINEKE, E. P. and HENNEMAN, H. A. (1961): Comparative indexes of thyroid function in sheep. Amer. J. Physiol., 201: 819-822.
HUTCHINSON, J. C. D. (1965): Photoperiodic control of the annual rhythm of wool growth. In: Biology of the Skin and Hair Growth. A. G. Lyne and B. F. Short (eds). Angus and Robertson, Sydney: 565-573.
HUTCHINSON, J. C. D. (1968): Deaths of sheep after shearing. Aust. J. exp. Agric. Anim. Husb., 8: 393-400.
JOLLY, M. and LYNE, A. G. (1970): The response of skin and wool growth to local subdermal temperature changes in the sheep. J. agric. Sci. (Camb.), 75: 501-515.
KLEMM, G. H. (1962): The reactions of unshorn and shorn sheep to hot wet and hot dry atmospheres. Aust. J. agric. Res., 13: 472-478.
LYNE, A. G., JOLLY, M. and HOLLIS, D. E. (1970): Effects of experimentally produced local subdermal temperature changes on skin temperature and wool growth in the sheep. J. agric. Sci. (Camb.), 74: 83-90.
MACFARLANE, W. V., MORRIS, R. J. H. and HOWARD, B. (1958): Heat and water in tropical Merino sheep. Aust. J. agric. Res., 9: 217-228.
MAQSOOD, M. (1955): Effects of the thyroid gland on fleece growth in sheep. Brit. vet. J., 111: 163-169.
MORRIS, L. R. (1961): Photoperiodicity of seasonal rhythm of wool growth in sheep. Nature (Lond.), 190: 102-103.
MULLANEY, P. D. (1967): The relation of birth coat and lamb survival. Aust. J. exp. Agric. Anim. Husb., 6: 84-87.
PANARETTO, B. A. and FERGUSON, K. A. (1969): Pituitary adrenal interactions in shorn sheep exposed to cold, wet conditions. Aust. J. agric. Res., 20: 99-113.
PANARETTO, B. A. and FERGUSON, K. A. (1969): Comparison of the effects of several stressing agents on the adrenal glands of normal and hypophysectomised sheep. Aust. J. agric. Res., 20: 115-124.
PARER, J. T. (1963): Wool length and radiant heating effects in sheep. J. agric. Sci., 60: 141-144.
PARER, J. T. (1963): The effects of wool length and nutrition on heat reactions of Merino sheep in the field. Aust. J. exp. Agric. Anim. Husb., 3: 243-248.
PHILLIPS, G. D. and RAGHAVAN, G. V. (1970): Response of unshorn and shorn sheep to thermal stress. J. Physiol. (Lond.), 208: 317-328.
PRIESTLEY, C. H. B. (1957): The heat balance of sheep standing in the sun. Aust. J. agric. Res., 8: 271-280.
ROUGEOT, J. (1957): Action de la variation saisonnière de la durée quotidienne d'eclairement sur la mue de certaines fibres de la toison de la race ovine Limousine. C.R. Soc. Biol. (Paris), 151: 834-837.
RUDALL, K. M. (1935): Growth changes in wool fibres: experiments on removing the heat conserving fleece. J. Text. Inst., 26: T358-370.
SLEN, S. B. and CONNELL, R. (1958): Wool growth in sheep as affected by administration of certain sex hormones. Canad. J. Anim. Sci., 38: 38-47.
SMITH, R. K. (1973): Seasonality of wool growth with particular reference to hormones. B. Rural Science Hons. Thesis, Univ. of New England, Armidale, Australia.

STEWART, A. M., MOIR, R. J. and SCHINCKEL, P. G. (1961): Seasonal fluctuations in wool growth in South-Western Australia. Aust. J. exp. Agric. Anim. Husb., 1: 85-91.
STORY, L. F. and ROSS, D. A. (1959): Effect cf shearing time on wool. IV. Effect on the fleeces. N.Z. J. agric. Res., 2: 1096-1103.
WAITES, G. M. H. (1961): Polypnoea evoked by heating the scrotum of the ram. Nature (Lond.), 190: 172-173.
WEBSTER, A. J. F. and BLAXTER, K. L. (1966): The thermal regulation of 2 breeds of sheep exposed to air temperatures below freezing point. Res. vet. Sci., 7: 466-479.

Chapter 1, Section 14: Effects of Temperature on Lactation of Cattle (H. D. JOHNSON)

ABILAY, T. A. and JOHNSON, H. D. (1973): Influence of high environmental temperature (33.5°C) on plasma progesterone and cortisol. J. Dairy Sci., 56: 642.
BERGMAN, R. K. (1963): Effects of a Prolonged High Environmental Temperature on Glucocorticoids in the Bovine. Ph.D. Thesis, Univ. Missouri.
BIANCA, W. (1965): Reviews of the progress of dairy science. Section A, Physiology: Cattle in a hot environment. J. Dairy Res., 32: 291.
BLACKWELL, R. E. and GUILLEMIN, R. (1973): Hypothalamic control of adenohypophysiol secretions. Ann. Rev. Physiol., 35: 357.
BROBECK, J. R. (1960): Food and temperature. Rec. Progr. Hormone Res., 16: 439.
BRODY, S. (1948): Physiological backgrounds. Mo. agric. exp. Stat. Res. Bull., 423.
BRODY, S., WORSTELL, D. M., RAGSDALE, A. C. and KIBLER, H. H. (1948): Heat production and cardiorespiratory activities during gestation and lactation in Holstein cattle. Mo. agric. exp. Stat. Res. Bull., 424.
CARGILL, B. F., STEWART, R. E. and JOHNSON, H. D. (1962): Effect of humidity of total room heat and vapor dissipation of Holstein cows at 65, 80 and 90°F. Mo. agric. exp. Stat. Res. Bull., 794.
CHRISTISON, G. I. and JOHNSON, H. D. (1972): Cortisol turnover in heat stressed cows. J. Anim. Sci., 35: 1005.
COBBLE, J. W. and HERMAN, H. A. (1951): The influence of environmental temperatures on the composition of milk of the dairy cow. Mo. agric. exp. Stat. Res. Bull., 485.
COWIE, A. T. (1961): The hormonal control of milk secretion. In: Milk, the Mammary Gland and its Secretion, I. S. K. Kon and A. T. Cowie (eds). Academic Press, New York: 163-203.
CROSS, B. A. (1961): The neural control of lactation. In: Milk, the Mammary Gland and its Secretion, I. S. K. Kon and A. T. Cowie (eds). Academic Press, New York: 229-250.
DAVIS, H. P. (1914): Effect of open-shed housing as compared with the closed stable for milk cows. Rep. Pa. agric. exp. Stat. Bull., 183.
DICE, J. R. (1935): The nutrients required by dairy cows kept in an open shed versus cows kept in a dairy barn. J. Dairy Sci., 18: 447.
DICE, J. R. (1940): The influence of stable temperature on production and feed requirement of dairy cows. J. Dairy Sci., 23: 61.
DICE, J. R. (1942): The ability of yearling heifers to withstand cold temperatures. J. Dairy Sci., 25: 678.
HARDY, J. D. (1961): Physiology of temperature regulation. Physiol. Rev., 41: 521.
HOLMES, C. W. (1971): Local cooling of the mammary gland and milk production in the cow. J. Dairy Res., 38: 3.
JOHNSON, H. D., RAGSDALE, A. C., BERRY, I. L. and SHANKLIN, M. D. (1962): Effects of various temperature-humidity combinations on milk production of Holstein cattle. Mo. agric. exp. Stat. Res. Bull., 791.
JOHNSON, H. D. and KIBLER, H. H. (1963): Temperature humidity effects on thyroxine I^{131} disappearance rates in cattle. J. appl. Physiol., 18: 73.
JOHNSON, H. D. (1965): Environmental temperature and lactation (with special reference to cattle). Int. J. Biometeor., 9, 2: 103-116.
JOHNSON, H. D., KIBLER, H. H., BERRY, I. L., WAYMAN, O. and MERILAN, C. P. (1966): Temperature and controlled feeding effects on lactation and related physiological reactions of cattle. Mo. agric. exp. Stat. Res. Bull., 202.

JOHNSON, H. D. (1967): Climatic effects on the physiology and productivity of cattle. In: Ground Level Climatology. R. H. Shaw (ed.). Amer. Ass. Advanc. Sci. (Wash.).
JOHNSON, H. D., HAHN, L., KIBLER, H. H., SHANKLIN, M. D. and EDMUNDSON, J. D. (1967). Heat and acclimation influences on lactation of Holstein cattle. Mo. agric. exp. Stat. Res. Bull., 916.
JOHNSON, H. D. (1972). Meteorological effects on milk production. In: Progress in Biometeorology, 1, III. S. W. Tromp (ed.). Swets & Zeitlinger B.V., Amsterdam.
KAMAL, T. H., ABRAHIM, I. A., SEIF, S. M. and JOHNSON, H. D. (1970): Plasma insulin and free amino acids changes with heat exposure in bovine. J. Dairy Sci., 53: 651.
KELLEY, M. A. R. and RUPEL, I. W. (1937): Relation of stable environment to milk production, U.S. Dept Agric. techn. Bull., 591.
KIBLER, H. H., JOHNSON, H. D., HAHN, L. and SHANKLIN, M. D. (1970): Thermal regulation in cattle at 2° to 35°C as influenced by controlled feeding, *ad libitum* feeding and fasting. Mo. agric. exp. Stat. Res. Bull., 975.
KOTBY, S. and JOHNSON, H. D. (1967): Rate of adrenal cortical activity during exposure to high (34°C) ambient temperature. Life Sci., 6: 1121.
LEE, D. H. K. (1951): Plant and animal nutrition in relation to soil and climate factors. Spec. Conf. Agric., Melbourne, (London, H.M.S.O., 1951).
MAUST, L. E., McDOWELL, R. E. and HOOVEN, N. W. (1972): The effect of weather on the performance of Holstein cows in three stages of lactation. J. Dairy Sci., 55: 1133.
MAWSON, W. F. Y. and WHITE, B. J. (1971): Climatic and biological limitations to dairy production in a tropical environment. Trop. Grasslands, 5: 145.
McDONALD, M. A. and BELL, J. M. (1958): The effects of low fluctuating temperatures on farm animals. IV. Influence of ambient air temperature on feed intake of lactating Holstein-Friesian cows. Canad. J. Anim. Sci., 38: 148.
MEITES, J., NICOLL, C. S. and TALWOLKER, P. K. (1963): The central nervous system and the secretion and release of prolactin. In: Advances in Neuro-Endocrinology. Univ. Illinois Press, Urbana, Illinois: 238.
MITRA, R., CHRISTISON, G. I. and JOHNSON, H. D. (1972): Effect of prolonged thermal exposure on Growth Hormone (GH) secretion in cattle. J. Anim. Sci., 34: 776.
MORGANE, P. J. (1962): Hypothalamic and rhinencephalic mechanisms in the regulation of caloric intake. 22nd Int. Congr. physiol. Sci. Excerpta med. (Amst.) 1, II: 660-677.
PAYNE, W. J. A. (1957): The problem of milk production in tropical countries. Trop. Agric., 33: 34.
POPOFF, J. S. (1927): Der Einfluss der Kälte auf die Produktivität der Milchkühe. Z. Tierz. Biol., 10: 285.
RAGSDALE, A. C., BRODY, S., THOMPSON, H. J. and WORSTELL, D. M. (1948): Influence of temperature 50° to 105°F on milk production and feed consumption in dairy cattle. Mo. agric. exp. Stat. Res. Bull., 425.
RAGSDALE, A. C., WORSTELL, D. M., THOMPSON, H. J. and BRODY, S. (1949): The influence of temperature, 50° to 0°F, and 50° to 95°F, on milk production, feed and water consumption and body weight in Jersey and Holstein cows. Mo. agric. exp. Stat. Res. Bull., 449.
RAGSDALE, A. C., THOMPSON, H. J., WORSTELL, D. M. and BRODY, S. (1950): Milk production and feed and water consumption responses of Brahman, Jersey and Holstein cows to changes in temperature 50 to 105°F and 50 to 8°F. Mo. agric. exp. Stat. Res. Bull., 460.
RAGSDALE, A. C., THOMPSON, H. J., WORSTELL, D. M. and BRODY, S. (1951): Influence of increasing temperature, 40 to 105°F on milk production in Brown Swiss cows, and on feed and water consumption and body weight in Brown Swiss and Brahman cows and heifers. Mo. agric. exp. Stat. Res. Bull., 471.
SEMENTOVSKAYA, N. M. and GARKAVI, O. V. (1950): Reference in: Milk, the Mammary Gland and Its Secretion. S. K. Kon and A. T. Cowie (eds). Academic Press, New York.
WITZEL, S. A. and HIEZER, E. E. (1946): A 5 year summary of dairy barn research. Agric. Eng., 27: 499.
WRIGHT, N. C. (1946): Report of the development of cattle breeding and milk production in Ceylon. Sessional Paper No. 20, Colombo, Ceylon Govt. Press.

Chapter 1, Section 15a: Breeding Management in Cattle (H. A. W. VENTER)

AMIR, S., KALI, J., VOLCANI, R. and PERLMAN, M. (1967): Early breeding of dairy heifers. Anim. Prod., 9: 268.
BONSMA, J. C. (1939): Breeding seasons on cattle ranches. Fing. S. Afr., June: 230.
BONSMA, J. C., SCHOLTZ, G. D. J. and BADENHORST, J. F. G. (1940): The influence of climate on cattle fertility and hardiners of certain breeds. Fing. S. Afr., 15: 7-12.
BONSMA, J. C. (1949): Breeding cattle for increased adaptability to tropical and sub-tropical environments. J. agric. Sci., 39: 204-221. [2]
BONSMA, J. C., VAN MARLE, J. and HORMEYR, J. H. (1953): Climatological research on animal husbandry and its significance in the development of beef cattle production in colonial terries. Emp. J. exp. Agric., 21, 83: 154-175. [1]
BONSMA, J. C. and JOUBERT, D. M. (1959): Gestation of cattle in the sub-tropics with special reference to the birth weight of calves. S. Afr. J. agric. Sci., 2, 2: 215-230.
BONSMA, J. C. and LOUW, G. N. (1966): Heat as a limiting factor in animal production. In: Biometeorology, 2. S. W. Tromp and W. H. Weihe (eds). Proc. 3rd Int. biometeor. Congr.: 371-382.
BONSMA, J. C. and SKINNER, J. D. (1969): Factors influencing birth date, birth weight and preweaning growth rate of Africander, Bonsmara and Hereford cattle in the sub-tropics. Proc. S. Afr. Anim. Prod., 8: 161-163.
BONSMA, J. C., BADENHORST, J. F. G. and SKINNER, J. D. (1972): The incidence of foetal dwarfism in Shorthorn cattle in the subtropics. S. Afr. J. Anim. Sci., 2: 19-21.
BOSMAN, D. J. and HARWIN, G. O. (1966): Genetic and environmental factors affecting preweaning traits in beef cattle under extensive ranching conditions. Proc. S. Afr. Anim. Prod., 5: 153-159.
BOSMAN, D. J. and HARWIN, G. O. (1969): The occurrence of postpartum oestrus in beef cows under ranching conditions. Proc. S. Afr. Soc. Anim. Prod., 8: 165-166.
DALE, H. E., RAGSDALE, A. C. and CHENG, C. S. (1959): Effect of constant environmental temperatures, 50° and 80°F, on appearance of puberty in beef calves. J. Anim. Sci., 18: 1363-1366.
DE ALBA, J. and RIERA, S. (1966): Sexual maturity and spermatogenesis under heat stress in the Bovine. Anim. Prod., 8: 137-144.
DEVENDRA, C. (1970): The comparative efficiency feed utilization of ruminants in the tropics. Trop. Sci., XIII, 2: 123-132. [3]
FERGUSON, F. (1970): Adaptive behaviour of cattle to tropical environments. Trop. Sci., XIII, 2: 113-122. [4]
FINDLAY, J. D. (1972): Climatological data needed to specify climatic stress. Wld Rev. Anim. Prod., 8, 2: 38-44. [5]
HAFEZ, E. S. E. (1967): Bioclimatological aspects of animal productivity. Wld Rev. Anim. Prod., 3, 14: 22-37. [6]
HALL, J. G., BRANTON, C. and STONE, E. J. (1959): Estrus, estrus cycles, ovulation time, time of service and fertility of dairy cattle in Louisiana. J. Dairy Sci., 42: 1086.
HAMMOND, J. C. (1949): Adaptation of livestock to new environments. Unsccur Proc., Land Resources. [7]
HARWIN, G. O., LAMB, R. D. and BIRSCHOP, J. H. R. (1967): Some factors affecting reproductive performance in beff females. Proc. S. Afr. Anim. Prod., 6: 171-177.
JOHNSTONE, J. E. (1972): Housing and management problems in cattle production in humid tropics. Wld Rev. Anim. Prod., 8, 1: 64-69. [8]
JOUBERT, D. M. (1954): The influence of high and low nutritional planes on the oestrus cycle and conception rate of heifers. J. agric. Sci. (Camb.), 15: 164-172.
JOUBERT, D. M. (1963): Puberty in female farm animals. Anim. Breeding Abstr., 31, 3: 295-306.
KIESEL, K. and GRIM, H. (1961): Is the ripening of the follicles dependent on temperature? Anim. Breeding Abstr., Annot. Bibl., No. 120.
KOGER, M., REYNOLDS, W. H., KIRK, W. G., PEACOCK, F. M. and WARWICK, A. C. (1962): Reproductive performance of crossbred and straightbred cattle on different pasture programs in Florida. J. Anim. Sci., 21: 14-19.
LIGGINS, G. C. (1968): Premature parturition after infusion of corticotrophin or corticol into foetal lambs. J. Endocr., 42: 323-329.

LIGGINS, G. C. (1969): Premature delivery of foetal lambs infused with glucocorticoid. J. Endocr., 45: 515-523.
MAREE, C. and HARWIN, G. O. (1971): Nutritional effects on ovarian activity, puberty and the early mating of beef heifers under extensive grazing conditions. Agroanimalia, 3: 103-106.
McCARTHY, W. O. and HAMILTON, C. P. (1964): Comparative performance of beef breeds in central Queensland. 1 Prod. trop. Agric., 41: 293-297.
McDOWELL, R. E. (1972): Improvement of livestock production in warm climates. W. H. Freeman and Company, San Francisco. [9]
MEREIER, E. and SALISBURY, G. W. (1947): Seasonal variations in hours of daylight associated with fertility level of cattle under natural breeding conditions. J. Dairy Sci., 30: 747-756.
MUKHTAR, A. M. S. (1972): Efficiency of feed conversion by some tropical ruminants. Wld Rev. Anim. Prod., 8, 3: 69-72. [10]
PENZHORN, E. J. and MEINTJIES, J. P. (1968): Preliminary results with regard to early reproduction in Africander heifers. Proc. S. Afr. Anim. Prod., 7: 155-156.
PINO, J. A. (1972): Animal breeding and climatology. Wld Rev. Anim. Prod., 8: 1: 45-47. [11]
PLASSE, D., WARWICK, A. C. and KOGER, M. (1969): Reproductive behaviour of *Bos indicus* females in a sub-tropical environment. Puberty and ovulation frequency in Brahman X British heifers. J. Anim. Sci., 27: 94-100.
RENDEL, J. M. (1972): Dairy cattle in hot climates. Wld Rev. Anim. Prod., 8, 1: 16-24. [12]
RENDEL, J. M. (1972): Breeding cattle for the Australian North. Wld Rev. Anim. Prod., 8, 1: 48-56. [13]
ROLLINSON, D. H. L. (1971): Further development of artificial insemination in tropical areas. A.B.A., 39, 3: 407-427. [14]
SHORT, R. E., BELLOWS, R. A., MOODY, E. H. and HOWLAND, B. E. (1972): Effects of suckling and mastectomy on bovine postpartum reproduction. J. Anim. Sci., 34: 70-74.
SKINNER, J. D. and JOUBERT, D. M. (1963): A further note on the duration of pregnancy and birth weight in beef cattle in the sub-tropics. Proc. S. Afr. Soc. Anim. Prod., 2: 104-106.
SKINNER, J. D. and LOUW, G. N. (1966): Heat stress and spermatogenesis in *Bos indicus* and *Bos taurus* cattle. J. appl. Physiol., 21, 6: 1784-1790.
STEWART, R. E. (1973): Environmental research with cattle since 1968. Presented Ann. Meeting A.S.A.E., Univ. Kentucky. [17]
STOTT, G. H., WIERSMA, F. and WOODS, J. M. (1972): Reproductive health program for cattle subjected to high environmental temperatures. J. Amer. vet. med. Ass., 161, 11: 1339-1344. [18]
SWEEMAN, W. J. (1950): Artificial breeding in Alaska and the effect of extra light during the short winter days. J. Dairy Sci., 33: 391-392.
TEMPLE, R. S. (1965): 14th Annual beef cattle short course. Flo. agric. exp. Stat., Univ. Florida.
TEMPLE, R. S. (1972): Temperate and indigenous beef cattle breeds in the tropics. Wld Anim. Prodc., 7: 39-44. [15]
VAN RENSBURG, S. J. (1967): Gestation in sheep after foetal adrenalectomy and cortical acetate administration. J. Endocr., 38-83.
VENTER, H. A. W. (1970): The influence of growth rate on the reproductive performance of young beef bulls. M.Sc. Thesis, Univ. Pretoria, S. Afr.
VENTER, H. A. W., BONSMA, J. C. and SKINNER, J. D. (1972): The influence of climate on the reproduction of cattle. Int. J. Biometeor., 17: 147-151. [19]
VINCENT, C. K. (1972): Effects of season and high environmental temperative on fertility in cattle. J. Amer. vet. med. Ass., 161, 11: 1334-1338. [20]
VORSTER, T. H. (1954): The influence of certain environmental factors and breed differences on cattle production. Rhod. agric. J., 51: 428-473.
YEATES, N. T. M. (1958): Foetal dwarfism in sheep, an effect of high atmospheric temperature during gestation. J. agric. Sci. (Camb.), 51: 34-89.
YEATES, N. T. M. (1965): Modern aspects of animal production Butterworths Publ. Co., London. [21]

Chapter 1, Section 15b (1): Effects of Heat on Reproduction of Sheep (C. W. ALLISTON)

ABILAY, T. A., JOHNSON, H. D. and SEIF, S. (1973): Heat effect on plasma steroid in Zebu and Highlands. J. Anim. Sci., 37: 298 (Abstract).
ALLISTON, C. W. and ULBERG, L. C. (1961): Early pregnancy loss in sheep at ambient temperatures of 70° and 90°F as determined by emtryo transfer. J. Anim. Sci., 20: 608-613.
ALLISTON, C. W., EGLI, G. E. and ULBERG, L. C. (1961): Loss of potential young in the ewe due to high ambient temperature. J. appl. Physiol., 16: 253-256.
ALLISTON, C. W., HOWARTH, B., Jr. and ULBERG, L. C. (1965): Embryonic mortality following culture in vitro of one- and two-cell rabbit eggs at elevated temperatures. J. Reprod. Fertil., 9: 337-341.
ASDELL, S. A. (1946): Patterns of Mammalian Reproduction. Comstock Publ. Co., Ithaca, New York.
BELLVÉ, A. R. (1972): Viability and survival of mouse embryos following parental exposure to high temperatures. J. Reprod. Fertil., 30: 71-81.
BLIGH, J. (1973): Temperature regulation in mammals and other vertebrates. In: Frontiers of Biology, 30. A. Neuberger and E. L. Tatum (eds), Elsevier Publ. Co., Inc., Amsterdam/New York.
BROOKS, J. R. and ROSS, C. W. (1962): Effect of ambient temperature and thyroxine therapy on semen quality of rams. J. Anim. Sci., 21: 700-705.
CASADY, R. B., MYERS, R. M. and LEGATES, J. E. (1953): The effect of exposure to high ambient temperatures on spermatogenesis in the dairy bull. J. Dairy Sci., 36: 14-23.
COLLINS, K. J. and WEINER, J. S. (1968): Endocrinological aspects of exposure to high environmental temperatures. Physiol. Rev., 48: 785-839.
CRIBLEZ, T. L. and ALLISTON, C. W. (1974): Uridine incorporation and pyruvate metabolism in rabbit eggs cultured in vitro at normal and elevated temperatures. J. Reprod. Fertil., 37, 387-391.
DUNN, T. G., et al. (1972): Progesterone levels in ewes during induced estrus and in hot environments. J. Anim. Sci., 34: 903 (Abstract).
DUTT, R. H. and HAMM, P. T. (1957): Effect of exposure of environmental temperature and shearing on semen production of rams in winter. J. Anim. Sci., 16: 328-334.
DUTT, R. H. and SIMPSON, E. C. (1957): Environmental temperature and fertility of Southdown rams early in the breeding season. J. Anim. Sci., 16: 136-143.
DUTT, R. H. (1960): Temperature and light as factors in reproduction among farm animals. J. Dairy Sci., 43: 123-144.
EDWARDS, M. J. (1969a): Congenital defects in guinea pigs: fetal resorptions, abortions and malformations following induced hyperthermia during early gestation. Teratology, 2: 313-328.
EDWARDS, M. J. (1969b): Congenital defects in guinea pigs: prenatal retardation of brain growth of guinea pigs following hyperthermia during gestation. Teratology, 2: 329-336.
EDWARDS, M. J. (1971): The experimental production of *Arthrogryposis multiplex congenita* in guinea-pigs by maternal hyperthermia during gestation. J. Path. Bact., 104: 221-229.
EGLI, G. E., ALLISTON, C. W. and ULBERG, L. C. (1956): The effect of environmental temperature upon fertility of sheep. In: Proc. Ass. Southern agric. Workers, 53: 80-81 (Abstract).
ELLIOTT, D. S. (1968): Developmental Alterations in Mouse Embryos Subjected to Temperature Stress in vitro During the First Cell Division. Ph.D. Thesis, N. C. State Univ., Michigan.
ENTWISTLE, K. W. (1973): Ram fertility and fertilization rates in the ewe in a semi-arid tropical environment. Int. J. Biometeor., 17: 109-113.
GORET, E. A., JOHNSON, H. D. and HAHN, L. (1973): Plasma TSH in lactating cattle at 10 and 30 C. J. Anim. Sci., 37: 313 (Abstract).
HOWARTH, B., Jr., ALLISTON, C. W. and ULBERG, L. C. (1965): Importance of uterine environment on rabbit sperm prior to fertilization. J. Anim. Sci., 24: 1027-1032.
HOWARTH, B., Jr. and HAWK, H. W. (1968): Effect of hydrocortisone on embryonic survival in sheep. J. Anim. Sci., 27: 117-121.
HULET, C. V., et al. (1956): The nature of early-season infertility in sheep. J. Anim. Sci., 15: 607-616.

KELLY, C. F. and BOND, T. E. (1971): Bioclimatic factors and their measurement. In: A Guide to Environmental Research on Animals. Proc. nat. Acad. Sci. (Wash.): 7-92.
McKENZIE, F. F. and BERLINER, V. (1937): The reproductive capacity of the ram. Mo. agric. exp. Stat. Res. Bull., 265.
MILLER, H. L. and ALLISTON, C. W. (1972): Corticoid levels of Angus heifers to two temperature regimes. J. Anim. Sci., 35: 1112 (Abstract).
MILLER, H. L. and ALLISTON, C. W. (1973): Bovine LH and progesterone at two temperature regimes. J. Anim. Sci. 37: 320 (Abstract).
MOULE, G. R. and WAITES, G. M. H. (1963): Seminal degeneration in the ram and its relation to the temperature of the scrotum. J. Reprod. Fertil., 5: 433-446.
MOULE, G. R. (1970): Australian research into reproduction in the ram. Anim. Breed. Abstr., 38: 185-202.
PEPELKO, W. E. and CLEGG, M. T. (1965): Influence of season of the year upon patterns of sexual behavior in male sheep. J. Anim. Sci., 24: 633-637.
RHYNES, W. E. and EWING, L. L. (1973): Plasma corticosteroids in Hereford bulls exposed to high ambient temperature. J. Anim. Sci., 36: 369-373.
RICH, T. D. and ALLISTON, C. W. (1970a): Influence of programmed circadian temperature changes on the reproductive performance of rabbits acclimated to two different temperatures. J. Anim. Sci., 30: 960-965.
RICH, T. D. and ALLISTON, C. W. (1970b): Influence of programmed circadian temperature changes on the reproductive performance of ewes. J. Anim. Sci., 30: 966-969.
RYLE, M. (1961): Early reproductive failure of ewes in a hot environment. I. Ovulation rate and embryonic mortality. J. agric. Sci. (Camb.), 57: 1-7.
SHEARER, R. W. and McCARTHY, B. J. (1967): Evidence of ribonucleic acid molecules restricted to the cell nucleus. Biochemistry, 6: 283-289.
SHEEAN, L. A., DURRANT, B. S. and ULBERG, L. C. (1973): Temperature induced alterations of ^{3}H-leucine incorporation into mouse embryos. J. Anim. Sci., 37: 329 (Abstract).
SHELTON, M. and HUSTON, J. E. (1968): Effects of high temperature stress during gestation on certain aspects of reproduction in the ewe. J. Anim. Sci., 27: 153-158.
SMITH, S. Y., YECK, R. G. and HAZEN, T. E. (1971): Development of research facilities. In: A Guide to Environmental Research on Amimals. Proc. nat. Acad. Sci. (Wash.): 267-305.
STEINBACH, J. (1973): Bioclimatic influences on the reproductive processes in swine in a humid tropical environment. Int. J. Biometeor., 17: 141-145.
STOTT, G. H. and WIERSMA, F. (1973): Climatic thermal stress, a cause of hormonal depression and low fertility in bovine. Int. J. Biometeor., 17: 115-122.
THWAITES, C. J. (1969): Embryo mortality in the heat stressed ewe. II. Application of hot-room results to field conditions. J. Reprod. Fertil., 19: 255-262.
TILTON, J. E. et al. (1972): Influence of adrenalectomy on embryonic mortality in thermally stressed ewes. J. Anim. Sci., 34: 605-608.
ULBERG, L. C. (1958): The influence of high temperature on reproduction. J. Hered., 49: 62-64.
VANJONACK, W. J., et al. (1973): Plasma glucocorticoids in lactating cattle. J. Anim. Sci., 37: 332 (Abstracts).
VENTER, H. A. W., BONSMA, J. C. and SKINNER, J. D. (1973): The influence of climate on the reproduction of cattle. Int. J. Biometeor., 17: 147-151.
WETTEMANN, R. P., et al. (1973): Effect of elevated ambient temperature on boars. J. Anim. Sci., 37: 332 (Abstract).
WHITEMAN, J. V. and BROWN, K. I. (1959): The effect of delayed shearing on ewes and daytime cooling of rams on late spring breeding performance. J. Anim. Sci., 18: 392-398.
WOODY, C. O. and ULBERG, L. C. (1964): Viability of one-cell sheep ova as affected by high environmental temperature. J. Reprod. Fertil., 7: 275-280.
YEATES, N. T. M. (1953): The effect of high air temperature on reproduction in the ewe. J. Agri. Sci., 43: 199-203.
YEATES, N. T. M. (1956): The effect of high air temperature on pregnancy and birth weight in Merino sheep. Aust. J. agric. Res., 7: 435-439.
YEATES, N. T. M. (1958): Foetal dwarfism in sheep – an effect of high atmospheric temperature during gestation. J. agric. Sci. (Camb.), 51: 84-89.
YOUNG, W. C. (1927): The influence of high temperatures on the guinea pig testis. J. exp. Zool., 49: 459-499.

Chapter 1, Section 15b (2): Breeding Management of Sheep (N. T. M. YEATES)

BRADEN, A. W. H. and MATTNER, P. E. (1970): The effects of scrotal heating in the ram on semen characteristics, fecundity, and embryonic mortality. Aust. J. agric. Res., 21: 509-518.

DUCKER, M. J. and BOWMAN, J. C. (1972): Photoperiodism in the ewe. 5. An attempt to induce sheep of three breeds to lamb every eight months by artificial daylength changes in a non-light-proofed building. Anim. Prod., 14: 323-334.

DUTT, R. H. and HAMM, P. T. (1957): Effect of exposure to high environmental temperature and shearing on semen production of rams in winter. J. anim. Sci., 16: 328-334.

ENTWISTLE, K. W. (1970): Fertilization rate and early embryonic mortality in ewes in north west Queensland. Proc. Aust. Soc. anim. Prod., 8: 353-357.

FOWLER, D. G. and KENNEDY, J. P. (1968): Skin folds and Merino breeding. 6. The effects of varying heat exposures and degree of skin fold on rectal, scrotal and testis temperatures. Aust. J. exp. Agric. Anim. Husb., 8: 133-141.

HAFEZ, E. S. E. (1952): Studies on the breeding season and reproduction of the ewe. J. agric. Sci. (Camb.), 42: 189-264.

HARRISON, R. G. (1949): The comparative anatomy of the blood supply of the mammalian testis. Proc. zool. Soc. Lond., 119: 325-344.

HULET, C. V., VOIGTLANDER, H. P., POPE, A. L. and CASIDA, L. E. (1956): The nature of early-season infertility in sheep. J. Anim. Sci., 15: 607-616.

LEE, D. H. K. (1950): Studies of heat regulation in the sheep with special reference to the Merino. Aust. J. agric. Res., 1: 200-216.

MACFARLANE, W. V., MORRIS, R. J. H. and HOWARD, B. (1958): Heat and water in tropical Merino sheep. Aust. J. agric. Res., 9: 217-228.

MASON, J. L. (1969): A Dictionary of Livestock Breeds. C.A.B., Farnham Royal, Buchs., England.

McGUIRK, B. J. (1969): Skin folds and Merino breeding. 8. Fertility of individual rams in flocks selected for high and low skin fold. Aust. J. exp. Agric. Anim. Husb. 9: 147-150.

MORGAN, P. D., ARNOLD, G. W. and BOUNDY, C. A. P. (1972): The behaviour of ewes and their new born lambs in a hot climate. Proc. Aust. Soc. Anim. Prod., 9: 371-375.

PARER, J. T. (1963): Wool length and radiant heating effects in sheep. J. agric. Sci. (Camb.), 60: 141-144.

PHILLIPS, R. W. and McKENZIE, F. F. (1934): The thermo-regulatory function and mechanism of the scrotum. Mo. Agric. exp. Stat. Res. Bull., 217.

RATHORE, A. K. and GEORGE, J. M. (1969): Scrotal wool cover and fertility in rams. Aust. vet. J., 45: 589.

RATHORE, A. K. (1970): Fertility of rams heated for 1, 2, 3 and 4 days, mated to superovulated ewes. Aust. J. agric. Res., 21: 355-358.

ROBINSON, T. J. (1974): Controlled intensive breeding of sheep in Europe. Proc. Aust. Soc. Anim. Prod., 10: 250-253.

SHELTON, M. and HUSTON, J. E. (1968): Effects of high temperatures stress during gestation on certain aspects of reproduction in the ewe. J. Anim. Sci., 27: 153-158.

SYMINGTON, R. B. (1960): Studies on the adaptability of three breeds of sheep to a tropical environment modified by altitude. J. agric. Sci., (Camb.), 55: 287-315.

THWAITES, C. J. (1965): Photoperiodic control of breeding activity in the Southdown ewe with particular reference to the effects of an equatorial light regime. J. agric. Sci. (Camb.), 65: 57-64.

THWAITES, C. J. (1971): Short term heat stress and embryo mortality in the ewe. Aust. J. exp. Agric. Anim. Husb., 11: 265-267.

YEATES, N. T. M. (1949): The breeding season of the sheep with particular reference to its modification by artificial means using light. J. agric. Sci (Camb.), 39: 1.

YEATES, N. T. M. (1956): The effect of high air temperature on pregnancy and birth weight in Merino sheep. Aust. J. agric. Res., 7: 435-439.

YEATES, N. T. M. (1958): Foetal dwarfism in sheep – an effect of high atmospheric temperature during gestation. J. agric. Sci. (Camb.), 51: 84-90.

Chapter 1, Section 15c: Effects of Heat on Reproduction of Swine (J. STEINBACH)

ABILAY, J. P. and ARGANOSA, V. G. (1962): Causes of death of pigs from birth to weaning age. Philipp. Agric., 46: 551-559.
AHLSCHWEDE, W. T. and ROBINSON, O. W. (1966): Influence of climatic conditions on litter size in swine. J. Anim. Sci., 25: 916 (Abstract).
ARGANOSA, V. G., PUYAOAN, R. B. and ALCANTARA, P. F. (1969): Reproductive performance of swine under commercial farming conditions. Philipp. Agric., 53: 246-253.
ARGANOSA, V. G., PUYAOAN, R. B. and DE RAMOS, M. B. (1970): Performance of pigs born in different years and months of the year. Philipp. J. Anim. Sci., 7: 29-37.
ARGANOSA, V. G. and PENALBA, F. F. (1971): The influence of sequence of birth and birth weight on pig performance. Philipp. Agric., 55: 119-124.
CHRISTENSON, R. K., TEAGUE, H. S., GRIFO, A. P. and ROLLER, W. L. (1972): The effect of high environmental temperature on the boar. Ohio Swine Res. Inform. Rep., R.S., 61-4: 19-23.
CORTEEL, J. M., SIGNORET, J. P. and DU MESNIL DU BUISSON, F. (1964): Variations saisonnières de la reproduction de la truie et facteurs favorisant l'anoestrus temporaire. Proc. Vth int. Congr. Anim. Reprod. and A.I., III: 536-540.
DUNCAN, D. L. and LODGE, G. A. (1960): Diet in Relation to Reproduction and the Viability of the Young, Pt. III, Pigs. Commonwlth Bureau Anim. Nutr. Rowett Res. Inst. techn. Communi, No. 21.
EDWARDS, R. L., OMTVEDT, I. T., TURMAN, E. J., STEPHENS, D. F. and MAHONEY, G. W. A. (1968): Reproductive performance of gilts following heat stress prior to breeding and in early gestation. J. Anim. Sci., 27: 1634-1637.
EGBUNIKE, G. N. and STEINBACH, J. (1972): Age changes in the testicular finction of boars reared in a tropical environment. Proc. VIIth int. Congr. Anim. Reprod. and A.I., III: 2087-2090.
FRASER, A. F. (1966a): Studies in piglet husbandry in Jamaica. I. The relationship of litter size to survival till weaning. Brit. vet. J., 122: 288-295.
FRASER, A. F. (1966b): Studies in piglet husbandry in Jamaica. II. Principle causes of loss between birth and weaning. Brit. vet. J., 122: 325-332.
GUPTA, S., BHATTACHARYA, S. and MOULICK, S. K. (1967): Studies on the weaning performance in Large White pigs. India J. vet. Sci. Anim. Husbandry, 37: 138-142.
HEITMAN, H., HUGHES, E. H. and KELLY, C. F. (1951): Effects of elevated ambient temperatures on pregnant sows. J. Anim. Sci., 10: 907-915.
JENSEN, A. H., YEN, J. T., GEHRING, M. M., BAKER, D. H., BECKER, D. E. and HARMON, B. G. (1970): Effects of space restriction and management on pre- and post-puberal response of female swine. J. Anim. Sci., 31: 745-750.
LAWRENCE, J. A., TURMAN, E. J., RICH, T., SHARP, A. and HILLIER, J. C. (1970): A study of seasonal changes in boar semen. Misc. Publ. Okla. agric. Exp. Stat., No. MP 84: 77-83.
LEGAULT, C. (1969): Etude statistique et génétique de performances d'élevage des etruies de la race Large White. 1. Effets du troupeau, de la période semestrielle, du numéro de portée et du mois de naissance. Ann. Génét. Sél. anim. 1: 281-298.
MAGNANI, G. (1966): Investigations into possible climatic influences on fertility in the sow. Atti Soc. Ital. Sci. Vet. 20: 315-317, ABA 35: 3864.
MAHONEY, G. W. A., OMTVEDT, I. T., EDWARDS, R. L., STEPHENS, D. F. and TURMAN, E. J. (1970): Heat stress effects on reproduction in gilts. Trans. ASAE 13: 832-834.
MATHAI, E. and RAMACHANDRAN, P. (1972): A study on litter size and sex ratio in pigs. Kerala J. vet. Sci., 3: 1-5, ABA 41, No. 3146.
MAZZARRI, G. (1971): Effecto de la elevacion de la temperatura testicular sobre la espermatogenesis del verraeo. Agron. trop., 21: 171-182.
McNITT, J. I. and FIRST, N. L. (1970): Effects of 72-hour heat stress on semen quality in boars. Intern. J. Biometeor., 14: 373-380.
McNITT, J. I., TANNER, C. B. and FIRST, N. L. (1972): Thermoregulation in the scrotal system of the boar. I. Temperature distribution. J. Anim. Sci., 34: 112-116.

MERKEL, J. A. and HAZEN, T. E. (1967): Zone cooling for lactating sows. Trans. ASAE 10: 444-447.

MITCHELL, H. H. (1962): Comparative Nutrition of Man and Domestic Animals, I. Academic Press, New York.

MOORE, C. P., DUTT, R. H., HAYS, V. W. and CROMWELL, G. L. (1971): One-day vs. 14-day flushing and seasonal effects on reproduction in gilts. J. Anim. Sci. 33: 261 (Abstract).

MOULICK, S. K., GUPTA, S. and BHATTACHARYA, S. (1968): Studies on litter size and birth weight in Large White pigs. India J. vet. Sci. Anim. Husbandry, 38: 149-159.

NIWA, T. and MIZUHO, A. (1954): Studies on spermatogenic function in swine. II. On the seasonal variation in spermatogenic function. Bull. nat. Inst. agric. Sci. (Chiba) G, 8: 31-42, ABA 23: No. 1812.

OLUTOGUN, O. and STEINBACH, J. (1973): Seasonal effects on spermatozoan abnormality in porcine ejaculates. Proc. 14th Ann. Conf. Sci. Assoc. Nig.: 21 (Abstract).

OMTVEDT, I. T., NELSON, R. E., EDWARDS, R. L., STEPHENS, D. F. and TURMAN, E. J. (1971): Influence of heat stress during early, mid and late pregnancy of gilts. J. Anim. Sci., 32: 312-317.

PUYAOAN, R. B. and CASTILLO, L. S. (1963): Production and chemical composition of milk in sows. Philipp. Agric., 47: 141-149.

RASBECH, N. O. (1969): A review of the causes of reproductive failure in swine. Brit. vet. J., 125: 599-616.

ROLLER, W. L., TEAGUE, H. S. and GRIFO, A. P. (1967): Reproductive performance of swine in controlled warm environments. Trans. ASAE 10: 517-518, 522, ABA No. 720.

ROLLER, W. L., TEAGUE, H. S., CHRISTENSON, R. K. and GRIFO, A. P. (1973): Effect of ambient heat exposure upon swine reproduction. ASAE Paper 73-416.

SIGNORET, J. P. and DU MESNIL DU BUISSON, F. (1968): Influence de conditions d'habitat du verrat sur la fécondance du sperme. Proc. VIe Congr. int. Reprod. Anim. Insém. Artif., I: 317-319.

SMITH, D. M. (1959): The yield and composition of milk from sows fed varying proportions of separated milk and concentrates. N.Z. J. agric. Res., 2: 1057-1070.

STEINBACH, J. (1968): Effects of ambient temperatures during the processes of follicular development, ovulation, and implantation on litter size in swine. Proc. VIth int. Congr. Anim. Reprod. and A.I., I: 325-327.

STEINBACH, J. (1970): Effect of wallows on diurnal and gestational variations in the tespiratory frequency of pregnant sows. Resp. Physiol., 10: 64-73.

STEINBACH, J. (1971): Effects of season and breed on sow performance in the seasonal-equatorial climate of Southern Nigeria. J. agric. Sci. (Camb.), 77: 331-336.

STEINBACH, J. (1972a): Bioclimatic influences on sexual activity in boars. Proc. VIIth int. Congr. Anim. Reprod. and A.I., III: 2081-2086.

STEINBACH, J. (1972b): Bioclimatic influences on the characteristics of porcine ejaculates. In: Biometeorology, 5. S. W. Tromp and J. J. Bouma (eds). Proc. 6th int. biometeor. Congr. Swets & Zeitlinger, B.V., Amsterdam: 111 (Abstract).

STEINBACH, J. (1972c): The oestrous cycle of gilts in a tropical environment. Proc. VIIth Int. Congr. Anim. Reprod. and A.I., III: 2075-2080.

STEINBACH, J. (1972d): Changes in the water content of the uterus during the oestrous cycle of the gilt. Proc. VIIth int. Congr. Anim. Reprod. and A.I., II: 821-824.

STEINBACH, J. (1973): Bioclimatic influences on the reproductive processes in swine in a humid-tropical environment. Int. J. Biometeor., 17: 141-145.

TANHUECO, S. G. and RIGOR, E. M. (1964): Factos affecting weaning weights and number of pigs weaned. Philipp. Agric., 47: 445-453.

TEAGUE, H. S., ROLLER, W. L. and GRIFO, A. P. (1968): Influence of high temperature and humidity on the reproductive performance of swine. J. Anim. Sci., 27: 408-411.

TEAGUE, H. S. (1970): Effect of temperature and humidity on reproduction. In: Symp. Proc. Effects of Disease and Stress on Reproductive Efficiency in Swine. Ext. Serv., Univ. Nebraska, Coll. Agricult.

THIBAULT, C., COUROT, M., MARTINET, L., MAULÉON, P., DU MESNIL DU BUISSON, F., ORTAVANT, R., PELLETIER, J. and SIGNORET, J. P. (1966): Regulation of breeding season and estrous cycles by light and external stimuli in some

mammals. In: Environmental on Reproductive Processes. W. Hansel and R. H. Dutt (eds). J. Anim. Sci., Suupl. 25: 119-139.
TOMPKINS, E. C., HEIDENREICH, C. J. and STOB, M. (1967): Effect of post-breeding thermal stress on embryonic mortality in swine. J. Anim. Sci., 26: 377-380.
VANDENBRANDEN, M. (1960): Prix de revient de l'engraissement du porc "Yorkshire Large White" dans la région d'Elisabethville en janvier 1959. Bull. Inf., INEAC 9: 107-114.
VERSTEGEN, M. W. A., VAN ES, A. J. H. and NIJKAMP, H. J. (1971): Some aspects of energy metabolism of the sow during pregnancy. Anim. Prod., 13: 677-683.
WAGELIE, E., GARCIA, G. and CAPINPIN, J. (1959): A study on some physical and chemical characteristics of Bershire boar semen including effects of atmospheric temperature at collecting time. Philipp. Agric., 43, 348-368.
WAITES, G. M. H. (1968): Temperature and fertility in mammals. Proc. VIth Int. Cong. Anim. Reprod. and A.I., I: 235-256.
WARNICK, A. C., WALLACE, H. D., PALMER, A. Z., SOSA, E., DUERRE, D. J. and CALDWELL, V. E. (1965): Effect of temperature on early embryo survival in gilts. J. Anim. Sci., 24: 89-92.
WETTEMANN, R. P., OMTVEDT, I. T., WELLS, M. E., POPE, C. E. and TURMAN, E. J. (1973): Effect of elevated ambient temperature on boars. J. Anim. Sci. 37: 332-333 (Abstract).
WHATLEY, J. A., PALMER, J. B., CHAMBERS, D. and STEPHENS, D. F. (1957): The value of water sprinklers for cooling pregnant sows during the summer. Okla. Agr. exp. Stat. MP, 48: 2-4.
WIGGINS, E. L., CASIDA, L. E. and GRAMMER, R. H. (1950): The effect of season of birth on sexual development in gilts. J. Anim. Sci., 9: 277-280.
WILLIAMS, T. A. (1971): Studies on lactation yield in sows. Student's Project Rep., Univ. Ibadan, unpublished.

Chapter 1, Section 15d: Effects of Heat on Reproduction of Rat (U. A. SOD-MORIAH and E. BEDRAK)

AMATAYAKUL, K., RYAN, R., UOZUMI, T. and ALBERT, A. (1971): A reinvestigation of testicular-anterior pituitary relationship in the rat. I. Effect of castration and cryptorchidism. Endocrinology, 72: 292-295.
BEDRAK, E., SAMOILOFF, V. and FINKELSTEIN, Z. (1971a): Testosterone biosynthesis in the desert mouse, Acomys cahirinus. J. Endocr., 51: 7-16.
BEDRAK, E., SAMOILOFF, V., SOD-MORIAH, U. A. and GOLDBERG, S. (1971b): Synthesis of androgens in vitro by testis of rats acclimatized to a hot environment. J. Endocr., 51: 489-498.
BEDRAK, E. (1972): Enzyme of androgen biosynthesis in the desert mouse Gerbillus dasyurus. Gen. comp. Endocr., 18: 524-533.
BEDRAK, E., SAMOILOFF, V. and SOD-MORIAH, U. A. (1973): Metabolism of 3β-hydroxy-5-ene steroids by testes of rats acclimatized to a hot environment. J. Endocr., 58-207-217.
BROBECK, J. R. (1960): Food and temperature. In: Recent Progress in Hormone Research. G. Pincus (ed.), Academic Press, New York, 16: 439-459.
CHANG, M. C. and FERNANDEZ-CANO, L. (1959): Effects of short changes of environmental temperature and low atmospheric pressure on the ovulation of rats. Amer. J. Physiol., 196: 653-655.
COLLINS, K. J. and WEINER, J. S. (1968): Endocrinological aspects of exposure to high environmental temperatures. Physiol. Rev., 48: 785-839.
DAVIS, J. R., MORRIS, R. N. and HOLLINGER, M. A. (1964): Incorporation of L-lysine-U-C^{14} into protein of cryptorchid testis slices. Amer. J. Physiol., 207: 50-54.
DAVIS, J. R., MORRIS, R. N. and HOLLINGER, M. A. (1965): Comparison of trsticular protein labelling in cryptorchidism induced in prepuberal and adult rats. J. Reprod. Fertil., 10: 149-152.
EDWARDS, M. J. (1968): Congenital malformations in the rat following induced hyperthermia during gestation. Teratology, 1: 173-178.

ELFVING, G. (1950): Effect of the local application of heat on the physiology of testis. An experimental study on rats. Thesis. T. A. Sshalan Kirjapaino Oz, Helsinki.

FERNANDEZ-CANO, L. (1958): Effect of increase or decrease of body temperature and hypoxia on pregnancy in the rat. Fertil. and Steril., 9: 455-459.

GELINEO, S. (1964): Organ systems in adaptation: the temperature regulating system. In: Handbook of Physiology. Sect. 4: Adaption to the Environment. D. B. Dill (ed.), Amer. physiol. Soc. (Wash.): 259-282.

GUPTA, S. R. and LACY, B. (1967): Effects of protein calorie deficiency on the reproductive performance of female rats. Indian J. med. Res., 55: 904-910.

HAFEZ, E. S. E. (1968): Environmental effects on animal productivity. In: Adaptation of Domestic Animals. E. S. E. Hafez (ed.), Lea and Febiger, Philadelphia; 74-93.

HAMILTON, C. L. (1963): Interactions of food intake and temperature regulation in the rat. J. comp. physiol. Psychol., 56: 476-488.

KARIEM, Z. and SAMUELS, L. T. (1974): Influenced of "FSH" and "FSH + LH" on steroidogenic enzymes in immature mouse ovaries. Endocrinology, 95: 660-668.

MACFARLANE, W. V., PENNYCUIK, P. R. and THRIFT, E. (1957): Resorption and loss of foetuses in rats living at 35°C. J. Physiol. (Lond.), 135: 451-459.

MACFARLANE, W. V., PENNYCUIK, P. R., YEATES, N. T. M. and THRIFT, E. (1959): Reproduction in hot environments. In: Recent Progress in the Endocrinology of Reproduction. C. W. Lloyd (ed.), Academic Press, New York; 81-96.

PENNYCUIK, P. R. (1964): The effects on rats of chronic exposure to 34°C. IV. Reproduction. Aust. J. biol. Sci., 17: 245-260.

RHYNES, W. E. and EWING, L. L. (1973): Testicular endocrine function in Hereford bulls exposed to high ambient temperature. Endocrinology, 92: 509-515.

SELYE, H. and HEUSER, G. (1955): Fifth Annual Report on Stress. Part I: The anterior lobe. H. Selye and G. Heuser (eds), M. D. Publ., New York; 53.

SHAH, M. K. (1956): Reciprocal egg transportation to study the embryo-uterine relationship in heat-induced failure of pregnancy in rabbits. Nature (Lond.), 177: 1134-1135.

SHIKITA, M. and HALL, P. F. (1967): The action of human chorionic gonadotrophin in vivo upon microsomal enzymes of immature rat testis. Biochim. biophys. Acta (Amst.), 136: 484-497.

SHORT, R. E., PETERS, J. B. and CASIDA, L. E. (1968): Effect of two levels of feeding an ovarian activity, embryo survival and ovarian compensatory hypertorophy in the rabbit. J. Anim. Sci., 27: 701-704.

SOD-MORIAH, U. A. and POLLACK, E. (1970): Food and water consumption and the reproductive performance of the heat-acclimated female rat. Int. J. Biometeor., 14: 293-300.

SOD-MORIAH, U. A. (1971a): Reproduction in the heat-acclimatized female rat as affected by high ambient temperature. J. Reprod. Fertil., 26: 209-212.

SOD-MORIAH, U. A. (1971b): Body weight and composition, oxygen and food consumption of female rats chronically exposed to 35°C. Int. J. Biometeor., 15: 92 (Abstract).

SOD-MORIAH, U. A. and BEDRAK, E. (1973): Corpora lutea activity in heat-acclimatized female rats. Acta Endocr. (Kbh.), Suppl. 177: 337 (Abstract).

SOD-MORIAH, U. A. and YAGIL, R. (1973): Hyperthermia in heat-adapted female rats. Comp. Biochem. Physiol., 46: 487-490.

SOD-MORIAH, U. A. and BEDRAK, E. (1974): FSH content of the anterior pituitary and the incorporation of L-lysine-U-C^{14} into testicular proteins of heat-acclimatized rats. Comp. Biochem. Physiol., 47A: 333-339.

SOD-MORIAH, U. A., GOLDBERG, G. M. and BEDRAK, E.: Intrascrotal temperature, testicular histology and fertility of heat-acclimatized rats. J. Reprod. Fertil., in press.

SUNDSTROEM, E. S. (1930): Contribution to tropical biochemistry and physiology. II. Supplementary experiments on rats adapted to graded levels of reduced cooling power. Univ. Calif. Publ. Physiol., 7: 103-195.

Chapter 1, Section 16: Effects of Temperature on Oviposition and Egg Formation in Poultry (W. O. WILSON)

BIELLIER, H. V. and OSTMANN, O. W. (1960): Effect of varying daylength on time of oviposition in domestic fowl. Mo. Agric. exp. Stat. Res. Bull., 747, 52 pp.

BYERLY, T. C. (1941): Feed and other cost of producing market eggs. Md Agric. exp. Stat. techn. Bull., A-1.

CAIN, J. R. (1941): Environmental effects of avian circadian rhythm. Ph.D. Thesis, Univ. Calif., U.S.A.

COMBS, G. F. (1968): Amino acid requirements of broilers and laying hens. Proc. Univ. Md Nutr. Conf.: 86-96.

CONRAD, R. M. (1939): The effect of high temperature on the blood calcium of the laying hen. Poultry Sci., 18: 327-329.

FRANCIS, D. W., REID, B. L. and WILSON, W. O. (1972): Effects of atmospheric stresses on the performance of poultry. Western Regional Project W-50. New Mexico Agric. exp. Stat. Bull., 601, 56 pp.

GILBERT, A. B. and WOOD-GUSH, D. G. M. (1971): Ovulatory and ovipositional cycles. In: Physiology and Biochemistry of the Domestic Fowl. D. J. Bell and B. M. Freeman (eds). Academic Press, London/New York, Chapt. 57.

GRAU, C. R. and KAMEI, M. (1949): Delayed oviposition observed in hens fed purified diets. Poultry Sci., 28: 469-471.

LEPKOVSKY, S. (1973): Newer concepts in the regulation of food intake. Amer. J. clin. Nutr., 26: 271-284.

LONGHOUSE, A. D., OTA, H. and ASHBY, W. (1960): Heat and moisture design data for poultry housing. J. agric. Eng., 41: 567-576.

MORRIS, T. R. and NALBANDOV, A. V. (1969): The induction of ovulation in starving pullets using mammalian and avian gonadotropins. Endocrinology, 68: 687-697.

MORRIS, T. R. (1973): The effects of ahemeral light and dark cycles on egg production in the fowl. Poultry Sci., 52: 423-445.

NORDSTROM, J. O. (1973): Duration of egg formation in chickens during heat stress. Poultry Sci., 52: 1687-1690.

OPEL, H. (1966): The timing of oviposition and ovulation in the quail (*Coturnix coturnix japonica*). Brit. Poultry Sci., 7: 29-38.

PAYNE, C. G., LINCOLN, D. W. and CHARLES, D. R. (1965): The influence of constant and fluctuating temperatures on time of oviposition under constant lighting. Brit. Poultry Sci., 6: 93-95.

PAYNE, C. G. (1967): The influence of environmental temperature on egg production. In: Environmental control in poultry production. T. C. Carter, (ed.). Oliver and Boyd, London: 40-54.

ROMIJN, C. and VREUGDENHIL, E. L. (1969): Energy balance and heat regulation in the White Leghorn fowl. Neth. vet. Sci., 2: 32-58.

SMITH, A. J. (1969): The effect of high ambient temperature on the productivity of laying pullets: their energy intake, egg production and weight of egg components. Proc. S. Afr. Soc. Anim. Prod., 8: 177-178.

SMITH, A. J. (1971): The productivity of laying pullets at high environmental temperatures. Feedstuffs, 43, 17: 26.

STURKIE, P. D. (1946): The effect of hypothermia upon the reproductive tract of the hen. Poultry Sci., 25: 369-372.

SUUKI, M. T., MANNEL, K. H., MATTI, S., FRIESECKE, H. K., AL-AZZAWI, K., ROLLINSON, D. H. L., HUSSEIN, A. N. and AL-RAWI, K. (1971): Studies on protein requirements of laying hens in relation to feed intake under dry heat stress in Iraq. UNFAO and Govt Iraq Ann. Husb. Res. and Training Project techn., Rep., 35.

SYKES, A. H. (1972): In: The energy cost of egg production in egg formation and production. B. M. Freeman and P. E. Lake (eds). Brit. Poultry Sci. Ltd.

VAN MIDELKOOP, J. H. (1972): The relationship between ovulation interval of White Plymouth Rock pullets and the laying of abnormal eggs. Arch. Geflügelk., 36: 223-230.

WARREN, D. C. (1939): Effect of temperature on size of eggs from pullets at different latitudes. J. agric. Res., 59: 441-452.

WARREN, D. C. and SCHNEPEL, R. W. (1940): Effect of air temperatures on egg-shell thickness in the fowl. Poultry Sci., 19: 67-72.
WILSON, W. O. (1949): High environmental temperatures affecting the reaction of laying hens to iodized casein. Poultry Sci., 28: 581-592.
WILSON, W. O., McNALLY, E. H. and HAJIME OTA (1957): Temperature and calorimeter study on hens in individual cages. Poultry Sci., 36: 1254-1261.
WILSON, W. O., MATHER, F. B. and TANAKA, K. (1965): Maintenance of egg production in coturnix following reduction in photoperiod. Poultry Sci., 44: 1299-1302.
WILSON, W. O., ITOH, S. and SIOPES, T. D. (1971): Production traits of Leghorn pullets in controlled temperatures. Poultry Sci., 51: 1014-1023.
WINGET, C. M., AVERKIN, E. G. and FRYER, T. B. (1965): Quantitative measurement by telemetry of ovulation and oviposition in the fowl. Amer. J. Physiol., 209: 853-858.
WOODARD, A. E. and WILSON, W. O. (1970): Behavioral patterns associated with oviposition in Japanese quail and chicken. J. interdiscipl. Cycle Res., 1: 173-180.

Chapter 1, Section 17: Effects of Temperature as a Function of Age in Mammals (W. BIANCA)

See References in P.H.B., Part IA, Chapter 6, Sect. 6a, pp. 644.

Chapter 1, Section 18a: Morphological and Anatomical Adaptations (W. G. BRADLEY and M. K. YOUSEF)

ALLEE, W. C. and SCHMIDT, K. P. (1951): Ecological Animal Geography. John Wiley and Sons, Inc., New York, 462-469.
ALLEN, J. A. (1877): The influence of physical conditions in the genesis of species. Radical Rev., 1: 108-140.
BABERO, B. B., YOUSEF, M. K., WAWERNA, J. C. and BRADLEY, W. G. (1973): Comparative histology of the respiratory apparatus of three desert rodents and the albino rat. Comp. Biochem. Physiol., 44A: 585-597.
BAKER, M. A. and HAYWARD, J. N. (1968): The influence of the nasal mucosa and the carotid rete upon hypothalmic temperature in sheep. J. Physiol. (Lond.), 198: 561-579.
BERGMANN, C. (1847): Über die Verhältnisse der Wärmeökonomie der Thiere zu ihrer Grösse. Göttinger. Studen. pt., 1: 595-708.
BLIGH, J. and JOHNSON, K. G. (1973): Glossary of terms for thermal physiology. J. appl. Physiol., 35: 941-961.
BRADLEY, W. G. and O'FARRELL, M. J. (1969): Temperature relationships of the western pipistrelle, *Pipistrellus hesperus*. In: Physiological Systems in Semiarid Environments. C. C. Hoff and M. L. Riedesel (eds). Univ. New Mexico Press, Albuquerque.
BRADLEY, W. G. and YOUSEF, M. K. (1972): Small mammals in the desert. In: Physiological Adaptations: Desert and Mountain. M. K. Yousef, S. M. Horvath and R. W. Bullard (eds). Academic Press, New York.
BRADLEY, W. G. and MAUER, R. A. (1973): Rodents of a creosote bush community in southern Nevada. Southwest. Nat., 17: 333-344.
BRADLEY, W. G., MILLER, J. S. and YOUSEF, M. K. (1974): Thermoregulatory patterns in pocket gophers: desert and mountain. Physiol. Zool., 47: 172-179.
BRADLEY, W. G. and YOUSEF, M. K. (1975): Thermoregulatory responses in the plains pocket gopher *Geomys bursarius*. Comp. Biochem. Physiol., 50A: 149-153.
DICE, L. R. (1947): Effectiveness of selection by owls of deer-mice (*Peromyscus maniculatus*) which contrast in color with their background. Contr. Lab. Vertebr. Biol., Univ. Mich., 34: 1-20.
FINCH, V. A. (1972): Energy exchanges with the environment of two East African antelopes, the eland and the hartebeest. Symp. zool. Soc. (Lond.), 31: 315-326.
FOLK, G. E., Jr. (1974): Textbook of Environmental Physiology. Lea and Febiger, Philadelphia, p. 200.
GLOGER, C. L. (1833): Das Abändern der Vögel durch Einfluss des Klimas. Breslau.

HAYWARD, J. N. (1968): Brain temperature regulation during sleep and arousal in the dog. Exp. Neurol., 21: 201-212.
HÉROUX, O. (1961): Climatic and temperature-induced changes in mammals. Rev. canad. Biol., 20: 55-68.
HOCK, R. J. (1965): An analysis of Glogers rule. Hvalrådets Skrifter, Nr. 48: 214-226.
JACKSON, D. C. and SCHMIDT-NIELSON, K. (1964): Counter-current heat exchange in the respiratory passages. Proc. nat. Acad. Sci. (Wash.), 51: 1192-1197.
KANWISHER, J. and SUNDNES, G. (1966): Thermal regulations in cetaceans. In: Whales dolphins and porpoises. K. S. Norris (ed.). Univ. Calif. Press, Berkeley.
KROG, J. O. and WIKA, M. (1974): The circulation in the growing reindeer antlers. In: Proc. 1st int. Reindeer/Caribou Symp. Inst. Arct. Biol., Univ. of Alaska, Fairbanks, in press.
KUHN, W. and RYFFEL, K. (1942): Herstellung Konzentrierter Lösungen aus verdünnten durch Membranwirkung. ein Modellversuch zur Funktion der Niere. Z. physiol. Chem., 276: 145.
MacARTHUR, R. H. (1965): Patterns of species diversity. Biol. Rev., 40: 510-533.
MAYR, E. (1963): Animal Species and Evolution. Harvard Univ. Press, Camb.: 304-235.
McNAB, B. K. (1966): The metabolism of fossorial rodents: A study of convergence. Ecology, 47: 712-733.
McNAB, B. K. (1971): On the ecological significance of Bergmann's Rule. Ecology, 52: 845-848.
MERRIAM, C. H. (1895): Monographic revision of the pocket gophers, family Geomyidae (exclusive of the species *Thomomys*). U.S. Dept Agric. North Amer. Fauna, No. 8: 1-258.
MORRISON, P. (1966): Insulative flexibility in the guanaco. J. Mamm., 47: 18-23.
NELSON, E. W. (1918): Wild Animals of North America. Nat. geogr. Soc. (Wash.).
PETTER, F. (1952): Note préliminaire sur l'éthologie et l'écologie de *Psammomys obesus* Cretzchmar. Mammalia, 16: 137-147.
REEDER, W. G. and COWLES, R. B. (1951): Aspects of thermoregulation in bats. J. Mamm., 32: 389-403.
RENSCH, B. (1936): Studien über klimatische Parallelität der Merkmalsausprägung bei Vögeln und Säugern. Arch. Naturg. (N. F.), 5: 317-363.
RENSCH, B. (1960): Evolution Above the Species Level. Columbia University Press, New York: 39-46.
RICKLEFS, R. E. (1973): Ecology. Chiron Press, Newton, Mass.: 179-215, 701-742.
ROSENZWEIG, M. L. (1968): The strategy of body size in mammalian carnivores. Amer. Midl. Nat., 80: 299-315.
RÜMMLER, H. (1938): Die Systematik und Verbreitung der Muriden Neuguineas. Mitt. zool. Mus. (Berl.), 23: 1-298.
SCHMIDT-NIELSEN, K. (1961): Concentrating mechanism of the kidney from a comparative viewpoint. Amer. Heart J., 62: 579-586.
SCHMIDT-NIELSEN, K. (1964): Desert Animals: Physiological Problems of Heat and Water. Oxford Univ. Press, London, 277 p.
SCHMIDT-NIELSEN, K. (1970): Animal Physiology. Prentice-Hall, Inc., Englewood Cliffs.
SCHOENER, T. W. (1969): Optimal size and specialization in constant and fluctuating environments: an energy time approach. Brookhaven Symp. Biol., 22: 103-114.
SCHOLANDER, P. F., WALTERS, V., HOCK, R. and IRVING, L. (1950): Body insulation in some arctic and tropical mammals and birds. Biol. Bull., 99: 225-236.
SCHOLANDER, P. F. (1955): Evolution of climatic adaptation in homeotherms. Evolution, 9: 15-26.
SCHOLANDER, P. F. and KROG, J. (1957): Countercurrent heat exchange and vascular bundles in sloths. J. appl. Physiol., 10: 405-411.
SEMPER, K. G. (1881): Animal life as affected by the natural conditions of existence. Appleton, New York.
SUMNER, F. B. (1925): Some biological problems of our south-western deserts. Ecology, 6: 352-371.
TAYLOR, C. R. (1966): The vascularity and possible thermoregulatory function of the horns of goats. Physiol. Zool., 39: 127-139.
TAYLOR, C. R. (1972): The desert gazelle: A paradox resolved. In: Comparative Physiology of Desert Animals, G. M. O. Maloiy (ed.). Academic Press, New York, p. 215.

Chapter 1, Section 18b: Adaptation to Climatic Zones and the Ecophysiology of Ruminants (W. V. MACFARLANE)

BLAIR-WEST, J. R., BROOK, A. H. and SIMPSON, P. A. (1972): Renin responses to water restriction and rehydration. J. Physiol. (Lond.), 226: 1-14.

BREWER, N. L. and MACFARLANE, W. V. (1967): Structure and secretion of the naso-labial glands of cattle. Aust. J. exp. Biol. med. Sci., 45: 37.

HOWARD, D. A., BURDIN, H. L. and LAMPKIN, G. H. (1962): Variation in the mineral and crude protein content of pastures at Muguga in the Kenya highlands. J. agric. Sci. (Camb.), 59: 251-262.

McEWAN, E. H. (1970): Energy metabolism of the barren-ground caribou (*Rangifer tarandus*). Canad. J. Zool., 48: 391-392.

MACFARLANE, W. V., MORRIS, R. J. H., HOWARD, BETH, McDONALD, JANET and BUDTZ-OLSEN, O. E. (1961): Water and electrolyte changes in tropical Merino sheep exposed to dehydration during summer. Aust. J. agric. Res., 12: 889-912.

MACFARLANE, W. V. (1964): Terrestrial animals in dry heat: ungulates. In: Handbook of Physiology. Sect. 4: Adaptation to the Environment. D. B. Dill, E. F. Adolph and C. G. Wilber (eds.). Amer. physiol. Soc. (Wash.).

MACFARLANE, W. V., HOWARD, B. and SIEBERT, B. D. (1967): Water metabolism of Merino and Border Leicester sheep grazing saltbush. Aust. J. agric. Res., 18: 947-958.

MACFARLANE, W. V., KINNE, R., WALMSLEY, C. M., SIEBERT, B. D. and PETER, D. (1967): Vasopressins and the increase of water and electrolyte excretion by sheep, cattle and camels. Nature (Lond.), 214, 979-981.

MACFARLANE, W. V. (1968): Comparative studies in the adaptation of ruminants to hot environments. In: The Adaptation of Domestic Animals. E. S. E. Hafez (ed.). Lea and Febiger, Philadelphia: 164-182.

MACFARLANE, W. V. and HOWARD, B. (1970): The ecophysiology of water. In: The Physiology of Digestion and Nutrition in Ruminants. A. T. Phillipson (ed.). Oriel. Press, Newcastle.

MACFARLANE, W. V. (1971): Salinity and the whole animal. In: Salinity and Water Use. T. Talsma and J. R. Philip (eds). MacMillan, London.

MACFARLANE, W. V., HOWARD, B., HAINES, H., KENNEDY, P. J. and SHARP, C. M. (1971): The hierarchy of water and energy turnover of mammals living in deserts. Nature (Lond.), 234, 483-484.

MACFARLANE, W. V. and HOWARD, BETH (1972): Comparative water and energy economy of wild and domestic mammals. Symp. zool. Soc. (Lond.), 31: 261-296.

MACFARLANE, W. V., HOWARD, B. and GOOD, B. F. (1974): Use of tracers in field measurements of water, milk and thyroxine metabolism of tropical ruminants. In: Tracer Techniques in Tropical Animal Production. IAEO/FAO Symposium, Djakarta. Int. Atomic Energy Agency, Vienna, 1-23.

MALOIY, G. M. O. (1972): Renal salt and water excretion in the camel (*Camelus dromedarius*). Symp. zool. Soc. (Lond.), 31: 243-260.

PEIRCE, A. W. (1968): Salt tolerance studies of sheep VII. The tolerance of ewes and their lambs in pens for drinking waters of types obtained from underground sources in Australia. Aust. J. agric. Res., 19: 577-587.

SHKOLNIK, A., BORUT, A. and CHOSHNIAK, J. (1972): Water economy of the beduin goat. In: Comparative Physiology of Desert Animals. G. M. O. Maloiy (ed.). Symp. zool. Soc. (Lond.), 31: 229-242.

SIEBERT, B. D. and MACFARLANE, W. V. (1969): Water turnover and renal function of dromedaries in the desert. Physiol. Zool., 44, 225-240.

SIMPSON, G. G. (1953): The major features of evolution. Clarion, New York.

UCKO, P. J. and DIMBLEBY, G. W. (1969): The Domestication and Exploitation of Plants and Animals. Duckworth Publ. Co., London.

ZEUNER, F. E. (1963): A History of Domesticated Animals. Hutchinson Publ. Co., London.

Chapter 1, Section 18c (1): Adaptability of Different Breeds of Cattle to Heat (J. E. VERCOE)

JOHNSON, K. G. (1970): Sweating rate and electrolyte content of skin secretions of *Bos taurus* and *Bos indicus* cross-bred cows. J. agric. Sci. (Camb.), 75: 397-402.

KENNEDY, J. F. and CHIRCHIR, G. I. K. (1971): A study of the growth rate of F_2 and F_3 Africander cross, Brahman cross and British cross cattle from birth to 18 months in a tropical environment. Aust. J. exp. Agric. Anim. Husb., 11: 593-598.

O'KELLY, J. C. (1973a): Changes in lipid metabolism in genetically different types of calves during chronic hyperthermia. Brit. J. Nutr., 30: 211-220.

O'KELLY, J. C. (1973b): Plasma lipid changes in genetically different types of cattle during hyperthermia. Comp. Biochem. Physiol., 44: 313-320.

SCHLEGER, A. V. and TURNER, H. G. (1965): Sweating rates of cattle in the field and their reaction to diurnal and seasonal changes. Aust. J. agric. Res., 16: 92-106.

SEEBECK, R. M. (1973): Sources of variation in the fertility of a herd of zebu, British and zebu x British cattle in Northern Australia. J. agric. Sci. (Camb.), 81: 253-262.

SPRINGELL, P. H. (1968): Water content and water turnover in beef cattle. Aust. J. agric. Res., 19: 129-144.

TURNER, H. G. (1962): Effect of clipping the coat on performance of calves in the field. Aust. J. agric. Res., 13: 180-192.

VERCOE, J. E. (1969): The effect of increased rectal temperature on nitrogen metabolism in Brahman cross and Shorthorn x Hereford steers fed on lucerne chaff. Aust. J. agric. Res., 20: 607-612.

VERCOE, J. E. (1970): In: Energy Metabolism of Farm Animals. A. Schürch and C. Wenk (eds). Juris Druck Verlag, Zurich: 85 (EAAP Publs. 13).

VERCOE, J. E. and FRISCH, J. E. (1970): The effect of increased rectal temperature on nitrogen metabolism in Brahman cross and Shorthorn x Hereford steers fed on a low nitrogen roughage. Aust. J. agric. Res., 21: 857-863.

VERCOE, J. E. (1971): The effect of water loading on urea metabolism in cattle. J. agric. Sci. (Camb.), 76: 487-493.

VERCOE, J. E., FRISCH, J. E. and MORAN, J. B. (1972): Apparent digestibility, nitrogen utilization, water metabolism and heat tolerance of Brahman cross, Africander cross and Shorthorn x Hereford steers. J. agric. Sci. (Camb.), 79: 71-74.

VERCOE, J. E. and FRISCH, J. E. (1974): Fasting metabolism, liveweight and voluntary feed intake of different breeds of cattle. In: Proc. 6th Symp. Energy Metab., Stuttgart. EAAP Publ. 14: 131-134.

Chapter 1, Section 18c (2): Adaptability and Thermoregulatory Behaviour of Pigs (D. L. INGRAM)

BALDWIN, B. A. and INGRAM, D. L. (1967a): Behavioural thermoregulation in pigs. Physiol. Behav., 2: 15-21.

BALDWIN, B. A. and INGRAM, D. L. (1967b): The effect of heating and cooling the hypothalamus on behavioural thermoregulation in the pig. J. Physiol. (Lond.), 191: 375-392.

BALDWIN, B. A. and INGRAM, D. L. (1968a): The role of behaviour in the regulation of body temperature. Vet. Ann., 9: 130-142.

BALDWIN, B. A. and INGRAM, D. L. (1968b): Factors influencing behavioural thermoregulation in pigs. Physiol. Behav., 3: 409-415.

BALDWIN, B. A. and INGRAM, D. L. (1968c): The effect of temperature acclimatization and food intake on behavioural thermoregulation in pigs and mice. Physiol. Behav., 3: 395-400.

BALDWIN, B. A. and LIPTON, J. M. (1973): Central and peripheral temperatures and EEG changes during behavioural thermoregulation in pigs. Acta Neurobiol. exp., 33: 433-447.

BRAY, C. I. and SINGLETARY, C. B. (1948): Effect of hog wallows on gains of fattening swine. J. Anim. Sci., 7: 521-522.

CARLISLE, H. J. and INGRAM, D. L. (1973): The effects of heating and cooling the spinal cord and hypothalamus on thermoregulatory behaviour in the pig. J. Physiol. (Lond.), 231: 353-364.

CLOSE, W. H., MOUNT, L. E. and START, I. B. (1971): The influence of environmental temperature and plane of nutrition on heat losses from groups of growing pigs. Anim. Prod., 13: 285-294.
HEITMAN, H. and HUGHES, E. H. (1949): The effects of air temperature and relative humidity on the physiological well being of swine. J. Anim. Sci., 8: 171-181.
HEITMAN, H., HAHN, L., BOND, T. E. and KELLY, C. F. (1962): The effects of modified summer environment on behaviour. Anim. Behaviour, 10: 15-19.
HOLMES, C. W. and MOUNT, L. E. (1967): Heat loss from groups of growing pigs under various conditions of environmental temperature and air movement. Anim. Prod., 9: 435-452.
INGRAM, D. L. (1965): Evaporative cooling in the pig. Nature (Lond.), 207: 415-416.
INGRAM, D. L. and LEGGE, K. F. (1970a): The thermoregulatory behaviour of young pigs in a natural environment. Physiol. Behav., 5: 981-987.
INGRAM, D. L. and LEGGE, K. F. (1970b): Variations in deep body temperature in the young unrestrained pig over the 24-hr period. J. Physiol. (Lond.), 210: 989-998.
MORRISON, S. R. and MOUNT, L. E. (1971): Adaptation of growing pigs to changes in environmental temperature. Anim. Prod., 13: 51-57.
MOUNT, L. E. (1960): The influence of huddling and body size on the metabolic rate of the young pig. J. agric. Sci. (Camb.), 55: 101-105.
MOUNT, L. E. (1963): The environmental temperature preferred by the young pig. Nature (Lond.), 199: 122-123.
MOUNT, L. E. (1964): Radiant and convective heat loss from the new-born pig. J. Physiol. (Lond.), 173: 96-113.
MOUNT, L. E. and INGRAM, D. L. (1965): The effects of ambient temperature and air movement on localized sensible heat loss from the pig. Res. vet. Sci., 6: 84-91.
MOUNT, L. E. (1968): The climatic physiology of the pig. Edward Arnold Publ. Co., London.
STEPHENS, D. B. (1971): The metabolic rates of new-born pigs in relation to floor insulation and ambient temperature. Anim. Prod., 13: 303-313.
VERSTEGEN, M. W. A. (1971): The influence of environmental temperature on energy metabolism of growing pigs housed individually and in groups. Meded. Landbouwhogeschool, Wageningen: 71-2.
VERSTEGEN, M. W. A., CLOSE, W. H., START, I. B. and MOUNT, L. E. (1973): The effects of environmental temperature and plane of nutrition on heat loss, energy retention and deposition of protein and fat in groups of growing pigs. Brit. J. Nutr., 30: 21-25.

Chapter 1, Section 18c (3): Genetic Differences in the Adaptability of Rodents to Temperature Extremes (Pamela R. PENNYCUIK)

BAKER, R. L. and COCKREM, F. R. M. (1970): Selection for body weight in the mouse at three temperatures and the correlated response in tail length. Genetics (Princeton), 65: 505-523.
BARNETT, S. A. (1959): The skin and hair of mice living at a low environmental temperature. Quant. J. exp. Physiol., 44: 35-42.
BARNETT, S. A. (1961): Some effects of breeding mice for many generations in a cold environment. Proc. roy Soc. B., 155: 115-135.
BARNETT, S. A. and SCOTT, S. G. (1963): Some effects of cold and hybridity on the growth of mice. J. Embryol. exp. Morph., 11: 35-51.
BARNETT, S. A. (1965): Adaptation of mice to cold. Biol. Rev., 40: 5-51.
BARNETT, S. A. and WIDDOWSON, E. H. (1965): Organ weight and body composition of mice bred for many generations at —3°C. Proc. roy. Soc. B., 162: 502-516.
BIGHAM, M. L. and COCKREM, F. (1969): Body weights, tail lengths, body temperatures, food intakes and some slaughter data for four strains of mice reared at three different environmental temperatures. N.Z. J. agric. Res., 12: 658-668.
BROWN, J. H. (1968): Adaptation to environmental temperature in two species of woodrats, *Neotoma cinerea* and *N. albigula*. Misc. Publ. Mus. Zool. (Univ. Mich.), 135: 5-48.
CASSUTO, Y. (1968): Metabolic adaptations to chronic heat exposure in the golden hamster. Amer. J. Physiol., 214: 1147-1151.

CASSUTO, Y., CHAYOTH, R. and RABI, T. (1970): Thyroid hormone in heat acclimated hamsters. Amer. J. Physiol., 218: 1287-1290.
CHAIM, A. and BORUT, A. (1973): Differences in heat production among populations of the golden spiny mouse. 5th Ann. Symp. Desert Physiology, 1973, Beer-Sheva, Israel (Abstract).
CLARK, R. V. (1971): Behavioural thermoregulation by the white rat at high ambient temperatures. J. exp. Zool., 178: 387-392.
COCKREM, F. (1959): Selection for relationships opposite to those predicted by the genetic correlation between two traits in the house mouse (*Mus musculus*). Nature (Lond.), 183: 342-343.
CRARY, D. D. and SAWIN, P. B. (1960): Genetic differences in growth rate and maturation of rabbits. Growth, 24: 111-130.
FALCONER, D. S. (1954): Validity of the theory of genetic correlation. An experimental test with mice. J. Hered., 45: 42-44.
FALCONER, D. S. (1960): Introduction to quantitative genetics. Oliver and Boyd Ltd., Edinburgh/London: 365 pp.
FOWLER, R. E. (1962): The efficiency of food utilization, digestibility of foodstuffs and energy expenditure of mice selected for large and small body size. Genet. Res., 3: 51-68.
GORECKI, A. and KRZANOWSKA, H. (1970): Oxygen consumption in two inbred mouse strains and F1 hybrids. Bull. Acad. pol. Sci., Cl.2, 18: 115-119.
HARRISON, G. A. (1958): The adaptability of mice to high environmental temperatures. J. exp. Biol., 35: 892-901.
HARRISON, G. A. (1963): Temperature adaptation as evidenced by growth of mice. Fed. Proc., 22: 691-698.
KINDER, E. F. (1927): A study of the nest-building activity of the albino rat. J. exp. Zool., 47: 117-161.
KING, J. A., MAAS, D. and WEISMAN, R. G. (1964): Geographic variation in nest size among species of *Peromyscus*. Evolution, 18: 230-234.
LAIRD, A. K. and HOWARD, A. (1967): Growth curves of inbred mice. Nature (Lond.), 213: 786-788.
LEE, C. T. (1970): Nest building behaviour of inbred mice: developmental and genetic studies. Diss. Abstr., 31B: 931.
LYNCH, C. B. and HEGMANN, J. P. (1972): Genetic differences influencing behaviourial temperature regulation in small mammals. I. Nesting by *Mus musculus*. Behavior Genet., 2: 43-53.
OGILVIE, D. M. (1970): Behavioural thermoregulation and cold adaptation in two strains of white mice. Int. J. Biometeor, 14: 285-291.
PENNYCUIK, P. R. (1967): A comparison of the effects of a variety of factors on the metabolic rate of the mouse. Aust. J. exp. Biol. med. Sci., 45: 331-346.
PENNYCUIK, P. R. (1969a): Reproductive performance and body weights of mice maintained for 12 generations at 34°C. Aust. J. biol. Sci., 22: 667-675.
PENNYCUIK, P. R. (1969b): Physiological and morphological characteristics of mice moved to 21°C after 10 generations at 34°C. Aust. J. biol. Sci., 22: 677-687.
PENNYCUIK, P. R. (1971): Effect of acclimatization to 33°C on the oxygen uptake, growth rate and reproductive productivity of hairless and naked mice. Aust. J. biol. Sci., 24: 301-310.
PENNYCUIK, P. R. (1972): Inheritance of patterns of oxygen consymption in mice. Aust. J. biol. Sci., 25: 1031-1037.
ROBERTS, R. C. (1965): Some contributions of the laboratory mouse to animal breeding research. Part I. Genetic analysis of growth and body weight. Anim. Breed. Abstr., 33: 339-353.
ROUBICEK, C. B. and RAY, D. E. (1971): Direct and transfer response of rats selected for growth at two environmental temperatures. Growth, 35: 1-10.
SCHLESINGER, K. and MORDKOFF, A. M. (1963): Locomotor activity and oxygen consumption variability in two inbred strains of mice. J. Hered., 54: 177-182.

Chapter 1, Section 18d (1): Adaptation of Mammals to Thermal Stresses (W.V. MAC FARLANE)

See References in P.H.B., Part IB, Chapter 6, Sect. 19a, pp. 694-695.

Chapter 1, Section 18d (2): Seasonal Adaptations of Birds to Cold (G. C. WEST)

ASCHOFF, J. and POHL, H. (1970a): Der Ruheumsatz von Vögeln als Funktion der Tageszeit und der Körpergrösse. J. Ornithol., 111: 28-47.
ASCHOFF, J. and POHL, H. (1970b): Rhythms variations in energy metabolism. Fed. Proc., 29: 1541-1552.
BARNETT, B. (1970): Seasonal changes in temperature acclimatization of the house sparrow, *Passer domesticus*. Comp. Biochem. Physiol., 33: 559-578.
BAUDINETTE, R. V., LOVERIDGE, J. P., MILLS, C. D., SCHMIDT-NIELSEN, K. and WILSON, K. J. (1974): Vascular responses to temperature in the Herring Gull (*Larus argentatus*). Fed. Proc., 33: 408.
BERGER, M. and HART, J. S. (1972): Die Atmung beim Kolibri *Amazilia fimbriata* während des Schwirrfluges bei verschiedenen Umgebungstemperaturen. J. comp. Physiol., 81: 363-380.
BLACKMORE, F. (1969): The effect of temperature, photoperiod and molt on the energy requirements of the house sparrow, *Passer domesticus*. Comp. Biochem. Physiol., 30: 433-444.
BRAUN, C. and ROGERS, G. E. (1971): The white-tailed ptarmigan in Colorado. Colo. Div. Game, Fish Parks techn. Publ., 27: 1-80.
BUDD, S. M. (1973): Thermoregulation, bioenergetics and endocrinology of cold acclimatization of black-capped chickadees, *Parus atricapillus*. Ph.D. Thesis, Cornell Univ., 123 pp. (Unpublished).
CALDER, W. A. (1974): Thermal and caloric relations of birds. In: Avian Biology. D. S. Farner and J. R. King (eds). Academic Press, New York: 259-413.
CHATFIELD, P. O., LYMAN, C. P. and IRVING, L. (1963): Physiological adaptations to cold of peripheral nerve in the leg of the Herring Gull (*Larus argentatus*). Amer. J. Physiol., 172: 639-644.
DARYDOV, A. F. (1971): Seasonal fluctuations of energy metabolism and thermoregulation in the black-headed gull at rest. Ekologiya, 2: 59-63. (In Russian).
DAVIS, E. A., Jr. (1955): Seasonal changes in the energy balance of the English sparrow. Auk., 72: 385-411.
DAWSON, W. R. (1958): Relation of oxygen consumption and evaporative water loss to temperature in the cardinal. Physiol. Zool., 31: 37-48.
DAWSON, W. R. and HUDSON, J. W. (1970): Birds. In: Comparative Physiology of Thermoregulation, I. G. C. Whittow (ed.). Avademic Press, N.Y.: 223-310.
DAWSON, W. R. and TORDOFF, H. B. (1959): Relation of oxygen consumption to temperature in the evening grosbeak. Condor, 61: 388-396.
DONTCHEFF, L. and KAYSER, C. (1934): Le rythme saisonnier du metabolisme de base chez le pigeon en fonction de la température moyenne du milieu. Ann. Physiol. Physicochim. biol., 10: 285-300.
FREEMAN, B. M. (1970): Thermoregulatory mechanisms in the neonate fowl. Comp. Biochem. Physiol., 33: 219-230.
FREEMAN, B. M. (1971): Non-shivering thermogenesis in birds. In: Non-shivering Thermogenesis. L. Jansky (ed.). Swets & Zeitlinger N. V., Amsterdam: 83-96.
GELINEO, S. (1969): Heat production in birds in summer and winter. Bull. Acad. serbe Sci. Cl. Sci. mathém., 46: 99-105.
GESSAMAN, J. A. (1972): Bioenergetics of the Snowy Owl (*Nyctea scandiaca*). J. J. Arct. Alpine Res., 4: 223-238.
GROSSMAN, A. (1974): Temperature regulation and metabolism in the Alaskan black-capped chickadee. M.S. Thesis, Univ. Alaska (Unpublished).
HART, J. S. (1962): Seasonal acclimatization in four species of small wild birds. Physiol. Zool., 35: 224-236.
HERREID, C. E. and KESSEL, B. (1967): Thermal conductance in birds and mammals. Comp. Biochem. Physiol., 21: 405-414.
HYRTL, J. (1863): Neue Wondernetze und Geflechte bei Vögeln und Säugetieren. Denkschr. Kaiserlichen Akad. Wissensch. Wien, Math.-Naturwissensch. Classe, 22: 113-152.
IRVING, L. and KROG, J. (1955): Temperature of skin in the arctic as a regulator of heat. J. appl. Physiol., 7: 355-364.
IRVING, L., KROG, J. and MONSON M. (1955): The metabolism of some Alaskan animals in winter and summer. Physiol. Zool., 28: 173-185.

IRVING, L., WEST, G. C., PEYTON, L. J. and PANEAK, S. (1967): Migration of willow ptarmigan in Arctic Alaska. Arctic, 20: 77-85.
IRVING, L. (1972): Arctic Life of Birds and Mammals. Zoophysiology and Ecology 2. Springer-Verlag, New York, 192 pp.
KAHL, M. P. (1963): Thermoregulation in the wood stork, with special reference to the role of the legs. Physiol. Zool., 36: 141-152.
KENDEIGH, S. C. (1949): Effect of temperature and season on the energy resources of the English Sparrow. Auk, 66: 113-127.
KENDEIGH, S. C. (1961): Energy of birds conserved by roosting in cavities. Wilson Bull., 73: 140-147.
KENDEIGH, S. C. (1974): Bioenergetics. In: The Role and Control of Granivorous Birds in the Ecosystem. Kendeigh and Pinowski (eds). Cambridge University Press, London, in press.
KING, J. R. and FARNER, D. S. (1961): Energy metabolism, thermoregulation and body temperature. In: Biology and Comparative Physiology of Birds, 2. A. J. Marshall (ed.). Academic Press, New York: 215-288.
KLEIBER, M. (1961): The Fire of Life. John Wiley and Sons, New York, 454 pp.
LIN, Y-C. and STURKIE, P. D. (1968): Effect of environmental temperature on the catecholamines of chickens. Amer. J. Physiol., 214: 237-240.
LUSTICK, S. (1969): Bird energetics: effects of artificial radiation. Science, 163: 387-390.
McNAB, B. K. (1966): An analysis of the body temperature of birds. Condor, 68: 47-55.
MOREHOUSE, K. (1974): Energetics and nutrition of Pacific Brant. Ph.D. dissertation. Univ. Alaska, 104 pp.
MORTON, M. L. (1966): The effects of insulation on the diurnal feeding pattern of white-crowned sparrows (*Zonotrichia leucophrys gambelii*). Ecology, 48: 690-694.
MOSS, R. (1973): The digestion and intake of winter foods by wild ptarmigan in Alaska. Condor, 75: 293-300.
MURRISH, D. E. and GUARD, C. L. (1973): Sympathetic control of non-shivering thermogenesis in south polar skua chicks. Antarctic J., 8: 197-198.
MUSSEHL, T. W. (1960): Blue grouse production, movements and populations in the Bridger Mountains, Montana. J. Wildl. Managem., 24: 60-68.
NORTON, D. W. (1973): Ecological energetics of calidridine sandpipers breeding in northern Alaska. Ph.D. dissertation. Univ. Alaska, Fairbanks, 163 pp. (Unpublished).
POHL, H. (1971): Seasonal variation in metabolic function of bramblings. Ibis, 113: 185-193.
POHL, H. (1972): Seasonal changes in light sensitivity in *Carduelis flammea*. Naturwissenschaften, 59: 518.
POHL, H. and WEST, G. C. (1973): Daily and seasonal variation in metabolic response to cold during rest and forced exercise in the common redpoll. Comp. Biochem. Physiol., 45A: 851-867.
RAUTENBERG, W. (1957): Vergleichende Untersuchungen über den Energiehaushalt der Bergfinken (*Fringilla montifringilla* L.) und des Haussperlings (*Passer domesticus* L.). J. Ornithol., 98: 36-64.
RIDDLE, F. G., SMITH, G. C. and BENEDICT, F. G. (1934): Seasonal and temperature factors and their determination in pigeons of percentage metabolism per degree of temperature change. Amer. J. Physiol., 107: 333-342.
RODBARD, S. (1950): Weight and body temperature. Science, 111: 465-466.
ROSENMANN, M. and MORRISON, P. (1974): Maximum oxygen consumption and heat loss facilitation in small homeotherms by He-O_2. Amer. J. Physiol., 226: 490-495.
SCHOLANDER, P. F., WALTERS, V., HOCK, R. and IRVING, L. (1950): Body insulation of some arctic and tropical mammals and birds. Biol. Bull., 99: 225-236.
SEIBERT, H. C. (1949): Differences between migrant and non-migrant birds in food and water intake at various temperatures and photoperiods. Auk, 66: 128-153.
STEEN, I. and STEEN, J. B. (1965): The importance of legs in the thermoregulation of birds. Acta physiol. scand., 63: 285-291.
STEEN, J. B. (1958): Climatic adaptations in some small northern birds. Ecology, 39:625-629.
STEEN, J. B. and ENGER, P. S. (1957): Muscular heat production in pigeons during exposure to cold. Amer. J. Physiol., 191: 157-158.
SULKAVA, S. (1969): On small birds spending the night in the snow. Aquilo. Ser. Zool., 7: 33-37.

TANNER, J. T. (1952): Black-capped and Carolina chickadees in the southern Appalachian Mountains. Auk, 69: 407-424.
VEGHTE, J. H. and HERREID, C. F. (1965): Radiometric determination of feather insulation and metabolism of arctic birds. Physiol. Zool., 38: 267-275.
WALLGREN, H. (1954): Energy metabolism of two species of the genus *Emberiza* as correlated with distribution and migration. Acta zool. fen., 84: 1-110.
WEEDEN, R. B. (1964): Spatial separation of sexes in rock and willow ptarmigan in winter. Auk, 81: 534-541.
WEKSTEIN, D. R. and ZOLMAN, J. R. (1968): Sympathetic control of homeothermy in the young chick. Amer. J. Physiol., 214: 908-912.
WEST, G. C. (1960): Seasonal variation in the energy balance of the tree sparrow in relation to migration. Auk, 77: 306-329.
WEST, G. C. (1962): Responses and adaptations of wild birds to environmental temperature. In: Comparative Physiology of Temperature Regulation. J. P. Hannon and E. G. Viereck (eds). Arct. aeromed. Lab., Ft. Wainwright, Alaska: 291-333.
WEST, G. C. (1965): Shivering and heat production in wild birds. Physiol. Zool., 38: 111-120.
WEST, G. C. and HART, J. S. (1966): Metabolic responses of evening grosbeaks to constant and to fluctuating temperature. Physiol. Zool., 39: 171-184.
WEST, G. C. (1968): Bioenergetics of captive willow ptarmigan under natural conditions. Ecology, 49: 1135-1145.
WEST, G. C. and MENG, M. S. (1968): Seasonal changes in bcdy weight and fat and the relation of fatty acid composition to diet in the willow ptarmigan. Wilson Bull., 80: 426-441.
WEST, G. C. (1972a): Seasonal differences in resting metabolic rate of Alaskan ptarmigan. Comp. Biochem. Physiol., 42A: 867-870.
WEST, G. C. (1972b): The effect of acclimation and acclimatization on the resting metabolic rate of the common redpoll. Comp. Biochem. Physiol., 43A: 293-310.
WEST, G. C. and NORTON, D. W. (1974): Metabolic adaptations of tundra birds. In: Physiological Adaptations to the Environment. F. J. Vernberg (ed.). Intext Educational Publ., New York.
WEST, G. C. and POHL, H. (1975): Energy demands of high latitude resident redpolls. Oecologia, in press.
WETMORE, A. (1921): A study of the body temperature of birds. Smithsonian Misc. Collect., 72: 1-51.
WILLSON, M. F. and HARMESON, J. C. (1973): Seed preference and digestive efficiency of cardinal and song sparrows. Condor, 75: 225-234.
ZIMMERMAN, J. L. (1965): Bioenergetics of the Dickcissel, *Spiza americana*. Physiol. Zool., 38: 370-389.

Chapter 1, Section 19: Indices for Heat Adaptability of Domestic Animals (T. H. KAMAL)

ABDEL-GHANI, W. and HATHOUT, M. K. (1966): Effect of month of calving on milk yield of an imported Friesian Herd. Agric. Res. Rev., (Cairo), 44: 24-30.
BENEZRA, M. V. (1954): A new index for measuring the adaptability of cattle to tropical conditions. J. Anim. Sci., 13: 1015.
BIANCA, W. (1963): Rectal temperature and respiratory rate as indicators of heat tolerance in cattle. J. agric. Sci. (Camb.), 60: 113-120.
BONSMA, J. C. (1949): Breeding cattle for increased adaptability to tropical and subtropical environments. J. agric. Sci. (Camb.), 39: 204-221.
BROWN, W. H., SHANKLIN, M. D., HAHA, G. L. and JOHNSON, H. D. (1969): Rate of rectal temperature rise as an index of heat sensitivity. Trans. ASAE, 12: 225-227 and 230.
CARTWRIGHT, T. C. (1955): Responses of beef cattle to high ambient temperatures. J. Anim. Sci., 14: 350-362.
EL-FOULY, H. A. (1974): The use of the marker techniques in nutritional studies in the rumen. Ph.D. Thesis, Cairo Univ., 103 pp.
JOHNSON, H. D. (1965): Response of animals to heat. Meteor. Monogr., 6: 109-122.
KAMAL, T. H., JOHNSON, H. D. and RAGSDALE, A. C. (1962): Metabolic reactions

during thermal stress (35° to 95°F) in dairy animals acclimated at 50° and 80°F. Mo. Agric. exp. Stat. Res. Bull., 785.
KAMAL, T. H. (1965): Physiological reactions of cows to hot environmental conditions. Proc. Symp. Use of Radioisotopes in Animal Nutrition and Physiol. (Prague 23-27 Nov. 1964) Proc. Series IAEA, Vienna: 767-792.
KAMAL, T. H. and SEIF, S. M. (1969): Effect of natural and controlled climates of the Sahara on virtual tritium space in Friesians and water Buffaloes. J. Dairy Sci., 52: 1657-1663.
KAMAL, T. H., CLARK, J. L. and JOHNSON, H. D. (1970): The effect of age on heat tolerance in cattle as determined by the whole body ^{40}K and nitrogen retention. Int. J. Biometeor., 14: 301-308.
KAMAL, T. H. and JOHNSON, H. D. (1970): Whole body ^{40}K loss as a predictor of heat tolerance in cattle. J. Dairy Sci., 53: 1734-1738.
KAMAL, T. H. and JOHNSON, H. D. (1971): Total body solids loss as a measure of a short-term heat stress in cattle. J. Anim. Sci., 32: 306-311.
KAMAL, T. H., SHEHATA, O. and ELBANNA, I. M. (1972): Effect of heat and water restriction on water metabolism and body fluid compartments in farm animals. Proc. Symp. Isotope Studies on the Physiology of Domestic Animals Proc. Series IAEA, Vienna: 95-102.
KIBLER, H. H. (1957): Energy metabolism and cardiorespiratory activities in Shorthorn, Santa Gertrudis and Brahman heifers during growth at 50 and 80°F temperatures. Mo. Agric. exp. Stat. Res. Bull., 643.
KIBLER, H. H. and YECK, R. G. (1959): Vaporization rates and heat tolerance in growing Shorthorn, Brahman and Santa Gertrudis calves raised at constant 50 and 80°F temperatures. Mo. Agric. exp. Stat. Res. Bull., 701.
LEE, D. H. K. and PHILLIPS, R. W. (1948): Assessment of the adaptability of livestock to climatic stress. J. Anim. Sci., 7: 391-425.
MACFARLANE, W. V. (1963): Endocrine functions in hot environments. In: Environmental Physiology and Psychology in Arid Conditions. Rev. Res., UNESCO (Paris): 153-222.
McDOWELL, R. E., LEE, D. H. K., FOHRMAN, M. H., SYKES, J. F. and ANDERSON, R. (1955): Rectal temperature and respiratory responses of Jersey and Sindhi-Jersey (F_1) crossbred females to a standard hot atmosphere. J. Dairy Sci., 38: 1037-1045.
McDOWELL, R. E. (1966): The role of physiology in animal production for tropical and subtropical areas. Wld Rev. Anim. Prod., 1: 39-46.
RHOAD, A. O. (1944): The Iberia heat tolerance test for cattle. Trop. Agric. (Trin.), 21: 162-164.
TURNER, H. G. and SCHLEGER, A. V. (1960): The significance of coat type in cattle. Aust. J. Agric. Res., 11: 645-663.
VERNON, E. H., DAMON, R. H., HARVEY, W. R., WARWICK, E. J. and KINCAID, C. M. (1959): Relation of heat tolerance determinations to productivity in beef cattle. J. Anim. Sci., 18: 91-94.
YECK, R. G. and KIBLER, H. H. (1958): Predicting heat tolerance from calf vaporization rates. J. Anim. Sci., 17: 1228-1229.
See also references in P.H.B., Part IA, Chapter 4, Sect. 2, pp. 138-145 and 620-621.

Chapter 1, Section 20a: Effects of Heat and Cold on Cattle Disease (R. W. DOUGHERTY)

BIANCA, W. (1959): Reaktionen des Rindes auf Kälte und Wärme. Züchtungsk., 31, 9/10: 476-489.
BINNS, W. et al. (1965): Chronologic evaluation of teratogenicity in sheep fed with *Veratrum californicum*. J. Amer. vet. med. Ass., 147, 8: 839-842.
COMMITTEE ON PHYSIOLOGICAL EFFECTS OF ENVIRONMENTAL FACTORS IN ANIMALS: A guide to environmental research on animals. Proc. nat. Acad. Sci. (Wash.).
EWBANK, R. (1968): An experimental demonstration of the effect of surface cooling upon the health of the bovine mammary gland. Vet. Res., 83: 685-686.

HUGES, D. E. et al. (1965): Ultraviolet radiation and *Moraxella bovis* in the etiology of bovine infectious kerato conjunctivitis. Amer. J. vet. Res., 26, 15: 1331-1338.
MIKAJLOVIC, B. et al. (1972): Effect of some microclimatic factors on artificial infections with bovine respiratory viruses. Acta vet. (Beogr.), 22, 6.
OLSEN, J. D. (1969): Physiologic response of cattle to controlled cold exposure: aortic blood pressure, heart rate and respiratory rate. Amer. J. vet. Res., 30, 1: 47-50.
OLSEN, J. D. and TRENKEL, A.: Exposure of cattle to controlled subzero temperature: growth hormone, glucose and free fatty acid concentrations in plasma. J. Pap. J-7072, Iowa Agric. Home Econ. exp. Stat., Ames, Iowa, Project 1790.
PUHAE et al. (1972): A study of the influence of high ammonia and humidity concentrations in the air on the course of artificial infection with IBR virus and glucose, alkaline phosphatase, cholinesterase in corticosteroids concentrations in the blood of steers. Acta vet. (Beogr.), 22, 4.
SHAW, K. E. and NICHOLS, R. E. (1964): Plasma 17-hydroxycorticosteroid in calves: the effects of shipping. Amer. J. vet. Res., 25: 252-254.
See also references in P.H.B., Part III, Chapter 10, Sections 6 and 7, pp 93-117 and 146-151.

Chapter 1, Section 20b: Effects of Heat and Cold on Diseases in Sheep (M. E. HUGH-JONES)

ALEXANDER, G., PETERSON, J. E. and WATSON, R. H. (1959): Neonatal mortality in lambs: intensive observations during lambing in a corriedale flock with a history of high lamb mortality. Aust. vet. J., 35: 433-441.
ALEXANDER, G. and PETERSON, J. E. (1961): Neonatal mortality in lambs. Intensive observation during lambing in a flock of maiden merino ewes. Aust. vet. J., 37: 371-381.
ALEXANDER, G. (1962): Temperature regulation in the newborn lamb IV the effect of wind and evaporation of water from the coat on metabolic rate and body temperature. Aust. J. agric. Res., 13: 82-99.
ALEXANDER, G. and WILLIAMS, D. (1966): Heat stress and growth of the conceptus in sheep. Proc. Aust. Soc. Anim. Prod., 6: 102-105.
ALEXANDER, G. and WILLIAMS, D. (1966): Teat seeking activity in newborn lambs: the effects of cold. J. agric. Sci. (Camb.), 67: 181-189.
ALEXANDER, G. (1970): Thermogenesis in young lambs. In: Physiology of Disgestion and Metabolism in the Ruminant. A. T. Phillipson (ed.). Oriel Press, Newcastle-upon-Tyne: 199-210.
ALEXANDER, G. and WILLIAMS, D. (1971): Heat stress and development of the conceptus in domestic sheep. J. agric. Sci. (Camb.), 76: 53-72.
ALEXANDER, G. (1972): Neonatal survival in sheep, cattle and pigs: a general approach. Victorian vet. Proceedings, 1971-72, 30: 70-73.
ALEXANDER, G. (1972): Survival of the newborn: nutritional factors, environmental and behavioural hazards. Victoria Vet. Proc., 1971-72, 30: 73-77.
ALEXANDER, G., THORBURN, G., NICOL, D. and BELL, A. W. (1972): Survival, growth and the metabolic response to cold in prematurely delivered lambs. Biol. Neonate, 20: 1-8.
ALLOWAY, B. J. (1973): Copper and molubdenum in swayback pastures. J. agric. Sci. (Camb.), 80: 521-524.
ARMSTRONG, D. G., BLAXTER, K. L., GRAHAM, N. McC. and WAINMAN, F. W. (1959): The effect of environmental conditions on food utilisation by sheep. Anim. Prod., 1: 1-12.
BENNETT, J. W. (1972): The maximum metabolic response of sheep to cold: effects of rectal temperature, shearing, feed consumption, body weight. Aust. J. agric. Res., 23: 1045-1058.
BLAXTER, K. L. (1962): The energy metabolism of ruminants. Hutchinson Publ. Co., London.
BLAXTER, K. L., CLAPPERTON, J. L. and WAINMAN, F. W. (1966): The extent of difference between six British breeds of sheep in their metabolism, feed intake and utilisation, and resistance to climatic stress. Brit. J. Nutr., 20: 283-294.
BRADEN, A. W. H. and MATTNER, P. E. (1970): The effect of scrotal heating in the ram on semen characteristics, Fecundity and embryonic mortality. Aust. J. agri. Res., 21: 509-518.

CARTWRIGHT, G. A. (1972): The influence of high ambient temperatures during gestation on the development of wool follicules in the foetal lamb. Aust. Soc. reprod. Biol., 28: 151-152.
COOPER, N. M. (1966): Winter housing of breeding ewes. Proc. BVA Congr. Vet. Rec. 79:
DAVIES, G. B. (1973): Effect of weather on conception rates of sheep. N.Z. Vet. J., 21: 54.
DUTT, R. H. (1963): Critical period for early embryonic mortality in ewes exposed to high ambient temperatures. J. Anim. Sci., 22: 713-719.
EDEY, T. N. (1969): Prenatal mortality in sheep: a review. Anim. Breed. Abstr., 37: 173-190.
EGAN, J. K. (1972): Death in newborn lambs. Exp. Agric. Anim. Husbandry, 12: 470-473.
ENTWHISTLE, K. W. (1973): Reproduction in sheep and cattle in the Australian arid zones. In: Studies of the Australian arid zone 2. Animal Production in the Arid Zone. A. D. Wilson (ed.). CSIRO, Melbourne.
GEMMELL, R. T., BELL, A. W. and ALEXANDER, G. (1972): Morphology of adipose cells in lambs at birth and during subsequent transition of brown to white adipose tissue in cold and in warm conditions. Amer. J. Anatomy, 133: 143-164.
GEYTENBEEK, P. E. (1963): Experimental Record. Dept of Agric., S. Aust., No. 1: 21 (quoted by Hutchinson, 1968).
HALES, J. R. S. and HUTCHINSON, J. C. D. (1971): Metabolic, respiratory and vasimotor responses to heating the scrotum of the ram. J. Physiol. (Lond.), 212: 353-375.
HALLIDAY, R., SYKES, A. R., SLEE, J., FIELD, A. L. and RUSSEL, A. J. F. (1969): Cold exposure of southdown and welsh mountain sheep. 4. Changes in concentration of free fatty acids, glucose, acetone, protein-bound iodine, and antibodies in the blood. Anim. Prod., 11: 479-491.
HUTCHINSON, J. C. D., BENNETT, J. W. and TOMASZEWSKA, M. W. (1960): Cold stress after shearing with a snow-comb. Proc. Aust. Soc. Anim. Prod., 3: 199-204.
HUTCHINSON, J. C. D. (1968): Deaths of sheep after shearing. Aust. J. exp. Agric. Anim. Husb., 8: 393-400.
HUTCHINSON, J. C. D. and BENNETT, J. W. (1969): Design and use of plastic coats to protect shorn sheep. CSIRO, Div. Anim. Physiol., Tech. Rep. No. 1/69.
HUTCHINSON, J. C. D. and BENNETT, J. W. (1970): Further tests of plastic coats for shorn sheep. CSIRO, Div. Anim. Physiol. techn. Rep. No. 2/10.
HUTCHINSON, K. J. (1966): Bodyweight criteria and post-shearing deaths. Proc. Aust. Soc. Anim. Prod., 6: 299.
HUTCHINSON, K. J. and MCRAE, B. H. (1969): Some factors associated with the behaviour and survival of newly shorn sheep. Aust. J. agric. Res., 20: 513-521.
INGLISS, J. S. S., WEIFERS, M. and PEARCE, P. J. (1969): Hypomagnesaemia in sheep. Vet. Rec., 71: 755-763.
JOHNSON, H. D. (1965): Environmental temperature and metabolism, Int. J. Biometeor., 9: 103-116.
MACFARLANE, W. V., MORRIS, R. J. H. and HOWARD, B. (1958): Heat and water in tropical merino sheep. Aust. J. agric. Res., 9: 217-228.
MACKAY, J. M. K., NISBET, D. I. and SLEE, J. (1971): A high incidence of Jaagsiekte following experimental cold exposures in housed sheep. Int. J. Biometeor., 15: 65-69.
MORGAN, P. D., ARNOLD, G. W. and BOUNDY, C. A. P. (1972): The behaviour of ewes and their newborn lambs in a hot climate. Proc. Aust. Soc. Anim. Prod., 9: 35-7.
MOULE, G. R. (1970): Australian research into reproduction in the ram. Anim. Breed. Abstr., 38: 185-202.
MUNRO, J. (1962): The use of natural shelter by hill sheep. Anim. Prod. 4: 343-349.
NICHOLS, J. E. (1960): Shelter needs in relation to land use. Forestry, 33: 8-12.
OBST, J. M. and DAY, H. R. (1968): The effect of inclement weather in mortality of merino and corriedale lambs on Kangaroo Island. Proc. Aust. Soc. Anim. Prod., 7: 239-242.
OBST, J. M. and EVANS, E. V. (1970): Genotype-environment interactions in lambs mortality with particulary reference to birth coat and haemoglobin type. Proc. Aust. Soc. Anim. Prod., 8: 149-158.
PANARETTO, B. A. and VICKERY, M. R. (1970): The rates of plasma cortisol entry and clearance in sheep before and during their exposure to a cold, wet environment. J. Endocr., 47: 273-285.
PEART, J. N. (1967): The effects of different levels of nutrition during late pregnancy on the subsequent milk production of Blackface ewes and on the growth of their lambs. J. agric. Sci. (Camb.)., 68: 365-371.

PURSER, A. F. and KARAM, H. A. (1967): Lamb survival, growth and fleece production in relation to birth coat type among welsh mountain sheep. Anim. Prod., 9: 75-85.
ROBINSON, J. J. (1973): Some aspects of ewe nutrition. Vet. Rec. 92: 602-606.
RUSSELL, A. J. F., DONEY, J. M. and REID, R. L. (1967): The use of biochemical parameters in controlling nutritional states in pregnant ewes, and the effect of undernourishment during pregnancy on lamb birthweights. J. agric. Sci. (Camb.), 68: 351-358.
RUSSELL, A. J. F., DONEY, J. M. and REID, R. C. (1967): Energy requirements of the ewe. J. agric. Sci. (Camb.), 68: 359-363.
RYLE, M. (1961): Early reproductive failure of ewes in a hot environment. I. Ovulation rates and embyonic mortality. J. agric. Sci. (Camb.), 57: 1-9.
SETCHELL, B. P. (1962): Poisoning of sheep with antihelminthic doses of carbon tetrachbride. A summary of the histories of 185 mortalities. Aust. Vet. J., 38: 487-494.
SHARMAN, G. A. M. (1973): The health of the sheep during investigations into the intensification of sheep production. Vet. Rec., 92: 607-608.
SHELTON, M. (1964): Relation of environmental temperature during gestation to birth weight and mortalitv of lambs. J. Anim. Sci., 23: 360-365.
SHELTON, M. and HOUSTON, J. E. (1968): Effects of high temperature stress during gestation on certain aspects of reproduction in the ewe. J. Anim. Sci., 27: 153-158.
SLEE, J. (1971): Physiological factors affecting the energy cost of cold exposures. Proc. Nutr. Soc., 30: 215-221.
SMITH, I. D. (1961): Thermoregulation in the newborn merino. Aust. Vet. J., 37: 205-210.
SMITH, L. P. (1970): Weather and animal disease. W.M.O., Geneva, Techn. Note, 113.
SYKES, A. R., FIELD, A. C. and SLEE, J. (1969): Cold exposure of southdown and welsh mountain sheep. 3. Changes in plasma calcium, phisphorus, magnesium, sodium and potassium levels. Anim. Prod., 11: 91-99.
THWAITES, C. J. (1967): Embryo mortality in the heat stressed ewe. I. The influence of breed. J. Reprod. Fertil., 14: 5-14.
THWAITES, C. J. (1969): Embryo mortality in the heat stressed ewe. II. Application of hot-room results to field conditions. J. Reprod. Fertil., 19: 255-261.
THWAITES, C. J. (1971): Short-term heat stress and embryo mortality in the ewe. Aust. J. exp. Agric. Anim. Husbandry, 11: 265-267.
VOISIN, A. (1963): Grass tetany. (Trans. by Catherine T. M. Herriott). Crosby Lockwood & Son Ltd., London.
WAITES, G. M. H. and ORTAVANT, R. (1968): Early effects of a short elevation of testicular temperature on spermatogenesis in the ram. Ann. Biol. Anim., Biochem. Biophys., 8: 323-331.
WAITES, G. M. H. (1970): Temperature regulation and the testes. In: The Testes, 1. Johnson, A. D., Gomes, W. R., Vandemark, N. K. (eds), Academic Press, New York: 241-279.
WATT, J. A. (1965): Abortion and neonatal losses in sheep. Vet. Rec., 77: 627-642.
WATT, J. A. (1966): Indoor sheep husbandry; associated conditions. B.V.A. Congr., Vet. Rec., 79: 750-756.
WEBSTER, W. M. (1939): Review of the past season's work upon the problem of infertility in Romney Rams. Proc. 8th Ann. Meeting of Sheep Farmers: 85+94 (quoted by Davies, 1973).
WHEELER, J. L., REARDEN, T. F. and LAMBOURNE, L. J. (1963): The effect of pasture availability and shearing stress on herbage intake of grazing sheep. Aust. J. agric. Res., 14: 364-372.
YEATES, N. T. M. (1958): Foetal dwarfism in sheep. An effect of high atmospheric temperature during gestation. J. agric. Sci. (Cambr.), 51: 84-89.

Chapter 1, Section 20c: Effects of Heat and Cold on Diseases in Birds (E. SIMENSEN and L. D. OLSON)

ALBERTS, J. O. and GRAHAM, R. (1948): Fowl cholera in turkeys. N. Amer. Vet., 29: 24-26.
BRANDLY, C. A., MOSES, H. E., JONES, E. E. and JUNGHERR, E. L. (1946): Epizootiology of Newcastle disease of poultry. Amer. J. vet. Res., 7: 243-249.
HANSON, R. P. (1972): Newcastle Disease. In: Diseases of Poultry, 6th Ed. Hofsted (ed.) Iowa State University Press, Ames.

HUDSON, D. B. (1931): The influence of environmental temperature on the mortality in chicks inoculated with the virus of infectious bronchitis. Poultry Sci., 10: 391.
JORDAN, F. T. W. (1966): A review of the literature on infectious laryngotracheitis (ILT). Avian Dis., 10: 1-26.
JUSZKIEWICZ, T. (1966): Hyperthermia and prednisolone acetate as provocative factors of *Pasteurella multocida* infection in chickens. Pol. Arch. Weterynar., 10: 141-151.
MAAS, H. J. L. and VOUTE, E. J. (1961): Een praktijkonderzoek over de blauwe kam ziekte bij hoenders in "het land van Weert". Vlaams diergeneesk. T., 30: 106.
OLESIUK, O. M. (1951): Influence of environmental temperature on viability of Newcastle disease virus. Amer. J. vet. Res., 12: 152-155.
SINHA, S. K., HANSON, R. P. and BRANDLY, C. A. (1957): Effect of environmental temperature upon facility of aerosol transmission of infection and severity of Newcastle disease among chickens. J. infect. Dis., 100: 162-168.
SINOVIC, B. (1970): The influence of environmental temperature, age, sex, and breed on the mortality from *infectious laryngotracheitis*. 14th. Wld Poultry Congr., 3: 285-290.
VAN NESS, G. (1953): Weather influence in blue comb. in chickens. Science, 118: 601-602.

Chapter 1, Section 20d: Effects of Heat and Cold on Diseases of Arthropods (W. O. HAUFE)

BEARD, R. L. (1945): Studies on the milky disease of Japanese beetle larvae. Conn. Agric. exp. Stat. Bull., 491: 505-581.
BURDETTE, W. J. (1959): Tumors in *Drosophila*. Texas Univ. Publ., 5914: 57-68.
BURNSIDE, C. E. and REVELL, I. L. (1948): Observations on nosema disease of honey bees. J. Econ. Entomol., 41: 603-607.
DUTKY, S. R. (1959): Insect microbiology. Advanc. appl. Microbiol., 1: 175-200.
DUTKY, S. R. (1963): The milky diseases. In: Insect Pathology, 2. E. A. Steinhaus (ed.), Academic Press, New York: 75-115.
FRANZ, J. M. (1961): Biologische Schädlingsbekämpfung. In: Handbuch der Pflanzenkrankheiten. H. Richter (ed.), 2nd Ed., 6, 3. Paul Parey Verlag, Berlin: 1-302.
GOETZE, G. and ZEUTZSCHEL, B. (1959): Nosema disease of honey bees and its control with drugs – a review of research work since 1954. Bee Wld, 40: 217-225.
HARSHBARGER, J. C. and TAYLOR, R. L. (1968): Neoplasms of insects. Ann. Rev. Entomol., 13: 159-190.
HAUFE, W. O. (1974): Weather and pest management. In: Progress in Plant Biometeorology, L. P. Smith (ed.). Swets & Zeitlinger, B.V., Amsterdam.
HEIMPEL, A. M. (1967): A critical review of *Bacillus thuringiensis* var. *thuringiensis* Berliner and other crystalliferous bacteria. Ann. Rev. Entomol., 12: 287-322.
JULIAN, ST. G. and HALL, H. H. (1968): Infection of *Popillia japonica* larvae with heat-activated spores of *Bacillus popilliae*. J. Invert. Pathol., 10: 48-53.
KRAMER, J. P. (1959): Observations on the seasonal incidence of microsporidiosis in European corn borer populations in Illinois. Entomophaga, 4: 37-42.
KROMAN, A. R. (1957): An investigation of a hereditary melanotic tumor in *Drosophila melanogaster*: tu^{51},. Ph.D. Thesis, Univ. Minnesota, U.S.A. (Unpublished).
MARTIGNONI, M. E. (1964): Pathophysiology in the insect. Ann. Rev. Entomol., 9: 179-206.
PAILLOT, A. (1933): L'infection chez les insectes. G. Patissier Trévoux, 535 pp.
SMITH, H. S. (1939): Insect populations in relation to biological control. Ecol. Monogr., 9: 311-320.
STEINHAUS, E. A. (1949): Principles of Insect Pathology. McGraw-Hill, New York, 757 pp.
TANADA, Y. (1954): A polyhedrosis virus of the imported cabbage worm and its relation to a polyhedrosis virus of the alfalfa caterpillar. Ann. entomol. Soc. Amer., 47: 553-574.
VAGO, C. (1951): Phénomènes de "Latentia" dans une maladie à ultravirus des insectes. Rev. canad. biol., 10: 299-308.
WEISER, J. (1956): Protozoäre Infektionen im Kampfe gegen Insekten. Z. Pflanzenkrankh. Pflanzenschutz, 63: 625-638.
WEISER, J. (1970): Recent advances in insect pathology. Ann. Rev. Entomol., 15: 245-256.
YOKOKAWA, S. and YAMAGUCHI, K. (1960): Induction of the cytoplasmic polyhedrosis

in silkworm larvae by feeding of mulberry leaves some agricultural chemicals. J. sericult. Sci. (Japan), 29: 133-136 (In Japanese).
ZERNOFF, V. and AJOLO, J. (1939): Chimiothérapie chez les insectes: Action du para-aminophenylsulphamide (1162F.) dans l'infection expérimentale chez la *Galleria mellonella.* C. R. Soc. Biol., 131: 232-234.

Chapter 1, Section 21a: Shelter Engineering for Cattle and Other Domestic Animals (G. L. HAHN)

ASAE DATA (1974): Effect of thermal environment on production, heat and moisture loss and feed and water requirements of farm livestock. Agric. Eng. Yrb, Amer. Soc. agric. Eng., St. Joseph, Michigan: 424-435.
BAXTER, S. H. (1969): The environmental complex in livestock housing – a review. In: Farm Building Report, 4. Scot. Farm Blds Invest. Unit., Aberdeen.
BERRY, I. L., SHANKLIN, M. D. and JOHNSON, H. D. (1964): Dairy shelter design based on milk production decline as affected by temperature and humidity. Trans. Amer. Soc. agric. Eng., 7: 329-331.
DUCKHAM, A. N. (1967): Weather and farm management decision. In: Weather and Agriculture. J. A. Taylor (ed.). Pergamon Press, Oxford/London: 69.
GATES, D. M. (1962): Energy Exchange in the Biosphere. Harper and Row, New York: 1-46.
HAHN, G. L. (1969): Predicted vs measured production differences using summer air conditioning for lactating dairy cows. J. Dairy Sci., 52: 800-802.
HAHN, G. L. and McQUIGG, J. D. (1970): Evaluation of climatological records for rational planning of livestock shelters. Agric. Meteor., 7: 131-141.
HAHN, G. L. (1974): Environmental effects on ruminant production – rational decisions based on current knowledge. In: Livestock Environment. Proc. int. Livestock Environm. Symp., SP-01-74. Amer. Soc. agric. Eng., St. Joseph, Michigan: 232-236.
HAHN, G. L., MEADOR, N. F., THOMPSON, G. B. and SHANKLIN, M. D. (1974): Compensatory growth of beef cattle in hot weather and its role in management decisions. In: Livestock Environment. Proc. int. Livestock Environm. Symp., SP-01-74. Amer. Soc. agric. Eng., St. Joseph, Michigan: 288-295.
HAHN, G. L. and OSBURN, D. D. (1969): Feasibility of summer environmental control for dairy cattle based on expected production losses. Trans. Amer. Soc. agric. Eng., 12: 448-451.
LILLIE, R. J. (1970): Air Pollutants Affecting the Performance of Domestic Animals, a Literature Review. Agriculture Handbook, No. 380. U.S. Dept. Agric., Washington.
McDOWELL, R. E. (1974): Effect of environment on the functional efficiency of ruminants. In: Livestock Environment: Proc. int. Livestock Environm. Symp., SP-01-74. Amer. Soc. Agric. Eng., St. Joseph, Michigan: 220-231.
MORRISON, S. R., BOND, T. E. and HEITMAN, H. (1968): Effect of humidity on swine at high temperature. Trans. Amer. Soc. Agric. Eng., 11: 526-528.
MORRISON, S. R., HAHN, G. L. and BOND, T. E. (1970): Predicting Summer Production Losses for Swine. Product. Res. Rep., 118, U.S. Dept. Agric., Washington.
NATIONAL RESEARCH COUNCIL (1971): A Guide to Environmental Research on Animals. Nat. Acad. Sci, (Wash.): pp. 63-69.
OTA, H., GARVER, H. L. and ASHBY, W. A. (1953): Heat and moisture production of laying hens. Agric. Eng., 34: 163-167.
PAINE, M. D., WITZ, J. A., BUTCHBAKER, A. F., McCROSKEY, L. E. and BACON, C. M. (1972): Mathematical simulation of energy metabolism in beef animals. ASAE Paper 72-510. Amer. Soc. agric. Eng., St. Joseph, Michigan.
SAINSBURY, D. W. B. (1967): Animal Health and Housing. Bailliene, Tindall and Cassell, London.
STOTT, G. H., WIERSMA, F. and LOUGH, O. (1972): Consider Cooling Possibilities: The Practical Aspects of Cooling Dairy Cattle. Report P-25, Coll. Agric., Univ. Arizona, Tucson.
STUBER, A. (1969): Landwirtschaftliche Betriebsgebäude (Farm buildings). CIGR Dokumentation 3 (Reports of Section II, 2-4): 97-108.
TAYLOR, J. A. (ed.) (1967): Weather and Agriculture. Pergamin Press, Oxford/London: 5.

TETER, N. C., DESHAZER, J. A. and THOMPSON, T. L. (1973): Operational characteristics of meat animals – Part I: Swine; Part II: Beef; Part III: Broilers. Trans. Amer. Soc. agric. Eng., 16: 157-159; 740-742; 1165-1167.

THATCHER, W. W., GWAZDAUSKAS, F. C., WILCOX, C. J., TOMS, J., HEAD, H. H., BUFFINGTON, D. E. and FREDRICKSSON, W. B. (1974): Milking performance and reproductive efficiency of dairy cows in an environmentally controlled structure. J. Dairy Sci., 57: 304-307.

WEISS, P. (1945): Biological research strategy and publication policy. Science, 101: 101-104.

WHARTON, C. R., Jr. (1968): Risk, uncertainty and the subsistence farmer. A/D/C Paper. Agric. Developm. Counc., New York.

Chapter 1, Section 21b: Management, Including Breeding, and Nutrition of Cattle (W. F. Y. MAWSON)

ALLEN, G. H. and COWDRY, W. A. R. (1961): Beef gains from irrigated pastures in the Burdekin Delta. Qd agric. J., 87: 175-179.

ALEXANDER, G. I. and CHESTER, R. D. (1956): Growth studies of beef cattle in Queensland. Qd J. agric. Sci., 13: 69-95.

ALEXANDER, G. I., BEATTIE, A. W. and SUTHERLAND, D. N. (1964): Studies on factors in beef cattle production in a subtropical environment. 2. Growth to weaning. Qd J. agric. Sci., 21: 25-32.

ALEXANDER, G. I. (1965): Management aspects in beef cattle production. Aust. vet. J., 41: 61-65.

ALEXANDER, G. I. (1968): Beef cattle studies in the dry tropics. Qd agric. J., 94: 298-306.

ALEXANDER, G. I., DALY, J. J. and BURNS, M. A. (1970): Nitrogen and energy supplements for grazing beef cattle. Proc. XIth Int. Grassland Congr., 793-796.

ALEXANDER, G. I. and CARRAILL, R. M. (1973): The beef cattle industry. In: The Grazing Industries of Australia. G. Alexander and O. B. Williams (eds). Sydney University Press: 143-170.

ANON (1969): New dairy cattle for the tropics, Rural Res. C.S.I.R.O., 65: 37-40.

ANON (1973): The cattle tick in Australia. Inquiry by Cattle Tick Control Comm., Dept Agric., Sydney.

ARBUCKLE, J. (1958): More beef from Brahman cross. Qd agric. J., 84: 427-431.

ARBUCKLE, J. (1959): Now much milk do beef cows give" Qd agric. J., 85: 173-178.

BROWN, G. D. and HUTCHINSON, J. C. D. (1973): Climate and animal production. In: The Pastoral Industries of Australia. G. Alexander and O. B. Williams (eds). Sydney University Press.

BURNS, M. A. (1964): Early weaning of beef calves in North Queensland. Qd agric. J., 90: 534-540.

CASSIDY, G. J. (1967 and 1968): Dairy Pastures in Queensland. Qd agric. J., 93: 656-669, 732-742 and 94: 43-52.

DALY, G. D. and HALL, W. T. K. (1955): A note on the susceptibility of British and some Zebu type cattle to tick fever (Babesiosis). Aust. vet. J., 31: 152.

DAVIDSON, P. C. (1967): Managing the breeding herd in the north, Qd agric. J., 93: 238-48 and 307-313.

EBERSOHN, J. P. (1970): Herbage production from native grasses and sown pastures in south-west Queensland. Trop. Grasslands, 4: 37-41.

EDGLEY, W. H. R. and TUCKER, V. C. (1973): Atherton dairymen lift solids-not-fat. Qd agric. J., 99: 120-127.

FRANCIS, J. and LITTLE, D. A. (1964): Resistance of Drought-master cattle to tick infestation and babesiosis. Aust. vet. J., 40: 247-253.

GEE, R. W., BAINBRIDGE, M. H. and HASLAM, J. Y (1971): The effect of cattle tick (*Boophilus microplus*) on beef production in the Northern Territory. Aust. vet. J., 47: 257-263.

HANCOCK, J. (1953): Grazing behaviour of cattle. Anim. Breed. Abstr., 21: 1-13.

HANCOCK, J. and PAYNE, W. (1955): The direct effect of tropical climate on the performance of European type cattle. 1. Growth. Empire J. exp. agric., 23: 55-74.

HART, B. and MITCHELL, G. L. (1965): Effect of phosphate supplementation on the fertility of an open range beef cattle herd on the Barkly Tableland. Aust. vet. J., 41: 305-309.

HOLDER, J. M. (1967): Milk production from tropical pastures. Add. to Aust. Soc. anim. Prod. (Qd branch) Brisbane, Australia.

JOHNSON, H. D. (1965): Meteorological Monogr., 6: 109-122.

KELLEY, R. B. (1943): Zebu-cross cattle in Northern Australia. An ecological experiment. Counc. sci. industr. Res., Bulletin No. 172. Melbourne, Australia.

LAMPKIN, G. H. and QUARTERMAIN, J. (1962): Observations on the grazing habits of grade and Zebu cattle. II. Their behaviour under favourable conditions in the tropics. J. agric. Sci. (Camb.), 58: 119-123.

LARKIN, R. M. (1954): Observations on the grazing behaviour of beef cattle in tropical Queensland. Qd J. agric. Sci., 11: 115-141.

LECKY, T. P. (1972): The development of the Jamaica Hope as a tropical-adopted dairy breed. 1962 Anim. Husb. Bull., No. 2. Min. of Agric., Jamaica.

McCARTHY, W. O. and HAMILTON, C. P. (1964): Comparative performance of beef breeds in Central Queensland. 1. Productivity. Trop. Agric., 41: 293-297.

McDANIEL, A. H. and ROARK, C. B. (1956): Performance and grazing habits of Hereford and Aberdeen Angus cows and calves on improved pastures as related to types of shade. J. anim. Sci., 15: 59-63.

McDOWELL, R. E. (1972): Improvement of livestock production in warm climates. W. H. Freeman and Co., San Francisco.

McINTYRE, K. H. (1967): A comparative study of the performance of dairy cows during summer and winter in a subtropical climate. Aust. J. exp. Agric. Anim. Husbandry, 7: 400-407.

MAWSON, W. F. (1956): Brahman cattle grow faster than British in North. Qd agric. J., 82: 173-179.

MAWSON, W. F. (1959): Feeding the dairy herd. Qd agric. J., 85: 363-375 and 431-437.

MAWSON, W. F. Y. and WHITE, B. J. (1971): Climatic and biological limitations to dairy production in a tropical environment. Trop. Grasslands, 5: 145-157.

MINSON, D. J. and McLEOD, M. N. (1970): The digestibility of temperate and tropical grasses. Proc. XI int. Grassland Congr., 719-722.

MINSON, D. J. (1971): The nutritive value of tropical pastures. Aust. Soc. anim. Prod. (Qld branch), Brisbane, Australia.

MURRAY, D. M. (1966): A comparison of cutaneous evaporation rates in cattle exposed to heat in a climate laboratory and in the field. J. agric. Sci. (Camb.), 66: 175-179.

NORMAN, K. L. (1971): 220 gal. a day from stall-fed Friesian herd. Qd agric. J., 97: 289-293.

NORMAN, M. J. T. (1965): Seasonal performance of beef cattle on native pasture at Katherine, Northern Territory. Aust. J. exp. Agric. Anim. Husb., 5: 227-231.

PAYNE, W. J. A. (1970): Cattle production in the tropics. 1. Breeds and breeding. Longman Group Ltd., London, 179 pp.

PEGG, S. E. (1955): Seasonal calving for dairy cows. Qd agric. J., 80: 287-294.

QUINLAN, T. J., EDGLEY, W. H. R. and SHAW, K. A. (1973): Nitrogen for tropical legumes. Qd agric. J., 99: 383-387.

SHAW, N. H. and 'T MANNETJE, L. (1970): Studies on a spear grass pasture in central coastal Queensland – the effect of fertiliser, stocking rate and over-sowing with *Stylosanthes humilis* on beef production and botanical composition. Trop. Grasslands, 4: 43-56.

SHEPPARD, A. J., BLASER, R. E. and KINCAID, C. M. (1957): The grazing habits of beef cattle on pasture. J. anim. Sci., 16: 681-687.

SMITH, C. A. (1959): Studies on the Northern Rhodesia Hyparrhenia Veld. 1. The grazing behavior of indigenous cattle grazed at light and heavy stocking rates. J. agric. Sci. (Camb.), 52: 369-375.

SPARKE, E. J. and LAMOND, D. R. (1968): The influence of supplementary feeding on growth and fertility of beef heifers grazing natural pastures. Aust. J. exp. Agric. Anim. Husb., 8: 425-433.

SUTHERLAND, D. N. (1959): Factors affecting the performance of beef cattle on unimproved pastures in Queensland. Aust. vet. J., 35: 129-134.

TEITZEL, J. K. (1969): Pastures for the wet tropical coast. Qd agric. J., 95: 304-11, 380-5, 464-7, 532-535.

TURNER, H. G. (1972): Selection of beef cattle for tropical Australia. Aust. vet. J., 48: 162-166.
VOHNOUT, K. and BATEMAN, J. V. (1972): Effects of crude fibre upon feeding efficiency of cattle in "warm" environments. J. agric. Sci. (Camb.): 78: 413-416.
WELLINGTON, K. E. (1966): The productive performance of the Jamaica Hope breed. Anim. Husb. Bull., No. 2 Min. Agric., Jamaica.
WHARTON, R. H. (1962): The current status and prospects for the control of ixodid ticks with special emphasis on Boophilus microplus. FAO/OIE Ad hoc Consultation on control of protozoal tick-borne diseases of cattle. Nairobi, Kenya.
YOUNG, N. D., FOX, W. F. and BURNS, M. A. (1959): A study of three important pasture mixtures in the Queensland subtropics. Qd J. agric. Sci., 16: 199-215.

Author Index

Subject Index